D0221172

MACHINE DESIGN FUNDAMENTALS

A Practical Approach

James Watt (1736–1819) of Scotland. Inventor of the modern steam engine, patented in 1769. Watt was a mechanical genius whose influence on technology continues unmatched to this day. (Courtesy of the Science Museum, London.)

MACHINE DESIGN FUNDAMENTALS

A Practical Approach

Uffe Hindhede
Black Hawk College
Coordinator and Principal Author

John R. Zimmerman
University of Delaware

R. Bruce Hopkins
Consulting Engineer
and Iowa State University

Ralph J. Erisman
International Harvester Company (Retired)

Wendell C. Hull
New Mexico State University

John D. Lang
Assistant Professor of English
Augustana College
Team Editor

Prentice Hall, Englewood Cliffs, NJ 07632

Library of Congress Cataloging in Publication Data:
Main entry under title:

Machine design fundamentals, a practical approach.
Includes index.
1. Machinery—Design I. Hindhede, Uffe
TJ230.M23 1983 621.8′15 82-16066

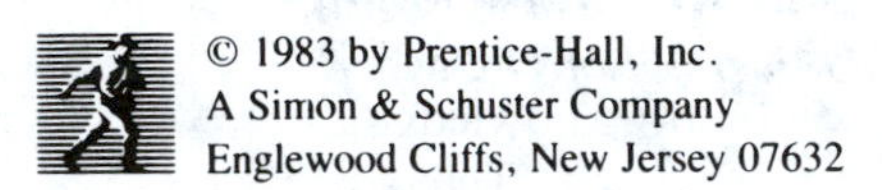

A Simon & Schuster Company
Englewood Cliffs, New Jersey 07632

Printed in the United States of America

10 9 8

ISBN 0-13-541764-3

Prentice-Hall International (UK) Limited, *London*
Prentice-Hall of Australia Pty. Limited, *Sydney*
Prentice-Hall Canada Inc., *Toronto*
Prentice-Hall Hispanoamericana, S.A., *Mexico*
Prentice-Hall of India Private Limited, *New Delhi*
Prentice-Hall of Japan, Inc., *Tokyo*
Simon & Schuster Asia Pte. Ltd., *Singapore*
Editora Prentice-Hall do Brasil, Ltda., *Rio de Janeiro*

Preface

This book evolved from material initially prepared for technicians in a six-credit machine design course at Black Hawk Community College.[1] The textbook for the course was an engineering text, since textbooks written specifically for technology students were not available. This created problems and, consequently, student handouts were prepared explaining in detail some of the more difficult topics. About the same time, the technology department began the transition to variable-entry courses. Explicit, well-written material became a necessity.

Then, within a year, local industries began a transition to SI units of measurement. Preparing metric material became a major assignment. Because my formal education took place in Denmark, I was familiar with the metric system. Additionally, I was able to draw on my knowledge of foreign languages, German in particular, to prepare metric handouts. The best German textbooks, I soon discovered, were all written by teams of experts. I therefore assembled a team of design engineers and educators to write a textbook using a ratio of 60% SI units to 40% English units and incorporating practical industrial knowledge at the technician's level. Dr. John Lang, Assistant Professor of English at Augustana College, became the team editor. His primary responsibility was to add clarity, concision, and uniformity to the manuscript, whose style varied because of the number of people involved in the book's preparation. In this capacity he transformed a rather rough manuscript into a readable text.

To use this book effectively, students should have a basic understanding of mechanical drawing, manufacturing processes, and strength of materials. Calculus is not a prerequisite for most of the material. Theory is presented when it promotes understanding of mechanical systems. Emphasis, however, is on practical industrial applications, since technology students will be employed in the application segment of industry. Whenever possible, a brief history or evolutionary development of machine elements and machines is given to promote an interest in and understanding of machines.

The textbook is aimed at both two- and four-year technology programs. There is adequate material for the four-year program, and the latter part of major chapters is often on a higher level and uses calculus. This text is unique in that it combines modern domestic industrial practices with the best information from metric European textbooks and handbooks. Because we draw on many areas of practical experience, there is a detailed treatment of all major topics. The text thus avoids the common pitfalls of exclusion and deemphasis. All topics are presented with the

---

[1] Moline, Illinois

broad, integrated approach common to current industrial practices because community colleges often serve specific needs determined by the nature of local industries.

The chapter sequence in this text is both logical and developmental. The material is presented under the three following headings.

1. Primary considerations; essentially introductory material concerned with the theory and fundamentals of machines and the design of machines.
2. Joining and fastening.
3. Machine members for transmission and control of power.

For clarity and pedagogy, we have treated individually as many machine members as possible. At the same time, we have been careful to tie each machine member in with other, often competitive, members, to see it in relation to an entire machine system.

We also present material and methods that are new to machine design. Topic highlights and a listing of new material follow.

Chapter 1 explains the nature and composition of machines. It includes an ingenious diagram for computing service factors.

Chapter 2 introduces nomographs and a quick method of implementing design changes.

Chapter 3 emphasizes design based on fatigue strength; it contains two exclusive charts to explain the nature of fatigue strength.

Chapter 4 discusses design based on rigidity, stability, and wear. The chapter stresses the growing importance of friction and wear in machine design.

Chapter 5 discusses machine frames and housings—a new topic in machine design textbooks.

Chapter 6 provides a comparative study of permanent fastenings, with an emphasis on arc welding.

Chapter 7 discusses unified and metric fasteners. Preloading and modes of failure are examined extensively.

Chapter 8 contains two unique charts for selecting coil springs.

Chapter 9 surveys the present art of power transmission and control in machinery—a new topic in machine design textbooks.

Chapter 10 emphasizes the growing importance of couplings in modern, high-speed machinery.

Chapters 11 to 14 discuss in simple terms power transmission by belts, chains, and gears. The selection of machine elements for industrial application is based on *Machinery's Handbook*.

Chapter 15 gives a much-needed simplified approach to the design of axles and shafts, followed by a more detailed presentation.

Chapter 16 examines the semipermanent fasteners most frequently used to transmit torque between a shaft and the machine parts assembled on it.

Chapter 17 presents a simplified approach to the selection of sliding element bearings and then offers a more advanced treatment of this topic.

Chapter 18 discusses rolling element bearings, a subject notorious for its poor presentation in domestic and foreign textbooks.

Chapter 19 analyzes the importance of clutches in terms of the safe and energy-efficient operation of machinery.

Chapter 20 discusses the function of brakes in the safe operation of vehicles and machinery; it emphasizes caliper disk brakes, which are rapidly replacing most other types of brakes.

Chapter 21 shows the importance of gaskets and seals in the safe operation of machinery utilizing fluids and gases under high pressure or high temperature.

The use of handbooks and industrial manuals is part of the learning experience in machine design. Because of widespread industrial use of *Machinery's Handbook,* we decided to use it as a reference manual in order to select interference fits, splines, eyebolts, flange couplings, belts and chains, and to calculate gears using AGMA standards.

The text contains examples relevant to all major principles demonstrated. Technology students, even more than engineers, need practical and explicit examples relevant to industry. Examples also bridge the gap between text or theory and the problems at the end of each chapter. The problems, like the text, progress from the simple to the complex and are realistic because of our extensive practical experience. In addition, numerous problems are taken from examinations given at technology schools and technical universities in Denmark and Sweden.

An instructor's manual is available that contains solutions to text problems and additional problems not found there. Tests for material comprehension are included in the instructor's manual, as is a list of appropriate industrial films.

We wish to express our appreciation to *Machine Design* (Penton Publication), Deere & Company, International Harvester Company, Caterpillar Tractor Company, General Motors Corporation, Mobil Oil Corporation, the American Chain Association, Black Hawk College, and the many companies and societies that furnished material and illustrations. The following individuals have contributed to this publication: Catherine Alexander, Ben Bean, David V. Hutton, Irving L. Kosow, Edwin C. McClintock, Lynn Schmelzer and O. J. Sorenson. The academic reviewers were: Ronald F. Amberger, Donald S. Bunk, John O. Pautz, Howard L. Paynter, Richard Rossignol, Howard Smith, and Ben C. Sparks. The principal reviewer was Dr. Stan Brodsky.

The following specialists volunteered to review appropriate sections: Louis M. Anderson, Charles M. Allaben, Jr., Richard B. Belford, Omer L. Blodgett, Arthur H. Breed, Richard Carrigan, William J. Derner, William Grube, Fred L. Heine, George Michalec, W. R. Miller, Larry L. Oliver, Robert O. Parmley, David W. Singley, Chon L. Tsai, Abraham I. Tucker, Richard L. VanEerden, and Richard J. Will. Their reviews greatly enhanced quality and assured that the material was current.

Special thanks to my wife, Lois, for her many contributions, to Linda Diedrick for typing much of the manuscript, to my editors, Judy Green, Susan Weiss and Vivan Kahane, and to the Wiley staff for a job well done.

Uffe Hindhede

Coordinator and Principal Author

To the Student

Knute Rockne, famous football coach at Notre Dame, once stated before a national audience:

> At Notre Dame we have a squad of about three hundred lads—both varsity veterans and newcomers. They keep practicing *fundamentals,* and keep it up, and keep it up, and keep it up, until these various fundamentals become as natural and subconscious as breathing. Then in the game they don't have to stop and wonder what to do next when the time comes for quick actions.

The same principles apply to machine design! If you want to become a first-rate designer, you must have your *fundamentals*—the ABCs of your job—firmly in mind. That is what this book is all about.

To the Industrial Practitioner of Mechanical Design

The material in this book has been obtained from sources considered reliable. If important designs are based on this material, the designer should nevertheless check all important data against those of available handbooks and other appropriate sources. Final judgment should be based on actual testing.

The authors of this text would welcome any comments or suggestions you might have on the book's contents. Such responses should be sent to my home address.

Uffe Hindhede
2800 25th Avenue A
Moline, IL 61265
309-762-2001

Contents

Note on the International System of Units

From a global point of view, this system, designated SI, is the dominant language of measurement in science, engineering, and technology. SI is the most recent and best version of the metric system. This system rose to prominence because of its simplicity and coherence. Because only decimal arithmetic is involved, fractions are avoided. Units of different sizes are formed by multiplying or dividing a single base unit by powers of 10. In mechanical design the millimeter is the fundamental unit of measurement. Because of its size (25.4 mm = 1 in.), it is much more convenient to use than the inch. Most dimensions are whole numbers, which greatly facilitates memorization. When decimals become necessary, one rarely needs more than two. The second decimal is often 0 or 5.

SI consists of a few *base* units and many *derived* units. Only those used in machine design will be mentioned.

BASE UNITS OF SI

The *meter* (m) is the unit of *length*.
The *kilogram* (kg) is the unit of *mass*.
The *second* (s) is the unit of *time*.

DERIVED UNITS OF SI

The *newton* (N) is the force that gives to a mass of 1 kg an acceleration of 1 m/s/s.

$$\text{Force (N)} = [\text{mass (kg)}]\,[\text{acceleration (m/s}^2)]$$

The *joule* (J) is the work done when the point of application of 1 N is displaced a distance of 1 m in the direction of the force. Therefore, $1\ \text{J} = 1\ \text{N} \cdot \text{m}$.

$$\text{Work (J)} = [(\text{force (N)}]\,[\text{distance (m)}]$$

The *watt* (W) is the power that produces energy at a rate of 1 J/s. $1\text{W} = 1\ \text{N} \cdot \text{m/s}$

$$\text{Power (W)} = [\text{force (N)}]\,[\text{velocity (m/s)}]$$

Although the SI unit of temperature is the *kelvin* (K), in technology it is common practice to use the Celsius temperature scale (equivalent to the older centigrade). The units are of the same magnitude, but the scales differ in zero point.

Once mastered, the rules for using SI units will be found less time consuming than those of the English system.

1. The decimal point is always preceded by a zero for numbers less than unity.

0.341 but not .341

2. Numbers having four or more digits are placed in groups of three (or four) and separated by a single-digit space instead of a comma.

1918 or 1 918 but not 1,918

43 198 but not 43,198

3. Parentheses, replacing the X, indicate multiplication.

$$(13.39)(73.5) = 984.165$$

When single letters are used, no parentheses are needed.

$$a(2.5 + 7.3) = 2.5a + 7.3a$$

4. Units should always be shown in any substitution, as follows.

$$F = m\,a$$

$$F = (10\text{kg})\ (10.3\ \text{m/s}^2)$$

$$F = 103\ \text{N}$$

For simplicity of operation, it has been recommended that preference be given to those derived units obtained by multiplying or dividing base units of 10^3. The preferred units of length are therefore the kilometer (1 km = 1000 m) and the millimeter (1 m = 1000 mm), but not the centimeter (1 cm = 10 mm).

One of the most important features of SI is that there is only one unit for each physical quantity—the meter for length, the kilogram for mass, and so forth. In both technical and nontechnical fields, however, the term "weight" is used with several meanings. In addition to the technically "correct" meaning such as force of gravity, *weight* is also used to mean mass. The best solution to this dilemma is to sharpen the distinction between *weight* and *mass* but not to exclude the word *weight*. In accordance with a resolution made by ASEE, the word *weight* meaning "force of gravity," but *never* "mass," will be retained.

The basic unit of length is the meter, but only the millimeter is used on mechanical engineering drawings. Hence it would be impractical to use the megapascal (MPa) as defined $10^6(\text{N/m}^2)$. Instead, we will use an equivalent unit: newton per millimeter squared ($1\ \text{N/mm}^2 = 1$ MPa). Thus the cumbersome conversion factor of 10^6 will be eliminated.

The sole purpose of different units is to facilitate measurements and calculations. Thus the meter is rarely used in mechanical drawings, but applies well to belt and chain speeds. Even though the use of the centimeter (0.01 m) is not recommended, there are numerous calculations involving moments of inertia that can be facilitated by such use.

A major aim of this book is to assist the transition to SI. In the mechanical industries this transition is taking place rapidly. Therefore, in this text, SI will be the dominant system of measurement. The ratio of SI to EU (English units) will be roughly 60 to 40%. This ratio will vary from chapter to chapter because some topics are more easily metricated than others.

General Symbols and Abbreviations

The following outline is a reference to symbols and abbreviations familiar to students. They will thus not be included in the nomenclature for each chapter.

Symbol	Meaning
a	linear acceleration
A	area of cross section
c	distance from neutral axis to fiber where stress is desired
C	constant
C	degrees Celsius
d, D	diameter of circle
E	modulus of elasticity in tension; energy
e	efficiency
F	force
F	degrees Fahrenheit
G	modulus of elasticity in shear
h	height
h	hour
I	area moment of inertia
J	polar moment of inertia
J	joule
k	radius of gyration; spring constant
L, l	length
M	moment of force; bending moment
m	mass
m	meter
mm	millimeter
N	Newton
N	normal force
n	revolutions per minute
p	pressure; force per unit area
P	power; axial force
r, R	radius
R	reaction or resultant force
R_C	Rockwell C hardness
S	strength
s	displacement; stress
s	second
S_{yc}	yield strength in compression
S_{yt}	yield strength in tension
S_{uc}	ultimate strength in compression
S_{ut}	ultimate strength in tension
T	torque
T	temperature
t	thickness; temperature
V	volume; shearing force in beam
v	velocity
W	weight; transverse force on beam
W	watt
Z	section modulus I/c based on area moment of inertia
Z_p	section modulus based on polar moment of inertia J/c

GREEK SYMBOLS

Symbol	Meaning
α (alpha)	an angle; coefficient of thermal expansion; angular acceleration
β (beta)	angle
γ (gamma)	angle; shearing strain unit

δ (delta)	elongation; beam deflection	ρ (rho)	density; variable radius
ϵ (epsilon)	normal unit strain	σ (sigma)	stress (tension and compression)
θ (theta)	angle; angular deflection	Σ (sigma)	summation
μ (mu)	micro (one millionth)	τ (tau)	shear stress
ν (nu)	Poisson's ratio	ω (omega)	angular velocity, rad/s
π (pi)	3.1416		

SUBSCRIPTS

avg	average	all	allowable
b	bending	c	compressive
e	elastic; effective	dyn	dynamic
i	internal; induced	max	maximum
min	minimum	nom	nominal
o	outer	r	radial
s	shear	st	static
t	tensile; tangential	tan	tangent
x	with respect to the x-axis	y	with respect to the y-axis

ABBREVIATIONS

AFBMA	Anti-Friction Bearing Manufacturers Association	B	Brinell hardness number
		ccw	counterclockwise
AGMA	American Gear Manufacturers Association	cfm	cubic feet per minute
		c.g.	center of gravity
AISC	American Institute of Steel Construction	CI	cast iron
		cw	clockwise
AISI	American Iron and Steel Institute	EU	English units
		cpm	cycles per minute
ASA	American Standards Association	cps	cycles per second
		fpm	feet per minute
ASEE	American Society for Engineering Education	fps	feet per second
		fs	factor of safety
ASLE	American Society of Lubrication Engineers	gpm	gallons per minute
		hp	horsepower
ASM	American Society for Metals	*ID*	inside diameter
		ips	inches per second
ASME	American Society of Mechanical Engineers	ips^2	inches per second-second
ASTM	American Society for Testing Materials	KE	kinetic energy
		kW	kilowatt
AWS	American Welding Society	kN	kilonewton
		kg	kilogram

km	kilometer
km/h	kilometers per hour
mph	miles per hour
m/s	meters per second
MA	mechanical advantage
OD	outside diameter
psi	pounds per square inch
rad	radians
rpm	revolutions per minute
rps	revolutions per second
SAE	Society of Automotive Engineers
SESA	Society for Experimental Stress Analysis
YP	yield point
TMA	theoretical mechanical advantage
μin.	microinch; equals 10^{-6} in.
μm	micrometer; equals 10^{-6} m

MACHINE DESIGN FUNDAMENTALS

A Practical Approach

PART 1

Primary Considerations

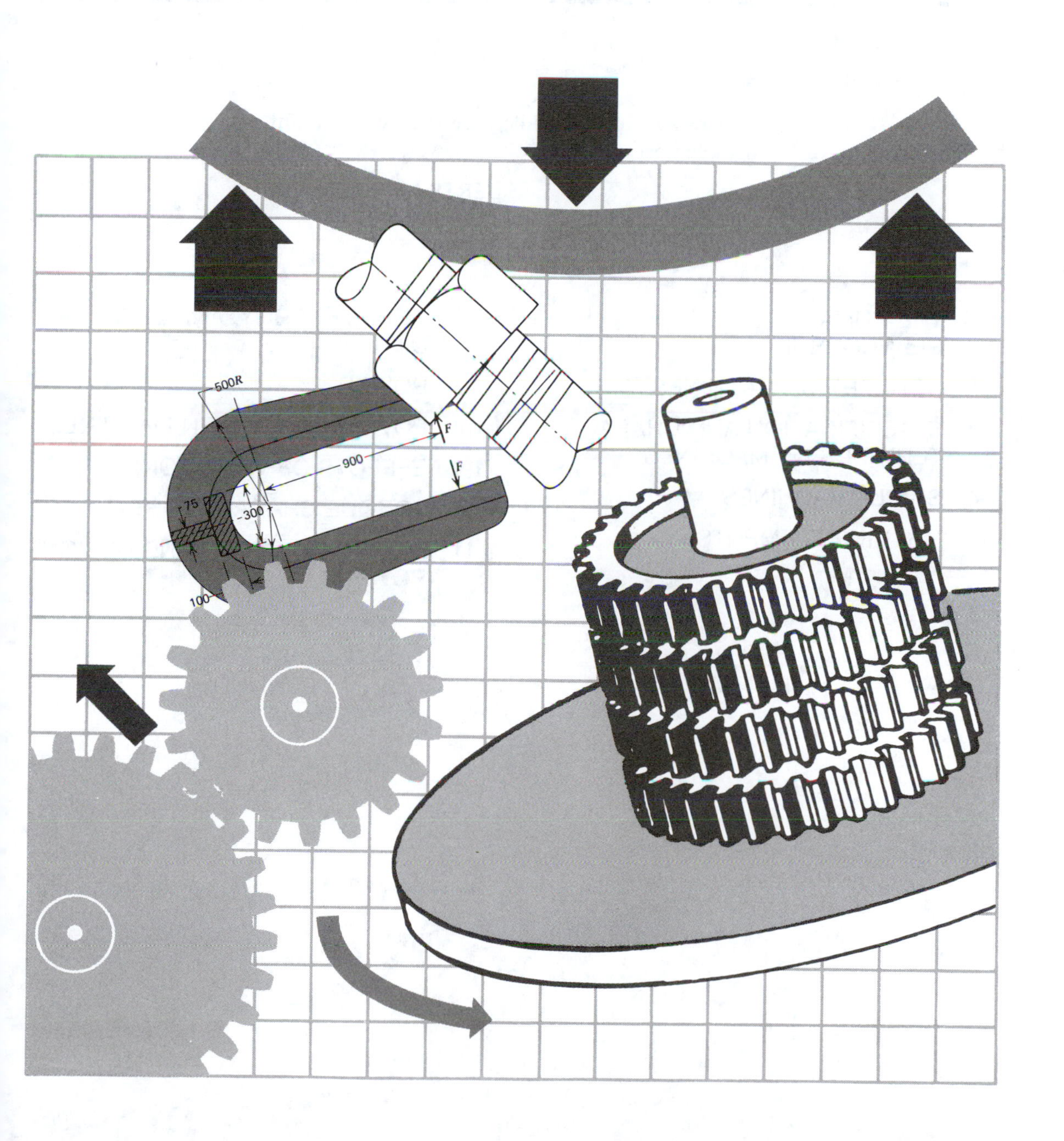

Chapter 1
Nature and Composition of Machines

Machines are worshipped because they are beautiful,
and valued because they confer power.

BERTRAND RUSSELL

In this chapter readers will meet machines as they relate to each other, to their operators, and to humanity. These machines will be seen as converters of energy and extensions of human power. The composition and characteristics of machines will be presented and underlying principles of mechanics demonstrated. Without this introductory knowledge students will not be ready to pursue the real goal of this book—design of machinery.

1-1 THE HUMAN-MACHINE RELATIONSHIP

A machine is a device for doing useful work, such as manufacturing and transporting goods. Depending on the nature of its useful function, a machine may be simple or complex, fragile or strong. Before the industrial age, machines were simple and low powered. By today's standards, they were crude devices designed to relieve people of backbreaking chores such as lifting heavy burdens, conducting mining operations, and milling grain. Animals, sails, windmills, and waterwheels provided power for these early machines. These sources were clearly inadequate for all but easy tasks. Thus the lack of mechanical power provided little incentive for anyone to invent machines that were faster or more powerful. Technical progress was slow except in the fields of mechanics and metallurgy. Development of the steam engine, however, provided the incentive needed for humans to improve vastly their standards of living.

Invention of the modern steam engine by James Watt[1] in 1769 gave people the mechanical power required for large-scale industry and mass transportation. Use of the steam engine for mine pumping and railroads led to other and much improved power sources such as the current internal combustion engines, the turbine, the jet engine, and the rocket. In 200 years (1769–1969) humanity advanced technically from a primitive steam engine to a powerful rocket engine that would carry people to the moon. The invention of the steam engine was one of the most important events in history.

The explosive nature of the industrial revolution underscores the fundamental fact that industrial production is much more dependent on available sources of power than on supplies of raw materials and production tools.

The most dramatic development in the history of the machine came, no doubt, with the introduction and rapid development of automatic operations—machines directed by machines instead of by humans. Termed *automation,* this interrelationship among machines initially depended on linkages and cams, then on punch cards and, finally, on tapes and electronic equipment to direct and control mechanical processes without the constant attendance of operators. At the heart of automation is the computer, a cybernetic[2] machine for processing, storing, and retrieving information.

1-2 CONCEPT OF MACHINES

A *machine* is any device that enables work to be done with greater ease or speed than that obtainable without its use. It is a term most commonly applied to mechanisms used in the industrial arts for shaping and joining materials. Machines are frequently named from their use (e.g., screw cutter) or from the product made (e.g., bolt maker). *Compound* machines are formed when two or more simple machines are combined. Tools are the simplest implements of the industrial art. Machines, on the other hand, are more complicated in structure. When machines act with great power, they generally take the name

[1] A revolutionary improvement of Newcomen's *atmospheric* steam engine.

[2] Cybernetics is the science of control and communication.

of engines (e.g., internal combustion, steam, or aircraft).

A machine is therefore essentially a structure consisting of a frame with various fixed and moving parts. The parts are rigid or resistant and are relatively constrained so that they can transmit power, modify force or motion, and do useful work. In short, a machine is a device that transmits and changes the application of energy.

Machinery is a derived term used to represent (1) the internal working parts of a complex assembly, usually of large size; and (2) a grouping, such as the machinery of a plant or mill.

1-3 CLASSIFICATION AND CHARACTERISTICS OF MACHINES

Machines can be classified broadly as basic or simple and complex or compound. A *simple* machine is a device with at least one mechanically actuated member (lever or screw). A *complex* machine is merely a combination of simple machines, as exemplified by a typewriter.

Complex machines may be classified as stationary, portable, or mobile. As the name implies, mobile machines, such as self-propelled combines or street sweepers, move themselves in doing useful work. This group includes all means of transportation, farm machinery, and construction equipment. Stationary machines include most factory production machinery. Portable machines are primarily items such as power tools, chain saws, or vacuum cleaners; that is, they can be carried by the user.

Complex machines are also classified as *prime* movers, *secondary* movers, and *power driven*. Any machine that utilizes a natural source of energy to produce power for other machines is a prime mover. An automobile engine is a prime mover that converts the chemical energy of the fuel into mechanical energy. This, in turn, is used to propel and direct the entire vehicle along some selected path. An electric motor, by contrast, is a secondary mover because it receives energy directly or indirectly from a generator driven by a prime mover, usually a steam turbine. A power-driven machine or power-absorption unit utilizes energy to do useful work. Typical power-driven machines are pumps, machine tools, and air compressors.

Machines are thus *dual* in their makeup; they all have a power source separate from the parts doing useful work. The power supplied to a machine is called *input* power, while the power delivered by a machine is called *output* power. Thus the output power of one machine becomes the input power of the driven machine. It is common to have a single place for power input and another for output power. Some machines, however (e.g., trucks and tractors), have two output shafts, one to propel the vehicle and another to drive accessories.

The dual makeup of machines is seen clearly in Fig. 1-1, which also shows many of the components common to nearly all machines.

1-4 SIMPLE MACHINES

Six simple machines have been known for centuries: lever, wheel and axle, pulley, inclined plane, wedge, and screw. Modern technology has added two: gear wheel and hydraulic press (Fig. 1-2). Simple machines provide a mechanical advantage (MA) by increasing force at the expense of speed; herein lies their primary importance. Increasingly, however, they are used in reverse to augment speed at the expense of force ($MA < 1$). The tremendous speeds of vehicles, for example, are obtained by using the wheel and axle in reverse.

Synthesis

From these eight simple machines, many complex machines originate. Because simple machines do not lose their identities within a complex machine, the forces required for their operation can be predicted accurately. Moreover, the total theoretical advantage (disregarding friction) of a complex machine is equal to the product of the individual theoretical me-

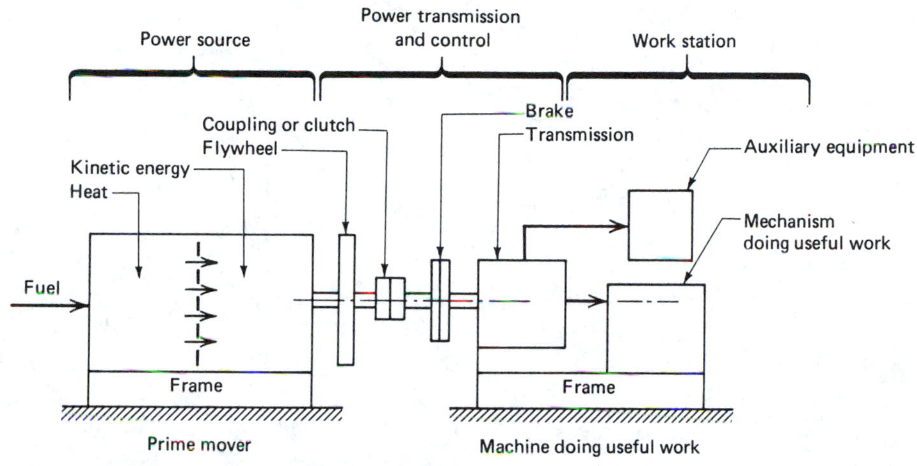

Figure 1-1 Basic components of a power-work unit. A machine is a device that transforms, transmits, and controls power. Power comes from a local source (primary or secondary mover) and is delivered to a work station. Everything in between is transmission and control. The power transmission elements shown are not all used simultaneously nor in the order shown.

chanical advantages for all simple machines that comprise the complex machine.

$$\text{TMA(total)} = (\text{TMA}_1)(\text{TMA}_2) \ldots (\text{TMA}_n) \quad (1\text{-}1)$$

Analysis

Conversely, complex machines are more easily understood when explained in terms of simple machines. For this reason, the topic of simple machines recurs throughout the text. Screw fasteners, for example, are a direct application of a simple machine, while worm gearing derives its characteristics from both the screw and gearing. A detailed discussion of simple machines is presented in Appendix A. For further details, consult *Machinery's Handbook,* pp. 299–303.[3]

1-5 MACHINE ELEMENTS

However simple, any machine is a combination of individual components generally referred to as machine elements or parts. Thus, if a machine is completely dismantled, a collection of simple parts remains such as nuts, bolts, springs, gears, cams, and shafts—the building blocks of all machinery. A machine element is, therefore, a single unit designed to perform a specific function and capable of combining with other elements (Table 1-1). Sometimes certain elements are associated in pairs, such as nuts and bolts or keys and shafts. In other instances, a group of elements is combined to form a subassembly, such as bearings, couplings, and clutches. These are often installed and serviced as a unit or pairs.

The individual reliability[4] of machine elements is well established and becomes the basis for estimating the overall life expectancy of a complete machine. Their simplicity satisfies the most stringent requirements for economic manufacturing.

Many machine elements are thoroughly standardized. Testing and practical experience have established the most suitable dimensions for

[3]All references are to the twenty-first edition.

[4]*Reliability* is the probability that a part will perform as intended under given circumstances.

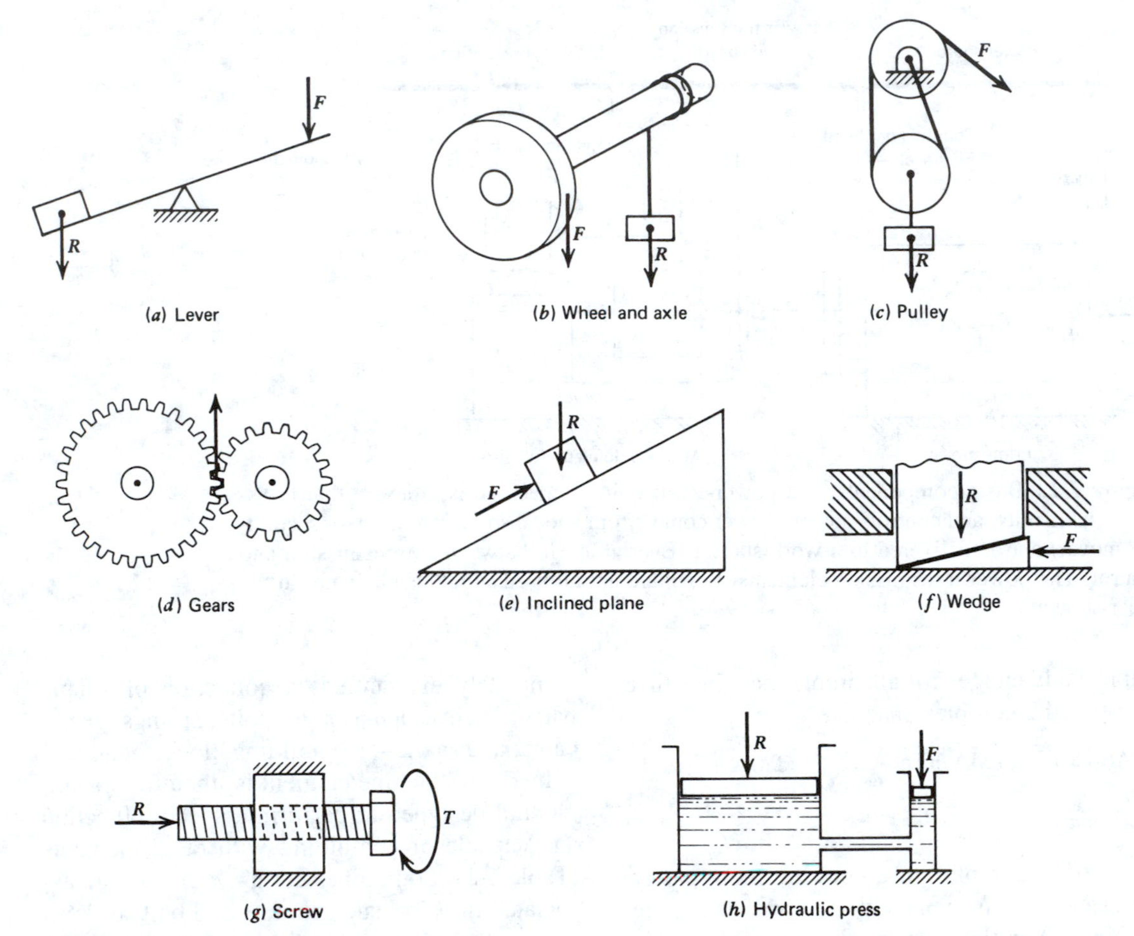

Figure 1-2 Simple machines.

common structural and mechanical parts. Through standardization, uniformity of practice and resulting economies are obtained. Not all machine parts in use are standardized, however. In the automotive industry only fasteners, bearings, bushings, chains, and belts are standardized. Crankshafts and connecting rods are not standardized.

1-6 WORK, ENERGY, AND EFFICIENCY

Simple machines facilitate work by changing force and speed without altering their product. They provide the huge forces needed to move heavy objects and mechanically shape and join materials for productive purposes.

In many machines the work stroke requires low speed, in contrast to the relatively high speed delivered at the output shaft of motors and engines. Because it is economical to design power units operating at high speeds, it is necessary to provide a speed reduction unit between driver and driven machines.

To do useful work, a machine must have a reliable source of energy. Energy is the capacity to do work or produce an effect, such as converting water to steam. Energy takes many forms; however, in machinery we are dealing

TABLE 1-1 Classification of Machine Elements

Function	Name
Fastening and joining	Rivets, screws, nuts, bolts, pins, snap rings, keys, washers
Storage of energy	Springs (coil, torsion bar, leaf, disk)
Power transmission and speed change	Belts, sheaves, pulleys, chains, coupling sprockets, gears, friction wheels, power screws, ball and roller bearing screws, linkages
Carrying and transmitting rotary motion	Axles, shafts, sliding and rolling bearings, linkages
Transmission and control of rotary power	Pistons, connecting rods, flywheels, cams, brakes, clutches
Sealing, enclosing, and controlling fluids	Seals, gaskets, valves, cocks, tubing, fittings, piston rings
Service elements	Handwheels, levers, steering wheels, cranks, linkages
Admitting, enclosing, restraining, or supporting integral machine parts or machines	Machine frames, housings, and foundations

primarily with mechanical energy, which includes potential[5] and kinetic energy.[6]

Although energy cannot be created or destroyed, it can be converted or transformed, always with some loss in the process. For example, when converting electrical energy to mechanical energy (motors), some energy is wasted in the form of heat. Usually this energy is not recoverable and represents a loss or inefficiency, as in Eq. 1-2.

$$\text{Work input} = \text{work output by machine} + \text{lost work} \qquad (1\text{-}2)$$

No device can be made to perform work unless a somewhat greater amount of energy—enough to make up for friction and other losses—can be supplied from some external source.

Equation 1-2 can be rearranged as follows

$$\frac{\text{Work input} - \text{lost work}}{\text{Work input}} = \frac{\text{work output}}{\text{work input}}$$

[5]Energy that exists by virtue of position; nonactive energy.
[6]The capacity for performing work possessed by a moving body by virtue of its momentum.

Clearly, this is a measure of performance or efficiency. Thus,

$$\text{Efficiency} = \frac{\text{work output}}{\text{work input}} = \frac{\text{energy delivered}}{\text{energy received}} \qquad (1\text{-}3)$$

Therefore the efficiency of a machine is the ratio of energy delivered by the machine to energy supplied. For example, the efficiency of an electric motor is the ratio between the energy delivered by the motor to the machinery it drives and the energy it receives from the generator or power line. The definition applies equally well to single machine members or mechanisms such as bearings, gears, belts, and chains. Efficiency will be designated by *e* and will always have a value less than or equal to one, unless given as a percentage.

Mechanical Efficiency

Loss of mechanical energy in machines is largely due to friction forces doing negative work. In good designs friction, vibration, and other such losses are purposely kept small because they contribute to inefficiency. They represent the energy required to maintain the motion of a machine or mechanism that is not doing useful work. Assume a certain amount of energy E_i is supplied to a machine and successively passes through three machine members with efficiencies e_1, e_2, e_3. By passing through the first member, E_i loses a small amount of energy and thereby reduces the output to $(e_1)(E_i)$. The output of the first member becomes the input of the second member. Again, energy is lost, and the output of the second member $(e_1)(e_2)(E_i)$ becomes the input of the third member. Total output E_o is therefore

$$E_o = (e_1)(e_2)(e_3)E_i \tag{1-4}$$

For n machine members with efficiencies e_1 $e_2 \ldots e_n$, total output is $E_o = (e_1)(e_2) \ldots (e_n)E_i$, and total efficiency becomes the *product*, not the sum, of the individual efficiencies.

$$e = (e_1)(e_2) \ldots (e_n) \tag{1-5}$$

Table 1-2 presents typical values of the efficiency for common machine elements.

TABLE 1-2 Mechanical Efficiency, Typical Values

Machine Member	Efficiency, e	
	High	Low
1. Ball bearings	0.999	0.99
2. Silent chains	0.99	0.97
3. Spur and helical gears (including bearings)	0.985	—
4. Roller bearings	0.98	—
5. Synchronous belts	0.98	—
6. Bevel gears	0.98	0.97
7. V-belts	0.98	0.95
8. Roller chains	0.97	0.95
9. Worm gearing	0.97	0.50
10. Ball bearing screws	0.93	0.90
11. Power screws (multithreads → single threads)	0.84	0.38
12. Roller bearing screws	0.80	—
13. Screw fasteners	0.38	0

EXAMPLE 1-1

An electric motor drives a machine by means of a gear train with three sets of spur gears and a roller chain drive. Find the efficiency of the drive system.

SOLUTION

Using values from Table 1-2 in Eq. 1-5, we obtain:

$$e = (e_g{}^3)(e_c)$$

$$e = (0.985^3)(0.96) = \mathbf{0.92}$$

Efficiency can be measured directly or calculated, but it can also be evaluated by means of noise or heat generated. Although not always recognized, noise represents an excellent medium for evaluating machine performance. Energy expended in unwanted machine motion or vibration reduces the useful output and results in noise. Generally, the condition of least noise corresponds to minimum loss in efficiency and minimum wear.

1-7 POWER

Power is the rate of doing work or transmitting energy. The SI power unit is the watt (W). Because the watt is such a small unit, the kilowatt (1000 W) is preferred in most technical calculations. With proper units, the power equations are:

$$P = \frac{Fv}{1000}\ \text{kW} \tag{1-6}$$

$$P = \frac{Tn}{9550}\ \text{kW} \tag{1-7}$$

$$P = \frac{T\omega}{1000}\ \text{kW} \tag{1-8}$$

$$P = \frac{Fv}{33{,}000}\ \text{hp} \tag{1-9}$$

$$P = \frac{Tn}{63{,}000}\ \text{hp} \tag{1-10}$$

where

P = power; kW, hp
F = force; N, lb

v = velocity; m/s, ft/min
T = torque; N · m, lb-in.
ω = angular velocity; rad/s
n = angular velocity; rpm

EXAMPLE 1-2

The belt speed of a simple drive is 15 m/s, and the net driving force is 1200 N. Calculate the belt power transmitted in kilowatts.

SOLUTION

Equation 1-6 is applicable for computing power in kilowatts. Thus,

$$P = \frac{Fv}{1000} = \frac{(1200 \text{ N})(15 \text{ m/s})}{1000} = \mathbf{18\ kW}$$

EXAMPLE 1-3

Find the torque capacity in newton-meters of a 10-hp, 1200-rpm electric motor.

SOLUTION

Equation 1-7 is applicable for computing the torque in newton-meters.

$$T = \frac{(9550)(P)}{n} = \frac{(9550)(10 \text{ hp})(0.746 \text{ kW/hp})}{1200 \text{ rpm}}$$

$$= \mathbf{59.4\ N \cdot m}$$

Power is a characteristic of motors and engines that facilitates comparison of one power unit with another of a similar type but different capacity. Rating, however, is not a simple matter because power varies with speed and numerous other factors. Currently, motors and engines are rated in horsepower but, with metrication, all ratings will be in kilowatts.

1-8 POWER FOR DRIVEN MACHINES

Power is generally provided for the purpose of carrying out some manufacturing process or moving people and goods. Increasing demand for standardized, low-cost products has led to the development of many machines for manufacturing and processing material. To drive these, a variety of engines and electric motors was developed to transform energy from available sources into mechanical energy. Large machine tools often have several motors installed to reduce the amount of shafting and gearing. The most important of these—electric motors, internal combustion engines, hydraulic motors, and air motors—are discussed next.

Electric Motors[7]

Primarily a constant speed device, the electric motor is by far the most widely used power source, and designers should be familiar with it. The primary advantage of electric power over other forms of energy is its great flexibility and cleanliness. Machines driven by electricity can be remotely located from the actual source of power, yet start at the flick of a switch. Furthermore, power can easily be increased or decreased to meet varying load needs. Many different motors are available for different applications, ranging from small fractional-horsepower sizes to several hundred horsepower. Generally, electric motors are very reliable, have a high efficiency, operate quietly, and are easy to install, use, and maintain. They have a much lower power-to-weight ratio than engines, however, and are therefore less suitable for vehicles. Electric units have been used almost exclusively for production machinery, hand tools, and appliances, although there is a good deal of research to develop economical automotive applications because of environmental and fuel considerations.

Extensively standardized with respect to speed, horsepower, and frame size, electric motors can be selected to suit a given design. Usually, the designer will specify the next larger size above the calculated value to allow a margin of

---

[7]In this book we will henceforth refer to electric motors simply as *motors*.

safety. Because of high initial cost, however, a great care must be exercised in selecting motors. A motor that is too large for its purpose will operate at low efficiency. A unit that is too small, on the other hand, will overheat, and its useful life will be greatly reduced.

Mechanical output available from the motor shaft at full speed is termed the *nominal* power rating. When solving problems involving power requirements, the designer must specify standard ratings and speeds according to Table 1-3. The tabulated speeds in Table 1-3 are nominal. Standards for alternating current electric motors allow as much as a 2% variation in speed from rated conditions. The designer must be aware of such limitations and take them into consideration.

Internal Combustion Engines

Engines are independent power plants with variable speed-torque characteristics and a high power-to-weight ratio. This is their main advantage. They are heat engines that burn their fuel internally. In this book we refer to internal combustion engines simply as engines.[8]

Of the more than 15 million engines produced annually in the United States, roughly 50% are automotive. The nonautomotive types are stationary industrial, marine, or small engines rated at less than 10 hp and used in lawn mowers, chain saws, small boats, motorcycles, and snowmobiles.

As a power source, the internal combustion engine is available in different types of cycles (two and four stroke) and cooling schemes (air, water), but is not actually standardized. Most engines have been designed for a specific, mass-produced mobile application. Although engines are available at low cost for other types of machinery, the driven machine must often be designed to suit the chosen power unit. Because the internal combustion engine is not self-starting, a clutching device must often be provided for disconnecting or lightening the load at low engine speeds to prevent stalling.

One of the disadvantages of engines as power sources is that they give rise to undesirable torsional vibrations. These vibrations are the result of power being delivered in pulses corresponding to the firing of individual cylinders instead of in a smooth, continuous power flow. Vibrations are particularly harmful if transmitted to the driven machine without suitable damping.

Chapters 10 and 19 on couplings and clutches will deal specifically with this problem as well as with starting and idling.

Air and Hydraulic Motors

These units are used where electric motors are impractical for lack of speed variation or because of wet or explosive environments, as in mechanical laundries and chemical plants. Air motors drive many power tools, particularly on mass production assembly lines. Hydraulic motors are used extensively in machine tools and lifting platforms and as the driving power to wheels in construction equipment.

These motors are either rotary or piston types and are powered by fluids (oil or air) under pressure. Compressed fluids provide a power medium somewhat similar to steam in performance but with one important difference—they can be used cold. Both types have very large starting torque and operate with little vibration, even at maximum speed. If stalled, both will resume turning without being restarted when the load is reduced within their respective power ranges.

Flywheels

The flywheel is unique in that it alternately functions as a power source and as a driven machine. It can be an integral part of power sources (engines) and production units. It can absorb tremendous amounts of energy at a very high rate—by increasing its rotational speed—and can give

[8]As used here, engines include reciprocating (piston-cylinder) spark ignition and compression-ignition (diesel) types.

TABLE 1-3 Standard Capacities and Speeds for Three-Phase, 60-Hz, Alternating-Current Induction Motors

hp	rpm	hp	rpm	hp	rpm	hp	rpm
0.25	1800	3	3600	25	3600	100	1800
	1200		1800		1800		1200
	900		1200		1200		900
			900		900		600
					600		450
0.50	1800	5	3600	30	1800	125	1800
	1200		1800		1200		1200
	900		1200		900		900
			900		600		720
							600
							450
0.75	1800	7.5	3600	40	1800	150	1800
	1200		1800		1200		1200
	900		1200		900		900
			900		600		720
							600
							450
1	3600	10	3600	50	1800	200	1800
	1800		1800		1200		1200
	1200		1200		900		900
	900		900		600		720
			600				600
							450
1.5	3600	15	3600	60	1800	250	1800
	1800		1800		1200		1200
	1200		1200		900		900
	900		900		600		720
			600				600
							450
							360
2	3600	20	3600	75	1800	300	1800
	1800		1800		1200		1200
	900		1200		900		900
	600		900		600		600
			600				450
							360

up this stored energy at an equally high rate, while slowing down.

1-9 EFFECTS OF VIBRATION

Mechanical vibration is motion that is usually unintentional and unwanted. A machine member is said to vibrate when it describes an oscillating motion about a reference point. Such oscillation occurs because of the dynamic effects of manufacturing tolerances, clearances, rolling and rubbing contact between machine parts, and out-of-balance forces in rotating and reciprocating members. Apparently insignificant vibrations can excite the resonant frequencies of some other structural parts and be amplified into major vibration and noise sources. In machine shafts resonant vibrations are known as critical speeds. Common remedies for resonant vibrations are altering speed, mass, or system rigidity.

1-10 TORQUE CHARACTERISTICS

A turning effort that causes, maintains, and changes rotation is called torque. Torque varies with speed, but rarely in a simple manner. When torque is plotted against speed, from start to full operating conditions, individual machines or groups of machines consistently display the same shape. Figures 1-3 to 1-5 show typical torque curves for common power sources and some commonly used driven machines. These unique curves provide a means for comparing torque output with required input and pairing drive units with driven units.

Generally, torque-speed curves are characterized by three levels: starting, accelerating, and running. *Starting torque* is that needed to start the driven machine or that delivered by the driving unit at start. *Accelerating torque* is that needed to bring the driven machine to running speed. *Running torque* is that needed to drive the machine continuously after attaining full speed.

Because of added friction at start plus large inertia forces to set the mass in motion, the starting torque of most driven machines exceeds the accelerating and running torque. The starting torques are therefore the ones to match when pairing driving and driven units. As shown in Fig. 1-3, commonly used power sources have either a high or low starting torque, as do most driven machines (Fig. 1-4). Power sources should therefore always be matched with driven machines of the same or lower characteristics.

Electric motors, with their high starting torque, generally match all types of driven machines, give quick starts, and allow a simple shaft connector or coupling without a clutch between driver and driven shafts. By contrast, the low starting torque of engines requires a more complicated shaft connector (a clutch) so that the engine can gain speed and therefore power prior to engagement. Figure 1-5 shows the time delay provided by a clutch when engines are used in vehicles. The starting torque, which may be 25 to 300% greater than the running torque, can be estimated as follows once the running torque is known.

$$T_s = k_r T_r \tag{1-11}$$

where

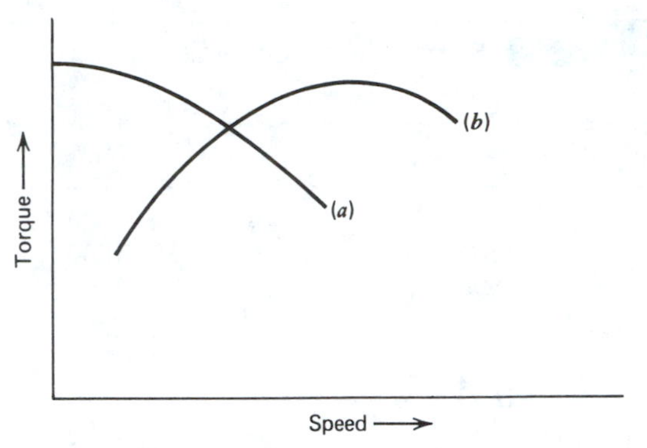

Figure 1-3 Speed-torque curves for power sources. (*a*) Typical for turbines and electric motors. (*b*) Typical for engines.

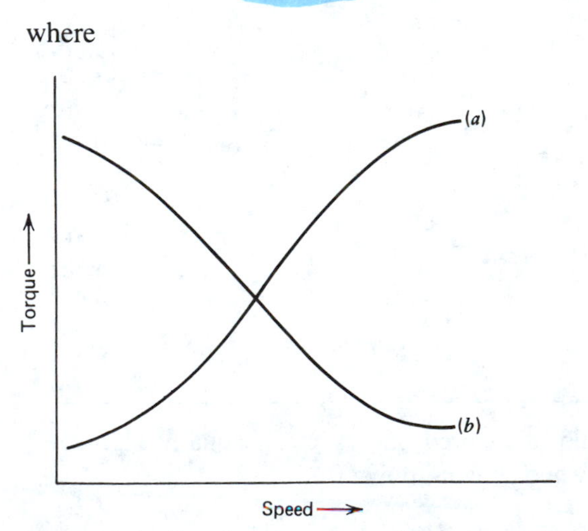

Figure 1-4 Speed-torque curves for various driven loads.
(*a*) Typical for fans, propellers, and pumps.
(*b*) Typical for trains, vehicles, cranes, and hoists.

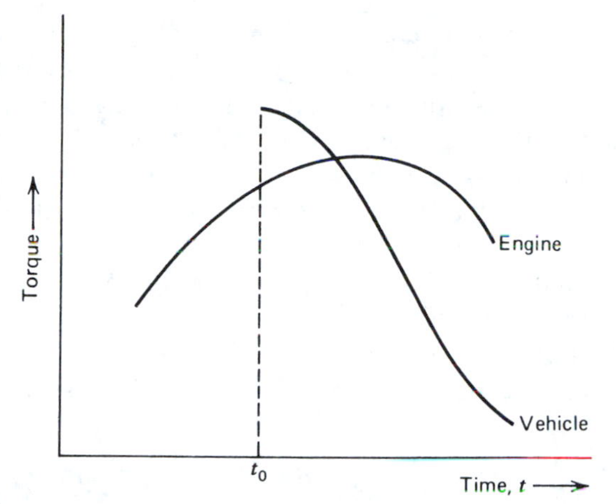

Figure 1-5 Time-torque curve for common vehicles; t_o = clutch engagement delay.

T_s = starting torque; N · m, ft-lb
k_r = running torque multiplier (obtained from Table 1-4)
T_r = running torque; N · m, ft-lb

The running torque multiplier is always greater than one and is determined from experience. Estimating the starting torque is important because it often determines the size of the power source.

If power and speed are known, torque can be determined by calculation. If several speeds are used, choose the lowest, because that is generally where the torque requirement will be greatest.

Most power sources will produce torque in excess of their nominal values, although not for long periods of time. For instance, the output torque of electric motors at start may momentarily reach values of three to five times nominal torque (Fig. 1-3). Although excess torque capabilities aid in overcoming inertia during periods of acceleration, they can also raise temperatures unduly and overload couplings, clutches, shafts, or gears. The designer must recognize these potential hazards and allow for them.

TABLE 1-4 Starting Torque Characteristics of Various Machines

Types of Applications	Running Torque Multiplier
General machines with ball and roller bearings	1.2–1.3
General machines with sleeve bearings	1.3–1.6
Conveyors and machines with excess sliding friction	1.6–2.5
Machines that have "high" load spots in their cycle (e.g. some printing and punch presses and machines with cam- or crank-operated mechanisms)	2.5–6.0

Source. "ZERO-MAX,® Variable Speed Drives," Zero-Max Industries, Inc., Minneapolis, 1973 (with permission).

EXAMPLE 1-4

The running torque of a stamping machine was measured at 100 N · m. Estimate the starting torque and the motor size.

SOLUTION

This machine has peak load zones and subsequent average torque multipliers of 2.5 to 6. Therefore,

$$\text{Minimum starting torque} = (2.5)(100) = \mathbf{250\ N \cdot m}$$

$$\text{Maximum starting torque} = (5)(100) = \mathbf{500\ N \cdot m}$$

Because output of electric motors at start may reach values of three to five times nominal torque, power requirements can be based on a torque value of 100 N · m. For a 1200-rpm motor, this corresponds to roughly 15 hp.

$$P = \frac{Tn}{9550}\ \text{kW} = \frac{(100\ \text{N}\cdot\text{m})(1200\ \text{rpm})}{9550}\ \text{kW}$$

$$= (12.57\ \text{kW})(1.34\ \text{hp/kW}) = \mathbf{16.8\ hp}$$

Reaction Torque

When a motor or engine produces torque, the reaction torque T_R tends to rotate the base or housing, which must therefore be constrained. The same, of course, holds true when a driven machine resists torque; it, too, must be constrained to prevent rotation. Thus reaction torque is an important design factor.

EXAMPLE 1-5

Estimate the maximum reaction torque exerted by a 10-hp, 1200-rpm motor driving a belt conveyor.

SOLUTION

The maximum reaction torque T_R is equal and opposite to the starting torque T_s. The starting torque is the running torque (T_r) times the maximum torque multiplier. Thus,

$$T_r = \frac{(10 \text{ hp})63{,}000}{1200 \text{ rpm}} = \mathbf{525 \text{ in.-lb}}$$

According to Table 1-4, the maximum torque multiplier for conveyors is 2.5. Hence,

$$T_R = -(525 \text{ in.-lb})2.5 = \mathbf{-1313 \text{ in.-lb}}$$

1-11 SERVICE FACTORS AND DESIGN POWER

Normal or nominal load demands of a driven machine are generally defined in terms of torque or power. However, the inherent characteristics of the power source, the driven machine, and the interaction of the two create conditions that generally make the actual load greater by 10 to 300% than that obtained from the power equations in Section 1-7. Extra loads may be due to vibratory torques, shocks, pulsations, or reversal of torque loads. Duration of service is also a factor.

Because reliable, on-the-spot measurements are difficult to conduct, component manufacturers have established *service factors* (experience factors) for estimating equivalent loads on common parts such as couplings, clutches, brakes, belts, chains, gears, or shafts. In each case the nominal power to be transmitted is multiplied by such a service factor to obtain an equivalent rating for the application. If the part is to be purchased, the value thus obtained must then be checked against manufacturers' catalog ratings to find a size rated above the calculated value.

$$P_d = k_s P_r \tag{1-12}$$

where

P_d = design power; kW, hp
k_s = service factor
P_r = power requirement; kW, hp

The power requirement or nominal power is obtained from the power equations.

Generally, service factors are numbers greater than 1 but less than 5. Service factors are deratings because they reduce the permissible capacity of the machine member being selected or calculated for strength. When specific service factors are not available, a reasonable approximation can be obtained through interpolation on the service factor diagram in Fig. 1-6.

This ingenious diagram covers broadly the entire field of major dynamic operational conditions encountered in machinery. It incorporates effects from all common power sources and 16 different operating conditions under four headings, thus covering hundreds of realistic industrial applications.

The diagram is a model of the different power sources driving a variety of machines. Once a starting point has been chosen according to type of power source, the four major effects are accounted for, in each case by following the general direction of the inclined lines. To get from the end point in one category to the starting point in the next category, one merely transfers the point horizontally.

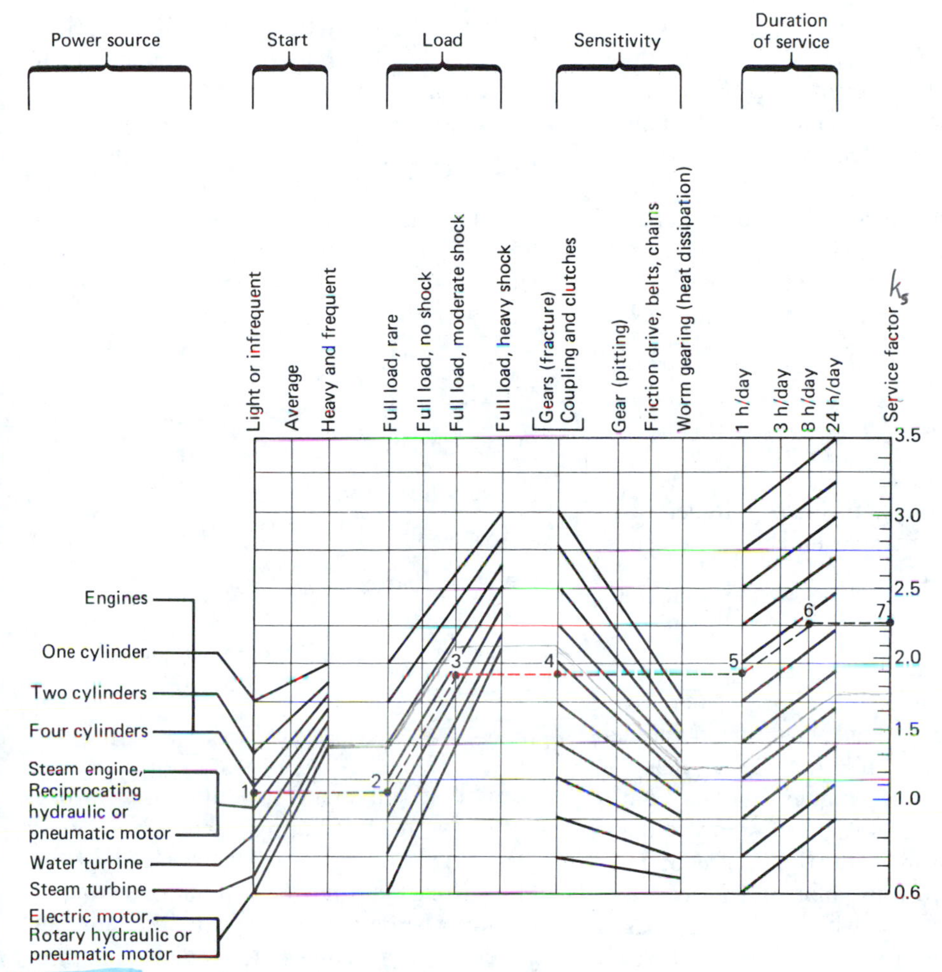

Figure 1-6 Service factor diagram. Service factors help to ensure optimal service life of machine components. (Developed by W. Richter and H. Ohlendorf.)

As might be expected, the service factor increases when:

1. Smoothly running power sources based on rotary motion are replaced by reciprocating engines.
2. The starting load becomes heavier and the number of starts increases.
3. The frequency of full-load conditions and shock phenomena increases.
4. The duration of services increases from 1 to 24 h/day.

Only the curves for sensitivities have a negative slope, thus effectively indicating relative smoothness of operation for the seven most common means of power transmission. Spur gears, for instance, require a service factor nearly twice that of worm gearing and 50 to 60% higher than those of belt and chain drives.

If the number of hours per day is not given, assume an 8-h day. Increasingly, however, the 8-h day for machinery is becoming a 16- or 24-h day. Automation and the high initial cost of large or complex machines often mandates around-the-clock operation.

The original service factor diagram (Fig. 1-6) did not contain references to hydraulic and pneumatic motors. When these are of the rotary type, they probably rate with electric motors; the piston types are more likely to rate with steam engines.

EXAMPLE 1-6

Find the service factor for a six-cylinder engine driving a generator in an amusement park.

SOLUTION

A six-cylinder engine runs more smoothly than a four-cylinder engine; hence the point of entrance, shown as 1 in Fig. 1-6, falls below that of a four-cylinder engine. Start is light and infrequent, so we do not follow any of the inclined lines upward to the right; instead, we move straight across to point 2. The load is judged to be full, but with moderate shock only. We therefore follow the general direction upward to the right until we intersect the vertical line representing "full load, moderate shock." We now have point 3. The engine is, no doubt, coupled directly (in-line) to the generator with a simple coupling, so we follow a horizontal line from point 3 until we intersect the vertical line for couplings. We now have point 4. Since couplings represent the first vertical line, there is no need to follow any of the inclined lines. Point 5 is thus reached by moving straight across from point 4. Assuming an 8-h day, we move upward to the right in the general direction indicated by the inclined lines. Where the dotted line intersects the vertical line for the 8-h day, we obtain point 6. The service factor, roughly 2.3, is obtained by going horizontally from point 6 to point 7, as shown.

EXAMPLE 1-7

A motor drives a piston-type air compressor by means of V-belts. The compressor delivers air to a shop. The shop operates 8-h/day, 5 days/wk. Find the service factor.

SOLUTION

Operating conditions are:

Start: heavy and frequent.

Load: full load, moderate shock.

Sensitivity: belts.

Duration of service: 8 h.

Service factor: **1.80** approximately.

SUMMARY

A machine is a device for doing useful work, such as manufacturing and transporting goods. Depending on the nature of its function, a machine may be simple or complex, fragile or strong. Regardless of function and structure, all machines are composed of simple elements.

Machine elements can be combined to form *simple* machines that are far more versatile than the individual element. These, in turn, can be combined to form systems that are more versatile than the individual machine could ever be. Complex machines can be analyzed by resolving them into simple machines first and then elements.

Machines designed to increase force at the expense of speed have a mechanical advantage (MA > 1). Machines designed to increase speed at the expense of force have a mechanical disadvantage (MA < 1). Machines are not 100% efficient. Some effort is always used to overcome friction. Output is thus always less than input.

Machines are dual in their makeup in that they have a power unit driving a power-absorption unit. Although power units are limited to a few basic types, variety in driven machines is virtually unlimited.

When a power source is coupled to a driven machine, the two torque curves should match; otherwise, starting will be adversely affected. When they do not match, a clutching device becomes necessary.

The starting torque for most machines is 25 to 300% greater than the running torque. From the running torque (easy to measure), the starting torque can be obtained by means of the running torque multiplier. Reaction torque often makes it necessary to anchor machine frames firmly to their base.

The operating loads in machinery are often much greater than the nominal loads due to vibratory torque, shock, pulsations, or reversed torque. They can be estimated with reasonable accuracy by means of service factors provided by equipment manufacturers. Thus the design power can be obtained by multiplying the power requirement by a suitable service factor. The service factor diagram by Richter and Ohlendorf is unique in that it covers practically the entire field of major dynamic operational conditions encountered in machinery.

The most dramatic improvement in machinery was the introduction and rapid development of automatic operations—machines directed by machines instead of humans. Termed automation, this development has recently reached new heights with the advent of the computer.

REFERENCES

1-1 Bosch, Robert, Inc. *Automotive Handbook*. Stuttgart, 1976. (Translated from German, this best-seller contains a wealth of useful information. Available from the Society of Automotive Engineers in the United States.)

1-2 Brooks, Eugene. "Air Motors Challenge Electrics," *Machine Design,* September 2, 1978. (This well-known publication is our most frequently used reference and will from now on be referred to merely as *Machine Design*. Its *Reference Issues* are published yearly and contain basic information as well as the latest developments.)

1-3 Bureau of Naval Personnel. *Basic Machines and How They Work*. New York: Dover, 1971.

1-4 Franks, D. G. K. "Wedge Analysis," *Machine Design,* October 12, 1978. (Franks' article goes beyond the typical textbook treatment by involving two or three friction pairs in real-life applications.)

1-5 *Machine Design. Electric Motor Reference Issue,* 1980.

1-6 *Machine Design. Mechanical Drives Reference Issue,* 1980.

1-7 Richter, W., and H. Ohlendorf. *Kurzberechnung Von Leistungsgetrieben*. Konstruktion in Maschinen-Apparate und Geratebau. Heft 11, 1959.

PROBLEMS

Problem solving constitutes one of the best means for clarifying the basic principles of machine design and fixing them in your mind. An effort has been made to present realistic problems in all chapters. However, many true engineering problems tend to be imprecisely or incompletely formulated, difficult to visualize, often have multiple answers, and are frequently beyond the scope of textbooks.

P1-1 Make a list of the machines that serve you directly and indirectly in the course of a day, a week, and a year. Classify them as either simple (basic) or complex (compound) machines.

P1-2 Which basic machines can be identified

in the following complex machines: (*a*) bicycle; (*b*) motorcycle; (*c*) chain saw; and (*d*) lawnmower?

P1-3 List at least six basic machines and several compound machines found in your automobile.

P1-4 The speed of a roller chain is 1000 ft/min. The net driving force is 99 lb. (*a*) Find the power transmitted in horsepower. (*b*) Change the given data to SI units and calculate the power transmitted in kilowatts. (*c*) Check your calculations by means of appropriate conversion factors.

P1-5 The starting torque of a small, general type of driven machine equipped with ball bearings was found by the "torque wrench method" to be roughly 25 N · m. What would be the running torque? For a speed of 1200 rpm, find the power required.

P1-6 Explain the purpose and function of the transmission and the differential in your automobile. Which basic machines are involved?

P1-7 Why are the pulse variations of electric motors and turbines less severe than those of engines? Why do engines require flywheels while electric, hydraulic, and pneumatic motors do not?

P1-8 Cars and propeller-driven aircraft are often powered by engines, yet only cars are equipped with a clutching device. Explain why. Show torque-speed curves.

P1-9 Estimate the starting torque of a conveyor when the running torque is 60 N · m. Estimate the reaction torque.

P1-10 Calculate the approximate efficiency of an ordinary bicycle, using data from Table 1-2. Make a simple sketch of a bicycle and indicate the parts affecting overall efficiency.

P1-11 A go-cart powered by a small engine has a belt at the engine and a chain at the rear axle. The drive sprocket and the driven sheave are both mounted on the same shaft (Fig. 1-7). Find the efficiency of this power train when the shaft and axle are mounted in roller bearings. Calculate the design power for a 1-hp requirement at the rear axle.

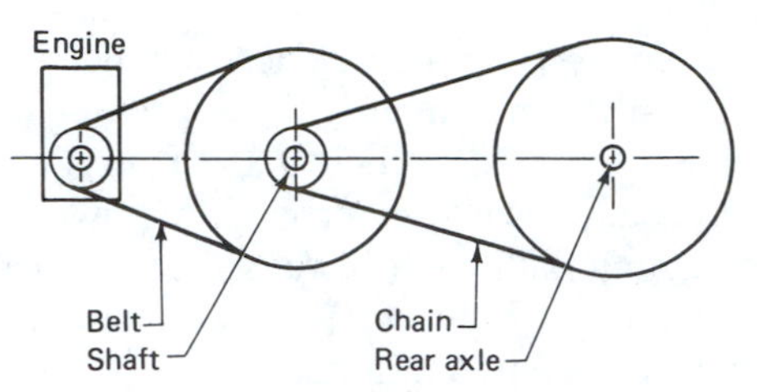

Figure 1-7 Go-cart. (Problem 1-11)

P1-12 See Fig. 1-1. Assume mechanical energy generated in the prime mover is roughly 100 kW. Find the power available for doing useful work when 10% of the available output energy is absorbed by auxiliary equipment. Because of damping, the coupling has an efficiency of 0.99. The transmission has two pairs of spur gears in mesh (assume two sets of sliding bearings in the transmission). (Fig. 1-8).

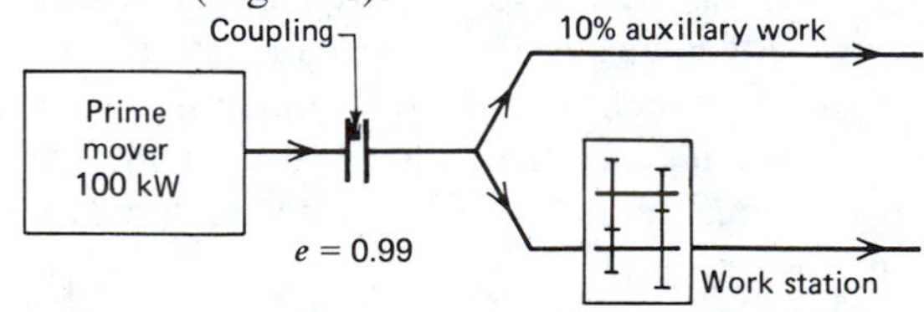

Figure 1-8 Power-work unit. (Problem 1-12)

P1-13 A passenger elevator for an office building can lift its maximum rated load of 7000 lb at a rate of 60 ft/min (Fig. 1-9). Find the service factor for the drive, which consists of an electric motor and a worm gear speed reducer. Select a motor. (*Hint:* Use Eq. 1-9, Fig. 1-6, and Tables 1-2 and 1-3.)

P1-14 An 1800-rpm electric motor drives a belt conveyor through a worm gear speed reducer and flexible coupling.

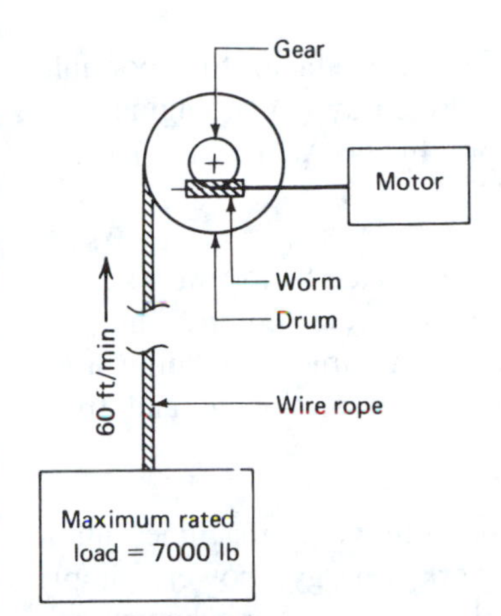

Figure 1-9 Elevator. (Problem 1-13)

The belt speed is 3 m/s, and the net driving force is 1200 N. If the combined efficiency of the speed reducer and coupling is 0.78, specify the standard-size motor required. The conveyor serves a limestone quarry (Fig. 1-10). Duration of service: 8 hr/day, 5 days/wk.

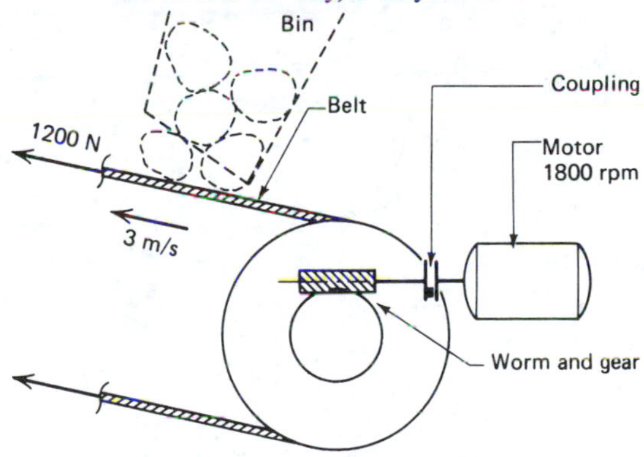

Figure 1-10 Belt conveyor. (Problem 1-14)

P1-15 The thermal (heat) efficiency of a 25-kW, 3000-rpm diesel engine on the test stand was 0.39. The mechanical efficiency was 0.78. (*a*) What is the total efficiency at 3000 rpm? (*b*) What is the thermal value in kw of the fuel consumed at 3000 rpm?

P1-16 The *imported* band saw shown in Fig. 1-11 is to be equipped with a *domestic* electric motor. This tool is characterized by having a taut, endless band of steel, with saw teeth on one edge, that passes over two large, flat pulleys or wheels, of which the lower is the driver. The manufacturer recommends an operating speed of 18 to 19 m/s and a minimum power requirement of 3.5 kW.

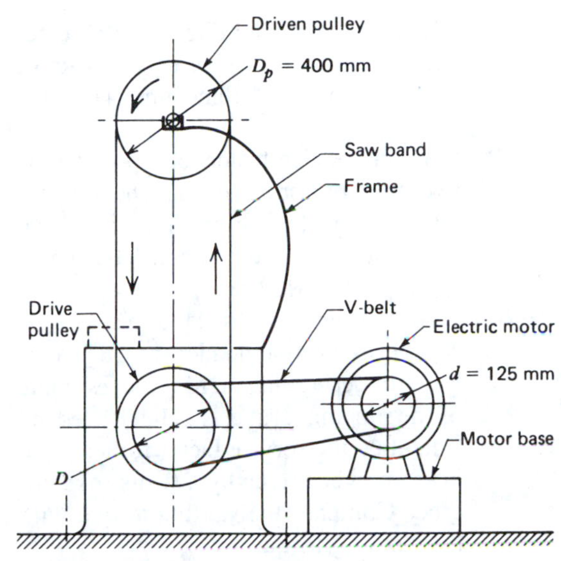

Figure 1-11 Band saw. (Problem 1-16)

(*a*) Calculate the speed of the drive pulley and diameter D of a driven V-belt pulley for a motor speed of 1200 rpm.

(*b*) Select a suitable motor based on intermittent operation and an 8-h working day.

(*c*) Calculate the cutting force F_c, which the saw blade exerts on the work piece. Assume 95% efficiency.

P1-17 An air motor of the piston type drives a conveyor belt in a mechanical laundry by means of a roller chain. The conveyor operates intermittently during an 8-h day. Estimate the service factor.

P1-18 A water turbine of the vertical type

(Kaplan) is coupled directly to an electric generator. Find the service factor, assuming that the turbine runs 24 hrs/day.

P1-19 Find the service factor for a piston-type hydraulic motor driving a planetary gear mounted inside a large wheel on a piece of construction equipment. The wheel has a pneumatic tire.

P1-20 A diesel engine drives a piston-type compressor with unloader valve 8 hr/day. Find the service factor. Make an estimate of the starting torque. Is a clutch needed?

P1-21 Calculate the service factor for a simple lawn mower driven by a two-stroke engine. Compare the starting torque with the running torque.

P1-22 Estimate the service factor for a portable chain saw driven by (*a*) an engine, and (*b*) a motor. 3 hr/day. Which design needs a clutch?

P1-23 Estimate the service factor for a single-cylinder motorcycle engine and its transmission. Assume it is being used primarily for commuting to and from work.

P1-24 Define the following: machine, automation, work, energy, power, simple and complex machines, mechanical advantage, torque, machine element, machinery, service factor, running torque factor, nominal power, design power.

Chapter 2
The Many Aspects of Machine Design

> There is no expedient to which man will not resort to avoid the real labor of thinking.
>
> THOMAS EDISON

This chapter introduces the various aspects of design and some of the procedures available to designers. Special attention will be given to computations and the use of nomographs.

2-1 THE BASIC ASPECTS OF DESIGN

Machine design is the art of envisioning, creating, and developing mechanical devices that will perform some desired function. It is the application of science and inventiveness to the conception, construction, and refinement of machinery. The basic concepts of machine design are:

1. Design concept.
   (a) Systems engineering (function, safety, reliability, maintainability).
   (b) Design of elements (size, shape, material, life).
2. Manufactureability.[1]
3. Cost (planning, materials, construction).

Systems engineering is concerned with proper functioning of the various parts of a machine within the total system. Machine parts are independent inasmuch as their properties can be defined without reference to other parts. A product composed of such separable parts is called a *system*. Systems engineering is thus concerned with the collective action of parts. It is not, however, concerned with the design of machine elements, but considers them as having certain specifiable properties. Because the main objective of this textbook is element design itself, systems engineering will be discussed only in terms of the function of major parts such as belts, chains, couplings, and gears.

Element design is concerned with the proper sizing of machine elements to perform a given function at some stated life criterion. Mechanical designers must also be familiar with properties of materials and machining processes to achieve optimal design. In addition, designers must always contend with the question of cost. The watchword should be simplicity, since a simple device is usually the least expensive.

Beyond these initial responsibilities, designers also have three critical responsibilities: safety, reliability, and maintainability. The automobile, for instance, must be designed with safety, reliability, and maintainability considered for every part and function if rising service costs and increased public safety standards are to be met. When a machine is down for repair, the owner loses production time. For some machines, including jet aircraft, the cost of lifetime maintenance may be two to three times the original production cost of the machine.

[1]Manufactureability is the ease or difficulty with which a part can be made.

2-2 HUMAN FACTORS ENGINEERING

Human factors engineering is concerned with all aspects of the man-machine relationship. Its purpose is to ensure that power tools, machinery, and manual controls are designed to accommodate the operator with safety, comfort, and efficiency.

The absence of human factors engineering explains why some power tools are cumbersome, ineffective, or dangerous. A close look may reveal that the better tool is more effective because it has low vibration, is light in weight, fits the operator's hand, or follows the natural movements of the body, especially the hands and arms. Human factors engineering has been applied extensively to all types of machinery and is a major reason why modern equipment is generally safer to operate than machinery of the past. Nowhere has this art been applied with greater success than in space vehicles, which literally are built around the astronauts. In short, by incorporating the human factor into a design, product efficiency can be improved and numerous potential operational problems can be avoided.

2-3 DEFINITION OF DESIGN

Design is the first step in manufacturing. Before a product can be manufactured, the engineering department must prepare a complete set of drawings and specifications for all component parts. Machine design is a means to this end.

Machine design is the process of creating something entirely new or significantly improved. It may result in a new component that becomes part of an existing machine to help the whole assembly perform more efficiently, or it may (rarely) lead to an entirely new machine. Therefore machine design is essentially a problem-solving process that may involve mathematical and graphical methods as well as imagination and three-dimensional visualization.

Design also requires familiarity with materials of all kinds and with basic manufacturing processes and machine shop practice. Machine design is an art, an activity, and a function, not merely a body of knowledge. As such, it is the main purpose of engineering.

2-4 THE DESIGN PROCESS

In designing machines one must first form a clear mental picture of the part, the machine, or the product to be made. The idea must then be conveyed to others through verbal and written instructions and through drawings. The engineering drawing is an indispensable aid to designers. It is a pictorial representation to which they confide their thoughts for later transmission to those who will build their machines. A technical drawing is an enormously condensed document and must therefore be precise.

Machine design differs from most of its supporting sciences, notably physics and mathematics, in that most problems generally have more than one solution. The automobile (primarily a device for transporting people) illustrates this principle. Many models are available on the market, and one may be as good a mechanical and aesthetic solution to the basic problem as another.

The solution to most design problems does not arise from a set of equations; instead, it is a compromise to satisfy a number of design requirements and practical limitations such as available tooling and servicing ease. Designs are often revised to introduce new features, but as much as possible of the old design is retained for economic reasons. Producing a revised design is usually not as difficult as producing a new design because the history of the original is available for evaluation. In a new design certain assumptions must first be made concerning operating conditions and requirements. Then a tentative design is developed based on these assumptions and available standards. From the tentative design, a prototype may be built and tested. Based on test data, the part or machine may be judged ready for production or may be redesigned and a second prototype built and tested. This process will be repeated until a satisfactory solution is obtained.

The following steps are necessary considerations in the analysis of a design problem.

1. A *kinematic* analysis to establish the necessary motion requirements.
2. A *force* analysis to establish the magnitude of the acting forces and moments.
3. A *strength* and *rigidity* analysis to establish the basic dimensions of machine members.

Kinematics is the study of motion requirements without regard to forces. Force analysis actually involves a dual consideration. It must include the nominal driving forces and the starting, stopping, and impact forces that might be encountered. Strength and rigidity draw heavily on the disciplines of statics and strength of materials as they apply to machine members.

Size and Space Occupied[2]

The art of design consists primarily of utilizing space effectively. Space is at a premium in almost any design, since space often represents money. In many new plants the cost per square foot may run as high as $40 to $50. Thus the machine that occupies the least floor space and still does the job is preferable. For mobile machines, road space, track space, or parking and

---

[2]See Fig. 9-6 and Table 9-1, column 6.

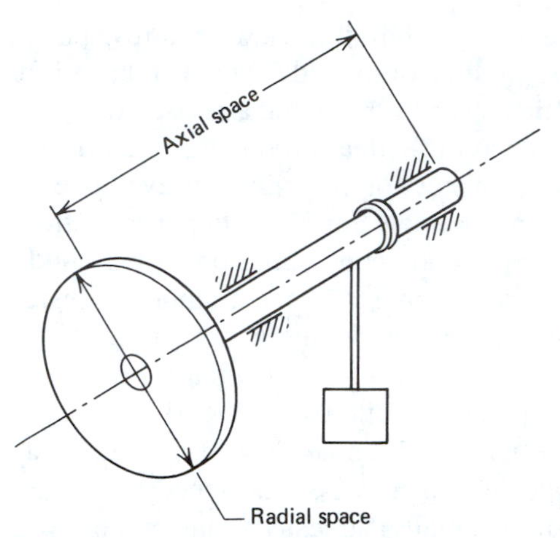

Figure 2-1 Axial and radial space.

storage space are all expensive and limited. Furthermore, a compact design yields maximum payload in any mobile machine. Utilizing the third dimension—height—is an important design consideration, since height generally is less restricted than area space (length and width). Limitations on height, however, may be imposed because of operator comfort or stability.

In discussions of various machine members or an entire design two additional expressions are often encountered: *axial* and *radial* space. A long, slender machine shaft, for instance, requires large axial space and little radial space. Most wheels, by contrast, need little axial space but large radial space (Fig. 2-1).

2-5 MODES OF FAILURE

Design for reliability involves many considerations because machine parts malfunction in many different modes. Failure does not necessarily mean that a part breaks; it simply means that it no longer functions as intended. For example, a gear may fail when a tooth breaks off; more often, however, failure is due to excessive wear. Wear, in turn, may be the cause of intolerable noise levels, damaging vibration, or loss of lubricants. Experienced designers consider all possibilities by asking:

1. How can this part fail?
2. Which mode or modes of failure are most likely to occur?
3. How can I design against these modes?

In gearing, for instance, the prevalent modes of failure are tooth breakage and wear (pitting). Gear calculations are aimed at preventing both. There are four basic modes of failure.

1. Lack of strength (rupture and surface destruction).
2. Lack of rigidity (excess elastic deflection).
3. Lack of stability (buckling or overturning).
4. Wear (removal of vital surface material).

Strength is the ability to resist loads (forces, bending moments, torques) and is expressed in terms of ultimate strength, yield strength, and fatigue strength.

Rigidity is the ability to resist change of form. In machinery it ensures accuracy and precision. Lack of rigidity leads to interference between parts and premature failure due to wear and fatigue.

Stability or steadiness is the ability of a part to resist displacement and, if displaced, to develop forces and moments that tend to restore the original condition. As a design criterion, stability relates primarily to machine columns, struts, coil springs, or pushrods. These compression members can all fail in buckling, characterized by excess transverse deflection or collapse.

Wear is the gradual, unintentional abrading of surfaces in contact as a result of relative motion between them.

Common modes of failure are shown in Fig. 2-2. The beam in Fig. 2-2*a* may deflect excessively or, if too highly stressed, the material may yield. Under cyclic loading the material may fail in fatigue (fracture). The tension member in Fig. 2-2*b* may stretch excessively; in case of overload, it will deform permanently or fracture.

The bar in Fig. 2-2*c*, loaded in torsion, may have excessive angular deflection. If it is over-

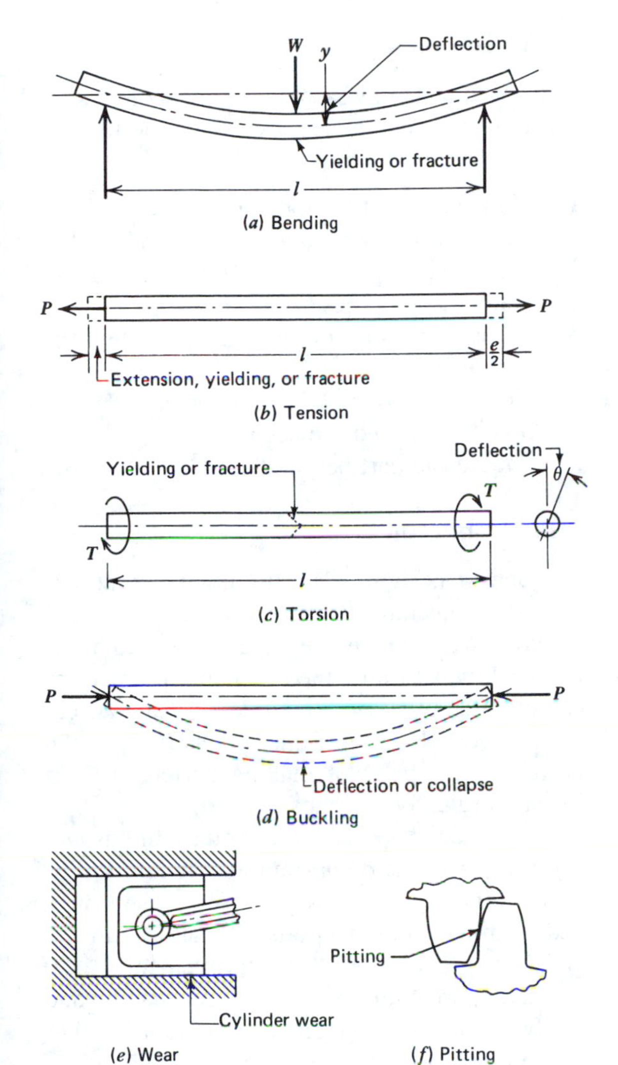

Figure 2-2 Common modes of failure of machine members.

loaded it will deform permanently or rupture. Cyclic overloading will cause fatigue fracture. A long compression member (Fig. 2-2*d*) will deflect transversely (buckle) if overloaded. When buckling occurs, the member can no longer carry a load.

Wear due to sliding is shown in Fig. 2-2*e*. Excess loss of material from both parts leads to reduced compression pressure and increased oil consumption. Figure 2-2*f* shows wear (pitting) due to rolling friction.

For safety and economy, machine members are often designed to fail in one mode in preference to another. Bolts, for instance, are designed to fail in tension, not in shear. Roller chains, in contrast, are designed to wear out instead of fail in tension, since such failure could have catastrophic consequences.

2-6 DESIGN CRITERIA

Even though machine parts fail in different modes, design against failure involves basically the same principal consideration.

A COMPARISON OF WORKING CONDITIONS WITH ALLOWABLE CONDITIONS

Strength: Maximum induced stress ≤ allowable or working stress.

Rigidity: Maximum operating deflection ≤ allowable elastic deflection.

Stability: Maximum load ≤ safe load.

Wear: Maximum wear ≤ permissible wear.

Induced, or calculated conditions to the left of the inequality often require the use of engineering disciplines, notably statics and dynamics. Field or laboratory testing may be necessary to establish actual load conditions from which stresses and deflections may be obtained. The limiting conditions to the right of the inequality depend on material properties, especially strength.

2-7 PROBLEM-SOLVING STRATEGY

No textbook can adequately cover the many individual situations of machine design. Designers must be knowledgeable about the basic principles and then develop their own methods and design philosophies to solve a specific problem.

Four-Step Approach

There is, however, a practical approach to problem solving that has been found to be most useful and quite flexible. Its simplest form involves four steps.

1. *Statement of Problem.* What is it all about? Is it a real problem? Is there a real need for a solution? A problem well stated is half solved.
2. *Analysis.* What is the cause of the problem? Improper development? A human error? A technical error? A need for greater performance?
3. *Illumination* (inspiration, imagination). What solutions are possible? What are the options?
4. *Decision.* Which solution is preferable? Which option has the most to offer?

Certain design problems require some hard thinking, in which case it becomes necessary to outline in writing the answers to each of the four steps.

In selecting the best solution in step 4, a balance sheet approach can be used. Divide a sheet of paper into two vertical columns. Place advantages on one side and disadvantages on the other. Do this for each solution or option. Faced with available facts, designers will find it easier to make decisions.

Value Engineering

The system of value engineering is designed to lower manufacturing costs without affecting product function, quality, or reliability. In a typical value engineering program a product is resolved to its smallest components to determine if cost-saving changes can be made in design, materials, and manufacturing. Value engineering is especially suitable for fastening and assembly because about 50% of all manufacturing costs are due to these two factors.

Checklist

A more direct approach to design consists of checking each component by asking the following questions.

1. How does it function within the system? Under what conditions must it function?
2. How well do we know loads, speeds, and material properties?
3. What is the mode of failure? How safe is the part against failure?
4. How can the part best be manufactured, maintained, and replaced?
5. Should the part be purchased?

Troubleshooting

In machine design, as in any other human endeavor, things rarely go as planned. The "bugs" are not always where one expects them to be. If the machine is still on the test stand, corrections can often be made with relative ease and at reasonable cost. However, once it has reached the customer, the price of making corrections has increased greatly (perhaps tenfold).

The remedial action is then referred to as *mechanical troubleshooting:* an exercise in symptom diagnosis. Since its purpose is to identify and eliminate malfunctions, it can be an invaluable source of product improvement.

The imperfections of new designs may be due partly to human negligence or ignorance, but they are primarily due to the enormous complexity of modern machines. The interacting factors are so numerous that the probability of covering all potential failures is small. Troubleshooting is therefore inevitable.

2-8 IMPORTANCE OF COMPUTATIONS

The purpose of calculations is primarily to confirm or predict:

1. The necessary *dimensions.*

2. The expected *effect* such as useful life, deflection, and so forth.
3. The *advantage* of one solution over another (e.g., the use of one material and not another).

An accurate record of the steps that led to a certain design can prove invaluable if it is ever necessary to backtrack to locate an error or support a patent application.

Engineering standards and codes enter into nearly all calculations. Standards evolved from a need to avoid repetitive calculations and simplify manufacturing of parts. Standard parts provide the savings inherent in high production. They guarantee the interchange of parts on the spot, with a minimum of downtime. A standardized machine member is also a known quantity. For instance, the load capacity of a standard bolt can be obtained from tables in engineering handbooks. For the neophyte in design, standards may seem to be a network of obstacles but, once understood, they are the designer's best friend.

The use of preferred numbers greatly facilitates standardization. Preferred numbers are a series of numbers used for purposes of standardization. Such series are geometric, not arithmetic. Therefore each size is larger than the preceding one by a fixed percentage. The range of a product should include the number of sizes that will adequately cover all possibilities or needs. For further details, consult *Machinery's Handbook,* p. 116.

2-9 ERRORS IN COMPUTATIONS

Designers should realize that most technical calculations are not exact. Many formulas used in calculating stress and/or deflection in machine members are merely approximations to actual physical conditions. They are mathematical models of typical conditions. A mathematical model is an idealized representation of the real mechanical system, simplified so that it can be analyzed more easily. The very figures used in these formulas may also contain errors—frequently errors of measurement.

Two types of computation errors must be considered: absolute and relative.

$$\text{Absolute error} = \text{measured value} - \text{exact value} \tag{2-1}$$

If the measured value is too large, the error is positive; if it is too small, the error is negative.

$$\text{Relative error} = \frac{\text{absolute error}}{\text{exact value}} \tag{2-2}$$

As an example, consider a round shaft of a uniform diameter d.

$$\begin{aligned} d\ (\text{nominal}) &= 50.00\ \text{mm} \\ d\ (\text{measured}) &= 50.10\ \text{mm} \\ \text{Absolute error (AR)} &= 50.10 - 50.00 = 0.10\ \text{mm} \\ \text{Relative error (RE)} &= \tfrac{0.10}{50} = \tfrac{10}{5000} \\ &= 0.0020 = 0.20\% \end{aligned}$$

In technology the relative error is frequently more important than the absolute error. The former provides a figure (in percent) that is easily compared with known or specified standards. Relative errors of 2 to 3% or less are generally acceptable in machine design.

2-10 REVISING COMPUTATIONS

Design is an evolutionary process with many changes involving size, material, and force diagrams. Often these changes are minor, repetitive, and involve only one or perhaps two variables out of many. Although electronic calculators have greatly eased the task of repetitious calculations, there are shortcuts, particularly when the changes are within 10% of the original values. The basic relation is:

$$\text{Relative change} = \frac{\text{new value} - \text{old value}}{\text{old value}} \tag{2-3}$$

$$\mathrm{RC}(Q) = \frac{Q_1 - Q_0}{Q_0} \qquad (2\text{-}3a)$$

where Q is a quantity subject to any of the six mathematical operations listed in Table 2-1. If Q contains only one variable A with a relative change RC(A) = a, then RC(Q) is simply a. Of greater practical importance is the more complicated case of Q being a function of two variables A and B with corresponding relative changes of a and/or b. This may involve addition, subtraction, multiplication, division, and powers of A and B. Corresponding relative changes of Q are shown in Table 2-1. Note the simplicity of relative changes when multiplication, division, and powers are involved. Fortunately, most practical applications involve these operational changes.

TABLE 2-1 Guide to Estimating Relative Changes

Q	RC(A)	RC(B)	RC(Q)
$A + B$	a	0	aA/Q
	a	b	$(aA + bB)/Q$
$A - B$	a	0	aA/Q
	a	b	$(aA - bB)/Q$
AB	a	0	a
	a	b	$a + b$
A/B	a	0	a
A^n, $n > 1$	a, $a < 10\%$		na
$A^{1/n}$, $n > 1$	a, $a < 10\%$		a/n

EXAMPLE 2-1

The centrifugal force F of a mass m moving in a circle of radius r with an angular speed ω (omega) is:

$$F = mr\omega^2 \qquad (2\text{-}4)$$

If the angular speed ω changes by 5%, how much does the centrifugal force F change?

SOLUTION

$$\mathrm{RC}(F) = \mathrm{RC}(mr\omega^2)$$

Since m and r do not change:

$$\mathrm{RC}(F) = \mathrm{RC}(\omega^2)$$

From Table 2-1 for $Q = A^n$,

$$\mathrm{RC}(Q) = na$$

$$n = 2$$

$$a = 5\%$$

$$\mathrm{RC}(F) = 2\,\mathrm{RC}(\omega)$$

$$\mathrm{RC}(F) = (2)(5\%) = \mathbf{10\%}$$

EXAMPLE 2-2

If the expression for F in Example 2-1 is solved for ω, one obtains:

$$\omega = \sqrt{\frac{F}{mr}}$$

If F changes by 5%, how much does ω change?

SOLUTION

$$\mathrm{RC}(\omega) = \mathrm{RC}\left(\sqrt{\frac{F}{mr}}\right)$$

From Table 2-1 for $Q = A^{1/n}$,

$$\mathrm{RC}(Q) = \frac{a}{n}$$

$$\mathrm{RC}(Q) = \frac{5\%}{2} = \mathbf{2.5\%}$$

2-11 NOMOGRAPHS AS A DESIGN AID

The nomograph (also called nomogram) is a visual device for solving equations by means of a straight line laid across three or more scales, each representing one of the variables in an equation. Nomographs make it possible to save time when the same equation is used repeatedly. They are well suited to quick, rough estimates in many preliminary computations and can be used by persons with limited mathematical training. Nomographs can provide a built-in safety device because the range of the scales can be limited to values for which the equation is valid. This prevents using them where the equation would not apply.

As an illustration, one may use a nomograph based on the equation

$$D = 68.47 \sqrt[3]{\frac{P}{n\tau}} \tag{2-5}$$

where

D = shaft diameter; in.
P = power transmitted; hp
n = speed of shaft; rpm
τ = torsional stress; psi

for computing the diameter of a shaft of uniform cross section that sustains pure torsion (Fig. 2-3).

EXAMPLE 2-3

Determine the shaft diameter that will transmit 500 hp at 800 rpm with a torsional stress of 2600 psi.

SOLUTION

Construct a line from 500 on the horsepower scale to 800 on the reference line, label point A. Construct a line from point A to 2600 on the τ scale; where this line intersects the D scale, read the answer of D = **4.25 in.**

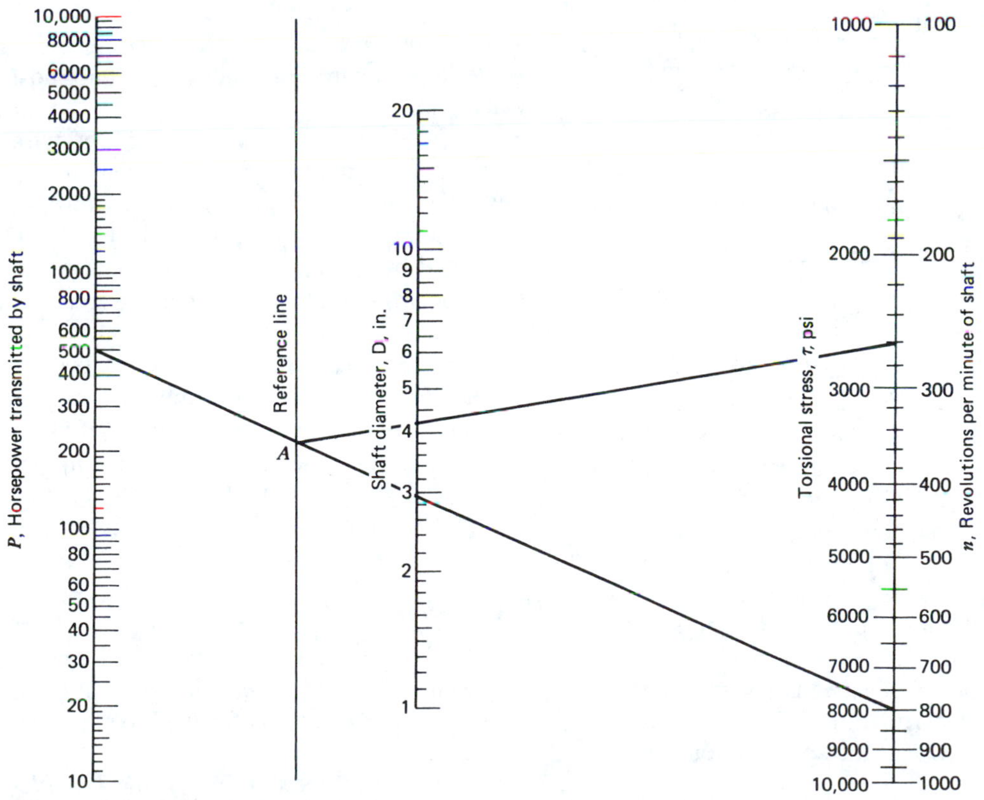

Figure 2-3 Nomograph for cylindrical shafts. (Courtesy *Design News*.)

EXAMPLE 2-4

Assume that the shaft in Example 2-3 is run at a higher speed of 875 rpm. (*a*) How would that affect the shaft diameter? (*b*) If the shaft diameter remains unchanged, how would the horsepower change?

SOLUTION

$$D = \text{constant} \left(\frac{\text{hp}}{\text{rpm}}\right)^{1/3}$$

(*a*) The relative change in speed is:

$$\text{RC}(n) = \frac{875 - 800}{800} = 9.375\%$$

$$D = \text{constant } (n)^{-1/3}$$

$$\text{RC}(D) = \text{RC}(n)^{-1/3}$$

$$= (-0.33)(9.375) = \mathbf{-3.12\%}$$

The shaft diameter can be reduced by 3.12%.

(*b*)

$$n = \text{constant hp}$$

$$\text{RC}(n) = \text{RC(hp)}$$

The horsepower would increase by **9.375%.**

SUMMARY

Machine design is primarily concerned with driven machines, of which there is an infinite variety. Within this group design focuses on improvements of existing machines, since there are few basically new machines. Proper sizing of machine elements to obtain a desired life criterion is therefore the essence of machine design. The legal aspects of design require close attention to product safety.

Design is the first step in manufacturing. Before a product can be manufactured, the engineering department must prepare a complete set of drawings and specifications for all component parts.

Design is also a compromise to satisfy a number of requirements and practical limitations, such as size and space occupied, tooling, servicing ease, and maintainability. Fortunately, most design problems have more than one satisfactory solution.

Design for reliability is dominated by the concept of strength. However, reliable operation coupled with a long service life often requires design for rigidity, stability, and resistance to wear. Since 60 to 90% of all machine members fail in fatigue, the laws governing fatigue assume the greatest importance.

The complexity of design is best encountered by means of computers and a variety of problem-solving strategies such as the four-step method, value engineering, checklists, and troubleshooting.

The purposes of calculations is primarily to confirm or predict (1) the necessary dimensions, (2) the expected effect, and (3) the advantage of one solution over another. Engineering standards and codes enter into most calculations.

Design is an evolutionary process with many changes involving size, material properties, and loads. Many of these relative changes are less than 10% and can thus be evaluated by means of simple formulas. Despite electronic calculators, nomographs are as useful to designers as ever.

Progress in machine design in the form of lower cost, longer useful life, and higher efficiency is due primarily to computerized optimization of systems and parts in conjunction with the application of new materials and manufacturing processes.

REFERENCES

2-1 Bronikowski, R. J. "Pareto's Law for Managers," *Machine Design,* July 24, 1975. (An article well suited to class discussion.)

2-2 Emerick, Robert Henderson. *Trouble-*

shooter's Handbook for Mechanical Systems. New York: McGraw-Hill, 1969. (An excellent and practical handbook.)

2-3 Ferguson, Eugene S. "The Mind's Eye: Nonverbal Thought in Technology," *Science,* August 26, 1977. (A fascinating but somewhat controversial subject well suited to class discussion.)

2-4 Greenwood, D. C. *Engineering Data for Product Design*. New York: McGraw-Hill, 1961. (Contains data for all topics covered in the following chapters. Although outdated in some areas, the book's topics are often fundamental in nature.)

2-5 Greenwood, D. C. *Mechanical Details for Product Design*. New York: McGraw-Hill, 1964. (Contains valuable information pertaining to fasteners, power transmission, springs, welding, and mechanical controls.)

2-6 Hilborn, E. H. *Handbook for Engineering Psychology*. Cambridge, Mass.: TAD, Inc., 1965. (An inexpensive volume containing a wealth of information in the form of graphs and tables.)

2-7 Leyer, Albert. *Machine Design*. Glasgow: Blackie & Sons, 1973. (Translated from German, this outstanding book by a Swiss professor complements the usual range of engineering texts and can be read to advantage by students of machine design as well as industrial practitioners. Available from major libraries.)

2-8 Matousek, R. *Engineering Design*. Glasgow: Blackie & Sons, 1966. (An outstanding book on machine design, translated from German. Available from major libraries.)

2-9 Middendorf, W. M. *Engineering Design*. Boston: Allyn & Bacon, 1969. (Emphasizes design techniques by means of examples and case studies; heavily oriented toward electromechanical engineering.)

2-10 Pitts, G. *Techniques in Engineering Design*. New York: Wiley, 1973. (A practical book on design philosophy with an emphasis on computers as a design aid.)

2-11 Siddal, James N. *Mechanical Design Reference Sources*. Toronto: University of Toronto Press, 1967. (Contains a wealth of useful information.)

2-12 Jensen, C. and J. Helsel. *Engineering Drawing and Design,* Second Edition. New York: McGraw-Hill, 1979. (Contains 700 pages of useful information. An excellent supplement to this textbook.)

PROBLEMS

P2-1 Discuss major components of a bicycle (chain, sprockets, wheels, brakes, etc.) by means of the five questions outlined in Section 2-7. Also discuss chain saw and lawn mower components.

P2-2 Determine the relative percentage error due to an absolute error of 5 g in (*a*) 1 kg, (*b*) 1.6 kg, and (*c*) 15 kg.

P2-3 The nominal measurements of a rectangular steel plate are 4 m (length) and 2.5 m (width), with errors not exceeding 1 cm. Determine the largest possible relative error in percentage in the area of the plate.

P2-4 The area A of a square is $A = 1/2d^2$, where d = the length of a diagonal. (*a*) If the diagonal is reduced by 3%, how much does this reduce the area? (b) How much should the diagonal be reduced in order to cut the area by 3%?

P2-5 The kinetic energy E_k of a moving mass is

$$E_k = 1/2mv^2$$

where

m = mass of body; kg
v = velocity; m/s

How much does the kinetic energy of an automobile rise when the speed is increased from 50 to 52 km/h?

P2-6 The maximum shearing stress τ in the wire of a coil spring can be calculated by the following formula:

$$\tau = KPD/(0.4\ d^3)\ \text{MPa}$$

where

P = axial load; N
D = mean diameter; mm
d = wire diameter; mm
K = Wahl stress factor

In the course of testing a certain spring, the axial load was discovered to be 10% higher than anticipated. How much should d be changed in order to retain the same values for τ and D?

P2-7 The total angle of twist, in degrees, for a circular shaft of uniform cross section is given by the following equation:

$$\theta = 584LT/Gd^4$$

where

θ = angle of torsion; deg
L = length of shaft; mm
T = torque on shaft; N · mm
G = torsional modulus of elasticity; MPa
d = shaft diameter; mm

(*a*) How does θ change for each of the following changes: (1) L by 5%; (2) T by 3%; (3) d by 4%; and (4) L and T by 2%?
(*b*) If T increased by 8%, how much must d be changed in order for θ to remain constant?

P2-8 The deflection δ of a simply supported circular shaft of diameter d, length L, and loaded at midpoint with a single load P is:

$$\delta = (4PL^3)/(3\pi Ed^4)$$

(*a*) If d is increased by 5%, how does that affect δ?
(*b*) If L is reduced by 5%, how does that affect δ?
(*c*) If L is increased by 7%, how must d be changed in order for δ to remain unchanged?

P2-9 A shaft is to transmit 150 hp at 750 rpm. The maximum allowable shear stress is 7500 psi.

(*a*) Determine required shaft diameter by means of the given nomograph (Fig. 2-3).
(*b*) If the horsepower is increased 8%, determine percent increase needed for shaft diameter.

P2-10 A circular shaft has a diameter of 25 mm. The allowable shear stress is 42 MPa. Determine at what speed the shaft will transmit 10 kW by (*a*) calculating the answer, (*b*) using a nomograph.

P2-11 Derive the formulas in Table 2-1.

P2-12 Write a paper on nomographs.

Chapter 3

Design for Strength

The test of a good plan is whether
it can be put to work.

AESOP (620–560 B.C.)

NOMENCLATURE

k_a = surface finish derating factor
k_b = size derating factor
k_c = reliability derating factor
k_d = temperature derating factor
k_e = stress concentration derating factor
k_f = impact derating factor
K_f = actual stress concentration factor
K_t = geometric stress concentration factor
N = number of cycles to failure
q = material notch sensitivity
S = stress amplitude
S_e = corrected machine member fatigue strength
S'_e = rotating beam fatigue strength
S_{se} = corrected machine member shear fatigue strength
S'_{se} = completely reversed torsional fatigue strength
S_{su} = shear ultimate strength
ν = Poisson's ratio
σ_a = amplitude stress
σ_i = induced normal stress
σ'_i = equivalent induced octahedral normal stress
σ_{ia} = induced amplitude stress
σ'_{ia} = equivalent induced octahedral amplitude normal stress
σ_{im} = induced mean stress
σ'_{im} = equivalent induced octahedral mean normal stress
σ_m = mean stress
σ_{max} = maximum normal stress
σ_{min} = minimum normal stress
σ_w = working normal stress
τ_a = amplitude shear stress
τ_i = induced shear stress
τ_m = mean shear stress
τ_w = working shear stress
θ = elastic stress-strain curve slope angle

The test of a good machine is its safe and reliable performance of useful work for a reasonable length of time without serious breakdown. Machine member design for strength seeks to attain, at minimum cost, a suitable combination of size, shape, and material to withstand service loadings for the expected life of the machine. It involves three basic types of information.

1. Load direction, magnitude, application frequency, and history.[1]
2. Mechanical properties of materials.
3. Methods of analysis to predict values of stress and strength.

Design for strength requires that the designer evaluate each mode in which a machine part can fail and then apply analytical and creative skills to devise a component that has strength enough to avoid *failure* (as discussed in Sections 2-5 and 2-6). In this chapter we will concentrate on two of the most common modes of failure: *yielding* due to *steadily applied loads* and *fatigue fracture* due to *cyclic loads*.

In failure from static or steady loading, the loads are high compared to available static strength, and the part fails by changing shape. Failure from cyclic loads, in contrast, can occur when loads are relatively low compared to static strength. Failure from permanent deformation is referred to as *yielding;* failure from cyclic loads is called *material fatigue*. The strength of mechanical contacts, an additional topic important in design for strength, will also be discussed.

Chapter 4 will consider buckling strength, friction, and wear.

[1]The history of a machine part includes load and environmental effects experienced by the part that have caused material changes in its ability to perform a given function. Examples of such changes are heat-affected strength variations, residual stresses, and corrosion.

3-1 LOADS AND THEIR EVALUATION

Definition and Origin

A load is any external force to which a machine member is subjected. Determination of external force magnitude, direction, point of application, and variation is the first step in the design of machine members, providing the *kinematic* requirements are met. In machinery, loads originate from several sources: actions of gas and fluids under pressure, thermal expansions, magnetic fields, gravity, inertial forces, and contacts between bodies.

Most machine member loads are contact forces. They arise as a natural consequence of power transmission—the primary function of machines. In machines, power flows from energy converters (motor, engine, hydraulic pump, etc.) to a work station (cutting tool, forming die, conveyor, etc.), where useful work is performed. Nearly all machine members participate in this activity, either by transmitting power or, through support action, by maintaining positional references for the parts that do.

Where loads are due to contact, a pair of equal and opposite forces occur (Fig. 3-1). One force acts as an external load on one contacting member, and the second force acts as an external load on the other member. This action-reaction force pairing is one of the basic natural laws put to practical use by engineers. Tracing these power transmission forces through connected machine linkages is an extremely useful visualization aid for identifying machine component loads.

Machine parts regularly sustain complex loadings. Fortunately, the equilibrium methods of statics usually allow such problems to be reduced to a simple set of basic loads. This is a convenience, since the well-known stress formulas from strength of materials may then be used to convert the basic loads to stress.

Machine member loads may differ in point of application, load magnitude, duration, frequency, and number of cycles. Thus the complete load description for machine members will require careful and thorough investigation to achieve reasonable certainty of adequate design strength.

Although every machine load is important, it

Figure 3-1 Simple types of loading.

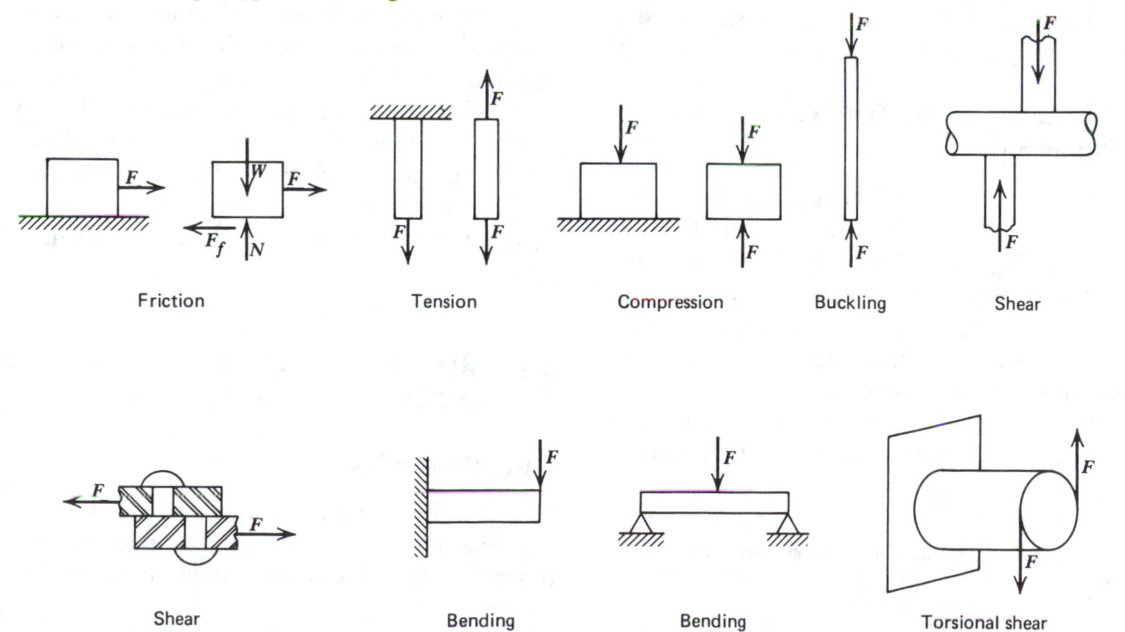

is estimated that 60 to 90% of all machine parts fail from cyclic loads and *material fatigue*. Many aspects of this important subject will be discussed in this chapter.

Basic Loads

Basic loads are resultant axial, bending, and shear loads. There are several simple load types from which the basic loads can be formed. Figure 3-1 illustrates tension, compression, buckling, direct shear, torsional shear, bending, and friction. The loads in this figure may be either static or cyclic.

As long as the maximum stress remains in the elastic range, superposition[2] is valid. The simple loads of Fig. 3-1 may then be summed as building blocks to "construct" analytically the combined basic loads at a point of interest in a machine member. The object of the calculation is to determine the resultant axial, bending, and shear basic loads at that point. Since it is not always possible to pick the weakest point in a member intuitively, the calculation may have to be made for several points. The results are compared to ascertain which point is the critical point where failure is most likely to originate. The first step in finding basic loads, then, is a simple load inventory.

Load Variation and Time-Load Relationships

A common load combination in machinery is a static load with a superimposed cyclic load, as shown in Fig. 3-2. Generally, the steady and repeated load magnitudes may also vary in complex ways during the load-life history of the part. This case usually requires special considerations and will not be discussed here.

Time-load relationships determine severity of loading: as the load-time interval decreases, severity of load increases. For the same magnitude, a shock load can be twice as severe as a static load. Likewise, many applications of relatively small repeated loads can be more severe than a single application of a high static load. Normally, machine components are made progressively heavier as loading goes from static through cyclic to *impact*. There are exceptions, however, since experience indicates that a few cycles of high stress can cause more equivalent damage than one or two relatively light impact loads.

Static Loading

Statics, which deals with bodies in equilibrium, is fundamental to the sizing of machine parts, yet relatively few parts are subjected to pure static loading. In some instances, nevertheless, it is important for a machine member to be relatively rigid and deform very little under load. Otherwise, it may fail to provide its primary kinematic and/or support function and can lead to machine stoppages and early failures. In other cases it is very important that the part be allowed, through spring action, to change shape or, through yielding, to deform permanently under load.

Experienced designers recognize the simplicity and versatility of statics and strength of materials. In their analyses they take full advantage of the equations of equilibrium, $\Sigma F = 0$ and $\Sigma M = 0$, and the stress formulas available for various load conditions. Use of these procedures and formulas greatly facilitates sizing of parts. Part rigidity will be discussed in Chapter 4.

3-2 MECHANICAL PROPERTIES OF ENGINEERING MATERIALS

Basic Properties

Design formulas for evaluating size and shape of machine members for strength include material properties. Of primary interest are properties in-

---

[2]Superposition is the addition and subtraction of forces and stresses.

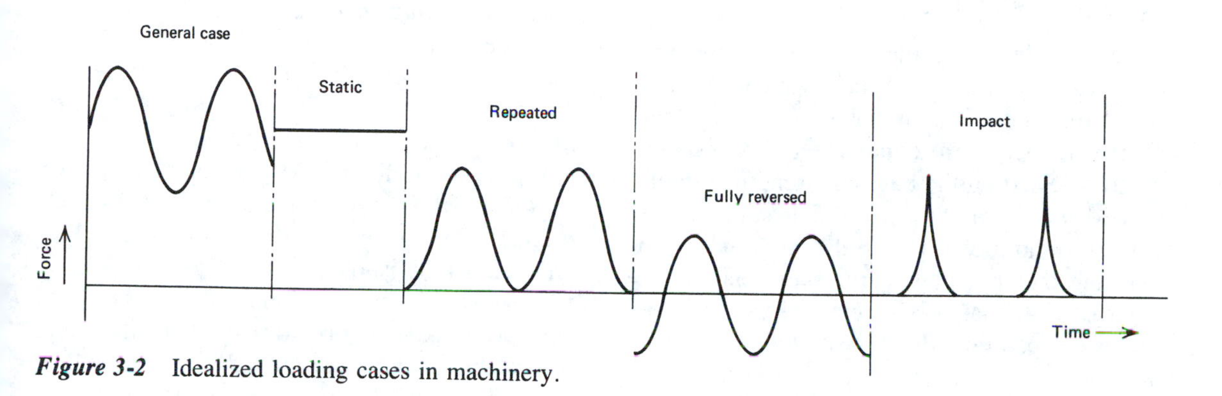

Figure 3-2 Idealized loading cases in machinery.

dicative of material behavior under loadings most frequently encountered in machinery. The properties commonly found in reference books and in the appendix of this text are:

1. Ultimate strength, S_u.
2. Yield strength, S_y.
3. Fatigue strength, S'_e.
4. Modulus of elasticity, E.
5. Shear modulus, G.
6. Brinell hardness, Bhn.

These properties give designers a basis for predicting material behavior. They are, however, data obtained under ideal laboratory conditions. Therefore they do not account for corrosive atmosphere, high temperatures, variable load history, load interactions, material strength variation, and special part features that affect strength. Consequently, designers must commonly face the problem of applying such data to situations considerably different from those in the laboratory. Methods that account for some of these effects will be introduced in this chapter.

To allow for lack of complete design information, design texts, handbooks, and engineering codes recommend use of design factors, referred to as factors of safety (fs), derating factors ($k_a \ldots k_f$), and service factors (k_s).[3] Such design factors are adequate in many cases. But designers are cautioned: whenever new machines are developed and failure may result in human injury, there is no substitute for prototype

[3]Optimal life is the main purpose of service factors. They are often used with purchased components.

test programs, manufacturing quality control, and continued field monitoring. The goals of these programs must be to establish accurate design factors, correct design deficiencies, and ensure human safety.

Materials Selection

Materials selection initially may seem simple because of past practices or the availability of information such as that in Table 3-1. It can, however, quickly become complicated. For instance, a part selected on the basis of strength, rigidity, or hardness may not be available in suitable standard configurations. Cases also arise where the machine member must survive unusual service environments that may require the use of an expensive superalloy. In this situation, designers may have to consult additional materials standards and experienced metallurgical or manufacturing engineers in an effort to achieve minimum cost consistent with quality design.

Carbon and Alloy Steels

Carbon steels which, for two centuries, were the backbone of machine design, continue to play a major role. However, if greater toughness,[4] strength, hardness, or high temperature and rust resistance is needed, expensive alloying elements such as nickel, molybdenum, chromium, and manganese become necessary.

Because carbon steel predominates, the discussion in this chapter will center on it. When alloy steels, cast iron, or nonferrous metals are discussed, they will be named specifically. Nonmetallic materials such as plastics, glass, composites, and concrete will not be considered here. Readers interested in them should consult reference books on engineering materials.

[4]Toughness is the ability of a material to absorb energy during deformation and fracture and to deform plastically before fracture.

3-3 ANALYSIS OF MATERIAL STRENGTH

Hooke's Law

A large part of material strength analysis is based on Hooke's law. According to this fundamental law, engineering materials undergo predictable linear deformations under the influence of loads. Materials that obey the law are called *elastic materials,* and their action under load is referred to as *elastic deformation*.

Elastic deformation has the desirable characteristic of allowing the material to recover its original shape when the load is removed. This type of action is observed in springs. Within their elastic range, machine members such as coil springs have deformations that are linear and predictable, and their original dimensions will be recovered when the load is removed.

Some materials (e.g., copper and plastics) have negligible elastic behavior, but most steels have a conveniently large elastic range. Thus steels enjoy wide usage in machine design whenever rigidity, flexibility, and low cost are desired. In transportation, however, aluminum plays a major role, especially in the air-frame industry. Aluminum and steel are both important engineering materials that obey Hooke's law.

Hooke's law states specifically that within a broad range of conditions, stress is proportional to strain. That many metals follow this simple relationship is of enormous practical importance. An additional manifestation of Hooke's law is the proportionality between load and deformation, which is analogous to the proportionality between stress and strain. Thus the force-time curves of Fig. 3-2 might equally well depict stress-time curves for simple loadings.

Stress-Strain Parameters

Stress and strain are terms used by designers to discuss the *action that takes place inside a machine member* when it is subjected to external load. Reasons for using these terms include the following.

TABLE 3-1 Steel Properties and Their Common Uses

Type of Steel	Symbol	Properties	Common Uses
Carbon steels			
Plain carbon	10XX		
Low carbon (0.06 to 0.20% carbon)	1006 to 1020	High toughness and lower strength	Shafts, rivets, chains, stampings, and sheet metal products
Medium carbon steel (0.20 to 0.50% carbon)	1020 to 1050	Medium toughness and strength	Machine parts, forgings, gears, axles, bolts, and nuts
High carbon steel (over 0.50% carbon)	1050 and over	Lower toughness and high hardness	Tools, drills, saws, knives, and music wire
Sulphurized (free cutting)	11XX	Improved machinability	Screw machine products
Phosphorized	12XX	Reduced ductility but increased hardness and strength	
Manganese steel	13XX	Improved surface finish	
Nickel steel Up to 5.00% nickel	2XXX to 25XX	Increased toughness and strength	Axles, connecting rods, and crankshafts
Nickel-chromium steel 0.70% nickel 0.70% chromium to 3.50% nickel 1.50% chromium	3XXX to 33XX	High toughness and strength	Gears, shafts, studs, chains, and screws
Molybdenum steel Small percentages chromium and nickel	40XX to 48XX	High strength and wear resistance	Cams, axles, forgings, and gears
Chromium steel	5XXX to 52XX	Great strength and toughness, high hardness	Gears, bearings, shafts, connecting rods, and springs
Chromium-vanadium steel	61XX	Strength and hardness	Piston rods, gears, axles, punches, and dies
Nickel-chromium -molybdenum steel	86XX	Hardness and strength, rust resistance	Surgical components and food containers
Silicone-manganese steel	92XX	Elasticity and flexibility	Springs

1. Stresses and strains can be expressed in units that are independent of machine member size or shape.

2. Stress and strain values characterize the localized internal or surface effects due to external loads on a machine member.

3. Material strength parameters are commonly expressed in terms of stress or strain.
4. Complex load systems can be combined (using superposition) into basic stress or strain parameters at each internal or surface point of a machine member.
5. Material failure generally starts at localized points in a machine member.

Stress and strain values are, therefore, extremely useful for predicting machine component performance and/or failure.

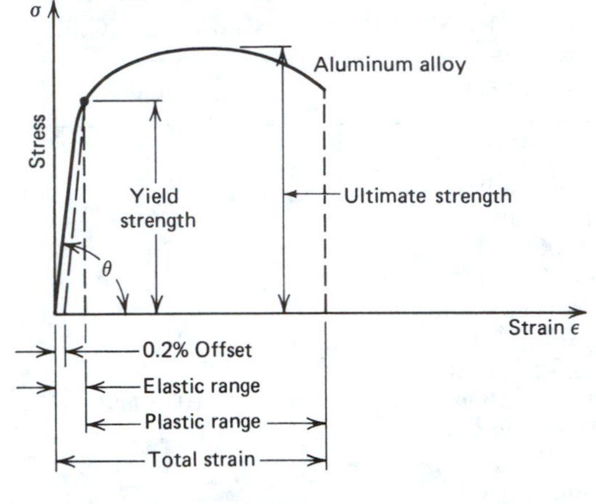

(a) Aluminum

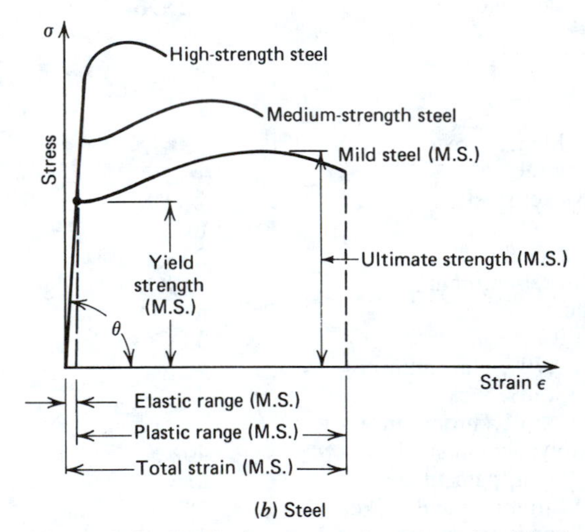

(b) Steel

Figure 3-3 Typical stress-strain curves for (a) aluminum and (b) steel.

The Stress-Strain Curve

The basic shape of the curve in Fig. 3-3 is typical of ductile materials. It defines yield strength, ultimate strength, and modulus of elasticity. Likewise, the curves illustrate the elastic range and Hooke's law. In mathematical terms, Hooke's law is:

$$\sigma = \pm E\epsilon \tag{3-1}$$

$$E = \tan\theta \tag{3-2}$$

where

σ = stress; MPa, psi
ϵ = strain; m/m, in./in.
E = modulus of elasticity; MPa, psi
θ = slope angle; deg

The importance of E may be summarized as follows.

1. Elastic range response to loading is represented by a single property E that remains constant through the life of the part.
2. E is virtually the same for all steels, E = 207 GPa (30×10^6 psi) and, for aluminum alloys, E = 71 GPa (10.3×10^6 psi).[5]
3. An inherent measure of rigidity and flexibility, E facilitates comparison of materials. For the same load, a large E yields a small deformation, and vice versa.

[5]Material properties for specific materials may vary ±10% from indicated values but, for many design calculations, these are sufficiently accurate.

Shear Modulus

When materials are loaded in shear, the shear stress is also proportional to shear strain. The proportionality factor is called the shear modulus (or modulus of rigidity). Therefore

$$\tau = G\gamma \tag{3-3}$$

where

τ = shear stress; MPa, psi
γ = shear strain; m/m, in./in.
G = shear modulus; MPa, psi

G is smaller than E. For steels, G = 79.3 GPa (11.5×10^6 psi) and, for aluminum alloys, G = 26.2 GPa (3.8×10^6 psi).

The relationship between the primary elastic constants is

$$G = \frac{E}{2(1 + \nu)} \quad (3\text{-}4)$$

ν is known as Poisson's ratio. For most metals it has a value of roughly 0.3 (0.2 to 0.4).

Poisson's ratio provides an indication of how a material will deform in directions perpendicular to an applied load. For instance, a cylindrical rod in tension elongates and becomes more slender. The ratio of strain in the transverse direction to strain in the axial direction is equal to Poisson's ratio.

$$\nu = \frac{\text{transverse strain}}{\text{axial strain}} \quad (3\text{-}5)$$

This relationship also holds for compression. A cylindrical rod, loaded in compression, will shorten and become thicker, thereby illustrating Poisson's ratio. An important use of Poisson's ratio is the design of interference fits. Values of Poisson's ratio are given in reference 3-1.

Hardening and Alloying

As previously noted, alloy steels are useful when high hardness, greater toughness, temperature resistance, and resistance to wear or corrosion are needed (see Table 3-1). However, to gain these advantages, designers must anticipate increased material and processing costs.

Metals generally display higher ultimate strength with increased hardness. Consequently, hardness measurements, which are easy to carry out, are an excellent shortcut for estimating ultimate strength. Moreover, hardness measurements can often be made somewhere on the actual part without affecting the usefulness of the part. Such measurements make small indentations in the surface and should thus be made only at points that experience low induced stresses.

The mean ultimate strength of steel can be approximated by using the following equation.

$$S_u\,(\text{MPa}) = 3.45\ \text{Bhn}$$

$$S_u\,(\text{psi}) = 500\ \text{Bhn} \quad (3\text{-}6)$$

Note that while the ultimate strength of steel varies over a wide range (depending on type of alloy, hardness, etc.), the moduli E and G are nearly constant. Consequently, contrary to popular belief, alloy and hardened carbon steels deflect the same amount per unit stress as mild steel. Hardening a steel part will not, therefore, make it less flexible, since E remains constant.

3-4 WORKING STRESSES

Design of machine members is dominated by the concept of an allowable working stress. This stress level, which should not be exceeded, is a fractional part of either the ultimate, yield, or fatigue strength of the material in question. The choice of which strength value to use depends on the material and loading. For static loading, yield or ultimate strength is needed; cyclic loading requires the use of fatigue strength. If the material is brittle (e.g., cast iron), it does not yield, so the ultimate strength or fatigue strength is used. The transition from strength to working stress is achieved by means of derating factors and factors of safety (*fs*).

Derating Factors

Derating factors are premultiplying factors (fractions; numbers between zero and one) applied to material strength values to reduce them in proportion to recognized service conditions that affect strength. Derating factors are of primary importance in fatigue loading, where they are used to account for the following conditions.

1. Surface finish derating factor (Section 3-11), k_a.
2. Size derating factor (Section 3-12), k_b.
3. Reliability derating factor (Section 3-13), k_c.
4. Temperature derating factor (Section 3-14), k_d.
5. Stress concentration derating factor (Section 3-15), k_e.
6. Impact derating factor (Section 3-16), k_f.

Derating factors are used as follows.

$$\text{Working stress} = (\text{derating factors})(\text{material strength}) \quad (3\text{-}7)$$

Factors of Safety

Factors of safety (numbers greater than one) are applied to material strength by division to reduce the material strength for two very important reasons.

1. *Lack of precise knowledge* of machine member strength and the service conditions.
2. The need to provide *insurance against failure*.

Preventing component failure means preventing machine downtime, costly repairs, and injury-causing accidents. Other considerations in selecting a factor of safety are specific design goals (low weight, fail-safe performance, etc.), availability of test data, the match between available test data and the case at hand, and prior experience with the subject machinery. Factors of safety ranging from 1.1 to 20 have been used.

An elevator manufacturer may use a safety factor of 14 because of the obvious dangers associated with failure. In another case where component weight is a major consideration and there is minimal risk of injury (economic or human), a factor of 1.1 or 1.2 might be selected. For some applications, machine design is restricted by codes established by professional groups and governmental bodies. In these cases, the factors of safety used by the writers of the code may not be specifically stated, but an allowable (or working) stress will be given. Designers of equipment affected by such codes must design components that conform to the working stress allowed by the standard.

Design factors (factors of safety and derating factors) represent an accumulation of experience with various types of machinery and machine components. However, the recommended values from references may be too high or too low for a given application. Therefore, in critical cases, the safety and derating factors should be verified through test programs and field use monitoring.

When the working stress is based on yielding, factors of safety between 1.5 and 4.0 are often used. The lower values are more appropriate when materials are reliable and the design and operating conditions are better known. For less reliable materials and uncertain service conditions, the higher values are safer. Similar factor-of-safety values have been applied for cyclic loading using the endurance limit as the strength parameter, although for brittle materials the factors are often doubled.

The relationship between working stress and factor of safety is written as follows.

$$\text{Working stress} = \frac{\text{material strength}}{\text{factor of safety}} \quad (3\text{-}8)$$

Simultaneous application of derating factors and factors of safety is illustrated in the following relationship.

$$\text{Working stress} = \frac{\text{derating factors}}{\text{factor of safety}}(\text{material strength}) \quad (3\text{-}9)$$

Cyclic Loading

Failure from cyclic loading is called *material fatigue*. Figure 3-4 portrays the transition in strength from static to cyclic loading. Note that the cyclic loading strength drops to 90% of the initial or static value after approximately 1000

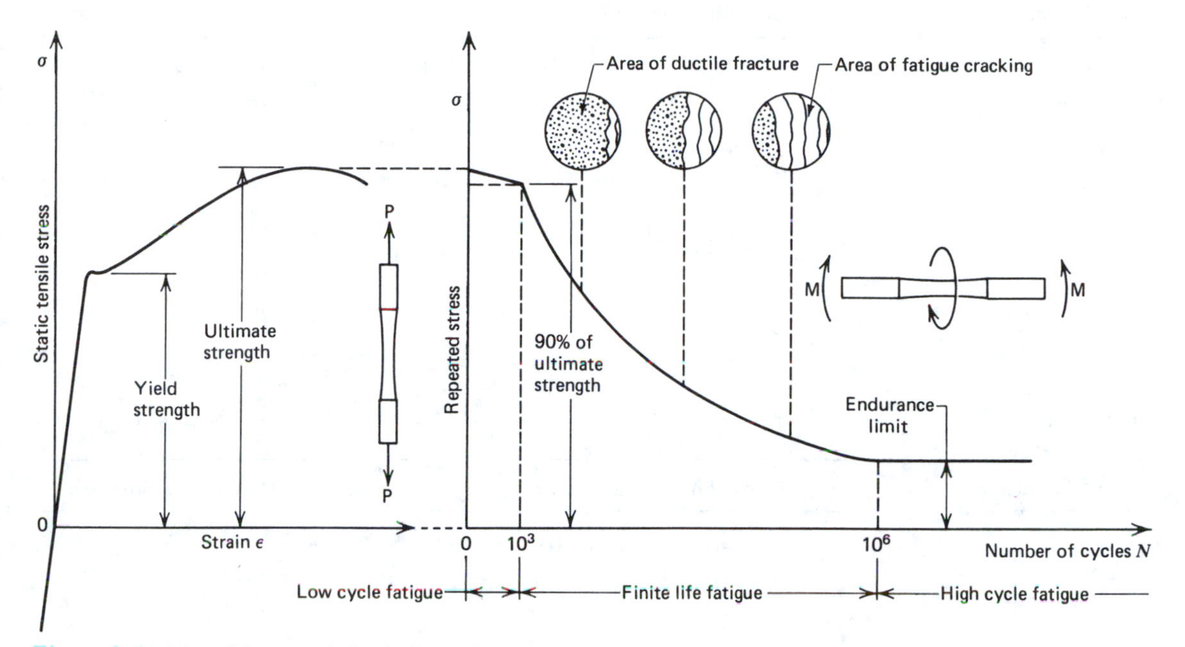

Figure 3-4 Material strength in design. (Developed by Uffe Hindhede and Wendell C. Hull.)

cycles and to roughly 50% after 1 million cycles and remains fairly constant thereafter. However, the flattening out of the curve after a large number of cycles is valid only for ferrous materials (iron and steel alloys).

Machine Member Specifications

Once the loading, induced stress, and working stress for a machine member are known, the designer can proceed to detailed part specifications. Machine member specification entails defining physical dimensions (shape), cross-sectional area (size), and area distribution (area moment of inertia). In addition, specifications of material type, fabrication method, heat treatment, and surface finish must be given. At each step in this process, designers must make choices from many different possibilities and, with each decision, they must weigh the consequences of that choice against a variety of often conflicting and competing criteria—function, safety, esthetics, economics, strength, availability, and so forth. In this chapter the emphasis is on design for strength and safety.

3-5 DIMENSIONING BASED ON STRENGTH

Generally, dimensioning involves four basic steps.

1. Calculating load-induced stresses σ_i or τ_i using the equations of statics and strength of materials (Table 3-2).
2. Estimating working stresses σ_w or τ_w using derating and/or factors of safety (Table 3-2).
3. Equating working and induced stresses.

$$\sigma_i = \sigma_w \qquad \text{or} \qquad \tau_i = \tau_w$$

4. Solving for the necessary cross-sectional dimensions.

Combined Stresses

Stresses in machine components (shafts in particular) are often a combination of normal stress (axial or flexure) and shear stress (direct or torsional), where the shear stress acts on the same cross section as the normal stress. In those cases, provided no other stresses act on the same section, the combined stress effect can be deter-

TABLE 3-2 Strength Formulas

Type of Loading	Basic Induced Stresses	Cross-sectional Dimensions
Tension compression	$\sigma = \frac{F}{A}$	$A = \frac{F}{\sigma}$
Bearing pressure	$\sigma = \frac{F}{A}$	$A = \frac{F}{\sigma}$
Bending	$\sigma = \frac{Mc}{I} = \frac{M}{Z}$	$\frac{I}{c} = \frac{M}{\sigma}$
Torsion	$\tau = \frac{Tr}{J} = \frac{T}{Z_p}$	$\frac{J}{r} = \frac{T}{\tau}$
Direct shear	$\tau = \frac{F}{A}$	$A = \frac{F}{\tau}$

Type of Failure	Normal Working Stresses	Shear Working Stresses
Brittle fracture	$\sigma = \frac{S_u}{\text{fs}}$	$\tau = \frac{S_{su}}{\text{fs}}$
Yield	$\sigma = \frac{S_y}{\text{fs}}$	$\tau = \frac{0.57S_y}{\text{fs}}$
Fatigue	$S_e = \frac{S'_e}{\text{fs}}$	$S_{se} = 0.57\frac{S'_e}{\text{fs}}$
Fatigue with derating factors	$S_e = k_a k_b \ldots k_f \frac{S'_e}{\text{fs}}$	$S_{se} = 0.57 k_a k_b \ldots k_f \frac{S'_e}{\text{fs}}$

Elements of Sections

Section	I	Z	J	Z_p
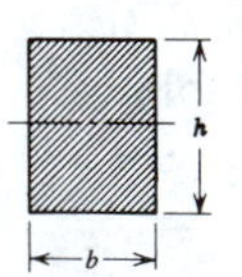	$\frac{bh^3}{12}$	$\frac{bh^2}{6}$	—	—
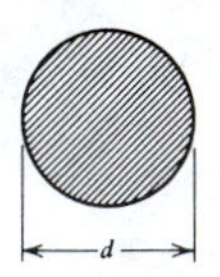	$\frac{\pi d^4}{64}$	$\frac{\pi d^3}{32}$	$\frac{\pi d^4}{32}$	$\frac{\pi d^3}{16}$
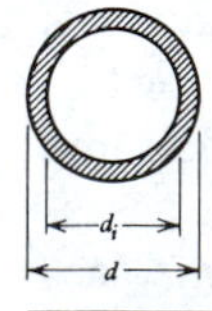	$\frac{\pi(d^4 - d_i^4)}{64}$	$\frac{\pi(d^4 - d_i^4)}{32d}$	$\frac{\pi(d^4 - d_i^4)}{32}$	$\frac{\pi(d^4 - d_i^4)}{16d}$

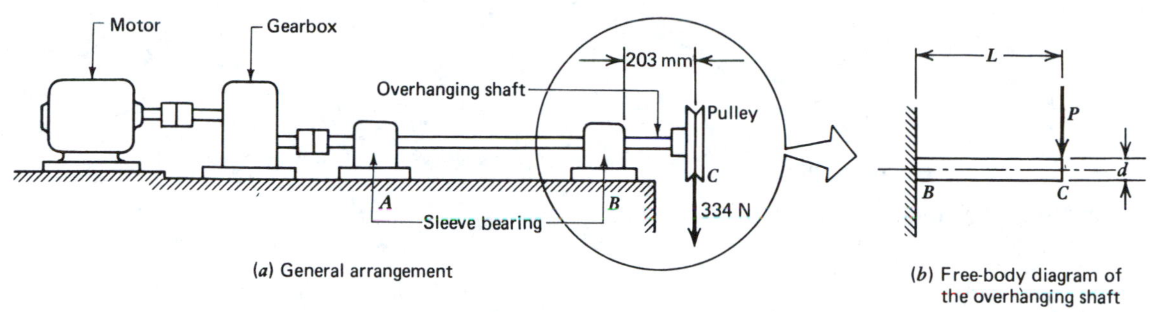

Figure 3-5 Cantilever shaft.

mined from the equivalent induced distortion energy or octahedral shear stress σ_i',[6] as follows.

$$\sigma_i' = \sqrt{\sigma_i^2 + 3\tau_i^2} \tag{3-10}$$

The equivalent induced normal stress may then be equated to the working stress. For yielding, the working stress is

$$\sigma_w = \frac{S_y}{fs} \tag{3-11}$$

EXAMPLE 3-1

A machine shaft carries a pulley at the end and is mounted in sleeve bearings, as shown in Fig. 3-5. The pulley is 203 mm from bearing *B*. Belt tension causes a load of 334 N at the pulley. The steel shaft is made from UNS G10500[7] drawn at 900°F with yield strength S_y = 896 MPa. A factor of safety for static yielding of 2.0 is desired. Neglecting torsional load, determine the shaft diameter d that will satisfy these requirements.

SOLUTION

Assume that the bearings resist shaft bending. Then the primary concern is limiting bending stress, which is maximum where the shaft enters the right-hand bearing in the accompanying figure. Using the bending (flexure) stress equation, the stress is

$$\sigma = \frac{Mc}{I} = \frac{(PL)(d/2)}{(\pi d^4/64)} = \frac{32PL}{\pi d^3}$$

where

$M = PL$ = bending moment at the bearing
P = load
L = cantilever shaft length
I = area moment of inertia of the shaft cross section

$$= \frac{\pi d^4}{64}, \text{ for a round shaft}$$

d = shaft diameter

$$c = \frac{d}{2} = \text{distance to extreme shaft fiber}$$

Induced Stress

$$\sigma_i = \left(\frac{32}{\pi}\right)\frac{(334 \text{ N})(203 \text{ mm})}{d^3}$$

$$= \frac{690\,600 \text{ N}\cdot\text{mm}}{d^3}$$

---

[6]A unit cube reference plane at the point of interest is referred to as the octahedral plane. The name comes from the fact that a set of eight such planes isolates an eight-sided element called an octahedron. The equilibriating shear stress on an octahedral plane is proportional to the associated distortion energy. Both concepts, the octahedral shear and the distortion energy, lead analytically to the same equivalent induced stress equation.

[7]Unified Numbering System for Metals and Alloys (UNS), Society of Automotive Engineers, Warrendale, Pa., 1975.

Working Stress

$$\sigma_w = \frac{S_y}{fs} = \frac{896\ \text{N/mm}^2}{2.0}$$

$$= 448\ \text{N/mm}^2$$

Equating the induced and working stress $\sigma_i = \sigma_w$,

$$\frac{690\ 600\ \text{N}\cdot\text{mm}}{d^3} = 448\ \text{N/mm}^2$$

$$d^3 = \frac{690\ 600\ \text{N}\cdot\text{mm}}{448\ \text{N/mm}^2} = 1542\ \text{mm}^3$$

$$d = \mathbf{11.55\ mm}$$

Therefore a shaft 11.6 mm in diameter will satisfy the problem requirements.

EXAMPLE 3-2

Assume that the shaft between bearings *A* and *B* in Example 3-1 transmits a torque of 11 300 N · mm. What shaft diameter *d* would be required in this section to avoid static yielding due to torsional shear stress?

SOLUTION

The torque is constant throughout the shaft section between *A* and *B* (Fig. 3-6). Torsional shear stress for this section is

$$\tau = \frac{Tr}{J} = \frac{T(d/2)}{(\pi d^4/32)} = \frac{16T}{\pi d^3}$$

where

T = torque

$r = \frac{d}{2}$ = shaft radius

d = shaft diameter

J = polar moment of inertia

$= \frac{\pi d^4}{32}$, for a round shaft

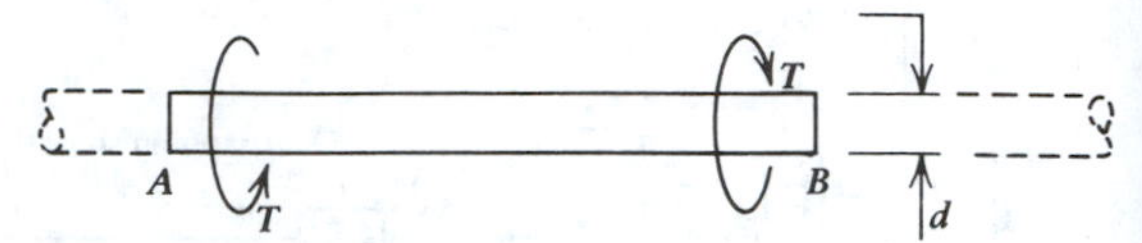

Figure 3-6 Shaft transmitting torque.

Induced Stress

$$\tau_i = \left(\frac{16}{\pi}\right)\frac{11\ 300\ \text{N}\cdot\text{mm}}{d^3}$$

$$= \frac{57\ 600}{d^3}\ \text{N}\cdot\text{mm}$$

Working Stress

$$\tau_w = \frac{0.57S_y}{fs} = \frac{(0.57)(896\ \text{N/mm}^2)}{2.0}$$

$$= 255\ \text{N/mm}^2$$

Set τ_i and τ_w equal.

$$\frac{57\ 600\ \text{N}\cdot\text{mm}}{d^3} = 255\ \text{N/mm}^2$$

$$d^3 = \frac{57\ 600\ \text{N}\cdot\text{mm}}{255\ \text{N/mm}^2} = 226\ \text{mm}^3$$

$$d = \mathbf{6.1\ mm}$$

Thus a 6.1-mm diameter shaft would be capable of carrying the torsional load on the shaft.

EXAMPLE 3-3

Since the torsional load of Example 3-2 would also act on the shaft section between bearing *B* and the pulley, the induced stresses of Examples 3-1 and 3-2 would combine in this region. Determine the shaft diameter *d* necessary to prevent yielding from the combined induced stresses.

SOLUTION

The equivalent normal stress for this case is

$$\sigma_i' = \sqrt{\sigma_i^2 + 3\tau_i^2} = \sqrt{\left[\frac{32\ PL}{\pi d^3}\right]^2 + 3\left[\frac{16T}{\pi d^3}\right]^2}$$

$$= \frac{16}{\pi d^3} \sqrt{(2PL)^2 + 3T^2}$$

$$= \frac{16}{\pi d^3} \sqrt{4P^2L^2 + 3T^2}$$

where the terms are as previously defined.

Induced Equivalent Normal Stress

$$\sigma_i' = \frac{16}{\pi d^3}$$

$$\sqrt{(4)(334\ \text{N})^2(203\ \text{mm})^2 + (3)(11\ 300\ \text{N}\cdot\text{mm})^2}$$

$$= \frac{697\ 800}{d^3}\ \text{N}\cdot\text{mm}$$

Working Stress

$$\sigma_w = \frac{S_y}{fs} = \frac{896\ \text{N/mm}^2}{2.0}$$

$$= 448\ \text{N/mm}^2$$

Equating these values $\sigma_i' = \sigma_w$,

$$\frac{697\ 800}{d^3}\ \text{N}\cdot\text{mm} = 448\ \text{N/mm}^2$$

$$d^3 = \frac{697\ 800\ \text{N}\cdot\text{mm}}{448\ \text{N/mm}^2} = 1558\ \text{mm}^3$$

$$d = \mathbf{11.55\ mm}$$

We find the resultant shaft diameter for the combined stress has changed very little over that for the bending load alone. However, depending on stress magnitudes, shear stresses can significantly influence machine member dimensions.

Brittle Materials

When designing for static loading of brittle materials, such as cast iron, the failure mode to be avoided is brittle fracture. Cast iron, for instance, has a smooth stress-strain curve from zero load to fracture, with no noticeable yielding; like many brittle materials, it is stronger in compression than in tension. The design criterion must, therefore, be based on ultimate, not yield strength, with consideration given to both tensile and compressive stresses. Thus, for these materials, failure is anticipated when the maximum induced normal stress becomes equal to the working stress for either tension or compression. Equivalent maximum and minimum induced normal stresses are calculated from the following formulas.

$$\sigma_{i1} = \frac{\sigma}{2} + \sqrt{\left(\frac{\sigma}{2}\right)^2 + \tau^2} \tag{3-12}$$

$$\sigma_{i2} = \frac{\sigma}{2} - \sqrt{\left(\frac{\sigma}{2}\right)^2 + \tau^2} \tag{3-13}$$

where

σ_{i1} = maximum induced normal stress

σ_{i2} = minimum induced normal stress

Note that σ_{i1} and σ_{i2} can be either tensile or compressive. When tensile, they are compared to the tensile ultimate working stress; when compressive, they are compared to the compressive ultimate working stress. Thus, for tensile stresses,

$$\sigma_{wt} = \frac{S_{ut}}{fs} \tag{3-14}$$

and for compressive stresses,

$$\sigma_{wc} = \frac{S_{uc}}{fs} \tag{3-15}$$

If either σ_{i1} or σ_{i2} becomes equal to σ_w in one of the last two equations, the brittle material design criterion is satisfied. If the criterion is satisfied with $fs = 1$, brittle fracture is anticipated.

EXAMPLE 3-4

An ASTM No. 30 cast iron machine component is subject to an axial stress $\sigma = 138$ N/mm^2 and a direct shear stress $\tau = 69$ N/mm^2. The ultimate strengths for this material are $S_{ut} = 214\ \text{N/mm}^2$ and $S_{uc} = 751$

N/mm². Determine the factor of safety against brittle fracture.

SOLUTION

The combined equivalent induced maximum and minimum normal stresses will be found from Eqs. 3-12 and 3-13, and the results will be compared to tensile and compressive working stresses.

Induced Stresses

$$\sigma_{i1} = \frac{\sigma}{2} + \sqrt{\left(\frac{\sigma}{2}\right)^2 + \tau^2}$$

$$= \frac{138}{2} + \sqrt{\left(\frac{138}{2}\right)^2 + (69)^2}$$

$$= 69 + 97.6 = 167 \text{ N/mm}^2$$

$$\sigma_{i2} = \frac{\sigma}{2} - \sqrt{\left(\frac{\sigma}{2}\right)^2 + \tau^2}$$

$$= 69 - 97.6 = -28.6 \text{ MPa}$$

Working Stresses

$$\sigma_{wt} = \frac{214}{fs}$$

$$\sigma_{wc} = \frac{-751}{fs}$$

Equating tensile and compressive working stresses with corresponding induced stresses gives

$$167 = \frac{214}{fs}$$

$$fs = \frac{214}{167} = 1.3 \text{ (tension)}$$

$$-28.6 = \frac{-751}{fs}$$

$$fs = \frac{-751}{-28.6} = 26.3 \text{ (compression)}$$

Therefore the factor of safety is 1.3, based on tensile brittle fracture.

3-6 FATIGUE IN RAILROAD AXLES

Repeated failures of railroad axles in Europe in the early 1800s caused August Wöhler, 1819–1914, to begin fatigue testing of materials in 1862. Wöhler realized that materials (like humans) were subject to fatigue when exposed to stresses repeated over many cycles.

Early railroad cars moved on wheels rigidly attached (shrunk) to a solid axle. The bearings were mounted outside the wheels (Fig. 3-7*a*). The corresponding free-body diagram shows the bearing supports of the beam shaft with vertical forces acting at each wheel (Fig. 3-7*b*). At any instant, the axle is loaded in bending with maximum stresses at top and bottom (Fig. 3-7*c*). However, because of rotation, the material at any point undergoes a complete stress cycle every revolution (Fig. 3-7*d*). During operation, stress cycles accumulate rapidly, and fracture may occur at either of the two bearings, coinciding with the maximum bending moment. Figs. 3-7*c* and 3-7*d* show the difference between ductile, uniaxial, tensile overload and fatigue fracture. Wöhler found the distinct pattern of lines and textures on the fracture surfaces to be an aid in explaining the difference between fatigue and simple ductile fracture. In short, Wöhler discovered that ductile materials subject to repeated loads generally behave elastically until suddenly, without external warning, they fracture.

Wöhler's work initiated the continuing struggle to comprehend the complex subject of material fatigue. Many investigators have followed Wöhler's pioneering example during the ensuing century, but there remain aspects of fatigue that are not well understood. Designers can, however, apply present-day knowledge to the machine components they design, with statistically reasonable assurance that such parts will perform satisfactorily for a predetermined length of time.

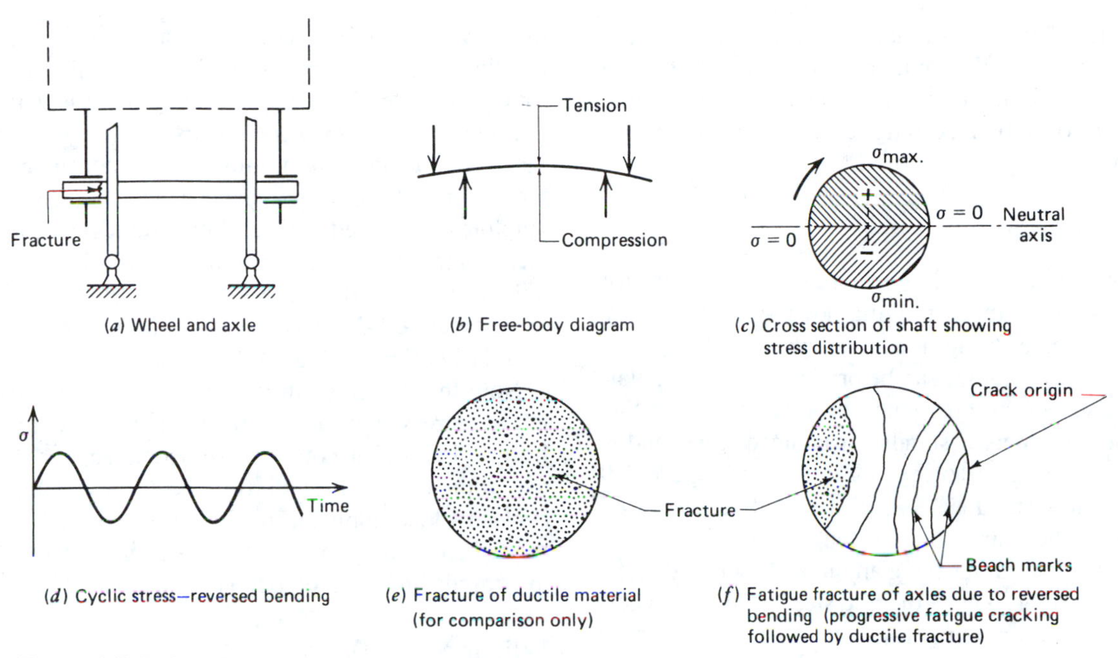

Figure 3-7 Fatigue failure of railroad axles (Wöhler).

3-7 STRESS-CYCLE DIAGRAMS

Rotary Beam Method

Since the load-stress conditions of railroad axles are easily simulated in the laboratory, materials were soon tested by what is known as the rotary beam method. In this method (Fig. 3-8), a motor spins a slender, round, solid, polished test specimen, supported at each end but loaded in pure bending. The configuration generates a cyclic bending stress, as shown in Figs. 3-7*c* and 3-7*d*.

Figure 3-8 Schematic diagram of rotating beam test machine. The majority of tabulated fatigue strength data was obtained using this method.

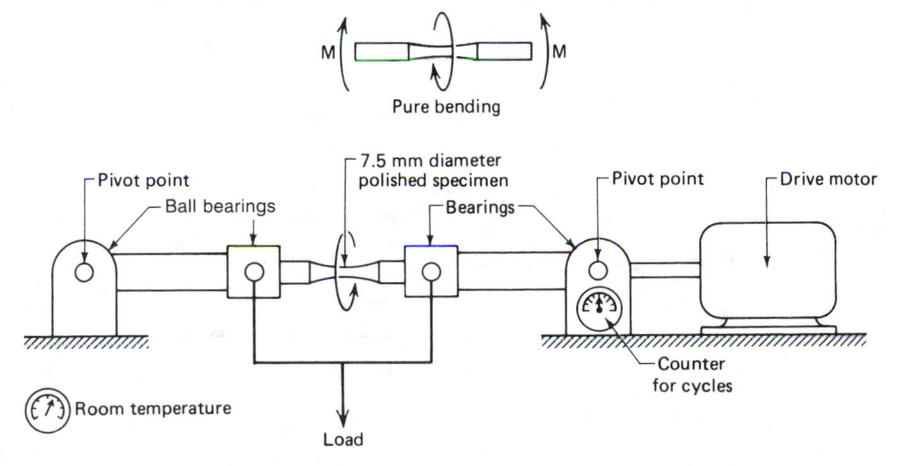

The rotary beam method is important because it simulates the dynamic stress conditions in many machine components and because the majority of published fatigue strength data was obtained using this method.

S-N Curves

By counting the number of cycles to failure for stress amplitudes ranging downward from 90% of ultimate strength, a stress cycles-to-failure (or fracture) diagram can be produced for any metal. These curves are generally referred to as *S-N* curves, where *S* stands for induced stress and *N* for the number of cycles to failure. Figure 3-9 shows a standard *S-N* curve, with *S* measured along the *y*-axis and *N* along the *x*-axis.

All *S-N* curves have an initial downward slope to the right. Those of mild steel (polished specimens at room temperature in air) have a horizontal asymptote, a feature of enormous practical importance. The asymptote indicates that statistically, below a certain stress level S_e', mild steels will endure a complete stress reversal indefinitely. The word "statistically" is significant because each curve is merely a statistical average of perhaps 50 specimens.

Unfortunately, aluminum and other nonferrous materials do not have a horizontal asymptote. Published fatigue limit values S_e' for these materials are for a specific statistical life to failure, usually 100 to 500 million cycles. A representative *S-N* curve for aluminum is shown as a dashed line in Fig. 3-9.

Note that the *S-N* curve starts at the extreme left at a stress amplitude of $0.9S_u$. This reduction from S_u helps to account for statistical scatter in the failure data and for starting the curve at 1000 cycles of load application. *S-N* curves are often plotted on semilog axes. Curves are thus reduced to straight lines and clarity is enhanced.

Fatigue Strength

Fatigue strength is associated with repeated loading. It is obtained by subjecting small specimens

Figure 3-9 Typical *S-N* curve for ductile materials. Note spread of the data points.

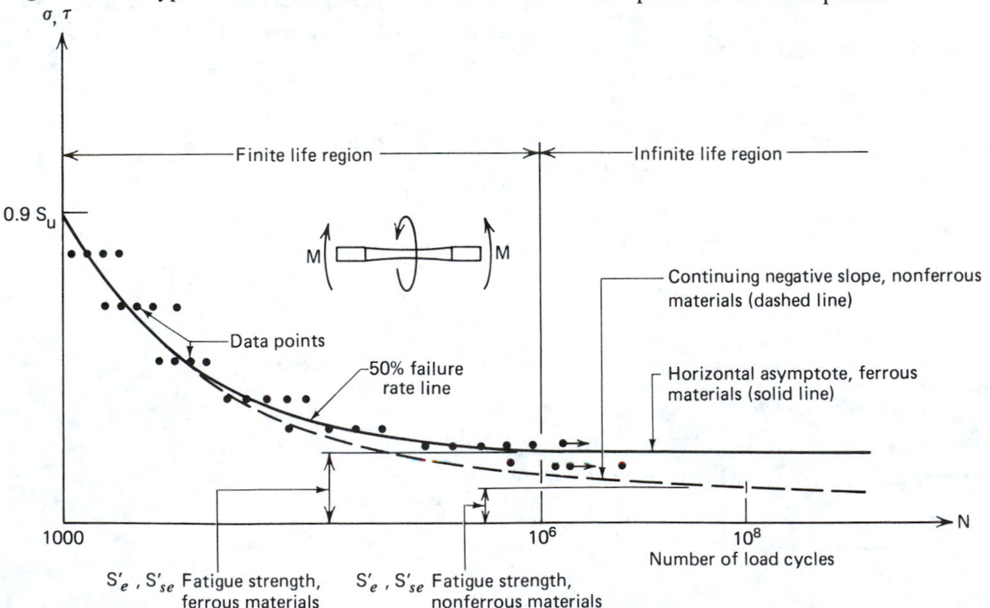

of a given material and shape to cyclic loading. An *S-N* plot of the resulting constant, fully reversed stress amplitude versus cycles to failure may then be used to establish the material's fatigue strength (or stress amplitude) for a given cycle life.

The asymptotic stress level S_e' (ductile ferrous materials only) is termed the fatigue limit. Since steel predominates in machine design, it is fortunate that a stress level exists below which a steel part can (in the absence of corrosion and high temperature) function almost indefinitely. Thus knowledge of fatigue limits has greatly enhanced the designer's ability to design for long life.

Rotary beam testing is a good simulation of the bending stress variation in shafts carrying gears, pulleys, and sprockets. With appropriate derating factors, the data may also be applied to machine members loaded in cyclic tension and shear. Published data are usually mean values of the statistical scatter (50% failed for the given stress amplitude S and number of cycles N), and were obtained using polished specimens of cylindrical shape ranging from 5 to 12 mm in diameter.

Cause and Appearance

Flaws in the material—scratches, minute invisible surface cracks, or microscopic nonmetallic inclusions—are a major cause of fatigue failure. With repeated loading, these flaws grow; eventually one (or a small number) shows significant propagation, resulting in cracks. The crack or cracks grow as the load cycles continue. Eventually, the remaining material can no longer support the load, and rupture occurs.

As can be observed, fatigue fracture surfaces often display two distinctly different zones. The one section, often discolored by corrosion, usually exhibits a pattern of lines or *beach marks* (Fig. 3-7*f*). At times, the beach marks (or fatigue striations) are so fine that they are visible only at great magnification (such as is possible with an electron microscope). Crack origin and direction of progression are often indicated by these markings, which thus give a clue to possible material flaws or inadequate design. The other zone of the fracture usually has the bright, grainy appearance of ductile rupture or fracture.

Fatigue is very sensitive to tensile stresses, as can be seen from their role in directional control of crack propagation. However, all types of cyclic loading can cause fatigue failure. In rolling bearings variable compressive stresses and shear dominate.

3-8 CLASSIFICATION OF COMMON FATIGUE FRACTURES

A major goal of designers is to obtain machine parts that do not fail. However, in the process of developing and testing new machines or evaluating broken parts from old machines, features of the fracture surface may help explain why the part failed. Figure 3-10 illustrates fracture surface features for the loading combinations of tension compression, torsion, and bending that can be used in determining the cause of failure of machine components.

Other investigative techniques for cause of failure include tensile testing of the material, hardness measurements on the broken part, microscopic examination of the fracture surface and of metallographic samples of the broken part, chemical analysis of the material, experimental strain measurement on a similar part under actual service conditions, and careful stress analysis.

In machine design, as in many other endeavors, experience is the best teacher; therefore the more experience designers have in discovering what causes machine members to fail, the better they will be able to prevent failure in the members they design. Students should, therefore, become familiar with failure analysis techniques and use them at every opportunity.

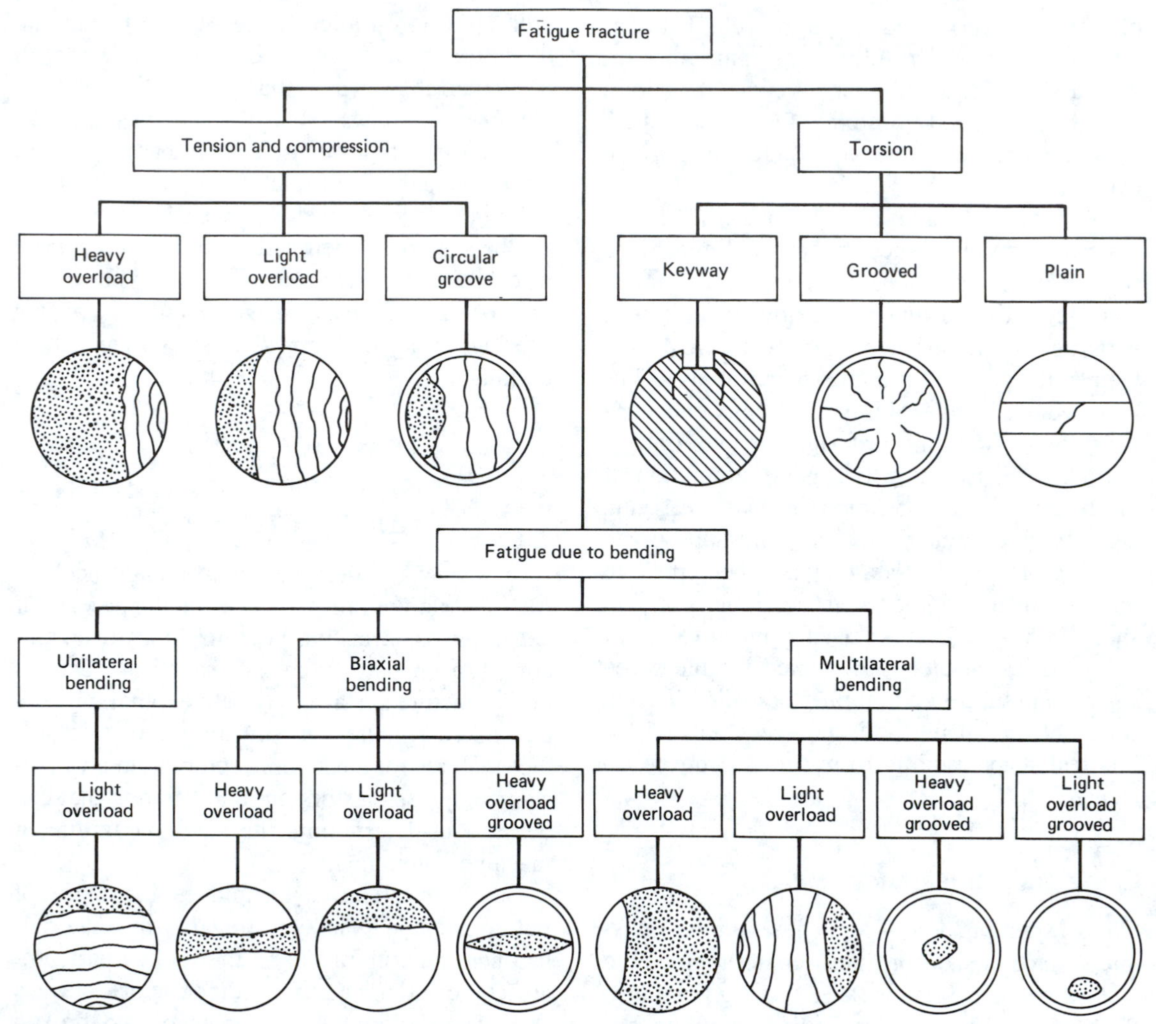

Figure 3-10 Classification of common fatigue fractures. The feature to note is that for tension and compression, the beach marks (lines) all curve in the same direction. For bending there is often a change in direction from one half to the other. (Adapted from Franz Findeisen, *Neuzeitliche Maschinenelemente,* Band 3, Abb. 65. Schweizer Drück-Verlagshaus AG, Zürich.)

3-9 FATIGUE STRENGTH APPROXIMATIONS

The availability of tabulated endurance limits and fatigue strength data for materials is steadily improving. However, in some cases designers may still have to design machine members with only tensile test data available. In these cases the approximate mean endurance limit can conveniently be estimated from formulas, providing the ultimate tensile strength S_{ut} is accurately known.

For steel,

$$S'_e = 0.5S_{ut}; \quad S_{ut} \leq 1400 \text{ MPa (200 Ksi)}$$
$$S'_e = 700 \text{ MPa (100 Ksi)}; \quad S_{ut} > 1400 \text{ MPa (200 Ksi)} \tag{3-16}$$

Note that S'_e is constant for ultimate tensile strengths above 1400 MPa (200 Ksi).

For cast iron,

$$S'_e = 0.4S_{ut} \tag{3-17}$$

The approximate mean fatigue strength of aluminum for the range 100 to 500 million cycles to failure is given by the following formulas. For wrought aluminum alloys,

$$S'_e = 0.4S_{ut} \tag{3-18}$$

For cast aluminum alloys,

$$S'_e = 0.3S_{ut} \tag{3-19}$$

The formulas in this section have been found to give reasonable estimates when compared to fatigue data, but designers are cautioned to seek actual test data in critical cases.

Torsional Fatigue Strength

Since tabulated fatigue strength data are predominantly rotating beam data, to analyze torsional fatigue, the data must be converted to corresponding shear fatigue strength. This is accomplished for ductile materials as follows.

$$S'_{se} = 0.57S'_e \tag{3-20}$$

where S'_{se} equals completely reversed torsional fatigue strength.

Torsional fatigue strength values S'_{se} could be generated by subjecting small laboratory specimens to completely reversed torsional shear stress and recording the amplitude of stress S_s versus number of cycles to failure N, yielding an S_s-N diagram. However, Eq. 3-20 has been found to give good approximations for S'_{se}. Designers may, therefore, estimate torsional or normal fatigue strengths from ultimate strength data using the equations of this section.

3-10 FATIGUE STRENGTH DERATING FACTORS

The primary use for derating factors is to determine the fatigue strength of a part subject to cyclic loading. These factors, as previously mentioned, are used to adjust the fatigue strength to account for loading effects, part features, and environmental conditions that are recognized to have a detrimental effect on the cycle life of the part. The object, then, is to take the laboratory specimen's rotating beam endurance limit S'_e and modify it using derating factors to yield the endurance limit S_e for the actual machine part under consideration. The calculations are made as follows.[8]

$$S_e = k_a k_b k_c k_d k_e k_f S'_e \tag{3-21}$$

where

S_e = endurance limit/fatigue limit of the machine member
S'_e = endurance limit/fatigue limit of a rotating beam specimen of the same material
k_a = surface finish derating factor
k_b = size derating factor
k_c = reliability derating factor
k_d = temperature derating factor
k_e = stress concentration derating factor
k_f = impact derating factor

3-11 SURFACE FINISH

Fatigue strength of steel parts increases rapidly as surfaces progressively improve through machining, grinding, and polishing. Polishing roughly doubles fatigue strength of low hardness, forged, and hot rolled parts, and gives larger increases for materials of greater hardness. The curves shown in Fig. 3-11 are, thus, of great practical importance, since they show the variation of fatigue limit with surface finish for steels.

Surface conditions affect fatigue strength for two reasons. First, the maximum induced stress appears on the surface of most of the parts. Second, minute cracks, fine scratches, or tiny, nonmetallic inclusions act as stress raisers that occur more often at the surface than within the part. The effect of machining, grinding, and polishing

---

[8]Joseph Marin, "Design for Fatigue Loading," *Machine Design*, Vol. 29, No. 2, p. 127, 1957.

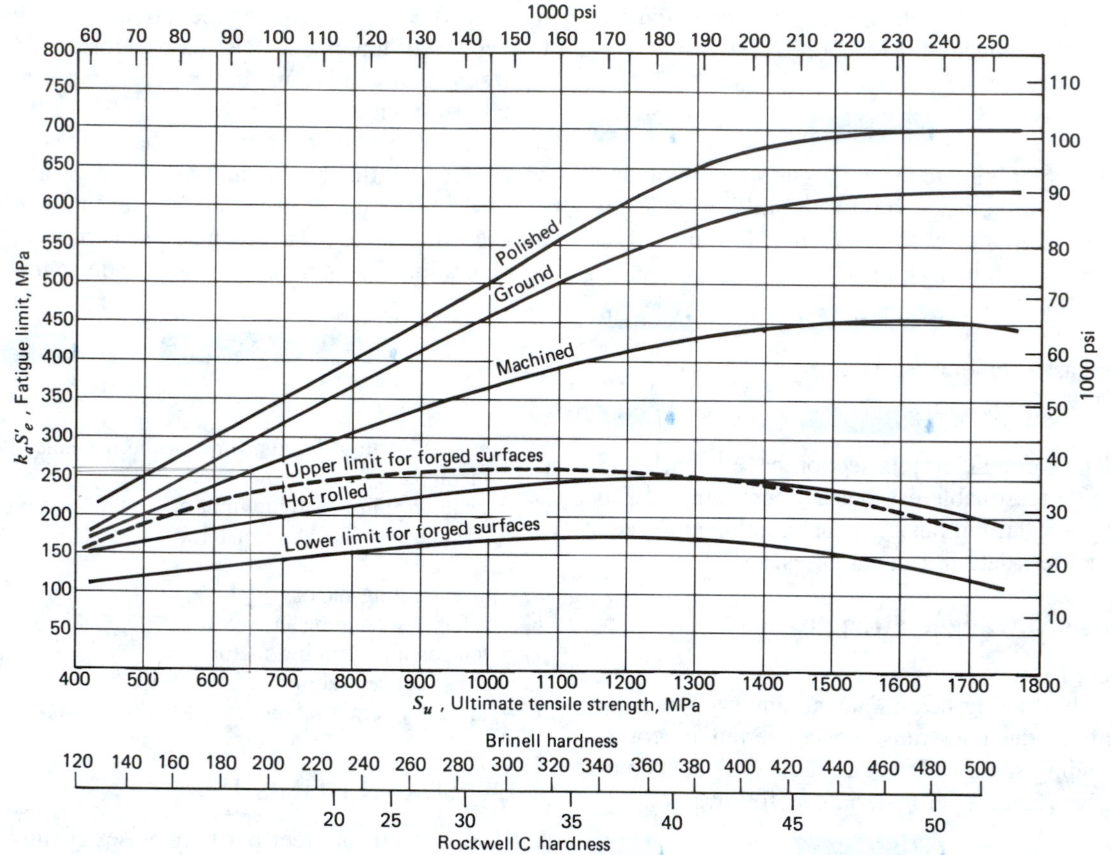

Figure 3-11 Effect of hardness and surface finish on the bending fatigue limit of steel. (Adapted from Joseph Shigley, *Machine Design,* First Edition, New York: McGraw-Hill, 1956.)

is to diminish or close surface irregularities, thereby reducing the probability of major cracks being opened by high surface stresses.

Surface Finish Derating Factor, k_a

In order to use the information contained in Fig. 3-11 to calculate a derating factor for surface finish, Eqs. 3-16 and 3-17 are used to estimate S'_e; then k_a is calculated as follows.

For steel,

$$k_a = \frac{k_a S'_e}{S'_e} = \frac{k_a S'_e}{0.5 S_{ut}}; \quad S_{ut} \leq 1400 \text{ MPa or } 200 \text{ Ksi}$$

$$= \frac{k_a S'_e}{700 \text{ MPa}}; \quad S_{ut} > 1400 \text{ MPa or } 200 \text{ Ksi} \tag{3-22}$$

For cast iron,

$$k_a = \frac{k_a S'_e}{0.4 S_{ut}} \tag{3-23}$$

The curves of Fig. 3-11 are for wrought steel but may also be used for cast steel and better grades of cast iron. To determine k_a from the figure, designers must first select the material's ultimate strength and the surface finish to be used for the machine part. Table 3-1 and similar reference materials are useful for making this selection. Once the material and surface finish are known, the k_a value can be calculated from Eqs. 3-22 and 3-23.

Tabulated endurance limits for aluminum and other nonferrous materials already include surface finish effects. Thus, for these materials, k_a is unity.

3-12 SIZE EFFECTS

Small machine parts exhibit greater fatigue strength than larger ones of the same surface finish, material, and configuration. This decreasing fatigue strength with increasing size results from the occurrence of higher stress gradients (flexure and torsion) and larger surface areas in all regions of high stress in larger parts. Larger surfaces have more defects totally; therefore the probability of failure is greater in large parts.

Size Derating Factor, k_b[9]

To account for the reduction in fatigue strength due to size effect, the following derating factors may be used.

$k_b = 1.0;\quad d \leq 7.6$ mm (0.30 in.)

$k_b = 0.85;\quad 7.6 \text{ mm} < d \leq 50$ mm (2.0 in.)

$k_b = 0.75;\quad d > 50$ mm (2.0 in.)

where

d = part diameter, section depth, or thickness of noncircular parts

For parts larger than 50 mm (2.0 in.), component testing is recommended. Since more machine parts of interest belong in the second category, a good starting point when the size is unknown is to set $k_b = 0.85$.

3-13 SURVIVAL RATES

Rotary beam data generally reflect a survival rate of 50%. This means that tabulated endurance limits are really the stress amplitude value that corresponds to the middle of the failure scatter band. In other words, for the same cycle life (number of constant amplitude stress cycles to failure), about 50% of the rotary beam specimens were able to sustain a higher stress amplitude, and about 50% a lower stress amplitude. Therefore, to insure that more than 50% of the parts will survive, the stress amplitude the part experiences must be lower than the tabulated endurance limit. The reliability[10] derating factor k_c is used to achieve this reduction.

TABLE 3-3 Reliability Derating Factors, k_c

Reliability, %	k_c
50.0	1.0
90.0	0.897
95.0	0.868
99.0	0.814
99.9	0.753
99.99	0.702

Source. Adapted from Joseph Shigley, *Mechanical Engineering Design,* Second Edition (New York: McGraw-Hill, 1963), p. 256.

Reliability Derating Factor, k_c

The greater the likelihood that a part will survive, the lower the derating factor k_c. Table 3-3 gives values of k_c corresponding to various reliability percentages. Note that the k_c values in Table 3-3 are approximations. For accurate survivability rates in critical cases, designers must rely on actual test data.

3-14 TEMPERATURE EFFECTS

Although material strengths are specified at room temperature, machine part operating temperatures are generally higher. Yet a decline in static and dynamic strengths occurs with increasing temperature. Whenever temperatures are increased significantly, metals usually re-

[9]Adapted from Joseph Shigley, *Mechanical Enginering Design,* Third Edition (New York: McGraw-Hill, 1977), p. 190.

[10]Reliability is defined as the probability that a system will perform satisfactorily for a given period of time when used under certain conditions.

spond by expanding in size and by softening, with accompanying reductions in strength.

A machine part may, therefore, lose strength with increased temperature and may also experience increased induced stress if its expansion is restricted by surrounding members. Temperature increases thus pose a potential problem whenever materials with different coefficients of expansion are assembled.

Some strength-reducing effects of temperature take place over long periods of time. These effects are termed *creep*. Creep is characterized by gradual relaxation of load and component elongation; it generally occurs when a machine member is held at high temperature for a long time. These effects can be very detrimental to close tolerance assemblies and to clamping-type connections under steady load conditions.

Another heat-sensitive component is the bearing. In both sliding and rolling bearings, overloads, inadequate lubrication, or both can effectively lower fatigue strength because of heating. It is very possible that some of the fatigue failures observed by Wöhler were precipitated by overheated journal bearings, a common occurrence in the early days of railroads.

Temperature Derating Factor, k_d

To account for reduction in the endurance limit of steels for temperatures above 71°C (160°F), the following expression for the temperature derating factor k_d is used.[11]

$$k_d = \frac{620}{460 + T} \text{ (Fahrenheit)}$$

$$k_d = \frac{344}{273 + T} \text{ (Celsius)} \qquad (3\text{-}24)$$

where T = the operating temperature of the machine part. Special alloys, other materials, and critical applications require actual test data to establish appropriate derating factors.

Note that k_d accounts for a reduction in available strength as the temperature of the material is increased above the ambient temperature at which the original strength data were obtained. It does not account for increased induced stresses that may be caused by higher temperatures.

[11]Adapted from Joseph Shigley, *Mechanical Engineering Design*, Second Edition (New York: McGraw-Hill, 1963), p. 257.

3-15 STRESS CONCENTRATIONS

When cross sections of machine members change abruptly due to holes, keyways, threads, shaft shoulders, and other essential discontinuities, stresses in those sections deviate from the nominal stress. The *net* effect of any discontinuity is to raise the stress in the immediate area by a factor as large as 20 but usually less than 5. Herein lies the danger of discontinuities (Fig. 3-12). Note that in Fig. 3-12 the load F and the cross-sectional area A under load are the same for each of the three members shown, but the maximum stress varies over wide extremes.

Discontinuities are sometimes referred to as *stress raisers*. Regions in which stress raisers occur are called *areas of stress concentration*. Geometric stress concentration occurs whether the stress is steady or variable and whether it is from axial loading, from bending, or from torsion. However, different loadings call for different stress concentration factors for the same part geometry. The transition from nominal to maximum stress is in each case accomplished by means of a geometric or theoretical *stress concentration factor* K_t. The maximum stress at a discontinuity is shown in Fig. 3-13 for three common loading configurations. For the various loadings, we have:

Direct load: $$\sigma = K_t \frac{F}{A} \qquad (3\text{-}25)$$

Bending: $$\sigma = K_t \frac{Mc}{I} \qquad (3\text{-}26)$$

Torsion: $$\tau = K_t \frac{Tr}{J} \qquad (3\text{-}27)$$

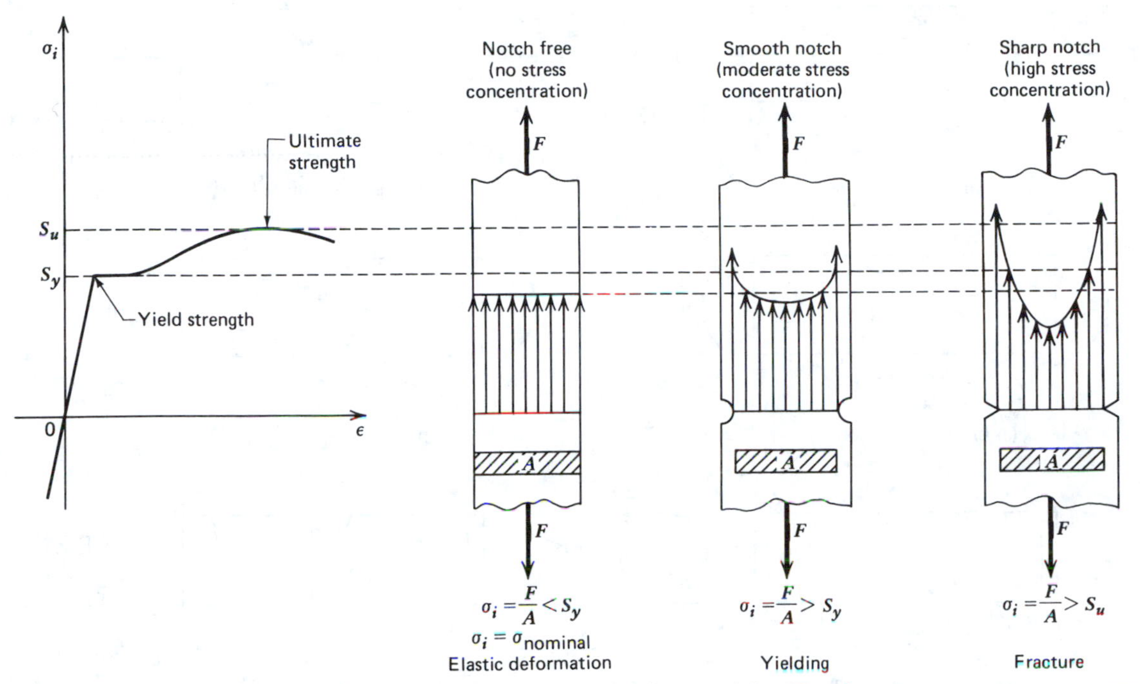

Figure 3-12 Effect of geometric stress concentration on ductile materials. (Developed by Uffe Hindhede and Wendell C. Hull.)

where A, I, and J are based on *net* section. Tables for K_t are available in the appendix and in *Machinery's Handbook.*

In the case of ductile materials, geometric stress concentration factors are seldom considered for static loads. The rationale is that yielding at the discontinuity redistributes internal stress until the entire cross section joins in sharing the load, which means that the maximum stress becomes approximately equal to the nominal stress and K_t is reduced to 1. However, this is not the case for static loads on brittle materials (cast iron, hardened tool steel, etc.), since they do not yield; hence full values of geometric stress concentrations K_t must be used. Likewise, K_t values cannot be neglected in fatigue loading.

K_t is accurate for static loads but often yields conservative stress values for repeated loads on ductile materials. Here, again, localized yielding at areas of high stress concentration will lower the stress (to a degree). K_t can be used for *preliminary* stress calculations. When greater accuracy is needed, however, fatigue stress calculations should include material *notch sensitivity*.

Stress raisers are present in all designs but, fortunately, diagrams and charts for K_t values are available to deal with many major cases. Figure 3-14 shows design techniques for reducing stress concentration and improving fatigue strength. One of these, labeled "Gear, Pulley, etc., Shrunk on Shaft," applies to axle wheel assemblies of railroad cars. Early designs observed by Wöhler were found to contain stress raisers due to excessive pressure at the wheel-axle interface and had a poorly designed hub.

Notch Sensitivity

It has been pointed out that, because of yielding, actual stress will drop below the values given by means of K_t. Additionally, some materials dem-

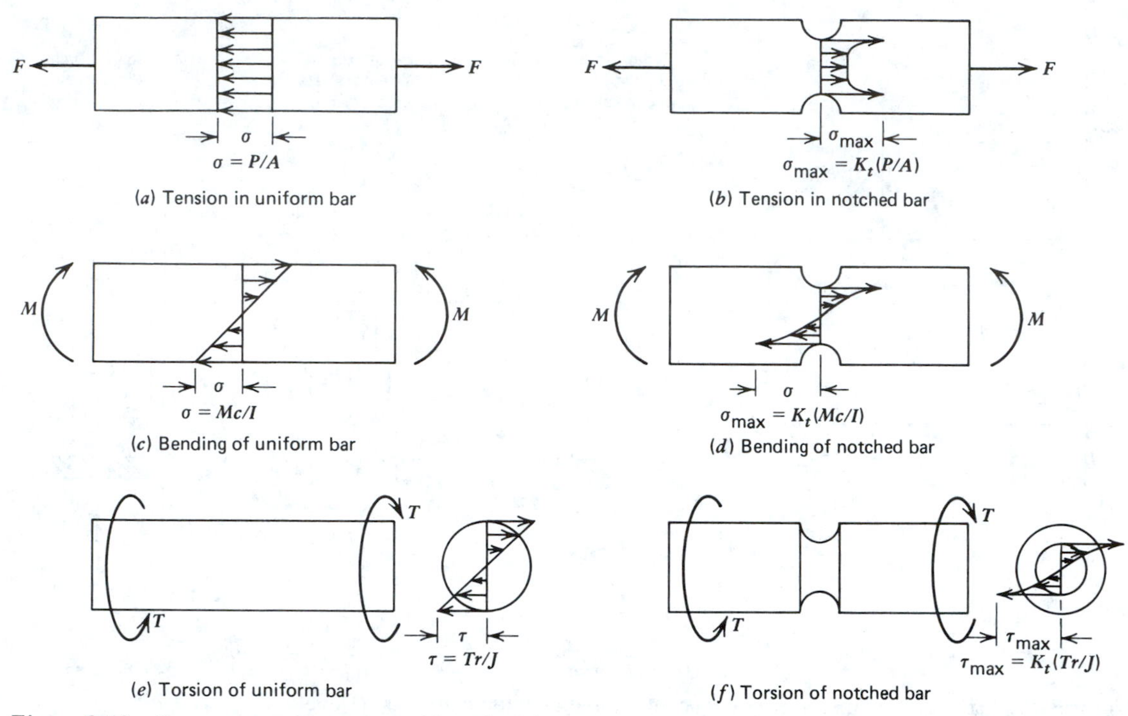

Figure 3-13 Comparison of stresses with and without stress concentration.

onstrate greater toughness and greater resistance to crack initiation in the presence of geometric stress concentration. To account for this, an important factor called *notch sensitivity* q is used. The relation between the *actual stress concentration factor* K_f and the *geometric* stress concentration factor K_t is

$$K_f = 1 + q(K_t - 1) \tag{3-28}$$

The magnitude of q depends on the material and the type of loading. Suitable values for carbon or alloy steels and aluminum can be obtained from Figs. 3-15 and 3-16. Note that r in Figs. 3-15 and 3-16 is the smallest part radius at the point where material strength is to be determined.

Cast iron has a unique internal structure (compared to other metals) that makes it less notch sensitive. A commonly accepted value of q for cast iron is 0.2.

Modifying Factor for Stress Concentration, k_e

In order to correct or modify the endurance limit for geometric stress concentration in the form of a derating factor, it is necessary to take the reciprocal of K_f. The calculation for k_e, the modifying factor for stress concentration, is thus given by the relationship

$$k_e = \frac{1}{K_f} \tag{3-29}$$

Figure 3-14 Design techniques for improving fatigue strength. (Adapted from Joseph Marin, *Machine Design*, 1961.)

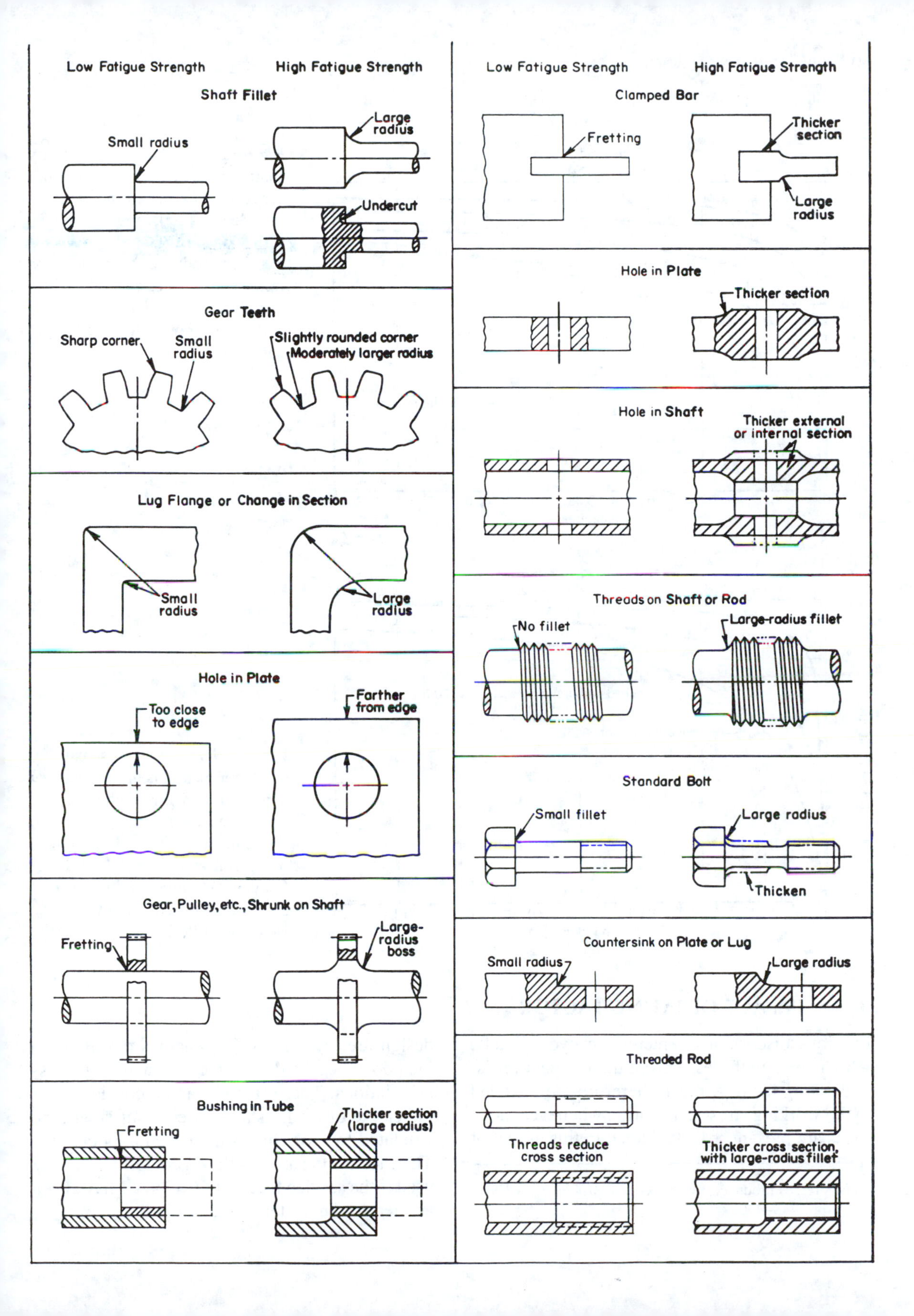

Low Fatigue Strength
High Fatigue Strength
Shaft Fillet
Small radius
Large radius
Undercut
Gear Teeth
Sharp corner
Small radius
Slightly rounded corner
Moderately larger radius
Lug Flange or Change in Section
Small radius
Large radius
Hole in Plate
Too close to edge
Farther from edge
Gear, Pulley, etc., Shrunk on Shaft
Fretting
Large-radius boss
Bushing in Tube
Fretting
Thicker section (large radius)
Low Fatigue Strength
High Fatigue Strength
Clamped Bar
Fretting
Thicker section
Large radius
Hole in Plate
Thicker section
Hole in Shaft
Thicker external or internal section
Threads on Shaft or Rod
No fillet
Large-radius fillet
Standard Bolt
Small fillet
Large radius
Thicken
Countersink on Plate or Lug
Small radius
Large radius
Threaded Rod
Threads reduce cross section
Thicker cross section, with large-radius fillet

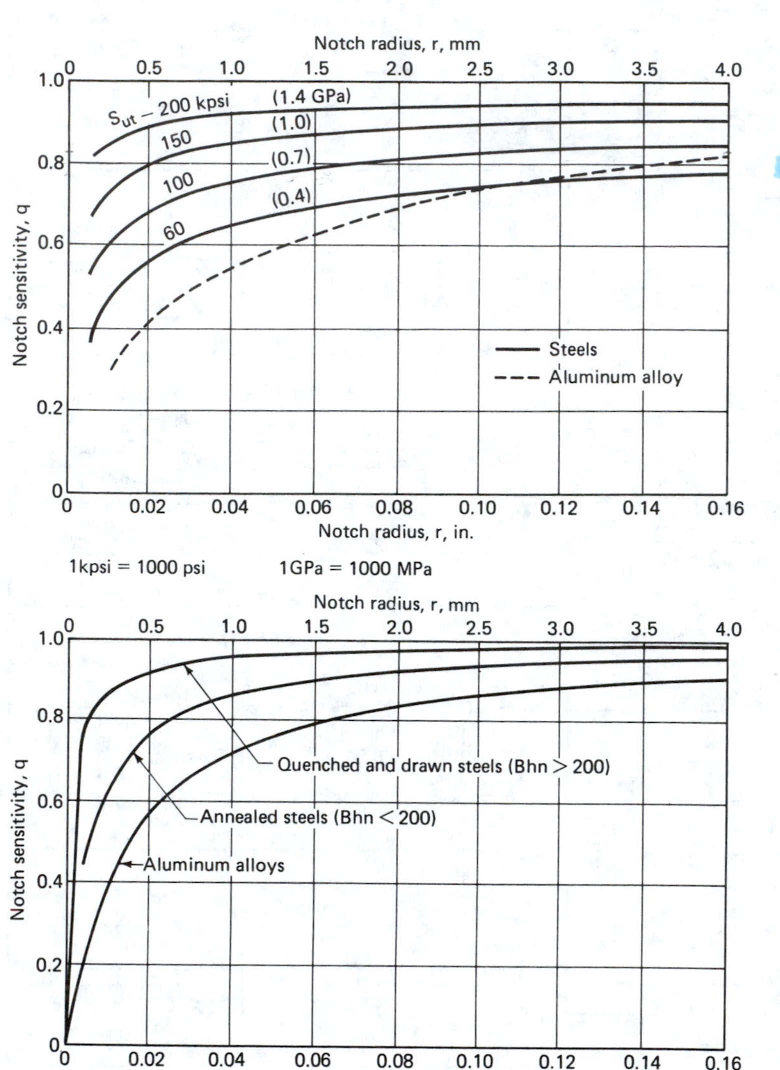

Figure 3-15 Notch sensitivity charts for steel and aluminum under *reversed bending* and reversed axial loads. For larger notch radii, use q values for $r = 4$ mm (0.16 in.). [Reproduced by permission from George Sines and J. L. Waisman (eds.), *Metal Fatigue* (New York: McGraw-Hill, 1959), pp. 256, 298.]

Figure 3-16 Notch sensitivity charts for materials under *reversed torsion*. For larger notch radii, use the values of q corresponding to $r = 4$ mm (0.16 in.). [Reproduced by permission from Joseph E. Shigley, *Mechanical Engineering Design*, 3rd ed. (New York: McGraw-Hill, 1977), p. 195.]

3-16 IMPACT DERATING FACTOR

Previous discussions centered on cyclic loads, which are usually less severe than impact loads. A derating factor k_f must, therefore, be included in the working stress calculation whenever shock loads are present.[12] Data related to this aspect of design are few. Table 3-4 enables designers to include shock loading effects in working stress calculations. These factors can be considered as indicative of cyclic shock effects but should be validated by experiment in critical applications. The factors in Table 3-4 are generally not to be used if the service factors of Chapter 1 have been applied to the system as a whole.

[12]For stresses produced by shocks, see *Machinery's Handbook*, pp. 448–450.

TABLE 3-4 Impact Derating Factor, k_f

Impact Load	k_f	Examples
Light	1.0–0.9	Machines based on pure rotary motion—turbines, centrifugal pumps, motors, etc.
Medium	0.8–0.7	Machines based on slider cranks, pumps, compressors, etc.
Heavy	0.6–0.5	Presses for tools and dies, shears, etc.
Very heavy	0.5–0.3	Hammers, rolling mills, crushers, etc.

Source. Adapted from G. Reitor and K. Hohmann, *Grundlagen des Konstruierens*, 3 auflage (Essen, West Germany: W. Girardet, 1977).

EXAMPLE 3-5

The machine element shown in Fig. 3-17 is made from A92024-T3 aluminum, which has a fatigue strength $S'_e = 20$ Ksi and ultimate strength $S_u = 70$ Ksi. The surface was cold drawn, then machined. A reliability of 99% is desired, while the machine operates at 21°C (70°F). The loading is axial and is classified as light impact. Determine the machine component fatigue strength S_e.

SOLUTION

Starting with the basic relationship for fatigue strength, Eq. 3-21,

$$S_e = k_a k_b k_c k_d k_e k_f S'_e$$

we must evaluate each of the derating factors k_a to k_f.

Surface Finish

$k_a = 1.0$, since the material is aluminum and the tabulated rotating beam fatigue strength S'_e includes surface effects.

Size Effect

$k_b = 0.85$ for section thickness greater than 7.6 mm and less than 50 mm.

Reliability

$k_c = 0.814$, since a reliability of 99% is called for.

Temperature Effect

$k_d = 1.0$, the part operates at room temperature.

Stress Concentration Effect

We must refer to the theoretical stress concentration figures in the appendix to find K_t before k_f can be calculated. From Figure F-2, we find, for $W/d = 0.37/1.0 = 0.37$, that $K_t = 2.3$. Then, from Fig. 3-15 for a notch radius $r = 0.37 = 0.185$ in., the notch sensitivity q is 0.82. Now we use Eq. 3-28 to calculate K_f (note that this is not the same as k_f, the impact derating factor), the reduced or actual stress concentration of the machine part.

$$K_f = 1 + q(K_t - 1)$$
$$= 1 + (0.82)(2.3 - 1)$$
$$= 2.07$$

Figure 3-17 Machine element in tension.

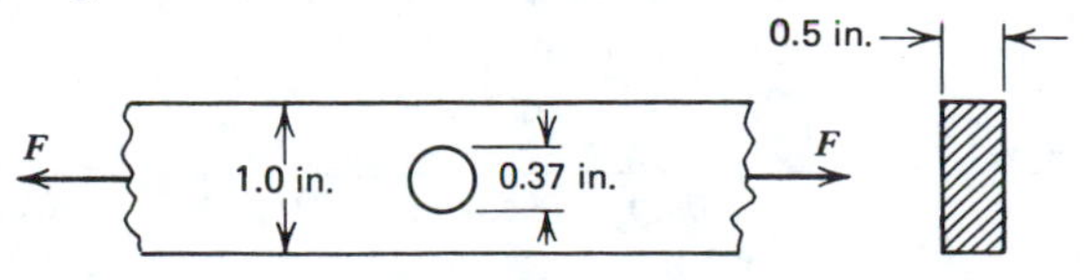

The fatigue strength stress concentration derating factor is found by using Eq. 3-29.

$$k_e = \frac{1}{K_f} = \frac{1}{2.07} = 0.48$$

Impact Loading Effect

$k_f = 0.9$, for light impact. If more were known about the specific application, this might be increased to unity.

Substituting the derating factors into the fatigue strength relationship gives

$$S_e = (1.0)(0.85)(0.814)(1.0)(0.48)(0.9)(20 \text{ Ksi})$$
$$= \mathbf{6\ Ksi}\ (6000 \text{ psi})$$

Therefore the fatigue strength of the machine part is 6 Ksi as compared to a rotating beam test specimen of the same material that had a fatigue strength of 20 Ksi. This example illustrates the strength-reducing effect of many of those factors affecting fatigue strength. Section 3-17 will discuss additional effects.

EXAMPLE 3-6

Assume that the machine component of Example 3-5 is subjected to a completely reversed force of $F = 500$ lb. Calculate the factor of safety for fatigue failure for a life of 10^8 cycles.

SOLUTION

Fatigue strengths given for aluminum alloys are for 10^8 cycles or more; thus we may use the machine member fatigue strength previously calculated in Example 3-5.

Induced Stress

The induced stress is the amplitude of stress caused by the cyclic load $F = 500$ lb. Thus

$$\sigma_{ia} = \frac{F}{A} = \frac{500 \text{ lb}}{(0.5 \text{ in.})(1.0 \text{ in.} - 0.37 \text{ in.})}$$
$$= \mathbf{1.6\ Ksi}\ (1600 \text{ psi})$$

Working Stress

$$\sigma_{wa} = \frac{S_e}{\text{fs}} = \frac{6 \text{ Ksi}}{\text{fs}}$$

Factor of Safety

Equating the induced and working stress amplitudes,

$$1.6 \text{ Ksi} = \frac{6 \text{ Ksi}}{\text{fs}}$$

$$\text{fs} = \frac{6}{1.6} = \mathbf{3.8}$$

Therefore the factor of safety is 3.8.

3-17 OTHER FACTORS AFFECTING FATIGUE STRENGTH

Fatigue strength of mechanical components depends on a great number of factors, many of which have been noted. Additional factors that are important because of their effect on the surface characteristics of machine members will be discussed in this section.

Surface Treatment

Fatigue strength can be improved by 25 to 30% through cold rolling, shot peening, hammer peening, and hardening of critical surfaces. These treatment effects are based on the fact that surface cracks originate and propagate in areas where the material is stressed in tension. Critical crack-initiating stresses may arise when induced tensile stresses are superimposed on *residual* tensile (or compressive) stresses at the surface. Residual stresses are "leftover" stresses (often tensile) from machining, forming, or welding, or they can be purposely implanted in the metal as beneficial compressive residual stresses by various surface treatments. Residual stresses exist independent of load-induced stresses. However, if the residual stresses are tensile, when they are summed with induced tensile stresses, the ultimate strength may be exceeded and a minute crack may appear (Fig. 3-12).

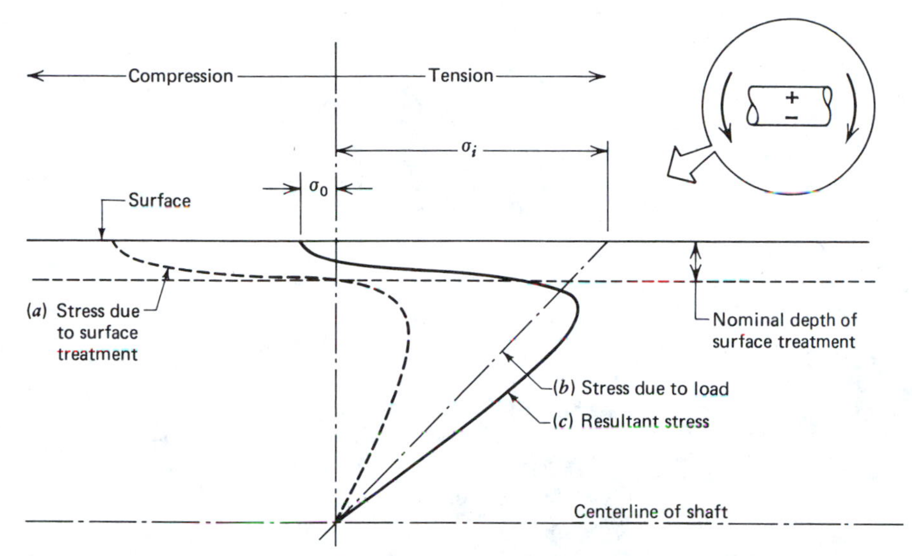

Figure 3-18 Schematic outline of stress distribution in a shaft that has undergone surface treatment—rolling, shot peening, or hardening—and is loaded in bending (see insert). Note how the load-induced stress σ_i has been changed to a compressive σ_0 at the surface. Only the upper half of the shaft is shown.

An obvious remedy is to induce through rolling, shot peening, hammer peening, or hardening permanent compressive surface stresses of such magnitude that even the largest combined stress remains below the tensile yield strength. Figure 3-18 shows successively (1) the stress due to surface treatment (compressive surface layer and tensile core), (2) induced bending stresses, and (3) the resultant stress, which is compressive at the surface. Note that surface treatment has caused the maximum stress to occur below the surface of the shaft, a place less likely to initiate cracks, and has reduced the magnitude compared to the maximum induced flexure stress.

Types of Surface Treatment

Shot peening consists of bombarding critical surfaces with small shot accelerated to high velocity in specially built machines. According to D. McCormick of *Design Engineering*:

> Hundreds of bird-shot-size pellets (usually of cast steel but sometimes of glass) hammer away at the metal surface of the workpiece. When the shot strikes, it deforms the top layer of metal, stretching it radially under tension to make a small, crater-like depression. When the shot rebounds, relieving the tension, the crater remains, along with some residual surface compressive stresses resulting from the metal's plastic deformation.[13]

Figure 3-19 illustrates the shot-peening process. Note from the figure that shot flung from a centrifugal wheel rebounds from the metal surface, leaving a shallow "crater" and wear-resistant compressive stresses. Figure 3-20 illustrates the magnitude of fatigue life improvement possible from shot peening.

Shot peening has the advantage, over some cold-work treatments, of reaching otherwise inaccessible areas that often are the most susceptible to stress raisers. The process is widely

---

[13]D. McCormick, "Shot Peening Gears for Longer Life," *Design Engineering*, July 1981, pp. 49–53.

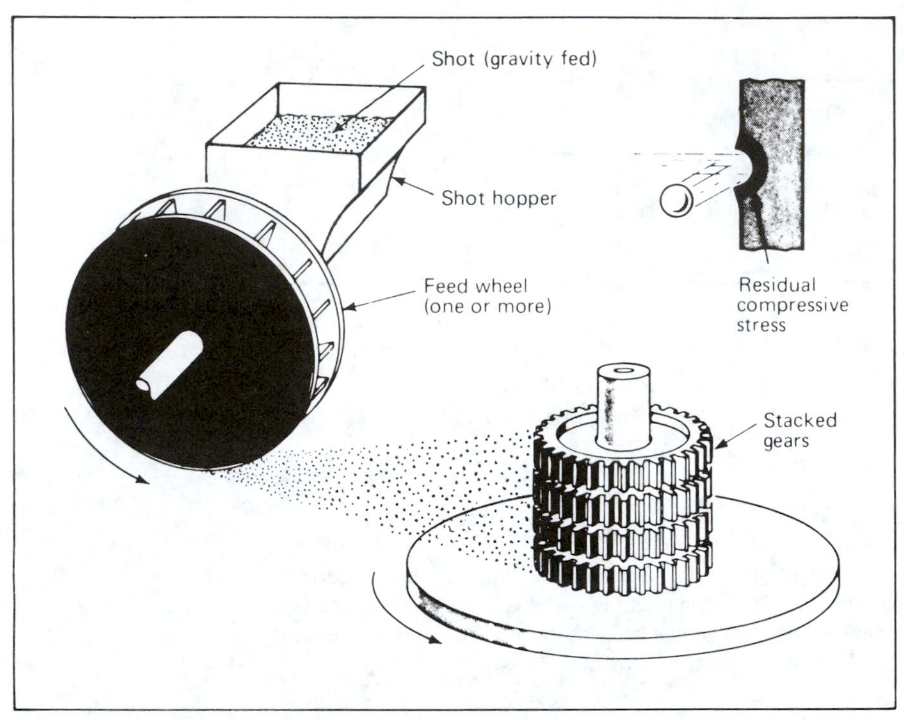

Figure 3-19 Illustration of the shot peening process. (Courtesy Doug McCormick, "Shot Peening Gears for Longer Life," *Design Engineering*, July 1981.)

used by automotive and aircraft industries as a means of prolonging the useful life of highly stressed parts. Shot peening is most beneficial to medium- and high-strength steel parts but is of little benefit to low-strength steels because of cyclic stress relaxation during fatigue.

Hammer peening is very similar to shot peening, except that a contoured tool is used repeat-

Figure 3-20 Fatigue strength improvements in carburized automotive gears. (Courtesy Doug McCormick, "Shot Peening Gears for Longer Life," *Design Engineering*, July 1981.)

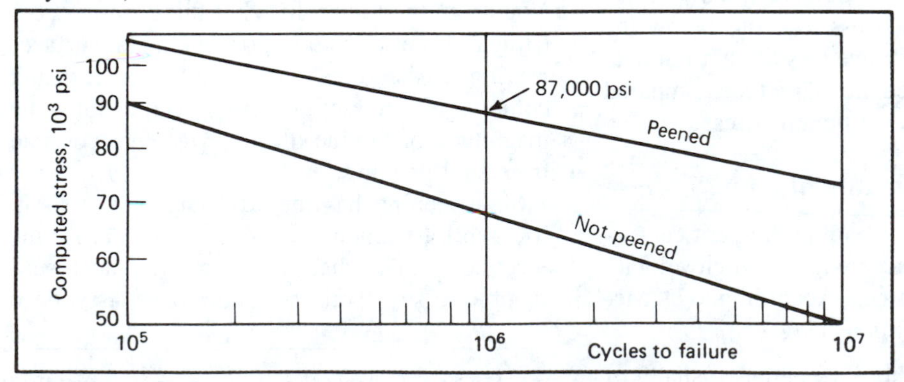

edly to impact the surface of a part. The tool is usually mounted in a reciprocating pneumatic hammer that has the advantage of being portable. With this method, however, inaccessible areas are more difficult to reach than with shot peening. Hammer peening is commonly used to prolong fatigue life of weldments (Section 6-10).

Surface hardening produces a thin, hard layer of superior strength that effectively arrests or delays crack propagation. Case hardening is advantageous for complex shapes because it penetrates readily into sharp corners and keyways where even shot peening cannot reach.

Surface rolling involves the use of contoured steel rollers to exert great localized pressure, thereby exceeding the yield strength and creating residual compressive stresses in the shaft surface. Commercially available cold-rolled steel shafts are referred to as *compressed shafts*. They are economical because they require no final surface machining.

Nearly all *plating* and *surface coating* processes tend to reduce the fatigue strength of materials. Common plating materials such as chromium, nickel, and cadmium have been known to reduce the fatigue strength of a part by 35% or more. In general, therefore, plating and surface coatings are not recommended for parts subject to fatigue loading.

Corrosion[14]

Machine components are exposed to corrosion whenever they are in contact with water (seawater in particular), oil with additives, acids, steam, or exhaust gas from engines. Depending on length of contact, scars may develop that have the effect of surface discontinuities. The discontinuities form stress raisers that are especially detrimental to materials sensitive to stress corrosion cracking. Thus heavy corrosion can quickly reduce fatigue strength.

Corrosion effects in fatigue tend to be complex and specific to the application. The subject should thus be given careful consideration whenever corrosion is likely to occur. This may mean that considerable testing would be required to validate long-term strength assumptions.

[14]Corrosion is the alteration of a material originating at the surface and caused by unintentional chemical or electrochemical attack.

3-18 FATIGUE STRESSES

General Case

A type of stress commonly encountered in machine parts is a cyclic stress superimposed on a steady or static stress. The regularly shaped curves shown in Fig. 3-21 only approximate real conditions. However, fatigue is relatively insensitive to the exact shape of load curve and very sensitive to stress amplitude. The following general formulas apply.

$$\sigma_m = 0.5(\sigma_{\max} + \sigma_{\min}) \tag{3-30}$$

$$\sigma_a = 0.5(\sigma_{\max} - \sigma_{\min}) \tag{3-31}$$

where

σ_m = mean or steady stress (due to a static load)
σ_a = alternating or amplitude stress (due to a cyclic load)
$\sigma_{\max}$ = maximum stress (corresponding to maximum load)
$\sigma_{\min}$ = minimum stress (corresponding to minimum load)

Figure 3-21 illustrates each of these stress parameters. An equivalent set of formulas holds for cyclic shear stresses.

The Soderberg Formula

Since the general load case is a combination of static and repeated load, most stress formulas that relate the induced fatigue stresses σ_a and σ_m to material strength contain the fatigue strength S_e and either the yield (S_y) or the ultimate (S_u) strength. The Soderberg formula uses the yield strength and is one of the simplest and perhaps most conservative formulas used.

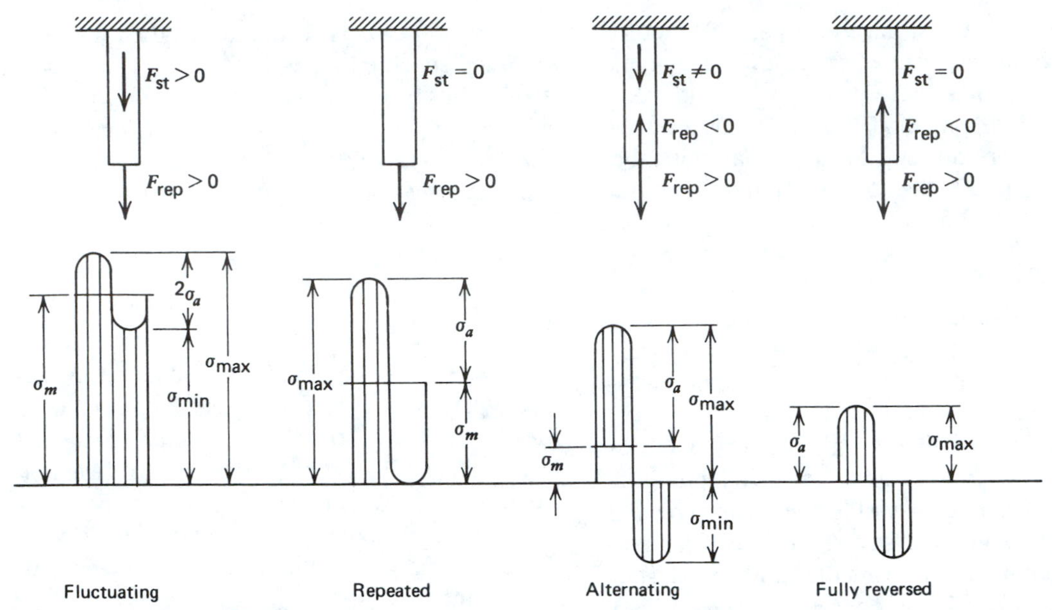

Figure 3-21 Fatigue stresses. [Adapted from G. Reitor and K. Hohmann, *Grundlagen des Konstruierens,* 3 auflage (Essen, West Germany: W. Girardet, 1977).]

$$\sigma_a = S_e\left(1 - \frac{\sigma_m}{S_{yt}}\right) \quad (3\text{-}32)$$

In this formula S_e is the endurance or fatigue limit of the machine part, which has previously been corrected by appropriate derating factors.

The Goodman Formula

Another of the most widely used stress formulas is the Goodman formula (Fig. 3-22). This formula is less conservative than the Soderberg formula and thus has closer agreement with some fatigue data. A plot of the Goodman formula (or line) seems very similar to that of the Soderberg line except that it passes through S_u instead of S_y.

$$\sigma_a = S_e\left(1 - \frac{\sigma_m}{S_{ut}}\right) \quad (3\text{-}33)$$

From Fig. 3-22 we deduce that the difference between the Goodman and the Soderberg formulas is more pronounced when the ratio of mean stress to stress amplitude is large, as, for example, in bolts.

The Amplitude-Mean Stress Diagram

This important diagram, also called the Soderberg diagram, is shown in Fig. 3-22. It is obtained by plotting alternating stress σ_a against the mean stress σ_m. Point $A(0, S_e)$ represents complete stress reversal, while points B and C correspond to the static yield strength for the Soderberg line (Eq. 3-32) and the static ultimate strength for the Goodman line (Eq. 3-33). The majority of data points (not shown) fall above and to the right of lines AB and AC. Values on the Soderberg line are thus generally quite conservative. For this reason the Soderberg line will be used for solving problems in this chapter.

Torsional Amplitude and Mean Stresses

When cyclic loading is pure torsion, metals respond somewhat differently than they do for cyclic axial or bending stress. In the case of cyclic axial and bending stress, metals become less durable as the mean stress σ_m increases, thus requiring the use of smaller and smaller values of

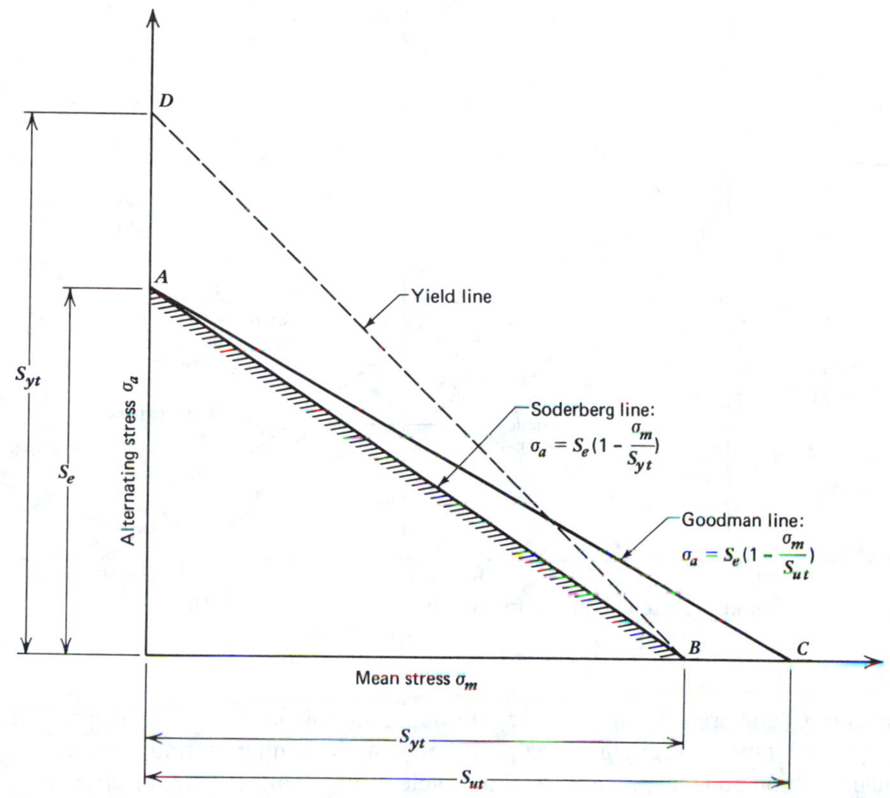

Figure 3-22 Alternating versus mean stress diagram. Soderberg and Goodman lines.

the amplitude stress σ_a as σ_m approaches the yield stress S_y. In contrast, pure torsional mean stress τ_m has no apparent effect on fatigue durability (horizontal Soderberg line) and may be neglected when calculating machine member fatigue strengths. However, for either torsional or normal (axial or bending) cyclic stress, cyclic elastic action stops, and the member fails by yielding if the sum of the amplitude stress and mean stress exceeds the yield strength S_y for normal stress or S_{sy} for shear stress. Therefore failure from fluctuating pure torsion is expected whenever the amplitude shear stress τ_a equals the shear fatigue strength.

$$\tau_a = S_{se} \tag{3-34}$$

Failure by yielding is expected when the sum of τ_a and τ_m equals S_{sy} (or $0.57S_y$).

$$\tau_a + \tau_m = 0.57S_y \tag{3-35}$$

where

$$\tau_m = 0.5(\tau_{max} + \tau_{min}) \tag{3-36}$$

$$\tau_a = 0.5(\tau_{max} - \tau_{min}) \tag{3-37}$$

Overview of Stresses

Figure 3-23 links three major diagrams: stress versus strain (static stresses), S versus N (cyclic stresses), and σ_a versus σ_m (static and cyclic stresses). Note that the two strength values, S_{yt} (or S_{ut}) from static loading and S_e from cyclic loading, combine to form the Soderberg (or Goodman) line.

Note particularly the difference between the solid and dashed cyclic stress curves. The solid curve is for small, polished, laboratory speci-

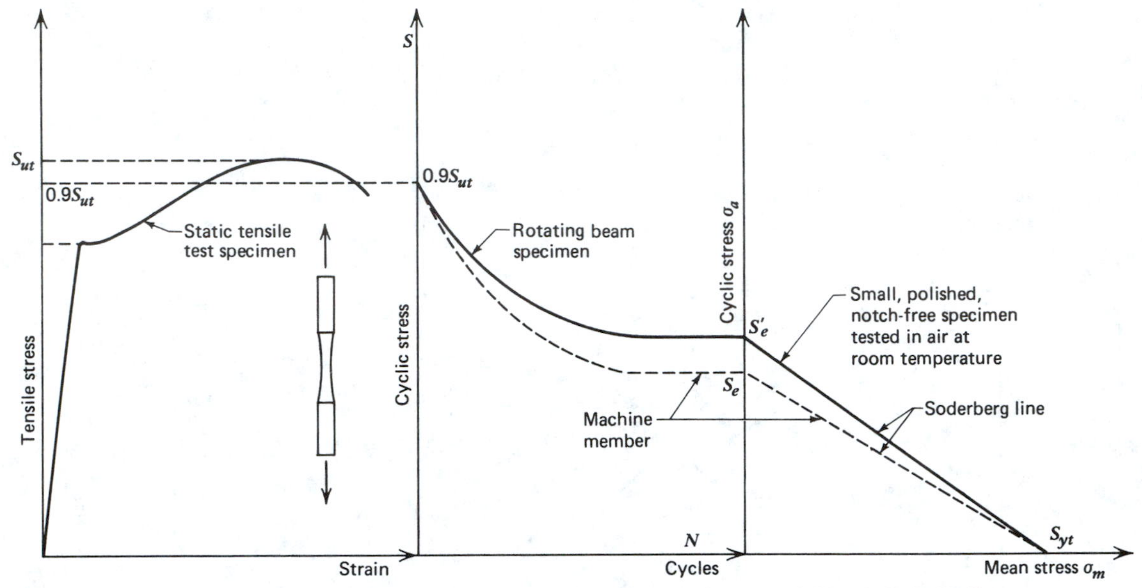

Figure 3-23 Compound stress diagram. (Developed by Uffe Hindhede and Wendell C. Hull.)

mens tested in air at room temperature; the dashed curve is the corrected fatigue strength curve for a specific machine component. Also note that the cyclic stress curve starts at $0.9S_u$ and ends at S'_e or S_e, depending on whether the curve is for a rotary beam specimen or an actual machine component. Both straight lines in the combined static and cyclic stress part of the diagram are Soderberg lines; again, the only difference between them is that one is for a specimen of the material and the other for a real machine part.

3-19 FLUCTUATING, REPEATED, AND FULLY REVERSED STRESSES

Fluctuating and Alternating

These terms are used in reference to nonzero mean stress (Fig. 3-21). Fluctuating stresses occur often from axial or longitudinal loads on bolts, pushrods, belts, and chains. In bolts, for instance, it is customary to prestress to $0.75S_{ut}$ when torquing during assembly. This preload helps to prevent loosening during subsequent loading cycles (parts tend to remain clamped). However, care must be exercised if loads are cyclic, since the external load is superimposed on the tightening prestress (Fig. 3-24).

For this type of loading, cyclic stress may be alternately smaller and larger than the static stress (Fig. 3-21). Consequently, negative stresses can occur. The stress cycle is representative of flexural (bending) stresses in shafts for belt and gear drives.

Repeated

Repeated is the special case obtained for $\sigma_m = \sigma_a$. Consequently, $\sigma_{\min} = 0$, and the curve touches the zero stress axis (Fig. 3-21). Repeated stresses may be found in camshafts, gear teeth, and chain drives (Fig. 3-25). A single gear tooth is essentially a short cantilever beam, stressed briefly in bending once during each revolution. In a chain drive the slack side carries no load and hence the tensile stress in a section

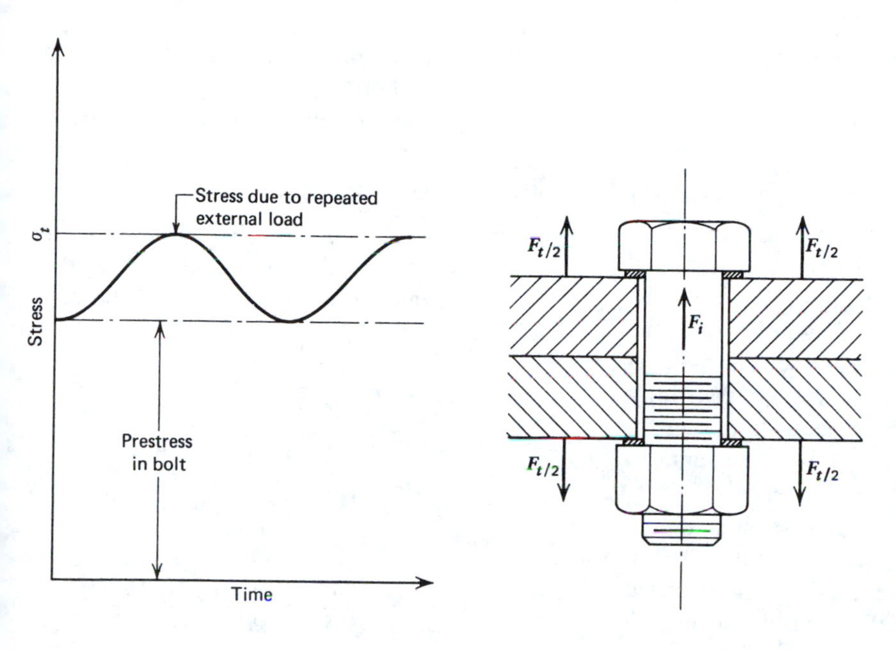

Figure 3-24 Fluctuating stress in bolt (simplified).

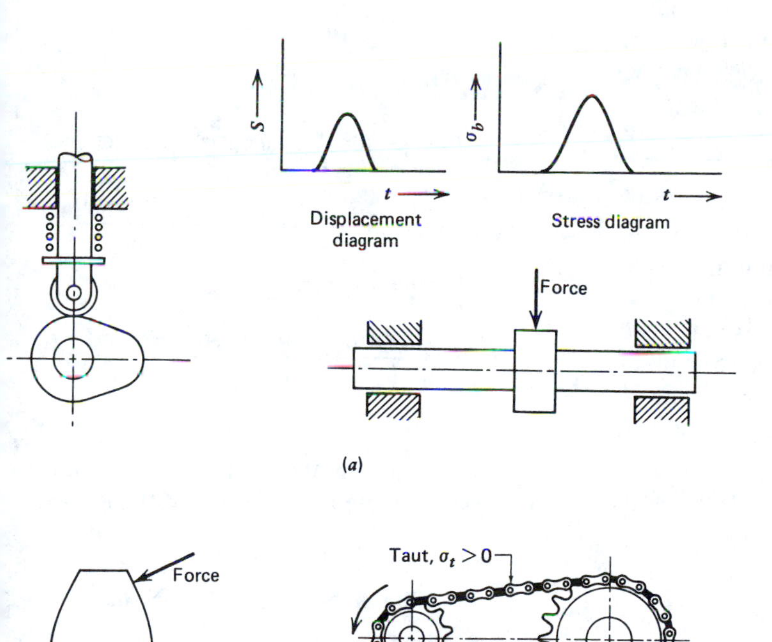

Figure 3-25 Repeated stresses occur in cam shafts, in individual gear teeth, and in chain drives.

reverts to zero each time this section traverses the slack side.

Fully Reversed

Fully reversed stress variation was the object of Wöhler's investigations and remains the universally used loading (*S-N* curves) for evaluating and comparing fatigue strength of materials. To this extent, it is cyclic loading for which the greatest amount of data is available. Thus endurance limits and fatigue strengths S'_e from fully reversed loading have become as important to designers as the static strength parameters, S_{yt} and S_{ut}.

When fully reversed loading occurs in machine components, the analysis is simplified in comparison to that of fluctuating and repeated loading, since σ_m is zero, which eliminates the need to use the Soderberg diagram. Thus, of the three major diagrams, only the stress-strain and *S-N* curves are needed in fully reversed loading. However, it will still be necessary to correct S'_e in order to obtain S_e for the machine part.

EXAMPLE 3-7

The machine member in Fig. 3-26 is subject to fluctuating bending with a maximum load of $F_{max} = 3738$ N, a minimum load of $F_{min} = 534$ N, and zero impact. The part is machined from steel alloy having an ultimate strength of $S_{ut} = 400$ MPa and a yield strength of $S_{yt} = 322$ MPa. The machine operates near a heat source, which results in a steady-state temperature of 121°C. Calculate the factor of safety for infinite fatigue life for the part.

Figure 3-26 Machine member subject to fluctuating bonding.

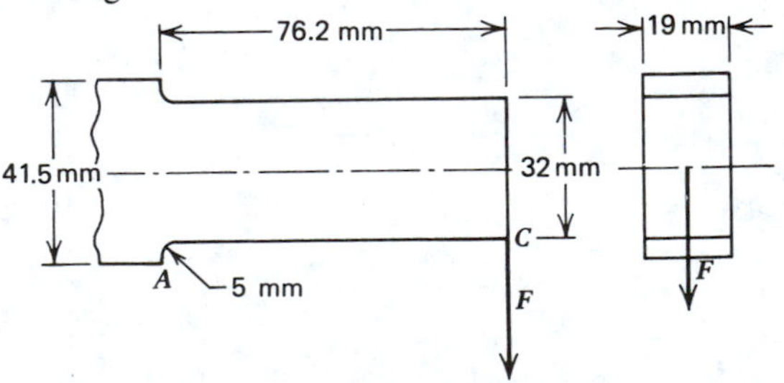

SOLUTION

Since this is a case of fluctuating bending stress, we will need to use the Soderberg diagram to evaluate fatigue strength of the part.

Rotating Beam Fatigue Strength

Using Eq. 3-16, we have

$$S'_e = 0.5S_{ut} = (0.5)(400 \text{ N/mm}^2)$$
$$= 200 \text{ N/mm}^2$$

Induced Stress at Section A

$$\sigma_i = \frac{Mc}{I} = \frac{(FL)(h/2)}{(bh^3/12)}$$
$$= \frac{6FL}{bh^2}$$

where

$$I = \frac{bh^3}{12} \text{ for a rectangular cross section}$$

$$\sigma_{imax} = \frac{6F_{max}L}{bh^2} = \frac{(6)(3738 \text{ N})(76.2 \text{ mm})}{(19 \text{ mm})(32 \text{ mm})^2}$$
$$= 87.8 \text{ N/mm}^2$$

$$\sigma_{imin} = \frac{6F_{min}L}{bh^2} = \frac{(6)(534 \text{ N})(76.2 \text{ mm})}{(19 \text{ mm})(32 \text{ mm})^2}$$
$$= 12.5 \text{ N/mm}^2$$

The mean bending and amplitude stress may be calculated using Eqs. 3-30 and 3-31.

$$\sigma_{im} = 0.5(\sigma_{imax} + \sigma_{imin})$$
$$= 0.5(87.8 \text{ N/mm}^2 + 12.5 \text{ N/mm}^2)$$
$$= 50.2 \text{ N/mm}^2$$

$$\sigma_{ia} = 0.5(\sigma_{imax} - \sigma_{imin})$$
$$= 0.5(87.8 \text{ N/mm}^2 - 12.5 \text{ N/mm}^2)$$
$$= 37.7 \text{ N/mm}^2$$

Fatigue Strength at Section A

We use Eq. 3-21 to calculate the fatigue strength at A. First, the derating factors must be determined.

(a) From Fig. 3-11 and Eq. 3-22,

$$k_a = \frac{(160 \text{ N/mm}^2)}{(0.5)(400 \text{ N/mm}^2)} = 0.80$$

(b) From Section 3-12, $k_b = 0.85$.

(c) From Table 3-3, $k_c = 1.0$. Note that in this case a reliability of 50% is assumed, since it was not specified in the problem statement.

(d) From Eq. 3-24,

$$k_d = \frac{344}{273 + 121} = 0.87$$

(e) Using the stress concentration tables in Fig. F-5 with $D/d = 41.5/32 = 1.3$ and $r/d = 5/32 = 0.16$, we find $K_t = 1.54$. From Fig. 3-15, $q = 0.78$; then Eq. 3-28 yields

$$K_f = 1 + q(K_t - 1)$$
$$= 1 + (0.75)(1.54 - 1)$$
$$= 1.4$$

Equation 3-29 is used next.

$$k_e = \frac{1}{K_f} = \frac{1}{1.4} = 0.71$$

(f) $k_f = 1.0$ for no impact.

Then

$$S_e = k_a k_b k_c k_d k_e k_f S'_e$$
$$= (0.80)(0.85)(1.0)(0.87)(0.71)(1.0)(200)$$
$$= 84 \text{ MPa}$$

Soderberg Diagram for Section A

In order to determine the factor of safety at section A, we will use the Soderberg line and a graphical technique. The Soderberg amplitude-mean stress diagram is plotted as shown in Fig. 3-27. Point D, corresponding to the induced stress at section A, is located on the diagram, and a line is drawn that passes through the point and the origin. An extension of this line intersects the Soderberg line at point E and defines the reference fatigue strength for the σ_a/σ_m ratio of point D. Either the ordinates or the abscissas of points D and E can be compared to determine the fatigue strength factor of safety at section A. Thus,

$$fs_A = \frac{82}{50.2} = \mathbf{1.63}$$

Combined Fatigue Stresses

In the case of combined cyclic normal and shear stress, both of which have alternating and mean components, equivalent alternating and mean octahedral induced normal stresses can be written that are similar to Eq. 3-10. The resulting expressions are as follows.

$$\sigma'_{im} = \sqrt{\sigma_{im}^2 + 3\tau_{im}^2} \qquad (3\text{-}38)$$

$$\sigma'_{ia} = \sqrt{\sigma_{ia}^2 + 3\tau_{ia}^2} \qquad (3\text{-}39)$$

where

σ'_{im} = equivalent normal octahedral mean stress
σ'_{ia} = equivalent normal octahedral alternating stress
σ_{im} = induced normal mean stress
σ_{ia} = induced normal amplitude stress
τ_{im} = induced shear mean stress
τ_{ia} = induced shear amplitude stress

Induced stresses σ_{im}, σ_{ia}, τ_{im}, and τ_{ia} are determined from the loading on the machine member. Then Eqs. 3-38 and 3-39 may be used to find equivalent octahedral stresses. After the octahedral amplitude and mean stress components are calculated, they are plotted on a Soderberg diagram to find the factor of safety for the machine part.

EXAMPLE 3-8

The shaft of Examples 3-1, 3-2, and 3-3 is driven at a constant speed of 300 rpm. This

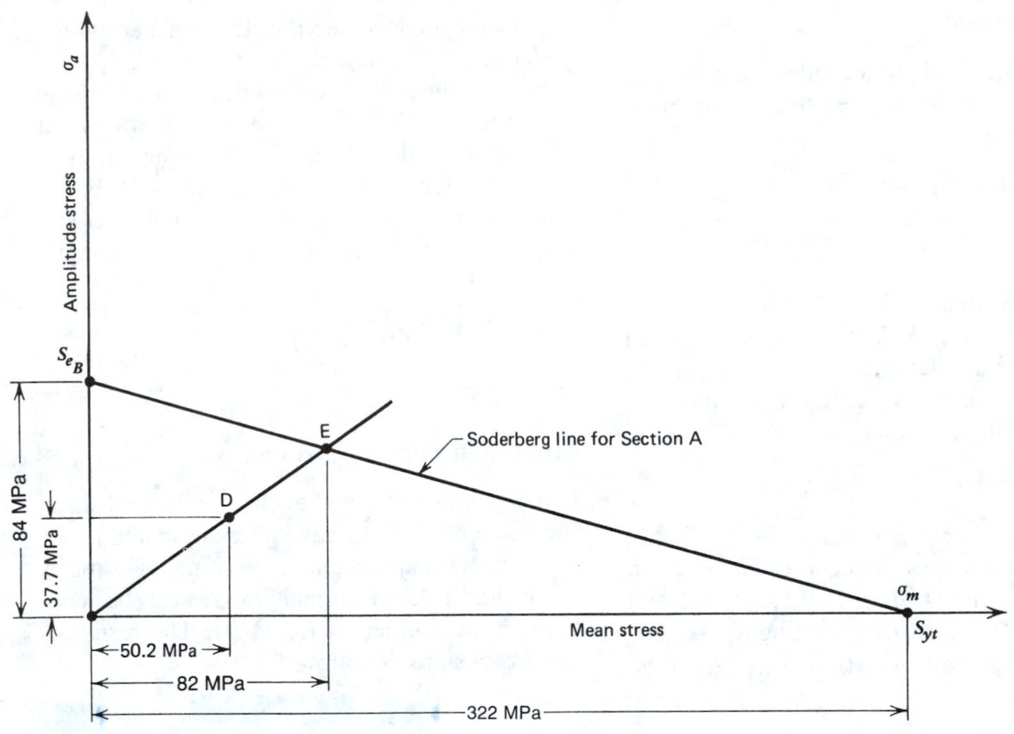

Figure 3-27 The Soderberg amplitude-mean stress diagram.

causes the shaft at bearing B to undergo constant torque and fully reversed bending. Assume that the pulley is attached to the shaft with a smooth collet so that no stress concentration exists between bearing A and the end C. Determine if the shaft will survive an infinite number of revolutions using a diameter of 11.6 mm and $S_{ut} = 1068$ MPa.

SOLUTION

Since this is a combined stress, normal and shear stress problem, we will use Eqs. 3-38 and 3-39 and the Soderberg diagram to find a solution.

Induced Stresses

The bending load varies from +334 N to −334 N as the shaft rotates; therefore,

$$\sigma_{imax} = \frac{690\ 600\ \text{N}\cdot\text{mm}}{(11.6\ \text{mm})^3} = 442\ \text{N/mm}^2$$

$$\sigma_{imin} = \frac{-690\ 600\ \text{N}\cdot\text{mm}}{(11.6\ \text{mm})^3} = -442\ \text{N/mm}^2$$

The corresponding amplitude and mean induced bending stresses from Eqs. 3-30 and 3-31 are

$$\sigma_{im} = 0.5(\sigma_{imax} + \sigma_{imin})$$
$$= 0.5(442\ \text{N/mm}^2 - 442\ \text{N/mm}^2)$$
$$= 0$$

$$\sigma_{ia} = 0.5(\sigma_{imax} - \sigma_{imin})$$
$$= 0.5(442\ \text{N/mm}^2 + 442\ \text{N/mm}^2)$$
$$= 442\ \text{N/mm}^2$$

The torsional stress is constant in direction and magnitude; thus,

$$\tau_{imax} = \frac{57\,600 \text{ N} \cdot \text{mm}}{(11.6 \text{ mm})^3} = 37 \text{ N/mm}^2$$

$$\tau_{imin} = \tau_{imax}$$

$$\tau_{im} = 0.5(\tau_{imax} + \tau_{imin})$$

$$= 0.5(37 \text{ N/mm}^2 + 37 \text{ N/mm}^2)$$

$$= 37 \text{ N/mm}^2$$

$$\tau_{ia} = 0.5(\tau_{imax} - \tau_{imin})$$

$$= 0.5(37 \text{ N/mm}^2 - 37 \text{ N/mm}^2) = 0$$

Substituting these values into Eqs. 3-38 and 3-39 yields, for the octahedral equivalent normal mean and amplitude stresses,

$$\sigma'_{im} = \sqrt{(0)^2 + 3(37)^2}$$

$$= 64 \text{ N/mm}^2$$

$$\sigma'_{ia} = \sqrt{(442)^2 + 3(0)^2}$$

$$= 442 \text{ N/mm}^2$$

Fatigue Strength

Before the Soderberg diagram can be drawn, the rotating beam endurance limit must be corrected by applying the appropriate fatigue derating factors. First, the rotating beam endurance limit S'_e is found by using Eq. 3-16.

$$S'_e = 0.5S_{ut} = (0.5)(1068 \text{ N/mm}^2)$$

$$= 534 \text{ N/mm}^2$$

The fatigue derating factors are as follows.

(a) $k_a = 0.68$ (Fig. 3-11).
(b) $k_b = 0.85$, $d > 7.6$ mm.
(c) $k_c = 1.0$, 50% reliability assumed.
(d) $k_d = 1.0$, assume room temperature.
(e) $k_e = 1.0$, no stress concentration.
(f) $k_f = 1.0$, assume no impact.

Therefore the corrected fatigue strength is

$$S_e = (0.68)(0.85)(1.0)(1.0)(1.0)(1.0)(534 \text{ N/mm}^2)$$

$$= \mathbf{309 \text{ MPa}}$$

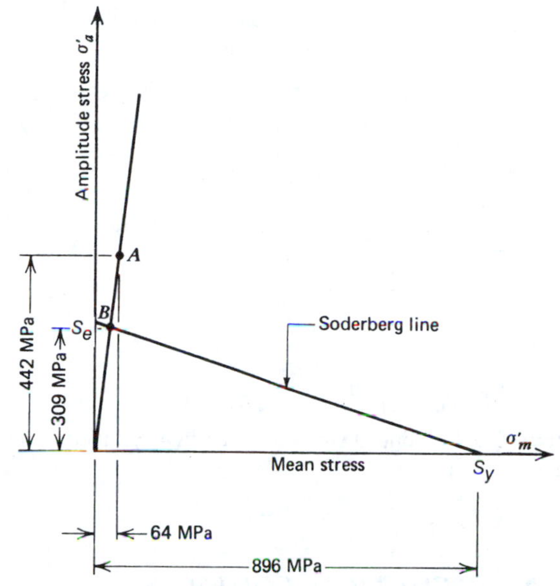

Figure 3-28 Soderberg diagram for shaft.

Soderberg Diagram

We are now ready to draw the Soderberg diagram (Fig. 3-28) and compare the induced amplitude and mean stress components with the strength of the shaft for this stress ratio. Again, a line is drawn through the induced stress point *A* and the origin, thus intersecting the Soderberg line at point *B*. In this case we find that point *A* falls outside the Soderberg line, which means that failure of the shaft could occur at a relatively short life and infinite life would not be expected. In order to achieve an infinite life design, designers would have to reduce the belt tension or increase the shaft size, or both, and then repeat the analysis until the induced stress point falls inside the Soderberg line with an appropriate factor of safety.

Note that a machine component strong enough to avoid yielding under static load may not be strong enough to sustain cyclic loads of similar magnitude. The subject of shaft design will be covered in greater detail in Chapter 15.

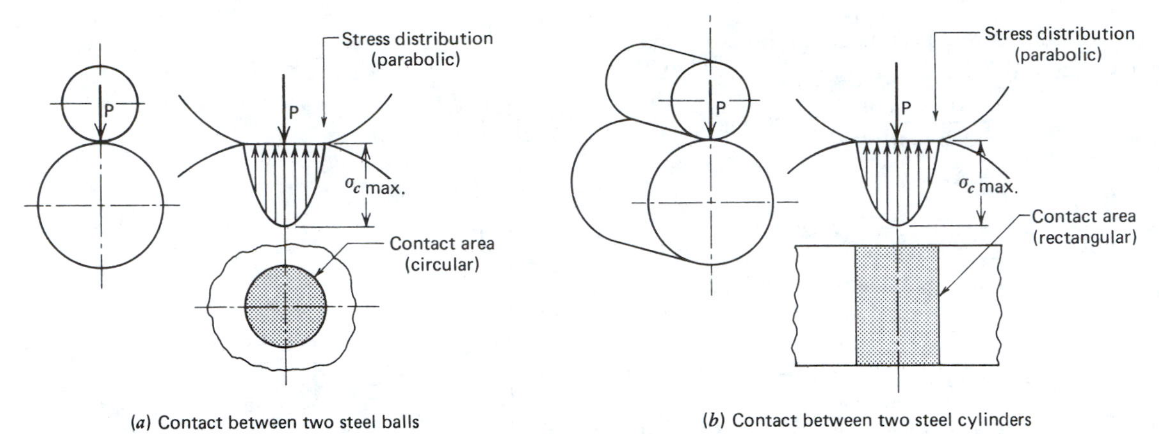

Figure 3-29 Contact stress between two cylinders: (*a*) circular contact area; (*b*) rectangular contact area.

3-20 MECHANICAL CONTACTS

One final source of stress—mechanical contacts—must be discussed. Mechanical contacts transmit force between rigid bodies. They are the simplest and most common means of transmitting forces in machinery.

Quite often contacting surfaces are curved. Typical contacts are those of meshing gear teeth, cam mechanisms, roller chains, metal wheels on rails, bearing journals, and roller-race (or ball-race) contacts in bearings (Fig. 3-25).

The contact area between two rigid, curved machine members is obviously small, leading to correspondingly high contact stresses. In roller bearings, they may reach magnitudes of 1500 to 3000 MPa. It is fortunate that such stresses can be tolerated because they facilitate transmission of large forces by small machine members.

Since even "rigid" bodies are slightly elastic, all contact areas deviate elastically from the basic surface curvatures. For instance, the contact area of two spheres of the same size and material is a minute circular plane. In general, the induced stresses are three dimensional but may be characterized by the stress perpendicular to a plane contact surface, which varies parabolically from zero at the periphery of the contact plane to maximum at the center (Fig. 3-29).

Contact stresses are functions of material properties, body geometry, and applied forces. Large forces generate large stresses, and vice versa. Sharp curvatures (small radii) lead to small contact areas and, consequently, to large stresses. For example, large-diameter wheels on a rail carry larger loads than small-diameter wheels because of larger contact areas.

The theoretical basis for contact stresses was laid down by H. Hertz nearly a century ago. The term *Hertz stress* is thus synonymous with contact stress. Hertz's stress formulas assume elastic bodies, continuous surfaces free of sharp edges, and large radii relative to those of the contact area.

The Hertz formulas are cumbersome, so nomograms have been developed for commonly used contacts. The advent of electronic calculators, however, has greatly facilitated analysis; hence both nomograms and formulas are used. Miller and Wright have developed many tables and curves for simplifying these calculations.[15]

Repeated loading of bodies in contact usually results in fatigue failure manifested by minute pockmarks or pits in the affected surfaces.

[15] W. R. Miller and D. K. Wright, Jr., "Contact Stresses," *Machine Design*, Vol. 35, No. 15, pp. 185–189, 1963.

TABLE 3-5 Allowable Contact Stresses for Gears

Metal	Minimum Surface Hardness	Allowable Contact Stress	
		1000 psi	MPa
Steel			
Through hardened	180 Bhn[a]	85–95	586–655
	240 Bhn	105–115	724–793
	300 Bhn	120–135	828–931
	360 Bhn	145–160	1000–1103
	440 Bhn	170–190	1172–1310
Case carburized	55 R_C[a]	180–200	1241–1379
	60 R_C	200–225	1379–1552
Flame or induction hardened	50 R_C	170–190	1172–1310
Gray cast iron			
AGMA[b] Grade 20	—	50–60	345–414
AGMA Grade 30	175 Bhn	65–75	448–517
AGMA Grade 40	200 Bhn	75–85	517–586
Nodular cast iron	165–300 Bhn	10% less than for steel of the same hardness	

Source. From E. J. Wellauer and H. R. Bergman, "Gear Materials—Ferrous Metals," *Machine Design*, 1968.

[a]Bhn = Brinell hardness number. R_C = Rockwell C scale.
[b]American Gear Manufacturers Association.

Termed pitting, this is the most common mode of failure in gearing and rolling bearings.

Table 3-5 provides values of contact stresses that experience has shown to be satisfactory for 10 million cycles of gear contact. They also apply to similar types of contact in cams and roller chains.

SUMMARY

Metals deform under the action of steady loads according to a set pattern. They first behave elastically; then, as the load increases, they behave plastically until fracture occurs.

Under cyclic loads, metals generally behave elastically until, suddenly, they fracture. Localized plastic deformation and rupture cause crack initiation and propagation, which often produce recognizable fracture surface features in the form of discoloration and beach marks.

Fundamental to strength calculations is the straight-line proportionality of stress and strain in the elastic range coupled with the existence of a fatigue limit under cyclic loading for many materials.

The strength of machine members may be significantly affected by a variety of factors, especially when loading is cyclic. Some of the more important factors that must be considered are:

1. Geometric stress concentrations.
2. Size effects.
3. Surface condition and treatment.
4. Reliability.
5. Temperature effects.
6. Corrosion.
7. Load history.
8. Type of loading (fluctuating, repeated, fully reversed, combined, shock, and contact).

Design for strength requires that designers investigate the function of each machine member in order to determine the loads it must sustain during normal operating cycles. To this information they add an estimate of special environmental effects (operating temperature, contact with corrosive substances), the variability of loading (operator error, unusual use patterns), frequency and magnitude of extreme loadings (malfunction, abuse, and other foreseeable circumstances), special design objectives (aesthetics, economics, scheduling), and the risk associated with part failure (human injury, property damage, machine downtime, warranty losses, customer goodwill, military readiness, etc.). Only when all this information is in hand are designers ready to select component dimensions, material, fabrication method, heat treatment, and surface finish and to define properly derating factors and factors of safety for the component. Even when these tasks have been accomplished as carefully as possible, complete knowledge of how the part will perform will be available only after it is tested under actual use conditions. Thus there is no alternative to testing whenever the risks include human injury from part failure.

REFERENCES

3-1 Baumeister, T. (ed.). *Marks' Standard Handbook for Mechanical Engineers,* Seventh Edition. New York: McGraw-Hill, 1967.

3-2 Boresi, A. P., et al. *Advanced Mechanics of Materials*. New York: Wiley, 1978.

3-3 Byars, E. F., and R. D. Snyder. *Engineering Mechanics of Deformable Bodies*. Scranton, Pa.: International Textbook Co., 1963.

3-4 Collins, J. A. *Failure of Materials in Mechanical Design*. New York: Wiley, 1981.

3-5 "Failure Analysis and Prevention," *Metals Handbook,* Eighth Edition. Metals Park, Ohio: American Society of Metals, 1975.

3-6 Fuchs, H. O. "Forecasting Fatigue Life of Peened Parts." *Metals Progress,* 1963.

3-7 Fuchs, H. O., and R. I. Stephens. *Metal Fatigue in Engineering*. New York: Wiley, 1980.

3-8 Kececioglu, D., et al. "Combined Bending-Torsion Fatigue Reliability of AISI 4340 Steel Shafting with $K_t = 2.34$." ASME paper 74-WA/DE-12, 1974.

3-9 *Metals Handbook,* Ninth Edition. Metals Park, Ohio: American Society of Metals, 1978.

3-10 Miller, W. R., and D. K. Wright, Jr. "Contact Stresses." *Machine Design,* 1963.

3-11 Mischke, C. "Designing to a Reliability Specification." SAE paper 740643, 1974.

3-12 Peterson, R. E. *Stress Concentration Factors*. New York: Wiley, 1978.

3-13 Popov, E. P. *Mechanics of Materials,* Second Edition. Englewood Cliffs, N.J.: Prentice-Hall, 1976.

3-14 Roark, R. J., and W. C. Young. *Formulas for Stress and Strain,* Fifth Edition. New York: McGraw-Hill, 1975.

3-15 *SAE Handbook*. Warren, Pa.: Society of Automotive Engineers, 1979.

3-16 Shigley, J. E. *Mechanical Engineering Design,* Third Edition. New York: McGraw-Hill, 1977.

3-17 Sines, G., and J. L. Waisman (eds.). *Metal Fatigue*. New York: McGraw-Hill, 1959.

PROBLEMS

External loads versus internal resistance is the essence of problem solving. Thus, all problems begin by stating that the induced stress must equal the working stress. Table 3-2 provides the appropriate formulas and should be consulted immediately.

P3-1 A 1-in. diameter steel shaft 24 in. long,

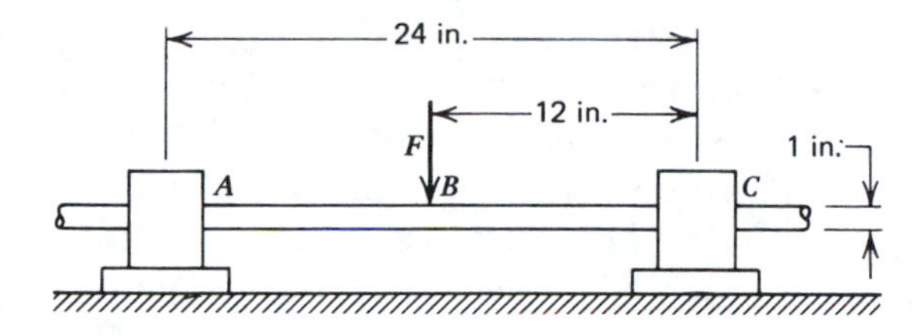

Figure 3-30 Illustration for Problem 3-1.

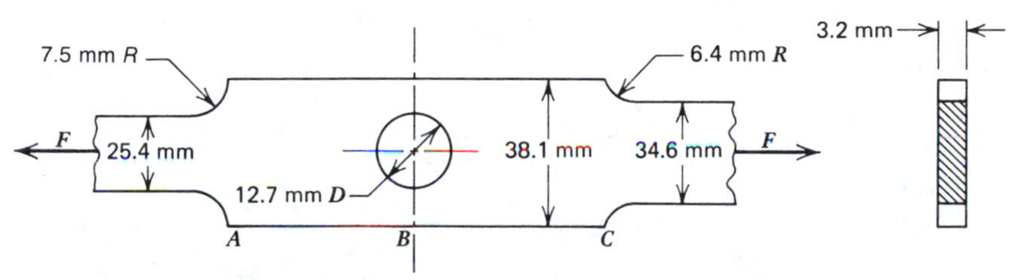

Figure 3-31 Illustration for Problem 3-2.

with yield strength $S_y = 67$ Ksi, has self-aligning bearings at either end (points A and B, Fig. 3-30). The only loading on the shaft is a steady force F at point B. Determine the force F that will give an fs of 2.0 based on yielding. Note that the maximum bending moment occurs at point B with magnitude $(FL)/4$, where $L = 24$ in. *Hint*: From Table 3-2: $Mc/I = S_{yt}/fs$.

P3-2 A flat bar is subject to completely reversed axial load, as shown in Fig. 3-31. Determine the theoretical stress concentration at A, B, and C. Which of these sections would be the most likely to fail from material fatigue?

P3-3 Figure 3-32*a* and 3-32*b* show a "dead" shaft and a "live" shaft, each carrying identical rollers with identical loads. The shafts are of the same length and the same material ($S_{ut} = 90{,}000$ psi, $S_{yt} = 50{,}000$ psi) and have ground surfaces. For simplicity, assume a single load acting at midpoint, and disregard stress concentration and shaft stiffening due to the roller. Assume a 50% reliability for

Figure 3-32 Illustration for Problem 3-3.

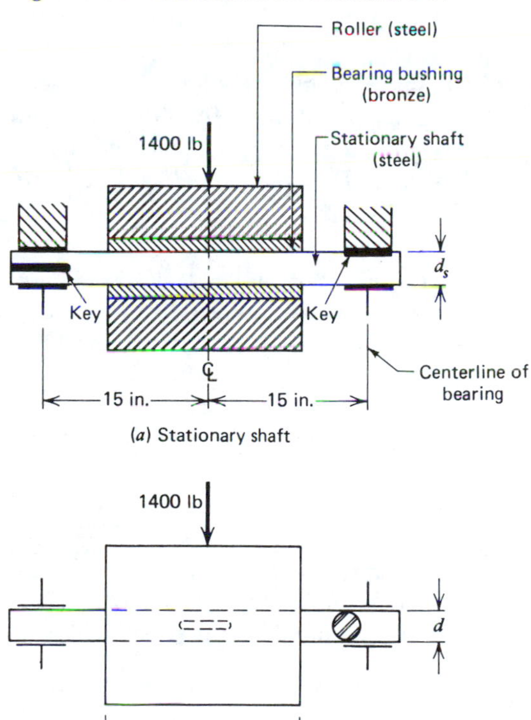

the rotating shaft. (*a*) Calculate the shaft diameter for a "dead" and a "live" load. *Hint*: Use Table 3-2. (*b*) Compare the change in section moduli. Give the answer as a percent.

P3-4 A circular shaft of an electric motor is 30 in. from bearing to bearing. The weight of the rotor is 1400 lb. For simplicity, we will assume the force of gravity concentrated at midpoint. The material is G10400 steel (S_{ut} = 90,000 psi). The surface is ground. Calculate the shaft diameter for light shock, a temperature of 70°C, no stress concentration, a reliability of 90% and fs = 1.5. *Hint*: Use Table 3-2 for bending and fatigue loading.

P3-5 A stepped shaft in circular cross section, as shown in Fig. 3-33, is made from steel (S_{ut} = 650 MPa), and the load (P = 22 kN) is repeated and fully reversed. Assuming $r/d = 0.25$, find the diameter d and the fillet r, corresponding to a reliability of 99.99%. The shaft is ground. Disregard notch sensitivity. Assume $D = d + 2r$. *Hint*: Use Table 3-2.

Figure 3-33 Illustration for Problem 3-5.

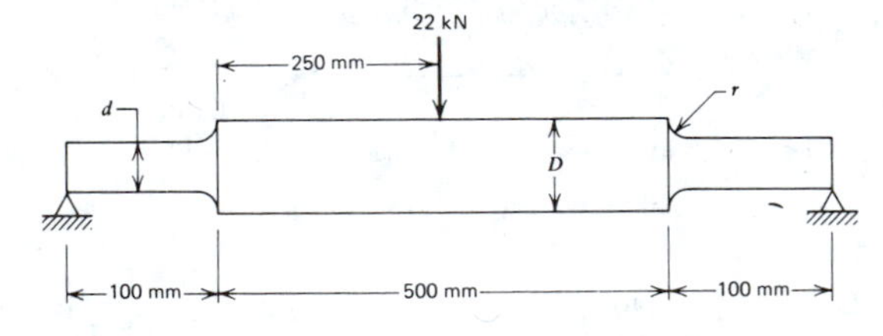

P3-6 A round steel bar is loaded in tension by a fully reversing force F = 6.6 *kN* (Fig. 3-34). The nominal cross-sectional area is 110 mm². K_t = 4, and q = 0.9. Calculate the factor of safety and make a sketch of the stress distribution. The bar material has an ultimate strength of 1516 MPa and a yield strength of 1240 MPa 50% reliability, machined surface, room temperature, no shock load. *Hint*: From Table 3-2, $F / A = S_e$.

Figure 3-34 Illustration for Problem 3-6.

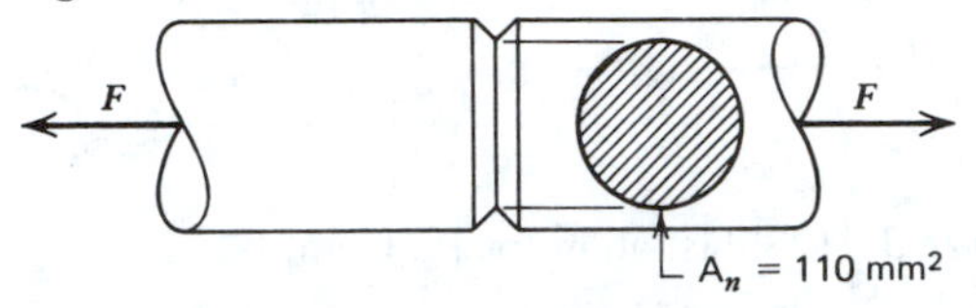

P3-7 Use the fillet radius r found in P3-5 as an approximation and recalculate r and d. This time include the notch sensitivity effect.

P3-8 Redesign the rotary shaft in P3-3 to minimize stress concentration caused by a light press-fit between shaft and roller, as shown in Fig. 3-35. Increase reliability to 90%, and design for light shock and an average temperature of 120°C. Use $r/d = 0.3$ and $D = d + 2r$. Neglect notch sensitivity. *Hint*: From Table 3-2, $M/Z = S_e$.

Figure 3-35 Illustration for Problem 3-8.

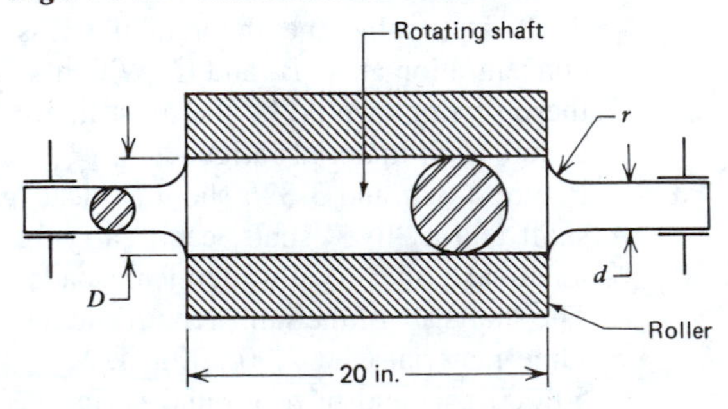

P3-9 A shaft is made from 37-mm SAE 1030 (S_{ut} = 520 MPa) bar stock and is not heat treated. The shaft has ground surfaces, and a portion of the shaft has the dimensions shown in Fig. 3-33. The fully reversed bending moment at the fillet is 157 000 N · mm. If a failure rate of 1% is acceptable in this application, is the stress too high? Note: d = 25 mm and r = 5 mm.

Figure 3-36 Illustration for Problem 3-10.

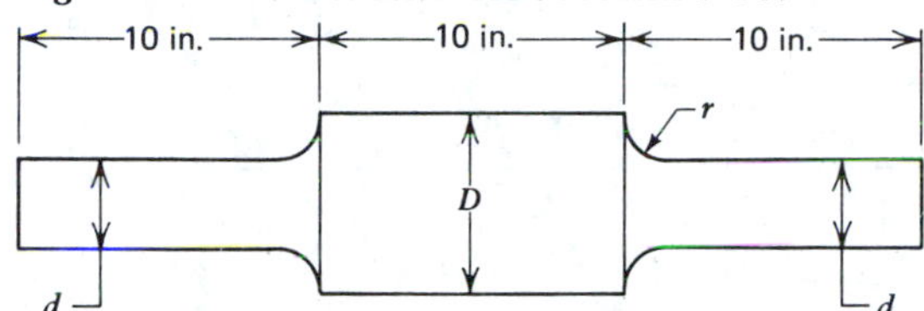

P3-10 Suppose the shaft in P3-5 was redesigned as shown in Fig. 3-36. D = 2.25 in., d = 1.75 in., r = 0.25 in. Would it have adequate strength? If so, what would be the factor of safety?

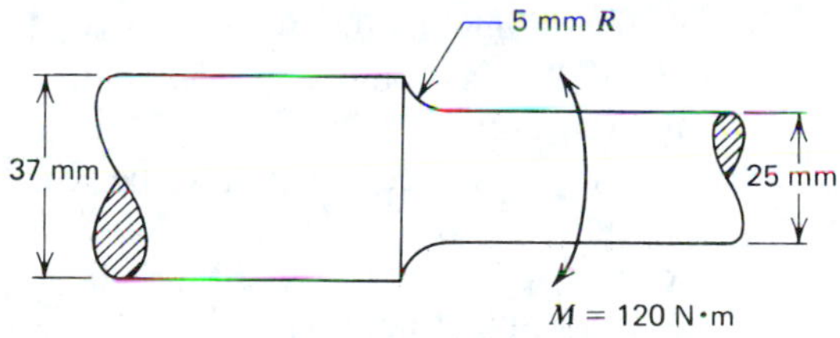

Figure 3-37 Illustration for Problem 3-11.

P3-11 A shaft is made from 38 mm diameter UNS G10350 CD bar stock having S_{ut} = 551 MPa and S_y = 462 MPa. The shaft has ground surfaces and a portion of the shaft has the dimensions shown in Fig. 3-37. The fully reversed bending moment at the fillet is 120 N · m. A reliability of 95% is desired, and the shaft must operate at 149°C. (*a*) What is the corrected fatigue strength S_e for this section of the shaft? (*b*) Is the stress at the fillet too high for infinite life of the shaft?

P3-12 An axle of circular cross section (Fig. 3-38) is made from hot-rolled, high-

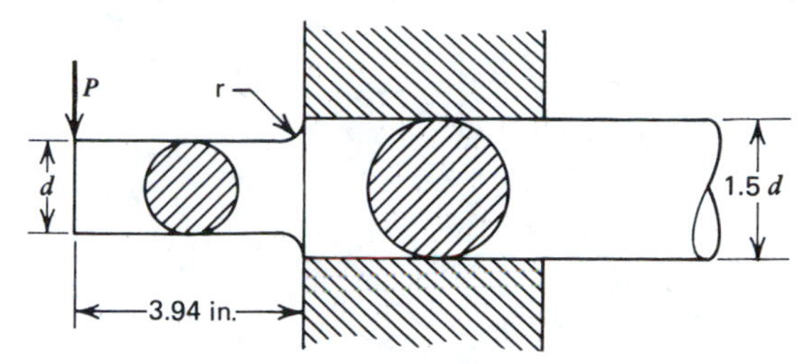

Figure 3-38 Illustration for Problem 3-12.

carbon steel with ground surface and an ultimate strength of 87 Ksi. The load P is completely reversed with a value of 2472 lb. Assuming r/d = 0.125, determine the diameter d and fillet radius r so that the maximum stress will be limited to a value corresponding to a factor of safety of 2.0.

P3-13 A round camshaft is 30 in. from bearing to bearing. The cam is placed at midpoint, as in Fig. 3-39. During one revolution, the transverse force goes from 0 to 1400 lb. and back again to 0 in a sinusoidal manner. The shaft made from annealed steel (S_{ut} = 90,000 psi and S_{yt} = 50,000 psi) is ground. Assume that the shaft has no discontinuities. Find the shaft diameter for 90% reliability under this loading, using (*a*) the Soderberg and (*b*) the Goodman equation. (*c*) What is the difference in percent?

Figure 3-39 Illustration for Problem 3-13.

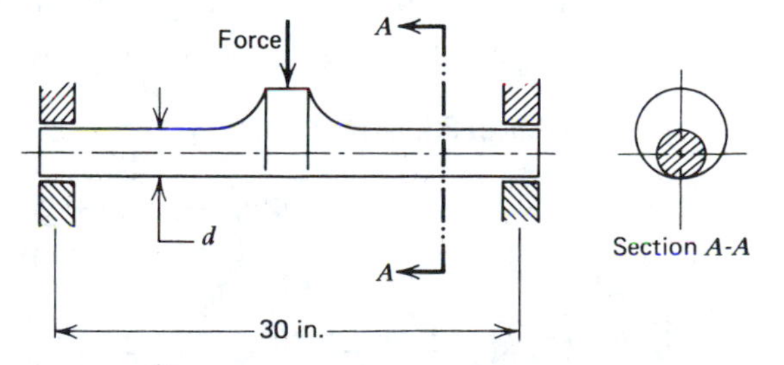

P3-14 The short, flat, hot-rolled bar in Fig. 3-40 carries an axial load F = 53 kN assumed evenly distributed across the

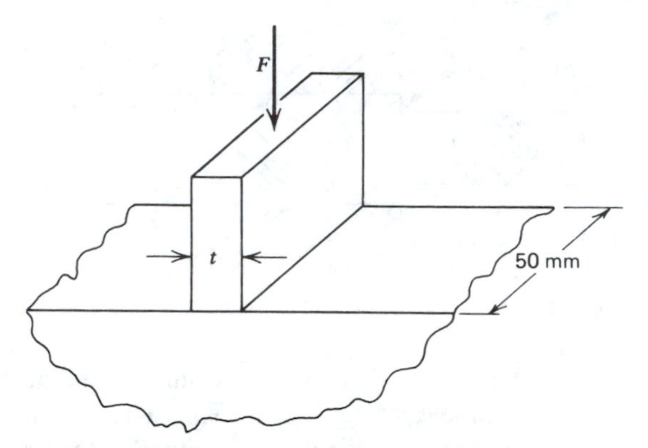

Figure 3-40 Illustration for Problem 3-14.

end of the bar. The material is steel (S_{ut} = 620 MPa, S_{yt} = 345 MPa). Find the plate thickness t for a reliability of 95% when (*a*) F is a static load. *Hint*: $F/A = S_{yt}$; (*b*) F is a fully reversed load. (Disregard stress concentration and notch effect.) *Hint*: $F/A = S_e$; and (*c*) Calculate the difference in percent.

P3-15 Solve P3-14 using the Soderberg and the Goodman equations when F is (*a*) a fluctuating load: F_{max} = 53 kN and F_{min} = 26.5 kN; (*b*) a repeated load; F_{max} = 53 kN and F_{min} = 0; and (*c*) an alternating load: F_{max} = 53 kN and F_{min} = −26.5 kN. Make a suitable sketch in each case as per Fig. 3-21.

P3-16 Draw a bar diagram as shown in Fig. 3-41 based on the values of t obtained in P3-14 and P3-15. Use a scale of 20:1 and place the diagram turned 90° from that shown in Fig. 3-41. What do you

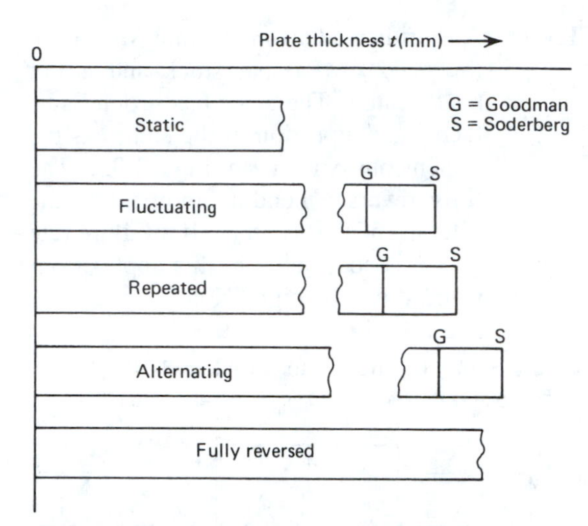

Figure 3-41 Bar diagram for Problem 3-16.

conclude from analyzing this bar diagram?

P3-17 A solid aluminum diving board is made of A95052-H32 with S'_e = 17.5 Ksi and S_{yt} = 27 Ksi. Assume that divers subject the diving board to fluctuating loads of 300 lb maximum and 150 lb minimum at point B (Fig. 3-42). The theoretical stress concentration for bending at point A is K_t = 1.3. Determine the factor of safety for the board for a board thickness of 0.75 in., a width of 18 in. and 10^8 cycles of load. *Hint:* Draw the Soderberg diagram.

P3-18 A coiled valve spring develops a shear stress of τ_{max} = 217 MPa when the valve is open and τ_{min} = 155 MPa when it is closed (Fig. 3-43). The spring material has a yield strength of S_{sy} = 496

Figure 3-42 Illustration for Problem 3-17.

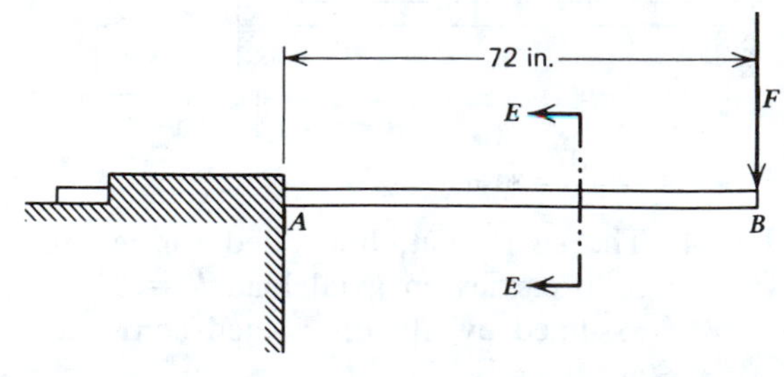

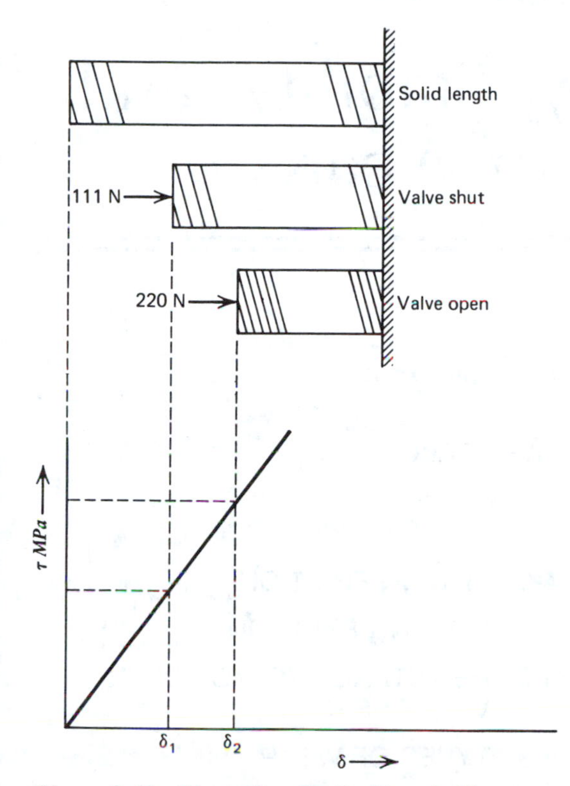

Figure 3-43 Illustration for Problem 3-18.

MPa in torsion and a corrected endurance strength in reversed torsion of S_{se} = 279 MPa. What is the *fs* for (*a*) static yielding and (*b*) infinite life?

P3-19 A cantilever spring (Fig. 3-44) is subject to fluctuating loading with F_{max} = 222 N and F_{min} = 111 N. The spring is made from heat-treated steel with a hardness of 390 Bhn. The theoretical stress concentration for bending is K_t = 1.5 at the support. Will the spring fail in fatigue if cycled for 10^7 cycles? The spring surface is ground. Use the Goodman criterion to evaluate failure.

Figure 3-44 Illustration for Problem 3-19.

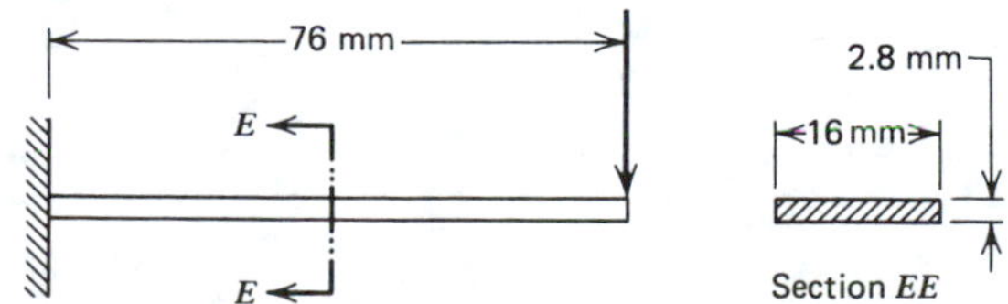

P3-20 If the shaft in P3-11 was subjected to a combined loading consisting of a fully reversed bending moment of 60 N · m and fluctuating torsion of T_{max} = 225 N · m and T_{min} = 60 N · m, would it have an infinite fatigue life? Use both Soderberg and Goodman equations. Compare the answers.

P3-21 Redesign the shaft in P3-4 so it will carry a synchronous sinusoidal torsional load in addition to the transverse load. The torsional load varies from 0 to 2000 in.-lb.

P3-22 Develop the equations for the Soderberg and Goodman formulas using Fig. 3-22.

P3-23 The flat machined bar shown in Fig. 3-45 carries an axial load of 50 kN. The material is steel (S_{ut} = 620 MPa; S_{yt} = 340 MPa). Assuming a factor of safety of 2 and 90% reliability, calculate the plate thickness for (*a*) static loading; (*b*) fully reversed loading (disregard notch sensitivity); (*c*) repeated loading, P_{max} = 50 kN; P_{min} = 0. Use Goodman and draw the corresponding diagram. *Scale:* abscissa, 1 mm = 50 MPa; ordinate, 1 mm = 5 MPa; and (*d*) Compare the answers in percent. Use the thickness for static loading as a base. Which statement in the text do these answers support?

Figure 3-45 Illustration for Problem 3-23.

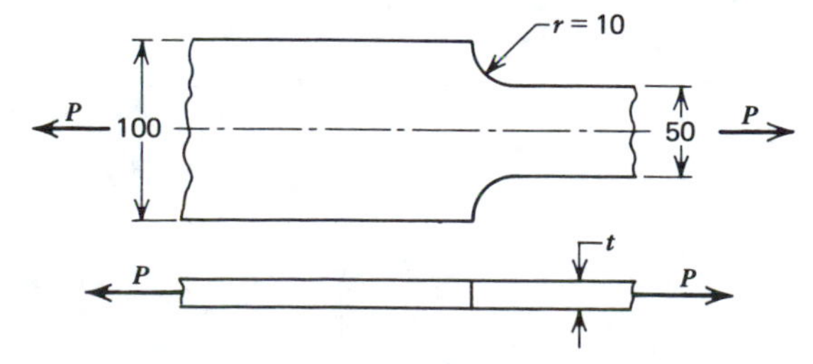

Chapter 4
Design for Rigidity, Stability, and Resistance to Wear

One good test is worth a thousand expert opinions.

ANONYMOUS

NOMENCLATURE

f = coefficient of sliding friction
f_R = coefficient of rolling friction
f_{st} = coefficient of static friction
f_v = coefficient of rolling friction of a vehicle
F_f = force of friction; N, lb
F_R = resistance to rolling; N, lb
F_{st} = force of static friction; N, lb
L = length of column; mm, in.
n = coefficient of end conditions
N = normal force; N, lb
P_{cr} = critical load; N, lb
r_g = radius of gyration; mm, in.
R = wheel radius; mm, in.
α = angle of inclination; deg
θ = friction angle; deg

Design of machine members is dominated by the concept of strength. Reliable operation coupled with a long service life, however, often requires additional considerations for rigidity, stability, and resistance to wear.

4-1 DESIGN FOR RIGIDITY

Rigidity is the ability to resist deformation. In machinery this is a very important quality for the frame and most members. This is especially true in machine tools for which normal deflection would otherwise affect the quality of the end products or reduce the life of cutting tools.

Rigidity becomes a design factor when:

1. A machine member is long relative to its width, depth, or diameter. Machine shafts, cantilever beams, and rollers for steel mills, printing presses, and paper machinery fall into this category.
2. Great stiffness is required as in machine tool frames.

Design is then based on allowable deflections established empirically.

Rigidity depends on several individual factors. A rigid shaft will not perform adequately unless its supporting bearings are firmly lodged in the frame. The frame, in turn, must have the rigidity necessary to maintain the alignment of the shaft and bearings adequately.

The adverse effect of deflection can often be controlled instead of eliminated, as demonstrated by the type of rollers used in steel mills (Fig. 4-1*a* and 4-1*b*). The tremendous pressure required for rolling sheets of metal causes even heavy rollers to deflect beyond acceptable limits. Thus cylindrical rollers would produce sheets thicker in the center than at the edges.

To eliminate the effect of deflection, rollers are contoured, that is, ground with a slightly curved surface. When a steel plate forces the rollers apart, both deflect until the opening has parallel sides. Therefore, all sheets will be of uniform thickness.

According to Hooke's law, the modulus of elasticity E is a measure of rigidity. In steel E is very high, providing a major reason for using steel parts. The rigidity of a part is also determined by its geometry, especially material distribution relative to force direction. In general, material should be as far removed as possible from the *neutral or centroidal axes,* as in I-beams or tubing. A measure of rigidity is found in the rectangular moment of inertia I and the polar moment of inertia J.

Example 4-1 will outline the type of reasoning used in assessing factors of rigidity.

EXAMPLE 4-1

Find an expression for the deflection δ of a simply supported round steel shaft of diameter d and length L loaded at midpoint by a single force P. How do the various parameters in the expression for δ affect the magnitude of δ?

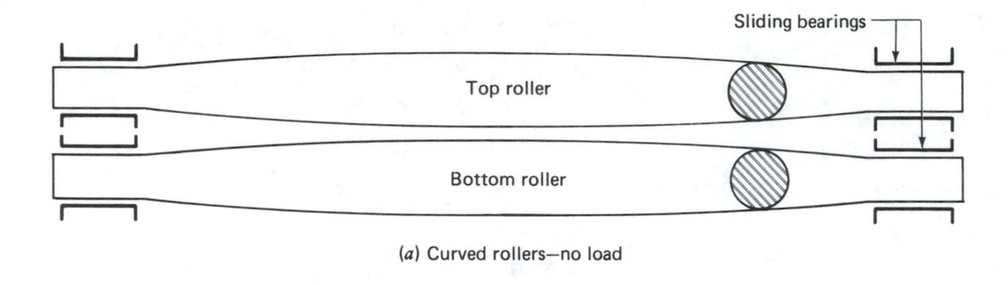

(a) Curved rollers—no load

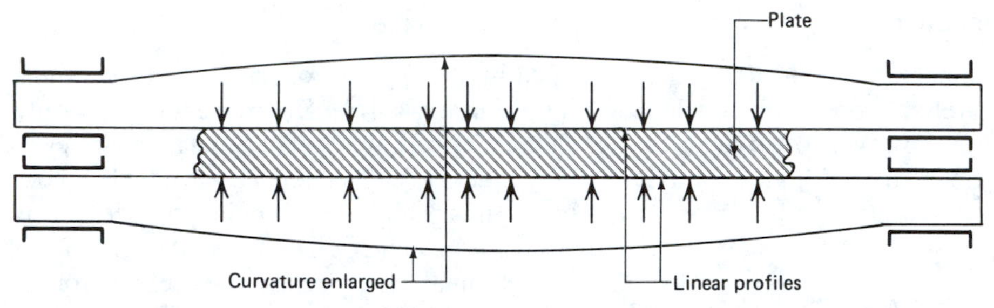

(b) Under load, both rollers deflect to a linear profile on the contact side. The initial curvature, a circular arc, is greatly exaggerated for reasons of clarity.

Figure 4-1 Contoured rollers eliminate the adverse effect of deflection when rolling of metal plates takes place.

SOLUTION

$$\delta = \frac{PL^3}{48EI}$$

Substituting,

$$I = \frac{\pi d^4}{64}$$

$$\delta = \frac{4}{3\pi}\frac{PL^3}{Ed^4}$$

For a given load P, deflection is determined by E, L^3, and d^4. Increasing E, for instance, by using tungsten parts instead of steel, is generally costly or impractical. Much better results are obtained by limiting shaft length L. This has the added benefit of reducing axial space (length) occupied. Most effective, however, is an increase in shaft diameter d. If, additionally, low weight is desirable, a hollow shaft should be specified.

Rigidity is a recurring topic that will be discussed again in Chapters 5, 15, 17, and 18.

4-2 STABILITY OF MACHINE COLUMNS (OR STRUTS)[1]

Machine columns, unlike those of civil engineering, are not necessarily vertical or stationary. Machine columns are the familiar links encountered in kinematics as the coupler of four-bar linkages or the connecting rod of slider cranks. These are the piston rods of hydraulic and pneumatic cylinders as well as the cylindrical body of power screws (Fig. 4-2). Machine columns are widely used because they will transmit large tensile and compressive forces in minimal space. Because they are weaker in compression, sizing is based on buckling.

[1]A strut is any part of a machine or structure whose principal use is to hold things apart; an inside brace.

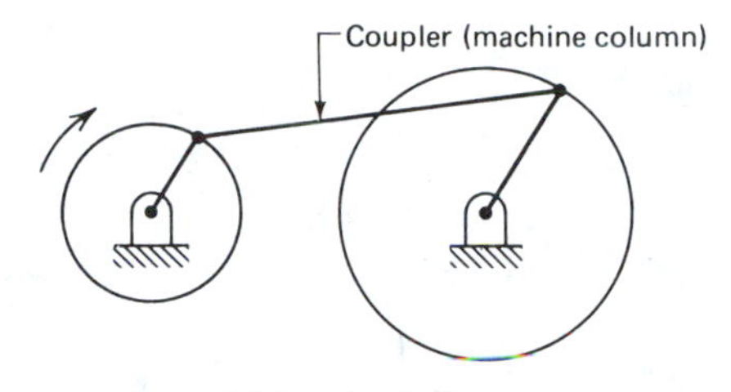

(*a*) Four-bar linkage

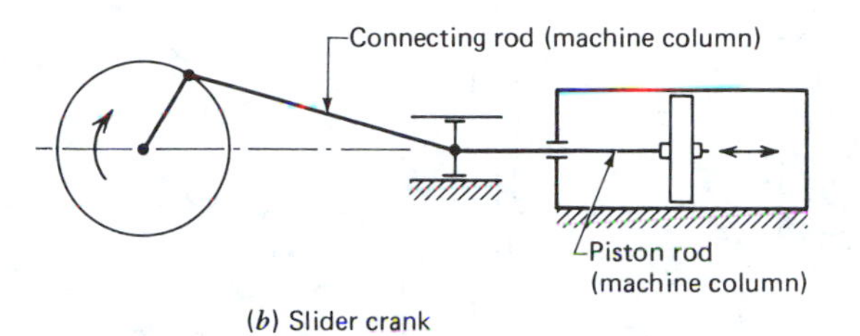

(*b*) Slider crank

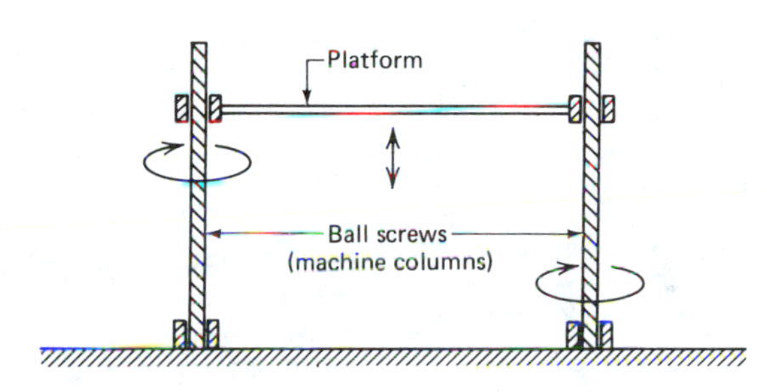

(*c*) Platform using ball screws

Figure 4-2 Typical machine columns.

Stability

A machine column or strut is a structural member which has considerable length in proportion to its width, depth, or diameter. The length of a column is the distance between points unsupported against lateral deflection. Because of a machine column's great length in comparison to its other dimensions, failure in longitudinal compression occurs by transverse deflection rather than by crushing (Fig. 4-3). Termed buckling, this mode of failure occurs rather abruptly—but only in members whose length exceeds the least dimension perpendicular to the centroidal axis by a factor of eight. The degree of transverse deflection accompanying buckling is immaterial. Even before visible deflection occurs, the member loses its load-carrying capacity and its structural usefulness; herein lies the vulnerability of columns.

Mode of Failure

Figure 4-3 shows three compression members, identically loaded and made from the same ductile material. The load, acting along the centroidal axis, is assumed uniformly distributed over identical cross sections (not shown for reasons of clarity). Furthermore, the load increases gradually from zero until failure occurs. During initial loading, all three members are compressed elastically according to Hooke's law. None of these minute axial deflections, however, affects the load-carrying capacity adversely.

Elastic Instability

As the load on all three members continues to increase at identical rates, a critical value is eventually reached for column 1, the longest and weakest member. At this point, the load becomes eccentric to the centerline of the member, introducing a bending component in addition to the compression. Deflection, still elastic, suddenly shifts from axial to transverse and increases exponentially as the load continues to increase. The longest column has now lost its usefulness, since further loading would cause collapse. The remaining two members are still carrying their load with no signs of failure by buckling or deformation.

Plastic Instability

The second member, being somewhat shorter, can obviously sustain a greater critical load than column 1. When it eventually buckles, the transverse deflection in critical surface areas will cause the material to exceed its yield strength ($\sigma_{c,\max} > S_y$). The line of demarcation between elastic and plastic deformation is clearly defined, as subsequent calculations will show.

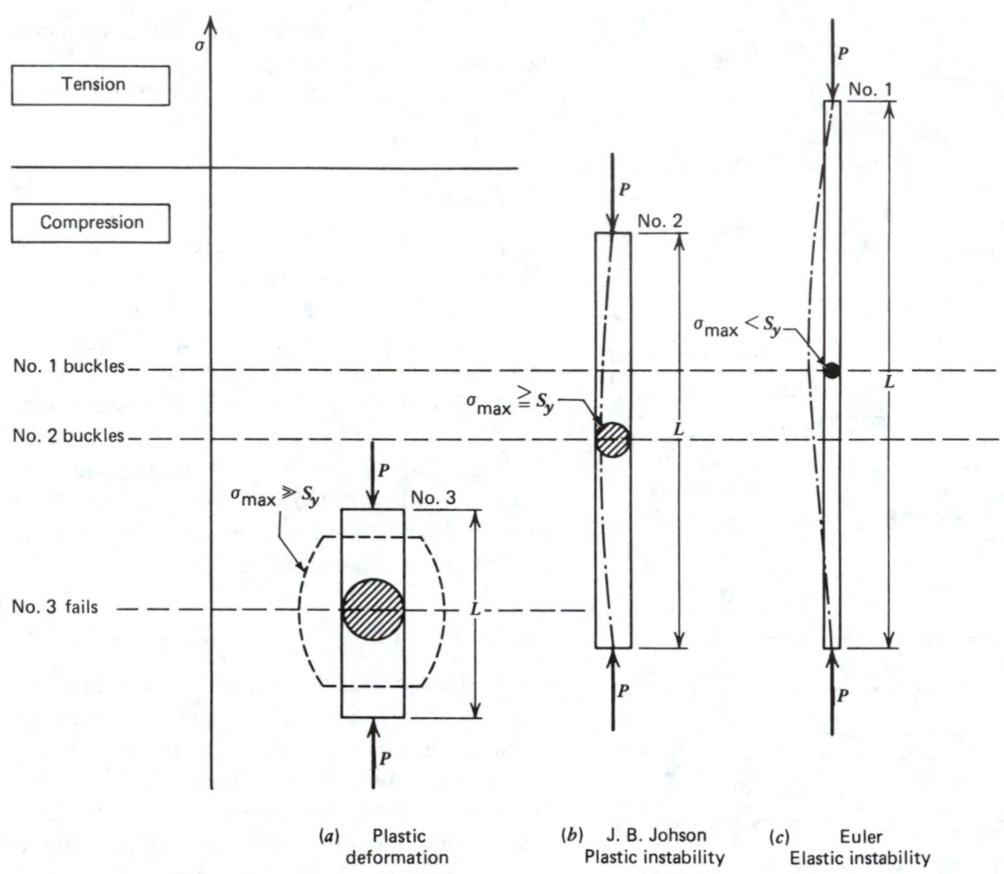

Figure 4-3 Modes of failure of compression members.

Plastic Deformation

The third member, too short to deflect transversely, will deform plastically as shown. The line of demarcation between a simple compression member and a short column is not clearly defined. Experience indicates that buckling prevails when the column length is eight or more times the smallest cross-sectional dimension.

Critical Load

The load that brings the column to the point of buckling is called the critical or buckling load, P_{cr}. The corresponding maximum stress is neither a simple nor a direct function of P_{cr}. Depending on column geometry, it may or may not exceed the material yield strength. In short columns, it is likely to do so; in long columns it will not. For this reason, column design is based on allowable or safe loads, not on allowable stresses. Consequently, derating factors are applied to critical loads.

End Conditions

The load at which a column may buckle is greatly influenced by the restraining conditions at the column's ends. The degree of restraint is generally expressed by a parameter n, termed the coefficient of end conditions.[2] The theoretical

[2]Also referred to as end-fixity or end-restraint coefficient.

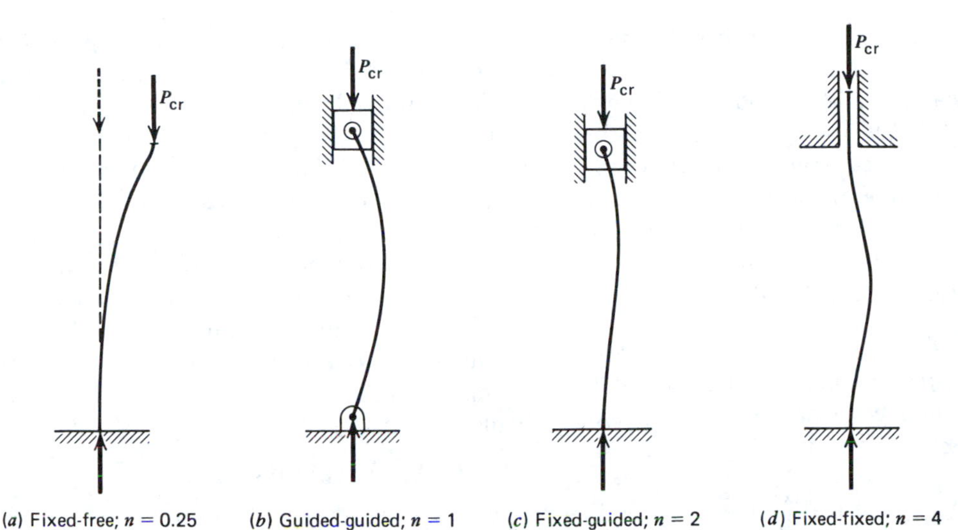

(*a*) Fixed-free; $n = 0.25$ (*b*) Guided-guided; $n = 1$ (*c*) Fixed-guided; $n = 2$ (*d*) Fixed-fixed; $n = 4$

Figure 4-4 Idealized end conditions for columns and coefficients for end conditions. (*b*) applies to connecting rods of slider-crank mechanisms. (*c*) depicts the behavior of cylindrical piston rods of hydraulic cylinders.

values of n vary from 0.25 to 4 and, as the value increases, so do the end restraints and the load capacity. The weakest case is when one end of the column is fixed and the other end is free of all restraint (Fig. 4-4*a*). Higher load capacity results when both ends are either round-ended and guided or hinged (Fig. 4-4*b*). Still more favorable is the case when one end is fixed and the other is round-ended or hinged (Fig. 4-4*c*). The increased load capacity is due to the fixed-end moment opposing any bending effort. Maximum load capacity is achieved when both ends are fixed rigidly (Fig. 4-4*d*). The term *fixed* implies that the tangent to the elastic curve at each end is parallel to the original axis of the column. The values shown for n are theoretical values, which tend to be high. Another way to visualize the effects of end conditions is shown in Fig. 4-5. Here, with loads and cross sections equal, the length, a measure of available space, varies with the end conditions.

Figure 4-5 The importance of end conditions as a means of obtaining a long column can be clearly seen in this array of columns that have identical cross sections and identical buckling loads but greatly varying lengths.

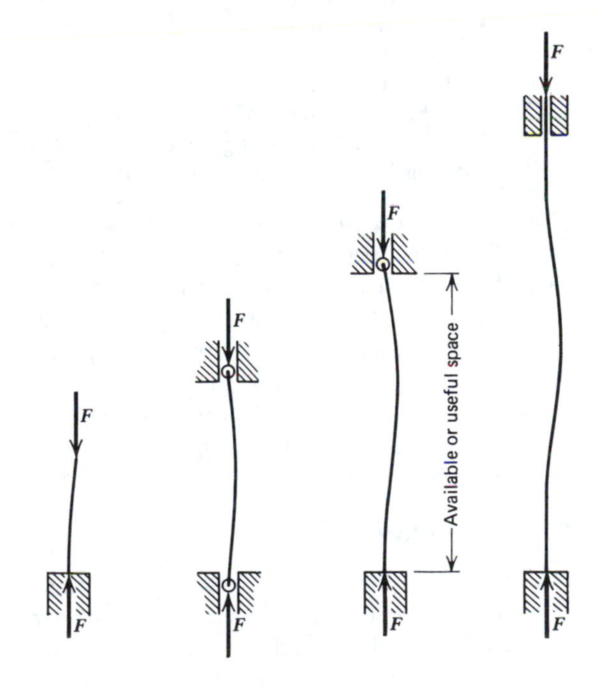

Eccentric Loading

When the column load is offset relative to the axis, the column is defined as eccentrically loaded. Because of the load-induced bending moment, such columns have a lower load capacity than equivalent, axially loaded columns. Their use is justified where space is limited. In all practical cases, a column has some eccentricity because of imperfections in the column itself or because of the load placed upon it.

4-3 CALCULATING MACHINE COLUMNS

The J. B. Johnson formula for short columns applies to most unbraced machine members and will therefore be presented first.

$$P_{cr} = AS_y\left(1 - \frac{Q}{4r^2}\right) \qquad (4\text{-}1)$$

This formula applies when Q/r^2 is less than 2, where

$$Q = \frac{S_y L^2}{n\pi^2 E}$$

where

P_{cr} = critical load; N, lb
A = area of section; mm^2, $in.^2$
S_y = yield strength; MPa, psi
L = length of column; mm, in.
n = coefficient of end conditions
E = modulus of elasticity; MPa, psi
r_g = $\sqrt{I/A}$ = least radius of gyration; mm, in.
I = moment of inertia of area; mm^4, $in.^4$

Q has no physical significance; it is a means of simplifying the expression for P_{cr} in the Johnson formula and later in the Euler formula. The expression for P_{cr} is valid only for $Q / r_g^2 < 2$. Since this expression contains both L and r, the two most common unknowns, it is rarely possible to place a value on Q / r_g^2 in advance. The procedure is therefore to assume $Q / r_g^2 < 2$ and find the quantities requested.

Plane of Failure

In determining the radius of gyration for noncircular cross sections, designers must consider the possibility of buckling in more than one plane. The plane of failure will be the plane for which the combination of bending resistance and coefficient for end conditions is minimum. (This is the plane for which the product nI achieves a minimum.) The I thus obtained is then used to calculate r_g. To illustrate this point, consider a column of rectangular cross section, as shown in Fig. 4-6. The column is pinned in such a manner that the plane of maximum bending resistance is the plane of minimum coefficient for the end condition, and vice versa. Thus the column has two distinct modes of failure in mutually perpendicular planes.

For buckling, as in Fig. 4-6*a*:

$$n_a I_a = 3\frac{ht^3}{12}$$

For the situation in Fig. 4-6*b*:

$$n_b I_b = \frac{th^3}{12}$$

The I to be used in calculating r is obtained from the smaller of the two values of nI. If equal strength is sought in both planes, the two values for nI must be *equated* and the proper relation

Figure 4-6 Rectangular column pinned at both ends. Assume buckling in the paper's plane.

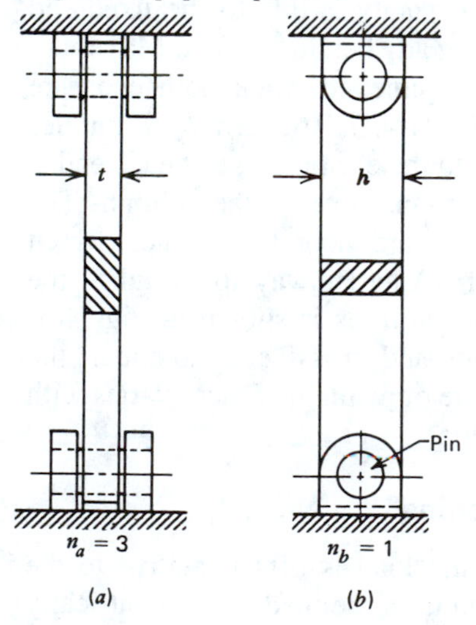

between t and h found. This relation is $h = t\sqrt{3}$. For $n_a = 4$, h becomes $2t$.

When the condition of loading and the physical qualities of the material used are accurately known, a factor of safety as low as 1.25 is sometimes used when minimum weight is important. Generally, a factor of safety of 2 to 2.5 is applied to steady loads. The factor of safety represents the ratio of the P_{cr} to the working load P_w.

The J. B. Johnson formula is most applicable to ductile materials such as steel. For brittle materials, such as cast iron, the formula yields somewhat high values. When a large strength-to-weight ratio is desirable in a column, the cross-sectional area should be tubular, or "I" shaped, and the depth should be larger in the midsection than at the ends.

Long Columns

When $Q/r^2 > 2$, the J. B. Johnson formula is not valid; the Euler formula takes its place.

$$P_{cr} = S_y A \frac{r^2}{Q} \tag{4-2}$$

This formula applies to medium and long columns, which are less common in machinery. The procedure is the same as for short columns. Note that S_y disappears because Q also contains S_y. This is to be expected, since the deflection is elastic.

EXAMPLE 4-2

A cylindrical steel column is to be made of steel ($S_y = 60{,}000$ psi). The column is 48 in. long and the end-restraint coefficient is 2.75. Find the diameter d for an applied load of 40,000 lb and a factor of safety of 2.

SOLUTION

Let us proceed on the assumption that the J. B. Johnson formula applies.

$$P_{cr} = AS_y\left(1 - \frac{Q}{4rg^2}\right)$$

$$P_{cr} = 2(40{,}000 \text{ lb}) = 80{,}000 \text{ lb}$$

$$A = \frac{\pi d^2}{4}$$

$$rg^2 = \frac{I}{A} = \frac{\pi d^4/64}{\pi d^2/4} = \frac{d^2}{16}$$

$$Q = \frac{S_y L^2}{n\pi^2 E} = \frac{(60{,}000 \text{ lb/in.}^2)(48 \text{ in.})^2}{2.75\pi^2(30)10^6 \text{ lb/in.}^2}$$

$$= 0.169 \text{ in.}^2$$

Substituting in the J. B. Johnson formula:

$$80{,}000 \text{ lb/in.}^2 = \frac{\pi d^2}{4}(60{,}000 \text{ lb/in.}^2)\left[1 - \frac{(0.169 \text{ in.}^2)(16)}{4d^2}\right]$$

$$1.695 = d^2 - 0.676$$

$$d^2 = 2.371 \text{ in.}^2$$

$d = 1.54$ in. Use d = $\mathbf{1\frac{9}{16}}$ **in.**

Check

$$\frac{Q}{rg^2} = \frac{(0.169)(16)}{2.371} = 1.14 < 2$$

The J. B. Johnson formula applies.

4-4 CYLINDRICAL PISTON RODS

Incredible power relative to mass, coupled with speed and smoothness of operation, accounts for the extensive industrial use of hydraulic cylinders—a basic machine. Manufacturers of cylinders have developed *piston rod selector charts*, as shown in Fig. 4-7. Three parameters—diameter, length, and load—are included. Thus any one parameter can be found if the other two are known. Although the rods seem short, the chart supposedly is based on Euler's equation applied to a column with both ends rounded. This corresponds to Fig. 4-4*b*. The other cases in

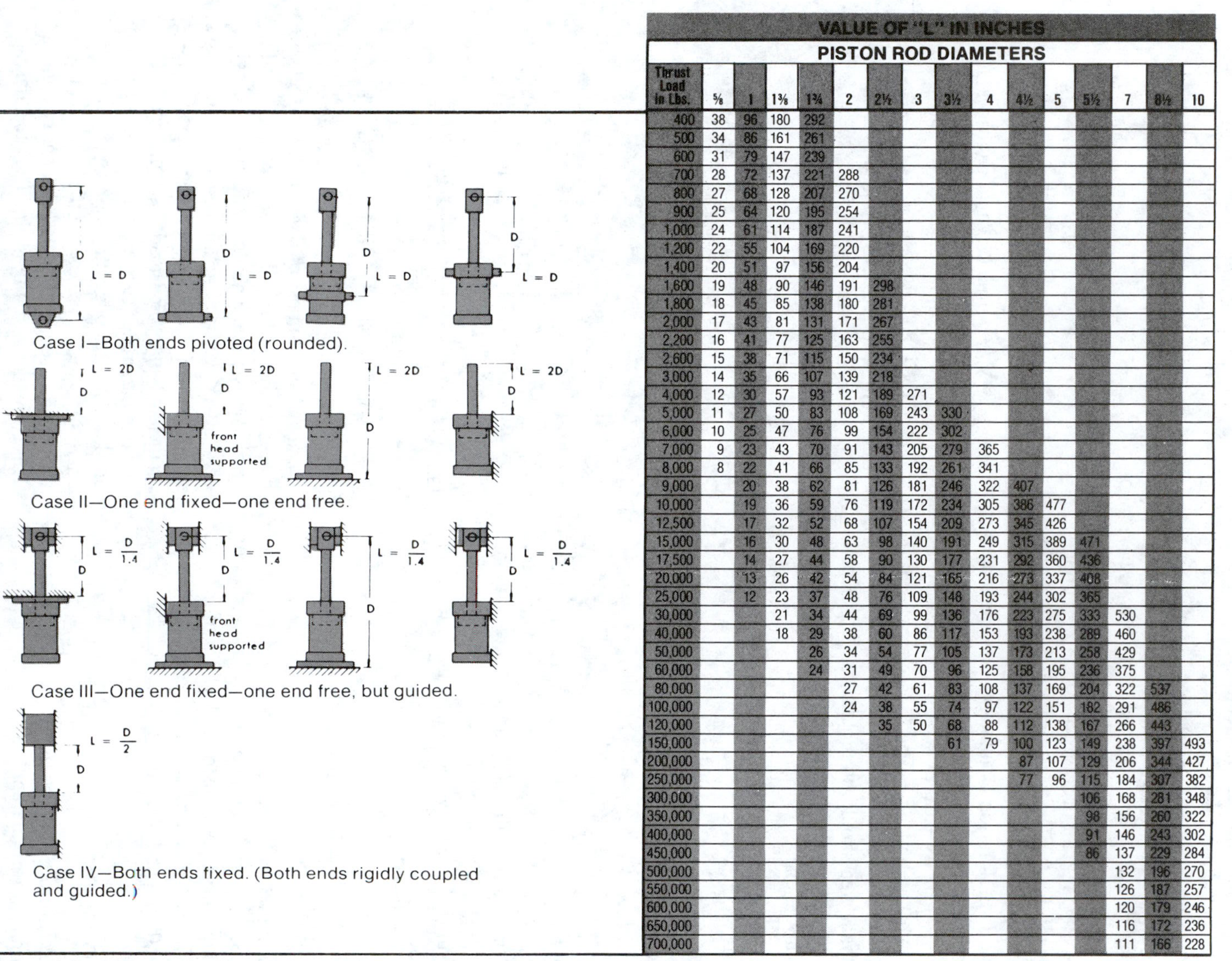

VALUE OF "L" IN INCHES

Thrust Load in Lbs.	PISTON ROD DIAMETERS														
	⅝	1	1⅜	1¾	2	2½	3	3½	4	4½	5	5½	7	8½	10
400	38	96	180	292											
500	34	86	161	261											
600	31	79	147	239											
700	28	72	137	221	288										
800	27	68	128	207	270										
900	25	64	120	195	254										
1,000	24	61	114	187	241										
1,200	22	55	104	169	220										
1,400	20	51	97	156	204										
1,600	19	48	90	146	191	298									
1,800	18	45	85	138	180	281									
2,000	17	43	81	131	171	267									
2,200	16	41	77	125	163	255									
2,600	15	38	71	115	150	234									
3,000	14	35	66	107	139	218									
4,000	12	30	57	93	121	189	271								
5,000	11	27	50	83	108	169	243	330							
6,000	10	25	47	76	99	154	222	302							
7,000	9	23	43	70	91	143	205	279	365						
8,000	8	22	41	66	85	133	192	261	341						
9,000		20	38	62	81	126	181	246	322	407					
10,000		19	36	59	76	119	172	234	305	386	477				
12,500		17	32	52	68	107	154	209	273	345	426				
15,000		16	30	48	63	98	140	191	249	315	389	471			
17,500		14	27	44	58	90	130	177	231	292	360	436			
20,000		13	26	42	54	84	121	165	216	273	337	408			
25,000		12	23	37	48	76	109	148	193	244	302	365			
30,000			21	34	44	69	99	136	176	223	275	333	530		
40,000			18	29	38	60	86	117	153	193	238	289	460		
50,000				26	34	54	77	105	137	173	213	258	429		
60,000				24	31	49	70	96	125	158	195	236	375		
80,000					27	42	61	83	108	137	169	204	322	537	
100,000					24	38	55	74	97	122	151	182	291	486	
120,000						35	50	68	88	112	138	167	266	443	
150,000								61	79	100	123	149	238	397	493
200,000										87	107	129	206	344	427
250,000										77	96	115	184	307	382
300,000												106	168	281	348
350,000												98	156	260	322
400,000												91	146	243	302
450,000												86	137	229	284
500,000													132	196	270
550,000													126	187	257
600,000													120	179	246
650,000													116	172	236
700,000													111	166	228

Figure 4-7 Piston rod selector chart. (Courtesy Ortman Fluid Power, Division of Garlock Inc.)

the chart correspond to those remaining in Fig. 4-4. Thus the chart shows four end conditions, 13 basic designs of cylinders, and 314 effective piston rod lengths at various thrust loads and rod diameters.

To determine the proper piston rod diameter for a given application, proceed as follows.

1. Determine the maximum thrust load required.
2. Relate the application to one of those illustrated as Case I, II, III, or IV.
3. Determine the value L with the piston rod fully extended.
4. Now, referring to the chart, select the thrust load figure that equals or exceeds your requirements.
5. Scan the chart on the right until the value of L equals or exceeds the L dimension on your application. The rod diameter sought is found at the top of this column.

The values shown are roughly 50% of the theoretical values obtained by Euler's equation. Service factors to match operating conditions must be added by designers.

The typical standard rod is made from medium carbon steel, case hardened to 54 Rockwell C, chrome plated, and polished. The yield strength is 90 to 100,000 psi. This rod is well protected against wear and corrosion. It also facilitates the use of high-pressure seals.

4-5 FRICTION AND WEAR CONSIDERATIONS IN MACHINE DESIGN

The soaring cost of natural energy has forced machinery builders to pay much closer attention to friction and wear control in the early stages of all designs. The economic importance of friction can be measured by the fact that 33 to 50% of the world's energy consumption is used to overcome friction. Wear, the constant companion of friction, represents one of the most significant operating costs of machinery. Every year thousands of otherwise useful machines are taken out of service because minute amounts of material have worn off vital contact surfaces.

Wear is the gradual, unintentional abrading of surfaces in contact as a result of relative motion between them. Because all machines contain parts moving against each other, wear is invariably present; under its influence, useful life will always be reduced. Wear can cause imbalance and subsequent vibration of rotating parts. It can adversely affect lubrication of parts designed with close-running tolerances. This, in turn, may lead to more vibration, more wear, and further loss in lubrication effectiveness. As a result, wear often progresses at an increasing rate; eventually, vibration becomes so violent that the machine ceases to function smoothly. These stages of deterioration can be observed in some automobiles after extended use without proper servicing.

Although friction and wear always appear together, there seems to be no consistent relationship between them. Some materials have a high coefficient of friction and a low wear rate, while others behave oppositely.

The laws of friction are fairly well substantiated, but those of wear are not. In fact, there are no satisfactory laws of wear, only rules of thumb. These are:

1. Wear increases with time of running.
2. With hard surfaces the wear is less than with softer surfaces.
3. Below a certain load wear is low (mild wear); above this load, it rises catastrophically to values that may be 1000 to 10,000 times greater (severe wear). This holds true for both clean and lubricated surfaces.

Wear, essentially a surface phenomenon, is affected by the environment, but not always in a predictable manner. Noncorrosive surface films, including the ever-present oxides, effectively lower friction and wear. The main effect of speed arises from increased surface temperatures that, in extreme cases, may cause the surface material to melt locally. When this happens, dry friction, which is high, is replaced by fluid friction, which is low. The results may be disastrous, as when brakes *fade*, or momentarily beneficial, as when a sliding bearing runs dry of lubricant oil.

4-6 SLIDING FRICTION

When two solid bodies rub against each other, the force opposing motion is known as solid or sliding friction. Solid friction is a complex phenomenon; fortunately, it can be expressed in a single formula, known as Coulomb's law, that is sufficiently accurate for most engineering applications.

$$F_f = fN \tag{4-3}$$

This formula states that the force of friction F_f varies directly with the normal load N but is independent of contact area and velocity (neither appears in the formula).

The constant f, defined as *coefficient of friction,* is the ratio of frictional force to normal force. It is higher for static than for dynamic conditions (Fig. 4-8) and assumes different values for differing materials in contact. In fact, coefficient tables list values of dissimilar materials often paired. Note that whereas friction denotes resistive force associated with a particular situation, the coefficient of friction defines frictional characteristics of materials. Herein lies its importance. It enables designers to compare frictional characteristics of different materials and pick the pair or combination that suits their design (Table 4-1). Since these characteristics vary within wide limits, however, which values does one change?

Even though f theoretically is independent of speed and surface finish, in reality it varies, as shown in Fig. 4-8. The minimum value of f, as a function of surface roughness, corresponds to a machined and ground surface. Thus, when surfaces are rough or velocity is high, one should choose the larger values.

The rapid increase in f, as surfaces become smoother, is really due to adhesion, which resists

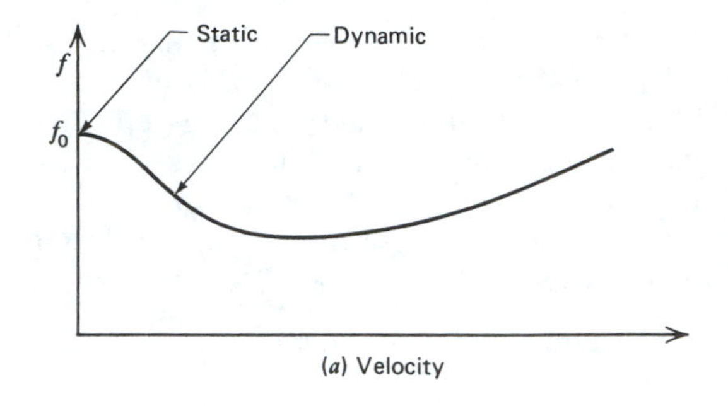

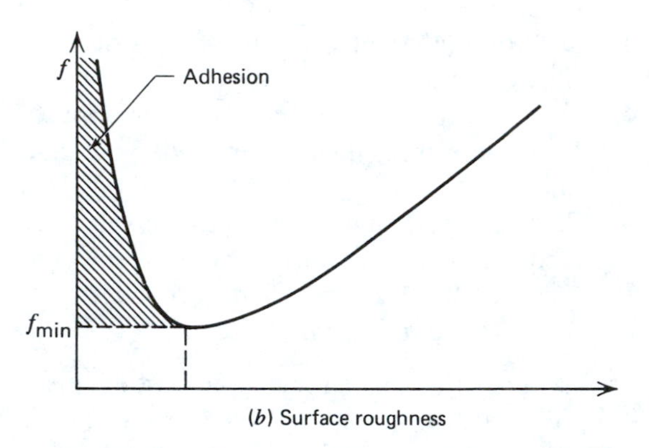

Figure 4-8 The coefficient of sliding friction as a function of velocity and surface roughness. $f_{\min}$ corresponds roughly to a machined and ground surface.

motion and tends to hold the two surfaces together so that they cannot be separated. Consequently, there is a practical lower limit on surface finish of machine members.

Rearranging terms in Eq. 4-3, we can write:

$$f = \frac{F_f}{N} = \tan\theta \tag{4-4}$$

where θ is the angle between the normal force N and the resultant of F_f and N. Termed friction angle, θ provides a practical equivalent to f in predicting motion in machine members based on the inclined plane, notably screws and worm gearing (Section A-3).

Of all the physical properties that designers have to use in calculating the behavior of mechanical systems, none is more difficult to specify than the coefficient of friction. Fortu-

TABLE 4-1 Coefficients of Friction for Selected Materials Under Varying Conditions

	Sliding Friction				Rolling Friction, f_R R is measured in mm
	Static f_{st}		Dynamic f_{dyn}		
Materials	Dry	Lubricated	Dry	Lubricated	
Steel–steel	0.11–0.33 (0.74)	0.10	0.10–0.11 (0.57)	0.01–0.05 (0.09)	$0.5/R$
Steel–cast iron	0.18–0.20	0.10 (0.183)	0.16–0.20	0.01–0.05	$0.3/R - 0.5/R$
Cast iron–cast iron	0.20–0.25 (1.10)	0.05–0.15	0.12–0.25	0.06–0.16	$0.5/R$
Steel–brakelining	—	0.10	0.50–0.60	0.20–0.50	—
Rubber–asphalt	0.50–0.80	0.30–0.50	—	—	—

Sources. *Dubbel Tashenbuch fur den Maschinenbau,* Bd I, 13. Aufl. Springer-Verlag, Berlin, 1970, p. 230. *Marks' Standard Handbook for Mechanical Engineers,* Eighth Edition, McGraw-Hill Book Company, New York, 1978, pp. 3-24 to 3-33. The figures in parentheses are from this reference. Used with permission of McGraw-Hill Book Company.

nately, friction forces generated between properly lubricated surfaces, as in most sliding bearings, are generally so small that they can be ignored. For mechanisms designed to utilize friction (screw fasteners, brakes, couplings), it is usually possible to make generous allowances for variations. In intermediate cases (e.g., translation screws), where friction is significant in relation to other forces, arriving at anything more than a reasonable estimate of the forces involved is often difficult. In these designs attempts should be made to replace sliding friction by rolling friction, which is much lower and more predictable.

Mechanical vibration effectively lowers friction by momentarily separating contacting surfaces. In screw fasteners this is a disadvantage because it tends to loosen nuts, bolts, and screws. Increasingly, however, controlled vibration is used to lower friction and facilitate motion in component feeders, rock drills, pile drivers, and soil-tilling machinery.

4-7 ROLLING FRICTION

The major advantage of rolling machine components, compared with sliding components, lies in the greatly reduced stresses generated at the mating surface. The result is low resistance to relative motion and thus reduced wear. Major examples include wheel on rail, rolling bearings (linear and circular), cams with roller followers, gear teeth, and friction wheels.

The basic equation for rolling friction parallels that for sliding friction.

$$F_R = f_R W \qquad (4\text{-}5)$$

where

F_R = resistance to rolling; N, lb
f_R = coefficient of rolling friction
W = weight of body; N, lb

The factor f_R depends on applied materials and surface conditions and also on wheel size. It varies inversely with wheel size, which explains the well-known advantage of using large-diameter wheels for low rolling resistance.

The advantage of rolling contact is greatest when the surfaces are high-friction types. Such surfaces ensure *pure rolling,*[3] a condition that arises when the force of static friction F_{st}, acting at the line of contact, is greater than the resistance to rolling F_R.

[3]In pure rolling action no one point of either member comes in contact with two successive points of the other.

4-8 RESISTANCE TO MOTION IN VEHICLES

When trains, trolleys, and mobile industrial equipment move laterally at constant speed, a simple empirical formula again provides a rough estimate of the frictional resistance. Because of standard components (wheel size, bearings, track gauge) and uniformity of materials (steel against steel), the only variable left is total weight. Thus,

$$F_R = f_v W \quad (4\text{-}6)$$

where

F_R = total resistance to rolling of an entire vehicle; N, lb

W = weight of vehicle and load; N, lb

f_v = coefficient of rolling friction for a specific vehicle operating under normal conditions

Railroad cars:

$$f_v = 0.0015 - 0.035$$

Vehicles on asphalt and concrete:

$$f_v = 0.015 - 0.025$$

Note, however, that data on rolling friction are scarce and highly variable (Table 20-5).

EXAMPLE 4-3

A railroad car of mass 10 000 kg carries a load of 80 kN. Calculate the force necessary to move the car at constant speed.

SOLUTION

$$\text{Total load} = (10\ 000\ kg)(9.81\ m/s^2) + 80\ 000\ N = 178\ 000\ N$$

Since F_R varies between extremes, we calculate maximum and minimum values for F.

$$F_{\min} = (178\ 000\ N)(0.0015) = \mathbf{267\ N}$$

$$F_{\max} = (178\ 000\ N)(0.0035) = \mathbf{623\ N}$$

4-9 TYPES OF WEAR

The principal types of wear are adhesive, abrasive, corrosive, and fatigue. Of these, abrasive action is the single greatest cause of wear in machinery. Often, however, two or more types operate simultaneously, one promoting the other.

Adhesive wear is the most common type and occurs whenever two solids are in sliding contact. Also called scoring or scuffing, wear of this type involves the transfer of small particles from one surface to the other (Fig. 4-9*a*).

Although modern tools are capable of producing parts with close tolerances and highly polished surfaces, many machine elements have surfaces that are too rough when newly machined to sustain the loads and speeds that they will ultimately carry. Frictional heat resulting from the initial roughness of mating parts may be sufficient to damage these parts, even to the point of failure. Thus a new machine, or a machine with new parts, must be operated below rated capacity until the opposing surfaces have been gradually worn to the required smoothness.

This process is known as run-in or break-in and takes place in all types of machinery. (In cars it occurs during the first 8000 km.) Under break-in conditions, friction is controlled by reduction of load. Sometimes a break-in lubricant fortified with load-carrying additives may be specified. Usually, this lubricant should be drained after a few hours of use, thus removing the initial debris from wear.

Abrasive wear is the only type of wear that arises from the penetration of one surface by another (Figs. 4-9*b* and 4-9*c*). Abrasive wear occurs when a rough surface slides against a softer one, plowing a series of grooves and removing material. This form of abrasive wear is often referred to as *gouging* and is exemplified by the teeth of a power shovel worn by handling rocks and other hard materials. Abrasive wear is also caused by abrasive particles carried by the oil between two rubbing surfaces, in which case it is referred to as *grinding* wear. Abrasive debris such as *tramp metal* in a lubricant can be very

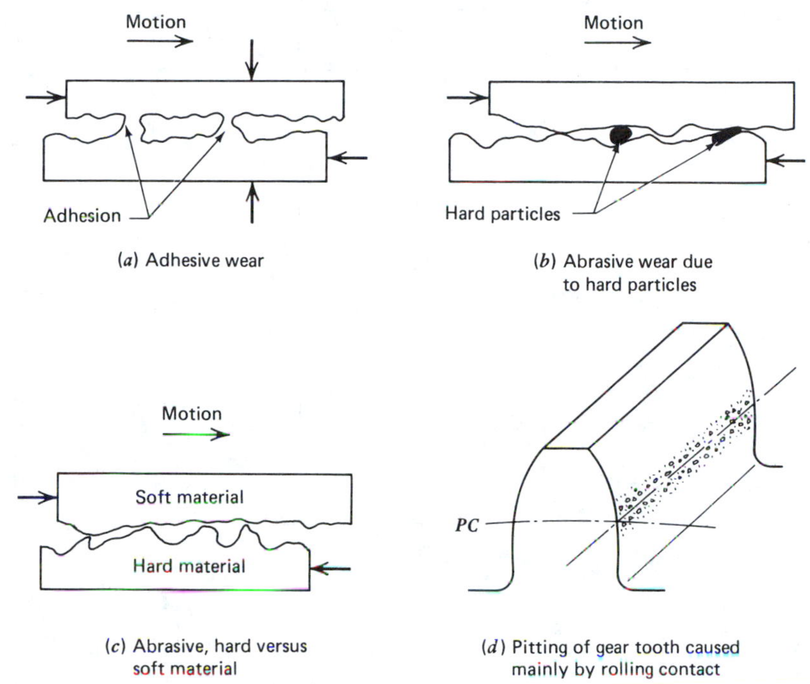

Figure 4-9 Basic wear mechanisms.

harmful because it grinds at precision parts, transmission screws, or cylinder walls. Often, scuffed bearings can be traced to contaminated lubricant.

Corrosive wear is a chemical action that causes destruction of metal surfaces by oxidation or rusting. In machinery the destructive chemicals are often fuel by-products or decomposed lubricant additives. Even though they are present in minute amounts only, their effect is relatively great because of interaction with mechanical wear, which constantly exposes fresh metal surfaces to attack.

Fatigue wear occurs when a surface is repeatedly stressed and unstressed (Fig. 4-9*d*). It is common for parts subjected to rolling contact, such as ball bearings and gears.

Cavitation, a combination of fatigue and corrosive wear, is often found in pumps, valves, and hydraulic turbines. It occurs on surfaces in contact with a liquid that is subjected to rapidly alternating changes of pressure.

The Three Stages of Wear (Fig. 4-10)

The wear process of new bearings can be divided into three stages. During the first stage, *running-in* takes place. In this stage minute peaks (asperities) left by machining operations on the mating surfaces engage each other and are partly sheared and partly plastically deformed; a relatively large amount of material is removed from the surface.

Running-in is followed by an extended working period of linear stabilized wear. Linear wear lasts the longest and corresponds to the normal working life of a sliding contact component. The last stage is destructive and is characterized by a rapidly increasing wear rate that leads to failure.

4-10 DESIGN FOR WEAR RESISTANCE

Because adhesive wear can be limited by specifying hard surfaces, this type of wear is inversely proportional to hardness. Carburizing, carbo-

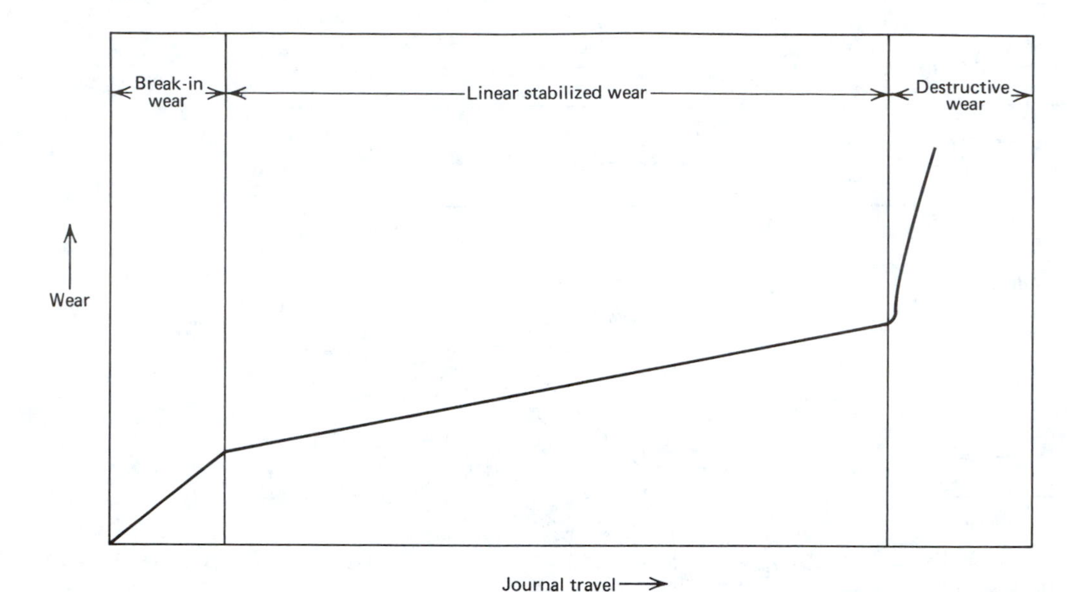

Figure 4-10 The three stages of wear. (From "Wear in Boundary-Lubricated Bearings," *Machine Design,* May 16, 1974.)

nitriding, and induction hardening are thus used to combat adhesive wear, as are inserts of extremely hard and wear-resistant materials such as tungsten carbide. Table 4-2 shows the nominal hardness for various machine elements.

Abrasive wear can be limited in numerous ways. Surfaces may be given a higher hardness than the abrasive particles. If a circulating lubricant is used, abrasive particles can be removed by filters. Magnetic filters are particularly useful because they attract and hold small steel particles torn from shaft journals. Such particles are very harmful because their hardness and cohesive strength exceed that of the material from which most sliding bearings are made. A combination of hard and soft surfaces can also be used so that the abrasive particles may embed themselves in the softer material. Cases where abrasion occurs without lubrication, as in soil-working tools, require both high carbon content and high hardness. White cast or chilled iron, although brittle, is especially resistant to abrasion.

Other known methods of reducing wear include the following.

1. *Replacement*. The most effective means of limiting the adverse effects of wear is to design for easy and inexpensive replacement of vital machine members, notably bearings.
2. *Dissimilar Materials*. Perhaps the most basic rule for reducing wear is to specify dissimilar materials for parts in sliding contact. Steel, for instance, does not run well with steel, but it is an excellent match for cast gray iron, bronze, brass, or plastics. Steel and nodular cast iron, however, frequently do not work well together.
3. *Compensation*. Adjustment and take-up provision for automatic compensation of wear is another effective method of fighting the adverse effects of wear in machinery. It may take the form of spring-loaded idlers or automatic take-up on machine slides subject to wear.
4. *Metallizing*. This term describes the spraying of molten metal onto the surface of a base metal. The coating metal, in the form of a wire, strip, or powder, is melted in a

spray gun by a high-pressure gas flame or in an arc. The metal is sprayed by means of a high-pressure gas, thereby producing a paper-thin coating. Because of oxidation these coatings are chemically and physically different from the metal used. They combine hardness with microscopic surface porosity, providing an excellent bearing surface that retains lubricants and is far more wear resistant than either the sprayed metal in original form or the underlying base metal.

5. *Hard-facing*. This process consists of welding a facing edge or point of hard, wear-resistant metal to surfaces that are subject to excessive wear, such as shovels on excavators. Metal surfaces that normally wear away rapidly in service thereby receive adequate protection through an added layer of wear-resistant alloy.
6. *Lubrication*. Wear can be greatly reduced by keeping sliding metals apart through the use of a lubricant—a substance that can be deformed or sheared without damage. The most versatile, reliable, and best-known lubricant is mineral oil. Its advantage over other fluids lies in its ability to maintain a strong film that separates contacting surfaces.

In some applications, fluid lubricants are inadequate because of the difficulty of retention, relubrication, or churning. In such

TABLE 4-2 Typical Steels and Nominal Hardness for Various Machine Elements

Hardness	Typical Steels	Application
64-66 R_C	1020 carburized	Cams, piston pins
	1115 carburized	Knuckle pins
62-64 R_C	4615, 4620, 3312	Gears, pinions, cams
	3120 carburized	Cups and cones for
		roller bearings
60-63 R_C	4615, 4620, 2515	Pinions, gears, cams
	3120 and 6120 carburized	Sprockets, stamps
		Aircraft crankshafts
59-61 R_C	1045 induction hardened	Gears, piston pins
	4615, 3120 carburized	Cams, sprockets
59-61 R_C	6152, 52100, 1095	Collets, feed fingers, etc.
55-58 R_C	3250, 6152	Chuck jaws, gears
	1045 induction hardened	Light shear blades
		Screwdriver blades
50-52 R_C	2345, 3145, 4342, etc.	Gears, pinions
45-50 R_C	6152, 1095	Light and medium springs
	9260	Arbors
43-45 R_C	3135, 2335	Locomotive drive gears
42-44 R_C	6152, 1095, 9260	Heavy springs, wrenches
38-40 R_C	3140, 2340, etc.	Wheel spindles, axles
350-375 Bhn		
300-350 Bhn	3135, 2330	Bolts, studs
		Crankshafts, connecting rods
250-300 Bhn	3135, 4130, 4337,	Studs, aircraft propellers
	2330	Crankshafts, connecting rods
225-275 Bhn	3135, 2330, etc.	Connecting rods, bolts, etc.

Source. *Metals Selection and Treatment*, by K. J. Trigger and D. L. Mykkanen.

cases greases are used. They consist of petroleum oil thickened by soap or inorganic matter.

Sometimes, however, solids can be used. Substances such as graphite or molybdenum sulfide, whose layers slide or flake against each other, act like liquids. Solid lubricants can be used alone or mixed with oil or grease. When used by themselves, they are referred to as dry lubricants.

7. *Seals and Gaskets*. These are effective means of reducing wear because they retain lubricants while excluding dust, dirt, and contaminants from vital machine parts.

Despite these many effective means of reducing wear, it remains the single largest operating cost of machinery. Science and industry have yet to develop adequate techniques for controlling wear and compensating for its effects.

SUMMARY

Design of machine members is dominated by the concept of strength. The designer's concern for reliable operation, coupled with long service life, requires design for rigidity, stability, and resistance to wear.

Rigidity is the ability to resist deformation. The rigidity of a part is determined by its geometry, especially material distribution relative to force direction and the modulus of elasticity of the material used. For single machine members rigidity becomes a design factor when the member is long relative to its width, depth, or diameter. For machine frames rigidity applies primarily to machine tools and production machinery. Design is based on allowable deflections established empirically.

Stability is the ability of a part to resist displacement and, if displaced, to develop forces and moments tending to restore the original condition. As a design criterion, stability relates primarily to machine columns. Machine columns are widely used because they will transmit large tensile and compressive forces in minimal space. They function in any position, whether stationary or mobile, and fail by deflecting transversely.

When columns are long relative to their width, depth, or diameter, they are prone to buckle elastically. Elastic instability is based on the Euler formula. Somewhat shorter columns can sustain greater critical loads. When they fail, the transverse deflection in critical surface areas will cause the material to exceed its yield strength. Plastic instability is based on the J. B. Johnson formula.

Friction plays a dual role in machinery. Friction is essential to rolling motion, screw fasteners, and friction drives, but it is also a major cause of wear on all contacting surfaces. Wear is the gradual, unintentional abrading of surfaces in sliding contact.

Although friction and wear always appear together, there is no consistent relationship between them. Furthermore, the laws of friction are fairly well established; those of wear are not.

Whenever sliding friction is undesirable, attempts should be made to replace it by rolling friction, which is much lower and more predictable. Friction and wear can be greatly reduced by keeping sliding metals apart through the use of a lubricant. Other known methods of reducing wear consist of replacing parts at regular intervals and using dissimilar materials and hardened surfaces. Despite the many effective means of reducing wear, it remains the single largest operating cost of machinery.

REFERENCES

4-1 Rabinowicz, Ernest. *Friction and Wear Materials*. New York: Wiley, 1965. (A basic reference book.)

PROBLEMS

P4-1 Using the Johnson formula, compute L as a function of n for the values of n

0.25, 1, 2, 3, 4. All other parameters remain constants. Assume $L = 50$ mm for $n = 1$. Plot L as a function of n in a bar diagram, with L horizontal. (*Hint:* L^2/n = constant.) Comment on the answers; what is their technical significance?

P4-2 Using the data from Example 4-1, plot a curve showing d as a function of S_y when S_y assumes values from 30,000 to 100,000 psi in increments of 10,000 psi. Plot a second curve for P_{cr} as a function of S_y for $d = 1.54$ in. (*Hint:* Use a minicomputer.)

P4-3 A 9000-lb load must be moved 50 in. by means of a hydraulic cylinder. Use Fig. 4-7 to specify a suitable rod diameter.

P4-4 Convert the data in Example 4-2 to SI. Use round numbers only. Calculate d for $n = 1.75$.

P4-5 Refer to Example 4-2. Find the necessary cross section for a hollow column in which the $OD = 2\ ID$. How much is the column mass reduced?

P4-6 Discuss the torsional rigidity of a circular shaft when the torsional deflection is given by the expression

$$\theta = \frac{32\ TL}{\pi GD^4}$$

where

θ = angular deflection; rad
T = applied torque; N•mm, lb-in.
L = length of shaft; mm, in.
G = shear modulus; MPa, psi
D = diameter; mm, in.

How do θ and mass change if the solid shaft is replaced by a hollow one with $ID = d = 0.5D$? What do you conclude from this?

P4-7 For the Johnson formula prove that buckling occurs for the plane in which nI achieves a minimum value.

P4-8 A cylindrical, hollow steel column of $ID = 0.5\ OD$ is made from SAE 1045 annealed steel, $S_y = 90{,}000$ psi. The length is 72 in., and the coefficient of end conditions is 2. Determine the cross-sectional area for a theoretical buckling load of 40,000 lb.

P4-9 The column in P4-8 is to be made from standard steel pipe. Use the appropriate tables in *Machinery's Handbook* (pp. 431-432) to specify a suitable pipe size. Indicate the various options. (Use the theoretical k value on p. 432.)

P4-10 The cylindrical piston rod of a reciprocating water plunger pump may be considered a column with one end fixed, the other hinged. The cylinder is 200 mm in diameter, and the maximum water pressure is 1.25 N/mm^2. The rod is 500 mm long and is made from hardened, polished mild steel ($S_y = 625$ MPa). Calculate the rod diameter for a safety factor of 2. How does this rod compare with one obtained from Fig. 4-7? Make a suitable sketch.

P4-11 A simple connecting rod of circular cross section is 150 mm long and is to carry a load of 22 kN. Determine the diameter using mild steel $S_y = 240$ MPa and a factor of safety of 2.

P4-12 A simple connecting rod of rectangular cross section is pinned as shown in Fig. 4-5. For $h = 125$ mm, calculate the other side t for equal buckling strength in both major planes.

P4-13 What is the allowable compressive load for a 50-mm × 100-mm steel bar that is 2 m long? Assume $S_y = 275$ MPa, $n = 1$, and $fs = 4$.

P4-14 A nut is mounted on a vertical screw thread. If the coefficient of friction is 0.10, what minimum value of helix angle will cause the nut to slide down on its own (Fig. 4-11)?

P4-15 A steel screw thread and nut have a helix angle of 6 deg. Is this screw con-

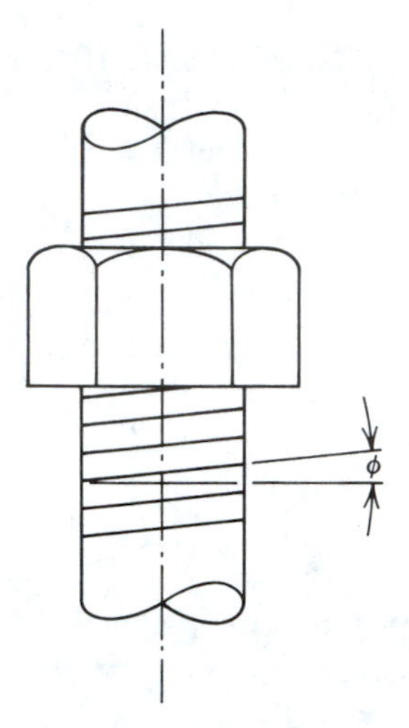

Figure 4-11 Nut on vertical screw thread (Problem 4-14).

nection self-locking when dry? What happens when a lubricant is applied?

P4-16 An overhead crane, moving on steel tracks, has a full load capacity of 442 kN and a mass of 65 000 kg. The four wheels, made from cast steel, are 300 mm in diameter. Make a suitable sketch and calculate the following.

(a) The resistance encountered at constant speed.
(b) The tractive force at each of the two drive wheels.
(c) The torque on the drive shaft.

(*Hint:* Use $F = f_R W$, Table 4-1, $f_R = 0.5/R$.)

P4-17 Find the power necessary to drive the crane in P4-16 at a travel speed of 1 m/s. Use Fig. 1-6 to estimate the service factor. The crane has a worm gear reduction unit and is driven by an electric motor at 1200 rpm. Specify motor size. (*Hint:* Use $P_n = Fv/1000$; $v = \pi Dn$.)

P4-18 Would it be advantageous to mount the crane in P4-16 on (*a*) automotive tires moving on concrete, or (*b*) solid rubber tires? Give reasons for your answer. Give motor size.

Chapter 5

Machine Frames and Housings

> The engine's even vibration shaking back through the fuselage's steel skeleton gives life to cockpit and controls . . . flowing up the stick to my hand, it's the pulse beat of the plane. Let a cylinder miss once and I'll feel it as clearly as though a human heart had skipped against my thumb.
>
> CHARLES LINDBERG

In nearly all machines there is one member that occupies a *fixed* position or carries the entire machine along with it during motion. This member is the frame. Design of frames and housings has, in recent years, undergone a revolutionary change due to the "finite element method," which utilizes the computer.

5-1 FUNCTION AND CHARACTERISTICS

The function of the frame is to admit, enclose, restrain, or support the integral parts of the machine. The frame is also the backbone of a structure whose essential parts have been put in their proper places and secured together. Generally, the role of a frame is to provide support or alignment to rotary bearings carrying machine shafts and axles or to linear bearings for reciprocating parts or machine tables (Fig. 5-1).

In mobile machines such as automobiles, the frame must also accommodate the means of locomotion. The vehicle chassis frame shown in Fig. 5-2 is an assembly of formed steel structural members that support and locate the vehicle body, front sheet metal structure, chassis components (wheel suspension, engine, steering

Figure 5-1 Frame for air motor. (Courtesy Gardner-Denver Company.)

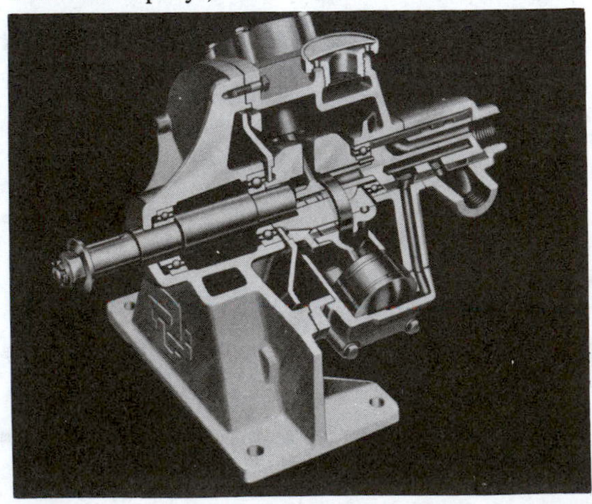

Figure 5-2 Open frame. Vehicle chassis frame for a convertible car. Note the assembly of formed steel structural member with heavy cross-bracing. (Courtesy Ford Motor Company.)

components, driveline, exhaust system, bumpers), and miscellaneous equipment. The chassis frame must provide accuracy of location as well as strength and rigidity of support for these components.

A well-designed frame is essential to vibration-free performance and extended machine life. Such a frame should be able to dampen vibrations directly or transfer them to its own support or foundation for absorption or distribution. The frame must also be able to cope with all acting forces. Mobile machines are generally designed with a somewhat flexible frame to minimize fatigue. Stationary machines, by contrast, are generally designed for maximum rigidity to ensure accurate constrained motion of all moving parts.

Human factors play a larger role in the design of frames than of most other machine members, and for obvious reasons. Frames are often built to accommodate human beings as operators, passengers, or both.

While the size of a machine frame depends largely on the acting forces, the shape depends primarily on the manner or direction in which those forces are brought to bear on it. Although the same principle applies to most machine members, a frame differs in that (1) it is the largest machine member, (2) it encounters several forces acting simultaneously in different directions, and (3) it is more likely to be affected by the force of gravity.

5-2 TYPES OF FRAMES

In its simplest form, a machine frame is merely a base plate on which major components such as bearings, pumps, or motors have been mounted (Fig. 5-3). In its most sophisticated form, it becomes the fixed member of a racing car, jet plane, or space vehicle.

Machine Bases

A machine base is the simplest type of frame. It is either a flat plate or a plate reinforced by bracings on one side. In either case, the depth is small relative to the length and width. Therefore the main problem is one of rigidity under bending and twisting loads. Optimum rigidity is achieved largely by controlling the direction and configuration of bracing.

A very broad classification of frames distinguishes between *open frames* (Fig. 5-2) and *closed frames* (Fig. 5-1). Open frames are essentially skeleton or beam structures; closed frames may be characterized as shell structures. Generally, open frames possess greater structural flexibility than closed frames and provide easier access to internal parts. When superior rigidity and a high strength-to-weight ratio are required, as in aircraft and space vehicles, a closed frame of lightweight metal is the best choice. Another type of frame, previously used in aircraft and now used in some racing cars and motorcycles, is the *space frame* (Fig. 5-4). A space frame is a multiple-tube frame deriving its name from the space a tubular structure encloses. A tube has a large section modulus (maximum material utilization) and is a widely used structural member. Since tubes have a high strength-to-weight ratio, space frames also have a very large strength-to-weight ratio.

The terms *air-frame* and *monocoque* are synonymous with shell structures. The word *monocoque* (French) literally means single frame and basically refers to a unit frame. A true *monocoque* is found only in rockets and spacecraft. It derives its strength from the outer stressed surface skin. The role of internal bulkheads and supports is mainly to prevent the skin surfaces from buckling and bending prematurely. The *monocoque* frame combined with the space frame led to a design that is well suited for vehicles and is known as the multicoque. Closed frames or shell frames are used almost exclusively for small- and medium-sized electric mo-

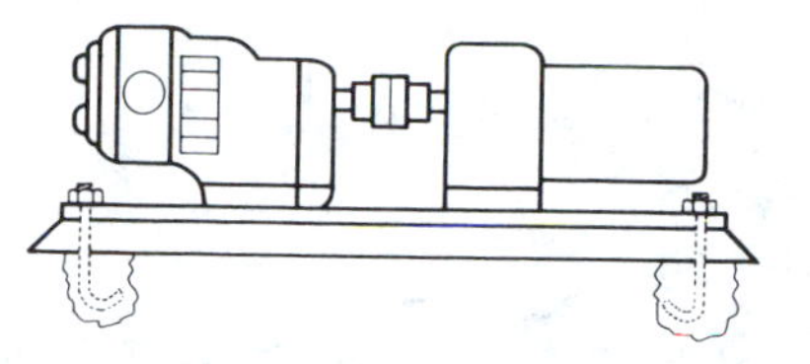

(a) Permanent installation with bolts and grout

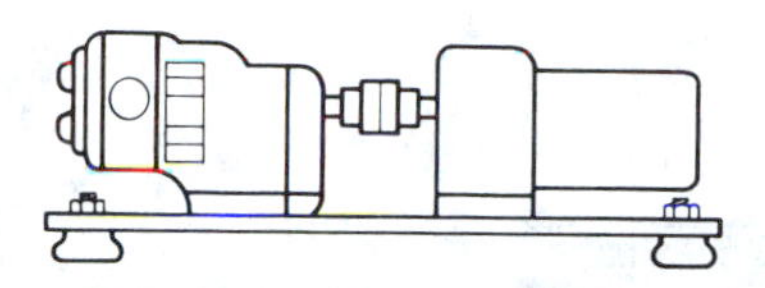

(b) Levelling and/or vibration pads

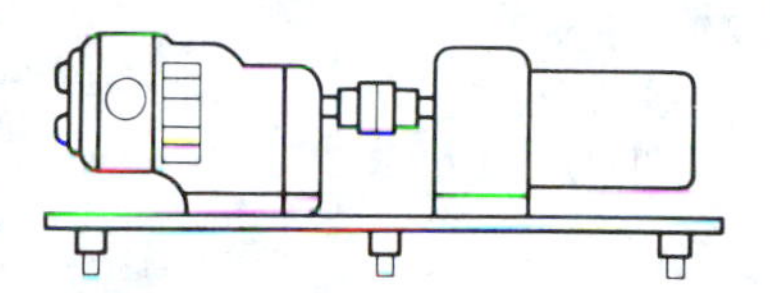

(c) Adjustable leg base can be easily moved or repositioned

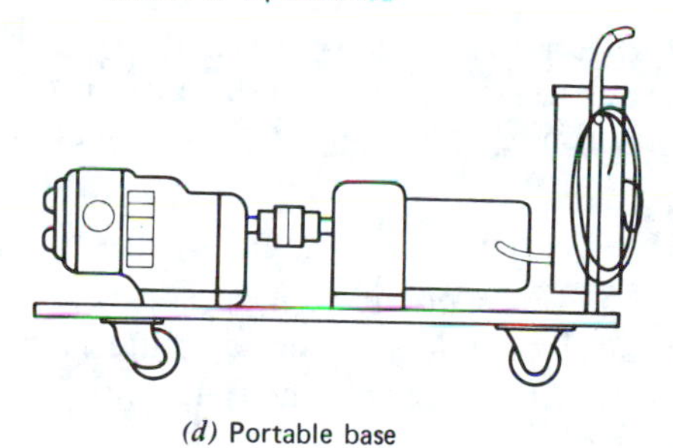

(d) Portable base

Figure 5-3 Base plate. (Courtesy Waukesha Pump.)

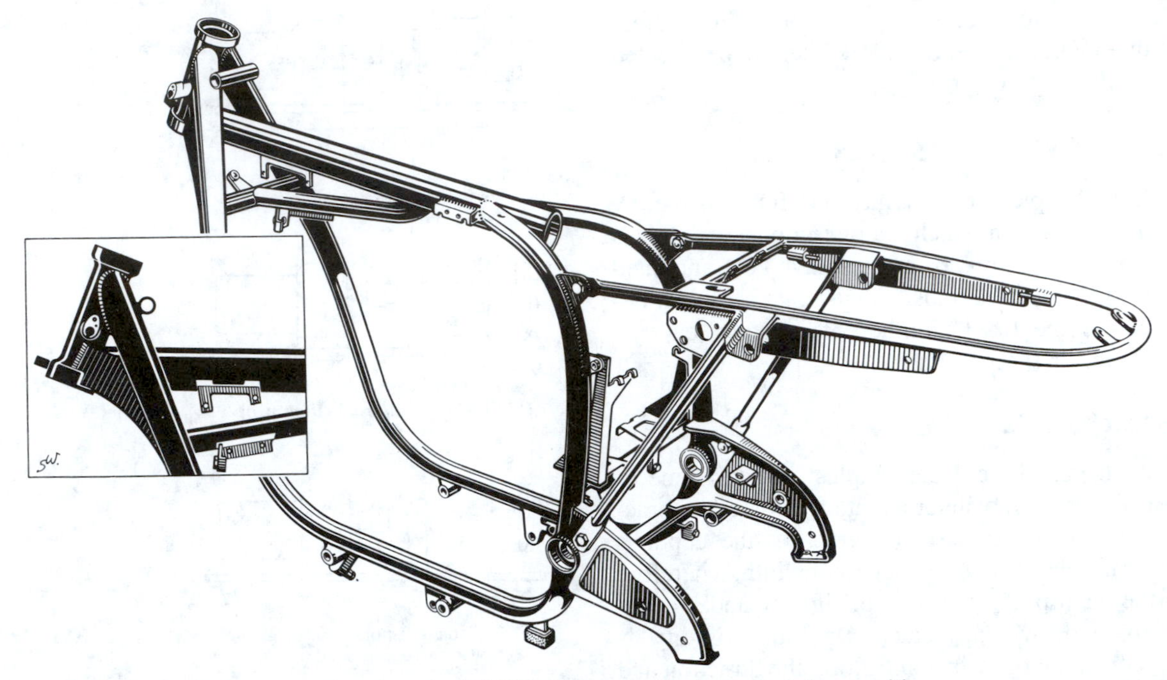

Figure 5-4 Space frame for motorcycle (BMW). It is a double loop steel frame with a forged steering head. For increased strength, the tubes are tapered and have oval cross sections in the appropriate locations. (Courtesy Bavarian Motor Works.)

tors (Fig. 5-5). They provide a rigid support for both rotor and stator and protect all vital parts against dust, dirt, foreign material, and impacting objects.

Figure 5-5 Closed frame used on small- and medium-sized electric motors. (Courtesy Bodine Electric Company.)

Unitized Body

Some modern cars have unitized bodies and no separate frames (Fig. 5-6). Unit bodies rely on box sections and gusseting for rigidity. The sheet metal used in some of these boxes and gussets is often the same gauge as the outer skins. Unitized bodies have a high strength-to-weight ratio. Their disadvantage is that the boxes tend to trap moisture and salt and are therefore subject to rust and corrosion. Unitized bodies can, and sometimes do, simply collapse from such corrosion.

C-Frames

The C-frame is probably the most widely used frame for stationary machinery such as drill presses, milling machines, and stamping machines. As shown in Fig. 5-7, the name is derived from its configuration. C-frames are easy

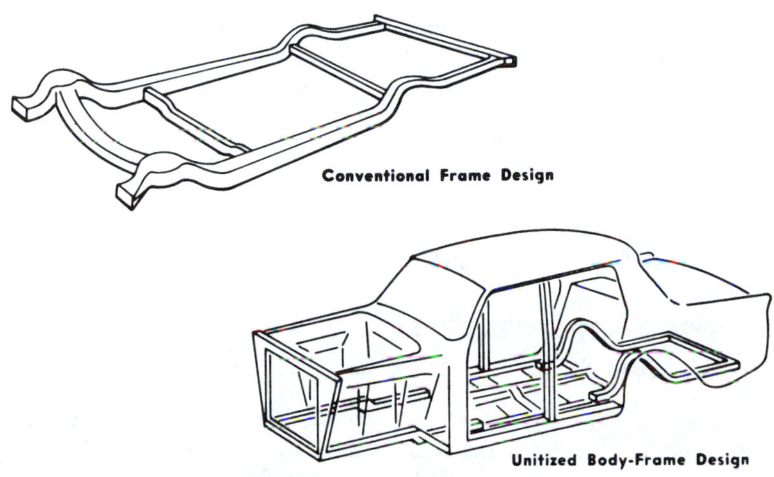

Figure 5-6 Conventional and unitized body-frame design. (Reprinted with permission from Applied Power, Inc.)

to design and manufacture and are usually made from gray cast iron. Simple C-frames are open frames; large C-frames are often hollow, curved beams. A major advantage of the C-frame is its accessibility for both operator and material being processed. Because of its opening, however, the C-frame lacks the rigidity necessary to resist large forces.

Figure 5-7 A typical C-frame, rigid yet accessible.

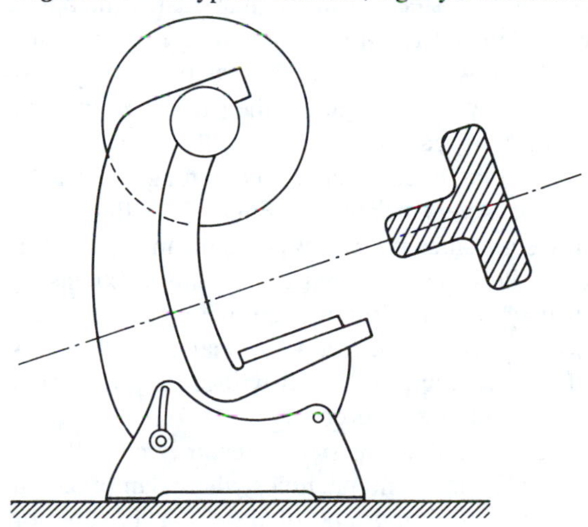

O-Frames (Fig. 5-8)

When large, vertical work forces are present, as in big hydraulic presses or rolling mills, a closed frame in the shape of the letter O may be advantageous. O-frames are stronger than C-frames, but accessibility is reduced.

Other Frames

Some machines do not have a specific frame in the ordinary sense of the word. A tractor's frame, for example, consists of boxlike housings bolted end to end to form a rigid structure. The differential is contained in a separate housing onto which the transmission is bolted. The engine, in turn, is bolted onto the transmission case.

5-3 CAST AND WELDED FRAMES

Machine frames can be either cast or welded. Gray cast iron is preferable to steel for mass-produced small- and medium-sized frames of simple configuration such as electric motors, pumps, compressors, and machine tools. Gray

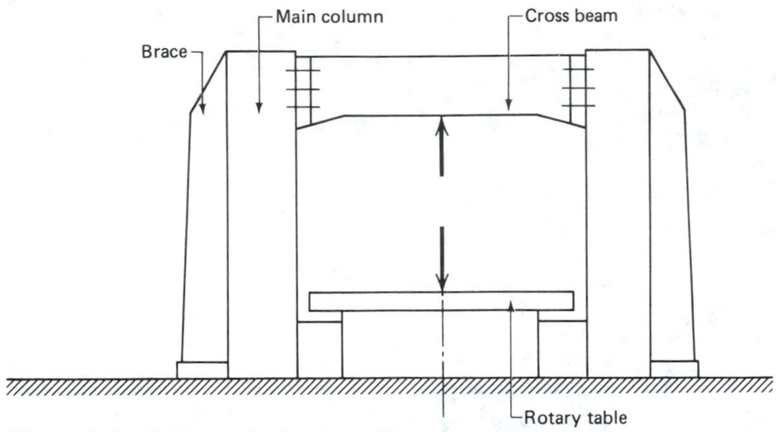

Figure 5-8 O-Frame for boring mill.

iron is easily cast to the proper shape, is inexpensive, and has a higher inherent damping capacity than steel. Once it has been stress relieved, gray iron also retains its shape. For these reasons, most precision machine tool frames of small or medium size are made from gray iron.

The casting process is a mass-production method for small- and medium-sized parts of simple configuration. Arc welding, by contrast, lends itself primarily to large- and medium-sized parts of more intricate shape, produced in limited quantities. Large frames are thus frequently welded from cast steel parts. The efficient use of welded steel frames is a practical alternative to the casting process. Welded frames provide uniform wall thickness with suitable ribbing to reduce high stress, thereby preventing warping in service. Such warping affects dimensional accuracy and is a serious drawback in welded frames and machine bases, one that must be adequately anticipated.

Figure 5-9 shows the housing of a large speed reducer made from gray cast iron. This housing is designed to maintain permanently rigid support for all components by using generous material cross sections and locating external ribs at points of highest stress. The upper half of the housing is removable for easy inspection and service without disturbing the base. Vertical chain or belt drives can be easily used, since the housing feet do not project beyond the bearing retainers. Mounting bolt holes extend beyond the housing walls for increased stability, easy access, and a choice of either bolts or studs.

This same speed reducer is also available with a steel housing, which should be considered when lightness of weight and cost are prime factors (Fig. 5-10). Such housings can be readily engineered for heavy shock loads and external impacts. Specifying high-strength steel, shot blasting, and stress relieving is adequate to ensure a quality product.

The trend in design of medium- to large-sized machine frames is clearly toward welded frames. Not only is steel stronger than cast iron, but it also enjoys a two-to-one advantage in modulus of elasticity. Thus a welded structure always has potentially less weight for the same strength and rigidity. A cast frame is usually so large and heavy that even a small percentage saved in weight is very important (Fig. 5-11). Reduction in weight does not necessarily mean a saving in cost, however, since steel is more expensive than gray iron. But reduction in weight is especially important to operational characteristics of mobile equipment and will always represent savings in shipping costs.

The advantage of steel over gray cast iron is less pronounced in machines where damping and rigidity are important qualities. The superior

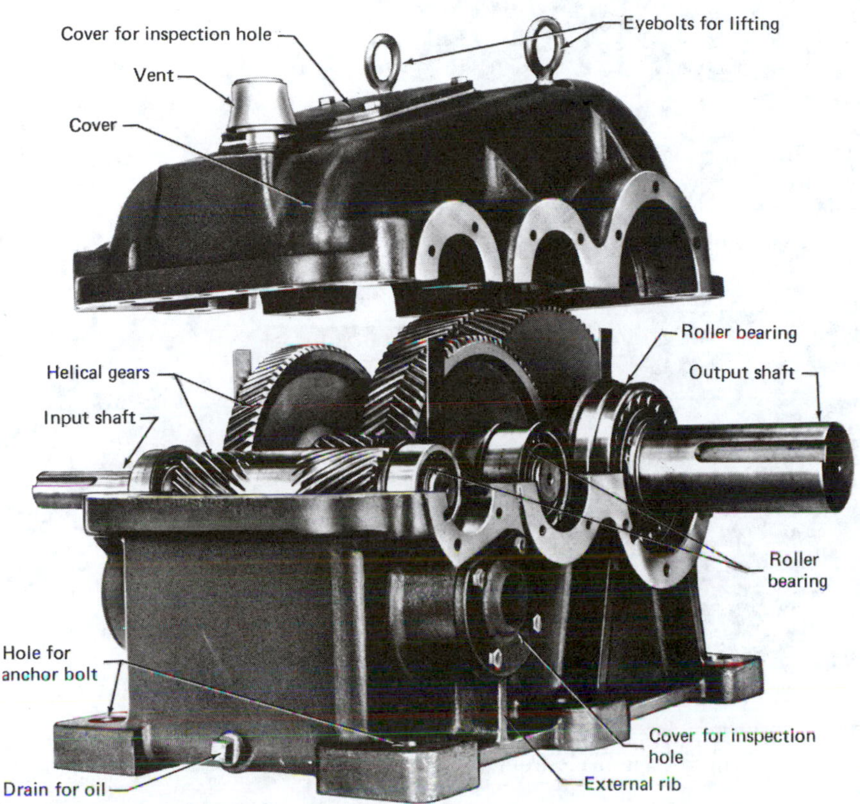

Figure 5-9 Housing of large-speed reducer made from gray cast iron. Note the heavy walls and external ribs. (Courtesy Link-Belt Drive Division.)

damping capacity of gray cast iron increases work quality. Moreover, in spite of stress relieving, welded machine tool frames are less reliable in the long run. Eventually they will undergo minute changes that affect accuracy.

5-4 DESIGN CONSIDERATIONS

Machine frames vary widely and so do many of the rules applying to their design. For instance, the rules that pertain to car frames are different from those for machine tools. Human factors engineering, however, applies to all. Moreover, specific design considerations apply within major groups such as mobile and stationary machines.

As an example of frame design, a lathe will be discussed in detail. Although constructed from the same basic elements and parts as other machines, machine tools differ in requiring greater precision and greater rigidity.

Forces

The forces acting on a frame may be external or internal. The internal forces are transmitted to the frame largely through bearings. Most bearings support rotary motion; however, machine ways, a type of guide bearing, support vertical or horizontal motion, as shown in Fig. 5-12.

The external forces on a lathe are primarily reactions both to the weight of the machine and

Figure 5-10 Welded housing of large-speed reducer. Note the thin walls and light external ribs compared with the same design made from cast iron. (Courtesy Link-Belt Drive Division.)

Figure 5-11 The broken-out section clearly shows the difference in wall thickness of cast and welded housings.

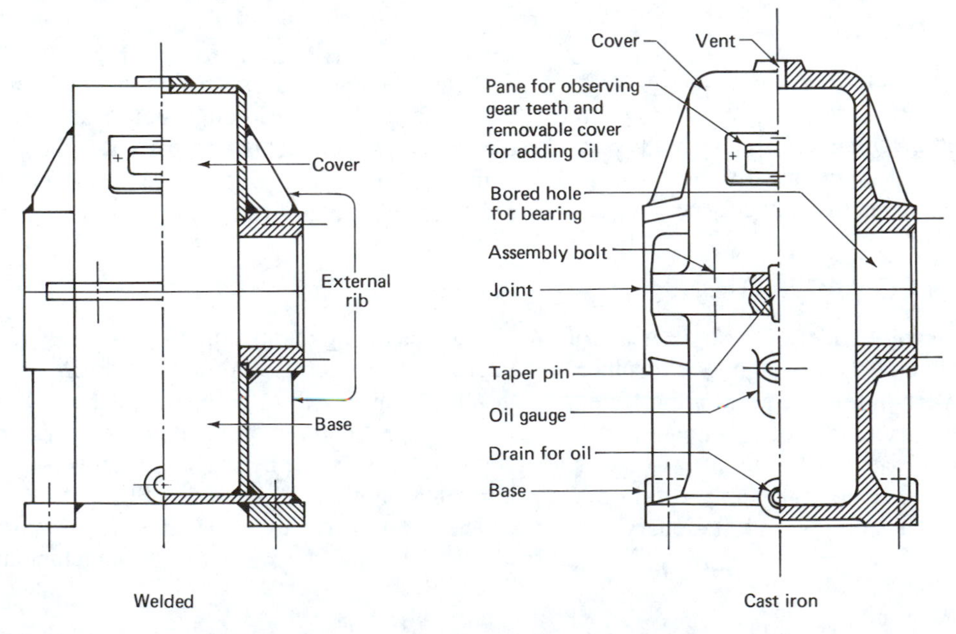

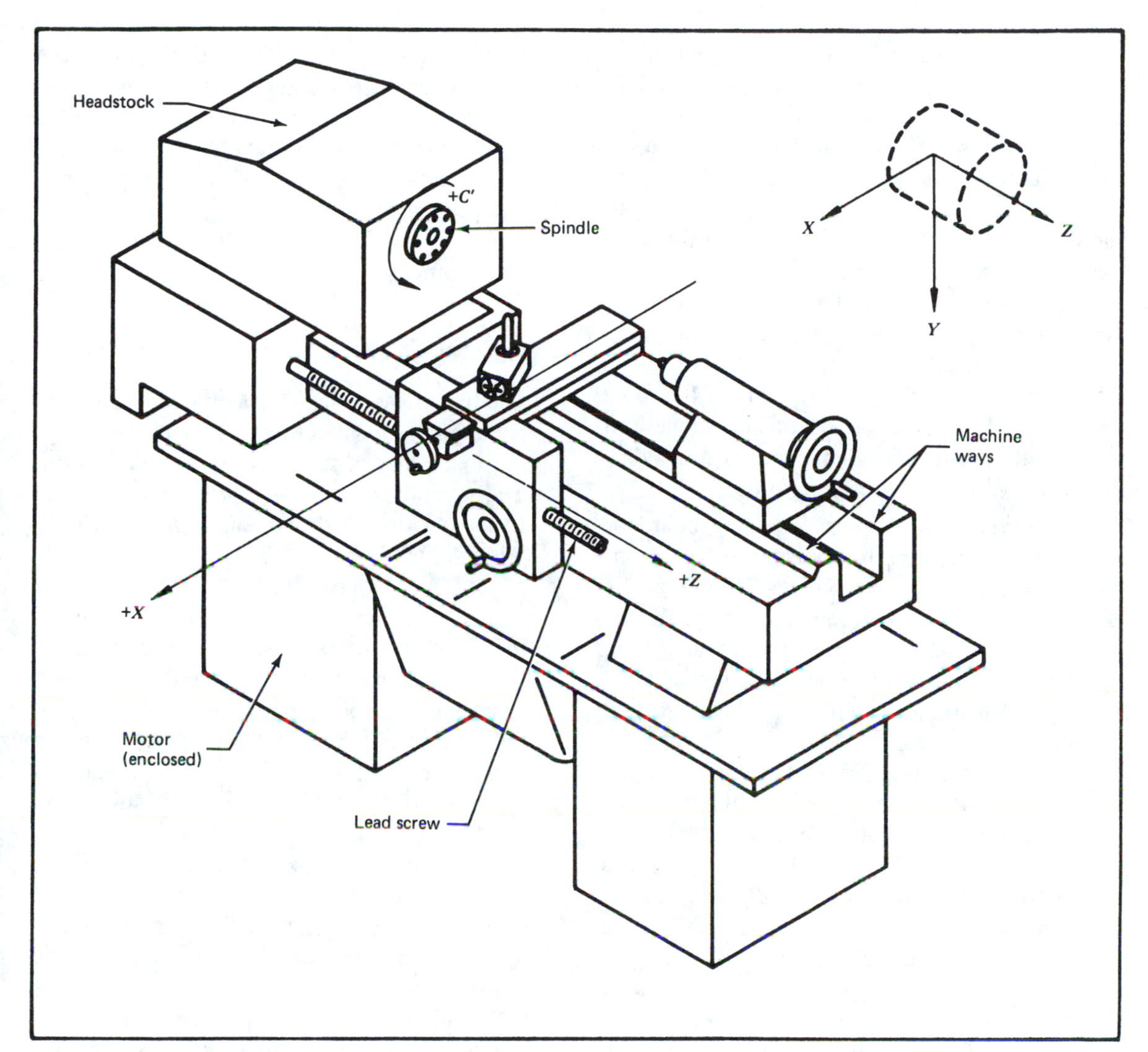

Figure 5-12 Typical lathe.

to the power input mode. The major internal loads acting on a lathe are due to cutting forces originating at the single point tool. These forces may be large and fluctuating, and their effects may be greatly increased by a form of vibration known as *chatter*. Chatter vibrations are often violent and can momentarily shake the entire machine, thereby adversely affecting the quality of finish. Other forces are generated by belt drives, gears, and reactions to these external forces.

Stability of the frame on the floor and rigidity against operational deflections are desirable characteristics of a machine tool. Both qualities are functions of material and configuration; but since only two materials, steel and cast iron, are used, designers are concerned primarily with configuration. Even configuration is not a major problem unless a new machine is being designed. Design changes usually occur as a result of vibration tests conducted during machining. Sections that fail to vibrate may be too heavy and thus may require a reduction of wall thickness. Sections that vibrate excessively need reinforcement.

The equations of static equilibrium and a force diagram provide a systematic means of studying the frame as a whole or a section of it. They will aid designers in optimum placement of forces to achieve maximum stability and rigidity. Maximum stability is realized when the forces fall within the support area of the base. Maximum rigidity is realized whenever forces and reactions oppose each other directly. A study of the force diagram will help to approach this ideal solution.

The lathe frame also demonstrates four other design principles or considerations.

1. A frame may act as a sounding board or transfer medium for noise and vibration. Remedial action or changes depend on the noise source and its location. Journal bearings, for example, are less noisy than rolling bearings and have greater damping capacity. Cast iron has a higher damping capacity than steel and is preferred.
2. Components that generate noise and vibration should be placed within the most rigid parts of a frame. Since motors are a source of noise and vibration, they are nearly always placed at or within the base of a machine; in some cases, they are placed directly on the machine foundation. In a lathe the electric motor is in the bottom of the left leg, directly under the main spindle.
3. The frame should be given the necessary height, length, and shape to accommodate the operator. To a lesser degree, the same consideration might be given to the individuals concerned with assembly, inspection, and repair. Recall how easily lathe operators can watch their work and manipulate the various levers and handwheels without undue fatigue.
4. Frames for light stationary machines (lathes, milling machines, and drill presses) generally are provided with two or four legs for support. Heavier machines, such as turbines, engines, and compressors, have frames that extend to the floor. Machines of high power-to-weight ratio (engines) and high operating speeds are generally bolted onto the floor or foundation. Machines of low power-to-weight ratio such as small lathes, drill presses, or grinders can be placed on the supporting floor without any attachment.

EXAMPLE 5-1

The backbone of a cast iron machine frame of rectangular cross section is subjected to varying loads that generate stress and deflection. The overall dimensions are approximately 15 in. × 5 in., with the length parallel to the principal load (Fig. 5-13).

(*a*) The designer wishes to obtain a size reduction by replacing cast iron with steel. If the allowable stress of steel S_{st} is three times that of gray iron S_{ci}, what dimensional reductions must take place assuming that the present ratio of height h to depth w (3:1) is to be maintained? Assume $E_{ci} = 15(10^6)$ psi and $E_{st} = 30(10^6)$ psi. Consider both stress and rigidity limitations. (1) Which condition (stress or rigidity) controls dimensioning? (2) What percent reduction in volume is realized with steel fabrication? (3) What stress ratio R gives the same answer for stress and rigidity considerations?

(*b*) In the preceding problem consider a welded box structure using 1 in. plates. Assume a working stress for weldments twice that of cast iron (Fig. 5-14).

Figure 5-13 Rectangular cross section of solid cast iron (15 in. × 5 in. approx.).

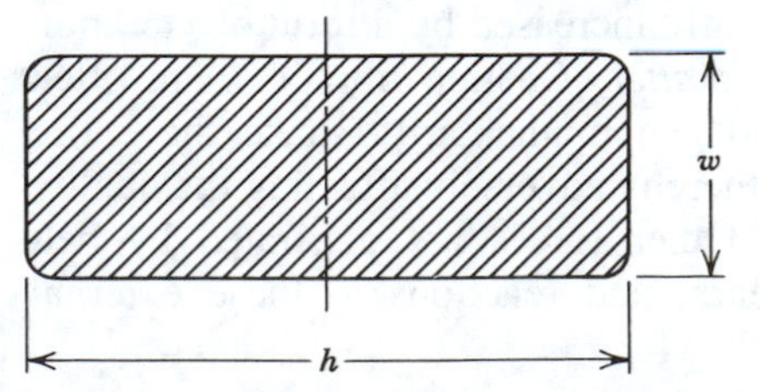

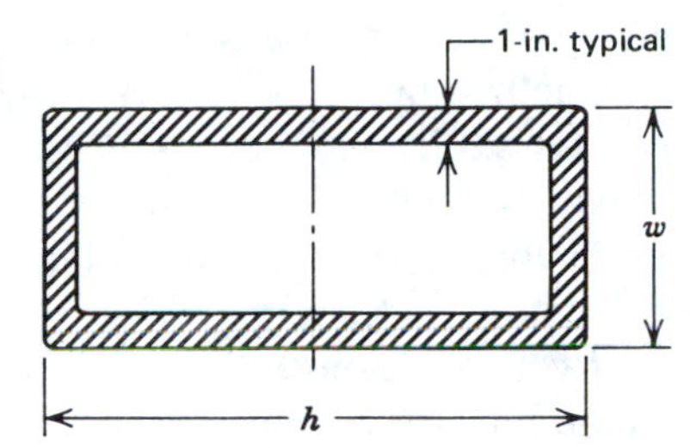

Figure 5-14 Rectangular cross section; a welded box structure using 1-in. steel plates.

SOLUTION

(a1) Calculations for width w and height h of the rectangular cross section will be based on strength and rigidity of a beamlike structure loaded in bending.

Strength

$$\sigma_b = M/Z \leq s_w$$

Cast iron: $\sigma_b = M/Z_{ci} \leq S_{ci}$ (strength of CI)

Steel: $\sigma_b = M/Z_{st} \leq S_{st}$ (strength of steel)

Since the induced bending moment is independent of choice of material,

$$\frac{S_{ci}}{S_{st}} = \frac{M/Z_{ci}}{M/Z_{st}} = \frac{Z_{st}}{Z_{ci}}$$

Because the strength of steel is three times that of cast iron,

$$\frac{S_{ci}}{3\,S_{ci}} = \frac{Z_{st}}{Z_{ci}}$$

$$Z_{ci} = 3\,Z_{st}$$

For a rectangular cross section: $Z = wh^2/6$

Cast iron: $Z_{ci} = (5\text{ in.})(15^2\text{ in.}^2)/6 = 187.5\text{ in.}^3$

Steel: $Z_{st} = (h/3)(h^2)/6 = h^3/18$

$$187.5 = 3(h^3/18)$$

$$h = 10.40\text{ in.} \qquad w = 3.47\text{ in.}$$

Rigidity

From Example 4-1: $\delta = \dfrac{PL^3}{48\,E\,I}$

Rectangular cross section: $I = wh^3/12$

For $w = h/3$: $I = h^4/36$

Since P and L are independent of the beam material,

$$\frac{S_{ci}}{S_{st}} = \frac{E_{st}I_{st}}{E_{ci}I_{ci}}$$

For equal strength,

$$E_{ci}I_{ci} = E_{st}I_{st}$$

Since $E_{st} = 2E_{ci}$ (given),

$$I_{ci} = 2I_{st}$$

$$(5\text{ in.})(15^3\text{ in.}^3)/12 = 2h^4/36$$

$$h = 12.61\text{ in.} \qquad w = 4.20\text{ in.}$$

The dimensions based on rigidity are the larger and the ones to use. Rigidity controls.

(a2) Assuming uniform cross section, we find that changes in volume are proportional to area changes so

$$\Delta V = \frac{5(15) - (4.20)12.61}{5(15)}100 = 29.4\%$$

Conclusion

$$\mathbf{h = 12.61\text{ in.} \qquad w = 4.20\text{ in.}}$$
$$\mathbf{\Delta V = 29.4\%}$$

(a3) For equal stress and rigidity,

$$\frac{15^2(5)}{6} = \frac{(12.6)^2 4.2}{6}R$$

R = stress ratio of steel to cast iron

$$\mathbf{R = 1.69}$$

Note that for this condition the stress ratio is nearly cut in half (1.69:1 versus 3:1).

(b) Dimensional comparisons must be based on area moments of inertia.

$$I_{ci} = \frac{15^3(5)}{12}$$

$$I_{st} = \frac{h^3(h)}{3(12)} - \frac{(h-2)^3}{12}\left(\frac{h}{3} - 2\right)(2)$$

$$\frac{15^3(5)}{12} = \left(\frac{h^4}{36} - \frac{h^4 - 12h^3 + 48h^2 - 80h + 48}{36}\right)2$$

$$25{,}313 = 12h^3 - 48h^2 + 80h - 48$$

$$h^3 - 4h^2 + 6.666h = 2113$$

This equation can be solved only through trial and error. Since h is less than 15, we may start with 14. This quickly leads to 14.12 and the following outside and inside dimensions.

Outside: $w \times h = 4.70 \text{ in.} \times 14.12 \text{ in.}$

Inside: $w_i \times h_i = 2.70 \text{ in.} \times 12.12 \text{ in.}$

Assuming uniform cross section, we find that changes in volume are proportional to changes in area. The volume ratio is therefore

$$R_v = \frac{(4.70)(14.12) - (2.70)(12.12)}{5(15)} \sim \mathbf{0.45}$$

or 55% savings in volume.

5-5 HOUSINGS

In machinery a housing is a case or enclosure to cover and protect vital parts such as gears, shafts, and bearings. Housings may be separate components fastened to a machine frame or foundation, or they may be an integral part of a machine. In the latter case, the frame is the bearing housing. The housing should provide a rigid support for the bearing. In addition to supporting the load, the housing protects the bearing and often provides a lubricant reservoir, a lubricant flow system, cooling, or seals.

Housings, whether standardized or custom designed, offer a relatively simple means of adapting bearing and shaft systems to a machine design. Primary consideration should be given to the loading, the stability and accuracy the housing must provide, and environmental conditions and servicing requirements.

Because a housing is only as stable as its supporting structure, the housing itself should rest on or within a stable foundation. If it is bolted to a weak machine frame that undergoes excessive deflection, it will be subject to bending and twisting loads for which it may not be designed. Even though such distortions may not harm the housing, they could cause undesirable misalignment of gears, shafts, and bearings.

The ultimate strength of any housing is the maximum load it can sustain without fracturing. Ultimate strength varies, depending on the direction of loading. The maximum capacity is usually attained with the load direction toward and perpendicular to the mounting surface. Where the load is parallel or oblique to the mounting surface on the frame, supplementary support is often required to assure a safe and reliable installation. Specific values for ultimate strength cannot be provided; however, the factor of safety should generally be at least 3. High-strength alloys or nonferrous metals are rarely needed for housings. Cast iron or steel is usually preferred due to low cost and compatibility with most frames.

One of the frequent errors in the planning of machines is failure to select a housing design that facilitates removal of bearings and shaft assemblies. Poor designs may require that bearings be burned off to be replaced. Designers must also consider sealing and lubrication, which often affect the size of a bearing housing. Special seals, for example, may require more space along the shaft. Similarly, lubrication and cooling may require oversize reservoirs. An alternative may be a compact housing with an external circulating oil system.

SUMMARY

In machinery the frame either occupies a fixed position or carries the entire machine along with it during motion. The frame supports, restrains, encloses, or admits the integral parts of a machine. It provides support and alignment to linear and rotary bearings. In self-propelled machines the frame must accommodate the means of locomotion.

Well-designed frames are essential to vibration-free performance and extended machine life. Mobile machines require flexible frames that will respond to changing road conditions. Stationary machines, in contrast, must have rigid frames to ensure continual accuracy of the product or service rendered. Human engineering applies to most frames.

A machine base is the simplest form of frame. More sophisticated frames are classified as open or closed. Cars often have unitized bodies, while machine tools and production machinery use C-frames and O-frames.

Machine frames are generally made from ferrous materials. As such, they may be castings or weldments. Frames of simple configuration produced in large quantities are generally castings. Since thermal stresses are less severe in castings than in weldments, cast frames are used almost exclusively in machine tools.

Arc welding is most economical when applied to large- and medium-sized frames of somewhat intricate shape that are produced in limited quantities. Large frames are frequently composed of large *steel* castings welded together, thus combining the best of both production methods.

Machine frames generally serve as sounding boards or transfer media for noise and vibration. Motors or components that generate noise and vibration should thus be placed within the most rigid part of the frame and near the base of the frame. Generally, frames should be given such height, length, and shape as accommodate the operator and facilitate the flow of material to and from the machine. In vehicles the frame must accommodate passengers as well as operators.

REFERENCES

5-1 Blodgett, O. W. *Design of Weldments.* Cleveland, Ohio: The James F. Lincoln Arc Welding Foundation, 1963. (This outstanding book, containing a wealth of information, is available to designers at a fraction of its real cost. Well suited for design projects.)

PROBLEMS

To obtain manageable figures, use centimeters instead of millimeters. (10 mm = 1 cm)

P5-1 A punch press with a C-frame similar to the one shown in Fig. 5-15 has a T-section as shown. The centroidal axis is located roughly 55 mm from the inside of the frame. The applied load is a single vertical force acting 245 mm from the inside of the frame. The frame material is built-up steel plate 5 mm thick. S_{ut} = 420 MPa. How large a force can safely be exerted by the punch press if fs = 2 and 99.99% reliability is required? *Hint:* Use eccentric fatigue loading. See Section B-12.

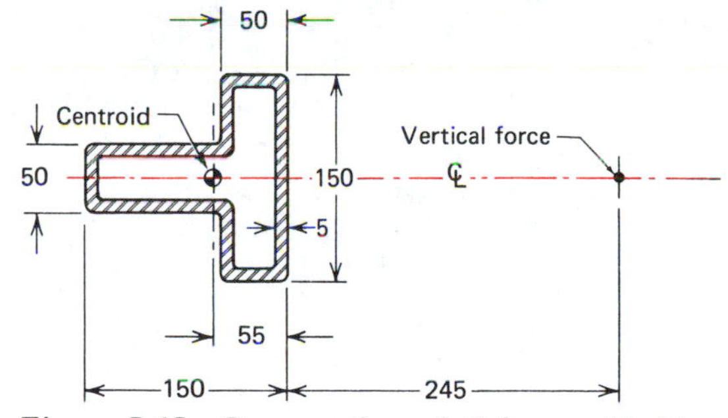

Figure 5-15 Cross section of C-frame. (Problem 5-1)

P5-2 The cast iron profile shown in Fig. 5-16 belongs to a machine frame loaded in bending about the x-x axis. The designer wishes to obtain a mass reduction while retaining the external dimensions by replacing cast iron with steel. Assume a working stress for steel three times that of cast iron and a modulus of elasticity for steel twice that of cast iron.

(a) Calculate the new wall for equal

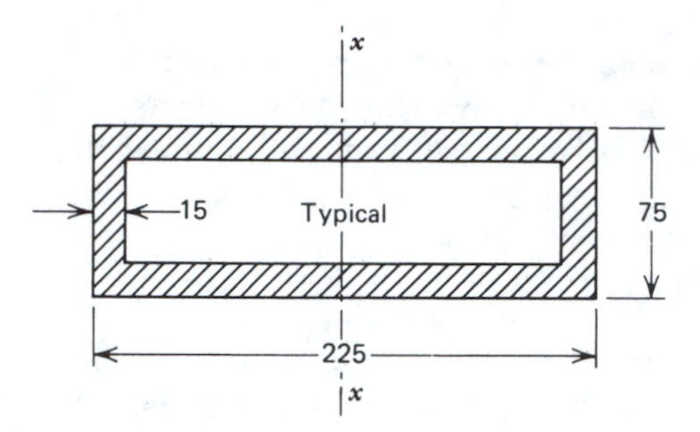

Figure 5-16 Box section. (Problem 5-2)

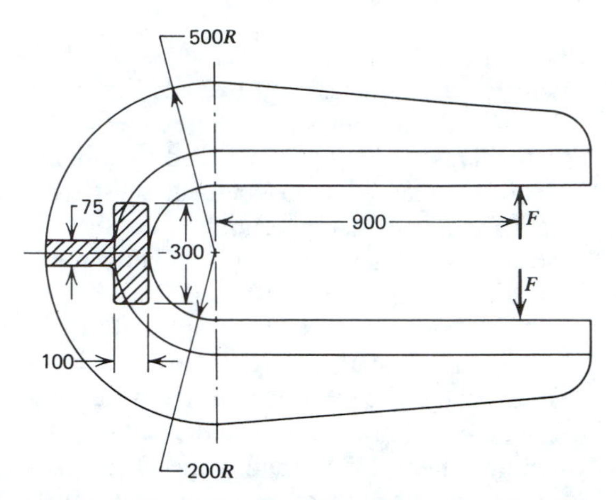

Figure 5-17 C-frame for punch press. (Problem 5-3)

strength and equal rigidity. Which condition controls dimensioning?

(b) Find the reduction in mass in each case.

(c) Which stress ratio R gives the same answer for stress and rigidity considerations?

P5-3 Find the maximum load P on the cast iron punch press frame shown in Fig. 5-17. Use eccentric cyclic loading. For high grade cast iron assume $S_{ut} = 175$ MPa. Use $fs = 2$ and assume a 99.99% reliability. Dimensions are in mm.

PART II
Rigid and Elastic Connections

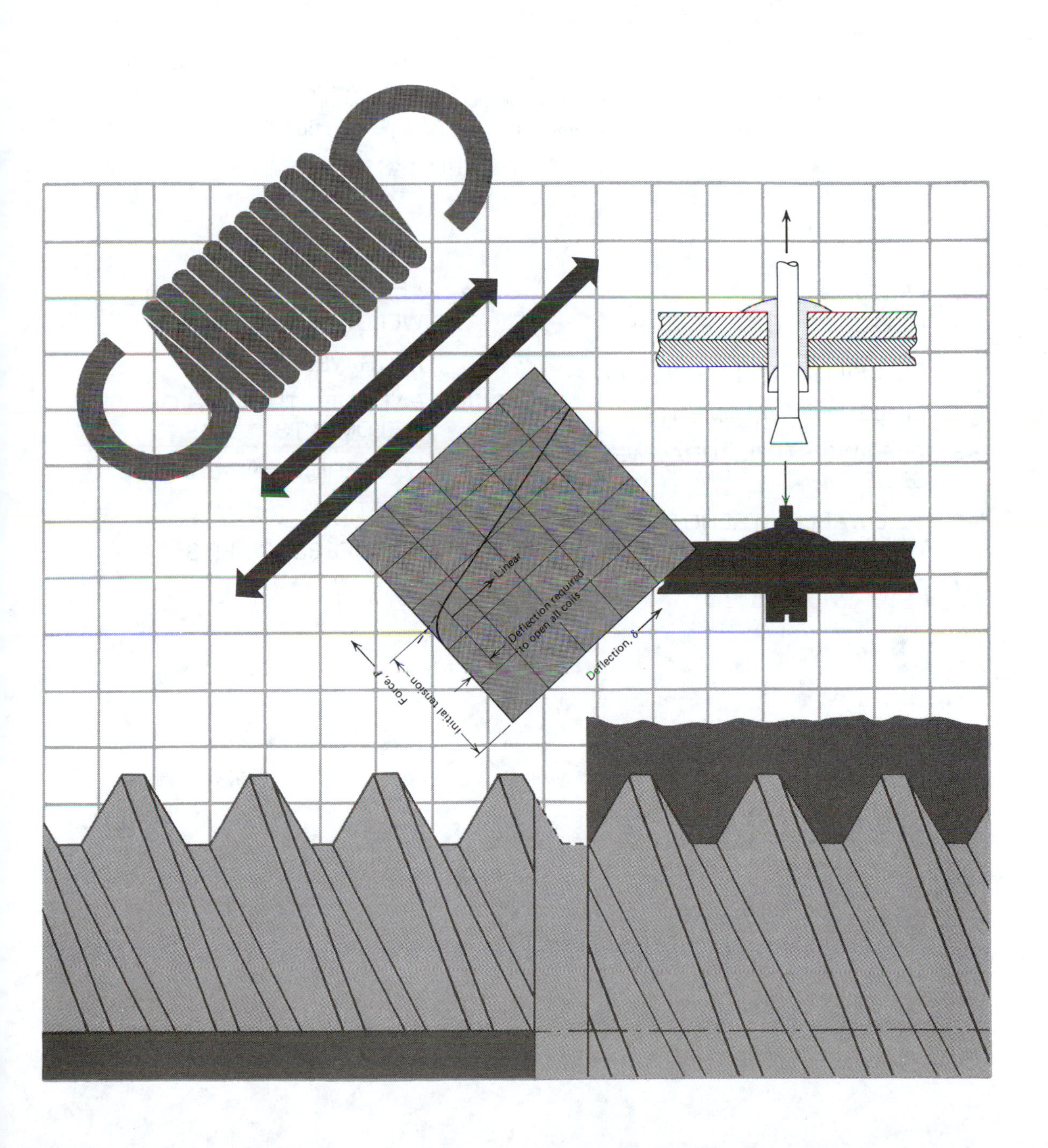

Chapter 6
Permanent Connections

Genius is one percent inspiration
and ninety-nine percent perspiration

THOMAS EDISON

NOMENCLATURE

a = allowance on interference fits
A_s = shear area
d = diameter of rod or shaft
f = allowable unit force on fillet weld
L = length of weld
r = distance from centroid to stress point for eccentric loading
S = strength of material
t = thickness of plate or tube wall
t_e = effective throat dimension
t' = depth of plug weld
w = width of hub or plate
ω = leg size of weld
σ_t = tangential stress
τ_1 = direct shear stress
τ_2 = torsional shear stress
ϕ = angle for locating stress point

As the name implies, permanent connections are those that, once assembled, cannot be readily disassembled. In general, disassembly of a permanent connection entails damage, if not destruction, of the related parts. Consequently, permanent connections should be used only where there is no foreseeable need for disassembly. (Interference fits are an exception; see Section 6-13.) The most commonly used techniques for permanently joining machine members are welding, brazing, soldering, riveting, pressing, and shrinking. Welding is the most popular because of its versatility, strength, and low relative cost. As a means of joining parts, it is rivaled in importance only by threaded fasteners. Table 6-1 gives the characteristics of riveting, welding, soldering, and brazing.

6-1 WELDING

Welding is the process of joining metallic parts by fusing them with heat at their junction, with or without pressure. The two pieces of metal are referred to as *base* metal or *parent* metal.

Welding without pressure is accomplished by melting the designated surfaces until they flow together; heat may be supplied by an electric arc or gas flame. Since there are several different methods of arc welding, each having certain advantages, arc welding should be compared with spot welding, projection welding, or riveting from a cost standpoint before a final decision is reached as to the method of assembly to be employed.

Welding with pressure is the process of making a weld by pressing the surfaces together while they are heated, generally to a plastic state. This process includes forge welding and resistance welding. Forge welding is done by hammering or rolling two surfaces together after they have been heated.

Arc Welding

In arc welding, the heat required for fusion is generated by an electric arc between the work piece and an electrode. This electrode may be composed of a filler metal and be consumed as welding progresses, or the electrode may be non-consumable. In the latter case, the filler metal is fed separately to the welding area. Figure 6-1 shows shielded metal arc welding in which the consumable electrode is coated with a flux material that vaporizes to form a gaseous shield around the molten metal. The primary purpose of the gaseous shield is to prevent the molten metal from absorbing oxygen, hydrogen, and nitrogen from the surrounding atmosphere. If not prevented, such absorption adversely affects the mechanical properties of the weld. The flux material, after serving its purpose of shielding the weld, will form a harmless slag deposit on the work piece. In other types of arc welding, shield-

TABLE 6-1 Characteristics of Permanent Fastenings

Characteristics	Riveting	Welding	Soldering Brazing
Strength of joint	High	High	Low
Weakening of the structure	Extensive (due to holes)	Slight	None
Stress flow at the joint	Unfavorable (overlapping)	Favorable	Unfavorable
Mass (weight) of structure	Large (overlapping edges)	Small	Small
Materials that can be joined	Dissimilar	Same metals only	Dissimilar metals
Changes in material	None	Change at joint	None
Structural changes	None	Slight warping	None
Resistance to temperature changes	Excellent	Excellent	Moderate
Cost of manufacturing	Average to low	Average	High to average
Necessary equipment	Simple	Substantial	Simple
Electrical conductivity across the joint	Excellent	Excellent	Excellent
Transfer of heat across the joint	Excellent	Excellent	Excellent
Tightness (for fluids)	Poor (unless specially sealed)	Excellent (after testing)	Excellent

Source. Georg Reitor and Klaus Hohmann, *Grundlagen Des Konstruieren*. Essen: Verlag W. Girardet. 1976.

Figure 6-1 Arc welding with coated electrode. (Courtesy American Welding Society.)

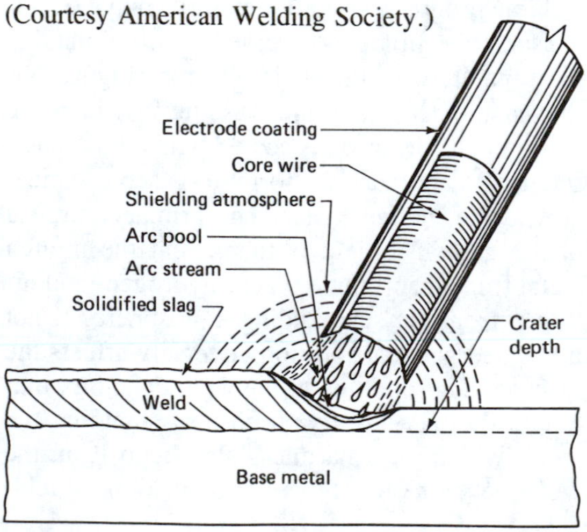

ing can also be accomplished by using an inert gas shield such as helium or argon around the welding area.

Gas Welding

Heat required to melt the metal is obtained from the combustion of a fuel gas such as acetylene in the presence of oxygen. If filler metal is used it is fed separately into the work area in the form of a wire or rod. For gas welding of steel, acetylene torches that produce a temperature of about 3300°C (6000°F) are commonly used. For aluminum, the lower temperature of an oxygen-hydrogen flame may be used. Various other gases, such as propane, are also employed for welding material with low melting temperatures.

Resistance Welding

Here the parts to be connected are clamped together by electrodes at the desired joint location, and a high-amperage current is passed through the work pieces (Fig. 6-2). Resistance of the work pieces to current flow creates the heat required for melting and fusion. Filler metal is not used. Resistance welding is readily automated and is thus often used as a mass-production technique, notably to produce spot welds for joining thin metal parts or sheets. It is a very common process in the automotive industry.

Two relatively new fusion processes of increasing value are *electron beam welding* and *friction welding*. In the first, heat required for fusion is obtained when the kinetic energy of a high-speed electron beam is converted to thermal energy as the beam strikes the work pieces. This process must occur in a vacuum, but it requires no filler metal and can produce deep, narrow welds in a variety of metals. In friction welding heat is produced as the work pieces are subjected to sliding contact under pressure. Usually, one part is held stationary while the other is rotated. When the proper temperature is reached, motion is stopped but contact pressure is maintained while fusion occurs. Friction-welding equipment is expensive but may be justified for mass production by time savings and reduced finish machining cost.

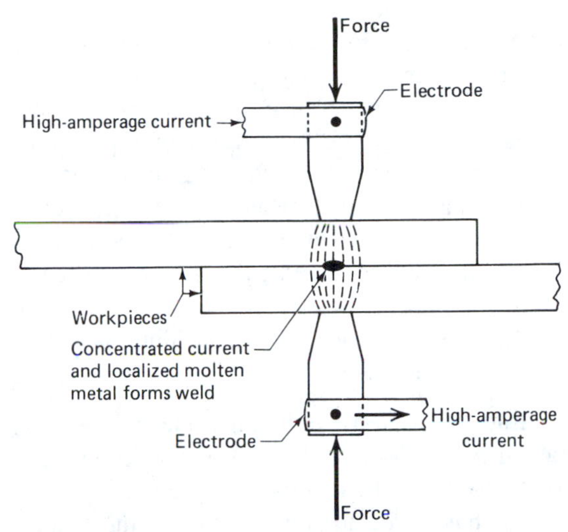

Figure 6-2 Resistance spot welding.

6-2 FIELD OF APPLICATION[1]

Welding applies primarily to medium- or large-sized machine members. Examples are gears, sprockets, sheaves, wheels, cranks, links push-rods, frames, housings, and support brackets. (Small parts can usually be made with greater accuracy and at lower cost as forgings and castings, and from sintered metal.)

Welding has proved particularly advantageous in two areas:

1. Replacement of rivets. (Large rivets have practically disappeared.)
2. As an alternate method for forging and casting (see Fig. 5-9 and Fig. 5-10).

Some of the advantages of welding are that it:

1. Simplifies fabrication and assembly. Standard shapes may be rolled or formed and joined with welding to cut cost by reducing material, machining, and finishing.
2. Lends flexibility to design. Changes are easily made without costly retooling. Material sizes (e.g., plate thickness) can be enlarged or reduced according to the need for strength and rigidity.
3. Reduces weight and size. For the same strength, welded machine parts may be up to 50% lighter than equivalent gray cast iron. (This saves on shipping.) This statement is *not* true for nodular cast iron, which is stronger than low-carbon steel.
4. Provides joints (caulk welds) tight to both liquids and gases (boilers and gas pipes).
5. Welded structures, in case of overloading, are often safer and more reliable than castings. A weldment will merely deform, whereas a brittle material casting may rupture.

[1]See *Machinery's Handbook,* 21st Ed., pp. 2211–2213.

6. Serves as a simple means of repair and salvage of machine parts.
7. Improves appearance because of smooth surfaces; there are no overlapping plates as in riveting and no protruding rivet heads.
8. Compared to riveting and forging, arc welding is a "silent" operation, so it contributes to the general efficiency and comfort of workers.
9. In mass production arc welding is often competitive with casting.

Disadvantages of welding include the need for trained personnel and the fact that welding cannot be used on a heat-sensitive assembly. Welding also has the disadvantage of reducing fatigue strength unless some form of mechanical surface treatment of the weld, such as rolling and hammer or shot peening, is included. Welding production shops must be ventilated extensively, an added expense to consider.

Unless stated otherwise, the discussion in the remaining sections on welding will be about arc welding of low-carbon steel parts.

6-3 CLASSIFICATION OF WELDS

Weds may be classified by type of joint, function of joint, and type of loading.

1. Classification by *type* of joint (Fig. 6-3).

Figure 6-3 Classification by type of joint. (Courtesy Chon L. Tsai.)

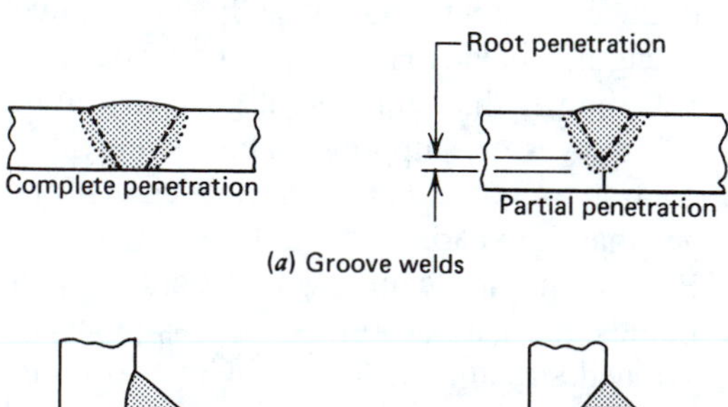

(*a*) Groove welds

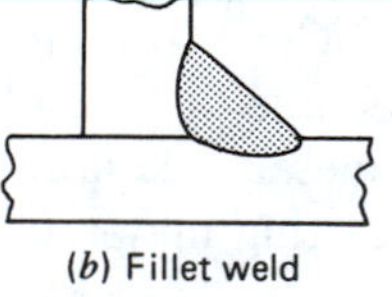
(*b*) Fillet weld

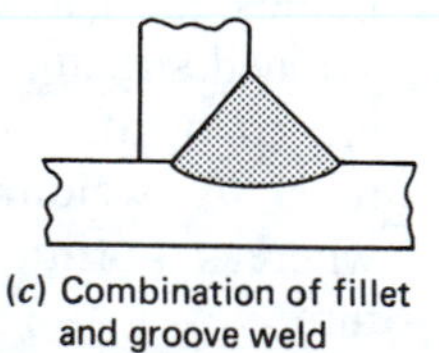
(*c*) Combination of fillet and groove weld

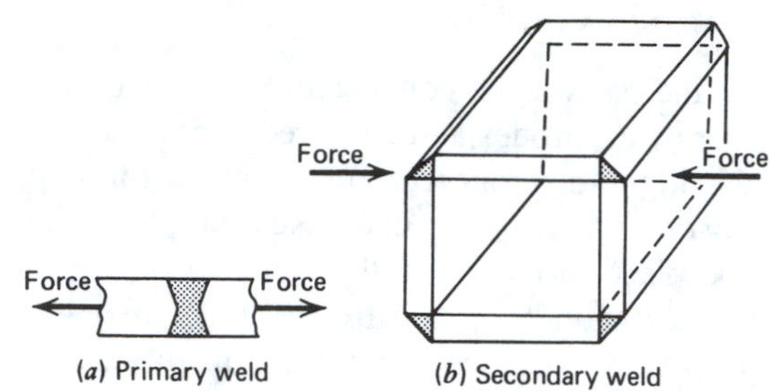

(*a*) Primary weld (*b*) Secondary weld

Figure 6-4 Classification by function. (Courtesy Chon L. Tsai.)

   (a) Groove welds: complete penetration; partial penetration.
   (b) Fillet welds.
   (c) Combination of partial joint penetration groove welds and fillet welds.
2. Classification by *function* of welds (Fig. 6-4).
   (a) Primary welds. Primary welds transfer the entire load at the joint. If the weld fails, the member fails. The weld becomes a part of the member.
   (b) Secondary welds. Secondary welds simply hold the parts together to form the member. The forces on the welds are low.
3. Classification by type of *loading* (Fig. 6-5).
   (a) Transversely loaded welds.
   (b) Longitudinally loaded welds.
   (c) Welds under bending.
   (d) Welds under torsion.
   (e) Welds under combined loading.

6-4 DESIGN OF WELDMENTS

Low-carbon steel, in addition to being inexpensive and readily available, has a third major advantage: it welds easily. For this reason, low-carbon steel (0.16 to 0.20% carbon) is the most widely used material for welded machine members. Alloy steel and carbon steel with a carbon equivalent above 0.3% tend to harden and develop cracks if welded. Preventative measures are:

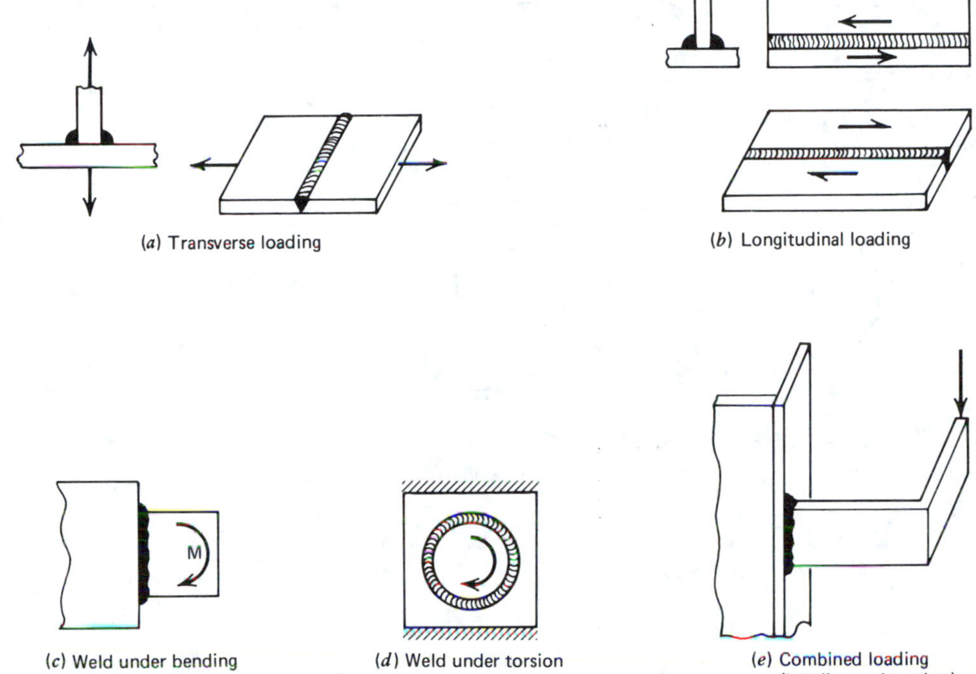

Figure 6-5 Classification by type of loading. (Courtesy Chon L. Tsai.)

1. Preheating of the weld area to retard cooling of the parent metal near the weld. American Welding Society (AWS) specifies proven preheat temperature for most metals.
2. Stress relieving of the weld area through proper heat treatment. A temperature range of 540 to 675°C (1000 to 1250°F) is widely used by industry.

An element of uncertainty is always present in any weldment because of the large (and often unknown), localized, thermal stresses that arise from *uneven* heating and cooling during fusion and subsequent solidification of material. As already observed, distortions can be reduced and thus an adverse effect on strength minimized through preheating and stress relieving. The following additional precautions should be observed.

1. Whenever possible, design the assembly so welds are *symmetrically* located about the neutral axis of the assembly. This will help ensure that any resultant distortion is uniformly distributed (Fig. 6-6).
2. Specify short *intermittent* welds instead of long continuous ones; this will reduce heat distortion and cut costs by reducing the amount of weld metal used (Fig. 6-6). Intermittent welds, however, introduce stress concentration.
3. Avoid stress changes in a weld. The entire cross section of a weld should preferably be loaded in tension, compression, or shear (Fig. 6-7).
4. Avoid joining several welds at a single location. Since each weld creates thermal stresses, several welds meeting will produce even larger and overlapping distortions (Fig. 6-8).
5. Reduce stress concentration by using a weld of sufficient size. Minimum sizes for fillet welds as a function of work piece thickness

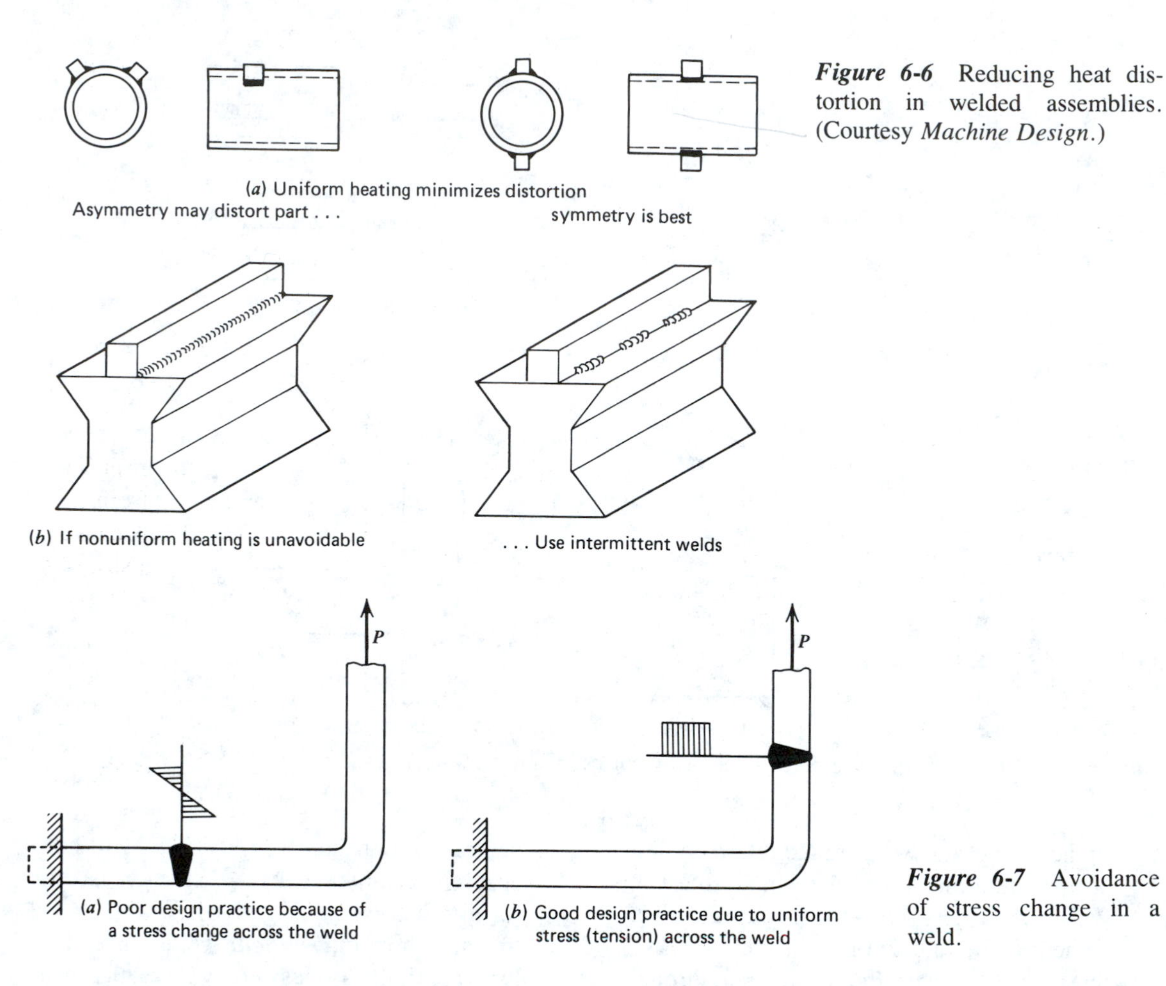

Figure 6-6 Reducing heat distortion in welded assemblies. (Courtesy *Machine Design*.)

Figure 6-7 Avoidance of stress change in a weld.

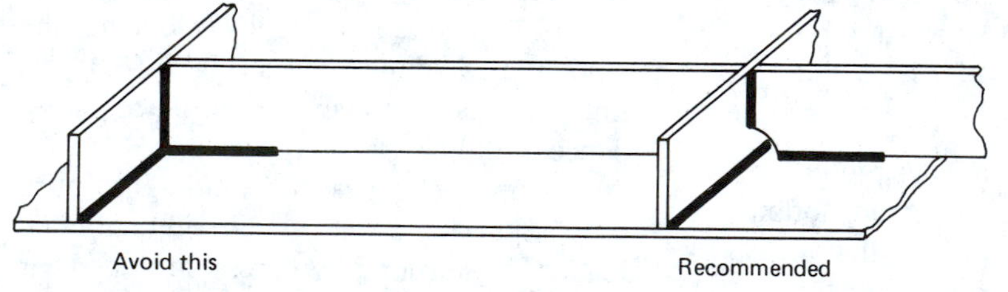

Figure 6-8 Reduction of thermal stress in welded assemblies. (Courtesy Lincoln Electric Company.)

are given in Table 6-2. Standard full fillet weld size is the leg length ω of the triangular cross section (Fig. 6-9). Maximum weld size is equal to the smaller of the thicknesses to be joined.

6. Minimize loads on the weld by locating the weld in low-stress areas whenever possible.

6-5 THE AWS STRUCTURAL WELDING CODE

The AWS Structural Welding Code offers designers of welded machine parts an opportunity for substantial cost reductions. Bulletin D412, *New Stress Allowables Affect Weldment Design*,

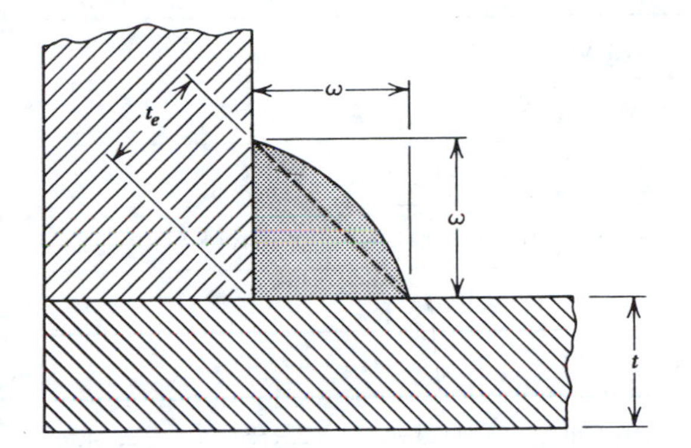

Figure 6-9 The strength of a fillet weld is determined by the throat size t_e.

by the James F. Lincoln Arc Welding Foundation explains the new code in sufficient detail. Additional sections in this chapter contain material derived from this bulletin.

6-6 ALLOWABLE STRENGTH OF WELDS UNDER STEADY LOADS

Table 6-3 gives allowable loads and shear stress for various fillet welds subject to a steady load. The seven columns in the table represent seven strength levels of the weld metal, ranging from 60 to 120 ksi (1 ksi = 1000 psi). These seven values are shown in the first row of figures. For butt welds in tension or compression, these values should match the strength of the parent metal near the weld.

The second row relates to fillet welds and partial penetration groove welds. The seven τ values are for the allowable shear stress on the throat (ksi). The values, as expected, are much lower than the corresponding values in the first row. The derating factor is $60/18 = 3.33$.

The third row and all the following rows contain values for the allowable unit force for various sizes of fillet welds ranging in size from 1 in. down to $\frac{1}{16}$ in. This is clearly a shortcut because it relates the induced load directly to the size and length of the weld.

TABLE 6-2 AWS Table of Minimum Fillet Weld Size (ω) or Minimum Throat of Partial Penetration Groove Weld (t_e)[a]

Plate Thickness (Thicker Plate)	Minimum Leg Size of Fillet
To $\frac{1}{4}$ in. inclusive	$\frac{1}{8}$ in.
Over $\frac{1}{4}$ to $\frac{1}{2}$ in.	$\frac{3}{16}$ in.
Over $\frac{1}{2}$ to $\frac{3}{4}$ in.	$\frac{1}{4}$ in.
Over $\frac{3}{4}$ to $1\frac{1}{2}$ in.	$\frac{5}{16}$ in.
Over $1\frac{1}{2}$ to $2\frac{1}{2}$ in.	$\frac{3}{8}$ in.
Over $2\frac{1}{2}$ to 6 in.	$\frac{1}{2}$ in.
Over 6 in.	$\frac{5}{8}$ in.

[a]Neither ω nor t_e should exceed the thickness of the thinner part.

6-7 PRIMARY WELDS: BUTT WELDS

Primary welds are welds that transfer the *entire* load at the point where it is located. The weld has the same property as the parent metal and, if the weld fails, the member fails.

The butt-welded joint is a typical primary weld (Fig. 6-10). It is a complete penetration weld or full-strength weld. For this butt weld the induced tensile stress is

$$\sigma_t = \frac{P}{wt} \tag{6-1}$$

Figure 6-10 Butt-welded joint. A primary weld.

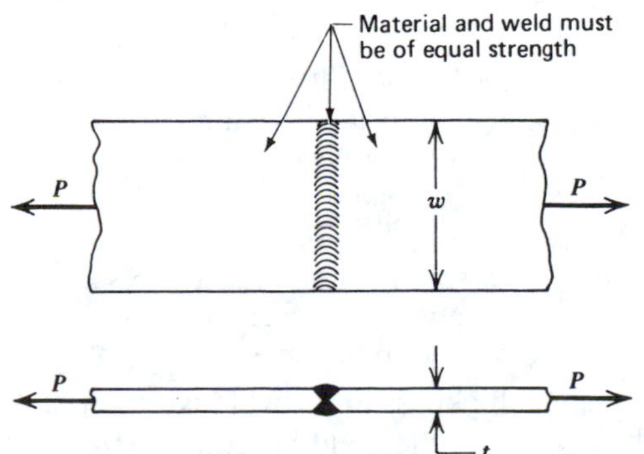

TABLE 6-3 Allowable Load for Various Sizes of Fillet Welds

	Strength Level of Weld Metal (EXX)						
	60[a]	70[a]	80	90[a]	100	110[a]	120
	Allowable Shear Stress on Throat ksi (1000 psi) of Fillet Weld or Partial Penetration Groove Weld						
$\tau =$	18.0	21.0	24.0	27.0	30.0	33.0	36.0
	Allowable Unit Force on Fillet Weld kips/linear in.						
$f =$	12.73ω	14.85ω	16.97ω	19.09ω	21.21ω	23.33ω	25.45ω
Leg Size ω, in.	Allowable Unit Force for Various Sizes of Fillet Welds kips/linear in.						
1	12.73	14.85	16.97	19.09	21.21	23.33	25.45
$\frac{7}{8}$	11.14	12.99	14.85	16.70	18.57	20.41	22.27
$\frac{3}{4}$	9.55	11.14	12.73	14.32	15.92	17.50	19.09
$\frac{5}{8}$	7.96	9.28	10.61	11.93	13.27	14.58	15.91
$\frac{1}{2}$	6.37	7.42	8.48	9.54	10.61	11.67	12.73
$\frac{7}{16}$	5.57	6.50	7.42	8.35	9.28	10.21	11.14
$\frac{3}{8}$	4.77	5.57	6.36	7.16	7.95	8.75	9.54
$\frac{5}{16}$	3.98	4.64	5.30	5.97	6.63	7.29	7.95
$\frac{1}{4}$	3.18	3.71	4.24	4.77	5.30	5.83	6.36
$\frac{3}{16}$	2.39	2.78	3.18	3.58	3.98	4.38	4.77
$\frac{1}{8}$	1.59	1.86	2.12	2.39	2.65	2.92	3.18
$\frac{1}{16}$	0.795	0.930	1.06	1.19	1.33	1.46	1.59

Source. Bulletin D412: *New Stress Allowables Affect Weldment Design*. The Lincoln Electric Company, with permission.

[a] Fillet welds actually tested by the joint AISC-AWS Task Committee.

where

P = tensile load; MPa, lb
w = width of plate; mm, in.
t = thickness of plate; mm, in.
$\sigma_w = \dfrac{S}{fs}$; MPa, psi
σ_w = the working or allowable stress
S = strength level of weld material obtained from Table 6-3

Note that the thickness of the weld is t regardless of whether or not the weld is reinforced.

6-8 SECONDARY WELDS: FILLET WELDS

Secondary welds simply hold parts together to form a rigid built-up member. They are the *non-full-strength* members, because of partial joint penetration. The joint may be a fillet weld or a partial penetration groove weld. In either case, Table 6-3 will provide the necessary allowable stresses and unit loads. Often rigidity is a major design criteria.

Fillet Weld: Direct Loading

Figure 6-11 shows two configurations for a fillet weld subjected to direct loading. The cross section of the weld appears in Fig. 6-9. For either parallel or transverse welds, the load is carried in shear. However, to find the magnitude of the shear stress in the weld, the appropriate stress area must first be determined. *Experience indicates that fillet weld failures most often occur across the minimum section of the weld.* Consequently, shear area is based on the *throat* dimension t_e (Fig. 6-9) which, for an equal-leg fillet weld, is $t_e = 0.707\omega$. Thus, for either the directly loaded parallel or for the transverse fillet weld, the nominal shear stress is

$$\tau = \frac{P}{0.707\omega(2L)} \tag{6-2}$$

Although strength calculations are dominated by the concept of allowable stress, the uniformity of welding seams permits a direct comparison of applied and allowable loads and immediate specification of weld size. The applied load must be equal to (or less than) the product of the weld length and a suitable unit load selected from Table 6-3. Thus

$$P = 1000fL \text{ lb} \tag{6-3}$$

where

f = allowable unit force for the particular size of weld; kips/linear in.
L = length of weld; in.

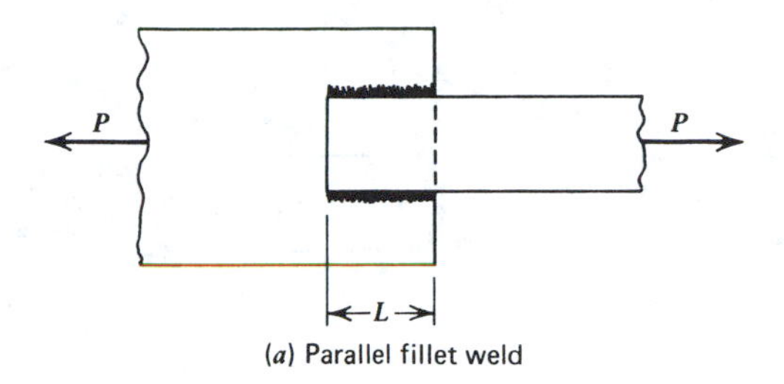

(*a*) Parallel fillet weld

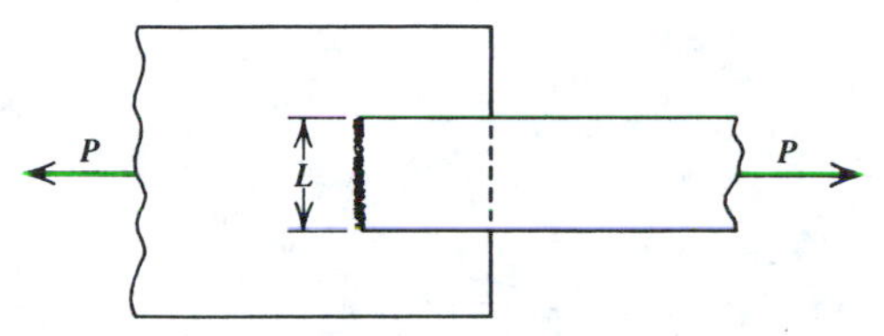

(*b*) Transverse fillet weld

Figure 6-11 Direct loading of fillet welds.

EXAMPLE 6-1

Determine the size and length of the fillet weld in Fig. 6-11*a* if P = 60,000 lb is a steady load, the working stress σ_w = 60 ksi, and the plate thickness $t = \frac{1}{4}$ in.

SOLUTION

For a plate thickness of $\frac{1}{4}$ in., a weld size of $\frac{3}{16}$ in. is adequate according to Table 6-2.

The corresponding unit force is f = 2.39 kips/in. according to Table 6-3. The weld length is therefore

$$2L = \frac{60 \text{ kip}}{2.39 \text{ kip/in.}} = 25.1 \text{ in.}$$

Use **L = 12.5 in.**

For comparison, L will now be computed by means of Eq. 6-2.

$$2L = \frac{60{,}000 \text{ lb}}{0.707(3/16 \text{ in.})(18{,}000 \text{ lb/in.}^2)} = \mathbf{25.1 \text{ in.}}$$

Stress calculations are clearly more involved and should be avoided whenever allowable unit forces apply.

Circular Fillet Weld: Axial Loading

A circular member such as a shaft or pipe that transmits an axial load to another member through a circumferential fillet weld is depicted in Fig. 6-12*a*. The load is supported primarily by shear stress in the weld. The length of the weld can be conservatively taken as the circumference of the shaft or pipe. The allowable unit force f is

$$f = \frac{P}{\pi d} \tag{6-4}$$

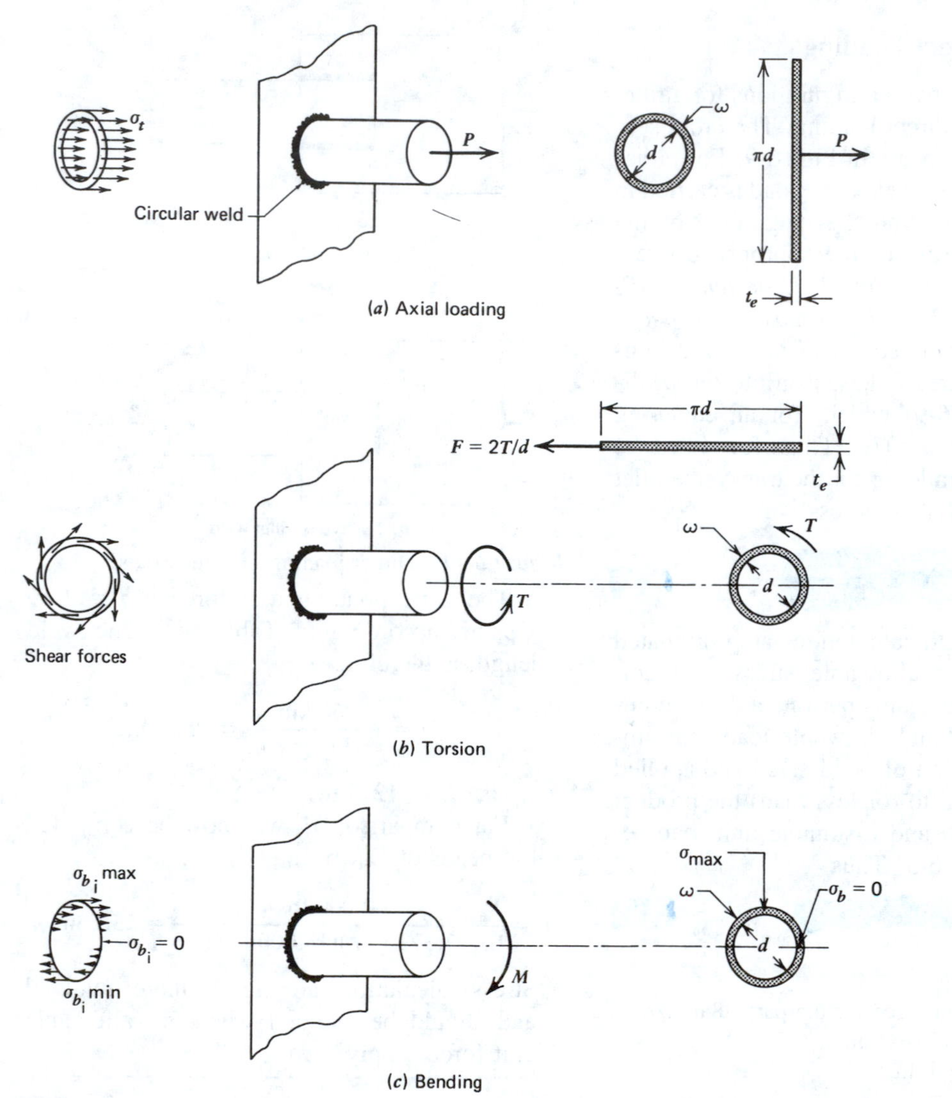

Figure 6-12 Circumferential fillet welds. Also shown are the shear areas and point of maximum stress for bending.

Given the f value, the corresponding leg size ω is obtained from Table 6-3.

Circular Fillet Weld: Torsion

The most common type of loading on a circular fillet weld is torsion (Fig. 6-12*b*). Again, the state of stress is shear and the weld length is πd. The force F is the result of applied torque T. Hence

$$F = \frac{2T}{d}$$

The unit force is

$$f = \frac{F}{\pi d} = \frac{2T}{\pi d^2} \tag{6-5}$$

Given the f value, the corresponding leg size ω is obtained from Table 6-3.

Circular Fillet Weld: Bending

Analysis of a circular fillet weld in bending is treated by using the standard bending stress formula $\sigma = Mc/I$, where section properties are for a hollow circular cylinder of inside diameter d and thickness t_e (Fig. 6-12*c*).

$$c = 0.5(d + 2t_e)$$

$$I = \frac{\pi[(d + 2t_e)^4 - d^4]}{64}$$

$$\sigma_b = \frac{32M(d + 2t_e)}{\pi[(d + 2t_e)^4 - d^4]} \qquad (6\text{-}6)$$

If $t_e < 0.1d$, this expression may be simplified to

$$\sigma_{\max} = \frac{4M}{\pi d^2 t_e} \qquad (6\text{-}7)$$

EXAMPLE 6-2

A shaft 2.5 in. in diameter is attached to a steel plate with a $\frac{3}{16}$ in. weld. Calculate the allowable bending moment for a strength level of the weld metal of 100 ksi and a factor of safety of 3.33.

SOLUTION

The allowable bending stress is 100,000/3.33 = 30,000 psi. Solve Eq. 6-7 for M and substitute the proper values.

$$M = 0.25\pi d^2(0.707)\omega\sigma_{\max}$$

$$= 0.25\pi(2.5 \text{ in.})^2(0.707)$$

$$(0.1875 \text{ in.})(30{,}000 \text{ lb/in.}^2)$$

$$= \mathbf{19{,}521 \text{ lb-in.}}$$

Fillet Weld: Eccentric Loading

As a final example of weld stress analysis, Fig. 6-13*a* shows a strut attached to a plate with fillet welds of length L and subjected to an eccentric load P. Such a condition induces both bending

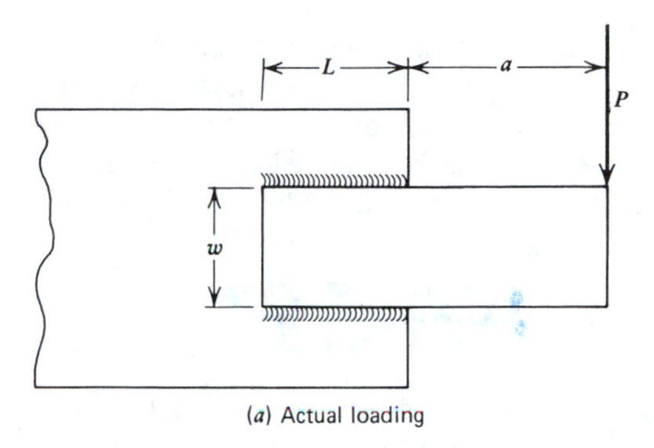

(*a*) Actual loading

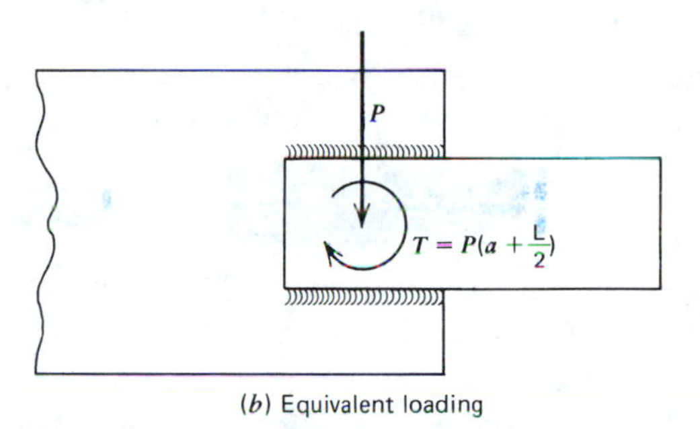

(*b*) Equivalent loading

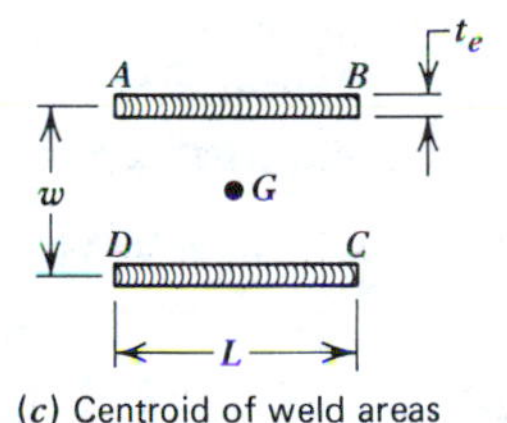

(*c*) Centroid of weld areas

Figure 6-13 Eccentrically loaded fillet welds.

and shear in the strut, but the welds will be considered as under shear only. The load is replaced by an equivalent force P and couple $T = P(a + 0.5L)$ acting through the geometric center or centroid of the welds (Fig. 6-13*b*). The welds are subjected to direct shear from the force and torsional shear from the twisting action of the couple.

Direct shear stress will be denoted by τ_1 and is given by

$$\tau_1 = \frac{P}{2t_e L} \qquad (6\text{-}8)$$

where t_e is the throat dimension. Torsional shear stress, τ_2, is approximated by analogy with the torsion equation for circular cylinders, $\tau = Tr/J$. Polar moment of inertia corresponds to throat areas of welds about their centroid (Fig. 6-13c) and can be approximated by

$$J = t_e\left(\frac{L^3 + 3Lw^2}{6}\right)$$

Maximum value of τ_2 will occur at points A, B, C, and D, at which

$$r = \sqrt{(0.5L)^2 + (0.5w)^2} = 0.5\sqrt{L^2 + w^2}$$

where r is the distance from the center G of a pattern to the point of stress, giving

$$\tau_2 = \frac{3T\sqrt{L^2 + w^2}}{t_e(L^3 + 3Lw^2)} \quad (6\text{-}9)$$

Because the direction of torsional shear stress is perpendicular to the radius from center G to the point at which stress acts, the state of stress at each corner is as shown in Fig. 6-14. Shear stress components τ_1 and τ_2 must be combined vectorially to determine total shear stress at any point. Clearly, maximum shear stress will occur at points B and C, and its magnitude is given by

$$\tau_{max} = \sqrt{\tau_1^2 + 2\tau_1\tau_2 \cos\phi + \tau_2^2} \quad (6\text{-}10)$$

where angle ϕ is given by

$$\phi = \arctan\frac{w}{L} \quad (6\text{-}11)$$

A similar analysis applies to load P acting at any angle to the axis of the strut provided the load acts in the plane of the strut.

EXAMPLE 6-3

A cable and pulley used for hoisting are mounted on a welded framework, as shown in Fig. 6-15. Four identical 12-mm fillet welds support the load. If the tension in the cable is 9 kN, calculate the maximum shear stress to which welds are subjected.

SOLUTION

We have L = 300 mm, a = 900 mm, w = 150 mm, and

$$t_e = 0.707(\omega)$$
$$= 0.707(12\text{ mm}) = 8.5\text{ mm}$$

Figure 6-14 Stresses at corners of eccentrically loaded fillet weld.

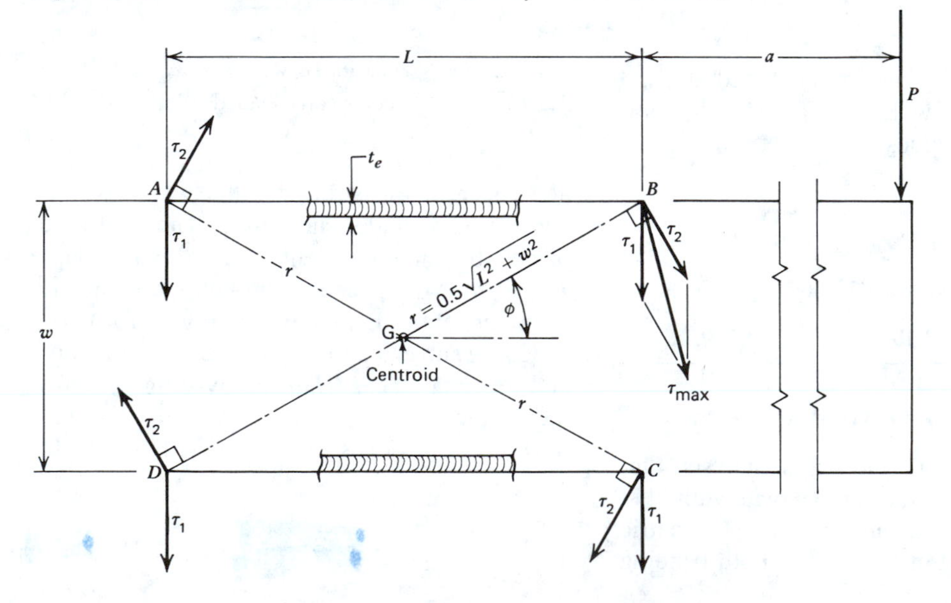

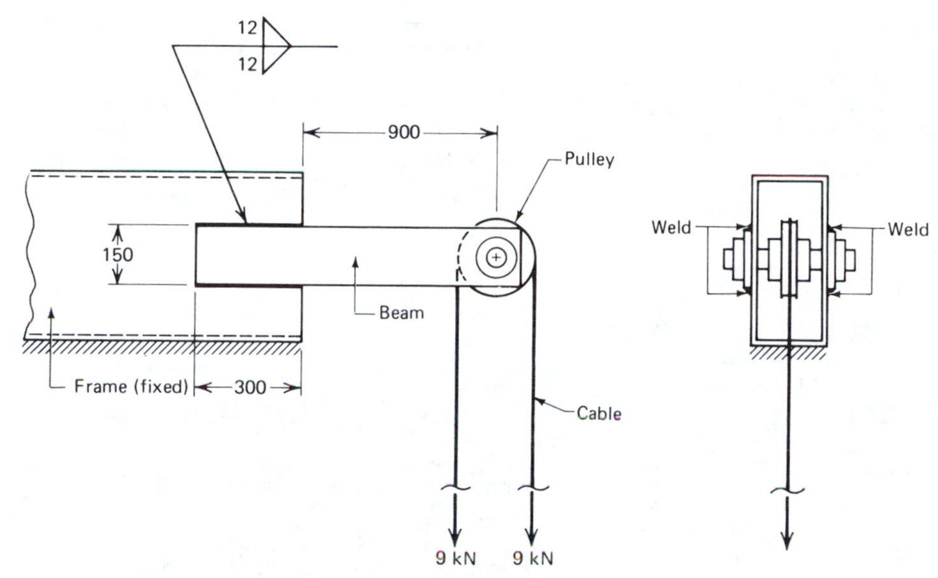

Figure 6-15 Simple hoist.

Because total load is twice the cable tension, we have $P = 2(9000) = 18\ 000$ N, which gives a torque of $T = 18\ 000(900 + 150) = 18.9(10^6)$ N · mm. Adapting Eq. 6-8 by doubling the area to account for both struts gives direct shear stress as

$$\tau_1 = \frac{P}{4t_eL}$$

$$= \frac{18\ 000 \text{ N}}{4(8.5 \text{ mm})(300 \text{ mm})} = \mathbf{1.76\ N/mm^2}$$

Similarly, since there are two struts, torsional shear stress is one-half the value given by Eq. 6-9. Thus,

$$\tau_2 = \frac{3T\sqrt{L^2 + w^2}}{2t_e(L^3 + 3Lw^2)}$$

$$= \frac{3(18.9)(10^6)\text{N} \cdot \text{mm}\sqrt{300^2 \text{ mm}^2 + 150^2 \text{ mm}^2}}{2(8.5 \text{ mm})[300^3 \text{ mm}^3 + 3(300)(150)^2 \text{ mm}^3]}$$

$$= 23.7 \text{ N/mm}^2$$

Applying Eqs. 6-11 and 6-10, respectively, gives

$$\phi = \arctan\left(\frac{150 \text{ mm}}{300 \text{ mm}}\right) = 26.6 \text{ deg}$$

and

$$\tau_{max} = [1.76^2 \text{ N/mm}^2 + 2(1.76 \text{ N/mm}^2)(23.7 \text{ N/mm}^2) \cos 26.6 \text{ deg} + (23.7 \text{ N/mm}^2)^2]^{1/2} = \mathbf{25.3\ MPa}$$

6-9 PLUG WELDS

A special type of fillet weld, the plug weld, is most often used to form a permanent connection between flat plates. This connection is made by welding inside a hole in one or both of the plates, as shown in Fig. 6-16*a*. Plugs have the advantage of requiring access to only one side of the work piece. At intermediate positions, where edge welding may be inadequate or inaccessible, plugs can be used as a supplement or replace-

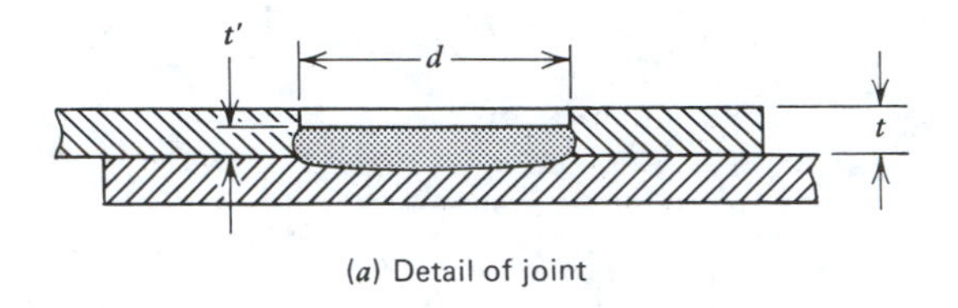

(a) Detail of joint

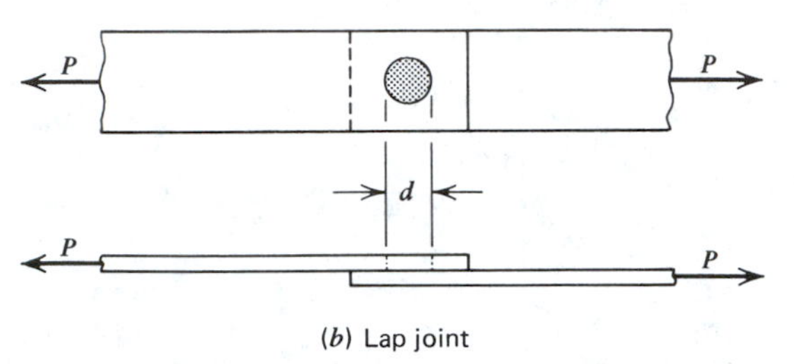

(b) Lap joint

Figure 6-16 Plug welding.

ment. The strength of a plug weld is based on load transfer in shear by the weld metal. Recommended shear stresses for steady loads are found in Table 6-3. Table 6-4 gives the size of plug welds as a function of plate thickness. The plug hole is most often completely filled with weld metal and ground flush for appearance and stress distribution. As indicated in Table 6-4, however, this is not required for thicker plates.

TABLE 6-4 Size of Plug Welds

t (mm)	d (mm)	t' (mm)	t (in.)	d (in.)	t' (in.)
6	19	6	$\frac{1}{4}$	$\frac{3}{4}$	$\frac{1}{4}$
10	25	10	$\frac{3}{8}$	1	$\frac{3}{8}$
13	29	11	$\frac{1}{2}$	$1\frac{1}{8}$	$\frac{7}{16}$
16	32	13	$\frac{5}{8}$	$1\frac{1}{4}$	$\frac{1}{2}$
19	35	14	$\frac{3}{4}$	$1\frac{3}{8}$	$\frac{9}{16}$
25	38	14	1	$1\frac{1}{2}$	$\frac{9}{16}$

Source. *Procedure Handbook of Arc Welding Design and Practice,* The Lincoln Electric Company, with permission.

Note. t = plate thickness, t' = depth of plug, and d = diameter of hole.

EXAMPLE 6-4

A lap joint fabricated by plug welding is shown in Fig. 6-16*b*. What is the allowable load P for this joint if d = 1 in., each plate is 0.375 in. thick, and the strength level is 70 ksi?

SOLUTION

Shear area is

$$A = \frac{\pi d^2}{4} = \frac{\pi(1^2 \text{ in.}^2)}{4} = 0.785 \text{ in.}^2$$

which, for a design shear stress of 21,000 psi, gives

$$P = \tau_w A = (21{,}000 \text{ lb/in.}^2)(0.785 \text{ in.}^2)$$
$$= \mathbf{16{,}485 \text{ lb}}$$

The previous examples of stress analysis of welds should acquaint readers with the basic concepts involved but do not, of course, cover all possibilities. For additional stress formulas for welded joints, see Table 6-5.

6-10 FATIGUE STRENGTH OF WELDMENTS

The performance of a weld subject to cyclic stress is clearly important. Although sound weld metal can have about the same strength as its parent metal, the fatigue strength is lower by a factor of 2 to 3. The reduction in strength is due primarily to stress concentration caused by the abrupt change in section, but minute cracks and rough surfaces also act as stress raisers. Grinding and hammer peening are common means of increasing fatigue strength of welded joints. Peening of a *hot* bead stretches the bead, thereby counteracting the tendency of the bead to contract as it cools. Residual stresses are thus minimized.

As fatigue strength calculations have become more exact, they have also become more compli-

TABLE 6-5 Stress Formulas for Welded Joints

Notation

- b = fillet height
- t_e = minimum dimension through throat of weld
- h = height of parent material or weld
- L = length of parent material or weld
- D = diameter of parent material
- P = load
- M = bending moment about neutral axis or torque about centroidal axis
- σ = tension and compression stress
- τ = shear stress

1. Any combination of welds with complete penetration (note $L > h$).

$$\tau = \frac{M(3L + 1.8h)}{h^2 L^2}$$

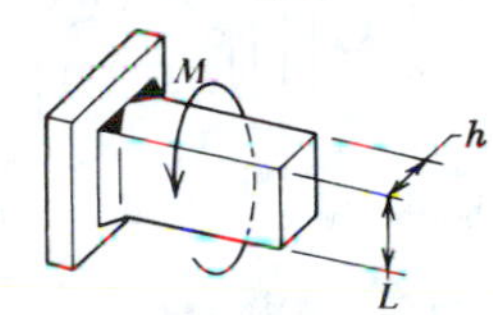

2. Any combination of welds with incomplete penetration (if $2t_e > h$, use formula 1).

$$f = (h - t_e)$$

$$\tau = \frac{3M(t_e^{\,2} + L^2)^{1/2}}{Lt_e(L^2 + 3t_e^{\,2})}$$

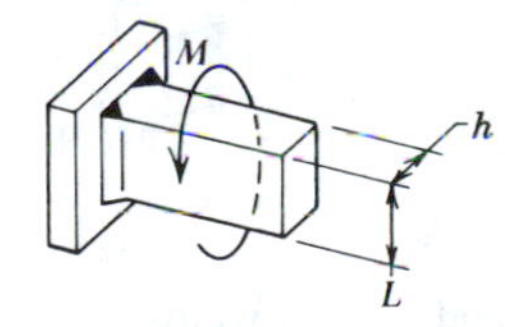

3. Equal-size fillet welds.

$$f = (h + b)$$

$$\tau = \frac{4.24M(t_e^{\,2} + L^2)^{1/2}}{Lt_e(L^2 + 3t_e^{\,2})}$$

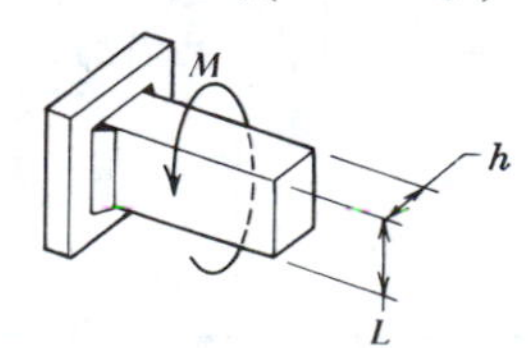

4. Fillet weld all around.

$$\tau = \frac{0.9M}{t_e D^2}$$

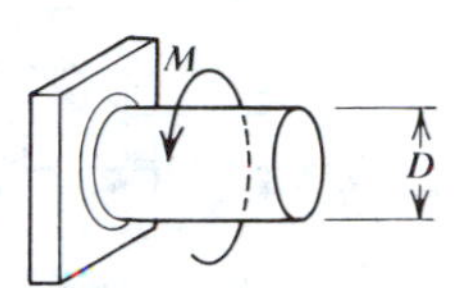

5. Fillet weld all around.

$$\tau = \frac{0.707M}{b(h + b)(L + b)}$$

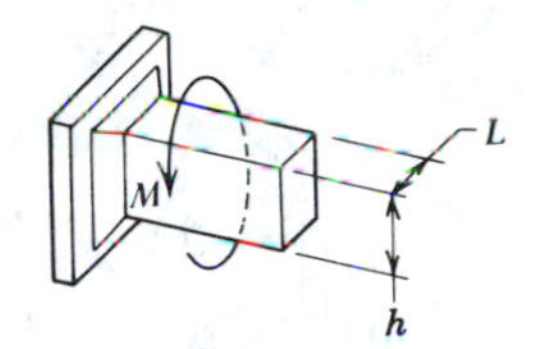

6. Any combination of welds with complete penetration.

$$\sigma = \frac{6M}{Lh^2}$$

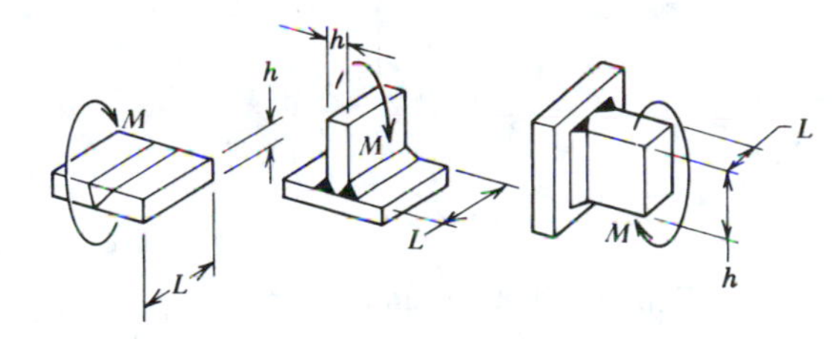

7. Any combination of welds with incomplete penetration (if $2t_e > h$, use formula 6).

$$\sigma = \frac{3Mh}{tL(3h^2 - 6ht_e + 4t_e^{\,2})}$$

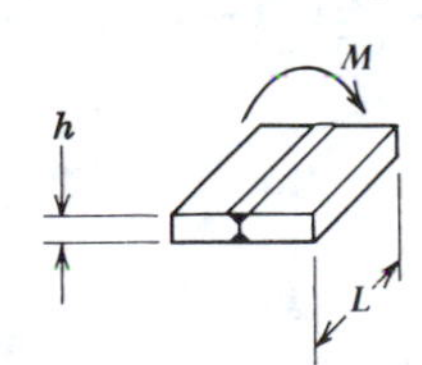

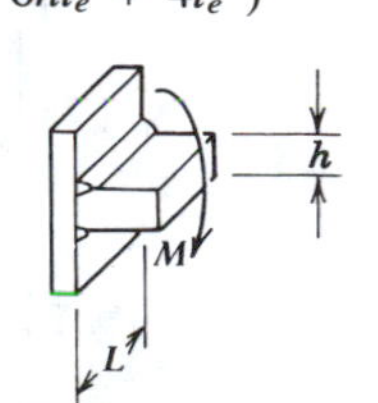

TABLE 6-5 (continued)

8. Equal-size fillet welds.

$$\sigma = \frac{1.414M}{bL(h + b)}$$

9. Any combination of weld with incomplete penetration (if $2t_e > h$, use formula 6).

$$\sigma = \frac{3M}{t_e L^2}$$

10. Equal-size fillet welds.

$$\sigma = \frac{4.24M}{bL^2}$$

11. Fillet weld all around.

$$\sigma = \frac{4.24M}{b[h^2 + 3L(h + b)]}$$

12. Fillet weld all around.

$$\sigma = \frac{1.80M}{bD^2}$$

13. Fillet weld—three sides.

$$\sigma = \frac{4.24(h + L)M}{bh^2(h + 2L)}$$

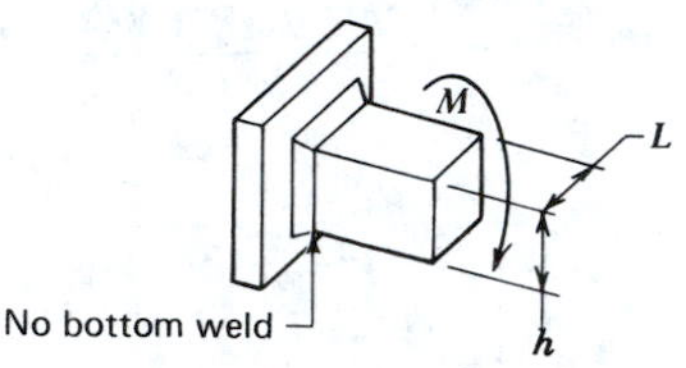

14. Fillet and groove welds.

$$b = 1.414t_e$$

$$\sigma = \frac{(h + b)M}{t_e L h^2}$$

15. Fillet and groove welds.

$$\sigma = \frac{6M}{L^2(t_e + 0.707b)}$$

Source. Courtesy *Design Engineering*, a Morgan-Grampian publication.

cated, as evidenced by Bulletin D412, referred to in Section 6-5. See this resource for detailed information about fatigue.

A *simplified* approach to fatigue loading is obtained by including a stress concentration factor in all calculations and in some cases a factor of safety of 2.0 to 3.0. This approach will make designers aware of the difference between static and dynamic loading. It will also demonstrate that some types of welds have greater fatigue strength than others.

Table 6-6 shows four major types of welds and the typical stress concentration factors. Butt welds, with a minimum of discontinuity, obviously have the lowest stress concentration factor.

EXAMPLE 6-5

Calculate Example 6-1 for a dynamic load.

SOLUTION

From Table 6-6 we find a derating factor $K_f = 1.5$. In this case it is better to apply K_f to the load instead of the working stress; hence

$$L = \frac{1.5\ (60\ \text{kip})}{2(2.39\ \text{ksi})} = \mathbf{18.8\ in.}$$

6-11 SOLDERING AND BRAZING

Soldering and brazing are processes that join metal parts by heating them and applying a nonferrous filler metal that has a *lower* melting temperature than either of the parent metals. Both are simple methods for joining sections in either mass production or low-volume production. The difference between soldering and brazing lies mainly in the type of filler alloy used. Soldering involves the use of lead- or tin-based alloys with melting points below 426°C (800°F). It provides a quick means of joining parts when low mechanical strength is acceptable. Soldering is often used to *seal* against leakage, as in automotive radiators, or to assure effective *electrical contacts*.

Brazing resembles soldering except that the filler metals—silver or brass alloys—have melting temperatures higher [above 426°C (800°F)] than soldering materials but lower than those of the parts. Because of the higher strength of the filler metal, brazing results in a stronger joint than soldering. This method accommodates dissimilar metals and requires little finishing. Brazing, however, often costs more than welding. Soldering and brazing are used extensively in mass production of electrical appliances and refrigeration equipment.

TABLE 6-6 Stress Concentration Factors for Welds[a]

Type of weld	K_f	
Reinforced butt weld	1.2	
Toe of transverse fillet weld	1.5	
End of parallel fillet weld	2.7	
T-butt joint with sharp corners	2.0	

[a]These factors are unnecessary if the AWS code is followed.

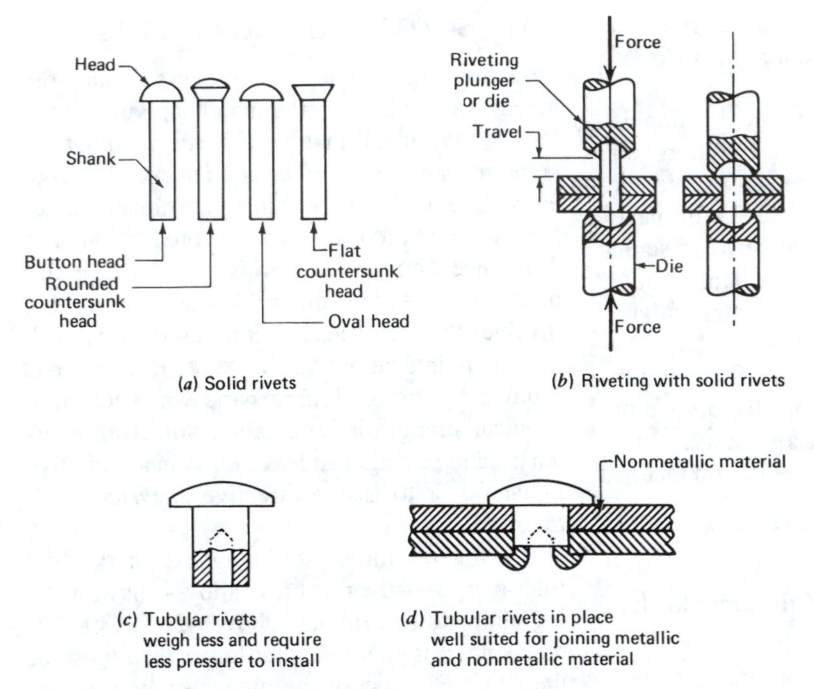

Figure 6-17 Common rivet types.

6-12 RIVETING

Riveting is popular because of its simplicity, dependability, and low cost. A rivet is a short metal pin, either solid or tubular. It has a preformed head at one end and a shank that can be worked into a second head following assembly (Fig. 6-17). Rivets are used to connect plates, structural steel shapes, sheet metal, and other relatively thin components. A joint is secured by means of rivets passing through the various thicknesses and clinched in place, often by high-speed automatic machinery. Initial cost of rivets is substantially lower than that of threaded fasteners because rivets are made in large volumes on high-speed heading machines with little scrap loss. To prevent unequal heat expansion and electrochemical corrosion, rivets and base metals should differ only with respect to hardness. A softer rivet material facilitates cold heading at assembly. A first-rate connection is best obtained when rivet holes are drilled in assembly rather than punched. Drilling provides a closer fit for the rivet shaft, even some interference.

Riveting has the following advantages over arc welding.

1. Initial cost, maintenance, and operation of equipment is low.
2. A minimum of skill is required to perform the operation of cold heading.
3. Dissimilar materials can be joined. Examples are brake linings—plastic or rubber sheets attached to metallic surfaces.
4. Parts of sheet metal too thin or too complicated in shape to be welded can be joined effectively by riveting.
5. There are no thermal stresses.
6. Rivets can be used as fasteners and as functional components such as pivots, electrical contacts, stops, or spacers.

Riveting has certain disadvantages compared with welding.

1. Riveting requires joint preparation in the form of punched or drilled holes, a procedure that adds to the cost while reducing strength.
2. Riveted assemblies weigh more because of overlapping edges.
3. Protruding rivet heads are a nuisance functionally and aesthetically. They also contribute to air and water current drag on streamlined shapes.

The use of large rivets requiring hot forming has sharply declined in recent decades due to competition from welding and high-strength bolts. The advent of blind rivets, however, has recently made riveting more competitive. A blind rivet has a self-contained unit that expands the end of the shank to form the second head. The chief advantage of a blind rivet is that access is required to only one side of the work piece. Two commonly used types of blind rivets are shown in Fig. 6-18.

Compressive loads associated with forming the second head result in high friction forces between contacting surfaces. Once friction is overcome by external loading, a riveted joint may fail in one of three common modes shown in Fig. 6-19. Design considerations for a riveted joint must include the crushing strength of the rivet-plate interface and the weakening effect of the rivet holes on the connected plates due to removal of material and shearing of the rivet. For

Figure 6-18 Blind rivets.

Force
Access required from this side only
Mandrel (removed)
Mandrel in its initial position
Vent

(*a*) Pullthrough mandrel and rivet in position for the final operation

(*b*) Finished rivet; mandrel removed

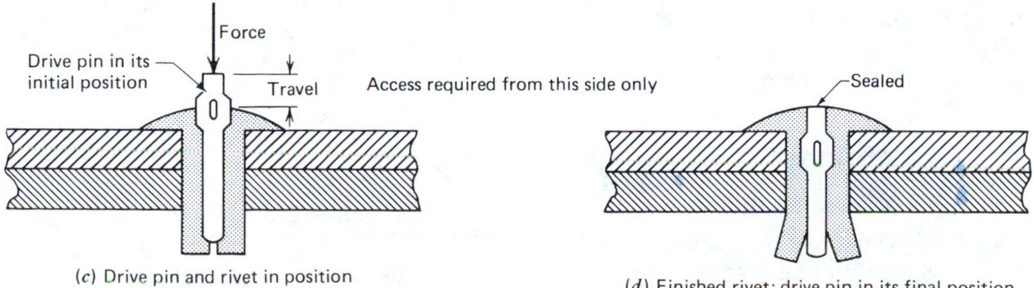

(*c*) Drive pin and rivet in position

(*d*) Finished rivet; drive pin in its final position

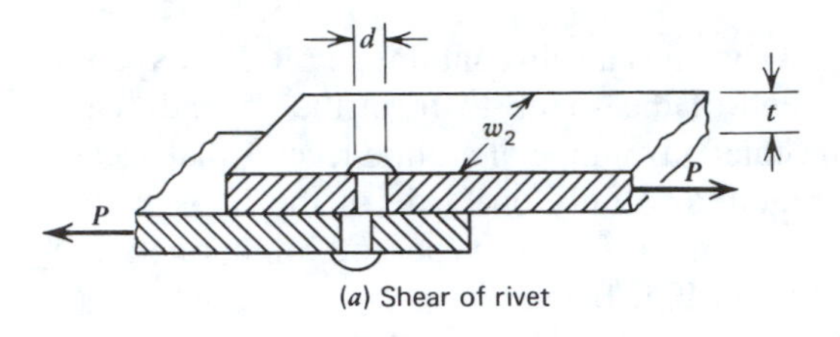

(*a*) Shear of rivet

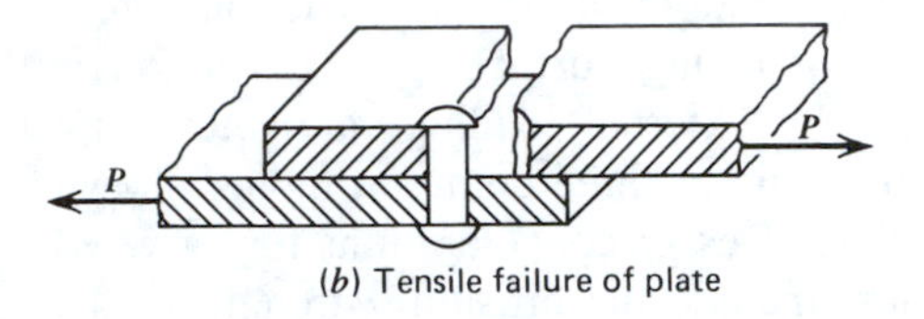

(*b*) Tensile failure of plate

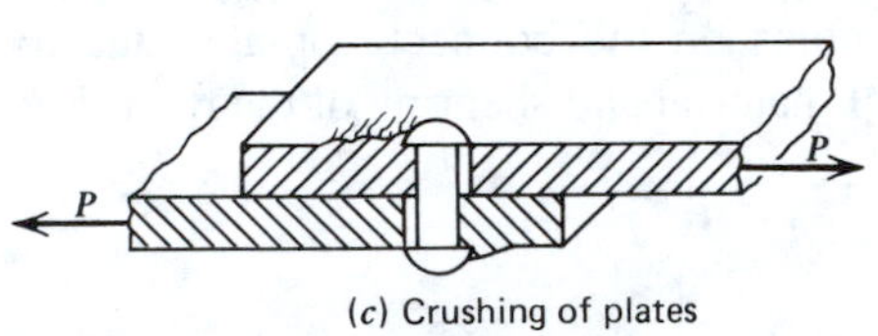

(*c*) Crushing of plates

Figure 6-19 Failure modes of riveted joints.

the joint with a single rivet, as in Fig. 6-19, shearing stress in the rivet is

$$\tau = \frac{P}{A_s} = \frac{4P}{\pi d^2} \tag{6-12}$$

where P is the tensile load on the joint and A_s, shear area, is $0.25\ \pi d^2$ for a solid rivet. If the rivet is subjected to double shear (as in Fig. 6-21), the shear area is doubled. Crushing stress due to load transfer at the contact between plate and rivet is

$$\sigma_c = \frac{P}{td} \tag{6-13}$$

with the stress area taken as the projected area of the rivet. The controlling tensile stress in the plate occurs at the section containing the rivet hole and is given by

$$\sigma_t = \frac{P}{(w - d)t} \tag{6-14}$$

where w is width of plate. Note that Eqs. 6-12 to 6-14 may also be applied to the case of a single transverse row of rivets if P is taken as *load per rivet* and w is *rivet spacing* or *rivet pitch*.

As with welded joints, design stresses for riveted joints are often set by code, depending on the application. For example, the boiler and pressure vessel code of the American Society of Mechanical Engineers lists the design stresses for riveted joints in steel pressure vessels as

60.7 MPa for rivets in *shear*
75.8 MPa for plate *tension*
131 MPa for *crushing*

These design stresses are based on ultimate strengths and a factor of safety of about 5. They are typical of riveted, low-carbon steel plates with rivets of similar material. For other materials, design stresses should also be based on corresponding ultimate strengths with an appropriate factor of safety.

One often used approach to the design of riveted joints is to select rivet diameter, rivet spacing, and plate thickness such that the joint is equally strong against the three common modes of failure. This approach, while it does have some merit, does not always yield a balanced design. Very often the size of the plates to be connected is known, but rivet diameter, number of rivets, and rivet spacing are to be determined. Rivet diameters most often fall in the range of $6\sqrt{t}$ to $7\sqrt{t}$, where t is plate thickness in millimeters. If rivet diameter is chosen accordingly, the procedure reduces to determination of the number of rivets and spacing required by ensuring that all stresses are below maximum design values. In order to avoid a fourth mode of failure, plate shearing (shown in Fig. 6-20), the center of a rivet hole should be no closer to the

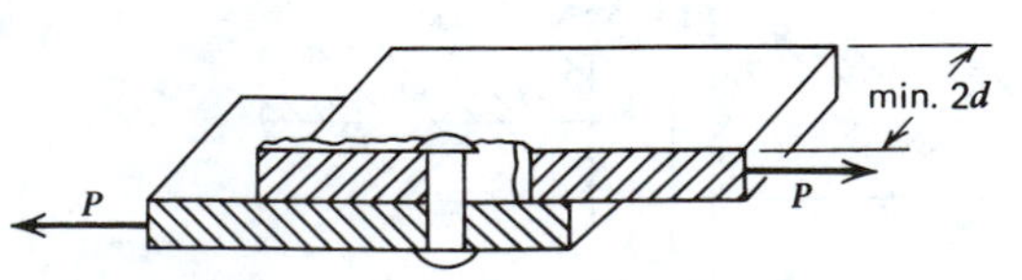

Figure 6-20 Shear of plate behind rivet.

edge of the plate than two times the rivet diameter.

When more than one rivet is needed, the spacing between rivets must be such that there is sufficient room for the driving tools. A pitch p of three times the rivet diameter is considered minimum spacing. For thin metal sheets, it is generally recommended that the spacing or pitch not exceed $24t$, where t is the sheet thickness.

EXAMPLE 6-6

Determine the maximum allowable tensile load P for the riveted joint in Fig. 6-21 if maximum allowable stresses are 103 N/mm² for shear, 275 N/mm² for crushing, and 138 N/mm² for tension. The thickness of each plate is 12 mm and the diameter of the rivets is 20 mm.

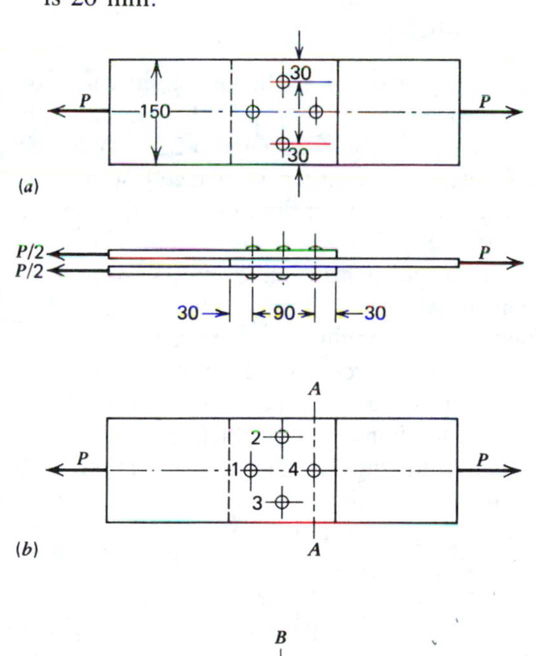

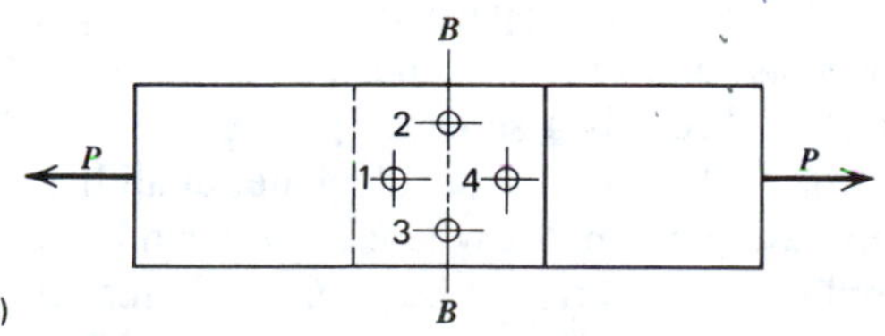

Figure 6-21 Riveted lap joint.

SOLUTION

First, consider shearing of all four rivets. Because each of the four rivets is in double shear, the shear area is

$$A_s = 2(4)(0.25\,\pi d^2) = 2\pi(20\text{ mm})^2$$
$$= 2513\text{ mm}^2$$

Allowable tensile load based on shear of rivets is then

$$P_1 = \tau A_s = (103\text{ N/mm}^2)(2513\text{ mm}^2)$$
$$= \mathbf{258\,839\ N}$$

Considering crushing of the plate, note that crushing would occur in the central plate since it carries full tensile load. For four rivets, crushing area is

$$A_c = 4td = 4(12\text{ mm})(20\text{ mm}) = 960\text{ mm}^2$$

so that allowable force becomes

$$P_2 = \sigma_c A_c = (275\text{ N/mm}^2)(960\text{ mm}^2)$$
$$= \mathbf{264\,000\ N}$$

In considering tensile stresses in the plate, two possibilities must be examined. The central plate could fail in tension across section A-A, as in Fig. 6-21b. In this case, tensile stress area is

$$A = (w - d)t = (150\text{ mm} - 20\text{ mm})(12\text{ mm})$$
$$= 1560\text{ mm}^2$$

and corresponding tensile force is

$$P_3 = \sigma A = (138\text{ N/mm}^2)(1560\text{ mm}^2)$$
$$= \mathbf{215\,280\ N}$$

A second mode of failure of the central plate across section B-B is depicted in Fig. 6-21c. Note that for this failure to occur, rivet 4 must either fail in shear or in crushing of the plate at that location. Thus we have our first glimpse of a combined failure mode. The tensile stress area at B-B is

$$A = (w - 2d)t = (150\text{ mm} - 40\text{ mm})(12\text{ mm})$$
$$= 1320\text{ mm}^2$$

Shear area of rivet 4 is

$$A_s = 0.25\pi d^2 = 0.25\pi(20^2\text{ mm}^2) = 314\text{ mm}^2$$

and crushing area at rivet 4 is

$$A_c = td = (12\text{ mm})(20\text{ mm}) = 240\text{ mm}^2$$

The load required for tensile failure at *B-B* and simultaneous shear of rivet 4 is then

$$P_4 = \sigma_t A + \tau A_s = (138\text{ N/mm}^2)(1320\text{ mm}^2) + (103\text{ N/mm}^2)(314\text{ mm}^2)$$
$$= \mathbf{214\,500\text{ N}}$$

For tensile failure at *B-B* and crushing at rivet 4, the tensile load is

$$P_5 = \sigma_t A + \tau_c A_c = (138\text{ N/mm}^2)(1320\text{ mm}^2) + 275\text{ N/mm}^2)(240\text{ mm}^2)$$
$$= \mathbf{248\,160\text{ N}}$$

Therefore maximum allowable load for the joint is

$$P = P_4 = \mathbf{214\,500\text{ N}}$$

based on combined tensile failure at section *B-B* and shear of rivet 4.

6-13 INTERFERENCE FITS

Fits between cylindrical parts govern the proper assembly and performance of countless rotating machine parts. Interference fits secure a certain amount of tightness between parts that are meant to remain permanently assembled. No fastening elements are used; instead, the *inherent elasticity* of the material is utilized to obtain a rigid connection (Fig. 6-22). Often these fits are used instead of costly splines, keyways, setscrews, and other fastening devices, although frequently keys and axial locating devices are added to prevent gradual creeping of the parts with press fits. Interference fits allow simple parts geometries and often more even stress distribution. Consequently, they offer great resistance to vibration. Four types are generally recognized: driving, forced, shrinkage, and expansion fits. Based on the method of assembly, driving and forced fits may be characterized as axial fits, whereas shrinkage and expansion fits are *radial* fits.

Driving Fits

When interference is such that the parts can be assembled "manually" by driving, this is known as a *driving fit*. Such fits are employed when small parts (plugs, pins, shafts, etc.) are to remain in a fixed position with their mating parts (Fig. 6-22*a* and 6-22*d*).

Forced Fits

Forced or *pressed* fit is the term used when the interference is such that assembly requires the use of a press (Fig. 6-22*a* and 6-22*d*). A forced fit therefore has a larger allowance (more interference) than a driving fit or the parts themselves may be larger. Cold press fits must be assembled slowly, at speeds less than 2 mm/s—otherwise the material will deform plastically, not elastically. The amount of interference for both "driving" and "forced" fits depends on the length of the bearing surface, the diameter of the hole, surface conditions, and the thickness and kind of metal surrounding the hole. Precise data for calculating forced fits are given in *Machinery's Handbook,* pp. 1531–1532.

Forced fits are limited to parts that will fit existing presses. This often excludes large-diameter wheellike parts such as ring gears, flywheels, and railroad wheels, where the outer rim (the "tire") is a separate part. Furthermore, forced fits do not provide adequate reliability for use in these examples where failure can be catastrophic. Such parts are best joined permanently by means of shrinkage fits.

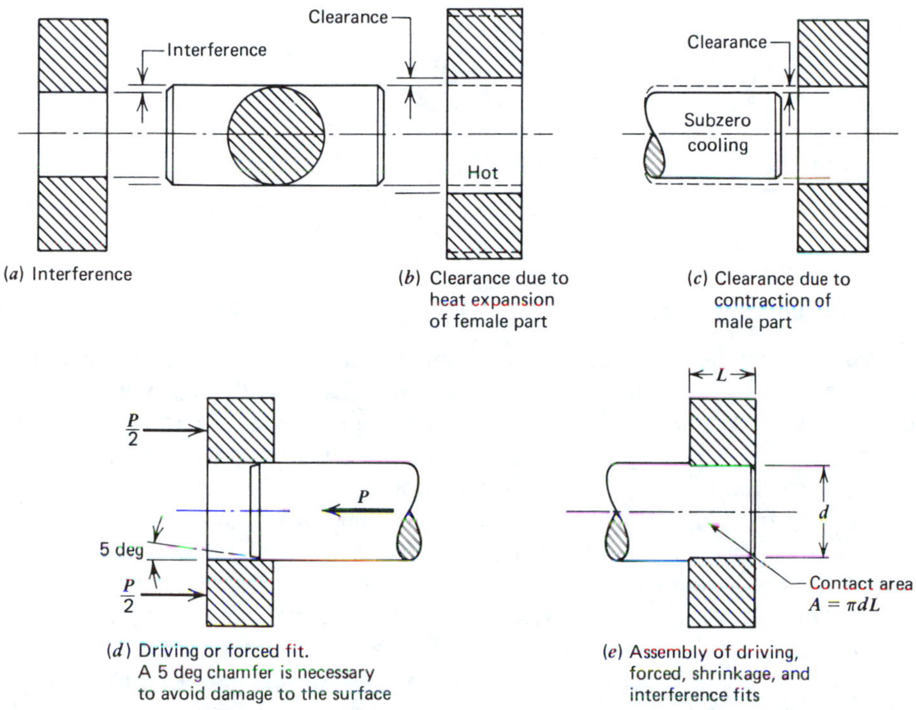

Figure 6-22 Interference fits. For clarity, interference and clearance are exaggerated.

Shrinkage Fits

This term is used when a ring-shaped outer member is mounted around a hub or inner member (Figs. 6-22*b* and 6-22*e* and 6-23). The ring is first heated to a temperature that will expand it *radially* slightly beyond the initial interference provided in machining. The ring is then slipped over the hub or wheel and allowed to cool in place. As the ring cools, it tries to shrink radially to its normal diameter, thereby producing a uniform pressure on its counterpart. Thus relative sliding at the interface is prevented by friction forces generated between the two parts. Shrinkage (or simply *shrink*) fits cost more than press fits but yield considerably more resistance to axial pull and relative rotation. For the same dimensions and tolerances, shrink fits provide at least three times as much resistance as press fits. Data and equations for calculating shrinkage fits are provided in *Machinery's Handbook*, pp. 1532–1535.

Expansion Fits

These are shrinkage fits in reverse (Fig. 6-22*c*). They are the result of modern cooling techniques that can generate subzero temperatures at low

Figure 6-23 Interference fits.

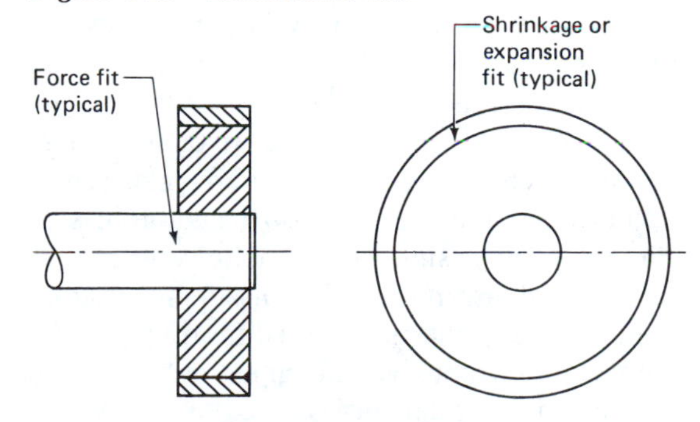

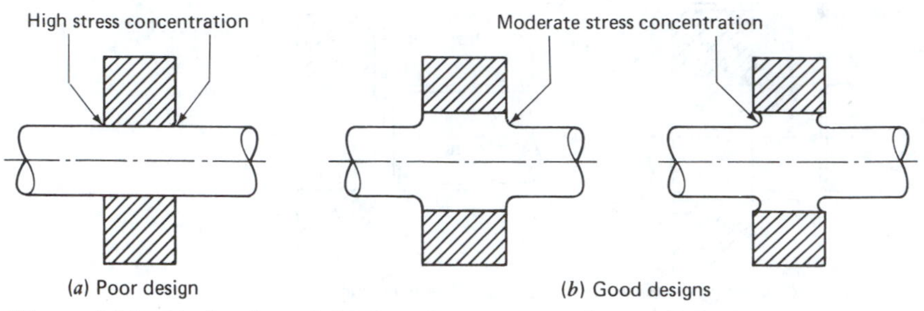

Figure 6-24 Design for minimum stress concentration.

cost. When a tight fit is required, the inner member is contracted by subzero cooling to permit insertion into the outer member. As the temperature rises, the inner part expands and a tight fit is obtained. A variety of temperatures can be obtained by means of a deep-freeze, −15°C (0°F); dry ice, −78°C (−109°F); and liquid nitrogen, −196°C (−320°F). This method has been used in assembling cast iron sleeves (liners) into engine cylinder blocks.

Interference fits are likely to fail in fatigue unless properly designed. Fatigue life of a shaft may be lowered by stress concentration at the end of the interferences. Figure 6-24 indicates two methods of increasing the fatigue life of shafts with interference fits.

SUMMARY

The most commonly used techniques for permanently joining machine members are welding, brazing, soldering, riveting, and shrinking. Welding is the most popular because of its versatility, strength, and low cost. Of the more than 40 different welding processes available, arc welding is the most useful in machine design. It is an alternative to casting and forging, a replacement for riveting, and a versatile repair medium. Major advantages include large reductions in size and/or mass, flexibility in design, and simplicity in fabrication and assembly. The disadvantages of welding are primarily the need for trained personnel and expensive equipment.

Design of weldments is based on the AWS structural welding code. The code specifies allowable stresses for static and dynamic loading. When applied to a variety of structural compositions, the code may cover as many as 200 separate structural details.

Soldering and brazing are methods for joining metal parts by heating them and applying a nonferrous filler metal. The filler metal has a lower melting temperature than that of the parts. Both methods will join parts of dissimilar metals, provide leak-proof joints, and assure effective electrical contacts.

Riveting is popular because of its simplicity, dependability, and low cost. There is no need for trained personnel, and the cost of equipment is low. Riveting is particularly advantageous where metallic and nonmetallic parts are joined. Sheet metal parts too thin or too intricate to be welded are easily joined by hollow rivets. Riveting is preferred to join aluminum alloys, which do not weld easily.

Interference fits are keyless stress connections that save material and reduce time for machining and assembly of cylindrical, concentric parts. Forced fits are cold press fits. Shrinkage fits rely on thermal expansion for assembly. Expansion fits are shrinkage fits in reverse.

REFERENCES

6-1 Blodgett, O. W. *Design of Welded Structures*. Cleveland: The James F. Lincoln Arc Welding Foundation, 1966.

6-2 *The Brazing Book*. New York: Handy & Harman, 1977. (This 50-page booklet contains a wealth of useful information on brazing.)

6-3 *General Motors Drafting Standards*. Warren, MI: GM Technical Center. (Available for subscription.)

6-4 Jefferson, T. B., and G. Woods. *Metals and How to Weld Them,* Second Edition. Cleveland: The James F. Lincoln Arc Welding Foundation, 1962.

6-5 Jordan, R. I. A. "Designing Interference Fits." *Machine Design,* Vol. 46, No. 23, pp. 68–72, 1974.

6-6 *Machine Design. Fastening and Joining Reference Issue,* 1976.

6-7 *Procedure Handbook of Arc Welding Design and Practice,* Eleventh Edition. Cleveland: The James F. Lincoln Arc Welding Foundation, 1957.

6-8 Sullivan, J. L. "Press-Fitted Shafts." *Machine Design,* Vol. 49, No. 13, pp. 102–106, 1977.

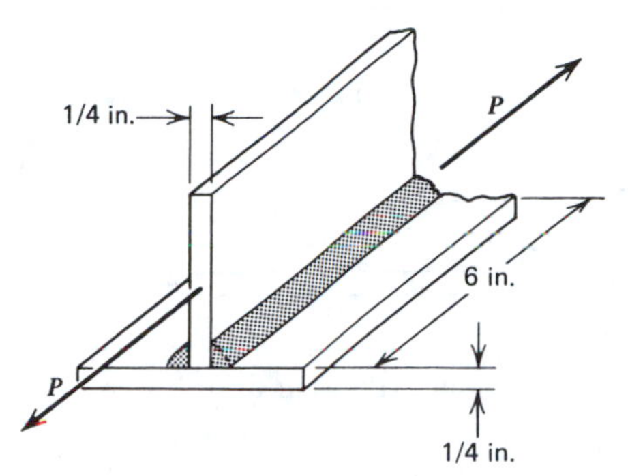

Figure 6-25 Bracket with fillet weld. (Problem 6-2)

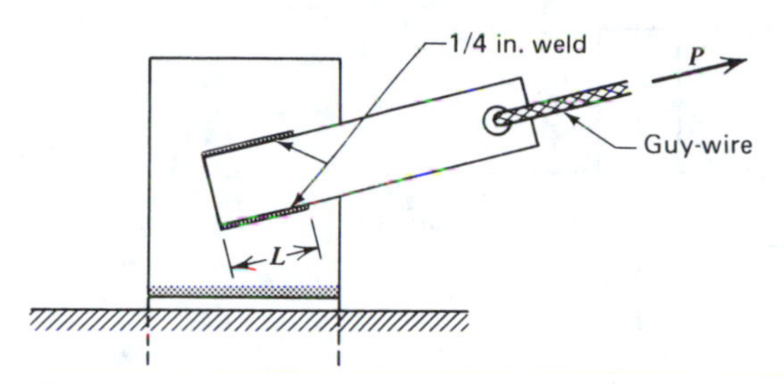

Figure 6-26 Support base. (Problem 6-3)

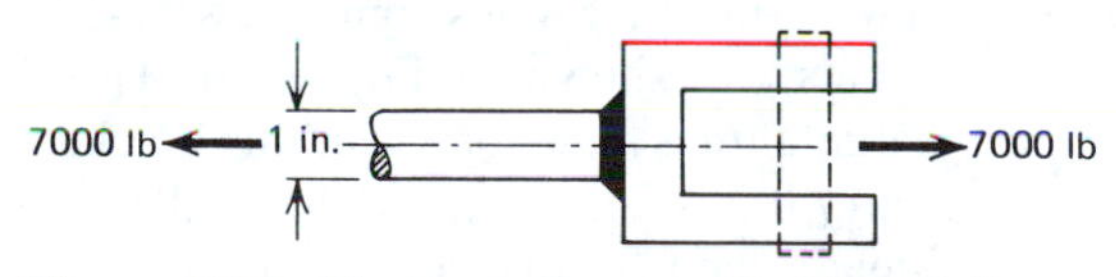

Figure 6-27 Clevis attachment. (Problem 6-4)

PROBLEMS

P6-1 Two low-carbon steel plates, each having a thickness of $\frac{3}{8}$ in. and a width w = 10 in., are butt welded, as shown in Fig. 6-10. If the plate material has a yield strength of 60,000 psi, what should be the strength level of the weld material? What is the maximum tensile, steady load P that the joint should be subjected to?

P6-2 The plates in Fig. 6-25 are both $\frac{1}{4}$ in. thick and joined by $\frac{3}{16}$-in. fillet welds 6 in. long, one on each side. Find the maximum, steady load P that the joint will withstand. The material has a strength level of 70,000 psi.

P6-3 Figure 6-26 shows the support base for an antenna guy-wire. What should be the length L of the $\frac{1}{4}$-in. weld if the maximum, steady, tensile load P is 40,000 lb? Assume a strength level of the material of 60,000 psi.

P6-4 A clevis (shackle) attachment for a pneumatic cylinder is to be attached to the actuating rod by a full-circumference fillet weld (Fig. 6-27). For a steady tensile load of 7000 lb, what weld size is required?

P6-5 A lap joint of the type shown in Fig. 6-28 has plates 0.5 in. thick, 16 in. wide, and a tensile load P. If the strength level is 70 ksi, what size load can be carried for (*a*) steady conditions, and (*b*) dynamic conditions? *Hint:* Use Table 6-6.

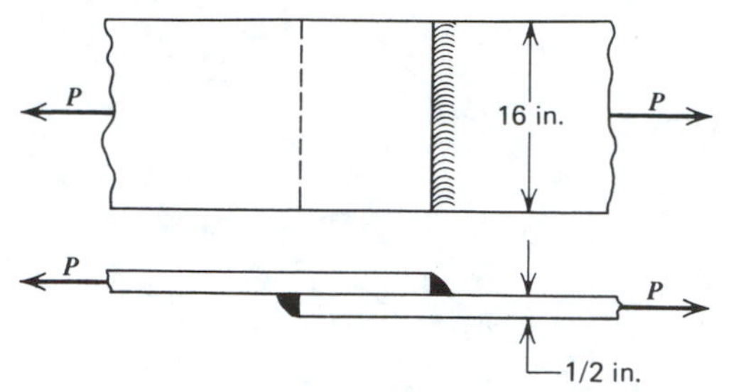

Figure 6-28 Lap joint. (Problem 6-5)

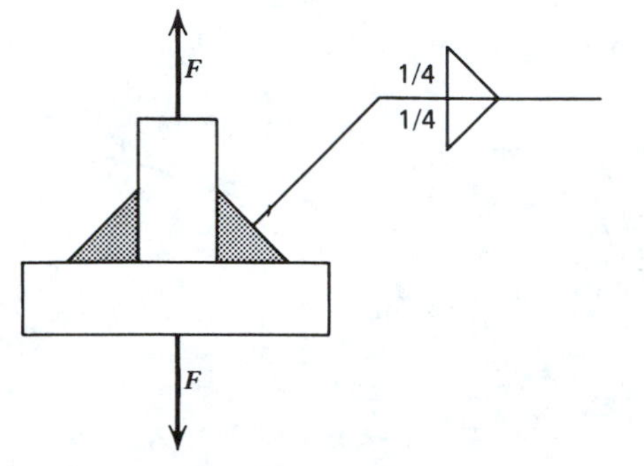

Figure 6-29 Normal weld. (Problem 6-6)

P6-6 Two $\frac{1}{4}$-in. fillet welds connect two $\frac{3}{8}$-in. plates, as shown in Fig. 6-29. How large a transverse force can this bracket carry if it is 2 in. long and the load is (*a*) steady, and (*b*) dynamic?

P6-7 Figure 6-30 shows a lever arm welded onto a shaft by means of a double $\frac{3}{8}$-in. fillet weld. How large a load can the lever arm endure for (*a*) a steady load, and (*b*) a dynamic load? The material has a strength of 80,000 psi. *Hint:* Use Eq. (6-5).

Figure 6-30 Lever arm. (Problem 6-7)

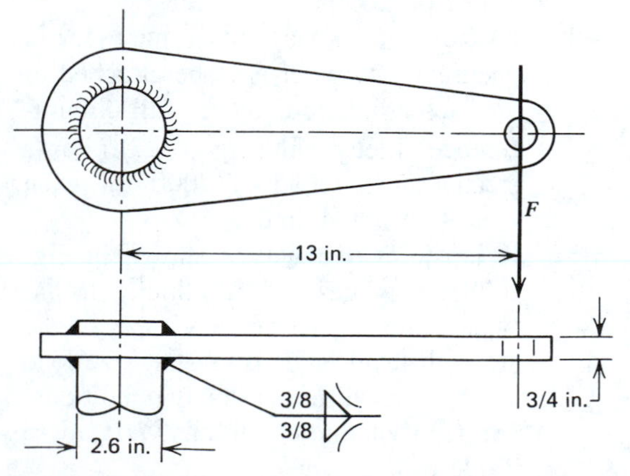

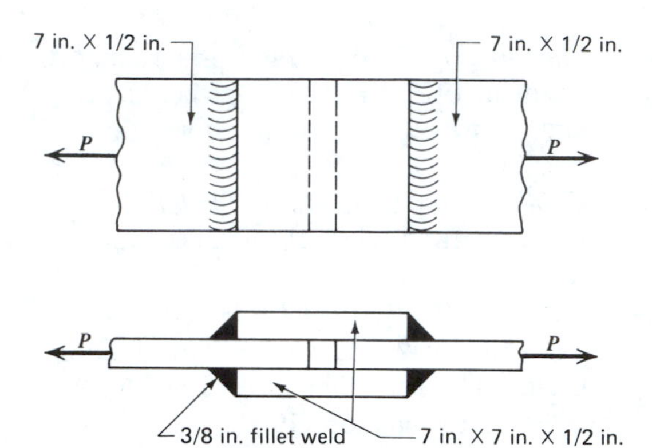

Figure 6-31 Double lap joint. (Problem 6-8)

P6-8 Find the force P that the double lap joint in Fig. 6-31 can endure for (*a*) a static load, and (*b*) a dynamic load. The plate material has a strength level of 60,000 psi.

P6-9 A large spur gear is welded from a hub, a web, and a rim, as shown in Fig. 6-32. The strength level of the material is 60 ksi. The torque is 24,000 lb-in. and steady. Determine the size of the welds to be recommended at the hub and the rim. Use a factor of safety of 2.0 to compensate for the light dynamic conditions, plus unknown factors. What type of weld is this—primary or secondary? *Hint:* Use intermittent welding.

Figure 6-32 Welded gear. (Problem 6-9)

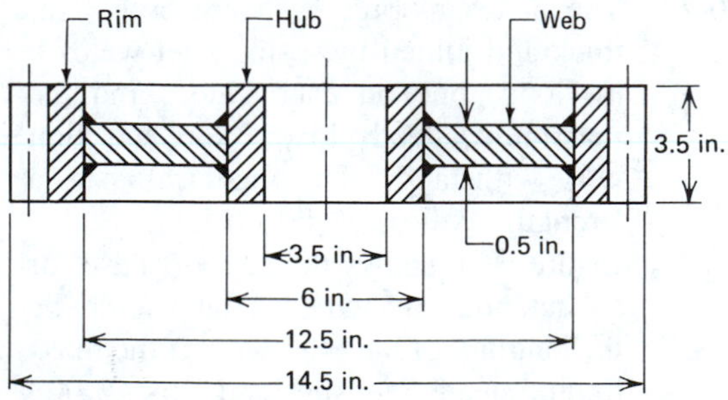

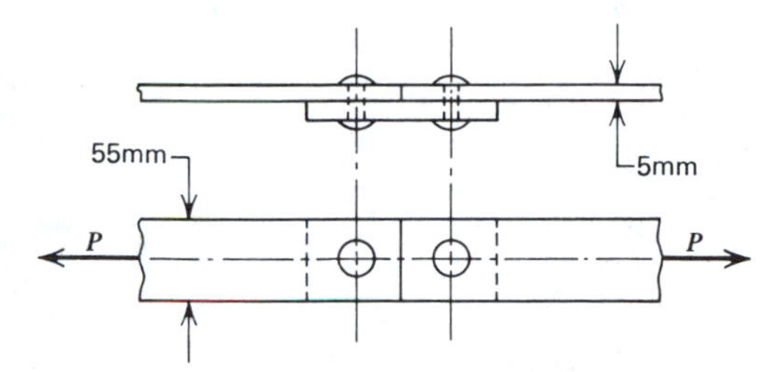

Figure 6-33 Rivet splice. (Problem 6-10)

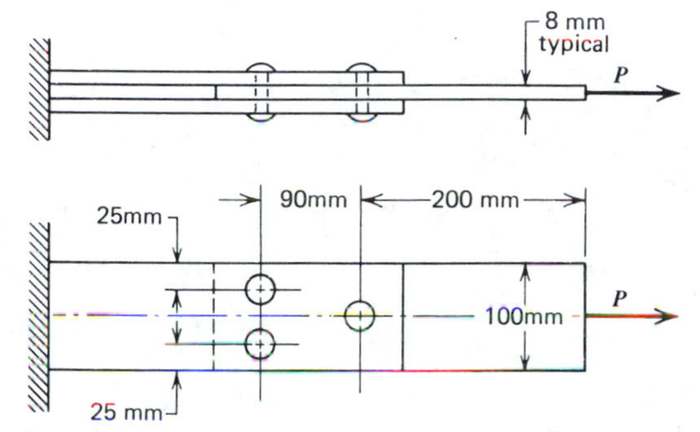

Figure 6-34 Double lap joint. (Problem 6-11)

P6-10 For the rivet splice shown in Fig. 6-33, determine the required rivet diameter using the ASME code design. The load is 6 kN.

P6-11 The diameter of each rivet in the double lap joint shown in Fig. 6-34 is 18 mm, and the thickness of each plate is 8 mm. Find the maximum load *P* that can be applied if the ASME code stresses are not exceeded. What is the primary mode of failure? How would you obtain a more balanced design?

P6-12 Figure 6-35 shows a brake band attached to its metal hinge by five rivets. Determine the size of the rivet needed for a load that can vary between zero and 12 kN. How wide must the band be for an allowable tensile strength $\sigma_{t,all}$ = 200 MPa? The maximum allowable strength of the rivets is 100 MPa in shear and 175 MPa for crushing.

P6-13 For the joint in P6-10, it was discovered that the load was actually 10% larger

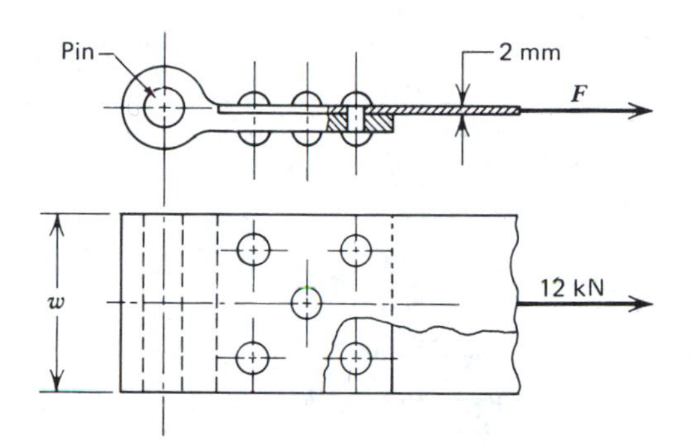

Figure 6-35 Riveted brake band. (Problem 6-12)

than indicated. By how large a percentage should the rivet diameter theoretically be increased? Is a change really necessary? *Hint:* Use Table 2-1, guide to estimating relative changes.

P6-14 Large sprockets for roller chains are often obtained by riveting a toothed steel disk onto a cast iron hub, as shown in Fig. 6-36. The roller chain, a single-strand RC80, is to transmit 0.50 kW at a minimum speed of 10 rpm in both directions.

(*a*) What are the advantages of this design compared with (1) a welded design, and (2) a unit made from a single piece of material?

(*b*) Find the diameter *d* of the six identical rivets needed to complete the assembly. For a fluctuating load the allowable rivet shear stress is 50 MPa, while the allowable compressive stress is 90 MPa. For the hub the allowable compressive stress is 20 MPa. *Hint:* The induced torque must equal the total shear torque in the rivets.

P6-15 Answer the following questions about forced fits using *Machinery's Handbook* as the main source of information.

(*a*) What is the allowance per inch of diameter?

(*b*) Explain the term *pressure factor*.

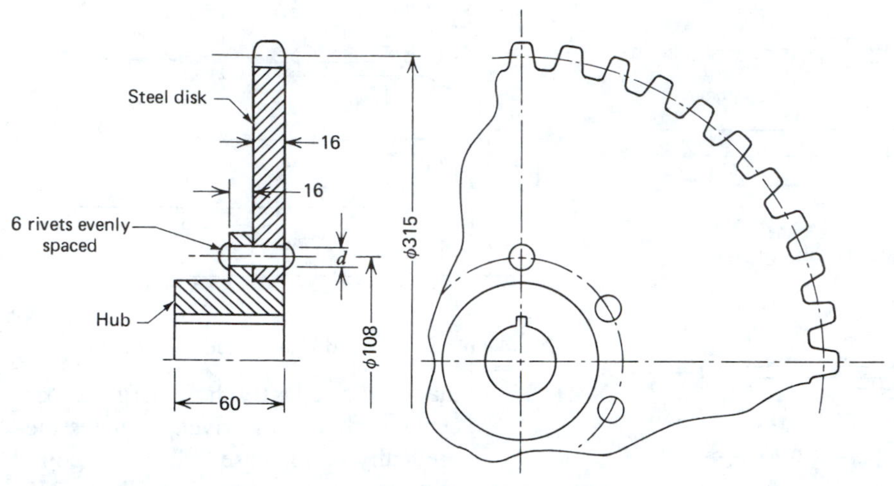

Figure 6-36 Riveted chain sprocket. (Problem 6-14)

(*c*) Give the expression for allowance for given pressure.

P6-16 Use *Machinery's Handbook* to answer the following questions about shrink fits.

(*a*) Plot and interpret a curve showing allowance as a function of the center diameter of a locomotive driving wheel tires.

(*b*) Give and interpret the formulas for calculating allowance and temperatures for proper assembly.

P6-17 An impeller for a centrifugal blower has a steel hub with an inside diameter of 1.5 in., an outside diameter of 3.75 in., and a length of 3.0 in. The impeller is pressed onto a steel shaft. What is the ultimate pressure required for assembly?

P6-18 A 35 in. diameter ring gear of steel is to be shrunk onto a 30 in. diameter rim of cast iron. Calculate (*a*) the allowance corresponding to an allowable true tangential stress of 4000 psi, and (*b*) the temperature to which the ring should be heated above room temperature for clearance at assembly.

Chapter 7
Detachable Fasteners

For want of a nail the shoe was lost.
For want of a shoe the horse was lost.
For want of a horse the battle was lost.

POOR RICHARD'S ALMANAC

NOMENCLATURE

A_s = tensile stress area (root area)

AS_s = thread shear area per unit of engaged length, for *external* threads

AS_n = thread shear area per unit of engaged length, for *internal* threads

C = torque-tension coefficient

D = basic major diameter

d = basic minor diameter

F_i = initial load or preload

f_t = friction coefficient at threads

h = length or depth of thread engagement

L = lead

n = number of threads per unit length

p = pitch

T = tightening torque

σ_p = proof stress

Threaded fasteners perform the function of locating, clamping, adjusting, and transmitting force from one machine member to another. They are thoroughly standardized and generally designed for use in mass production of machines. The use of screw-thread fasteners remains the basic assembly method in the design and construction of machines despite advances in other methods of joining. To be effective, each application must be properly engineered and installed, since failure of a single fastener can be destructive or even catastrophic. Thus the designers must select standard fasteners of the type and size that will most adequately suit the application at hand.

7-1 FUNCTION AND DESCRIPTION

A threaded fastener, examples of which appear in Fig. 7-1, is a device that can effectively exert and maintain a large force in one direction (axially) through the application of a small force in another direction (tangentially). All are based on the single-threaded screw, a simple machine, that yields a large mechanical advantage in minimal space and theoretically is self-locking.[1] Effective use, however, requires the aid of two other simple machines: the lever and the wheel-and-axle machine. The wrench is basically a lever; the screwdriver is a wheel-and-axle machine.

Threaded fasteners are basically small, intricate, highly stressed tensile components. Threads are helical ridges formed by cutting or cold forming a groove onto the surface of a cylindrical bar, thus producing what is known as a screw, bolt, or stud. Threads are also formed internally in cylindrical holes and, when produced individually in symmetrical shapes, constitute what is known as a nut. Matching external and internal threads so that they may be assembled is the key to all threaded fasteners. The rotary motion of a nut against a stationary screw first imparts an axial movement along the screw by reason of the matching helices. When resistance is encountered, the threads generate an axial force as dictated by the principle of the wedge. Further rotation demands increased effort (torque), with a resulting increase in axial force. Thus the connection remains tight unless some external influence such as vibration or temperature change overrides the initial condition.

The following definitions, which apply to essentially all types of screw threads, are made in reference to Fig. 7-2, which shows portions of mating external and internal threads.

Basic Major Diameter—D. On straight threads, external or internal, it is the diameter over the thread crests that is furthest from the axial centerline.

Minor Diameter—d. On straight threads, ex-

[1] Self-locking means that an axial force, no matter how large, cannot generate any relative motion and thus cause the screw to back-drive.

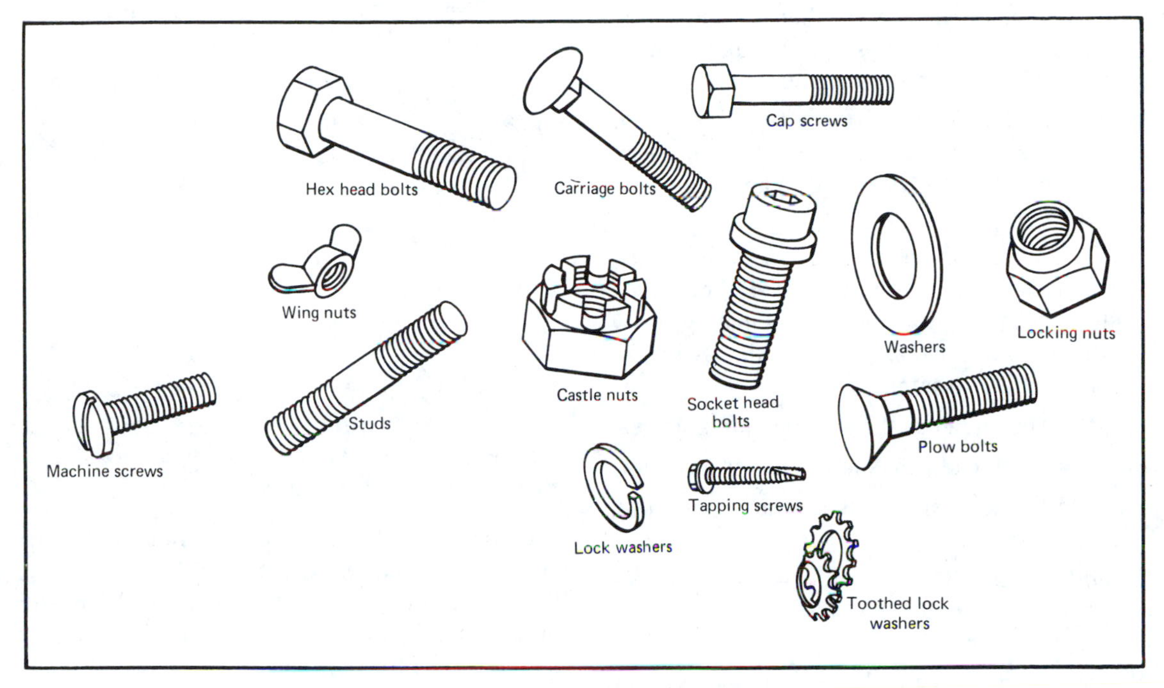

Figure 7-1 Typical fasteners. (Courtesy Deere and Company Technical Services.)

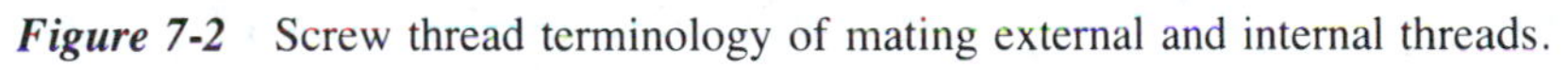

Figure 7-2 Screw thread terminology of mating external and internal threads.

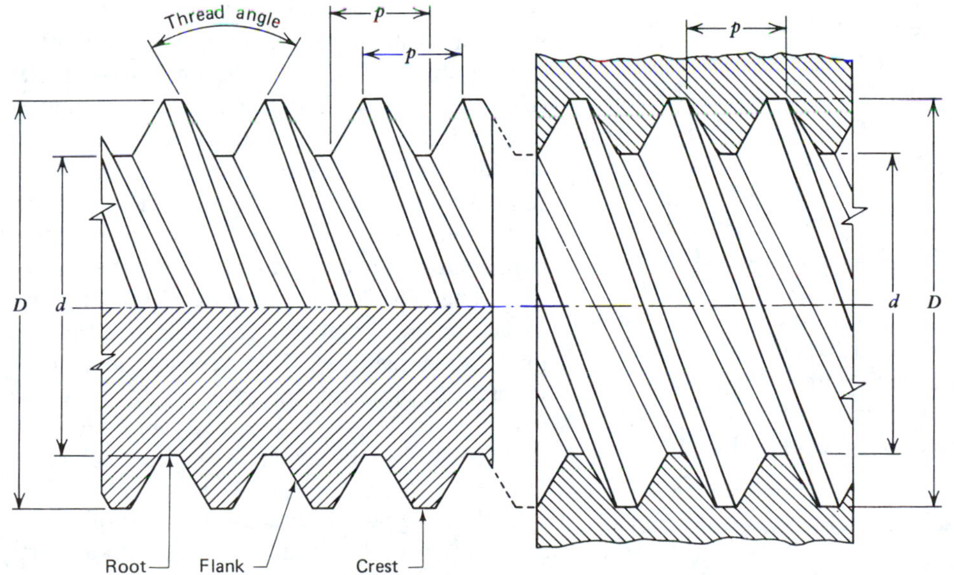

ternal or internal, it is the diameter under the thread crests that is closest to the axial centerline. Also called *root* diameter.

Pitch—p. This is the distance from a point on one thread to the corresponding point on an adjacent thread as measured in an axial direction.

Lead—L. This is the axial distance the thread advances in one revolution. Lead, then, is the distance a nut will advance on a mating screw in one revolution. On a single-thread screw, lead is equal to pitch.

Crest. The thread peaks at the major diameter.

Root. The thread peaks at the minor diameter.

Flank. Flank (or side) is the surface connecting crest and root.

Thread Angle. Thread angle is the angle between flanks of adjacent threads measured in an axial plane.

Nominal Size. This is the designation used for general identification. Nominal size is the same as basic major diameter. Actual size will vary according to manufacturing tolerances.

7-2 SCREW THREAD SYSTEMS[2]

Of the various screw-thread forms developed, those most widely used have symmetric sides inclined at equal angles to the axis. Early production methods were limited to simple cutting processes. As a result, fasteners with a single, sharp, V-thread were the only style available (Fig. 7-3*a*). The simple V-thread design is likely to fail in fatigue at the sharp corner of the root with high-strength materials. A Whitworth thread (Fig. 7-3*b*) utilizes rounded crests and roots and was the British standard for many years. American national thread (Fig. 7-3*c*) was an outgrowth of the sharp V-thread in which a 60-deg thread angle was retained but the crest

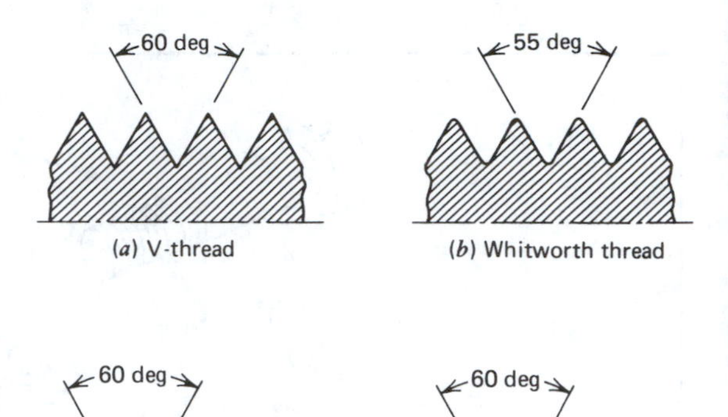

Figure 7-3 Screw thread systems.

and root were flattened. By eliminating sharp crests, both Whitworth and American national threads became stronger in fatigue. In 1948, the unified standard thread (Fig. 7-3*d*) was adopted by the United States, Canada, and Great Britain. In addition, a unified tolerance and gauging system was introduced that greatly enhanced interchangeability.

Unified standard thread is available in three basic series of diameter-pitch combinations. The coarse-thread series (UNC) is most common and is recommended for general assembly use where vibration is not a problem and where frequent disassembly is required. The fine-thread series (UNF) is somewhat stronger in tension and is more suitable where fine adjustment may be required. This series is often used in aircraft assemblies. Because of the smallness of the threads, the design of the nut is very important in high load applications to prevent thread stripping. The extra-fine-thread series (UNEF) is employed where the mating external thread is in a thin-walled member. This series is also more resistant to vibration and provides for very fine adjustments. For each thread series, the unified standard thread is available in three tolerance classes, as shown in Table 7-1.

Unified standard threads are identified on drawings, parts lists, and so forth, with a short-

---

[2]For further details see *Machinery's Handbook,* 21st Ed., p. 1256.

TABLE 7-1 Classes of Fit and Tolerance for Threaded Fasteners

	Tolerance Class				
	Unified		Metric		
Class of Fit	External	Internal	External	Internal	Remarks
Loose	1A	1B	8g	7H	Used for joints requiring frequent disassembly
Standard	2A	2B	6g	6H	General assembly work
Close	3A	3B	4g	5H	Used where accuracy and fit requirements are demanding

hand notation that includes size, thread series, class of fit, and the hand of the thread. For example, the designation

$$\tfrac{1}{4} - 20\text{UNC} - 2\text{A} - \text{RH}$$

identifies an externally threaded part having a basic major diameter of $\frac{1}{4}$ in., unified coarse thread with 20 threads per inch, tolerance class 2A, and right-hand threads. Usually the hand designation is omitted for right-hand, since these are standard. Another example is

$$\tfrac{3}{4} - 16\text{UNF} - 2\text{B} - \text{LH}$$

which designates an internal, left-hand thread having a basic major diameter of $\frac{3}{4}$ in. and 16 unified fine threads per inch of length.

The basic profile of general-purpose metric screw threads, as prescribed by the International Organization for Standardization (ISO), is shown in Fig. 7-4. Thread height is $0.47978p$ measured toward the axis from basic major diameter, and thread angle is 60 deg. On external threads, crests are truncated so that a flat length of $0.125p$ coincides with maximum major diameter. On internal threads, crests are truncated so that a flat length of $0.321p$ coincides with minimum minor diameter. ISO standards include many diameter-pitch combinations, but proposed U.S. standards specify a single diameter-pitch series. Metric threads are designated by the letter *M* followed by basic major diameter in millimeters, which is then followed by the pitch in millimeters separated by the symbol "×." For example, *M4 × 0.7* specifies a metric thread with basic major diameter of 4 mm and pitch of 0.7 mm.

7-3 MATERIALS FOR THREADED FASTENERS

Threaded fasteners often fail in fatigue; hence carbon steel and steel alloys are the most commonly used materials. The new ISO standards are now being added to the current ASTM and SAE standard steel specifications. In ISO, strength grades are known as property classes;

Figure 7-4 Basic profile for ISO general-purpose metric screw threads.

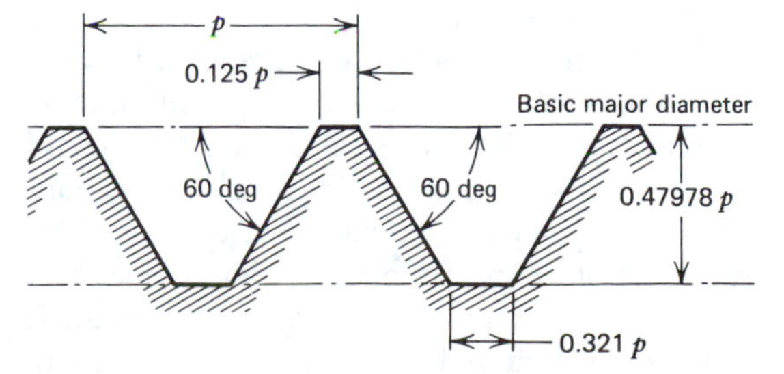

TABLE 7-2 Mechanical Properties of Fasteners—Bolts, Screws, and Studs

Property Class	Proof Stress, MPa	Minimum Tensile Strength, MPa	Minimum Yield Strength, MPa	Rockwell Hardness	
				Minimum	Maximum
4.6	225	400	240	B67	B95
4.8	310	420	340	B71	B95
5.8	380	520	420	B82	B95
8.8	600	830	660	C23	C34
9.8	650	900	720	C27	C36
10.9	830	1040	940	C33	C39
12.9	970	1220	1100	C38	C44

Source. ISO 898/1. Courtesy Industrial Fastener Institute.

this terminology will also be used by ASTM and SAE.

Table 7-2 details seven (out of nine) property classes of strength grading systems for metric bolts, screws, and studs. Property classes are identified by a two- or three-digit number. The first numeral of a two-digit number (or the first two digits of a three-digit number) are approximate minimum tensile strength (in megapascals) divided by 100. For example, property class 4.8 is a carbon steel having a minimum tensile strength of about 400 MPa (4 × 100) and a minimum strength of approximately 320 MPa (0.8 × 400). Note that a proof load stress, in addition to yield and tensile strengths, is specified for the various property classes. This is defined as the stress that a fastener must withstand without any significant deformation or failure in a specified test. Proof load stress is usually 90 to 96% of the yield strength. Property classes and proof load values for steel metric nuts are shown in Table 7-3.

The American Society for Testing Materials (ASTM) and the Society of Automotive Engineers (SAE) have established standard specifications for materials used in unified standard and metric threaded fasteners. SAE specifications classify materials into grades that must meet minimum tensile strength, proof load, and composition and treatment requirements. In addition, SAE grades are identified by a system of markings found on the bolt heads. Specifications for commonly used SAE grades of steel fasteners are given in Table 7-4. A particular SAE grade specification can be met by a number of different steels. Thus fastener manufacturers have some latitude in selecting the materials best suited to their specific production facilities.

In addition to carbon steels, many stainless steel alloys are used for threaded fasteners where improved corrosion resistance is required. Austenitic stainless steels such as Types 304 and 316 offer the best corrosion resistance, although ferritic alloys such as Type 430 are often used for economic reasons. Nickel-based superalloys such as Monel, Inconel, and Hastelloy are used for fasteners where strength at high temperatures as well as corrosion resistance is required. Aluminum, bronze, and brass are also used for threaded fasteners. In corrosive environments where great strength is not required, fasteners made of nylon and other plastics are both suitable and economical.

7-4 COMMON TYPES OF THREADED FASTENERS

There are many fastener configurations available, but the major difference among them is the

TABLE 7-3 Nut Proof Load Values, kN

Nominal Diameter and Thread Pitch	Thread Stress Area, mm^2	Property Class			
		5	9	10	12
M4 × 0.7	8.78	4.57	7.90	9.13	
M5 × 0.8	14.2	8.23	13.0	14.8	16.3
M6 × 1	20.1	11.7	18.4	20.9	23.1
M8 × 1.25	36.6	21.6	34.4	38.1	42.5
M10 × 1.5	58.0	34.2	54.5	60.3	67.3
M12 × 1.75	84.3	51.4	80.1	88.5	100
M14 × 2	115	70.2	109	121	137
M16 × 2	157	95.8	149	165	187
M20 × 2.5	245	154	225	260	294
M22 × 2.5	303	—	—	—	—
M24 × 3	353	222	325	374	424
M27 × 3	459	—	—	—	—
M30 × 3.5	561	353	516	595	673
M36 × 4	817	515	752	866	980
M42 × 4.5	1120	706	1030	—	1340
M48 × 5	1470	920	1350	—	1760
M56 × 5.5	2030	1280	1870	—	2440

Source. ASTM, A563M-80. American Society for Testing Materials. Reprinted with permission.

TABLE 7-4 Mechanical Requirements and Identification Marking for Bolts, Screws, and Studs

		Full-Size Bolts, Screws, and Studs				
Grade Designation	Product	Nominal Size, Diameter in.	Proof Load (Stress) psi	Tensile Strength, Minimum psi	Material	Grade Marking
2	Bolts Screws Studs	$\frac{1}{4}-\frac{3}{4}$	55,000	74,000	Low-carbon steel	None
		Over $\frac{3}{4}-1\frac{1}{2}$	33,000	60,000		
5	Bolts Screws Studs	$\frac{1}{4}-1$	85,000	120,000	Medium-carbon steel, quenched and tempered	
		Over $1-1\frac{1}{2}$	74,000	105,000		
7	Bolts Screws	$\frac{1}{4}-1\frac{1}{2}$	105,000	133,000	Medium-carbon alloy steel, quenched and tempered. Rolled threads after heat treatment	
8	Bolts Screws Studs	$\frac{1}{4}-1\frac{1}{2}$	120,000	150,000	Medium-carbon alloy steel, quenched and tempered	

Source. SAE J429k. Reprinted with permission, copyright © 1981, Society of Automotive Engineers, Inc.

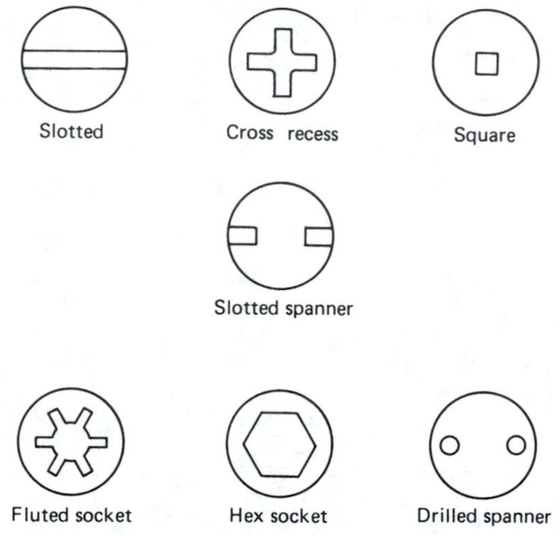

Figure 7-5 Various fastener head styles.

size and shape of the fastener head. The question "Why do the various threaded fasteners have different heads?" is often raised. The answer involves considerations such as torque application, appearance, and safety. Many of the various fastener head types are depicted in Fig. 7-5. Small fasteners can be tightened sufficiently with a screwdriver and thus need only a slotted head. Driving slots such as a cross recess (Phillips head) were devised to help retain a screwdriver in the slotted head during torque application. Larger fasteners require a square or hexagonal head to develop higher torques through the use of wrenches. A combination of high strength and small head size was realized with the development of the 12-point head bolt (Fig. 7-6).

Protruding heads may affect appearance or safety and are often replaced by flat head or socket head screws that can be placed flush with the surface. The preponderance of fasteners with square or hexagonal heads is due to manufacturing techniques, not design. Only in recent years has progress in cold heading and metallurgical skills led to the development of the varied and more functional head designs.

The following discussion of major types of threaded fasteners is not complete, but it should lead to a better understanding of designers' options.

Figure 7-6 Twelve-point head bolt. (Courtesy Deere and Company Technical Services.)

Bolts

A bolt is a headed, threaded fastener that can be used with or without a nut. A nut is a small, symmetrically shaped piece containing matching internal threads. Bolts are available with square or hexagonal heads and in standard lengths. The length of the threaded portion of bolts has been standardized (Fig. 7-7). For unified standard fasteners up to 6 in. in total length, the thread length is twice the basic major diameter plus $\frac{1}{4}$ in. For metric fasteners up to 125 mm in total length, the thread length is twice the basic major diameter plus 6 mm.

When a bolt passes through two or more parts and clamps them between the bolt head and a nut, it is referred to as a through bolt (Fig. 7-8*a*). Since access is required on both sides, through bolts are suited to symmetric connections and are employed whenever possible because it costs less to use a nut than to drill and tap a hole.

Figure 7-7 Standard thread length for bolts.
$L < 6$ in. : $x = \frac{1}{4}$ in.
$L > 6$ in. : $x = \frac{1}{2}$ in.
When bolts are too short for formula thread length, thread will extend as close to neck as possible.

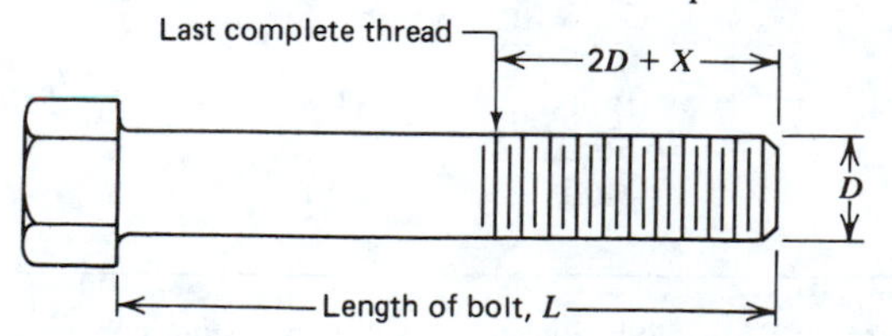

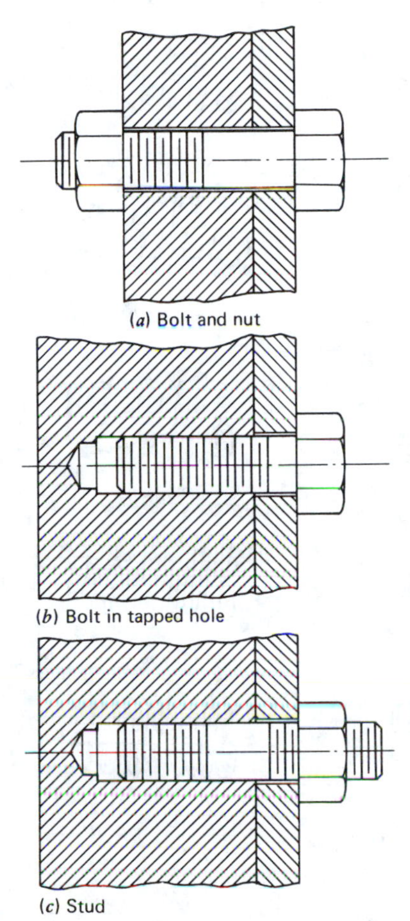

Figure 7-8 Most applications of screw fasteners fall into one of the three categories shown. The latter two are merely modifications of the bolt and nut principle.

By definition, a *tap bolt* (Fig. 7-8*b*) is not retained by a nut. It develops a clamping force between the bolt head and a threaded hole in one of the connected parts.

In contrast to through bolts and tap bolts, a stud (Fig. 7-8*c*) does not have a head but is simply a rod with threads at both ends. One end is retained in a tapped hole, while the other end is used with a nut. The primary advantage of studs over bolts is their ability to compensate partially for misalignment of the tapped hole and the clamped surface. Studs also act as pilots to facilitate the assembly of heavy parts such as engine cylinder heads. Studs are well suited for automated assemblies because they permit quick and easy stacking of gaskets, shims, and other components.

Screws

A screw is a tensile fastening device that allows proper tightening by torquing the head. Screws generally have a multiplicity of head and tip configurations, as outlined next.

Cap Screws. A cap screw is the same as a tap bolt except that it is available with various types of heads for specific uses. In addition to hexagonal heads, there are filister, button, flat, and socket heads. Filister, button, and flat head styles are provided with a screwdriver slot, thus limiting the tightening torque that can be applied. Originally designed to replace studs for small covers on steam engines, they were simply designated as cap screws.

Machine Screws. Machine screws are small cap screws that can be tightened exclusively by a screwdriver. The slotted head assumes a wide variety of shapes to suit various design conditions (Fig. 7-9).

Tapping Screws.[3] A tapping screw is essentially a tap and a screw in one piece, the outer end being the tap (Fig. 7-10). It cuts and forms a mating internal thread when screwed into a hole. The mating thread fits closely and keeps the screw tight, even against vibration. Tapping screws are used primarily in sheet metal, where little tapping is required.

Completing a screw joint requires three distinctly different operations: drilling or punching a clearance hole; tapping an internal thread; and torquing the screw itself. Since it is in-place cost that counts and not the fastener cost, there are obvious advantages to combining two, or per-

[3]Often referred to as self-tapping screws.

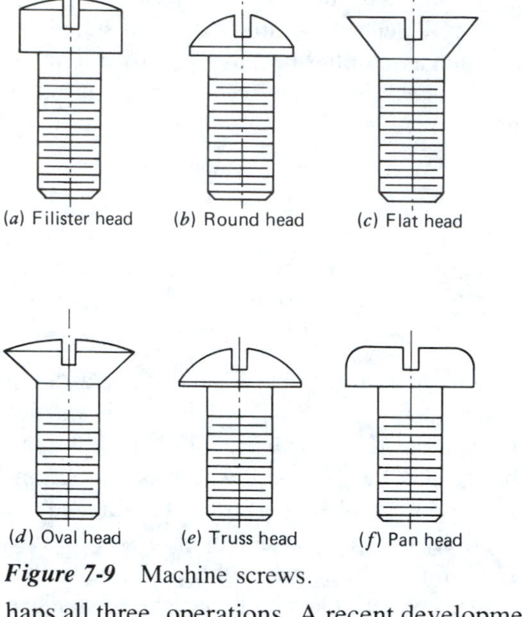

Figure 7-9 Machine screws.

haps all three, operations. A recent development is thus the *self-drilling and self-tapping screw.* These fasteners have a drilling point on the end, tapping threads on the body, and may have locking serrations under the head (Fig. 7-11).

7-5 NUTS AND LOCKING DEVICES

Nuts for threaded fasteners are available in a variety of standard styles to suit specific applications (Figs. 7-12 and 7-13). Whenever possible, use of special designs should be avoided in favor of commercially available styles.

The most common design is the *hexagon nut,* which is available in plain, slotted, and castellated styles, with or without an annular bearing surface on one end. Slotted and castellated hex nuts are designed for use with some types of retaining devices that will be discussed later. *Square nuts* are rough, unfinished nuts most often used with square head bolts or carriage bolts. A *cap nut* is a variation of the hex nut in which one end is enclosed by a crown that covers the tip of the bolt. Cap nuts are used to enhance appearance and safety and to protect the portion of external thread that would otherwise be exposed.

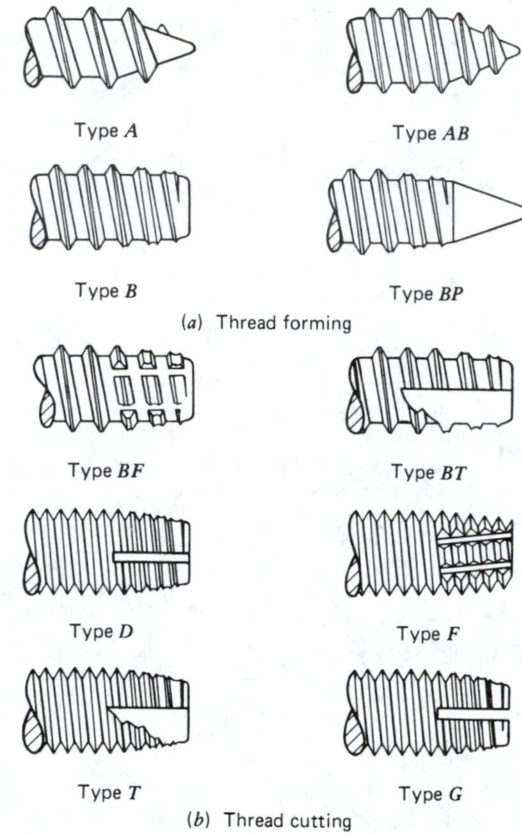

Figure 7-10 Self-tapping screw point types.

Figure 7-11 A self-drilling and self-tapping screw. (Courtesy Elco Industries, Inc.)

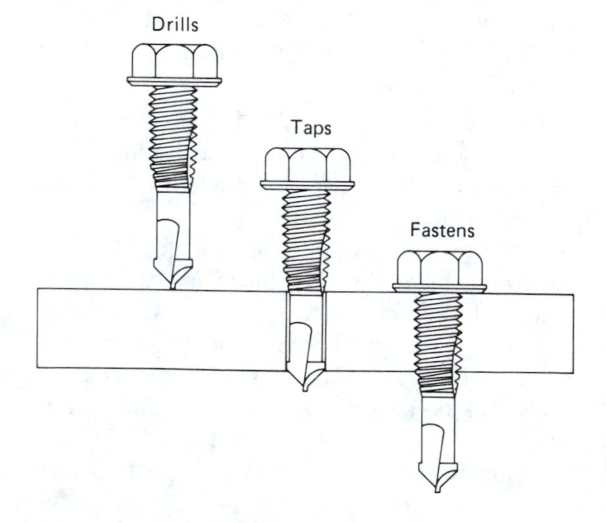

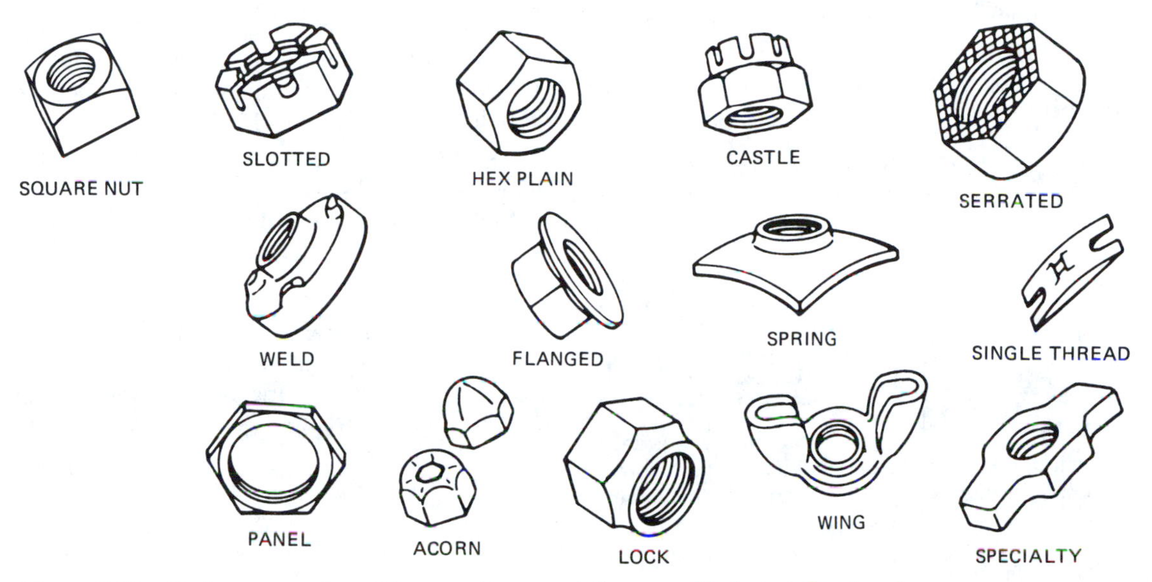

Figure 7-12 Various nut styles. (Courtesy Deere and Company Technical Services.)

Wing nuts are designed for tightening and loosening by hand. They are used where high tightening torque is not required and frequent assembly and reassembly are necessary, such as on access covers and safety guards.

Screw fasteners, despite their inherent self-locking design, cannot remain tight for long when subjected to dynamic loads. Nuts are very prone to loosening unless secured by some additional locking device. Friction locking devices resist rotation by gripping the mating thread or material of the connected parts. Interlocking devices, referred to as positive locking devices, are also used. With these fasteners a material failure is a necessary prelude to loosening.

Possibly the simplest locking method is the use of two standard nuts, with locking action provided by tightening one against the other. This method is highly efficient; in fact, studs are often withdrawn in this manner after having rusted in place for a number of years. In practice, a thin *jam nut* (Fig. 7-14) is used with a nut of

Figure 7-13 Specialty nuts. (Courtesy Deere and Company Technical Services.)

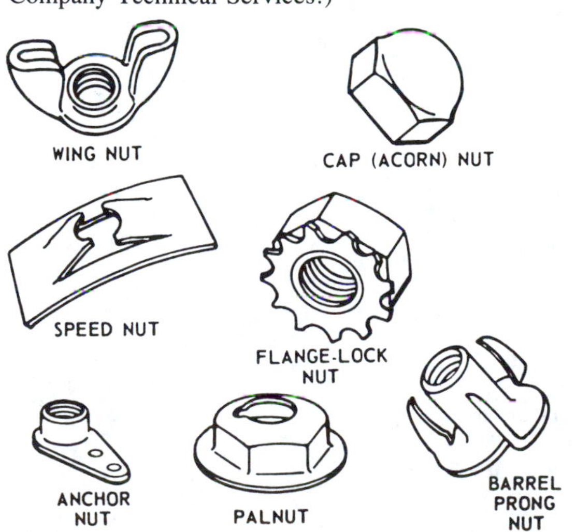

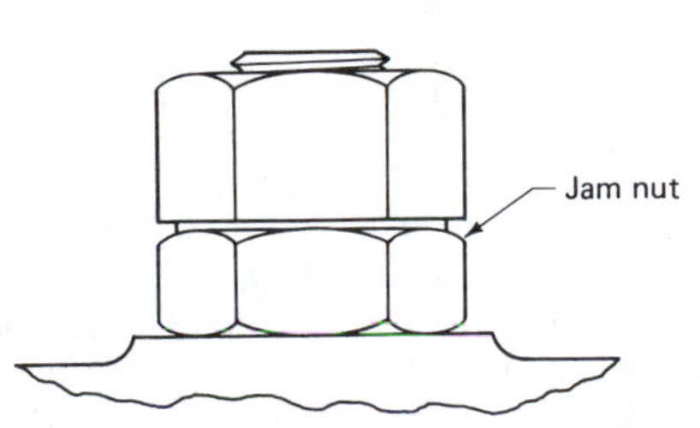

Figure 7-14 Nut locking with a jam nut.

regular thickness to reduce the connection's overall height.

Locknuts

Prevailing torque nuts employ a self-contained locking feature such as deformed or undersize threads, variable lead angles, plastic or fiber washers, or plug inserts. This type of nut resists screwing on or unscrewing and does not depend on bolt load for locking (Fig. 7-15).

Free-spinning torque nuts develop their locking action *after* the nut has been seated by reactive spring force against the threads or by friction

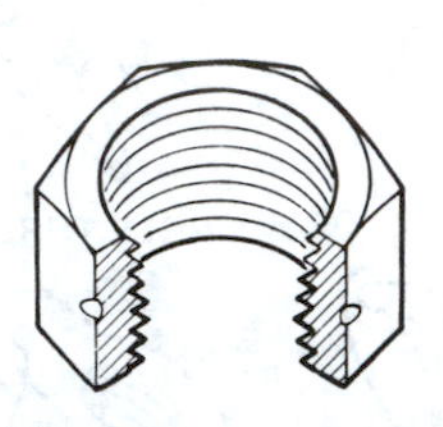

Distorted portion of nut thread produces an interference fit. Center dimple allows nut to be assembled with either side up

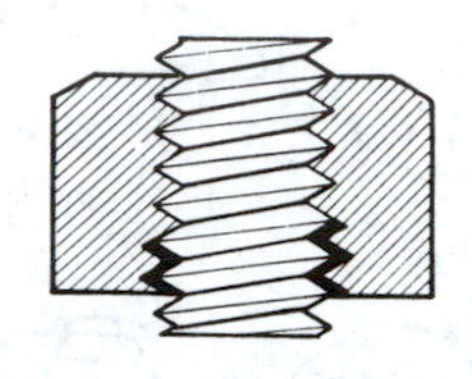

Top-crimp nut is easy to start but must be properly oriented before assembly

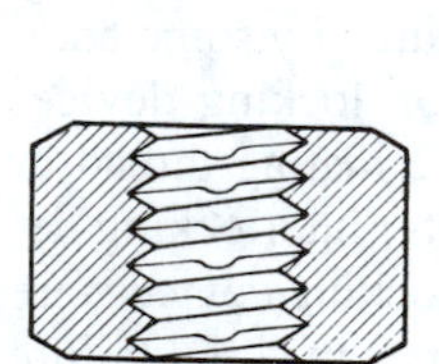

Thread profile is distorted, increasing interference. Complex manufacturing process increases cost

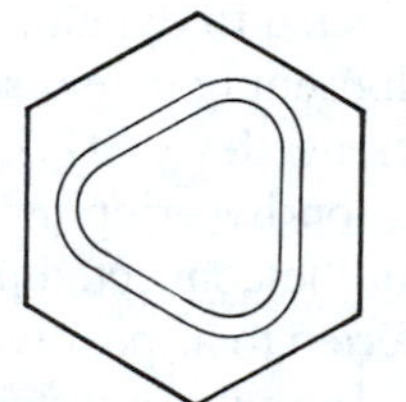

Nut hole, forced into an out-of-round shape after initial forming, produces spring action that maintains an interference fit after assembly

(a) **DISTORTED SHAPE**

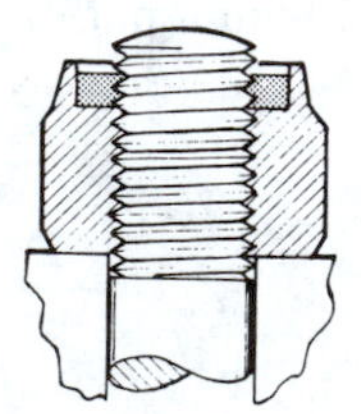

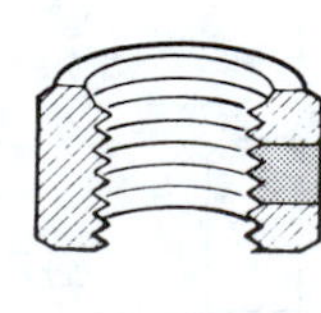

Plastic or metal section is added to the nut to increase mating-thread friction

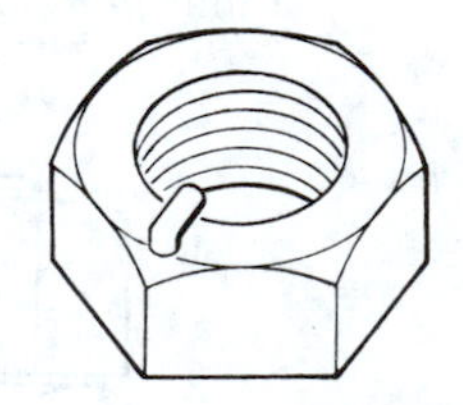

Metal-insert locknut has a projecting hardened wire or pin built in to provide a ratchet-like locking action. Reuse is limited by wear of the pin tip.

(b) **INSERT**

Figure 7-15 Prevailing-torque locknuts incorporate features that increase friction between mating threads. (Courtesy *Machine Design*.)

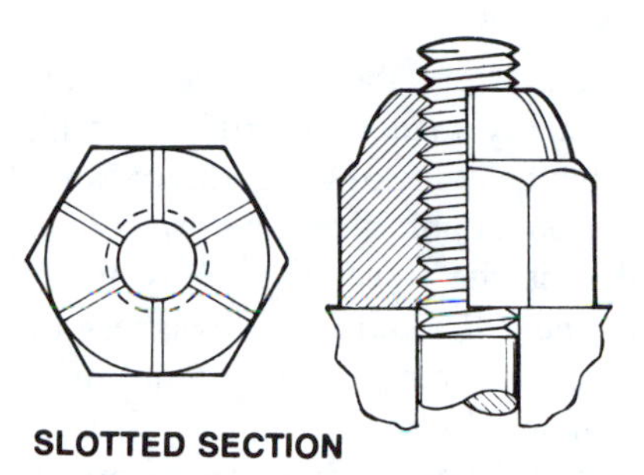

SLOTTED SECTION

Spring arms formed on the domed nut top are deflected inward. When the nut is threaded on, these arms grip the bolt threads.

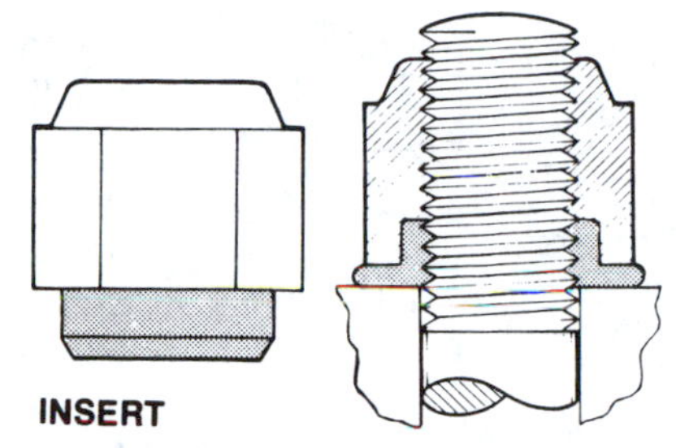

INSERT

A plastic or metal washer built into the nut base is permanently deformed and grips the bolt threads when the nut is seated.

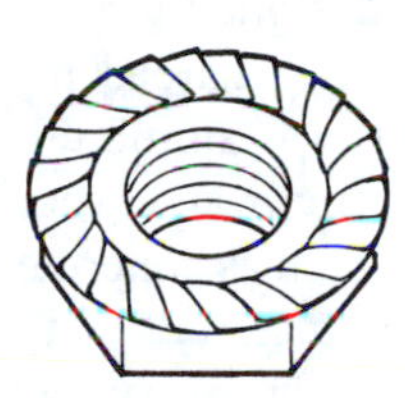

SERRATED FACE

Serrated or grooved face of nut digs into the bearing surface during final tightening.

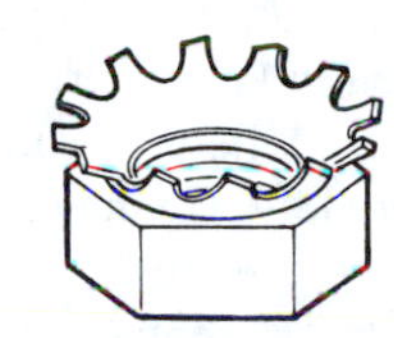

CAPTIVE WASHER

Toothed or spring washer attached to the bearing face of the nut increases friction between the bearing surface.

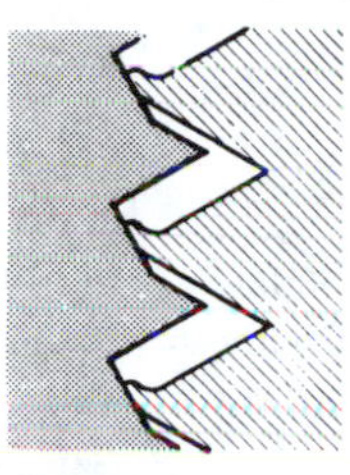

MODIFIED THREAD

Nut threads have a modified cross section which crimps the bolt-thread crests when clamp load is applied.

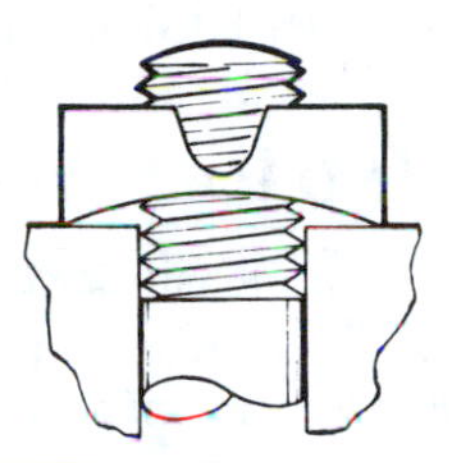

SPRING HEAD

When fully seated, the concave portion of the nut is forced inward and clamps against the bolt threads.

Figure 7-16 Free-spinning locknuts turn freely on the bolt until seated. Additional turning activates the locking mechanism. Free-spinning nuts improve holding power by creating an interference fit, increasing the friction at the bearing face, or by a combination of the two effects. (Courtesy *Machine Design*.)

against the bearing surface (Fig. 7-16). Slotted and castellated nuts become locknuts when used with a *cotter pin* or *wire* that passes through a hole in the shank of the externally threaded member (Fig. 7-17). The nut cannot then rotate unless a shear failure of the pin or wire occurs. This arrangement is found on most automobile front wheels and steering linkages.

Nuts and bolt heads can also be locked with a *locking plate* (Fig. 7-18) when bolt spacing is close. After the bolt or nut is tightened, the edges or corners of the plate are bent up against flats of

Figure 7-17 How a cotter pin is used to secure a castle or slotted nut. (Courtesy Deere and Company Technical Services.)

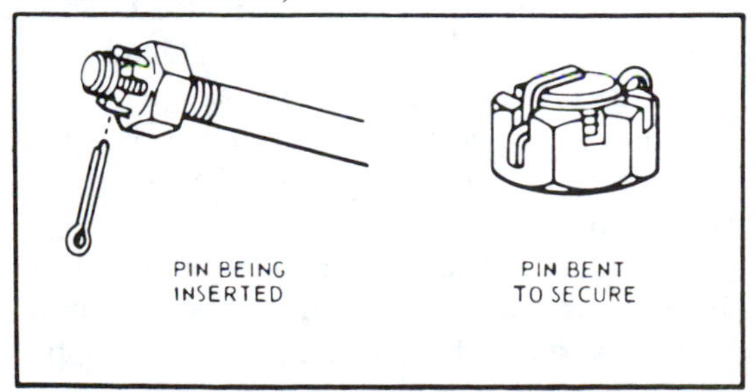

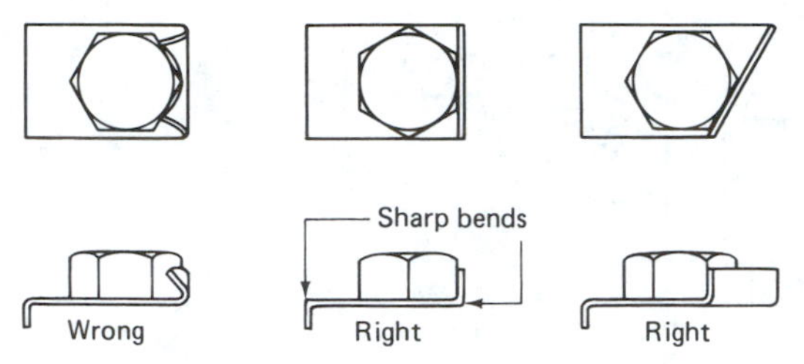

Figure 7-18 Application of flat metal locks.

the hexagon to prevent rotation. Locking plates have a disadvantage in that the soft material required for bending will also yield when the fastener is seated. Thus it may be difficult to develop sufficient fastener tension.

7-6 WASHERS AND LOCK WASHERS

Most screw fasteners are tensile devices sensitive to bending loads. *Washers* (Fig. 7-19) are often used with threaded fasteners to provide a better bearing surface for nuts and bolt heads, to provide a bearing surface over large clearance holes or slots, to distribute the load over a larger area, to prevent marring of parts during assembly, to improve torque-tension ratio (by reducing friction), and to provide locking, in some cases through spring action. *Flat washers* are thin, annular-shaped disks used primarily for bearing surface and load distribution and have no locking capability. *Conical spring washers* are made of hardened and tempered steel that is slightly dished. Conical washers deform when the bolt is tightened and act as springs that compensate for small losses in bolt tension due to thermal expansion or compression set of the gaskets. *Helical spring lockwashers* are essentially single-coil helical springs that flatten under load. Spring action assists in maintaining the bolt load, while split edges provide some locking action by biting into the bearing surfaces. These lockwashers are generally made of hardened steel or bronze and aluminum alloys.

Toothed lockwashers provide locking action by biting into the material of the bolt head or nut and the surface of the connected part. Gripping action results from the teeth being deformed axially. Usually made of hardened steel, they provide the best locking action of any lockwasher of comparable size. The *external toothed* lockwasher is the most commonly used type, although *internal toothed* lockwashers may be used where appearance is important. When the clearance hole is large, *internal-external toothed* lockwashers may be used; internal teeth grip the bolt head or nut and external teeth span the clearance hole to grip the work piece.

7-7 THREAD INSERTS

A thread insert is a special nut designed to perform the function of a tapped hole. Thread inserts are used in (1) materials such as aluminum or magnesium where thread strength and wear resistance are low, (2) materials such as concrete that cannot be threaded, and (3) materials with damaged internal threads, where they serve as repair elements. A variety of thread inserts are commercially available, and a few examples of the most common types are depicted in Fig. 7-20. The solid bushing is designed to be pressed into a drilled hole in soft materials such as wood and plastic. The expansion type (often referred to as an anchor) is most commonly used in concrete. The tapered portion containing threads is steel; the outer shell is lead. A series of blows applied with a setting tool expands the lead shell, which then grips the inside surface of the hole tightly. Helical coil wire inserts and self-tapping bushings are used in soft metals to provide more durable threads.

7-8 PRELOAD, FATIGUE, AND RESILIENCY

One of the most critical factors in joint reliability is preload—the force a tightened fastener exerts on an assembly. A preload ensures optimum performance if it prevents the clamped parts from separating in service. Thus a preload should always exceed any external load or payload. In

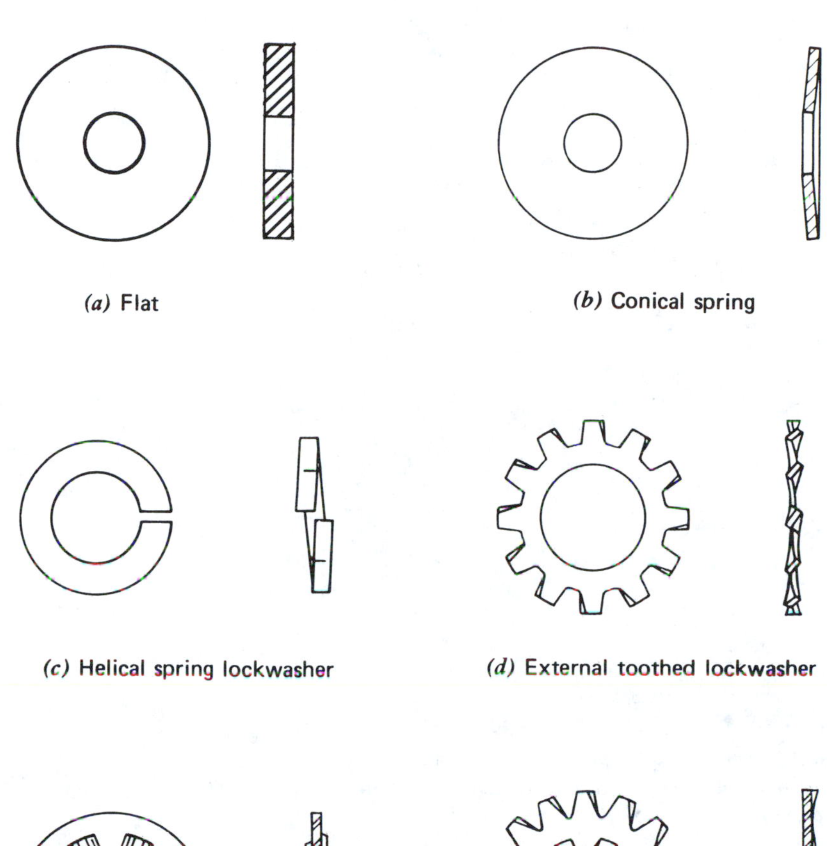

(a) Flat

(b) Conical spring

(c) Helical spring lockwasher

(d) External toothed lockwasher

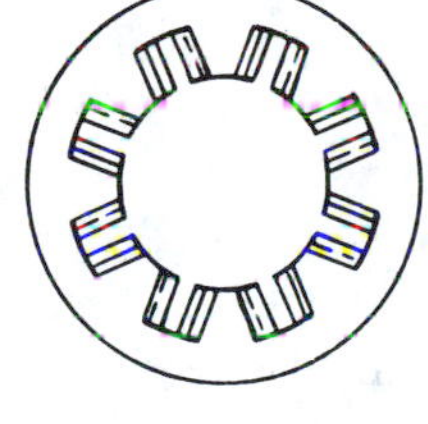

(e) Internal toothed lockwasher

(f) Internal—external toothed lockwasher

Figure 7-19 Washers.

Figure 7-20 Thread inserts.

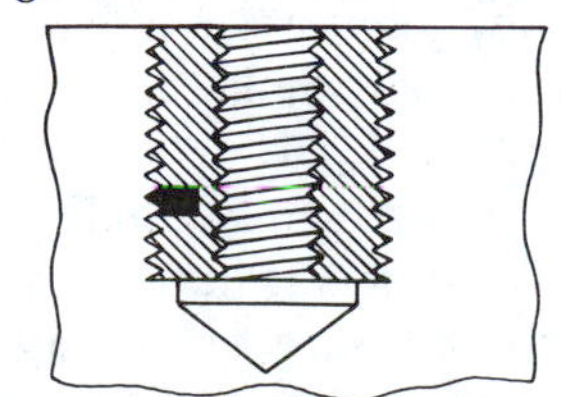

(*a*) Solid bushing with plastic inserts

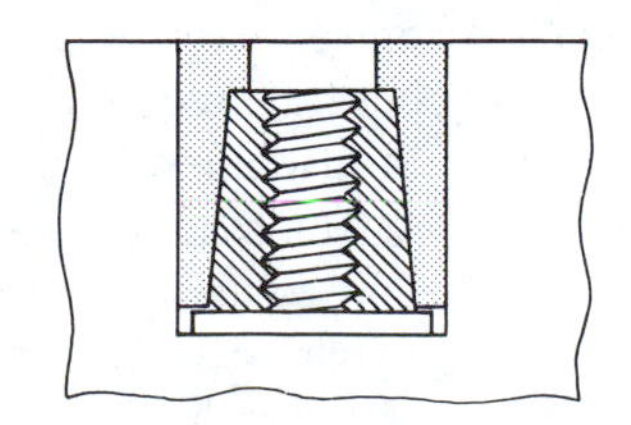

(*b*) Expansion anchor

(*c*) Helical coil

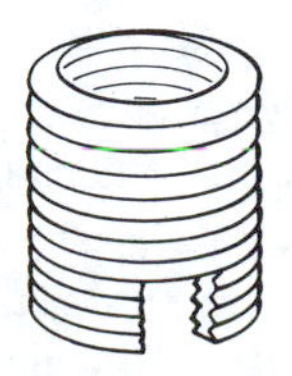

(*d*) Self-tapping

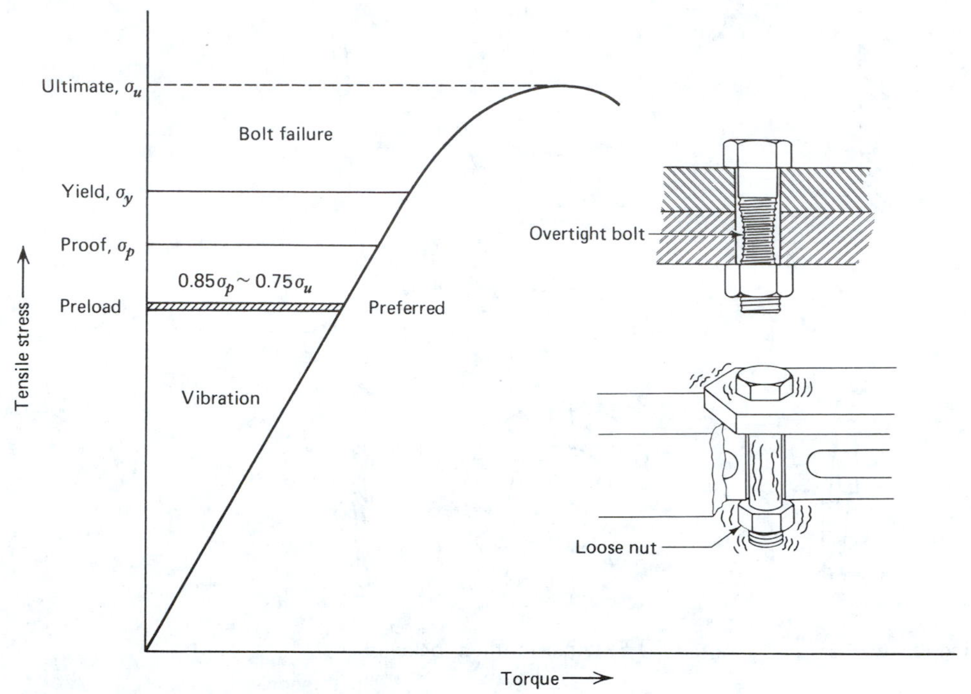

Figure 7-21 Tension-tightening torque graph for bolts. Improper use is a constant source of trouble with bolts and screws. Undertorqued bolts will shake loose, while overtorqued bolts will become permanently damaged and weakened. (Courtesy Deere and Company Technical Services.)

fact, recommended practice in most manufacturing industries, except aircraft, calls for a fastener to be tightened to 85% of proof load (or 75% of yield strength). Proof loads are 5 to 10% below the yield point of the fastener and 35% below tensile strength (Fig. 7-21).

Seemingly, this is a poor utilization of any material, since only 15% of total load capacity is left for payload. However, the obvious is not always true because, generally, the bolt load remains unchanged until the external load exceeds the preload. Therefore the higher the preload, the greater potential there is for withstanding larger external loads. These remarks apply to perfectly rigid joints, which solid, metal-to-metal joints approximate. A gasketed joint reacts differently, but then preload is minimized to prevent damage to the gasket material. Failure of such a joint is leakage, not bolt breakage.

The other extreme, a small preload perhaps no larger than the payload, may easily become more damaging. It may cause leaks at joints that were supposed to be tight. Worse, it can induce vibrations that may effectively reduce the preload or cause fatigue failure.

A high preload helps retain friction at the interface, which is important when shear loads are present. Friction at screw joints is an effective but little recognized means of damping vibration in machinery.

Machinery in action is a dynamic condition; consequently, most screw fasteners have a small dynamic load superimposed on a much larger static preload. This is a type of fatigue loading

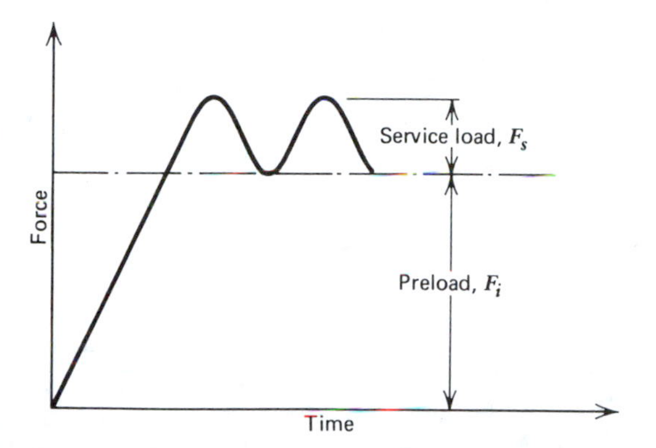

Figure 7-22 Typical fatigue load. This is a simplified approach. The actual load is less than the combined load ($F_i + F_s$). The preload is due to the elastic reaction of the clamped parts. The effect of the service load is partially to relieve the compressive load thereby making the total load less than the combined load.

known as fluctuating (Fig. 7-22). Although fluctuating loads are the least severe, they are obviously augmented by stress concentration and, to a smaller degree, by bending. The geometry of all screw-thread systems is such that the stress concentration factor of rolled threads is roughly 3.0. For cut threads it is around 3.8 or 25% higher. Most bolts fail in fatigue at or near the last thread because this thread does not receive full benefit of stress-field interference from other threads. Hence it is considered good design practice to reduce the body diameter of the bolt to slightly less than the root diameter of the thread. Since nearly all commercially available fasteners have cold-rolled threads, fatigue strength has been much improved by (1) improving grain flow, and (2) closing all microcracks (Fig. 7-23).

Fatigue strength in short bolts can be increased by adding resilience.[4] Usually this is done by reducing the diameter of the nonthreaded section or by drilling a hole axially (Fig. 7-24). Normally, bolt elongation is concentrated largely in

[4]*Resilience* is the ability of a strained body to recover its form and size after deformation.

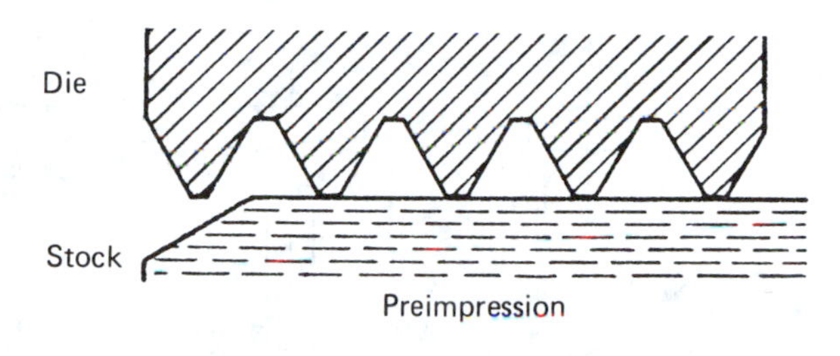

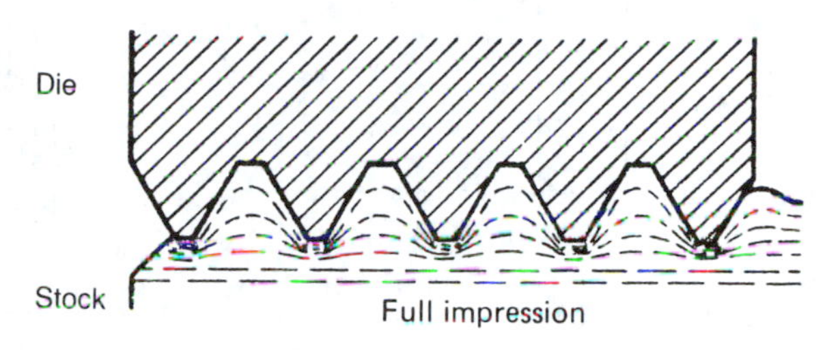

(a) Rolling screw threads is essentially a forging process producing the finished part by displacing metal to conform with the contour of the dies. The root of the thread work-hardens to a greater extent than the threads, as indicated by the density of the lines.

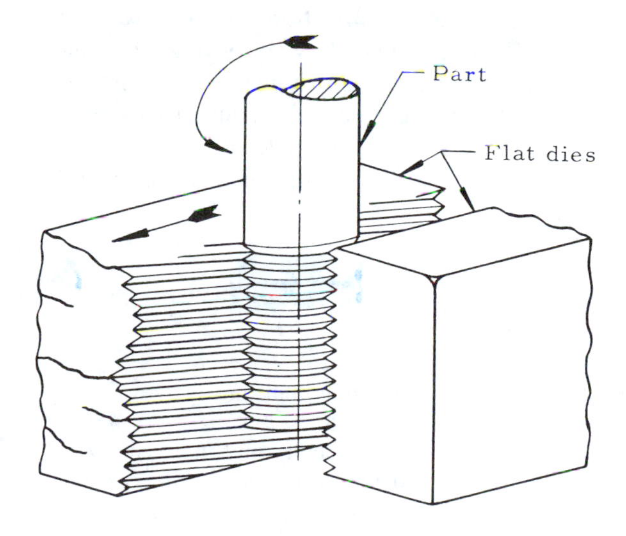

(b) As one die is moved past the other die, the thread profile is impressed into the surface of the rod.

Figure 7-23 Rolled threads. (Courtesy General Motors Corporation.)

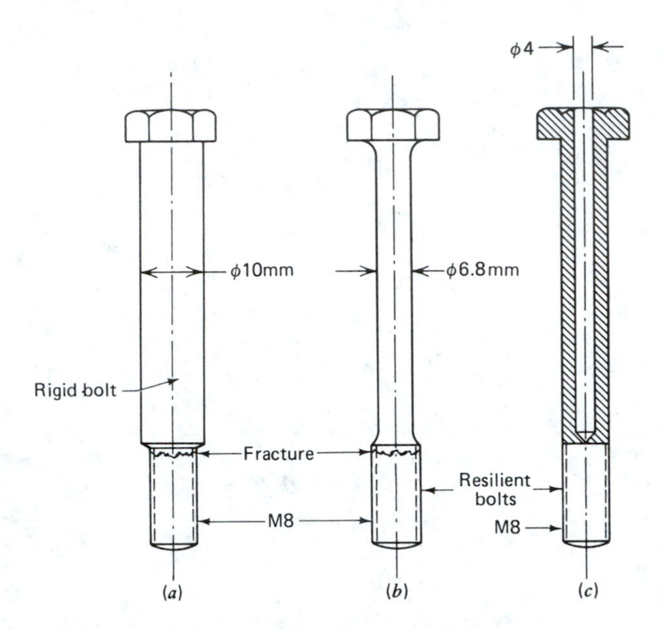

Figure 7-24 Fatigue strength of fasteners. The two resilient bolts have a fatigue strength of more than three times that of the rigid bolt.

the threaded section. When body area is reduced so that it equals stress area, total elongation becomes evenly distributed over the entire length and is greater as a whole. Bolts of this type are used in large engines to provide a flexible means of bolting the main sections together. Resilience can also be increased by specifying a bolt material of greater strength (higher elastic limit). Bolt length can be extended artificially by means of a tubular section placed under the bolt head.

Under load, a bolt will stretch according to the formula

$$\delta = \frac{FL}{EA_s} \tag{7-1}$$

where

δ = elongation; mm, in.
F = axial load; N, lb
L = length of bolt; mm, in.
E = modulus of elasticity; MPa, psi
A_s = root area; mm^2, in.^{2}

Knowing the exact amount of bolt elongation during service is important in the selection of gaskets, as we shall see in Chapter 21, Gaskets and Seals.

7-9 STRESS CONSIDERATIONS FOR THREADED FASTENERS

A threaded fastener may fail in three distinctive modes (Fig. 7-25): the fastener may fail in tension; the external threads may fail in shear by being stripped; or the internal threads may fail in shear by being stripped. The actual mode of failure depends on relative material strengths and height of the nut h. For connections that use an internally threaded component other than a nut, h will be understood to represent length of thread engagement of internal and external threads.

The tensile stress area for standard metric fasteners is given by

$$A_s = 0.25\pi(D - 0.9382p)^2 \text{ mm}^2 \tag{7-2}$$

where D is basic major diameter and p is pitch. Values of A_s corresponding to standard thread sizes are given in Tables 7-5 and 7-6.

For unified standard threads,

$$A_s = 0.25\pi(D - 0.9743p)^2 \text{ in.}^2 \tag{7-3}$$

Tensile stress is then given by

$$\sigma_t = \frac{F}{A_s} \tag{7-4}$$

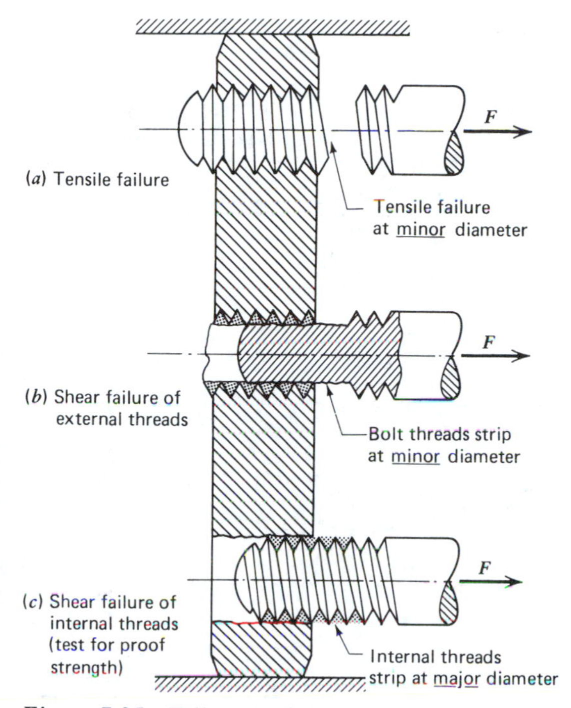

Figure 7-25 Failure modes of threaded fasteners.

where F is *total* tensile load on the bolt, including both preload and a portion of the external load (Fig. 7-25*a*).

Stripping of external thread occurs when bolt threads fail in shear at the minor diameter of internal thread (Fig. 7-25*b*). Similarly, stripping of internal thread is a shear failure at the major diameter of external thread (Fig. 7-25*c*).[5] Corresponding thread shear areas per unit of engaged thread are also given in Tables 7-5 and 7-6.

The minimum thread shear area for *external* threads is AS_s; the minimum thread shear area for *internal* threads is AS_n, where subscripts denote screw and nut, respectively. Shear stress in bolt thread is

$$\tau = \frac{F}{(AS_s)\,h} \tag{7-5}$$

[5]This stripping test, with a hardened-threaded steel mandrel, is a measure of thread strength in a nut and is known as the *proof strength*.

while shear stress in nut (internal thread) is

$$\tau = \frac{F}{(AS_n)\,h} \tag{7-6}$$

Since the shear area of internal thread is greater than that of the external thread, stripping of bolt thread will usually precede stripping of nut thread. For this reason, nut strength and hardness are purposely made lower than those of the bolt. This affords some cost savings in material and processing and, at the same time, guards against a failure of the bolt. This is seen to be true when considering the bolt as a tension member with large stress concentrations and the nut as a compression member where the stress concentrations are not as significant. By causing the nut to deform first, the bolt is protected.

Standard threaded fasteners are designed such that the mode of failure will be that of tensile failure of the bolt (preferential mode of failure). This is accomplished by making nut height large enough (80 to 90% of basic major diameter) to ensure that bolt threads are stronger in shear than the bolt cross section is in tension. In connections where the nut is replaced by an internally threaded hole, Eq. 7-4 will determine the required length of engagement. In general, length of thread engagement for steel in steel should be no less than the basic major diameter. For brass, bronze, and cast iron, length of engagement should be 1.5 times the basic major diameter. Aluminum, zinc, or plastic should have a thread engagement of at least twice the basic major diameter. For the latter materials it is often safer, but more expensive, to use thread inserts.

EXAMPLE 7-1

An M10 × 1.5 bolt is fitted with a standard hex nut and used in a bolted connection. Bolt material satisfies the requirements of property class 4.6, and nut material corresponds to property class 5. What is the maximum total tensile load that the connection can withstand without failure?

TABLE 7-5 Thread Stress Areas

Basic Major Diameter and Pitch	Tensile Stress Area, mm^2 A_s	Thread Shear Area per mm of Engaged Threads, mm^2	
		AS_s (screw)	AS_n (nut)
M4 × 0.7	8.78	5.47	7.77
M5 × 0.8	14.2	7.08	9.99
M6 × 1	20.1	8.65	12.2
M8 × 1.25	36.6	12.2	16.8
M10 × 1.5	58.0	15.6	21.5
M12 × 1.75	84.3	19.0	26.1
M14 × 2	115	22.4	31.0
M16 × 2	157	26.1	35.6
M20 × 2.5	245	33.3	45.4
M22 × 2.5	303	37.0	50.0
M24 × 3	353	40.5	55.0
M27 × 3	459	46.2	62.0
M30 × 3.5	561	51.6	69.6
M36 × 4	817	61.3	84.1
M42 × 4.5	1120	74.3	99.2
M48 × 5	1470	85.8	114
M56 × 5.5	2030	101	134

TABLE 7-6 Stress Areas for Unified Standard Fasteners

Size, Threads per Inch, and Series	Tensile Stress Area, in.^{2} A_s	Thread Shear Area per in. of Engaged Threads, in.^{2}	
		AS_s (screw)	AS_n (nut)
$\frac{1}{4}$–20 UNC	0.0318	0.4615	0.6743
$\frac{5}{16}$–18 UNC	0.0524	0.5946	0.8435
$\frac{3}{8}$–16 UNC	0.0775	0.7242	1.0132
$\frac{7}{16}$–14 UNC	0.1063	0.8486	1.1833
$\frac{1}{2}$–13 UNC	0.1419	0.9819	1.3527
$\frac{9}{16}$–12 UNC	0.1819	1.1128	1.5223
$\frac{5}{8}$–11 UNC	0.2260	1.2407	1.6938
$\frac{3}{4}$–10 UNC	0.3345	1.5121	2.0323
$\frac{7}{8}$–9 UNC	0.4617	1.7783	2.3730
1–8 UNC	0.6057	2.0374	2.7144
$1\frac{1}{8}$–7 UNC	0.7633	2.2863	3.0551
$1\frac{1}{4}$–7 UNC	0.9691	2.5809	3.3952
$1\frac{3}{8}$–6 UNC	1.1549	2.8147	3.7373
$1\frac{1}{2}$–6 UNC	1.4052	3.1092	4.0776
$1\frac{3}{4}$–5 UNC	1.8995	3.6132	4.7604
2–$4\frac{1}{2}$ UNC	2.4230	3.9920	5.1864

SOLUTION

The maximum load that the bolt can withstand is the product of the proof stress and the stress area. For M10 × 1.5,

$$A_s = 58.0 \text{ mm}^2 \qquad \text{(see Table 7-5)}$$

For property class 4.6, the proof stress is

$$225 \text{ MPa (see Table 7-2)}$$

The proof load is

$$F = (58 \text{ mm}^2)(225 \text{ N/mm}^2)$$

$$= \mathbf{13.05\ kN}$$

The nut proof load, according to Table 7-3, is 34.2 kN, which is greater than the bolt proof load by 162%. This verifies that screw fasteners are designed with tension as a preferential mode of failure.

EXAMPLE 7-2

A 1/2–13UNC–2A bolt is used as part of a cylinder head to an engine block assembly. Internal thread is tapped in the engine block, which is made of an aluminum alloy having yield strength in shear of 6600 psi. Bolt material is SAE Grade 2 steel. Determine the required length of thread engagement that will ensure that tensile failure occurs prior to thread stripping.

SOLUTION

From Table 7-4, minimum tensile strength of SAE Grade 2 material is 74,000 psi for this size fastener. Tensile load at failure is then

$$F = S_{ut}A_s = (74{,}000 \text{ lb/in.}^2)(0.1419 \text{ in.}^2)$$

$$= 10{,}500 \text{ lb}$$

where A_s is obtained from Table 7-6.

Rewriting Eq. 7-6 in the form

$$h = \frac{F}{\tau(AS_n)}$$

substituting F = 10,500 lb, obtaining AS_n from Table 7-6, and setting τ equal to the yield strength of internal thread gives

$$h = \frac{10{,}500 \text{ lb}}{(6600 \text{ psi})(1.3527 \text{ in.})} = 1.18 \text{ in.}$$

This procedure gives a length of engagement for which tensile failure of the bolt and stripping of internal thread would occur simultaneously. Thus h should be increased to provide a margin of safety. For this example, h = **1.5 in.** should be adequate.

In addition to tensile and shear stresses on the bolt and threads, consideration should be given to *compressive* stress that arises at the bearing surface. The tensile load on the bolt is transferred to connected parts through the bolt head and nut. Thus the so-called bearing stress is a function of tensile load and contact area of the bolt head or nut. In order to prevent permanent deformation of bearing surfaces of connected parts, bearing stress should be less than the yield strength of the material in compression.

7-10 THE TORQUE-TENSION RELATIONSHIP

A nut produces bolt tension by rotation that wedges mating threads together. Wedging action and the associated production of tension cannot occur until the nut is seated against a bearing surface. Figure 7-26*a* shows a nut and bolt assembly in which the nut is not yet seated. Only the "leading" flanks of threads are in contact. As the nut reaches the bearing surface (Fig. 7-26*b*), resistance to axial motion is encountered and thread contact shifts to "trailing" flanks. As increased torque is applied to rotate the nut further, the nut and its bearing surface experience compression, and tension is developed in the bolt. The flanks of both nut and bolt threads are in compression, while the threads' base experiences bending, not unlike a stubby cantilever beam. Thus torque produces the wedging action

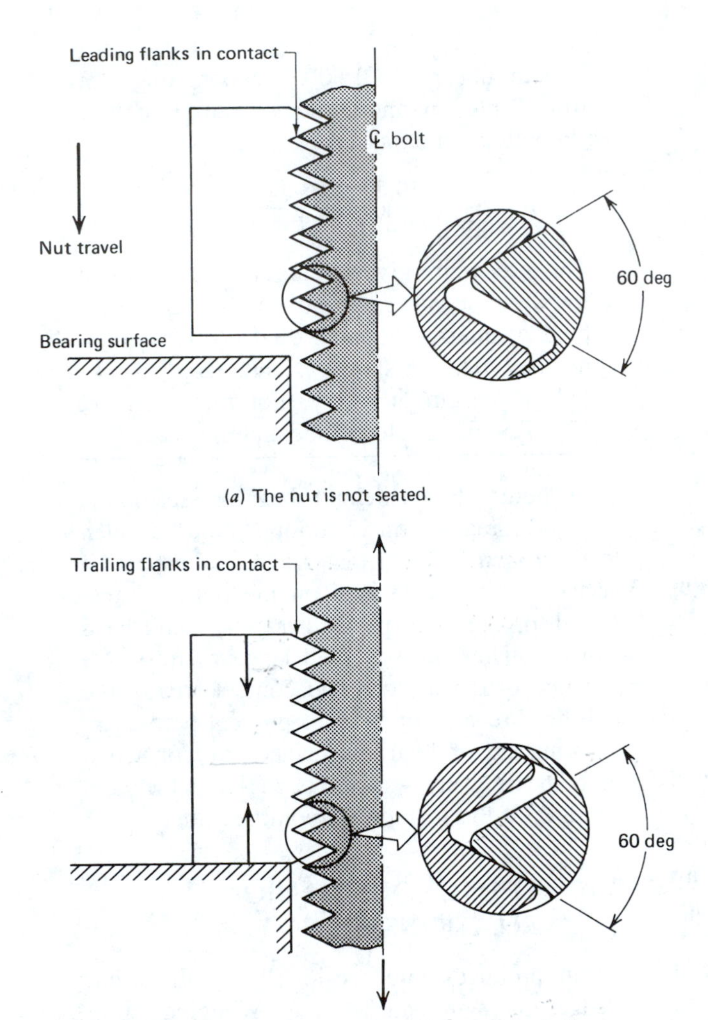

Figure 7-26 Development of bolt tension.

of threads that elongates the bolt to produce tension.

In any fastening situation the basic aim is to determine as accurately as possible the least expensive fastener that, when properly tightened, will secure a joint during product life. Tension induced in a fastener at assembly should always be greater than any external load the joint will experience in service. One of the anomalies of fastener engineering is that fastener preload is usually obtained from a torque measurement, not from a tension measurement. The required amount of torque to produce desired tension must be determined based on a torque-tension relationship. This relationship varies with fastener size and degree of lubrication but can be determined by means of a simple device called a torque-tension tester. Using such equipment, standard tables have been developed that list commonly encountered torque-tension coefficients. Proper bolt tension on a job, then, is achieved by means of a torque wrench.

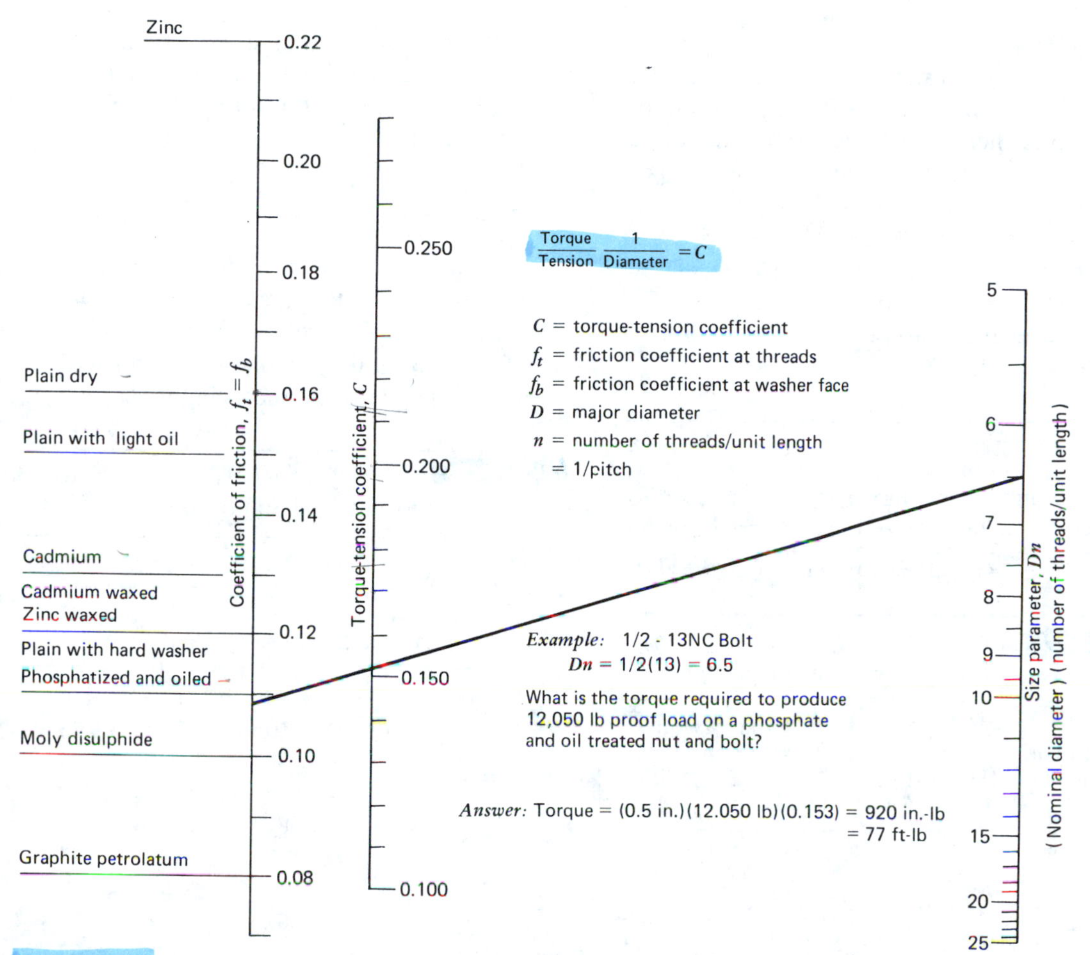

Figure 7-27 Approximate torque-tension relationship. (Developed by R. J. Erisman.)

The torque-tension coefficient C is defined as

$$C = \frac{T}{F_i D} \tag{7-7}$$

where T is tightening torque, F_i is preload, and D is basic major diameter. The value of C is often determined experimentally, but there is enough information available in the literature to obtain a reasonably accurate value without a test program. The critical factor is the coefficient of friction at both the mating threads and the bearing interface. Variations in coefficient of friction can change the C values by as much as 250%. There is also a slight interdependence with lead angle of the thread that can produce variations of about 5%. The nomograph in Fig. 7-27 may be used to obtain torque-tension coefficients for various fastener treatments and sizes. As would be expected, the nomograph shows that as surface friction increases, so does the torque-tension coefficient, since more torque is required to overcome friction.

It was stated earlier that bolt preload should be greater than any applied external load. Good design practice is to stress a bolt at preload to 85% of proof stress. This allows up to 15% additional stress for shear stress while tightening. If this procedure is followed, it is a simple matter to calculate bolt size from the given load and a strength level consistent with the type of bolt contemplated for use. The torque-tension relationship is then used to determine the torque necessary at assembly to obtain the desired preload. For dry thread surfaces, Tables 7-7 and 7-8 may be used to obtain the required torque. These torque values were calculated so as to result in a preload stress of 85% of the proof stress. For most fastener materials, this approximates 75% of yield strength.

EXAMPLE 7-3

What would be the maximum external load for an M8 × 1.25 fastener if the material is property class 5.8? Assume a 20% margin of safety. Calculate the assembly torque if the threads are lubricated with molybdenum disulfide. Why does this torque differ from the one given in Table 7-7?

SOLUTION

The clamped parts should never separate in service; hence, the maximum load should never exceed the preload. With a margin of safety of 20%, the payload should not exceed 80% of the preload. According to Table 7-2,

TABLE 7-7 Recommended Tightening Torque for Metric Standard Fasteners

Basic Major Diameter, mm	Recommended Torque, N · m				
	ISO 4.8	ISO 5.8	ISO 8.8	ISO 10.9	ISO 12.9
5	4	5	7	10	12
6	7	8	14	19	24
8	14	18	29	40	48
10	29	35	57	80	96
12	50	62	99	140	167
14	80	98	157	222	265
16	123	153	245	345	414
20	241	300	479	674	808
24	417	518	828	1166	1397
30	830	1030	1645	2318	2775
36	1450	1800	2875	4050	4850
39	1980	2458	3927	5531	6624

The torque values in this table were calculated from

$$T = \frac{0.85 f D \sigma_p A_s}{1000}$$

where

T = torque; N · m
f = 0.2 = torque coefficient
D = basic major diameter; mm
σ_p = proof stress; N/mm^2
A_s = tensile stress area; mm^2

TABLE 7-8 Recommended Tightening Torque for Unified Standard Fasteners

Basic Major Diameter, in.	Recommended Torque, ft-lb			
	SAE Grade 2	SAE Grade 5	SAE Grade 7	SAE Grade 8
$\frac{1}{4}$	6	10	12	14
$\frac{5}{16}$	13	20	24	28
$\frac{3}{8}$	23	35	43	49
$\frac{7}{16}$	36	56	69	79
$\frac{1}{2}$	55	85	106	121
$\frac{9}{16}$	80	123	152	174
$\frac{5}{8}$	110	170	210	240
$\frac{3}{4}$	195	302	373	426
$\frac{7}{8}$	189	486	601	687
1	283	729	901	1030
$1\frac{1}{8}$	401	900	1277	1460
$1\frac{1}{4}$	566	1270	1802	2060
$1\frac{1}{2}$	985	2210	3135	3583
$1\frac{3}{4}$	1554	3485	4945	5651
2	2266	5080	7208	8238

The torque values in this table were calculated from

$$T = \frac{0.85 f D \sigma_p A_s}{12}$$

where

T = torque; ft-lb
f = 0.2 = torque coefficient
D = basic major diameter; in.
σ_p = proof stress; psi
A_s = tensile stress area; in.^{2}

the proof stress for property class 5.8 is 380 MPa. The tensile stress area for M8 × 1.25 fastener is 36.6 mm² (Table 7-5).

$$\text{Preload:} \quad F_i = 0.85\ (380\ \text{N/mm}^2)(36.6\ \text{mm}^2)$$

$$= 11.8\ \text{kN}$$

$$\text{Payload:} \quad F = 0.80\ (11.8\ \text{kN}) = 9.44\ \text{kN}$$

To determine the assembly torque, the torque-tension coefficient must be found using Fig. 7-27. The size parameter is

$$Dn = D/p = 8\ \text{mm}/1.25\ \text{mm} = 6.4$$

The corresponding torque-tension coefficient is 0.142. Solving Eq. 7-7 for the tightening torque gives

$$T = F_i\,CD$$

$$= (11.8\ \text{kN})(0.142)(8\ \text{mm}) = \mathbf{13.4\ N \cdot m}$$

The recommended tightening torque for M8, ISO 5.8 in Table 7-7 is 18 N · m.

The 34% higher value is due to a coefficient of friction of 0.20, on which the table is based. Lubrication has lowered the torque by about 26%.

7-11 OTHER LOAD CONDITIONS

No Preload

Screws without preloading are simply loaded in tension or compression. The necessary cross section is thus obtained by dividing the load by allowable stresses. A factor of safety of 1.50 applied to yield strength is generally adequate. Figure 7-28 shows a fastener loaded in compression. If the extended length of the bolt is large, the bolt should be checked for buckling.

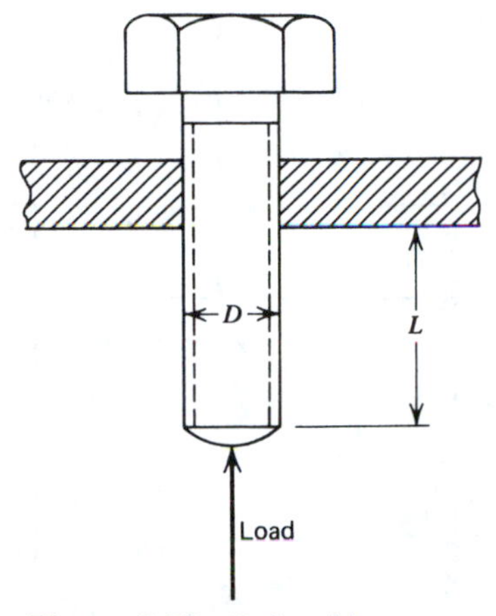

Figure 7-28 Bolt without preloading.

Eccentric Loading

Threaded fasteners are tensile devices that should not be loaded transversely. When a transverse load is present, as with an eccentrically loaded bracket, the shear force should be carried by a shear pin instead of by friction forces at the interface of the bracket and wall. Figure 7-29 shows this common type of loading.

When the bolts are preloaded during mounting, the base of the bracket will be subject to compression. In fact, as long as compression exists between the base and the wall, the bracket and the wall will act as a continuous structure. Thus, when the offset force P is next applied, the bracket will tend to rotate about the centroidal axis of the contact area.

If each bolt is preloaded with a force F_i, as shown in Fig. 7-29, the uniform compressive stress in the base plate is

$$\sigma_c = \frac{2F_i}{b2c} = \frac{F_i}{bc} \qquad (7\text{-}8)$$

The eccentric load P superimposes a tensile stress that, at the top, is

$$\sigma_t = \frac{Mc}{I} = \frac{12Prc}{b(2c)^3} = \frac{3Pr}{2bc^2} \qquad (7\text{-}9)$$

where I is the moment of inertia about the centroidal axis. The resulting stress at A is

$$\sigma_A = \frac{F_i}{bc} - \frac{3Pr}{2bc^2} \qquad (7\text{-}10)$$

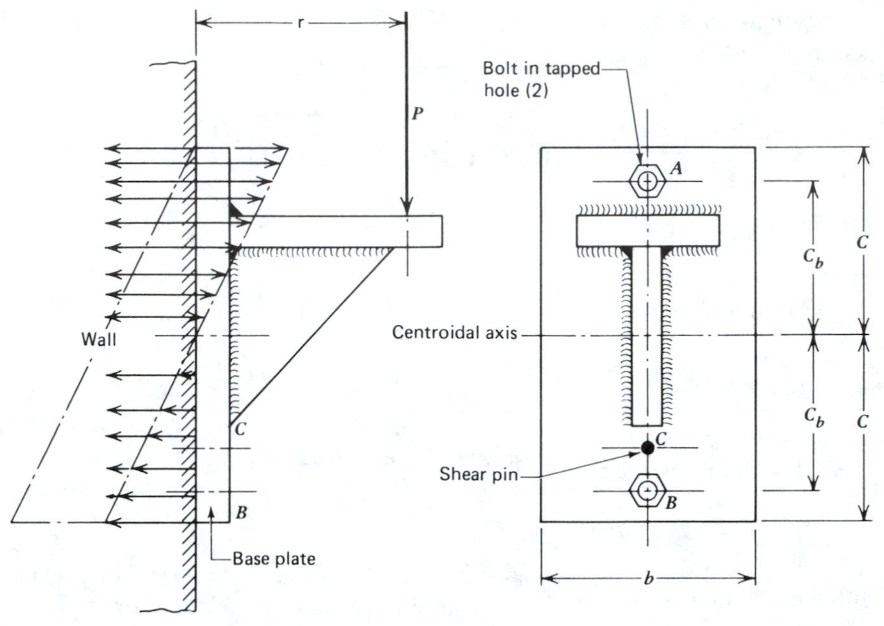

Figure 7-29 Eccentric loading of bracket. Preloading of the bolts places the base plate in compression. Following application of the load p, only part of the base plate may remain in compression.

To avoid separation of the bracket and wall at A, $\sigma_A > 0$, or

$$F_i > \frac{3Pr}{2c} \tag{7-11}$$

Maximum tensile stress occurs in the upper bolt and is due to preloading superimposed by the eccentric loading.

$$\sigma_t = \frac{F_i}{A_s} + \frac{Prc_b}{I} \tag{7-12}$$

A 25% margin on the preload is sufficient to account for overload. Thus

$$F_i = \frac{15Pr}{8c}$$

Substituting F_i in Eq. 7-10 yields

$$\sigma_t = \frac{15Pr}{8cA_s} - \frac{3Prc_b}{2bc^3} \tag{7-13}$$

Since total bolt stress should not exceed the proof stress, we may substitute σ_p for σ_t for any suitable grade material and solve for A_s to obtain the bolt size.

Combined Loading (Tension and Torsion)

This type of loading is present in turnbuckles. A turnbuckle is a loop or eye of metal with a nut or internal thread at each end; one is a lefthand thread and the other is a righthand thread (Fig. 7-30). It is used to tighten stayrods or similar tension members in many structures. Consequently, the threaded part is loaded in both tension and torsion. Because of combined loading, the threaded part should be calculated by the methods outlined in Chapter 3.

Transverse Loading

When a bolt connection is loaded transversely, the load P may be taken up by the friction force F_f at the interface. As long as the friction force exceeds the transverse load, the bolts will not be

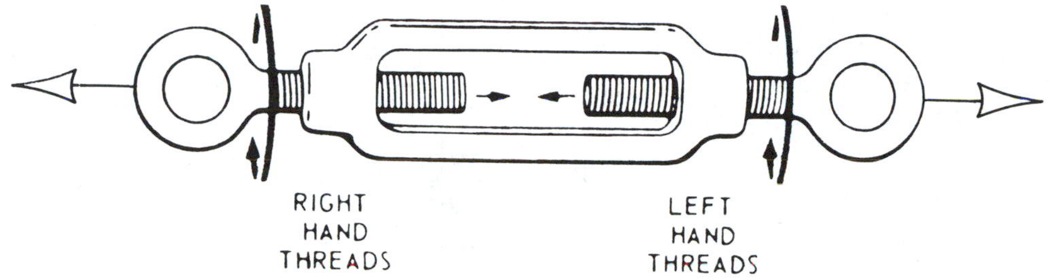

Figure 7-30 A turnbuckle has both right and left-hand threads. (Courtesy Deere and Company Technical Services.)

subject to shear, which is to be avoided (Fig. 7-31).

$$F_f = fF_i > P \tag{7-14}$$

SUMMARY

Screw-thread fasteners are tensile devices thoroughly standardized for the purpose of efficiently locating, clamping, adjusting, and transmitting force to and within machinery. They are usually small, mass-produced, intricate, highly stressed components, all requiring the use of some tool for torquing. Nearly all of them have a head, a shank, and some form of nut. The shanks are very much alike, but the nuts and heads vary widely to accommodate a multiplicity of design and manufacturing considerations. Nevertheless, most fasteners fall into one of three categories: bolt and nut, bolt in tapped hole, and stud.

Threaded fasteners fail primarily in fatigue so carbon steel and steel alloys are the most commonly used materials. They are specified by strength grades. In the ISO system strength grades are known as property classes.

Threaded fasteners are devices that generally should not be loaded in bending or in compression. Transverse loads should be taken up by shear pins unless the friction forces (due to clamping) provide ample resistance to shear.

The primary aim of designers is to select the least expensive fastener that, when tightened, will secure a joint. The torque-tension relationship for fasteners, however, is subject to variations within wide margins. The preload obtained is therefore at best a statistical average of the one specified. Generally, the preload is specified at 85% of proof load or the near equivalent 75% of minimum tensile strength.

Screw fasteners cannot remain tight for long when subjected to dynamic loads. A locking device is therefore necessary. A locking device based on friction resists rotation by gripping the mating thread or material of the connected parts. Interlocking (positive) devices are more reliable than those based on friction, but they cost more. Washers are a simple but indispensable supplement to screw connections. They distribute the load more evenly and over a larger bearing surface. Thus bending of the bolt shank is minimized.

The transition to metric fasteners in the

Figure 7-31 Transverse loading.

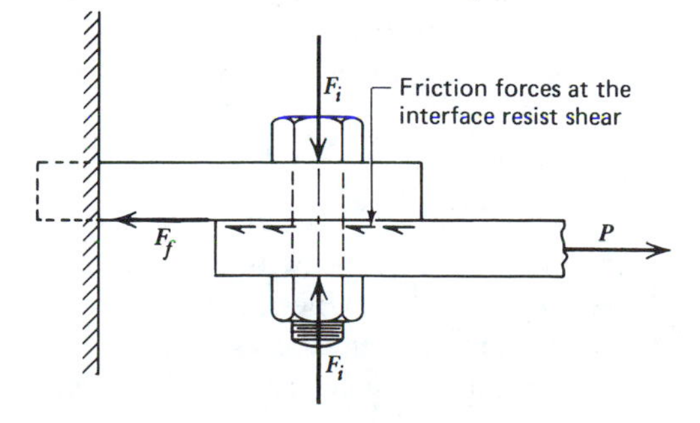

United States is well underway. The economic incentives are many, including worldwide interchangeability, greater strength relative to mass, fewer standards, and smaller inventories.

REFERENCES

7-1 *Fundamentals of Service: Fasteners*. Moline, Ill.: John Deere Service Publications, 1975. (The accompanying slide collection is well suited for teaching.)

7-2 *Machine Design. Fastening and Joining Reference Issue,* 1976.

7-3 *Metric Fastener Standards*. Cleveland: Industrial Fasteners Institute, 1981.

PROBLEMS

P7-1 How long a wrench is needed to tighten a 1-in. bolt if the maximum force (pull) that can be exerted by the mechanic is roughly 100 lb? Assume SAE Grade 2 material. *Hint:* Use Table 7-8.

P7-2 Find the pull that must be applied to a wrench with a lever arm of 450 mm when tightening an M30 × 3.5 bolt. Assume ISO class 4.8. What is the corresponding stress in the bolt? *Hint:* Use Tables 7-2 and 7-7.

P7-3 A bolt is loaded in tension by 45 kN. The property class is 4.8.

(*a*) What is the proof stress?

(*b*) How much root area is required for an allowable stress of 85% of proof stress? *Hint:* Use Eq. 7-4.

(*c*) Select a suitable thread from Table 7-5.

(*d*) What is the safety factor based on this thread?

P7-4 Builders of heavy machinery prefer to use size $\frac{5}{8}$-in. bolts and larger to prevent overtorquing. The average mechanic, with available standard wrenches that have 6-, 9-, 12-, and 15-in. handles, generally cannot exert the pull necessary for overtorquing. Calculate the pull required for the above-mentioned standard handles applied to a $\frac{5}{8}$-in. bolt of SAE Grade 2 material (the weakest). Comment on the answers.

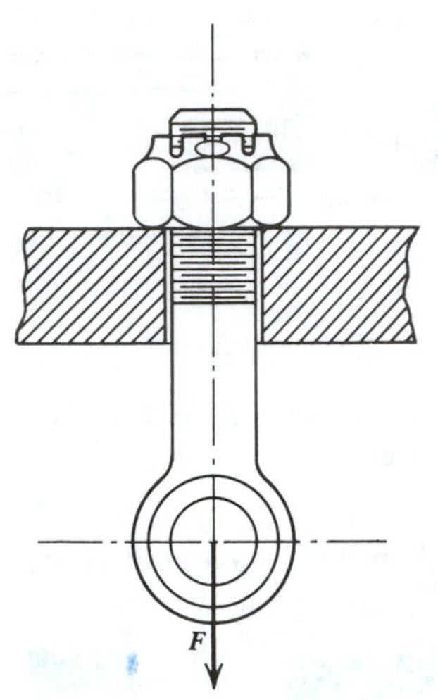

Figure 7-32 Eye bolt. (Problem 7-5)

P7-5 An eye bolt of the type shown in Fig. 7-32 is used in assembly work to carry a steady load of 29 kN. Select a suitable thread for property class 5.8. For a steady load, use a factor of safety of 1.25. What is the real factor of safety?

P7-6 The bolt shown in Fig. 7-33 is of the type often used in large engines to provide a flexible means of bolting the main components together. By reducing body area to equal the thread stress area, the bolt can absorb impact energy. The bolt is an M10 × 1.5 of ISO 8.8 material. The cylindrical body is 667 mm long.

Figure 7-33 Engine bolt. (Problem 7-6)

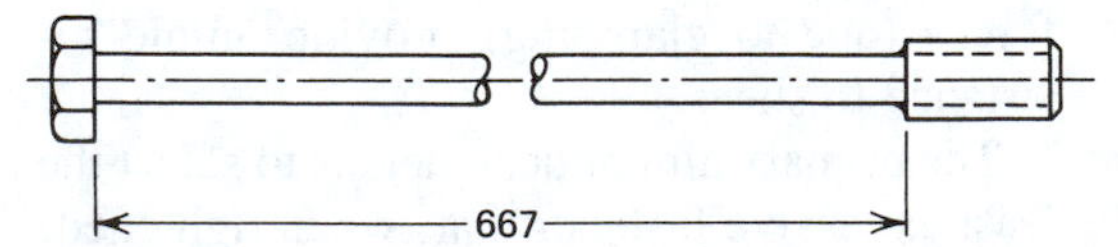

The nut has been run up snugly, so that there is neither slack nor stress in the bolt.

(*a*) How much stress is generated if the nut is rotated one revolution (E = 207 000 MPa)? *Hint:* $\sigma = \epsilon E$ and $\epsilon = \Delta L/L$.

(*b*) How many degrees must the nut be rotated to generate a stress equal to 90% of yield strength?

P7-7 A manufacturer of electric motors lists the net weight of a motor as 1356 lb. This motor has an eye bolt screwed into its cast-iron frame for lifting purposes. The eye bolt is located directly over the center of gravity. (See *Machinery's Handbook,* 21st edition, pp. 1103-1104.)

(*a*) What are the modes of failure of this screw connection? For instance, how can the bolt fail? How can the thread in the casting fail? Which mode of failure is the least likely to occur?

(*b*) Which size standard threads should be used, assuming an allowable tensile stress of 6000 psi?

(*c*) How far should the thread of the eye bolt be extended into the casting in order to carry the load of 1356 lb? Assume an allowable tensile stress of 2000 psi for cast iron.

(*d*) Make a full-size sketch of the area of engagement.

P7-8 Solve P7-5 by specifying an eye bolt from *Machinery's Handbook*. Make a full-scale drawing of the selected eye bolt seated in the frame.

P7-9 Repeat P7-7 for a generator of mass 1360 kg using an eye bolt with a metric thread. Use equivalent material strength in MPa.

P7-10 Calculate the approximate preload in an M14 × 2 fastener after applying a 100-N · m tightening torque if the threads are: (*a*) plain dry, (*b*) cadmium plated, and (*c*) phosphatized and oiled. *Hint:* Use Eq. 7-7.

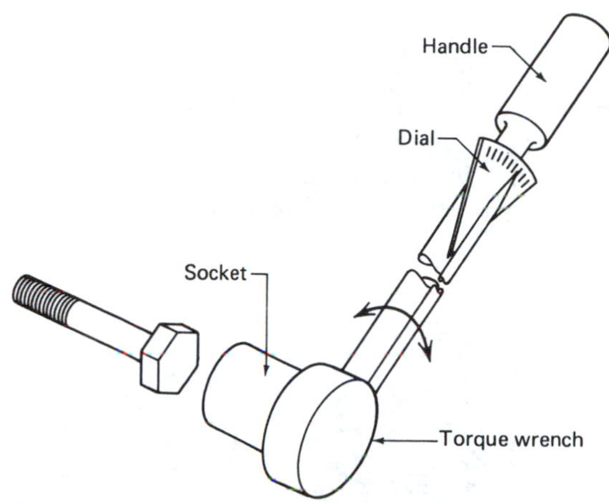

Figure 7-34 Wrench. (Problem 7-12)

P7-11 This problem is the same as P7-10, except that the fastener is $\frac{1}{2}-13$ UNC and the tightening torque is 75 ft · lb.

P7-12 Calculate the force P that must be applied to a wrench with a torque arm of 225 mm when tightening an M30 × 3.5 oil-lubricated bolt to produce a tensile load of 8700 N (Fig. 7-34).

P7-13 A pipe hanger (Fig. 7-35) is to support a total load of 2.8 kN. Determine the

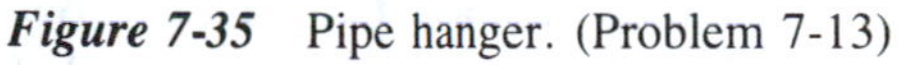

Figure 7-35 Pipe hanger. (Problem 7-13)

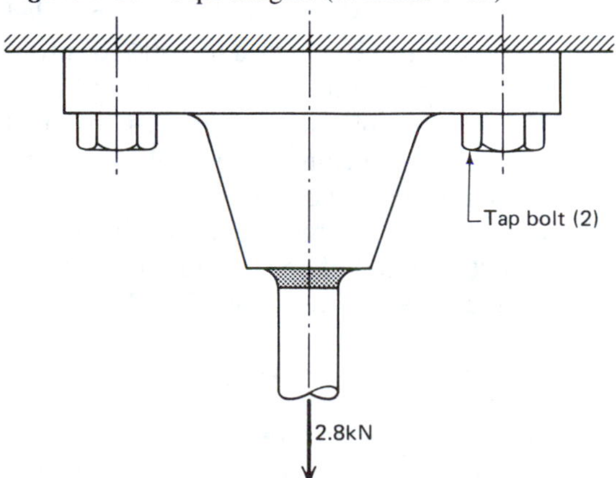

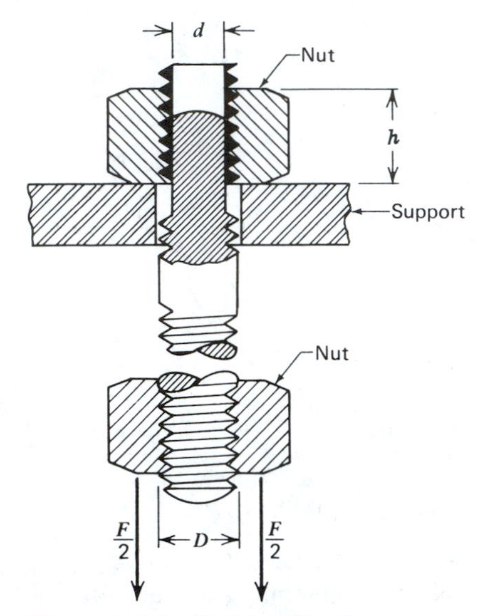

Figure 7-36 (Problem 7-16)

required size and tightening torque of the standard metric fasteners to be used. The material class is ISO 4.6, and the threads will be lightly oiled.

P7-14 A $\frac{3}{4}-10$ UNC, SAE Grade 2 fastener is assembled into a threaded hole in a nodular cast-iron housing. The yield strength of the housing material is 55,000 psi. What maximum depth of tapped hole will ensure tensile bolt failure prior to thread stripping?

P7-15 A certain application calls for a nonlubricated M36 × 4 ISO 8.8 threaded fastener to be torqued to 100% of proof load. Find this torque. At assembly, a cadmium-plated fastener was installed using the specified torque. What are the likely consequences of this error?

P7-16 Calculate the theoretical height h of a nut for a bolt if equal strength in tension and shear are required (Fig. 7-36). Assume (*a*) nut and bolt are made from the same material; (*b*) load is equally distributed over all threads; (*c*) no stress concentration is present; (*d*) root diameter d is 80% of the nominal diameter D; and (*e*) material yield strength in shear is roughly 50% of yield strength in tension. *Hint:* Substitute in the formula $\sigma_t = 2\tau$.

P7-17 For the engine bolt shown in Fig. 7-33 and described in P7-6, calculate

(*a*) the preload based on 85% proof stress.
(*b*) the corresponding torque when a light oil is applied to the threads.
(*c*) the total elongation generated by the preload.
(*d*) the corresponding rotation in degrees of the nut.

Are a washer and a locking device needed? What would you recommend? Give reasons.

P7-18 The following data apply to the bracket shown in Fig. 7-29: $P = 1500$ lb; $r = 8$ in.; $b = 10$ in.; $C = 7$ in.; $C_b = 6$ in.; and $f = 2.0$.

(*a*) Find a suitable tap bolt. Assume a 25% margin on the preload.
(*b*) Estimate the preload on the tap bolt.
(*c*) Select a steel pin to carry the shear load. Assume an allowable shear stress of 6000 psi.

P7-19 A bolt connection of the type shown in Fig. 7-31 is to resist a load P of 2 kN. The dry surfaces have a coefficient of friction of 0.20. Find a suitable low-cost bolt.

Chapter 8
Springs

I, wisdom, dwell with prudence
and find out knowledge
of witty inventions.

PROVERBS

NOMENCLATURE

a = outside radius, Belleville spring
b, b_o = width of flat stock
C = spring index
d = wire diameter; inside diameter, Belleville spring
D = mean coil diameter (pitch); outside diameter, Belleville spring
E_p = potential energy
E'_p = potential energy per unit volume
f = elastic limit deflection per coil, coil springs
h = coned height, Belleville spring
IT = initial tension, extension coil spring
k = spring rate
K = Wahl correction factor; parameter for tapered leaf spring
L = total wire length, coil springs; span length, leaf springs
$L_1 L_2$ = loaded lengths, coil springs
L_o = free length, coil springs
L_s = solid height, coil springs
n = number active coils
N = number total coils
P = spring load
P_e = elastic limit load
S = shear force
t = material thickness
α = slope of load deflection curve
β = Belleville spring load parameter
δ = spring deflection
δ_s = spring deflection, to solid length
ν = Poisson's ratio
$\emptyset$ = Belleville spring stress parameter
σ = unit stress, bending
σ_e = unit stress, tensile elastic limit
τ = unit stress, torsional
τ_1 = unit stress, combined torsional and shear
τ_2 = unit stress, torsional with Wahl correction factor
τ_e = unit stress, torsional elastic limit
τ_p = unit stress, shear
θ = angle of twist

Springs are simple machine elements that play an important role in the operation of many mechanical and electrical devices. Their primary function, unlike that of most other components, is to introduce controlled flexibility by deflecting under applied loads. Spring deflection in machine design is utilized to absorb the energy of suddenly applied loads and to store energy for subsequent release.

Springs vary in size from small ones found in watches to giant ones used to support and control heavy industrial equipment. The simplicity of springs frequently makes it advantageous to combine a springlike element with other mechanical members. Examples of the use of these special purpose springs are spring washers, safety pins (torsion spring), and farm implement tools (chisel plow).

8-1 FUNCTION AND DESIGN

A spring can be described as an elastic body having a predetermined relationship between the force that can be applied to it or exerted by it and the accompanying change in length. Typical engineering applications where springs perform desired functions according to this principle are as follows.

1. *Control* of forces due to impact, shock loading, or vibration, as in vehicle suspension.
2. *Restraint* of motion, as in valves of internal combustion engines (Fig. A-2).

3. *Exertion* of forces to actuate mechanisms, as in brakes or clutches (Fig. 19-3).
4. *Storage* of mechanical energy, as in timing devices, switches, and starters.
5. *Measurement* of forces, as in scales.
6. *Compensation* for heat expansion of machine parts.

The design of a successful spring involves determining the proper geometric proportions to meet the following objectives.

1. Attainment of a required load in acceptable spatial confines.
2. A given or acceptable rate of load increase per unit of deflection.
3. A satisfactory stress level.

Spring designers must, therefore, be conversant with available materials and heat treatment practices as well as with the basic design techniques for performance and stress analysis.

8-2 CLASSIFICATION

Springs may be classified into two main categories: springs made from wire stock and springs made from flat stock. Within these two divisions there is much overlapping because of the diverse nature of configurations, usage, materials employed, and principal stresses induced. Wire springs include helical compression and extension springs, both of which stress the material in torsion. Also included are torsion springs, so named because they produce a torque about an axis but stress the material in bending. This category further includes torsion bars, which are essentially straight and provide torsional deflection from torsional stresses.

Flat springs include springs made of flat or strip materials having a principal stress in bending. In this category are single-leaf and multileaf springs. These types are often termed elliptical or cantilever springs and are, in fact, simply supported or cantilever beams.

Perhaps a third classification should be added to include some of the special springs that are becoming popular. Belleville springs, volute springs, ring springs, rubber springs, and gas springs should be placed in this category.

Helical coil and leaf springs are the most commonly used, probably because they are readily mass produced at nominal costs. These and additional types of springs are shown in Fig. 8-1.

8-3 MATERIALS

Various materials are available to meet the often conflicting needs of strength, cost, supply, surface condition, and other specialized properties. Generally, the preferred material will possess high strength and flexibility. These requirements are best achieved by using steel with a high-carbon content or steel alloys containing silicon, chromium, manganese, vanadium, or tungsten. For statically loaded springs, hard-drawn or oil-tempered carbon spring wire is adequate. When the loading is cyclic, however, a higher-quality material such as music or valve spring wire or a steel alloy with restricted surface conditions is preferred. Life of springs is often increased by shot-peening.

High-carbon spring steels in wire form are the most common because they are the least expensive, easily coiled, and readily available. They are inadequate, however, for springs operating at extremely high or low temperatures or for extremely severe impact loading.

Alloy spring steels, including stainless steels, have physical properties that greatly reduce the preceding shortcomings, but obviously at higher cost. These steels also offer corrosion and creep resistance.

Nonferrous spring materials are often specified when good electrical conductivity or nonmagnetic properties are desired. They also have good corrosion resistance. Phosphor bronze, beryllium copper, and monel are representative materials in this category.

Typical physical properties for these materials are shown in Figs. 8-2 and 8-3. These graphical presentations show the inverse proportionality of

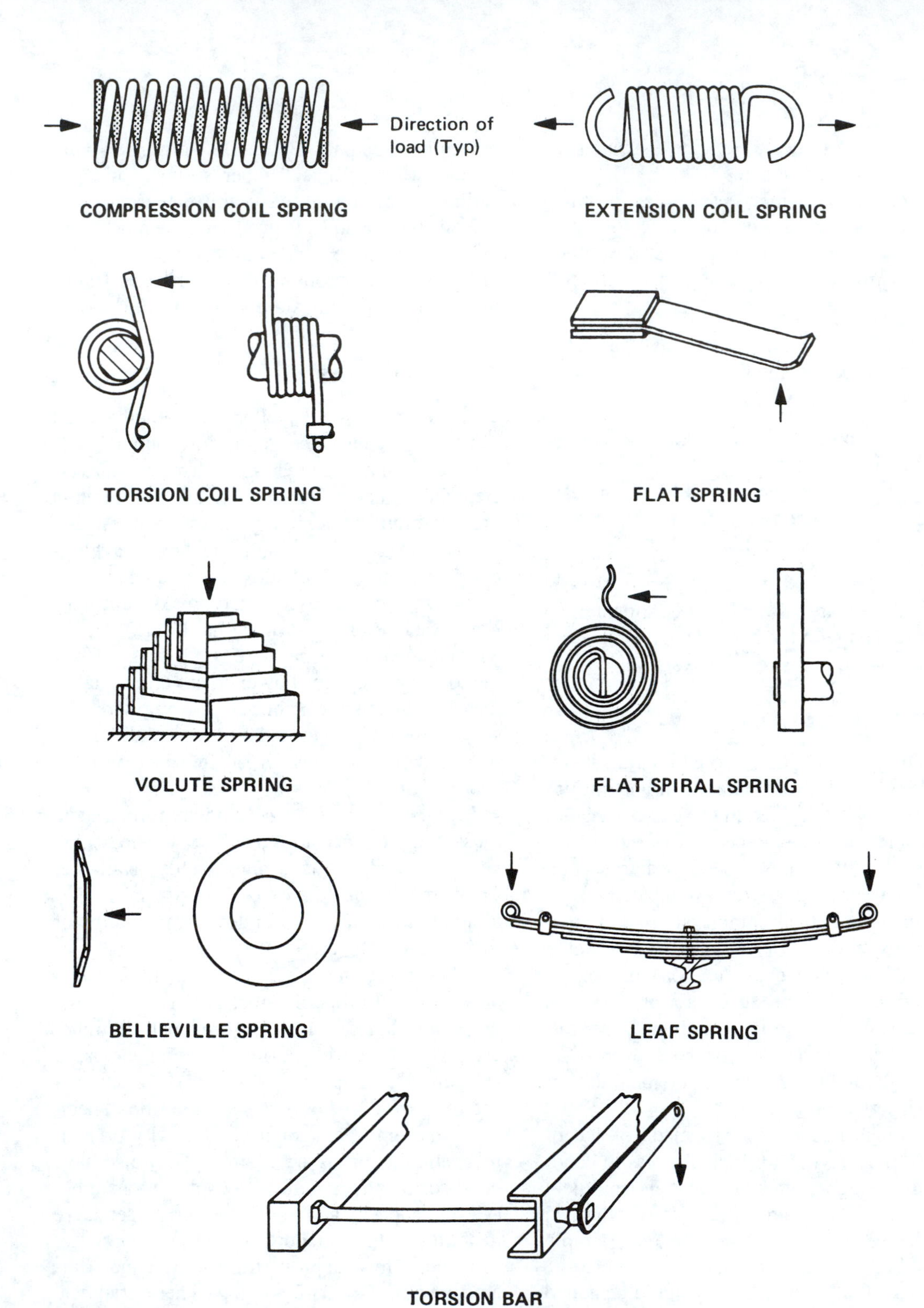

Figure 8-1 Types of mechanical springs. (Courtesy General Motors.)

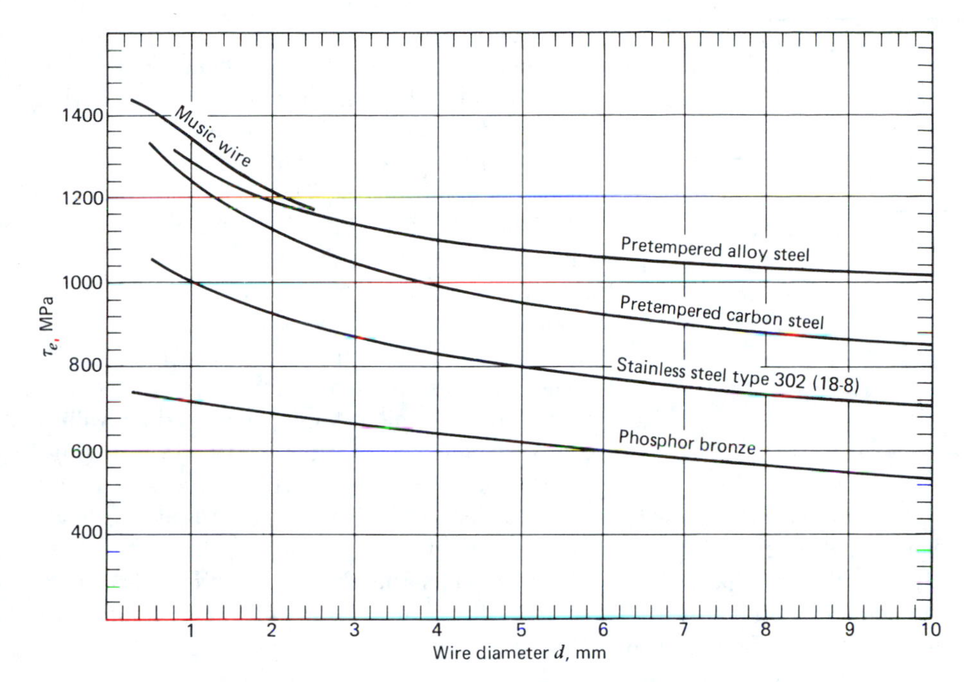

Figure 8-2 Torsional elastic limit strength of spring materials.

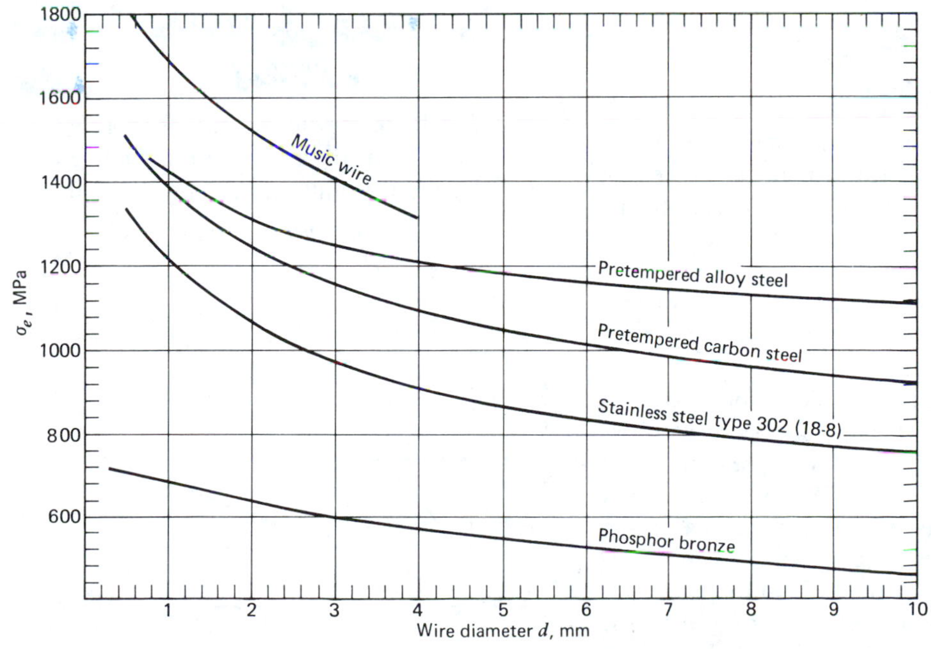

Figure 8-3 Tensile elastic limit strength of spring materials.

TABLE 8-1 Moduli of Elasticity for Commonly Used Spring Materials

Material	*E*		*G*	
	10^3MP$_a$	10^6psi	10^3MP$_a$	10^6psi
Steel (music wire)	207	30	79.3	11.5
Steel (carbon or alloy)	207	30	79.3	11.5
Steel (stainless 18-8)	193	28	68.9	10
Phosphor bronze	103	15	43.1	6.25

wire diameter to both torsional and tensile strengths, a factor that should be considered in the solution of design problems. Spring deflections are based on another material property: modulus of elasticity (Table 8-1). Both tensile and torsional moduli are used in spring design, depending on whether the principal stress is bending or torsion. Table 8-1 shows commonly used values.

8-4 LOAD-DEFLECTION RELATIONSHIP

Springs are unique among machine members because (1) the applied load is simple in nature (not compound) and (2) the deflections caused by the load are large, visible, and important. In most machine members the deflection is very small and is often ignored. Springs, on the other hand, have appreciable deflections that must be considered in the design process.

The deflection capabilities of a spring are surprising, considering the space it occupies. Consider, for example, a helical compression spring of 25 mm diameter and a length of 65 mm. A deflection of 25 mm is typical for such a spring. A solid piece of steel occupying the same space would deflect only about 0.25 mm. How is it possible to increase the deflection by 100:1? The answer lies in the *configuration* of the spring. Instead of using a compact piece of material, the designer substitutes a wire of about 1000 mm in length with a diameter of only 3 or 4 mm. This wire is loaded in torsion with a radius arm of about 11 mm. A spring thus becomes a very efficient deflecting device.

Total deflection depends on the magnitude of the applied load and the stiffness characteristic. A stiff spring has very little deflection, even under a large load. Conversely, a flexible spring has a large deflection, under even light loads. This relationship can be seen graphically by plotting, on rectangular graph paper, various values of load and the corresponding deflection (Fig. 8-4). When the load-deflection curve is a straight line, deflection is directly proportional to load, as predicted by *Hooke's law*, a fundamental relationship in spring design. From Fig. 8-4 a useful equation can be derived from the similar triangles shown.

$$\tan \alpha = \frac{P_1}{\delta_1} = \frac{P_2}{\delta_2} = \frac{P_2 - P_1}{\delta_2 - \delta_1} = \text{constant} \qquad (8\text{-}1)$$

From this we obtain

$$P_2 - P_1 = k(\delta_2 - \delta_1) \qquad (8\text{-}2)$$

$$P = k\delta \qquad (8\text{-}3)$$

where k is the spring characteristic or spring rate. The characteristic or rate is a useful parameter in the design and testing of springs. Its units are

Figure 8-4 Linear deflection characteristics.

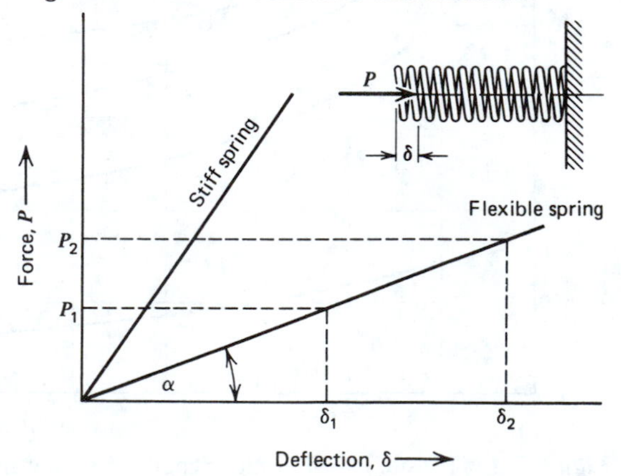

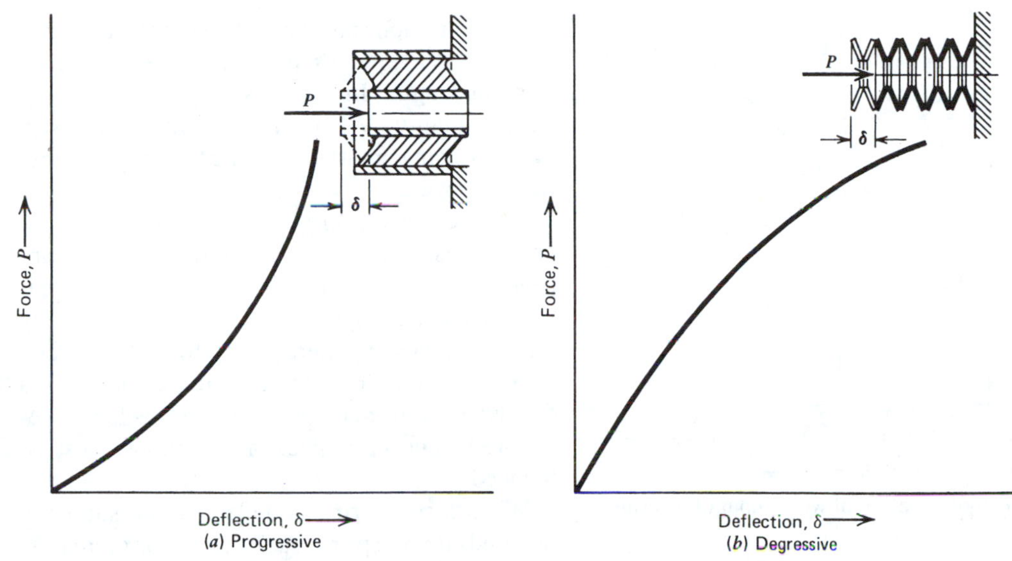

Figure 8-5 Nonlinear deflection characteristics.

obviously force per unit of length deflected and will be given in units of newtons per millimeter or pounds per inch throughout this chapter.

Not all spring characteristics are straight lines. In fact, it is sometimes desirable to obtain a nonlinear characteristic relationship. Two types of curves are encountered in spring design: *progressive* and *degressive* curves. These are shown in Fig. 8-5. Figure 8-5*a* shows a progressive characteristic typical of rubber springs stressed in shear. The curve shows an exponential increase in resistance as deflection increases. Consequently, maximum deflection can be effectively controlled. This type of characteristic is useful in damping harmful vibrations and is often found in vehicle or railway car suspension systems. Progressive characteristics can be obtained by winding a compression spring with variable coil spacing. With this arrangement, the number of active coils is reduced as the load increases, thus causing an increase in the spring characteristic. Progressive characteristics are also typical of multileaf springs.

A degressive spring characteristic displays a leveling off of the load as the deflection increases (Fig. 8-5*b*). This characteristic is typical of some Belleville spring washers and offers the advantage of maintaining a nearly constant load, regardless of the spring's height in its mounted position.

8-5 ENERGY STORAGE AND ENERGY DISSIPATION

When a spring deflects under load, the work performed by the force is stored as elastic or strain energy, a form of potential energy that exists by virtue of position. For a given characteristic curve, the work input is equal to the area under the curve. This is shown in Fig. 8-6 for a linear characteristic curve, and it follows that

$$\text{Work} = 0.5P\delta$$

since

$$P = k\delta$$

$$E_p = 0.5k\delta^2 \qquad (8\text{-}4)$$

This equation shows that energy stored increases exponentially with the deflection. A twofold increase in deflection results in the quadrupling of the potential energy.

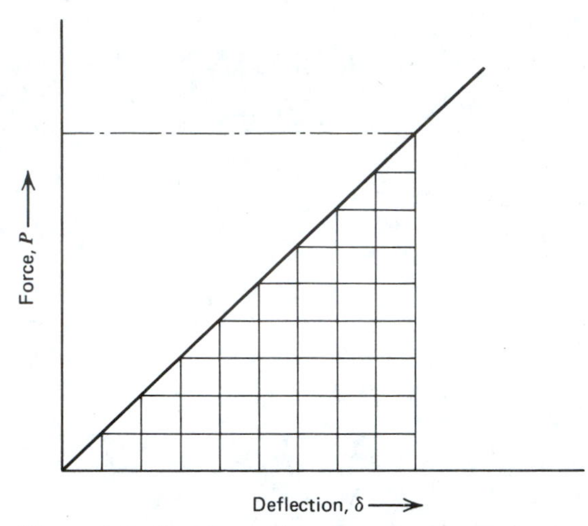

Figure 8-6 Representation of work input of a spring system.

Because the primary purpose of springs is energy absorption, a major design goal is maximum energy capacity for a given space. The amount of energy stored depends on the volume of the elastically deformed material. In turn, volume required is proportional to the modulus of elasticity but inversely proportional to the square of the elastic limit stress. The capability of stored potential energy in a helical coil spring for a given unit of volume is

$$E_p' = \frac{\tau^2}{4G}$$

As an example of applying this concept, assume that the maximum allowable stress could be increased by 5%; then energy capacity would be increased by 10% (according to the rules in Section 2-10). In terms of material, this would also mean that the mass or volume of the spring could be reduced by 10% for the same energy capacity.

Maximum load capacity in a spring is thus achieved by:

1. Using materials and manufacturing methods that will permit high allowable stresses.
2. Choosing a configuration that will stress all of the material evenly.

When friction forces are absent, stored energy is fully recovered as the spring returns to its original length. In motor springs this is a desirable quality. When present, however, as in systems using multileaf springs, friction forces must be overcome with part of the applied load so that the stored energy is not completely recoverable. (Usually the unrecovered energy will appear as heat, which may be an undesirable feature in itself.) Figure 8-7 shows schematically the mechanics of energy transformation when some energy is lost. Friction forces are thus a disadvantage in many cases but are actually desirable whenever impact or vibrations must be reduced.

When a coil spring is deflected and suddenly released, it will spring back past its original position and go on oscillating for some time. Each swing will be of smaller amplitude than the one before while taking the same amount of time. The oscillation does not continue indefinitely because of internal molecular friction (hysteresis) within the spring material. A well-made steel spring has low hysteresis; therefore, coil springs are often combined with dampers, the most common application being in automotive suspension systems. A hydraulic damper offers resistance to movement by the simple expedient of squeezing a liquid through a small hole. Thus, potential

Figure 8-7 Representation of work input versus work output of a spring system.

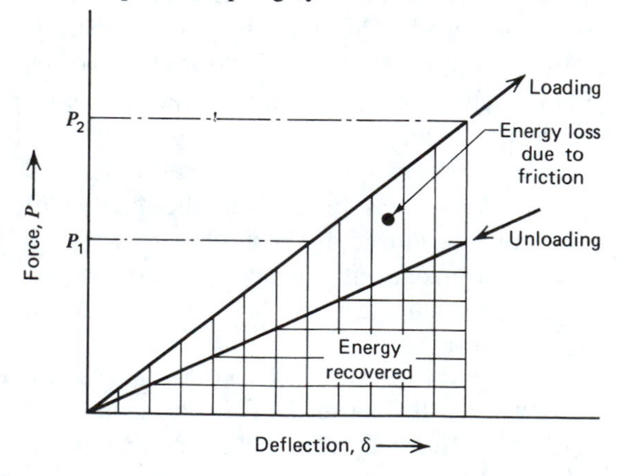

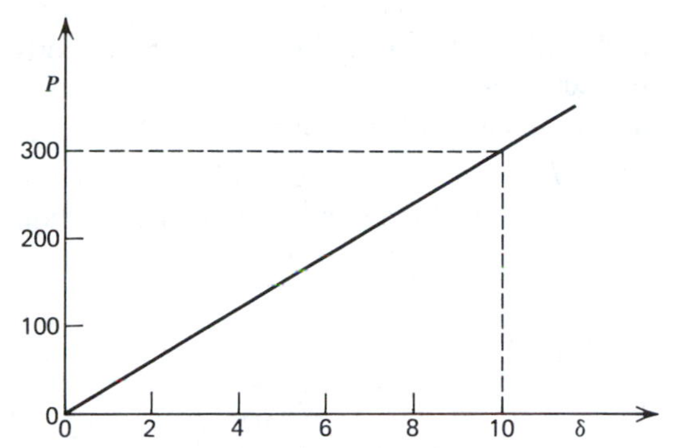

Figure 8-8 Spring characteristic, Example 8-1.

energy is converted into heat that is readily dissipated. A good damper stops the oscillation in less than two cycles.

EXAMPLE 8-1

A helical valve spring has a rate of 30 N/mm; maximum allowable load is 300 N. Construct a diagram showing the spring characteristic. Calculate both the maximum allowable deflection and the maximum energy that can safely be stored in this spring (see Fig. 8-8).

SOLUTION

$$\delta = \frac{P}{k} = \frac{300 \text{ N}}{30 \text{ N/mm}} = \mathbf{10 \text{ mm}}$$

$$E_p = 0.5k\delta^2$$

$$= 0.5(30 \text{ N/mm})(10 \text{ mm})^2 = \mathbf{1.5 \text{ N} \cdot \text{m}}$$

or

$$E_p = 0.5P\delta$$

$$= 0.5(300 \text{ N})10 \text{ mm} = \mathbf{1.5 \text{ N} \cdot \text{m}}$$

8-6 ALLOWABLE STRESSES

Springs are designed to operate at high stress levels that permit, as we have seen, the maximum storage of energy. Why can springs tolerate these high levels of stress? This phenomenon can be explained in several ways. First, the material itself is prepared under the most stringent controls to ensure freedom from impurities that often lead to failure. Second, by its nature, wire is very uniform in size, a trait that optimizes its response to heat treatment. Third, the wire, especially in smaller sizes, is obtained by drawing it through a sizing die that produces excellent surface conditions and, at the same time, work-hardens the material. Fourth, springs respond favorably to the application of residual prestressing. Helical compression springs, for instance, are sometimes wound with a greater free length than specified. This procedure results in a solid height stress in excess of the elastic limit. The springs are then compressed solid several times, causing the free length to decrease and thereby introducing beneficial residual stresses. This operation is known as *setting* or *presetting* and can be controlled to produce the desired free length of the spring. Presetting is beneficial when operating stresses are in the same direction as the induced residual stresses and when no reversals occur, as is the case for helical compression springs.

Thus the conventional spring is a highly favored machine element capable of operating very near the elastic limit of the material. Its surface finish must, therefore, be carefully protected; the slightest nick or abrasion will serve to concentrate stresses and produce a premature failure. Hydrogen embrittlement in finishing can seriously weaken the hook at the end of the spring, unless proper precautions are taken.

8-7 NOMENCLATURE, HELICAL COMPRESSION SPRINGS

A straight-sided (constant diameter) helical compression spring is the most common of springs. It is easy to specify, simple to manufacture, and readily adapts to a multitude of design situations. When compressed from its original length, it will exert an axially outward force. Figure 8-6 shows a force-deflection diagram for this type of spring. The following terms are used to describe the various factors associated with the design of a helical compression spring (Fig. 8-9).

Outside Diameter (OD). Diameter of the coils measured on the outside of the wire perpendicular to the axis of the spring.

Inside Diameter (ID). Diameter of the coils measured on the inside of the wire perpendicular to the axis of the spring.

Pitch Diameter (D). Mean diameter of the coils measured to the center of the wire section

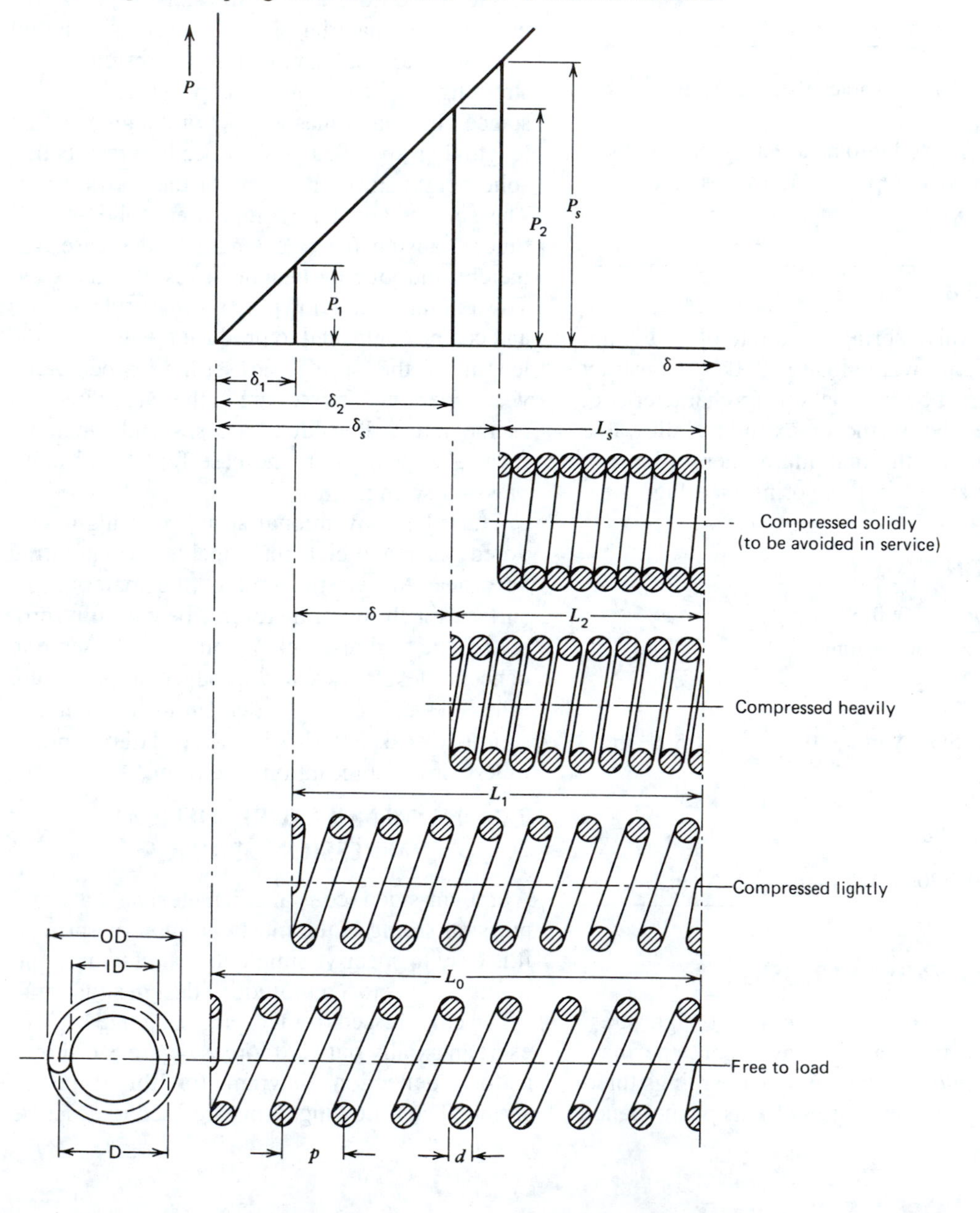

Figure 8-9 Force-deflection diagram and nomenclature for a constant diameter helical compression spring made from round wire of uniform thickness.

perpendicular to the axis of the spring. $D = OD - d$. This diameter is used in the equations for stress and deflection.

Wire Diameter (d). Cross-sectional diameter of the wire used in forming the spring. Wire sizes have been standardized. Table 8-2, Spring Design Chart, Section 8-10, shows typical metric sizes.

Spring Index (C). Ratio of pitch diameter to wire diameter. $C = D/d$. It is a measure of the relative severity of coil curvature.

Free Length (L_0). Overall length measured parallel to the axis in the unloaded or free condition.

Compressed Length (L_1, L_2, . . .). Axial length when compressed by an external load.

Solid Length (L_s). Compressed length with all coils touching adjacent ones. Also called solid height.

Deflection (δ). Reduction in axial length, when compressed by an external load.

Solid Deflection (δ_s). The deflection causing solid length (L_s).

Coil. One complete turn or convolution of the wire about the axis of the spring.

Direction of Helix. The hand of the helix, right or left, corresponding to screw-thread nomenclature.

Active Coils (n). Coils that are free to deflect under the influence of an external load. Also called working coils.

Inactive Coils. End coils rendered inactive by contact with the supporting structure.

Total Coils (N). Total number of complete turns. Total coils = active coils + inactive coils (Fig. 8-10).

Pitch (p). Distance from one active coil to the next measured in the axial direction.

Rate (k). Ratio of change in load to the corresponding change in deflection.

$$k = \frac{P_2 - P_1}{L_1 - L_2}$$

Figure 8-10 Relationship between Active Coils, Total Coils and Solid Height.

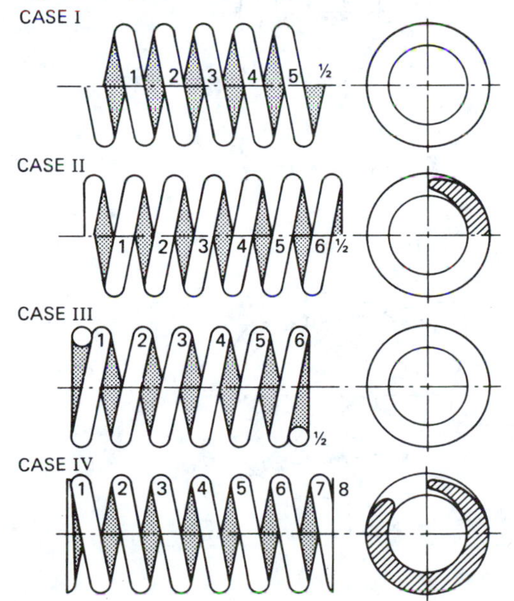

I–BOTH ENDS OPEN
Working coils = 5½
Total coils, N = 5½
Solid Height = (N + 1) d

II–BOTH ENDS OPEN-GROUND
Working coils = 6
Total coils, N = 6½
Solid Height = Nd

III–BOTH ENDS CLOSED
Working coils = 4½
Total coils, N = 6½
Solid Height = (N + 1) d

IV–BOTH ENDS CLOSED-GROUND
Working coils = 6
Total coils, N = 8
Solid Height = Nd

8-8 DESIGN EQUATIONS, HELICAL COMPRESSION SPRINGS

There are *two* basic design equations, with others relating to them. The *first* is the *stress* equation derived from the free-body diagram shown in Fig. 8-11*a*. The partial spring is shown in equilibrium under the action of the external axial force *P*, the shear force *S*, and the torsional moment *T*. Summing the axial forces, *S* is found equal in magnitude to *P*. Summing the moments with respect to the wire center *0* yields

$$T = P(0.5D)$$

The torsional shearing stress due to the torque *T* is (Fig. 8-11*b*)

$$\tau = \frac{Tc}{J} = \frac{0.5PD(0.5d)}{\pi d^4/32} = \frac{8PD}{\pi d^3} \tag{8-5}$$

If shearing stress due to *S* (equivalent to *P*) is τ_P (Fig. 8-11*c*), then

$$\tau_P = \frac{P}{A} = \frac{4P}{\pi d^2} \tag{8-6}$$

The maximum *combined* shear stress is therefore (Fig. 8-11*d*)

$$\tau_1 = \frac{8PD}{\pi d^3} + \frac{4P}{\pi d^2}$$

$$= \frac{8PD}{\pi d^3} + \frac{4PD}{\pi C d^3} = \frac{8PD}{\pi d^3}\left(1 + \frac{1}{2C}\right) \tag{8-7}$$

Equation 8-7 is adequate for *static* and *low-cycle* loading. For springs subjected to many (greater than 10^4) cycles of loading, it is inadequate. The calculated stress for this condition must include a concentration factor that takes into account the curvature at the inside of each coil. This factor, *K*, named for its originator, is the *Wahl correction factor*. It is a dimensionless multiplier of the basic torsional stress, τ. In equation form,

Figure 8-11 Loading and stress distribution for helical compression springs.

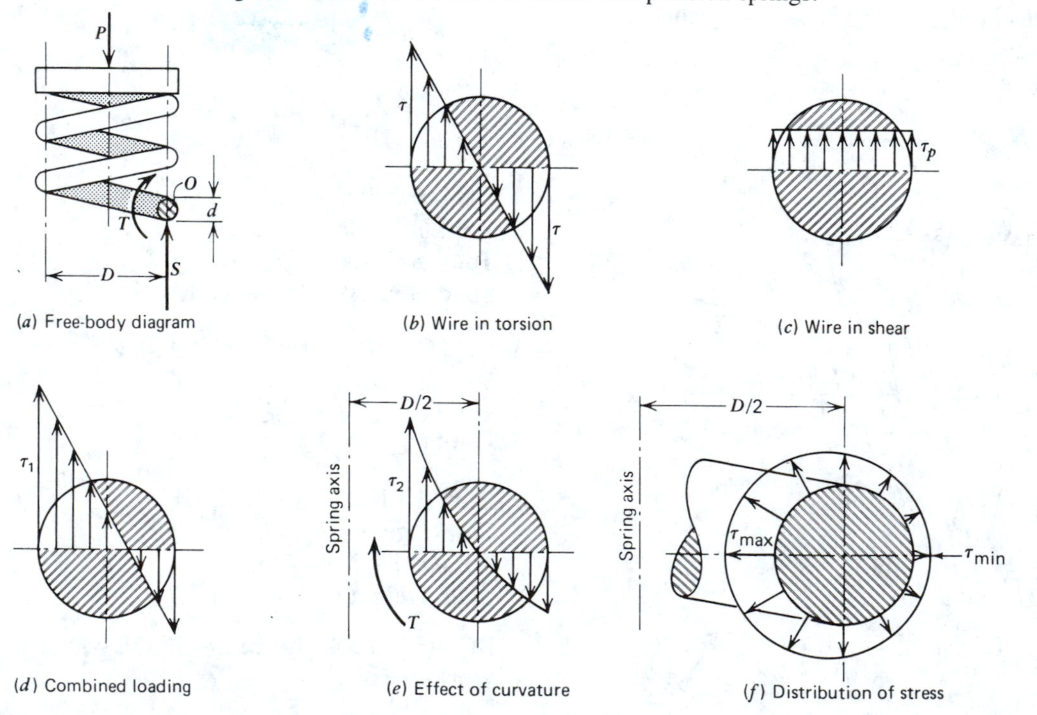

the Wahl factor is expressed as follows (Fig. 8-12 shows a graphical representation):

$$K = \frac{4C - 1}{4C - 4} + \frac{0.615}{C} \qquad (8\text{-}8)$$

The combined shear stress with the Wahl correction factor is (Fig. 8-11*e* and 8-11*f*)

$$\text{Stress, for cyclic loads, } \tau_2 = \frac{8PD}{\pi d^3}K \qquad (8\text{-}9)$$

The *second* basic design equation relates to *deflection* and is derived from *energy* considerations. Assume a spring, initially unloaded, under the action of an external force. As the force is applied, the spring deflects. The expression for the work done is

$$\text{Work} = 0.5P\delta$$

Under the influence of the induced torque, the wire will twist. This strain energy is $0.5T\theta$. These two forms of energy are equal; therefore

$$0.5P\delta = 0.5T\theta$$

$$\theta = \frac{TL}{GJ}$$

$L = \pi Dn$ (accurate for helical angles less than 10 deg)

$$T = 0.5PD$$

$$J = \frac{\pi d^4}{32}$$

By substitution and rearrangement, we have

$$P\delta = T\theta = \left(\frac{PD}{2}\right)^2 \frac{L}{GJ}$$

$$\delta = \frac{8PD^3 n}{Gd^4} \qquad (8\text{-}10)$$

Additional design relationships can be derived from these basic equations.

$$\delta = \frac{\pi \tau_2 D^2 n}{KGd} \qquad (8\text{-}11)$$

$$\delta = \frac{P}{k}$$

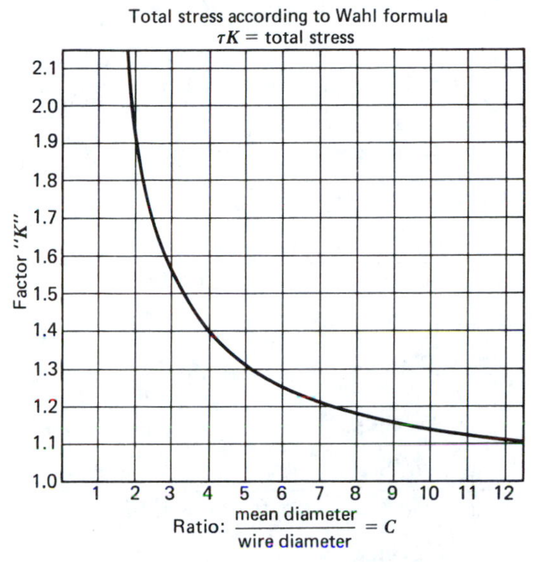

Figure 8-12 Wahl stress correction factor.

$$\text{Rate, } k = \frac{Gd^4}{8D^3 n} \qquad (8\text{-}12)$$

$$\text{Number of active coils, } n = \frac{Gd^4}{8D^3 k} \qquad (8\text{-}13)$$

EXAMPLE 8-2

A helical compression spring has the following dimensions: $D = 1.55$ in., $d = 0.192$ in. (number 6 gauge), $n = 8$. Assuming steel wire ($G = 11{,}500{,}000$ psi) and a maximum load of 145 lb, what are the basic torsional stress τ and the combined stresses τ_1 and τ_2 for static and cyclic loading? What distance does the spring deflect to resist the 145 lb load? What is the spring rate?

SOLUTION

$$\tau = \frac{8PD}{\pi d^3}$$

$$= \frac{8(145\text{ lb})(1.55\text{ in.})}{\pi(0.192\text{ in.})^3} = \mathbf{80{,}860\ psi}$$

$$C = \frac{D}{d} = \frac{1.55\text{ in.}}{0.192\text{ in.}} = 8.07$$

Static:

$$\tau_1 = \tau\left[1 + \frac{1}{2C}\right] = 80{,}860\left[1 + \frac{1}{2(8.07)}\right]$$

$$= \mathbf{85{,}870\ psi}$$

Cyclic:

$$K = \frac{4C - 1}{4C - 4} + \frac{0.615}{C} = \frac{4(8.07) - 1}{4(8.07) - 4} + \frac{0.615}{8.07}$$

$$= 1.182$$

$$\tau_2 = K\tau = (1.182)(80{,}860\ \text{psi})$$

$$= \mathbf{95{,}580\ psi}$$

$$\delta = \frac{8PD^3n}{Gd^4} = \frac{8(145\ \text{lb})(1.55\ \text{in.})^3 8}{(11{,}500{,}000\ \text{psi})(0.192\ \text{in.})^4}$$

$$= \mathbf{2.21\ in.}$$

$$\text{Rate, } k = \frac{P}{\delta} = \frac{145\ \text{lb}}{2.21\ \text{in.}} = \mathbf{65.6\ lb/in.}$$

or

$$\text{Rate, } k = \frac{Gd^4}{8D^3n} = \frac{(11{,}500{,}000\ \text{psi})(0.192\ \text{in.})^4}{8(1.55\ \text{in.})^3 8}$$

$$= \mathbf{65.6\ lb/in.}$$

8-9 COMPUTATIONS FOR HELICAL COMPRESSION SPRINGS

The design of a helical compression spring involves *four basic design parameters* that will meet the load objective at a satisfactory stress level. These four parameters are *outside diameter*, *wire diameter*, *free length*, and *number of active coils*. If the design objective is simply to attain a given load, there are many solutions. But when additional requirements such as loaded length, load-deflection rate, or a specified solid height are imposed, the number of solutions is reduced. In fact, the design may be restricted to such an extent that no solution is possible. Designers, however, are usually faced with the problem of selecting the best solution from several possible ones.

The design equations are such that load, rate, and stress are readily obtainable when the physical dimensions of the spring are known. It is quite another matter to calculate the physical dimensions of the one best spring when load, rate, and stress are the known quantities. As a result, *spring design is based on trial and error, with the final selection based on designers' judgment*. In terms of mathematics, there are *more unknowns than available equations*, so a single rigorous mathematical solution is *not* possible. A computer program for spring design calculations is often used to generate several possible solutions from which designers make the final selection.[1]

A common approach is:

1. Assume a reasonable outside diameter, if not prescribed.
2. Assume a wire type and size (wire diameter should be limited to available commercial sizes).
3. Assume the number of active coils to yield a reasonable solid height and/or load-deflection rate. (*Note:* One or both of these items may be a requirement.)
4. Calculate the stress and free length.

If these assumptions do not yield a satisfactory spring, the computations are repeated with new values for wire size, outside diameter, and number of coils. Thus, through trial and error, one can obtain the desired spring proportions for the prescribed conditions.

EXAMPLE 8-3

Design a simple hold-down spring (no fatigue) to exert a force of 500 N. Stress should be about 635 MPa. Coil ends are closed but not ground. Wire material is steel (G = 79 300 MPa).

---

[1] "Home Computers Aid Spring Design," *Design Engineering*, June 1981.

SOLUTION

Assume $OD = 18$ mm and $d = 3.2$ mm. Then $D = 14.8$ mm and $C = 4.625$.

$$\tau_1 = \frac{8PD}{\pi d^3}\left(1 + \frac{1}{2C}\right)$$

$$\tau_1 = \frac{8(500\text{ N})(14.8\text{ mm})}{\pi(3.2\text{ mm})^3}\left[1 + \frac{1}{2(4.625)}\right]$$

$$= (575.1)(1.108) = \mathbf{637\ N/mm^2}$$

This stress is satisfactory. Now assume 15 working coils in spring.

$$n = 15; \qquad N = 17$$

$$\text{Rate } (k) = \frac{Gd^4}{8nD^3} = \frac{(79\ 300\text{ N/mm})(3.2\text{ mm})^4}{8(15)(14.8\text{ mm})^3}$$

$$= \mathbf{21.375\ N/mm}$$

From Fig. 8-10, Case III (both ends closed)

$$\text{Solid height } (L_s) = (N + 1)d$$

$$= (17 + 1)(3.2\text{ mm})$$

$$= \mathbf{57.6\ mm}$$

$$\text{Deflection } (\delta) = \frac{P}{k} = \frac{500\text{ N}}{21.375\text{ N/mm}}$$

$$= \mathbf{23.4\ mm}$$

$$\text{Free length } (L_0) = L_s + \delta$$

$$= 57.6\text{ mm} + 23.4\text{ mm}$$

$$= \mathbf{81\ mm}$$

Say **85 mm** to allow some coil clearance.

The required spring specifications are:

***OD* = 18 mm**   ***d* = 3.2 mm**

L_0 = 85 mm   ***N* = 17 coils**

There are other springs that will satisfy the load-stress requirements. For example, a spring with the following specifications is functionally equivalent.

***OD* = 25 mm**   ***d* = 3.6 mm**

L_0 = 68 mm   ***N* = 10 coils**

The designer must now decide which of these two springs or others are best suited to the design. The student should verify that design load, stress, and rate are indeed equivalent.

8-10 SPRING DESIGN CHART

The trial-and-error approach to spring design was noted in the preceding section. Much of this tedium can be alleviated by the use of a spring design chart (Table 8-2). Such a chart can be of considerable help because the various parameters are displayed for the popular size ranges, thus permitting the elimination of several trials by inspection. Commercially available wire sizes are shown along the left side, together with the torsional elastic limit τ_e strength for each size. Along the top is a sequence of outside diameters that might be considered in any design problem. (Outside diameter is not restricted to the values shown, as is the case for wire diameters, because it is not based on fixed manufacturing processes or standards.) In the body of the chart the three important parameters for solving spring problems are listed.

P_e = elastic load limit; N (load that a spring of these proportions, d and OD, would attain if stressed to the elastic limit, τ_e, without considering any correction factors). Note:

$$P_e = \frac{\pi d^3 \tau_e}{8D}$$

K = Wahl correction factor

f = elastic limit deflection per coil; mm (deflection for one coil if P_e is applied)

A much simplified expression for the spring rate results from using the chart

$$k = \frac{P_e}{fn} \qquad (8\text{-}14)$$

where n is the number of active coils.

TABLE 8-2 Spring Design Chart

		Outside Diameter (D + d), mm													
Wire Diam. d, mm	τ_e MPa	7.0	8.0	9.0	10.0	11.0	12.0	13.0	14.0	15.0	16.0	17.0	18.0	19.0	20.0
		68.59	58.79	51.44	45.73	41.15	37.41	34.30	31.66	29.40					
1.0	1048	1.252	1.213	1.184	1.162	1.145	1.131	1.119	1.110	1.102					
		1.495	2.034	2.657	3.363	4.152	5.024	5.979	7.017	8.138					
		134.3	114.4	99.66	88.27	79.22	71.85	65.73	60.58	56.17	52.36	49.04			
1.25	1007	1.342	1.284	1.243	1.213	1.189	1.170	1.155	1.142	1.131	1.122	1.113			
		1.055	1.454	1.917	2.443	3.034	3.688	4.406	5.188	6.034	6.944	7.917			
		191.1	162.1	140.8	124.4	111.5	100.9	92.24	84.92	78.68	73.29	68.59	64.46	60.80	
1.4	993	1.404	1.332	1.283	1.246	1.218	1.195	1.177	1.162	1.149	1.139	1.129	1.121	1.114	
		.8812	1.224	1.623	2.078	2.590	3.157	3.781	4.461	5.197	5.990	6.838	7.743	8.704	
		289.5	244.3	211.3	186.1	166.3	150.3	137.1	126.1	116.7	108.6	101.5	95.33	89.85	84.97
1.6	972	1.498	1.404	1.340	1.294	1.259	1.231	1.209	1.190	1.175	1.162	1.151	1.141	1.133	1.125
		.7018	.9858	1.318	1.698	2.127	2.603	3.128	3.701	4.321	4.991	5.708	6.473	7.287	8.148
		416.2	349.1	300.6	263.9	235.2	212.2	193.2	177.4	164.0	152.4	142.4	133.6	125.8	118.9
1.8	945	1.610	1.485	1.404	1.346	1.303	1.269	1.242	1.221	1.202	1.187	1.174	1.162	1.152	1.143
		.5624	.7995	1.078	1.399	1.760	2.164	2.609	3.096	3.624	4.194	4.805	5.458	6.153	6.889
			483.8	414.7	362.9	322.5	290.3	263.9	241.9	223.3	207.3	193.5	181.4	170.8	161.3
2.0	924		1.580	1.476	1.404	1.351	1.310	1.278	1.252	1.231	1.213	1.197	1.184	1.172	1.162
			.6589	.8968	1.171	1.483	1.830	2.215	2.636	3.093	3.587	4.118	4.686	5.290	5.930
				598.4	521.2	461.6	414.3	375.7	343.8	316.8	293.8	273.8	256.5	241.1	227.6
2.25	903			1.580	1.485	1.418	1.367	1.327	1.295	1.269	1.247	1.229	1.213	1.199	1.187
				.7244	.9550	1.217	1.511	1.837	2.195	2.585	3.006	3.459	3.944	4.461	5.009
					722.4	637.4	570.3	516.0	471.1	433.4	401.3	373.7	349.5	328.4	309.6
2.5	883				1.580	1.493	1.430	1.381	1.342	1.310	1.284	1.262	1.243	1.227	1.213
					.7871	1.011	1.263	1.543	1.851	2.186	2.550	2.942	3.362	3.809	4.285
							807.7	728.5	663.5	609.1	562.9	523.3	488.9	458.7	432.0
2.8	862						1.515	1.453	1.404	1.365	1.332	1.305	1.283	1.263	1.246
							1.032	1.269	1.530	1.815	2.125	2.459	2.818	3.201	3.608
								1095	993.7	909.5	838.4	777.7	725.1	679.2	638.8
3.2	834							1.564	1.498	1.446	1.404	1.369	1.340	1.315	1.294
								.9916	1.204	1.438	1.692	1.966	2.262	2.578	2.914
										1308	1203	1113	1036	968.4	909.4
3.6	814									1.540	1.485	1.441	1.404	1.373	1.346
										1.164	1.377	1.608	1.857	2.124	2.409
												1533	1424	1329	1246
4.0	793											1.523	1.476	1.437	1.404
												1.327	1.539	1.767	2.011
														1905	1782
4.5	772													1.528	1.485
														1.429	1.633
5.0	752														
5.6	738														
6.3	717														
7.0	703														
8.0	679														
9.0	665														
10.0	652														

P_e elastic limit load; N
K Wahl correction factor
f elastic limit deflection per coil; mm

Pretempered carbon steel wire

$$P_e = \frac{\pi d^3 \tau_e}{8D}$$

$$f = \frac{\pi D^2 \tau_e}{Gd}$$

G = 79 300 MPa

TABLE 8-2 Spring Design Chart (*continued*)

Wire Diam. d, mm	τ_e MPa	Outside Diameter (D + d), mm 22.0	25.0	28.0	30.0	32.0	35.0	38.0	40.0	45.0	50.0	55.0	60.0	65.0
1.0	1048													
1.25	1007													
1.4	993													
1.6	972	76.64 1.112 10.02												
1.8	945	107.1 1.128 8.487	93.29 1.111 11.19											
2.0	924	145.1 1.145 7.321	126.2 1.125 9.682	111.6 1.110 12.37										
2.25	903	204.5 1.166 6.202	177.5 1.143 8.229	156.9 1.126 10.54	145.6 1.116 12.24									
2.5	883	277.8 1.189 5.321	240.8 1.162 7.084	212.5 1.142 9.099	197.0 1.131 10.58	183.7 1.122 12.18	166.7 1.110 14.78							
2.8	862	387.0 1.218 4.496	334.7 1.186 6.011	294.9 1.162 7.745	273.2 1.149 9.023	254.5 1.139 10.40	230.8 1.125 12.65	211.1 1.114 15.11						
3.2	834	570.8 1.259 3.649	492.3 1.219 4.907	432.7 1.190 6.350	400.4 1.175 7.416	372.6 1.162 8.564	337.5 1.146 10.44	308.4 1.133 12.50	291.6 1.125 13.98					
3.6	814	810.5 1.303 3.033	696.9 1.255 4.102	611.2 1.221 5.333	564.9 1.202 6.243	525.1 1.187 7.225	475.0 1.168 8.832	433.5 1.152 10.60	409.7 1.143 11.87	360.2 1.125 15.35				
4.0	793	1107 1.351 2.545	949.1 1.294 3.464	830.4 1.252 4.524	766.5 1.231 5.309	711.8 1.213 6.158	642.9 1.190 7.548	586.2 1.172 9.079	553.6 1.162 10.18	486.1 1.141 13.20	433.3 1.125 16.62	390.8 1.112 20.43		
4.5	772	1579 1.418 2.081	1348 1.346 2.856	1176 1.295 3.753	1083 1.269 4.419	1005 1.247 5.140	905.8 1.221 6.322	824.7 1.199 7.627	778.2 1.187 8.565	682.1 1.162 11.15	607.2 1.143 14.07	547.0 1.128 17.33	497.8 1.116 20.93	
5.0	752	2171 1.493 1.722	1846 1.404 2.383	1605 1.342 3.152	1477 1.310 3.724	1367 1.284 4.344	1230 1.252 5.362	1119 1.227 6.489	1055 1.213 7.299	922.8 1.184 9.533	820.3 1.162 12.07	738.3 1.145 14.90	671.2 1.131 18.02	615.2 1.119 21.45
5.6	738		2623 1.482 1.965	2272 1.404 2.620	2086 1.365 3.108	1928 1.332 3.639	1731 1.294 4.513	1571 1.263 5.481	1480 1.246 6.178	1292 1.212 8.105	1146 1.186 10.29	1030 1.166 12.74	935.6 1.149 15.45	856.8 1.136 18.42
6.3	717			3244 1.485 2.123	2971 1.435 2.533	2739 1.394 2.978	2453 1.346 3.714	2221 1.308 4.531	2089 1.287 5.121	1819 1.246 6.753	1611 1.215 8.610	1446 1.191 10.69	1311 1.172 13.0	1199 1.156 15.54
7.0	703				4117 1.515 2.105	3788 1.464 2.487	3382 1.404 3.119	3055 1.358 3.823	2869 1.332 4.333	2492 1.283 5.745	2202 1.246 7.356	1973 1.218 9.167	1787 1.195 11.18	1633 1.177 13.38
8.0	679						5056 1.498 2.451	4551 1.437 3.026	4266 1.404 3.443	3690 1.340 4.603	3251 1.294 5.931	2905 1.259 7.428	2625 1.231 9.092	2395 1.209 10.92
9.0	665							6565 1.528 2.462	6141 1.485 2.813	5288 1.404 3.794	4643 1.346 4.921	4139 1.303 6.194	3733 1.269 7.614	3400 1.242 9.180
10.0	652								8535 1.580 2.325	7315 1.476 3.164	6401 1.404 4.133	5690 1.351 5.231	5121 1.310 6.457	4655 1.278 7.814

Maximum Initial Tension
for Extension Springs
Max $IT = P_e \dfrac{d}{D}$

To use this chart, a designer would consider the maximum load demanded for the application, increase it by a trial factor, and select several combinations of wire diameter and outside diameter that provide this value for P_e, the elastic limit load. The choice of outside diameter will probably be made in accordance with the space availability dictated by surrounding parts. For economy, wire diameter is selected on the basis of the smallest size compatible with the load. To determine rate, divide the P_e value by both the f value and n, the number of active coils. When a maximum solid height is stipulated, the number of total coils is first determined for a particular wire diameter. The number of active coils is then found based on the type of spring ends to be used. If the Wahl factor is to be used in determining the stress, it can be taken into account by dividing the P_e value by the K value to obtain the maximum permissible load for this consideration.

EXAMPLE 8-4

Interpret the spring parameters shown for an outside diameter of 15 mm and a wire size of 2 mm. What load can be specified for a safety factor of 1.35? What number of coils should be specified for a rate of 8 N/mm?

The three parameters listed in the chart are:

$$P_e = 223.3 \text{ N}$$

$$K = 1.231$$

$$f = 3.093 \text{ mm}$$

The elastic limit stress τ_e is 924 MPa.

SOLUTION

A spring of these proportions having a maximum load of 223.3 N/1.35, or 165.4 N, will produce a stress of 924 N/mm²/1.35, or 684 N/mm², without considering the effects of direct shear or stress concentration. To verify this, use Eq. 8-5.

$$\tau = \frac{8PD}{\pi d^3}$$

$$D = 15 \text{ mm} - 2 \text{ mm} = 13 \text{ mm}$$

$$\tau = \frac{8(165.4 \text{ N})(13 \text{ mm})}{\pi(2 \text{ mm})^3} = 684 \text{ N/mm}^2$$

Similarly, with the Wahl correction factor, the load should be limited to 223.3 N/1.231, or 181.4 N. The simple stress would then be 924 N/mm²/1.231, or 751 N/mm².

To find the number of coils required for a rate of 8 N/mm, proceed as follows.

$$n = \frac{P_e}{fk} = \frac{223.3 \text{ N}}{(3.093 \text{ mm})(8 \text{ N/mm})} = \mathbf{9.02 \text{ coils}}$$

This can be verified by using Eq. 8-13.

$$n = \frac{Gd^4}{8D^3k} = \frac{(79\,300 \text{ N/mm}^2)(2 \text{ mm})^4}{8(13 \text{ mm})^3(8 \text{ N/mm})}$$

$$= \mathbf{9.02 \text{ coils}}$$

Considerable labor has been eliminated from the calculations by the use of the chart values. It is important, however, to remember the limitations of the chart: a fixed value for the torsional modulus of elasticity and a given stress-size relationship for the wire. In this particular case, only metric values are shown. The torsional modulus of elasticity is taken as 79 300 MPa, while the stress-size relationship is for pretempered carbon steel wire, taken from Fig. 8-2.

8-11 GOOD DESIGN PRACTICES

A well-designed spring is one that functions properly and is easily manufactured. Adherence to the following guidelines will help accomplish this goal.

Spring Index (C). The practical limits for good design and manufacturing considerations are C values between 5 and 12. Some applications, however, use a C value as low as 3.

Buckling. Since compression springs are essentially columns, their resistance to buckling

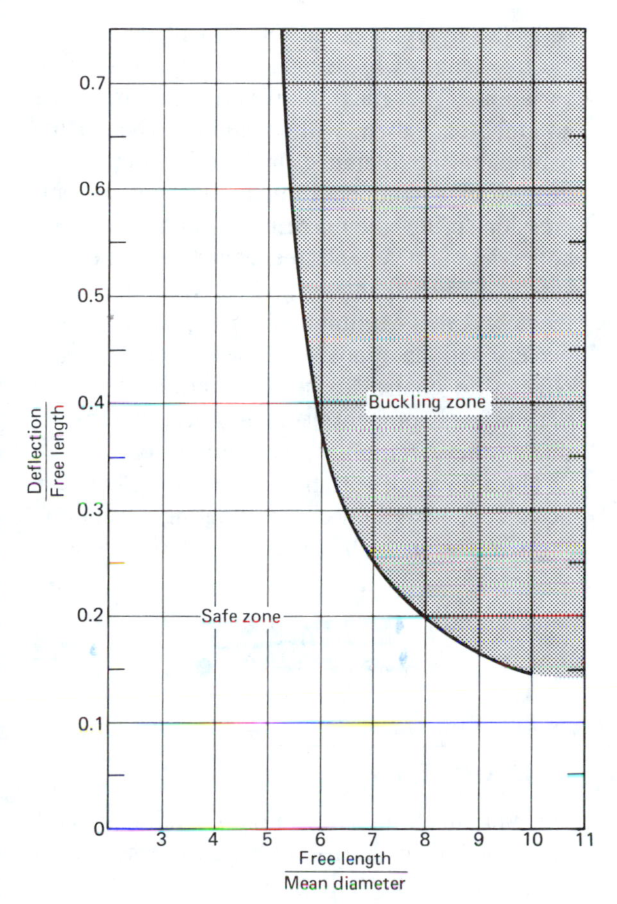

Figure 8-13 Buckling characteristics of compression springs with closed and ground ends compressed between parallel plates.

must be verified. If the free length is less than five times the mean diameter, the spring will be safe from buckling at any deflection. If the free length is more than five times the mean diameter, the spring may still be free from buckling if the deflection is not too great. The amount of deflection that can be tolerated before buckling ensues can be determined from the curve in Fig. 8-13.

Active Coils. The number of active coils specified should be greater than three. Springs with a small number of active coils are very susceptible to large variations in the rate.

Solid height versus minimum working length of spring. Springs should not be designed so that solid height is reached in the normal working range.

Spring life versus stress. There is an inverse proportionality between the number of load cycles a spring can accumulate before failure and the stress range imposed during each cycle; the higher the stress, the lower the number of cycles.

A design guide based on this concept is shown in Fig. 8-14, which indicates the required safety factor for a given number of cycles. The stress range considered is from zero to maximum. Factor of safety is defined as the ratio of elastic limit stress to Wahl corrected stress.

Figure 8-14 Life cycles versus factor of safety. (Developed by R. B. Hopkins and R. J. Erisman.)

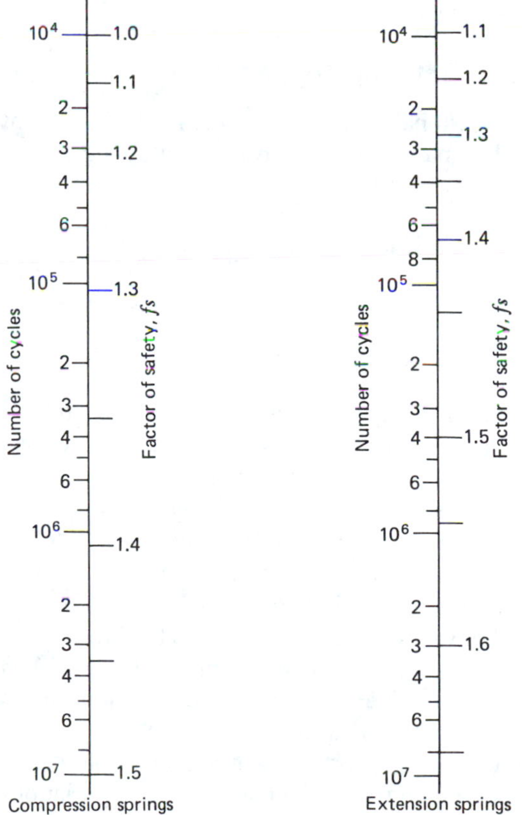

For less than 10 000 cycles, conditions are *static,* hence the stress should be calculated on the basis of the torsional and direct shear relationship.

Stability and axis alignment. The type of end configuration will affect the stability and axis alignment of helical compression springs. The open end, in which the wire is cut off without changing the spacing of end coils, is unstable. Partial grinding of the end coils perpendicular to the spring axis will improve the stability of open-ended springs. The amount of grinding, however, should never produce a thin, "feather" edge.

Stability is greatly enhanced with the practice of closing the end coils. Optimum alignment is achieved by also specifying a grinding operation. Normally a 3-deg alignment precision can be obtained with only a nominal cost penalty.

Manufacturing Tolerances

Dimensional tolerances specified and met by spring manufacturers have the following values.

$$d \pm 1.5\%$$

$$D \pm 2\%$$

$$n \pm 5\%$$

$$L_0 \pm 3\%$$

With these manufacturing tolerances, the resulting variation in load can actually exceed ±20%. Fortunately, the spring manufacturer can compensate for some of the variations so that load tolerance can be reduced to ±10% for springs with six or more coils. This will be understood when it is remembered that the load that a spring can develop is related to both the characteristic or rate and the distance it is deflected ($P = k\delta$). The mathematical expression for rate (Eq. 8-12) contains wire diameter, pitch diameter, and number of active coils. These quantities are all subject to manufacturing variations. Similarly, the actual distance that a spring is deflected to its installed length will depend on the free length which, again, is subject to some variation. The maximum variation from the nominal load, then, will depend on whether the manufacturing tolerances are all in a reinforcing instead of a compensating mode. For a compression spring, the maximum load occurs when both wire diameter and free length are to the high side *and* pitch diameter and active coils are on the low side. Minimum load, of course, will occur when the opposite is true. A spring maker will try to control those four variables so that their effects compensate each other, thereby reducing the variation in the nominal load.

The amount of variation in the spring load can be calculated as a ratio by evaluating the following expression with a specific set of tolerance values.

$$\frac{P_{max}}{P_{nom}} = \frac{k_{max}\ \delta_{max}}{k_{nom}\ \delta_{nom}} \tag{8-15}$$

EXAMPLE 8-5

What maximum deviation in loading could be expected in the following spring because of manufacturing tolerances?

$D = 0.865$ in. $\quad L_0 = 5.75$ in. $\quad n = 20$ coils

$d = 0.135$ in. $\quad L_1 = 3.60$ in.

SOLUTION

$$k_{nom} = \frac{Gd^4}{8D^3 n}$$

$$k_{max} = \frac{G(d_{max})^4}{8(D_{min})^3 n_{min}}$$

$$\frac{k_{max}}{k_{nom}} = \frac{(1.015)^4}{(0.98)^3(0.95)} = 1.187$$

$$\frac{\delta_{max}}{\delta_{nom}} = \frac{L_{0(max)} - L_1}{L_0 - L_1}$$

$$= \frac{(1.03)(5.75\text{ in.}) - 3.6\text{ in.}}{5.75\text{ in.} - 3.6\text{ in.}} = 1.08$$

$$\frac{P_{max}}{P_{nom}} = (1.187)(1.08) = \mathbf{1.282}$$

P_{max} could exceed P_{nom} by **28%**.

EXAMPLE 8-6

A compression spring of steel wire is required to support a load of 550 N at 160 mm length and a load of 1050 N at 125 mm. A life of 10^7 cycles is desired. The ends are to be closed and ground. What are the specifications for the required spring? Draw the corresponding force-deflection diagram.

SOLUTION

(*a*) Assume $P_1 = 550$ N, $L_1 = 160$ mm, $P_2 = 1050$ N and $L_2 = 125$ mm and draw part of the force-deflection diagram (Fig. 8-15).

(*b*) Calculate the required elastic limit load for spring chart comparison. The Wahl correction factor is not known, so an average value of $K = 1.2$ is assumed. The factor of safety for 10^7 cycles is 1.5.

$$P_e = P_2 K\, fs = (1050 \text{ N})1.2(1.5) = \mathbf{1890\ N}$$

(*c*) From the spring chart, select an outside diameter and wire combination with P_e equaling about 1900 N. Try $OD = 55$ mm and $d = 7.0$ mm. The table values are:

$$P_e = 1973 \text{ N};\quad K = 1.218;\quad f = 9.167 \text{ mm}$$

(*d*) Calculate the rate k.

$$k = \frac{P_2 - P_1}{L_1 - L_2}$$

$$= \frac{1050 \text{ N} - 550 \text{ N}}{160 \text{ mm} - 125 \text{ mm}}$$

$$= \frac{500 \text{ N}}{35 \text{ mm}} = \mathbf{14.3\ N/mm}$$

(*e*) Calculate the number of active coils n.

$$n = \frac{Gd^4}{8kD^3}$$

$$= \frac{(79\,300 \text{ N/mm}^2)(7.0 \text{ mm})^4}{8(14.3 \text{ N/mm})(55 \text{ mm} - 7 \text{ mm})^3}$$

$$= 15.04 \text{ coils} \qquad \textbf{Use 15 coils}$$

(*f*) Recalculate rate on the basis of 15 coils.

$$k = \frac{Gd^4}{8nD^3}$$

$$= \frac{(79\,300 \text{ N/mm}^2)(7.0 \text{ mm})^4}{(8)15(48 \text{ mm})^3}$$

$$= \mathbf{14.3\ N/mm}$$

(*g*) Calculate free length L_0.

$$L_0 = L_2 + \delta_2$$

$$= L_2 + \frac{P_2}{k} = 125 \text{ mm} + \frac{1050 \text{ N}}{14.3 \text{ N/mm}}$$

$$= 198.4 \text{ mm} \qquad \textbf{Use 198 mm}$$

(*h*) Calculate solid height L_s.

$$L_s = (n + 2)d = (15 + 2)7 \text{ mm} = \mathbf{119\ mm}$$

This value is less than minimum loaded length of 125 mm and is therefore satisfactory.

(*i*) Calculate actual loads at designated lengths.

$$P_1 = \delta_1 k = (L_0 - L_1)k$$

$$= (198 \text{ mm} - 160 \text{ mm})(14.3 \text{ N/mm})$$

$$= \mathbf{543\ N}$$

$$P_2 = \delta_2 k = (L_0 - L_2)k$$

$$= (198 \text{ mm} - 125 \text{ mm})(14.3 \text{ N/mm})$$

$$= \mathbf{1044\ N}$$

(*j*) Calculate fs at maximum load P_2.

$$fs = \frac{P_e}{P_2 K} = \frac{1973 \text{ N}}{(1044 \text{ N})1.218} = \mathbf{1.552}$$

The loads are acceptably close to the desired design loads; it is important to remember that a load tolerance of ±10% of the calculated loads is required by the manufacturer. The required specifications are:

Outside diameter: **55 mm**

Wire diameter: **7.0 mm**

Total number of coils: **17**

Free length: **198 mm**

8-12 EXTENSION COIL SPRINGS

Extension springs are formed by winding wire into either an open or closed helix to which a pull or tension may be applied to increase its length. When extended, this spring tends to draw two objects together (Fig. 8-16*a*). Such springs are usually *closely wound,* with all coils in tight contact. Close coil-to-coil contact, obtainable on spring-making equipment, imposes a *preload* on the spring. The effect of preload (sometimes called initial tension) is an internal force tending to hold the coils together. This force must be overcome by an external load before the coils will begin to separate. Consequently, the desired spring load can be achieved with less deflection, making a more compact design.

Designers obviously must know how much initial tension (*IT*) is practical to complete the design and so must specify a realistic value. Based on experience, a maximum value obtainable is given by

$$IT = P_e \frac{d}{D} \tag{8-16}$$

After a load great enough to overcome initial tension has been applied, an extension spring will deflect in a theoretical straight-line relationship with a uniform rate. For example, suppose the initial tension is 35 N and the spring rate is 2 N/mm; then, at a deflection of 25 mm, the load is

$$35 \text{ N} + (2 \text{ N/mm})(25 \text{ mm}) = 85 \text{ N}$$

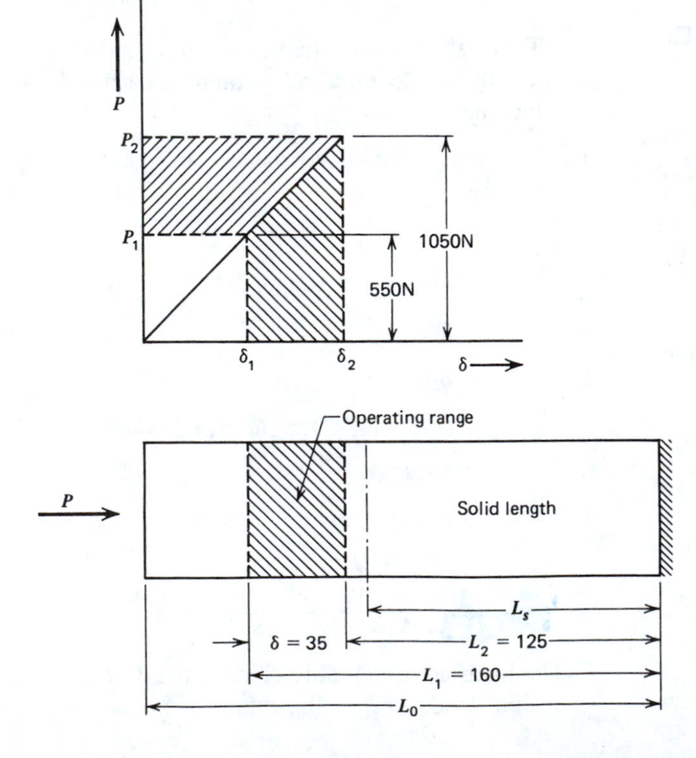

Figure 8-15 Force deflection diagram for Example 8-6.

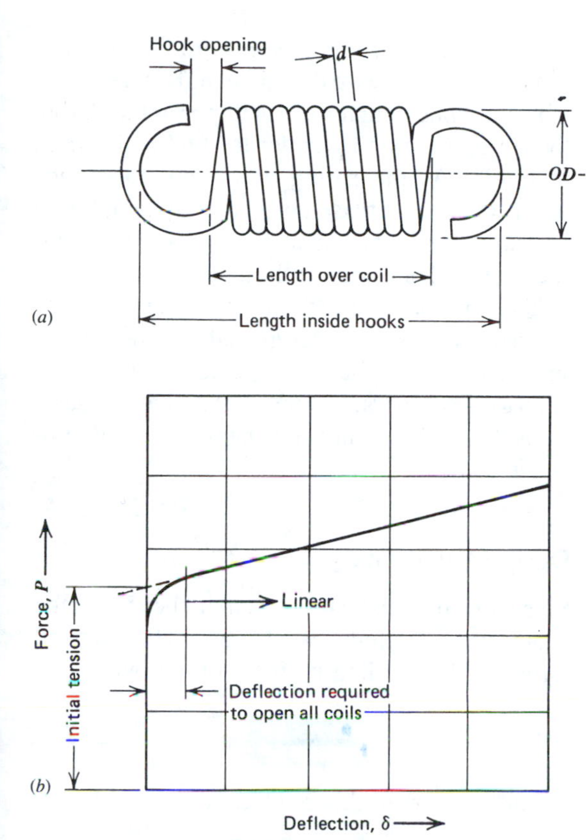

Figure 8-16 Helical extension coil spring. (*a*) nomenclature (*b*) characteristic with initial tension.

Without initial tension the spring load would be only

$$(2 \text{ N/mm})(25 \text{ mm}) = 50 \text{ N}$$

Figure 8-16*b* shows the characteristic of an extension spring with built-in initial tension.

Except for the initial tension consideration, the same formulas apply as for compression springs. Types of ends, however, have considerable influence on design, not only controlling methods of attaching but also limiting loads and space. An extension spring cannot be loaded to the same extent as a similar compression spring because of the added stress concentration at the hook juncture. This is taken into account in the design procedure by requiring a larger (by about

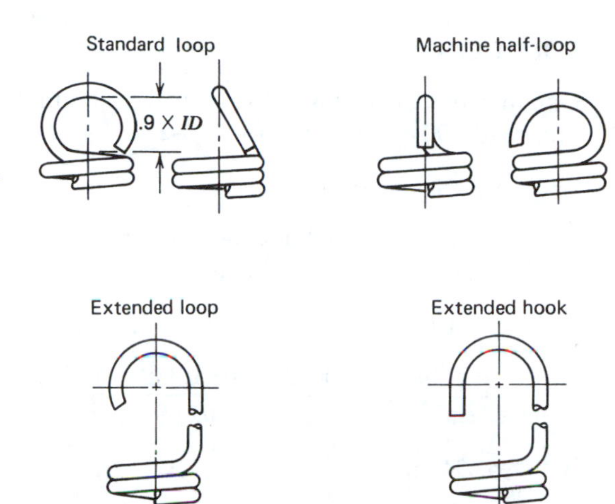

Figure 8-17 Extension spring ends.

10%) factor of safety for the same life cycles (Fig. 8-14).

The simplest hook is formed by distorting end coils in such a way as to direct the applied and reactive force along the center line of the spring. This and other possible configurations are shown in Fig. 8-17.

All formed hooks have a certain amount of *innate flexibility* that, if not taken into account, leads to an apparent discrepancy between a spring's calculated and actual rate. To avoid this confusion, designers sometimes calculate rate on a larger number of coils than specified. The increase is on the order of $\frac{1}{2}$ coil.

Free length of an extension spring is defined as hook-to-hook length, measuring from the inside of one hook to the other. Thus the type of hooks must be specified before the design can be completed.

EXAMPLE 8-7

What is the expected life of the following extension spring? Assume pretempered carbon steel wire.

$OD = 1.5$ in.	$L_0 = 4.25$ in.
$d = 0.162$ in.	$IT = 10$ lb.
$N = 12\frac{1}{2}$	$L_1 = 6.875$ in.

Assume $n = N + \frac{1}{2}$ for hook flexibility.

SOLUTION

Obtain τ_e by converting 0.162 in. to 4.115 mm and read 990 N/mm² from Fig. 8-2.

$$\tau_e = 990 \text{ N/mm}^2 = 143{,}550 \text{ psi}$$

$$P_e = \frac{\pi d^3 \tau_e}{8D} = \frac{\pi(0.162 \text{ in.})^3(143{,}550 \text{ psi})}{8(1.338 \text{ in.})}$$

$$= 179.1 \text{ lb}$$

$$\delta = \frac{\pi \tau_e D^2 n}{Gd} = \frac{\pi(114{,}200 \text{ psi})(1.338 \text{ in.})^2(13)}{(11{,}500{,}000 \text{ psi})(0.162 \text{ in.})}$$

$$= 5.634 \text{ in.}$$

$$k = \frac{P_e}{\delta} = \frac{179.1 \text{ lb}}{5.634 \text{ in.}} = 31.8 \text{ lb/in.}$$

$$P_1 = IT + k(L_1 - L_0)$$

$$= 10 \text{ lb} + (31.8 \text{ lb/in.})(6.875 \text{ in.} - 4.25 \text{ in.})$$

$$= 93.5 \text{ lb}$$

$$K = \frac{4C - 1}{4C - 4} + \frac{0.615}{C}$$

$$C = \frac{1.5 \text{ in.} - 0.162 \text{ in.}}{0.162 \text{ in.}} = 8.26$$

$$K = 1.178$$

$$\tau_2 = \frac{8P_1 D}{\pi d^3} K = \frac{8(93.5 \text{ lb})(1.338 \text{ in.})}{\pi(0.162 \text{ in.})^3} 1.178$$

$$= 88{,}270 \text{ psi}$$

$$fs = \frac{\tau_e}{\tau_2} = \frac{143{,}550 \text{ psi}}{88{,}270 \text{ psi}} = 1.63$$

Life = 3.5 × 10⁶ cycles

8-13 TORSION COIL SPRINGS

When a coil spring is adapted to provide a torque about the helical axis, it is called a torsion spring. The adaptation is a matter of forming the ends so that load application and reaction subject the spring to torsion. For this reason, there are no standard ends and each design will have its own unique features.

Torsion springs stress the material in *bending*. Since coiling the wire during manufacture is also a bending phenomenon, it is important to design the spring to wind up under load to avoid stress reversals. As the spring winds up, the inside diameter will decrease and the body length will increase. Designers must provide for these dimensional changes in each application.

Torsion spring use varies so widely that no standard forms are recognized. Some of the more common applications include spring hinges used to close doors, ordinary safety pins, mousetraps, and the return mechanism in window shade rollers.

Design Equations

Since the material is stressed in bending, the beam stress formula can be used; the applied torque is the bending moment on the wire.

$$\sigma_b = \frac{Tc}{I}$$

For round wire this reduces to

$$\sigma = \frac{0.5dT}{\pi d^4/64}$$

$$= \frac{10.2T}{d^3} \qquad (8\text{-}17)$$

The angular deflection required to develop a given torque can be derived from an approach based on conservation of energy. The design formula becomes

$$T = \frac{Ed^4}{10.2Dn} \quad \text{for 360 deg deflection} \qquad (8\text{-}18)$$

Designers must recognize that in many instances the spring ends will deflect under the application of the torque, thus giving rise to a more flexible condition than the formula indicates. Some judgment factor should be applied to compensate for this, a factor usually determined by arbitrarily calculating with a larger number of coils than will actually be wound into the spring.

EXAMPLE 8-8

Design a torsion spring to produce a torque of 45 lb-in. when deflecting through an angle of 60 deg. Allowable stress is 80,000 psi, E = 30 million psi. Allow a 5% correction factor for the flexibility of the ends. The spring must fit over a 2.25-in. diameter pin. Assume that the clearance between spring and pin is 5% of the pin diameter.

SOLUTION

From Eq. 8-17,

$$d^3 = \frac{10.2T}{\sigma} = \frac{(10.2)(45 \text{ lb-in.})}{80{,}000 \text{ psi}} = 0.00574 \text{ in.}^3$$

$$d = \mathbf{0.179 \text{ in.}}$$

Use 0.177 in. (a commercial gauge). Assume

$$D = \text{pin diameter} + \text{wire diameter} + \text{clearance}$$

$$D = 2.25 \text{ in.} + 0.177 \text{ in.} + 0.123 \text{ in.}$$

$$= \mathbf{2.55 \text{ in.}}$$

From Eq. 8-18, the theoretical number of coils is:

$$n = \frac{Ed^4}{10.2TD} \frac{\text{deg travel}}{360 \text{ deg}}$$

$$= \frac{(30)(10^6 \text{ lb/in.}^2)(0.177 \text{ in.})^4}{(10.2)(45 \text{ lb-in.})(2.55 \text{ in.})} \frac{60 \text{ deg}}{360 \text{ deg}}$$

$$= 4.2 \text{ coils}$$

The actual number of coils is:

$$n_a = \frac{4.2}{1.05} = \mathbf{4.0 \text{ coils}}$$

8-14 LEAF SPRINGS

Leaf springs are made from essentially flat stock and are stressed in bending because of their beamlike configuration. They are used in appliances and many industrial products and have been commonly used as a part of automotive suspension systems. These springs are classified in the same manner as beams: cantilever, elliptical, and semielliptical (Fig. 8-18). The cantilever designation implies a single clamped support at one end with loading imparted at the free end. Elliptical and semielliptical springs are really simple beams suitably curved at midspan to allow greater deflections. One such spring by itself is designated semielliptical. Two springs mounted with curvature reversed (approximating an ellipse in appearance) are designated elliptical.

All categories of leaf springs can be designed as single leaf or multileaf to suit the application. The advantages of lamination are a more compact design and greater allowable deflections for a given load. By shortening successive leaves (Fig. 8-19), the spring becomes a modification of a beam of uniform strength. Also, it is possible to group the leaves to form a spring with a progressive rate. This is desirable in automotive suspension systems, where a soft ride is preferred over minor road disturbances but protection against bottoming is necessary for severe road irregularities.

Design Equations

Design procedures will be shown for the cantilever-type spring. Equations for stress and

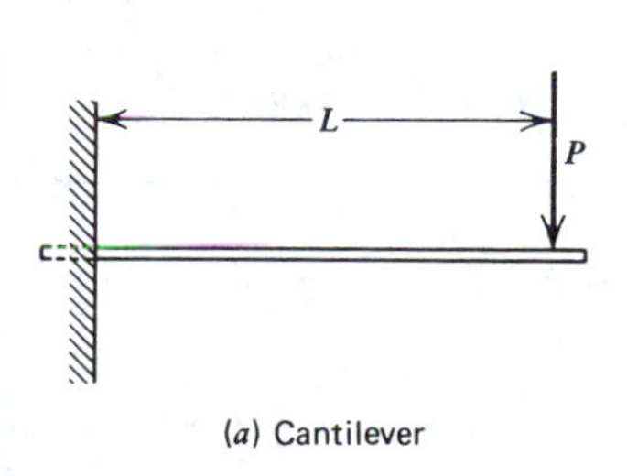

(*a*) Cantilever

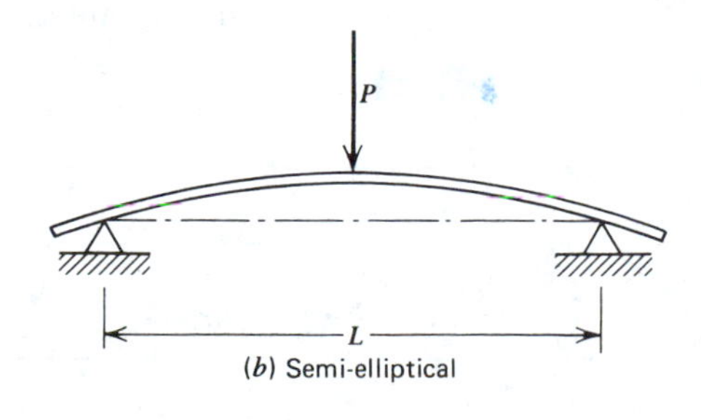

(*b*) Semi-elliptical

Figure 8-18 Single-leaf springs.

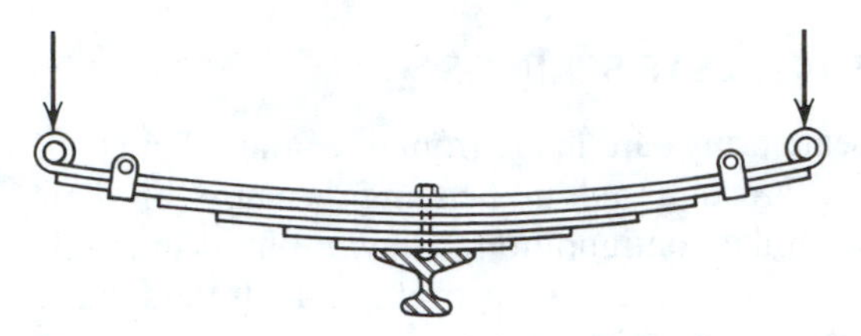

Figure 8-19 Typical multileaf spring.

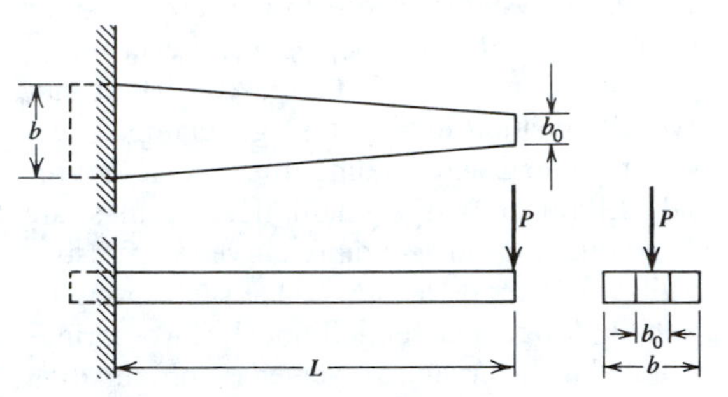

Figure 8-20 Tapered cantilever spring.

deflection are adapted from those for a constant section cantilever beam.

Stress: $$\sigma_b = \frac{PLc}{I}$$

Deflection: $$\delta = \frac{PL^3}{3EI}$$

Rate: $$k = \frac{P}{\delta} = \frac{3EI}{L^3}$$

The deflection can be increased without changing the maximum stress by tapering the spring width from the support to the free end, as shown in Fig. 8-20. The amount of increase is in accordance with the ratio of free-end width to fixed-end width of the beam. Representative values are given in Table 8-3. The design equations for spring analysis thus become:

$$\sigma_b = \frac{6PL}{bt^2} \tag{8-19}$$

$$\delta = \frac{4PL^3K}{Ebt^3} \tag{8-20}$$

$$k = \frac{Ebt^3}{4L^3K} \tag{8-21}$$

When deflection and stress (as well as load and length) are considered the *known* values, the dimensional relationships can be determined directly by using the following adaptation of Eqs. 8-19 and 8-20.

$$t = \frac{L^2\sigma_b K}{1.5\delta E} \tag{8-22}$$

$$b = \frac{6PL}{\sigma_b t^2} \tag{8-23}$$

Maximum deflection is achieved when $K = 1.5$, in which case the leaf assumes a *triangular* configuration. This arrangement produces the greatest shear stress at the tip of the spring directly under the load and is impractical from the standpoint of load application. Minimum deflection occurs when $K = 1.0$, in which case the leaf has a rectangular shape. The tapered shape permits better material utilization than the simple rectangular configuration. This can be seen by comparing the volumes of material in equivalent springs of each shape. Equivalent springs are defined as those having equal stress and deflection values when the loads and lengths are the same.

EXAMPLE 8-9

Compare the volume of material of two cantilever springs that will deflect 40 mm under a 20 N load. Each beam is 150 mm long and has a stress limitation of 500 MPa. One is of constant cross section and the other has a b_0/b ratio of 0.2. Assume $E = 207\ 000$ MPa.

TABLE 8-3 *K* Values

b_0/b	1.0	0.8	0.6	0.5	0.4	0.2	0
K	1.0	1.071	1.154	1.2	1.25	1.364	1.5

SOLUTION

Substitute the given values into Eqs. 8-22 and 8-23. For the case of $K = 1.0$:

$$t = \frac{(150\text{ mm})^2(500\text{ N/mm}^2)(1.0)}{(1.5)(40\text{ mm})(207\,000\text{ N/mm}^2)} = \mathbf{0.906\text{ mm}}$$

$$b = \frac{6(20\text{ N})(150\text{ mm})}{(500\text{ N/mm}^2)(0.906\text{ mm})^2} = \mathbf{43.9\text{ mm}}$$

The volume of material in the beam is as follows.

$$\text{Volume} = btL$$
$$= (43.9\text{ mm})(0.906\text{ mm})(150\text{ mm})$$
$$= \mathbf{5966\text{ mm}^3}$$

For the case of $b_0/b = 0.2$, $K = 1.364$:

$$t = (0.906\text{ mm})(1.364) = \mathbf{1.236\text{ mm}}$$

$$b = \frac{6(20\text{ N})(150\text{ mm})}{(500\text{ N/mm}^2)(1.236\text{ mm})^2} = \mathbf{23.56\text{ mm}}$$

$$b_0 = b(0.2)$$
$$= (23.56\text{ mm})(0.2) = \mathbf{4.712\text{ mm}}$$

$$\text{Volume} = (0.5)(b + b_0)tL$$
$$= (0.5)(23.56\text{ mm} + 4.71\text{ mm})$$
$$(1.236\text{ mm})(150\text{ mm})$$
$$= \mathbf{2621\text{ mm}^3}$$

$$\text{Reduction} = \frac{5966\text{ mm}^3 - 2621\text{ mm}^3}{5966\text{ mm}^3}100$$
$$= \mathbf{56\%}$$

Thus the two springs are functionally equivalent but the tapered one uses only 44% as much material as the straight-sided one.

8-15 BELLEVILLE SPRINGS

Belleville springs have gained wide acceptance for applications requiring *large* loads with *small* deflections. Their geometry can be tailored to produce variously shaped load-deflection

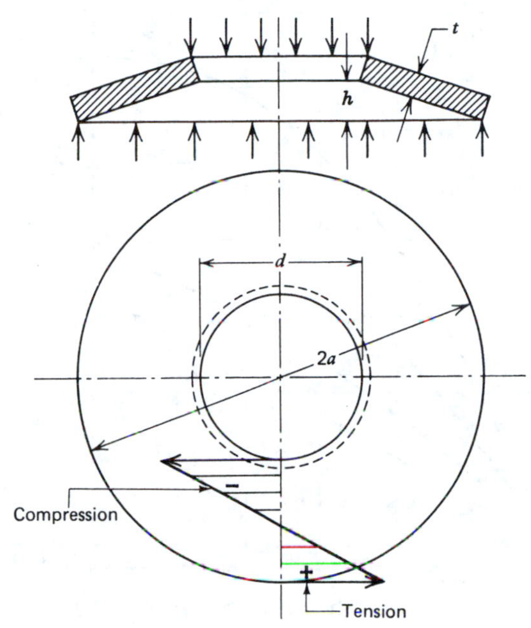

Figure 8-21 Belleville spring.

curves, making them advantageous when this characteristic is desired.

In its simplest form, a Belleville spring consists of a single *coned* disk or washer of the configuration shown in Fig. 8-21. A force applied axially will load the disk in bending. The maximum stress occurs on the upper surface at the inside diameter and is compressive. Tensile stresses of lower magnitude occur at the outside diameter. Stress levels for static loads can safely reach 2100 MPa. This exceeds the yield strength for most steels and simply indicates that while some localized yielding occurs, a compensating residual stress is produced that lowers the actual stress at this point. This redistribution of stresses is possible when high stresses are confined to a small area and are surrounded by a large volume of material at lower stresses. If the spring must endure many repeated load cycles, lower design stresses are mandatory.

The characteristic load-deflection curves for Belleville springs are shown in Fig. 8-22. Note that the curves are *nonlinear*, varying from a weak to a strong *degressive* characteristic de-

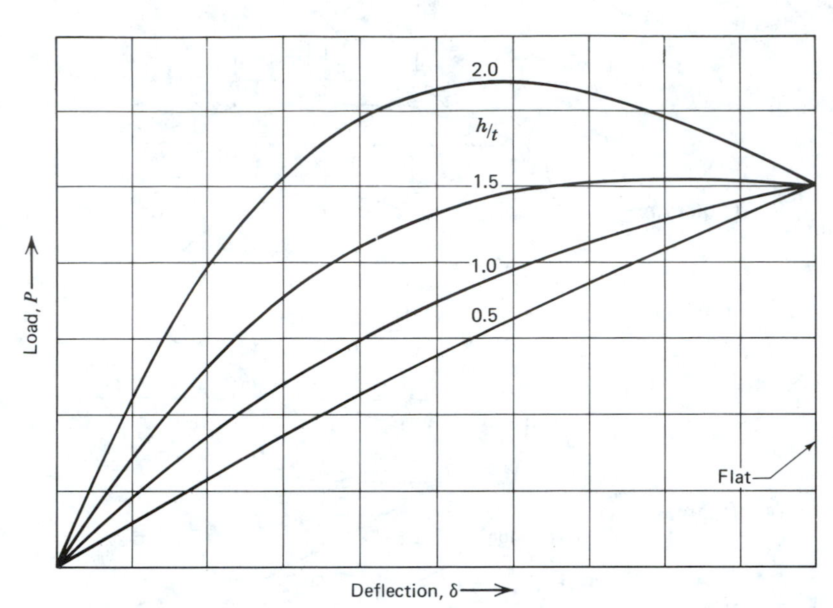

Figure 8-22 Belleville springs load-deflection curves.

pending on whether the geometric factor of h/t is small or large. For most applications the ratio of coned height h to thickness t will have a value between 0.5 and 2.0. If the ratio exceeds 2.83, the load will become negative at some point beyond the flat position and will require loading in the opposite direction to return it to its unloaded position.

Since the deflection relative to load is small, it is at times desirable to stack several washers in series to increase the overall deflection (Fig. 8-23). An increase in load capacity results when stacking is in parallel. Stacking in parallel series will provide an increase in both deflection and load.

The theory of Belleville springs is very complex, making the equations for calculating loads, deflections, and stresses intricate and cumbersome. It is possible, however, to reduce much of the theory to dimensionless relationships, and we will further facilitate solutions by considering only loads and stresses at the *flat* position. Design charts are presented for this purpose in Figs. 8-24 and 8-25. Quantities $\emptyset$ and β have no physical significance. They are a means of simplifying the expressions for σ and P. ν (nu) is Poisson's ratio (0.30 for steel) and a the outside radius.

$$\sigma_c = \frac{-E}{(1 - \nu^2)} \emptyset \left(\frac{t}{a}\right)^2 \qquad (8\text{-}24)$$

Note that maximum stress is compressive.

$$P = \frac{-\sigma t^2}{\beta} \qquad (8\text{-}25)$$

EXAMPLE 8-10

Find the dimensions of a steel Belleville spring to resist 2.6 kN (flat position) with a material stress of 1400 MPa (compression). A

Figure 8-23 Belleville spring combinations.

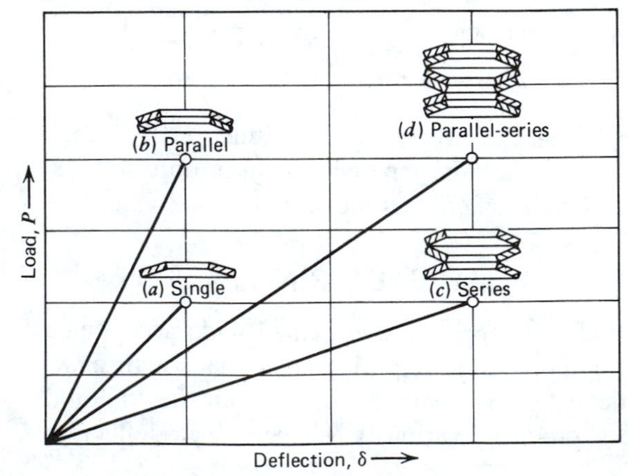

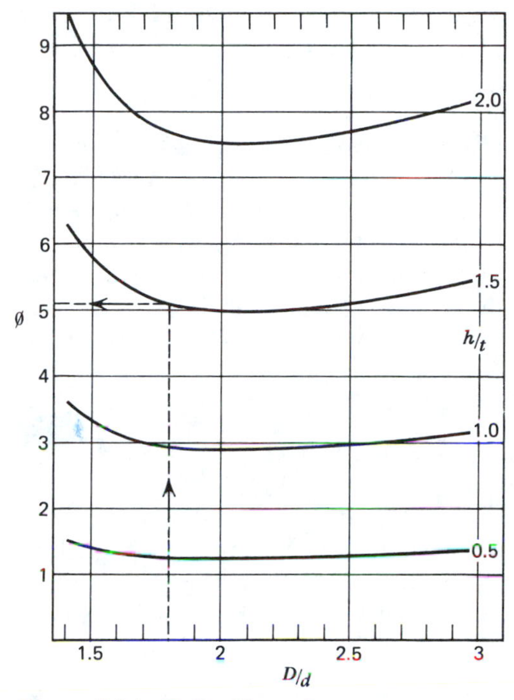

Figure 8-24 Belleville springs stress parameter.

Figure 8-25 Belleville springs load parameter.

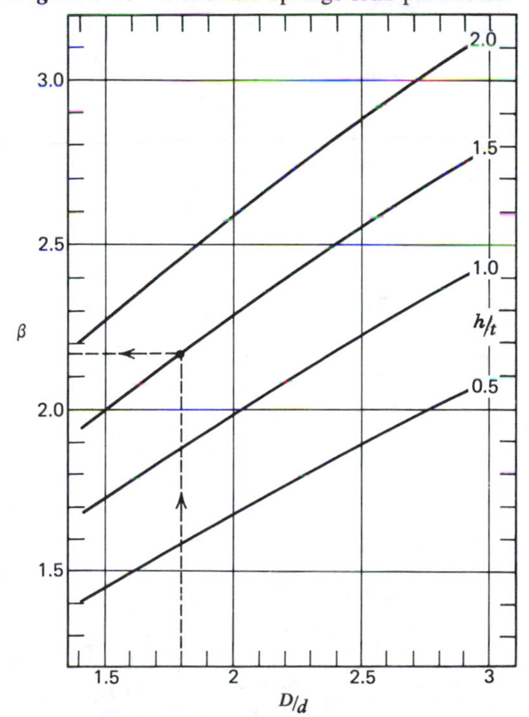

load-deflection characteristic of $h/t = 1.5$ is desired. Let $D/d = 1.8$.

SOLUTION

From Fig. 8-24 $\quad \emptyset = 5.09$

From Fig. 8-25 $\quad \beta = 2.175$

From Eq. 8-25 $\quad t^2 = \dfrac{-P}{\sigma}\beta$

$$t^2 = \frac{(-2600 \text{ N})(2.175)}{-1400 \text{ N/mm}^2} = 4.03 \text{ mm}^2$$

$t = 2.010$ mm $\quad$ **Use 2.0 mm**

From Eq. 8-24 $\quad a^2 = \dfrac{-E}{(1 - \nu^2)} \dfrac{t^2}{\sigma_c} \emptyset$

$$a^2 = -\frac{207\ 000 \text{ N/mm}^2}{1 - 0.3^2} \frac{4 \text{ mm}^2}{-1400 \text{ N/mm}^2}(5.09)$$

$a^2 = 3308$ mm^2 $\quad$ **a = 57.5 mm**

The required dimensions are:

$D = 2(57.5$ mm$) =$ **115 mm**

$d = 115$ mm$/1.8 =$ **63.9 mm**

$t =$ **2.0 mm**

$h = 1.5(2.0$ mm$) =$ **3.0 mm**

8-16 COUPLING OF SPRINGS

Springs are often used in combination because of space limitations and because a combination may be more efficient than a single equivalent spring. Often a spring combination will yield a load-deflection curve not possible with a single spring.

Parallel

Two or more springs are said to be coupled in *parallel* when (1) each spring is in direct contact with the load and a common support member so that they deflect the same distance, regardless of total load (Fig. 8-26), and (2) total deflection is consequently less than if one spring were used to

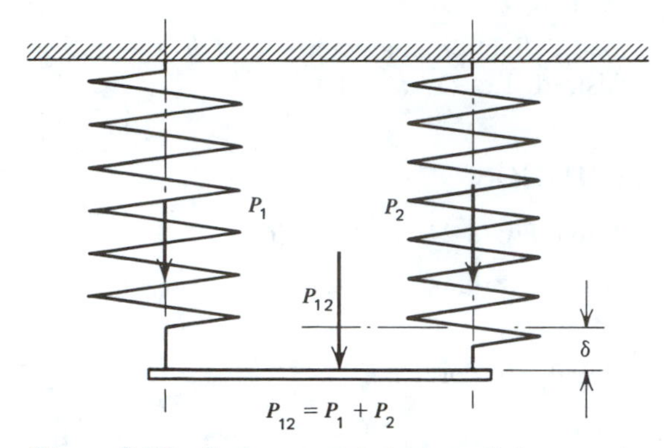

Figure 8-26 Springs coupled in parallel.

support the identical load. A compound system based on parallel coupling therefore has a spring rate that is the algebraic sum of the individual spring rates, or

$$k = k_1 + k_2 + k_3 + \cdots + k_n \quad (8\text{-}26)$$

When helical compression springs are coupled in parallel, the axes must be parallel. If, additionally, the coils are mounted *concentrically,* they form a *nested* spring. This combination is useful when a heavy load must be carried in a restricted space and a single spring would be overstressed (Fig. 8-27). Nested springs should have some circumferential *clearance* and should be *wound alternately* left and right to avoid pinching adjacent coils.

Figure 8-27 Nested springs.

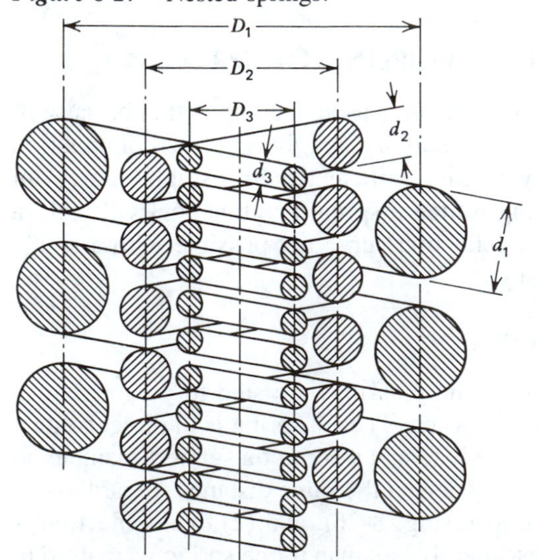

Series

For increased overall deflection a set of springs can be coupled in *series,* as shown in Fig. 8-28. *The total force is then sustained by each spring.* As a result, the total deflection is made up of the individual deflections of each spring reacting to total force. For springs with linear characteristics, the resulting rate k can be found by means of the following formula.

$$\frac{1}{k} = \frac{1}{k_1} + \frac{1}{k_2} + \frac{1}{k_3} + \cdots + \frac{1}{k_n} \quad (8\text{-}27)$$

Parallel series

When a larger deflection as well as a larger load capacity is needed in a design, a set of springs may be combined in *parallel series* (see Example 8-11).

EXAMPLE 8-11

What is the rate k of the spring system shown in Fig. 8-29?

$k_1 = 30$ lb/in. $\quad k_2 = 40$ lb/in.

$k_3 = 40$ lb/in. $\quad k_4 = 50$ lb/in.

$k_5 = 50$ lb/in. $\quad k_6 = 50$ lb/in.

SOLUTION

k_2 and k_3 are in parallel.

From Eq. 8-26,

$$k_{23} = k_2 + k_3$$
$$= 40 \text{ lb/in.} + 40 \text{ lb/in.}$$
$$= 80 \text{ lb/in.}$$

k_4, k_5, and k_6 are in parallel.

$$k_{456} = 50 \text{ lb/in.} + 50 \text{ lb/in.} + 50 \text{ lb/in.}$$
$$= 150 \text{ lb/in.}$$

k_1, k_{23}, and k_{456} are in series.

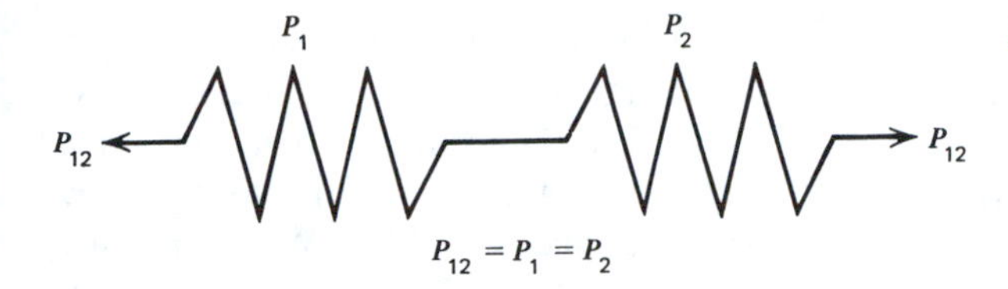

Figure 8-28 Springs coupled in series.

From Eq. 8-27,

$$\frac{1}{k} = \frac{1}{k_1} + \frac{1}{k_{23}} + \frac{1}{k_{456}}$$

$$\frac{1}{k} = \frac{1}{30 \text{ lb/in.}} + \frac{1}{80 \text{ lb/in.}} + \frac{1}{150 \text{ lb/in.}}$$

$$= 0.0333 \text{ in./lb.} + 0.0125 \text{ in./lb}$$

$$+ 0.00667 \text{ in./lb}$$

$$= 0.0525 \text{ in./lb}$$

$$\boldsymbol{k = 19.05 \text{ lb/in.}}$$

SUMMARY

Springs are elastic machine members that can undergo large deflections without being permanently deformed. *The primary purpose of springs is energy absorption during deflection and energy release during recoil.* Most springs do not possess damping capacity and therefore deflect proportionally to the imposed load. Such springs are said to have linear characteristics or a constant spring rate.

Figure 8-29 Springs in parallel series. (Example 8-11)

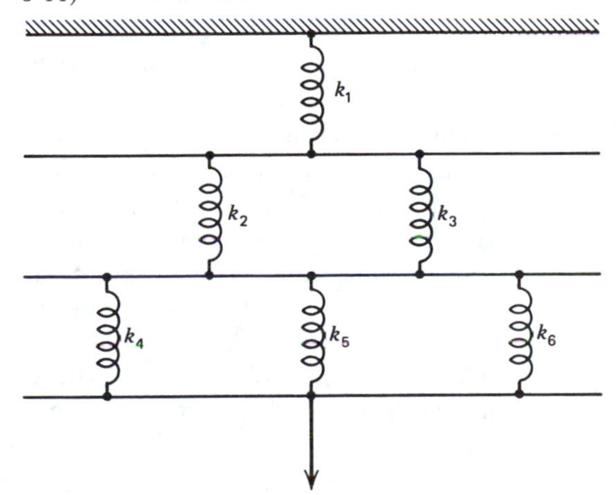

Mechanical spring design is based on the deflection characteristics of highly flexible metallic shapes such as wires and flat strips. Most springs are coiled, since this configuration yields maximum material and space utilization for a multitude of designs. Designers are usually trying to find the spring proportions that will permit an acceptable or stated deflection within an allowable stress condition. As is often the case, these two qualities are in opposition. In a coil spring, with a constant pitch diameter, decreasing the wire diameter will increase deflection but raise the stress. With a constant wire diameter, increasing the pitch diameter will increase deflection but, again, the stress is also raised. Designers are further limited by the amount of space available to mount the spring. The calculations to satisfy all of these requirements will usually involve trial and error. A spring table has been included to alleviate this mathematical burden.

Techniques are developed in this chapter to illustrate the design of (1) wire springs (compression, tension, and torsion), (2) flat stock springs, and (3) Belleville washer springs. The helical compression spring is probably the most common spring encountered in industrial products. The procedures can be adapted to include the design of specialty devices having a springlike member.

Nominal spring stresses are very high in comparison to other machine elements. This is necessarily true, since deflections are also comparatively larger. Fortunately, the spring material is structurally uniform and can be processed to allow high operating stresses.

High-carbon steel is the most commonly used spring material. It is inexpensive, has a large modulus of elasticity, is easily worked, and its endurance can be improved through hardening. The most important consideration in spring engineering is to impart and maintain a smooth surface condition. Any surface imperfection coupled with high operating stresses will produce early fatigue failure.

Spring making is a specialty, and springs are generally mass produced. With special processing tools and high production runs, springs are relatively inexpensive. Tooling can be adapted so that special-purpose, nonstandard springs can also be produced economically.

REFERENCES

8-1 *Handbook of Mechanical Spring Design*. Bristol, Conn.: Associated Spring Corporation, 1964.

8-2 Mather, G. E., et al. *Mechanical Spring Design Guide*. Clawson, Mich.: North American Rockwell, Mechanical Spring Operations, 1972.

8-3 *Spring Design in Brief*. Bristol, Conn.: Associated Spring Corporation, 1969.

8-4 "Spring Selection Simplified," *Machine Design*, September 16, 1965.

PROBLEMS

P8-1 A helical spring must absorb 3.0 N·m while deflecting 10 mm. Find the corresponding load. Draw the characteristic curve.

P8-2 A helical spring is made from 2.8 mm phosphor bronze wire. The outside diameter is 18 mm. How many active coils are required to obtain a spring rate of 6.5 N/mm?

P8-3 A helical compression spring is made from 0.1483 in. (number 9 gauge) pretempered carbon steel wire. Its outside diameter is 0.906 in., and it has 15 active coils. Find the maximum stress (assume a high fatigue condition) for a deflection of 1.375 in. The spring has a short service life. Why? Will it buckle?

P8-4 A stainless steel helical compression spring has an outside diameter and a wire diameter of 32 mm and 4.5 mm, respectively. It must sustain a static load of 725 N after deflecting 60 mm. What are the maximum stress and the number of active coils? Will it buckle?

P8-5 Two helical coil springs are mounted in tandem (series). Their rates are 70 and 45 lb/in. What force is required to produce a deflection of 2.4 in.?

P8-6 A compression spring has an outside diameter of 35 mm, a wire diameter of 4 mm, and 12 coils. The free length is 120 mm; the ends are closed and ground. Material is steel. Use the design chart to find:

(*a*) Rate of spring.
(*b*) Load at solid height.
(*c*) Stress at solid height.
(*d*) Length and load at which the spring could operate for 10^7 cycles.

P8-7 Design a simple hold-down spring (compression) to exert 250 N at a length of 30 mm and fitting inside an 18 mm diameter hole. Use pretempered carbon steel wire. Assume spring ends closed and ground. Specify appropriate tolerances.

P8-8 A governor spring requires a maximum force of 70 N and a rate of 2 N/mm. The spring should be able to fit over a 10 mm diameter pin. Length is not critical. Assume ends closed and ground, and design for $(5)(10^6)$ cycles. What are the required spring dimensions? Material is steel. Set tolerances at 50% of normal because of the critical nature of the spring.

P8-9 An extension spring has an outside diameter of 1.00 in., a wire diameter of 0.1205 in. (number 11 gauge), and 21 coils. The free length of the spring is 3.75 in. Material is steel. Assume no hook flexibility. Find:

(*a*) Load required to separate coils (specify maximum).
(*b*) Spring rate.
(*c*) Load sustained by the spring when stretched to a length of 5.9 in.
(*d*) Torsional stress at this load.
(*e*) Combined torsional and shear stress.
(*f*) Torsional stress with curvature correction.
(*g*) Expected number of fatigue cycles.

P8-10 Design an extension spring to meet the following requirements. Material is steel.

(*a*) Load at 110 mm = 160 N.
(*b*) Load at 135 mm = 410 N.
(*c*) Outside diameter = 32 mm.
(*d*) Design for 100,000 cycles.
(*e*) Hooks to mount over pins.
(*f*) Allow $\frac{1}{2}$ coil extra for hook flexibility.

P8-11 Bumper springs are required in the bottom of a freight elevator shaft to provide a cushioned stop in the event of overtravel. The mass of the loaded elevator is 1800 kg, and its speed of descent is 1.3 m/s. What are the spring dimensions needed to bring the elevator cage to a stop in 150 mm? Base the calculations on four springs, one in each corner of the shaft. Springs will have ends closed and ground. Allowable stress for this infrequent operation may be assumed to be 620 MPa. If outside diameter is 100 mm, what are the required wire diameter, number of coils, and free length? Material is steel.

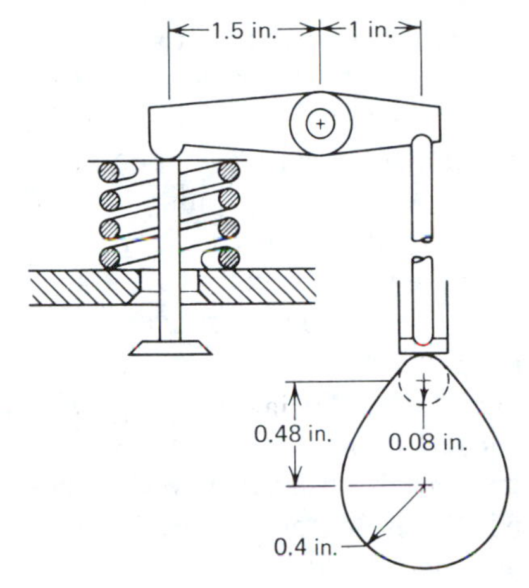

Figure 8-30 Illustration for Problem 8-12.

P8-12 The valve and cam arrangement for a four-cycle engine is shown in Fig. 8-30. During the compression cycle, the spring holds the valve against its seat with a force of 90 lb. In the valve open position (shown), the cam has forced the spring to its maximum load, minimum length position. From the geometry of the cam and rocker arm, calculate the amount of spring compression between valve closed and valve open positions. Design a spring that will operate under these conditions and have a life of at least $(5)(10^6)$ cycles. Assume that $k = 60$ lb/in., the spring outside diameter is 1.5 in., the spring length with valve open is 2.4 in., material is pretempered carbon steel, and spring ends are closed and ground.

P8-13 A compression spring for a toggle mechanism has an outside diameter of 32 mm and a wire diameter of 3.6 mm. The required force in the engaged positions is 200 N. When the toggle goes over center, the spring is deflected an additional 25 mm. How many coils must be specified to achieve 10^5 cycles? What should the spring length be at maximum load, and what is free length? Assume pretempered carbon steel wire and ends closed and not ground.

P8-14 An automotive clutch has a pressure plate that is loaded by nine equally spaced compression coil springs (Fig. 8-31). The pressure plate has a 250 mm outside diameter and an inside diameter of 200 mm. The desired engaged unit pressure is 250 000 Pa. What force should each spring contribute? What is the rate of each spring if they are compressed 25 mm on assembly? What is the combined spring rate?

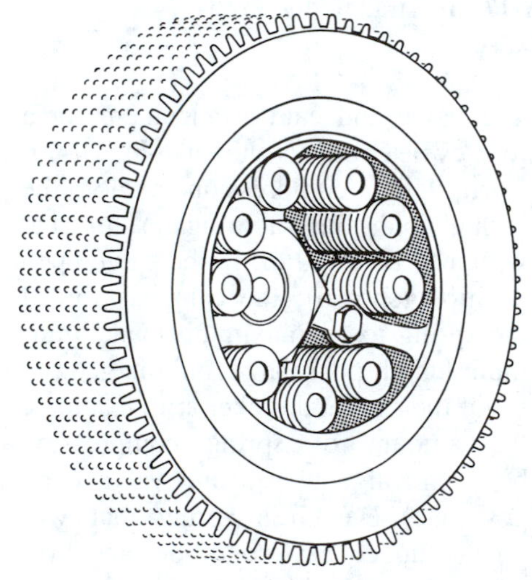

Figure 8-31 Illustration for Problem 8-14.

P8-15 A manufacturer makes a cantilever spring from 5 mm stock. The width of the spring is 30 mm (constant). After the 5 mm stock is depleted, the manufacturer wants to know if 6 mm stock can be used and still get the same characteristics. What would you think?

P8-16 The static loading of a spring is to be increased 20%. The specifications are currently $d = 0.207$ in.; $OD = 1.25$ in.; $n = 13$; free length = 4.5 in.; and loaded length = 3.25 in. What are the new specifications in metric units? The stress should be about the same for the two springs.

P8-17 A torsional helical spring is used to lift an oven door that raises vertically a distance of 0.75 m. The mean diameter of the helix is 150 mm (free), and the gravity force acting on the door is 300 N. The spring is contained inside a reel whose diameter is 200 mm, on which the lifting cable is wound. How many coils, and what wire diameter, should be specified to limit the closing force to 60 N? Assume steel wire with a maximum permissible stress of 750 MPa. What diameter tube should the spring be mounted on?

P8-18 The control for a furnace fan uses a strip of phosphor bronze mounted as a simple beam. The strip is 2.00 in. long and 0.25 in. wide. What thickness is required to obtain a 0.125 in. deflection with a force of 0.5 lb located in the center of the beam? What is the working stress?

P8-19 A steel compression spring of the proportions $d = 0.162$ in., $OD = 1.5$ in., and $n = 9$ is made to the standard manufacturing tolerances. What extreme values of rate may be expected? If the nominal free length is 4 in., what variation in load can be expected at a specified length of 2 in.?

P8-20 What are the dimensions of a steel Belleville spring that will develop a load of 3.15 kN at a stress of 1400 MPa in the flat position? Use $D/d = 1.6$ and $h/t = 1.0$.

P8-21 Two coil springs are used to resist an imposed force through a connecting cross bar. The springs have rates of 20 N/mm and 25 N/mm, respectively, and are placed 300 mm apart. Where should the force be applied to keep the cross bar horizontal? The springs are of equal length in their unloaded condition.

PART 3

Machine Elements for Torque-Speed Change and Rotary Power Transmission

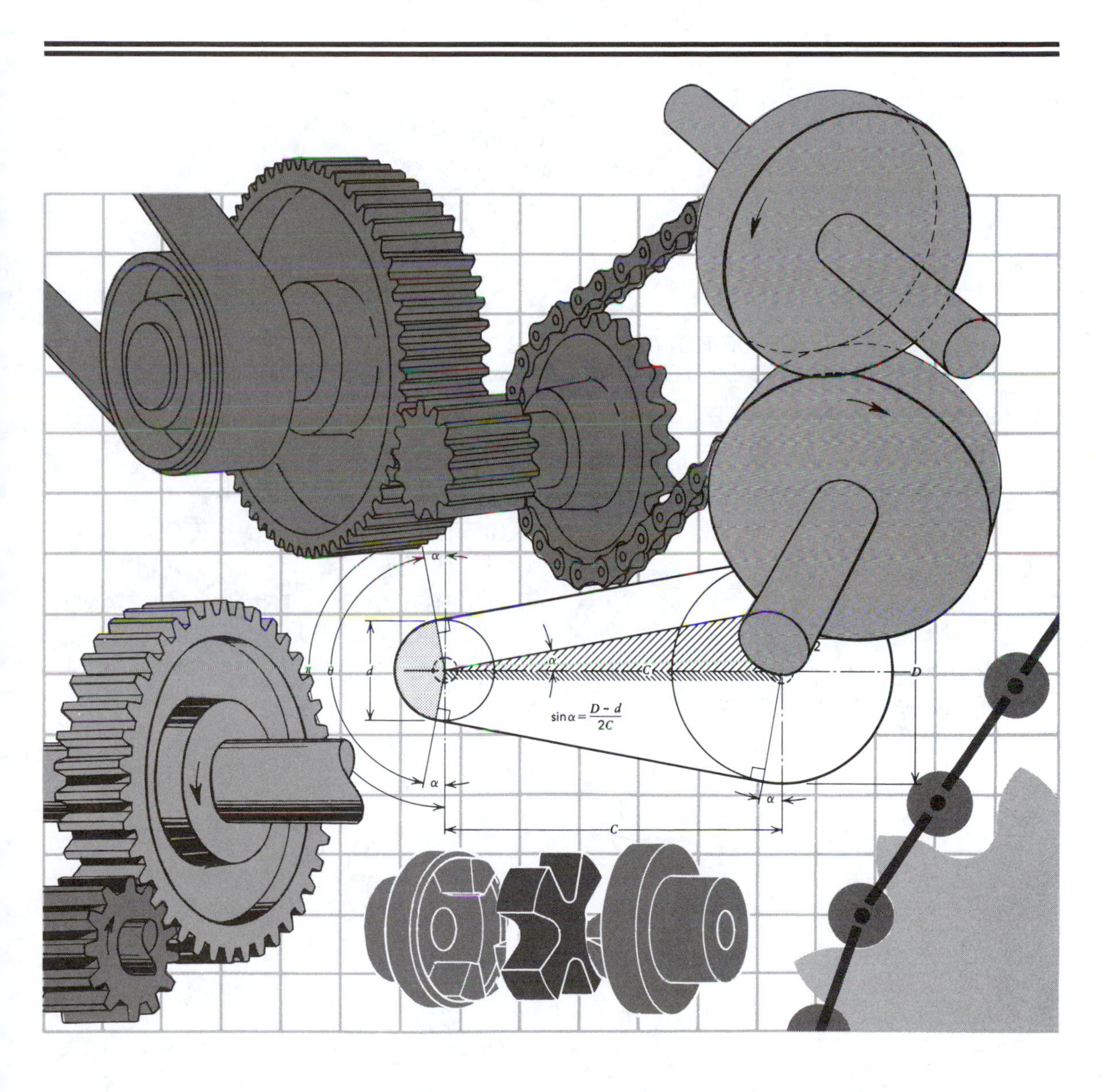

Chapter 9
Fundamentals of Transmission and Control of Power

I hear and I forget.
I see and I remember.
I do and I understand.

CONFUCIUS

This chapter surveys the differences and similarities among common machine systems and/or machine members for transmission and control of power and motion. A knowledge of how they differ enables designers to select from among several systems or members the ones most useful to a specific job at minimum cost.

A survey of similarities enables us to discuss in one place common selection procedures and common traits, and thus avoid repetition later.

This chapter also discusses briefly the systems and machine members to which an entire chapter cannot be devoted. These include mechanical and hydraulic systems, power screws, friction drives, and flywheels.

9-1 THE SIGNIFICANCE OF POWER TRANSMISSION AND CONTROL

A machine is a device that transforms, transmits, and controls power. It can thus be no surprise that machine designers devote much of their effort to selecting and designing components directly responsible for transmission and control of power. Indeed, these components are the largest single group of machine elements—a group that includes belt, chain, and gear drives; couplings; clutches; shafts; and bearings.

Logically, discussion begins with couplings, which transfer power directly, followed by the speed-torque multipliers (belts, chains, and gears); last come shafts and their related members (bearings, clutches, and brakes). Controlling power for useful work is essential to high productivity. Control of power involves starting, stopping, and changing speed, torque, and power flowing to or within a machine. To a large extent this is the task of couplings, clutches, and brakes, which can be operated separately or in unison.

With few exceptions, these machine members are part of a speed-reduction system. The same basic principles and equations thus apply to them. All are subject to vibration, inertia forces, and some sort of waste load.[1]

9-2 PRINCIPLES OF POWER TRANSMISSION

Too often new designs are simply minor variations of old ones. To avoid this pitfall, designers should remember that there are usually several different physical principles that will produce a desired effect. Figure 9-1 shows schematically how power can be transmitted and controlled. Thus, in any contemplated design all avenues should be explored and all but one discarded by following the problem-solving strategy outlined in Section 2-7.

Positive Transmission[2]

Any drive or transmission that is unyielding or dependent on the positive contact of intermeshing teeth or other parts rather than frictional resistance is classed as positive. Intermeshing gears illustrate a positive drive; friction drives such as belts represent a transmission that is not positive. In general, slippage between driving and driven members of a positive drive could not occur without breakage or excessive distortion or displacement of the driving and driven members. In contrast, any transmission that is not positive might slip without injury to the mechanism whenever the load exceeded the frictional resistance between driving and driven surfaces. Most mechanisms use a positive drive, either because the power to be transmitted would be excessive for a nonpositive drive or because a definite relation must be maintained between the driving and driven members. In many cases, a positive drive is needed to meet both these requirements.

---

[1]Commonly called parasitic loads.

[2]Reprinted with permission from F. D. Jones and P. B. Schubert, *Engineering Encyclopedia*, New York: Industrial Press, Inc., 1963.

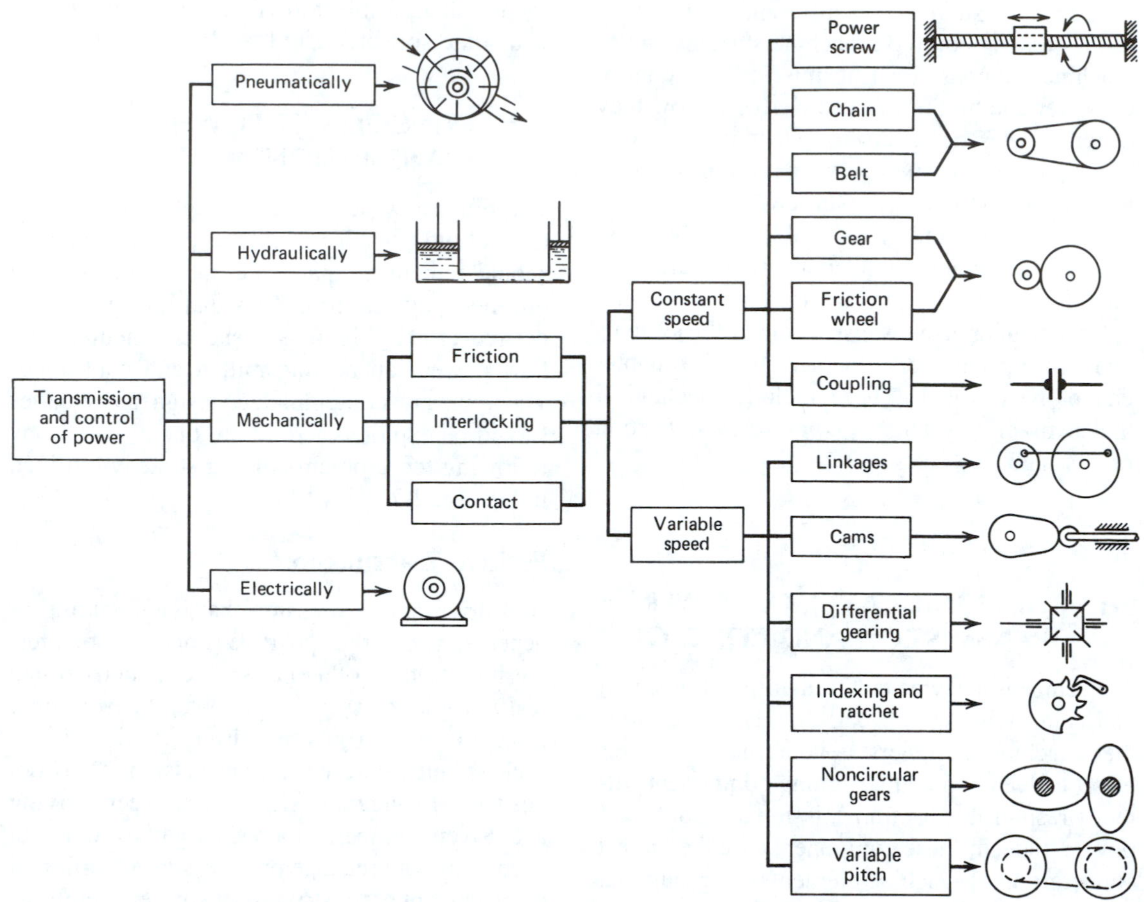

Figure 9-1 Principal means of power transmission. (Developed by Uffe Hindhede.)

Data for Power Transmission

To select or design a drive, the following data should be known.

1. Type of power source (motor or engine).
2. Amount of power required (kW, hp).
3. Type of operation (continuous or intermittent).
4. Operating conditions (start, load duration).
5. Magnitude of speed (input or output).
6. Speed modification (output to input).
7. Environmental conditions (ambient temperature, corrosive elements, etc.).

As will be seen in the next 10 chapters, these basic data are required for the selection or design of belts, chains, gears, shafts, bearings, couplings, clutches, and brakes.

In exercising control over machine operations there are only two variables to consider: force and manner of application. By controlling these variables, designers also control speed, work, and power. The third variable, time, is a basic reference; it can be measured but not controlled.

9-3 MECHANICAL POWER TRANSMISSION

Mechanical power transmission is generally associated with shaft-mounted rotating members such as belts, chains, gears, and couplings. Fig-

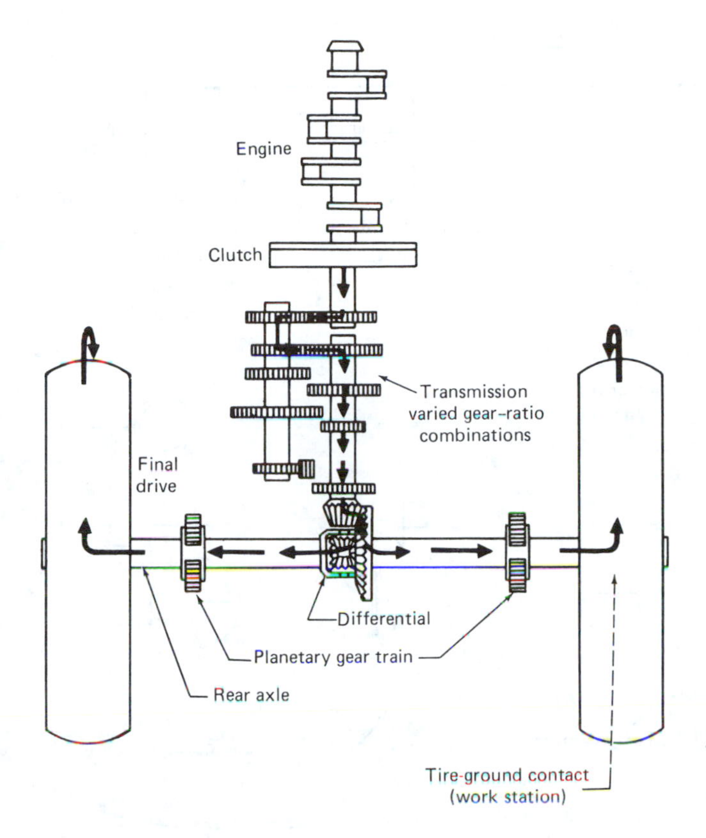

Figure 9-2 Basic power train of tractor. (Courtesy Deere and Company Technical Services.)

ure 9-2 shows the basic parts of a common power train. The driveline shown illustrates well the two functions of transmission and control of power. For example, the differential at the rear axle transmits power from the drive shaft to the two wheel axles. It also exercises control. During cornering the outer wheel always speeds up by the same ratio that the inner wheel slows down.

The choice of machine member depends first on the relative shaft position, but speed, torque, and shaft distance are also important. Figure 9-3 provides a schematic outline of common means of mechanical power transmission. The outline is based on the four basic relative shaft positions: (1) collinear (in line), (2) parallel, (3) intersecting, and (4) nonintersecting, or skew. Corresponding means of power transmission are outlined next and shown schematically. For parallel shafts, the division due to large or small shaft center distance is also shown. The size and composition of these machine members is determined primarily by speed, power to be transmitted, useful life, and environment. Figure 9-4 gives the approximate breakdown based on industrial usage. Note the predominance of gear drives and couplings.

When motor speed is the same as shaft speed, a simple flexible coupling is often utilized. For speed reduction and a smooth flow of power, however, gears, chains, or belts are needed (Fig. 9-5). Gears predominate because of their versatility, compactness, and high efficiency. Unlike most belts and all chains, gears (bevel, worm) permit change in the plane of rotation, a trait that greatly increases their usefulness. Gears provide power transmission over a wide range of speeds but are generally used for short center distances such as those found in vehicle transmissions.

When shaft distances are large enough to render gears impractical, chains or belts are normally used. Drive chains, as on motorcycles,

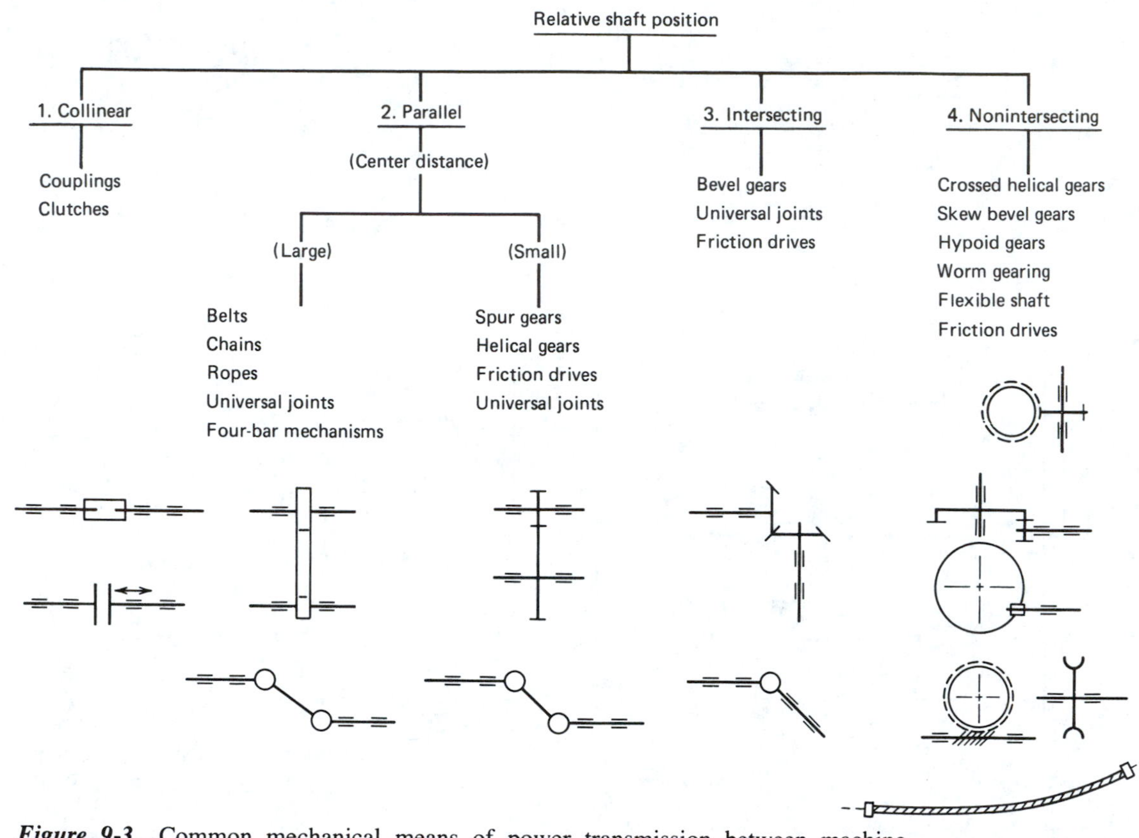

Figure 9-3 Common mechanical means of power transmission between machine shafts.

Figure 9-4 How industry uses mechanical components. Approximate breakdown based on sales. [Data by Frost and Sullivan (*Machine Design,* 1975 reference issue).]

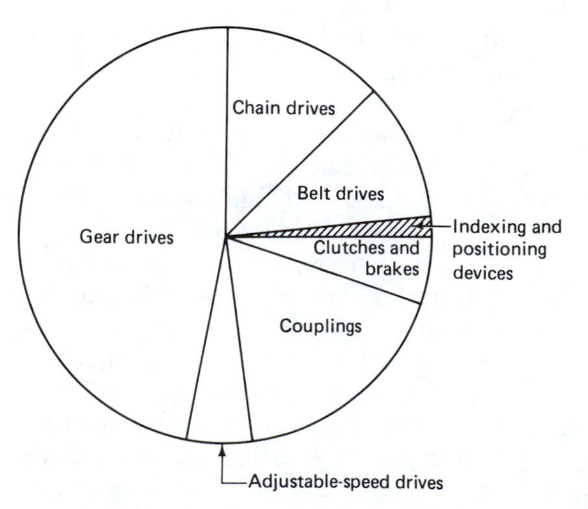

help solve the weight problem, but they tend to wear and elongate. Wear can be minimized with adequate lubrication and the use of protective housings to keep out abrasive material. A principal advantage of chains is their high load capacity, which results from their being loaded in simple tension and having many teeth engaged on the sprocket. This is in contrast to gears, where only one or two pair of teeth are engaged at one time.

For higher speeds and lower torques, belts are more suitable. Belts do not need lubrication but require more frequent tightening and are prone to slip under heavy loads. Flat and V-belts have a lower load capacity, primarily because they must rely on friction. Synchronous belts, which are fabricated from rubber or plastics and steel cable, have molded gear teeth on their inside surface. They therefore share features of both

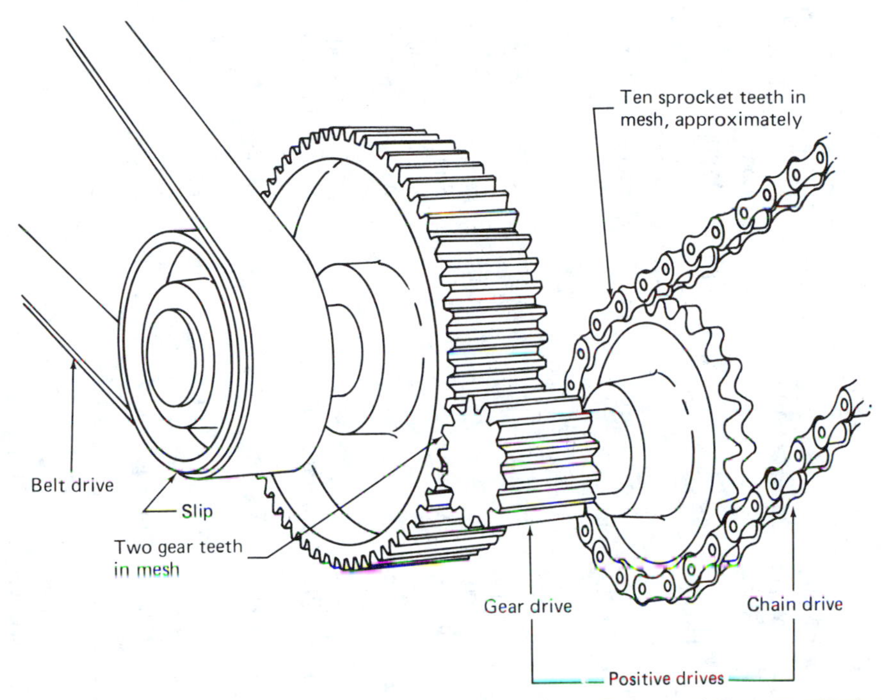

Figure 9-5 The three major types of drives. (Courtesy Deere and Company Technical Services.)

belts and chains. With respect to efficiency, chains are somewhat inferior to gears but somewhat superior to belts. Gears, synchronous belts, and chains are used almost exclusively when a positive drive (no slip) is used. In other cases, hydraulic drives may be a possible alternative.

To select the drive best suited to a specific job, designers should compare the characteristics of available drives with those required. Table 9-1 shows the limits and characteristics of constant-speed drives. Primary characteristics are usually speed ratio, power, maximum speed, and efficiency. Note that the upper limits on speed and power are constantly rising because of new developments. Other considerations are mass-to-power ratio, initial cost, useful life, and space occupied. Space is a major design consideration; Fig. 9-6 is presented to illustrate the relative space requirements for different constant-speed drives for the same speed reduction and power delivered.

Figure 9-6 Dimensional comparison of different mechanical drives. Reduction ratio: $i = 4.3$, $P = 136$ kW. (From *Konstruieren von Getrieben*, G. Reitor and K. Hohmann, Essen: Verlag Girardet, 1976.)

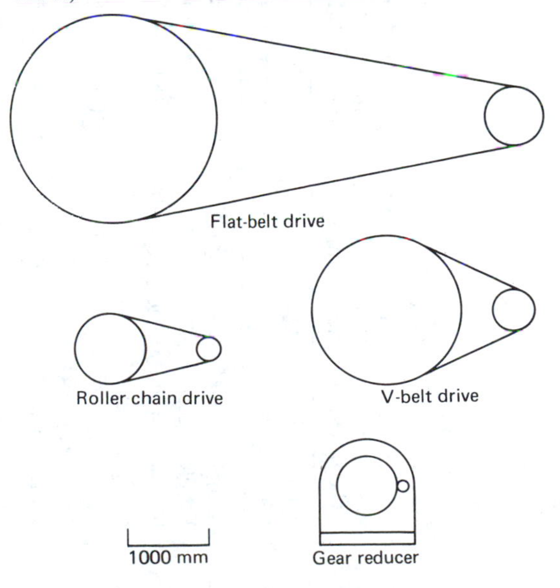

TABLE 9-1 Limits and Characteristics of Constant-Speed Drives

Type of Drive	Speed Ratio (one step)	Power (per mesh)	Maximum Speed	Peripheral Speed	Efficiency (including bearings)	Shaft Center Distance (a measure of space occupied)	Relative Shaft Position (reducer)	Mass Relative to Power	Initial Cost (spur gear =100%)	Noise Level	Life (relative)	Damping
Nomenclature rate, etc.	m_w	P, kW	n_{max}, rpm	v, m/s	e, %	C, m			%			
Straight Spur	<5	3,000		<50			Parallel			High	Average	
Parallel gear	<8			<100	99–97	0.1–0.63	Collinear	Average	100			Poor
helical drive	(20)[a]	(20,000)	100,000	(200)						Average	Long	
Straight Bevel	<8			<25	99–97	—				High	Average	
gear	<8	4,000	15,000	<50			Inter-	Small	150			Poor
Spiral drive	(15+)			(200)	99–97	—	secting			Average	Long	
Crossed helical gearing	<100	<8	25,000	<50	95–50	0.1–0.4	Non-parallel,	Very large:	100	Low	Moderate	Poor
Worm gearing	<60 (100)	<150 (1,000)	30,000	<70	97–50	0.05–0.5	Non-intersecting	Small	<100	Low	Short	Moderate
Friction wheel drive	6 (10)	20 (150)	10,000	<20	98–95	0.1–0.5	All positions	Average	50	Very low	Short	Moderate to good
Flat	5 (20)	300 (1,600)	18,000	<100	98–96	$(0.5–3)x (d_1+d_2)$	Arbitrary	Average	65	Low	Average	Moderate to good
V-type Belt drives	8 (40)	220 (1,100)	10,000	<60	97–95	$>1_i(2–3)$	Parallel	Average	65	Very low	Moderate	Good
Synchronous	8 (15)	200 (1,200)	5,000	<60	98–95	$>1_i(2–3)$	Parallel	Average	65	Low	Moderate	Good
Roller	<6	700						Average	85	High	Moderate	Poor
Chain	(14)	(4,000)	5,000	<17	97–95							
Silent drives	<8					0.3–3	Parallel	to large	125	Low	Average	Poor
	(12)	(900)		<30	99–97							

[a]( ) Refers to maximum value.

Source. Adapted from G. Reitor and K. Hohmann, *Grundlagen des Konstruieren*. Essen: Verlag W. Girardet, 1976.

9-4 FRICTION DRIVES

In friction drives power is transmitted from the drive wheel to the driven wheel by means of friction. Being quite simple, they are usually low-cost drives. The amount of power that can be transmitted is determined by the surface condition of the two materials, the normal force at the interface, and the coefficient of friction. The capacity of friction drives is very limited because of the relatively small areas in contact. For this reason, the material strength is important.

Although belt drives rely on friction for power transmission, they are not classified as friction drives by the trade. Friction drives are normally those in which power is transmitted through contacting wheels. Two types of friction wheels are normally recognized: *constant-* and *variable-speed* ratios.

There is a wide variety of constant-speed drives. The type shown in Fig. 9-7*a* illustrates the simplicity of design that accounts for low cost and long life.

Figure 9-7*b* shows a simple drive for continuous-speed variation. Low cost and stepless variation within wide limits are its main advantages.

Friction drives are more common in Europe than in the United States. The best domestic source of information is *Mechanical Design and Systems Handbook* by H. A. Rothbart, published by McGraw-Hill.

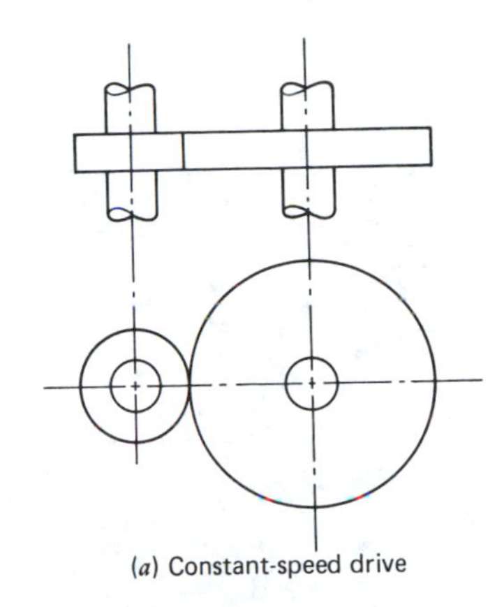

(*a*) Constant-speed drive

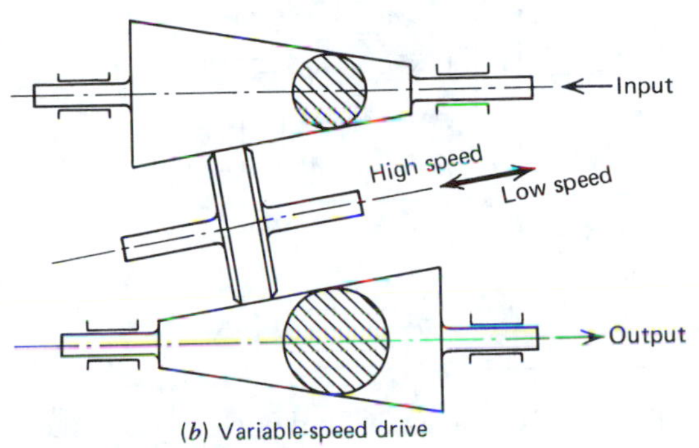

(*b*) Variable-speed drive

Figure 9-7 Friction drives.

9-5 HYDRAULIC POWER TRANSMISSION

Hydraulic drives are used in preference to mechanical systems when (1) power is to be transmitted between points too far apart for chains or belts; (2) high torque at low speed is required; (3) a very compact unit is needed; (4) a smooth transmission, free of vibration, is required; (5) easy control of speed and direction is necessary; or (6) output speed must be varied steplessly.

These advantages can be obtained by using fluids, usually oil, as the power media. Although nearly incompressible, fluids are infinitely flexible; they readily change shape, and their flow can be easily divided. Because of low frictional resistance, fluids move rapidly from one location to another and transmit force in any direction. As opposed to solids, fluids provide no mechanical lockup; however, once in motion, they resist shear and possess kinetic energy that can be transferred from the drive part of a fluid clutch or coupling to the driven part. Fluids, therefore, have *torsional* flexibility at low speeds and are advantageously used with engines, where they prevent stalling at low speed and slipping during load surges.

Control of fluids is easy; the turn of a valve may quickly reverse motion with minimal shock. Moreover, the presence of relief or control valves can guard against overloading.

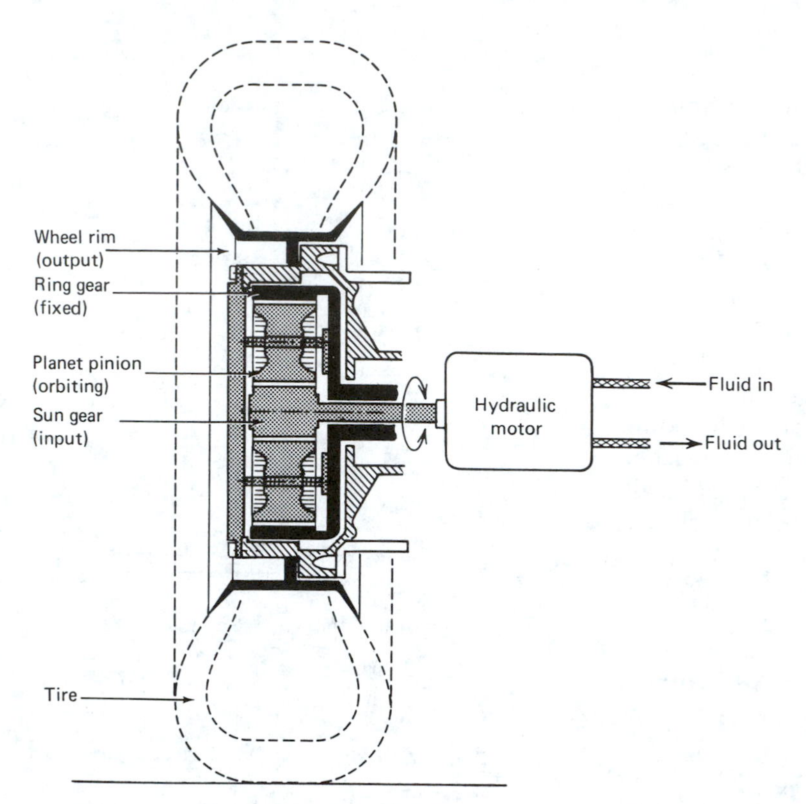

Figure 9-8 Combined planetary and hydrostatic transmission for vehicles.

The disadvantages of fluid power systems are the need for a separate power unit (pump, sump, drive motor and, in some cases, coolant), higher initial cost, and lower efficiency (0.70 – 0.80). However, in many applications, operating conditions can overshadow efficiency as the primary selection factor. The limitations of both mechanical and fluid systems may often be overcome by combining the best of both systems.

High torque at low speeds is a characteristic of fluid motors, but hydrostatic[3] drives have torque limitations because of pressure limitations (35 MPa or 5000 psi). Planetary gear reduction units, unlike some mechanical drives, have a very high torque capacity at low speeds. A combined hydrostatic and planetary transmission in which the planetery part (output) is coupled to the hydrostatic motor vastly increases output torque while reducing overall cost and space requirements (Fig. 9-8).

Hydrostatic drives consist of a power-driven pump unit and one or more hydraulic motors connected to it by means of pressure lines (Fig. 9-9). They are compact, adaptable, and easily controlled, but their size is strongly influenced by current limitations on pressure. Because pressure lines can (within reason) be almost any length and size, power can be transmitted over long distances, around corners, and into confined spaces.

Since hydraulic motors are totally enclosed, they can operate in harsh environments. Unlike engines, they provide a constant torque over a wide range of speeds. The average hydraulic motor provides 80 to 90% of running torque at start-up and can be briefly overloaded without harm. Because of their low speed output and high power-to-weight ratio, hydraulic motors

[3]Hydrostatics is the branch of physics that treats the laws governing fluids at rest, as distinguished from hydraulics, which refers to fluids in motion.

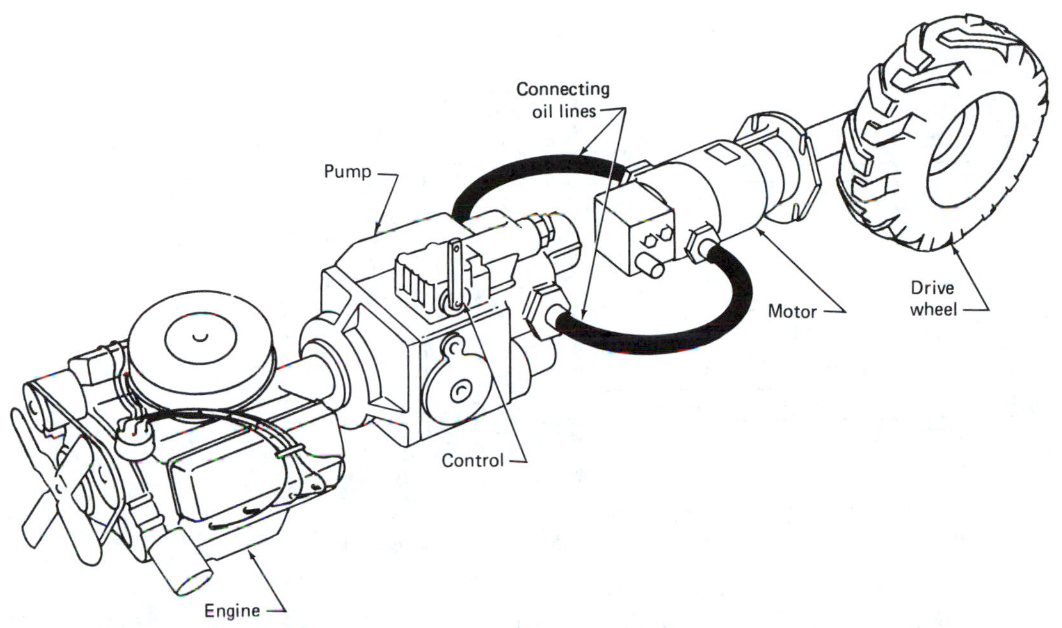

Figure 9-9 Hydrostatic drive in a complete power train. The hydrostatic drive is an automatic fluid drive that uses fluid under pressure to transmit engine power to the drive wheels of the machine. (Courtesy Deere and Company Technical Services.)

usually can be coupled directly to driven mechanisms. Thus universal joints, couplings, belts, pulleys, and chains can be eliminated, thereby producing a more compact and more reliable drive package.

Hydrostatic drives are presently used in machine tools, slow-moving vehicles, farm machinery, and cranes. Low efficiency tends to limit their application to intermittent operations.

Mobile Application

Figure 9-10 shows the application of a hydrostatic drive to a vehicle. The key to greater freedom in vehicle design lies with the four *hydraulically powered, high-torque wheels.* Mounted inside each wheel is a high-ratio, multi-planetary gear drive powered by a relatively expensive high-speed, low-torque hydraulic motor. Smooth, infinitely variable power is thus available for both forward and reverse and for individual wheels according to need. Hydraulic power is generated by a central variable-flow pump driven by the engine. Power output varies with engine speed. This system consequently eliminates the need for conventional drive shafts, clutches, transmissions, axles, differentials and, often, brakes.

9-6 SCREWS FOR POWER TRANSMISSION (TRANSLATION SCREWS)[4]

Generally, power transmission is based on continuous rotary motion, as in belt, chain, and gear drives. When slow, uniform, linear, intermittent motion is occasionally required for lifting and actuating purposes, design options are usually limited to screws and fluid power cylinders.

[4]For further detail consult H. A. Rothbart, *Mechanical Design and Systems Handbook,* section 26, "Power Screws"; New York: McGraw-Hill, 1964.

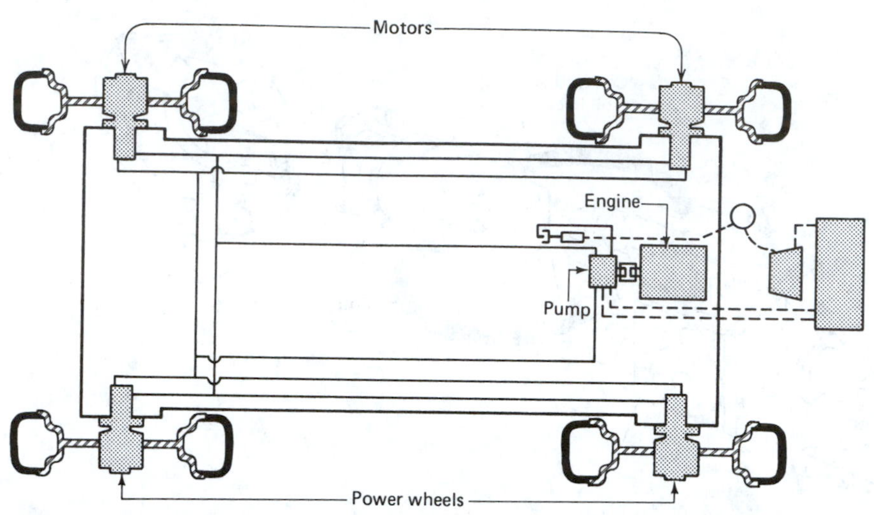

Figure 9-10 Vehicle with hydraulic drive. (Courtesy Warner Gear, Division Borg-Warner Corporation.)

When, as in machine tools, precisely controlled motion is also required, only screws will do. The screw is a unique way to convert rotating motion to linear travel and does so with great precision. Three types are presently available: conventional screws (the simple machine) and two sophisticated, recent developments—ball and roller bearing screws.

A power screw is essentially a threaded shaft carrying one or two nuts. It can function in two distinctly different ways. In one mode, the nut has axial motion against a resisting force, while the shaft rotates in its bearings. This is the lead-screw type used in lathes. The second mode has a nut, a form of bearing, forcing the screw shaft to move axially. This is the power screw used in both manually and power-operated jacks, presses, and valves (Fig. 9-11).

Screws for power transmission are purposely designed for a mechanical advantage. The threads must therefore be strong. For lead-screws the threads must also be very accurate, and compensation for wear should be provided. Friction should be minimized through careful machining at manufacture.

Of suitable thread profiles—Acme, Buttress, and Square—Acme is the best compromise (Fig. 9-12). It is strong, has reasonably low frictional resistance, and allows compensation for wear by means of a split nut. As wear progresses, looseness is prevented by "taking up" the nut.

Screws for power transmission, as opposed to screw fasteners, often have two or three parallel single threads. They have the advantage of increasing the lead of a screw without weakening it by cutting a coarse single thread. Increased efficiency plus greater travel speed result from multiple threads. The travel distance resulting from one shaft revolution is termed lead, L. For single threads the lead is equal to the pitch, p: $L = p$. For double threads, $L = 2p$, and so forth.

The exceedingly useful principle of the screw is greatly hindered by inherently high frictional resistance—resulting in excessive wear and low efficiency—on the order of 0.40 (Fig. 9-13). Adding recirculating balls as in ball screws (Fig. 9-14) or rollers as in roller screws (Fig. 9-15) will double efficiency and greatly reduce wear. Stick-slip,[5] a serious drawback to conventional

[5]Stick-slip, a phenomenon inherent in sliding friction, is the tendency to start-stop and stutter when slow linear motion is desired.

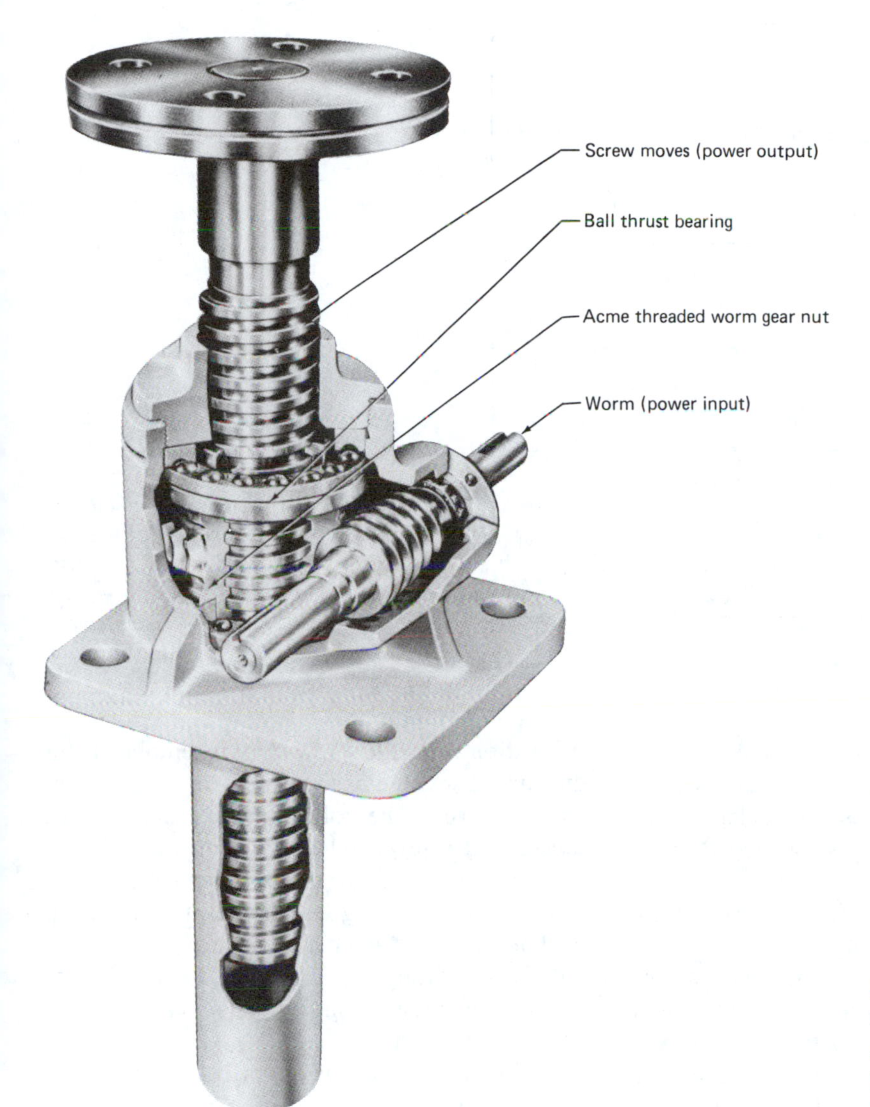

Figure 9-11 Machine screw actuator driven by a worm gear. The nut is an integral part of the gear. (Courtesy Duff-Norton Company.)

screws, is nearly eliminated in the ball screw system. Conventional screws must be adjusted periodically for gradual wear, but the ball screw system is virtually wear-free with adequate lubrication and protection from dirt contamination. The actual life of a ball screw is limited by metal fatigue, not metal wear. Usually preloading the screw (in tension) can reduce end play and backlash to nearly zero.

Roller screws employ a number of planetary

Figure 9-12 Common translation screw threads. Thread forms deviating from the standard V-thread are used primarily on translation screws where high strength and low frictional resistance are desirable.

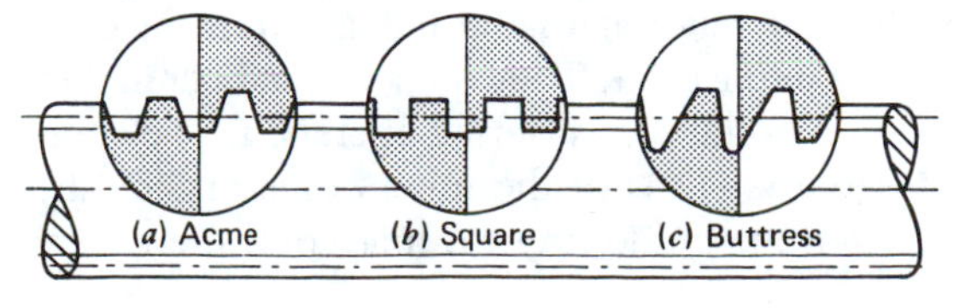

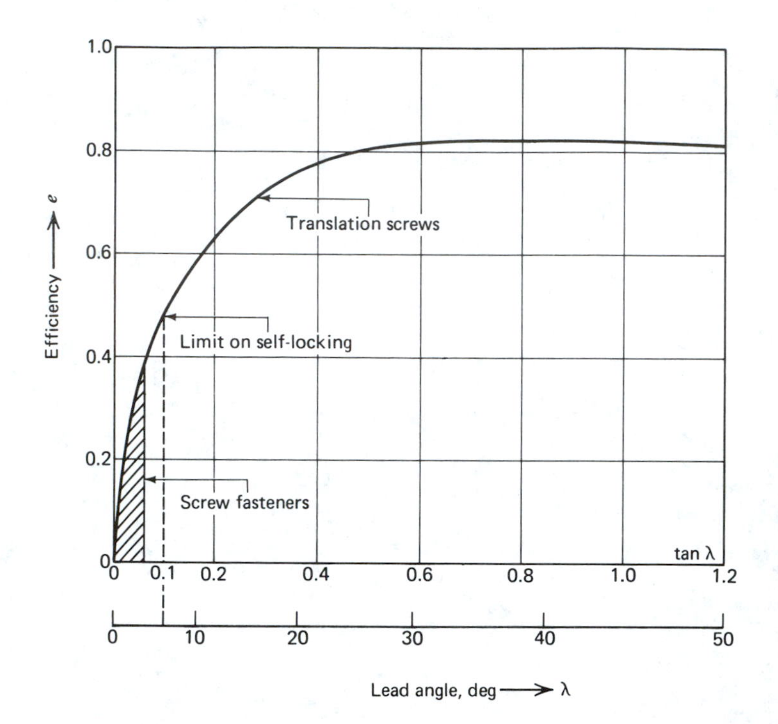

Figure 9-13 Efficiency of a standard metric screw, as a function of the lead angle, for a coefficient of friction $f = \tan = \theta = 0.1$; $\theta = 5.71$ deg. (From G. Nieman, *Maschinenelemente, I.*, p. 203, Berlin: Springer-Verlag, 1975.)

rollers in order to increase their mechanical advantage and load capacity. They are used to move heavy loads only. Presently roller screws (LTI-Transrol) capable of carrying 2500 kN each with an efficiency of 0.80 are in use.

Type SR Transrol screws (Fig. 9-15) employ planetary rollers. They consist of a threaded screw shaft (1), an internally threaded nut (2), and a number of threaded rollers (3). The rollers are positioned in the space between the screw shaft and the nut and mesh with the threads of both. When the screw shaft or the nut rotates in relation with the other element, the rollers follow a planetary path around the screw shaft. The ends of the rollers are toothed (8) and terminate in a cylindrical spigot (4). The roller teeth mesh with the toothed rings (9) located in the nut by pins (7); thus the rollers cannot be displaced by sliding in relation to the nut and remain constantly in position. They are also held parallel to the axis of the screw during operation. The spigots (journals) (4) of the rollers rotate inside the holes in the guide ring (5) that holds the rollers in position when the screw shaft is removed from the nut. These guides are held in position by snaprings (6). The rollers have a single start thread. The screw and the nut, on the other hand, have the same number of starts, generally 4, 5, 6, or 8. The resulting lead is the product of the lead of the rollers and the number of starts on the nut. This resulting lead is also constant and independent of possible sliding between the screw shaft and the rollers.

Ball and roller screws are often selected from a manufacturer's catalog by means of nomograms. The various design criteria are buckling, critical speed, life span, input power, and stiffness. Both types are often preloaded axially in tension in order to increase load capacity.

9-7 EFFECTS OF CENTRIFUGAL AND INERTIA FORCE

Any discussion of power transmission and control of power must consider the adverse effects of centrifugal and inertia force.

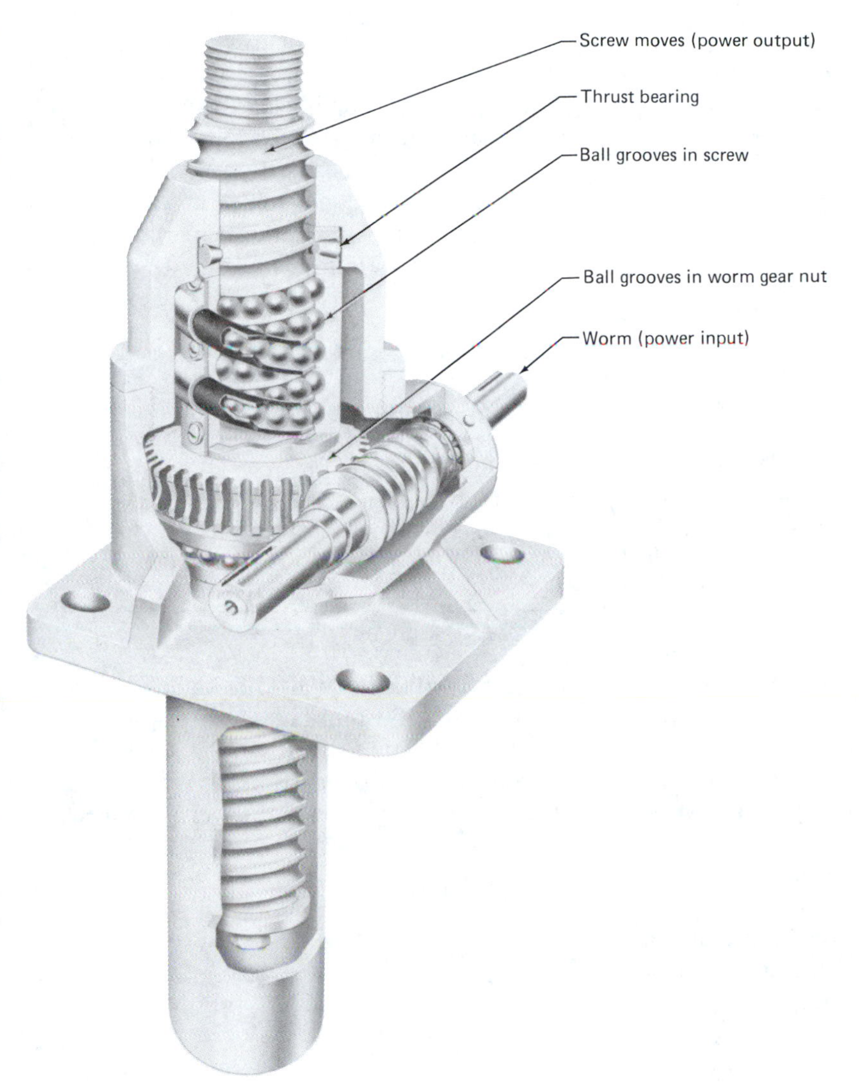

Figure 9-14 Ball screw actuator driven by a worm gear. The nut, mounted on an axial ball bearing, is an integral part of the gear. (Courtesy Duff-Norton Company.)

Centrifugal force acts on any rotating machine part by tending to force its mass farther away from the axis, thereby creating undesirable internal stresses and imposing an extra load on the supporting bearings. However, centrifugal force is used to advantage in centrifugal pumps, speed regulators, and starter clutches.

Centrifugal force increases directly with mass, radius of rotation, and velocity squared. Because this undesirable form of load increases with velocity squared, centrifugal force inevitably places an upper limit on the velocity of all rotating machine members. Any reduction of mass, speed, or radius of rotation will reduce the adverse effects of centrifugal force.

Inertia is a term for the resistance of a station-

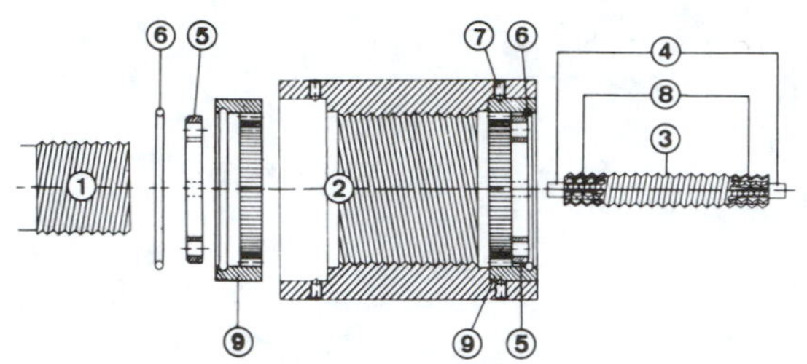

Figure 9-15 Roller screw with planetary rollers. (Courtesy La Technique Integrale.)

ary object to being moved and for the tendency of a moving body to continue moving in a straight line or to spin around its axis of rotation. Inertia can be overcome only by the application of force.

Inertia is directly proportional to changes in mass or moment of inertia and velocity, but it acts in the opposite direction. Heavy or rapidly accelerating machine parts are thus often subject to sizable inertia forces that cause large material stresses. Few machine parts are exposed to greater inertia forces than engine pistons, where motion may be reversed more than 100 times per second. In this case, the connecting rod must sustain the powerful alternating tension and compression cycles.

Mass reduction is one way of alleviating undesirable or harmful inertia forces. Pistons, for example, are frequently made from aluminum alloys. Damping is another principle used to advantage in combating inertia forces. When used in vehicle suspension systems, inertia forces are reduced because velocity changes are reduced.

Inertia forces produced by purely rotary mechanisms, such as the rotor of an electric motor, demonstrate the *flywheel effect.*[6] By virtue of this effect a driven machine, equipped with a flywheel, may operate with a 10-kW motor, yet produce work through a portion of its operating cycle at a 20-kW rate. Energy builds up during most of the work cycle, only to be consumed during a fraction of it. The resulting velocity variation, characterized by a "dip" during *energy takeoff,* forms the basis for calculating the energy transfer. With an assumed or known speed change and a known rotational inertia, a calculation is made as follows.

$$E = 0.5k^2m(\omega_1^2 - \omega_2^2) \qquad (9\text{-}1)$$

where

E = transfer energy; N · m, ft-lb
k = radius of gyration; m, ft
m = mass; kg, slug
ω_1 = maximum angular velocity; rad/s
ω_2 = minimum angular velocity; rad/s
$\omega = 2\pi(n/60)$, where n = rpm

[6]For detailed information on flywheels, consult *Machinery's Handbook,* pp. 333-345, and *Mark's Standard Handbook for Mechanical Engineers,* by Theodore Baumeister, McGraw-Hill Book Company, Eighth Edition, 1978, pp. 3-11; 8-49 to 8-51; and 9-170.

EXAMPLE 9-1

A rotor with a mass of 4500 kg has a nominal speed of 1800 rpm but slows down to 1600 rpm in 1 min during energy takeoff. The radius of gyration is 610 mm. What is the (*a*) energy take-off, and (*b*) power generated?

SOLUTION

(*a*)

$$\omega_1 = \frac{2\pi 1800}{60} = 188.5 \text{ rad/s}$$

$$\omega_2 = \frac{2\pi 1600}{60} = 167.6 \text{ rad/s}$$

$$E = 0.5k^2m(\omega_1^2 - \omega_2^2)$$
$$= 0.5(0.61 \text{ m})^2(4500 \text{ kg})[(188.5 \text{ rad/s})^2 - (167.6 \text{ rad/s})^2]$$
$$= \mathbf{6231\,000\ N \cdot m/min}$$

(*b*)

$$P = \frac{6231\,000 \text{ N} \cdot \text{m/min}}{1000(60 \text{ s/min})} = \mathbf{104\ kW} \text{ (139 hp)}$$

Momentarily, this flywheel has a power capacity equivalent to an engine with 139 hp.

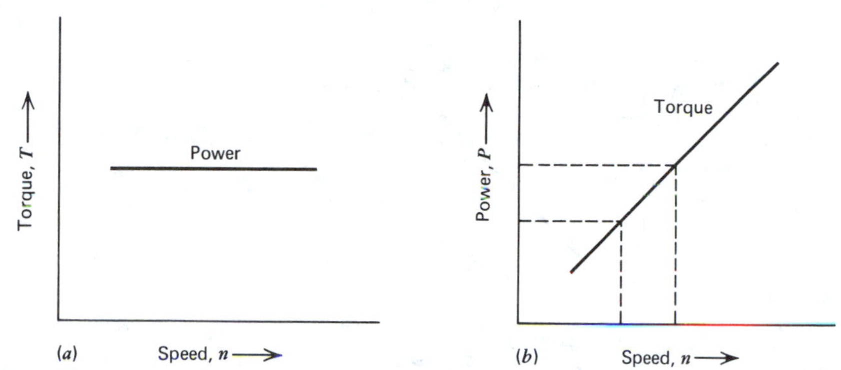

Figure 9-16 Constant-torque machines. For contant torque, each speed increase is followed by a specific power increase.

9-8 OPTIMUM POWER

The power equation for rotary motion has three variables: torque, power, and speed. In many applications one of these variables remains constant.

For *constant torque*: $P = Kn$

Power is proportional to speed (Fig. 9-16).

Most production machines (perhaps 90%) require a constant torque for each specific operation. In such cases power is proportional to speed. Because different jobs, such as drilling operations, require different torque, a means for varying the speed is necessary.

For *constant power*: $Tn = K$

Torque and speed then have a hyperbolic relationship (Fig. 9-17). If we design for high operating speeds, we can tolerate parts with lower torque capacity and achieve notable savings in size, mass, and initial cost. Speed is thus often the key to more power per unit of mass. Again, varying speed or torque is desirable, even though limitations on speed are generally imposed by centrifugal forces, excess vibration, or impact. The need for constant power levels is characteristic of machine tools.

For *constant speed*: $P = KT$

Figure 9-18 depicts the linear relationship between power and torque. Synchronous motors are constant-speed devices. Other power

Figure 9-17 Constant-power machines. Equal speed changes yield different torque changes.

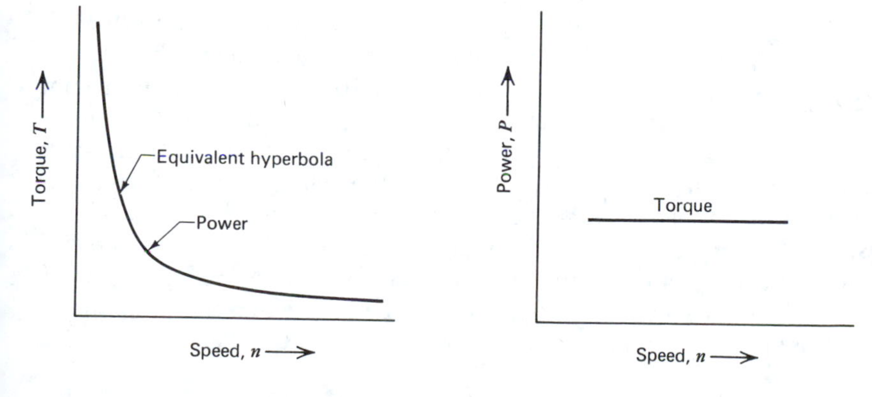

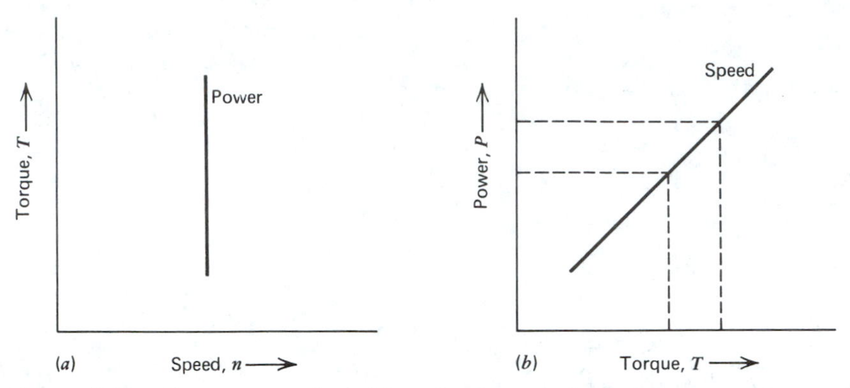

Figure 9-18 Constant-speed machines. Each torque increase corresponds to a specific power increase.

sources—steam, water, or gas turbines and gas or diesel engines—can be held at a constant speed for optimum performance.

Waste Loads (Parasitic Loads)

Whenever power is transmitted there is a loss to consider. Usually friction is the largest contributor. Other loads are static in nature and do no useful work at all. Chief among these are shaft and bearing loads. If loads cause deflection of parts, a net loss in useful work of the machine will result because work must be expended to deflect the part.

Centrifugal forces in belts and chains create tensile forces that add to tension, but not to power transmission. Dynamic loads in gearing, superimposed on the power-transmitting loads, may easily be twice the useful loads. Effective design therefore optimizes useful loads and minimizes waste loads.

9-9 MECHANICAL ADJUSTABLE-SPEED DRIVES

These are transmissions that provide various output speeds for given input speeds. Available as standard units, they are usually less expensive than equivalent electrical and hydraulic units. Two basic types—stepped (usually termed variable) and stepless drives—are used.

Variable-Speed Drives

Despite their wide application, engines are not inherently suited to perform as vehicle power sources. The basic problem lies in their torque-speed characteristics, which do not match the torque-speed requirements at the rear axle. A transmission, essentially a variable torque multiplier placed between engine and rear axle, will match the two.

Engines deliver a nearly constant but *low* torque over most of their speed range. Generally there is a slight (10%) rise in the operating range before dropping off. Power thus increases only as engine speed increases (Fig. 9-19*a*). Because of low torque, a directly driven vehicle would provide inadequate acceleration and exhibit unsatisfactory climbing ability. The function of the transmission is to provide the maximum power, which is present only at one engine speed, at any driving speed. At low speed, when a vehicle is accelerating and/or climbing, a very high torque is required, often 10 times that of the engine torque. At cruising speed, a direct drive will do. For greater torque multiplication, a rear-axle gear reducer and even planetary wheel drives can be included in the drive system.

Ideally, engines should deliver constant power for optimum fuel efficiency. For constant power, torque and speed are inversely proportional ($T \cdot n$ = constant). Thus the ideal torque-speed

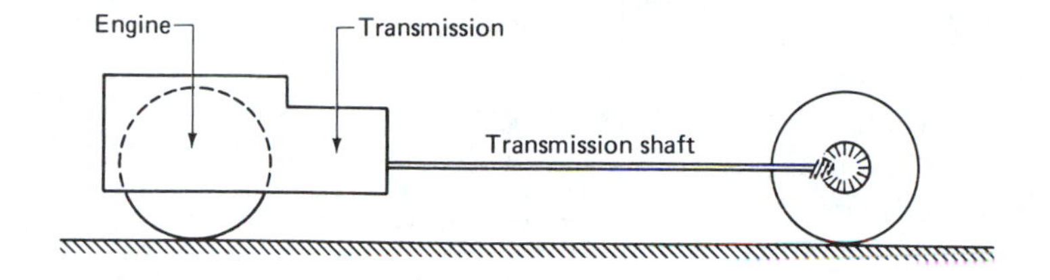

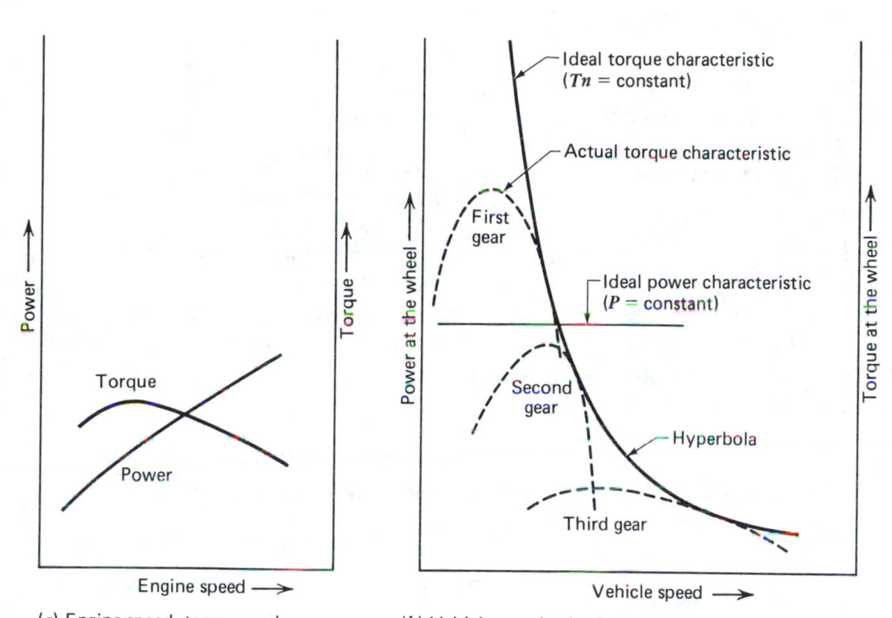

(*a*) Engine speed, torque, and power (*b*) Vehicle speed, wheel torque, and power at the wheels

Figure 9-19 Fundamentals of power transmission in vehicles. (Courtesy *Machine Design*.)

curve is hyperbolic, as shown in Fig. 9-19*b*. To match this curve, as road and operating conditions vary, a transmission should be continuously variable. Such transmissions are available for motor scooters, snowmobiles, and the like, but not for vehicles and large machine tools requiring a wide speed range and large power.

A practical alternative is the conventional, multistep transmission in which engines accelerate through a series of fixed gear ratios as the vehicle picks up speed. Figure 9-19*b* shows this arrangement for a three-speed, manually shifted transmission. Such a device will provide ideal torque at only three points in the vehicle speed range. Increasing the number of speeds will match the individual curves more closely to the ideal curve. Trucks often have a five-speed main transmission and a four-speed auxiliary transmission—20 speeds altogether. Suppose the truck is operating on a level road at about optimum engine speed, but a downshift is needed to negotiate a slight grade. If there are too few speeds available, the change in driveline ratio is so great that it creates a large change in engine speed and a loss in performance. Unless the engine can remain at optimum speed, vehicle performance is not optimum. Generally, the number of speed ratios is limited to minimize size and cost of the transmission and keep gear shifting from becoming burdensome.

Stepless Speed Variation

A stepless transmission is the answer. This has been the goal of designers for nearly a century and, within certain limits, it has been achieved. The limits imposed relate to power, speed, and control. Available drives are mechanical, hydraulic, and electrical.

Figure 9-20 shows current operating ranges of major adjustable-speed drives based on speed and power. Least desirable are the variable-stroke drives, most of which are based on the one-way clutch principle. In comparison, traction drives offer both higher speed and higher power. Basically, they are friction wheel drives. For still greater power and speed, the variable-pitch group is superior. It includes belts, chains, and woodblock belt drives. Most powerful are the fluid drives. Note that in the power range above 100 hp, only fluid drives and *stepped* gear transmissions (not shown) are available for optimum performance.

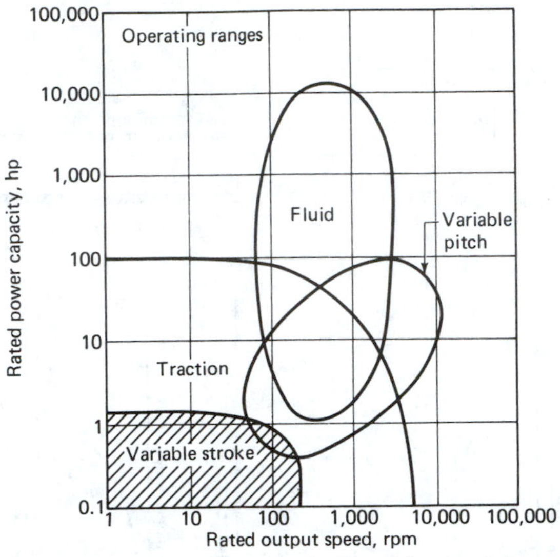

Figure 9-20 Comparison of adjustable-speed drives based on speed and power. (Courtesy *Machine Design*.)

EXAMPLE 9-2

A small tractor semitrailer fully loaded has a gross vehicle mass of 1000 kg. It uses 8.25 – 20/10 PR tires that make 343 rev/km. At a top speed of 90 km/h on level highway, the aerodynamic drag and tire rolling resistance total 4250 N. The driveline efficiency is 86%.

(*a*) What should be the engine power? What is the engine torque, if the engine speed is 2800 rpm?

(*b*) If a diesel engine is used that provides maximum power at 2800 rpm, what should be the overall driveline ratio of engine speed to wheel speed?

SOLUTION

The power needed at the driving wheels is found by using $P = Fv/1000$.

$$(a) \quad P = \frac{(4250\text{ N})(90\text{ km/h})(1000\text{ m/km})}{(1000)(60\text{ min/h})(60\text{ s/min})}\text{ kW}$$

$$\text{wheel power } P = 106\text{ kW}$$

The power needed at the engine is

$$P_{\text{eff}} = \frac{106}{0.86} = \mathbf{123\ kW}$$

The engine torque is found by means of $P = Tn/9550$.

$$T = \frac{(123\text{ kW})(9550)}{2800\text{ rpm}} = \mathbf{419\ N \cdot m}$$

(*b*) The vehicle speed in kilometers per minute is

$$v = (90\text{ km/h})(1\text{ h}/60\text{min}) = 1.5\text{ km/min}$$

The wheel speed is

$$n = (1.5\text{ km/min})(343\text{ rev/km}) = 514\text{ rpm}$$

So the driveline ratio should be

$$\frac{2800\text{ rpm}}{514\text{ rpm}} = \mathbf{5.45}$$

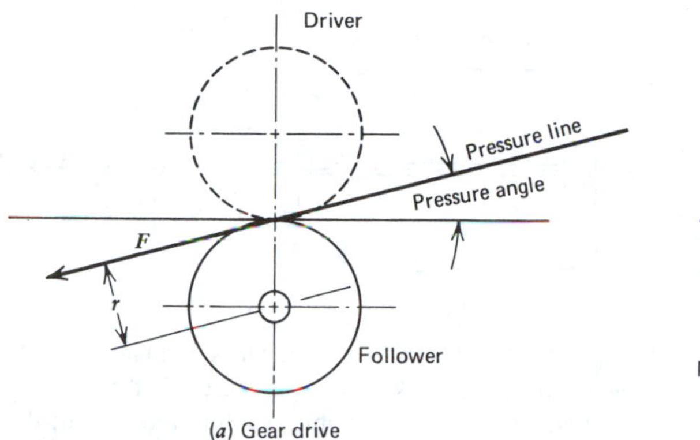

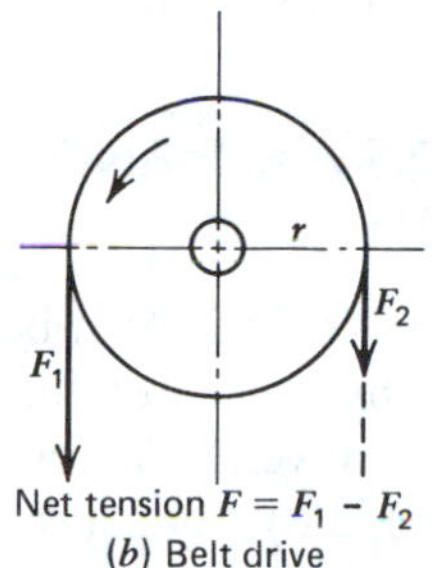

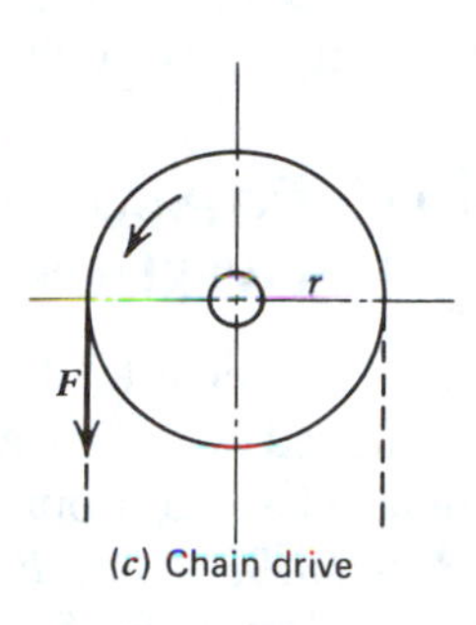

Figure 9-21 Gear, belt, and chain drives.

If, at top speed, the transmission ratio is equal to 1, the rear axle ratio is made equal to 5.45:1.

9-10 TRANSVERSE FORCE DUE TO TORQUE

Power transmission by gears, belts, and chains accounts for roughly 70% of all power transmitted within machines and perhaps half of all power transmitted between machines. In all three cases, the net effect is a combination of torsional and transverse shaft loading. Torsion is an obvious consequence of transmitting rotary energy; transverse loading is not. However, a transverse force can be produced by a torque carried between two shafts by means of gears, belts, and chains. Because this phenomenon may at first seem contradictory, it requires further discussion.

Figures 9-21*a*, 9-21*b*, and 9-21*c* show forces acting on a set of spur gears, a belt, and a chain. All three force systems can be reduced to a single force acting at a distance r from the shaft center (Fig. 9-22*a*, 9-22*b*, and 9-22*c*). This simplified approach is adequate in all preliminary calculations. What we have, then, is offset loading, as shown in Fig. 9-22*a*. We now add two equal but opposite forces of magnitude F at the shaft center O, as shown in Fig. 9-22*b*. Because the two forces added are both equal and opposite, however, shaft loading has not been altered. We now have acting on the shaft a "couple" plus a transverse force. The net result is a shaft loaded in torsion and transversely (Fig. 9-22*c*).

The transverse force, which causes bending of the shaft and loading of bearings, is clearly a

Figure 9-22 Tangential loading of circular machine member.

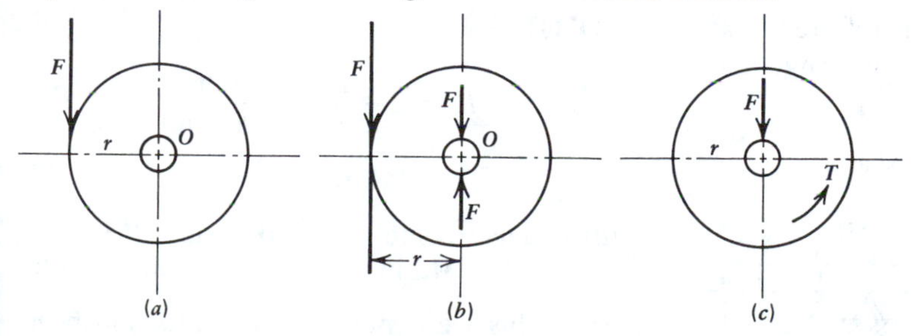

"waste load," since it provides no useful work. The torque, by contrast, transmits power.

9-11 ESTIMATING TRANSVERSE AND OVERHUNG LOADS

The transverse load shown in Fig. 9-23 can be calculated with fair accuracy by means of formulas derived from the two power equations ($P = Tn/9550$ and $P = Tn/63{,}000$). By substituting $T = F_t\,(0.5D)$ and solving for F_t, one obtains

$$F_t = k_t \frac{19\,100P}{Dn}\ \text{N} \tag{9-2}$$

$$F_t = k_t \frac{126{,}000P}{Dn}\ \text{lb} \tag{9-3}$$

where

F_t = transverse or tangential force; N, lb
P = power transmitted; kW, hp
D = pitch diameter; m, in.
n = shaft speed; rpm
k_t = overhung load factor

Approximate values for k_t are as follows:

Single-strand chain drives: $k_t = 1.$
Double-strand chain drives: $k_t = 1.25.$
Gears: $k_t = 1.25.$
V-belts: $k_t = 1.5.$
Flat belts. $k_t = 2.5.$

Note that F_t increases with power transmitted and decreases with higher speeds and larger pitch diameters. Since P and n usually are given, only the size of D can be used to keep F_t below any stipulated upper limit.

Depending on application, F_t is referred to as chain pull or net belt tension; in gearing, it is referred to as the transmitted load. They all generate an overhung load on output shafts of motors and speed reducers. An overhung load is thus the radial load imposed by sprockets, pinions, pulleys, or sheaves (Fig. 9-24). It should not exceed the *rated* load stipulated by manufacturers of motors and gear reducers.

If the estimated overhung load exceeds the rating, the designer may (1) increase the pitch diameter, (2) use an outboard bearing (expensive), or (3) go to a larger unit (even more expensive).

Figure 9-23 Combined loading of shaft.

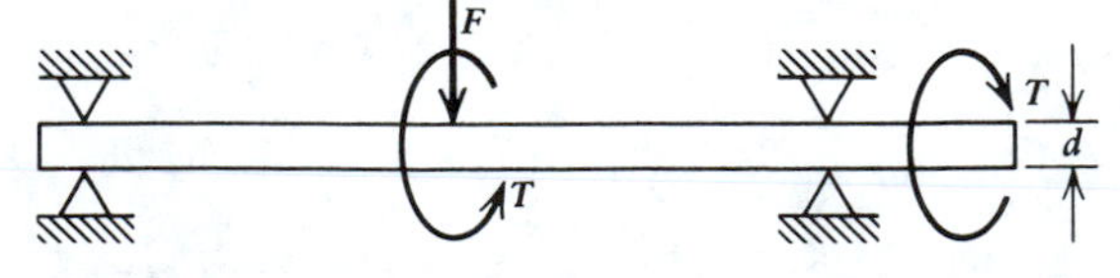

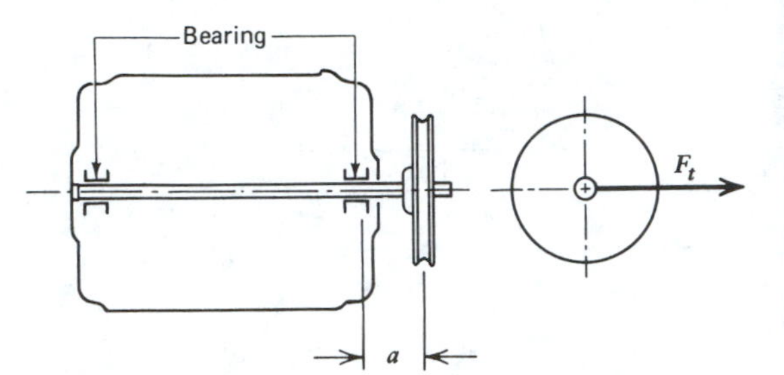

Figure 9-24 Overhung load. The safe load is one for which the moment $F_t a < M_w$, where a is the moment arm and M_w is the allowable bending moment of the shaft. Furthermore, F_t should not exceed the allowable bearing load.

EXAMPLE 9-3

Calculate the transverse load of a 10-hp, 1200-rpm electric motor equipped with a (*a*) sprocket, (*b*) gear, (*c*) V-belt sheave, and (*d*) flat belt pulley, assuming all four members have a pitch diameter of 5 in.

SOLUTION

$$F_t = k_t \frac{126{,}000\,(10\ \text{hp})}{(5\ \text{in.})\,(1200\ \text{rpm})}$$

$$= k_t\ 210\ \text{lb}$$

(*a*) Chain: **210 lb** (*b*) Gear: **263 lb**
(c) V-belt: **315 lb** (*d*) Flat belt: **525 lb**

From this we conclude that shafts and bearings must increase in size as we go from chains to flat belts.

SUMMARY

This chapter surveys the differences and similarities among common machine members. A knowledge of how they differ enables designers to select from several systems or members the ones most useful to a specific job at minimum cost. Most are part of a speed-reduction system. Thus the same basic principles and equations apply to the majority. All are subject to vibration, inertia forces, and parasitic loads.

Power can be transmitted and controlled mechanically, pneumatically, hydraulically, and electrically. In mechanical systems power is transmitted by means of friction, interlocking, and contact. Within these three categories transmission may take place at constant or variable speed. In each category there are six basic means of power transmission, so power can be transmitted mechanically in 12 different ways.

Power transmission is necessary because available forces can rarely be used effectively without being transferred to a suitable work station, adjusted and, in general, magnified. Even though power can be transmitted mechanically, pneumatically, hydraulically, and electrically within compact machinery, the mechanical means dominate. Gears are the most important means in this group, followed by belts and chains. To adjust speed and torque to the prevailing but often changing loads, stepless variation is ideal. Whenever this is achieved, however, it is usually at the expense of capacity or efficiency.

Hydraulic, friction, belt, and chain drives can all be designed to provide stepless speed variation over a wide range but, due to inherent shortcomings, are very limited in their use. Multistep gear transmissions are therefore the most widely used means of adjusting torque and speed to prevailing operating conditions. As a drive, the power screw provides unparalleled precision for slow, uniform, linear, intermittent motion of heavy platforms and tool carriages.

Power transmission, when subject to extreme torque variation, usually employs a flywheel for smooth running and as a source of instant power. Besides torque variation, there are other undesirable side effects from power transmission such as centrifugal force, inertia force, and transverse shaft loads that place an additional load on vital parts without doing useful work. Thus power technology is far from its goal of transmitting power at any capacity and speed with high efficiency.

REFERENCES

9-1 Carson, R. W. "New and Better Traction Drives Are Here," *Machine Design,* April 18, 1974.

9-2 Jaeschke, R. L. *Controlling Power Transmission Systems.* Cleveland, Ohio: Penton, 1978. (An outstanding book, easy to read.)

9-3 "Mechanical Adjustable Speed Drives," *Machine Design,* September 12, 1968.

9-4 Metzger, Jack. "Hydraulic or Pneumatic: Which Fluid Power System?" *Machine Design,* June 26, 1969.

9-5 Schulthorpe, H. J., and R. Lemon. "Gears Put Muscle in Hydrostatic Drives," *Machine Design,* October 26, 1977.

PROBLEMS

P9-1 Use Table 9-1 to answer the following.

(*a*) What are the speed and ratio limitations on spur gears, helical gears, V-belts, and roller chains?

(*b*) Which of these drives have:

(1) The highest rate of power to mass?

(2) The lowest noise level?
(3) The lowest initial cost?
(4) Longest life (relatively)?
(5) The lowest and highest efficiency?

P9-2 What are the advantages and disadvantages of using friction, interlocking, and hydrostatics for power transmission?

P9-3 Use the pie diagram in Fig. 9-4 to calculate the percentage use of mechanical components, assuming that use roughly corresponds to sales. List the components according to sales and comment on the answer.

P9-4 A speed reducer steps down a motor speed to a lower load speed. The shaft is long, because the motor and work station are far apart. Where should the reducer be located for optimum design—close to the motor, close to the work station, or perhaps midway between the two (Fig. 9-25)? *Hint:* Use the power equation. For shafts, $d^3 = 16T/(\tau\pi)$.

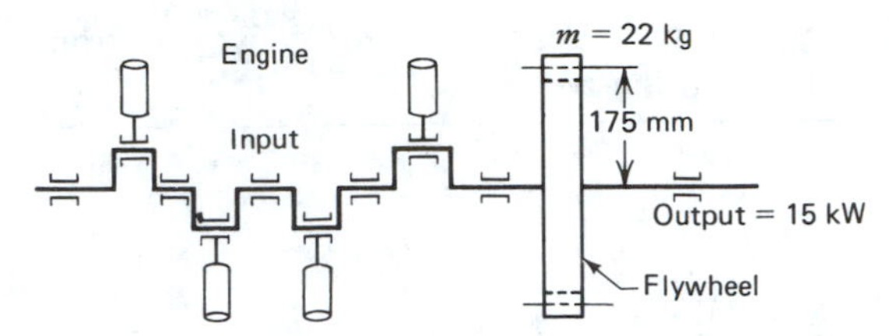

Figure 9-27 Engine and flywheel. (Problem 9-6)

P9-6 A four-cylinder engine has a flywheel of mass 22 kg concentrated at a radius of 175 mm. The energy, which momentarily makes the flywheel accelerate from minimum to maximum speed, is 20% of the engine output per revolution. At an output of 15 kW, the mean speed is 1800 rpm. Find the total speed variation percentage. (Fig. 9-27). *Hint:* $E = 0.5k^2m(\omega_1 + \omega_2)(\omega_1 - \omega_2)$.

P9-7 An industrial speed reducer has a pinion mounted on its output shaft. This shaft rotates at 90 rpm, transmits a torque of 200 N · m, and has an overhung load

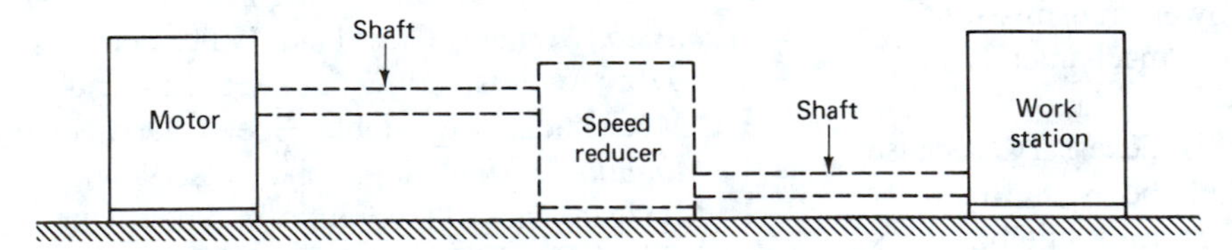

Figure 9-25 Location of speed reducer. (Problem 9-4)

P9-5 A flat belt conveyor is required to transport parts at a rate of 450 fpm with a 5-in. diameter head shaft pulley. A 10-hp, 1800-rpm motor is to be used. Service time is 4 h/day, 5 days/wk, 50 w/yr. Select one or several suitable drives. How many of these drives qualify if space is at a premium (Fig. 9-26)?

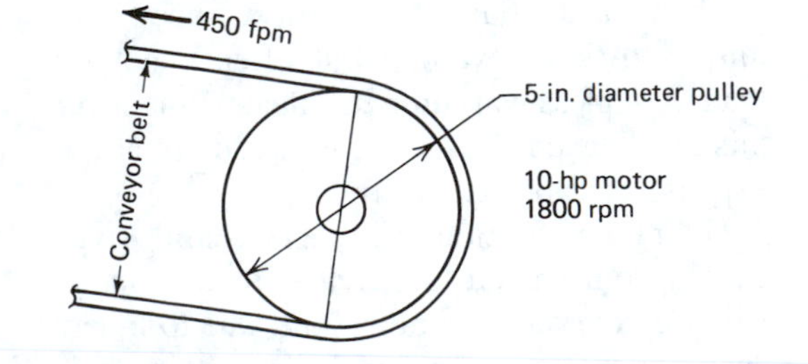

Figure 9-26 Conveyor. (Problem 9-5)

capacity of 5.8 kN. Estimate the minimum pitch diameter D for the pinion. Disregard power losses in estimating D. *Hint:* Use $F_t < 5.8$ kN to find D.

P9-8 Calculate the transverse force F_t on a shaft transmitting 10 kW at 900 rpm by means of a wheel of pitch diameter 100 mm if the wheel successively is part of a (*a*) single-strand chain drive, (*b*) gear drive, (*c*) V-belt drive, and (*d*) flat belt drive. What is the increase of F_t in percent?

P9-9 At a speed of 72 km/h, the transmission shaft of a small vehicle (1500 kg) delivers a torque of 315 N · m to the differential. This torque control device has a reduction ratio of 4.12:1 and an efficiency of 98%. The transmission efficiency is 0.92. The car has rear wheel drive (Fig. 9-28).

(*a*) What is the torque on the rear axle (use $P = Tn/9550$)? Except for the 2% loss, power is the same on both sides of the differential.

(*b*) For a wheel diameter of 762 mm, calculate the traction F_t at each rear wheel (use $T = Fr$).

(*c*) At 72 km/h, what is the wheel speed (use $v = \pi D n_w$)?

(*d*) Estimate the engine power at 72 km/h.

(*e*) What is the resistance to rolling F on asphalt (use $F_R = f_v W$. $f_v = 0.025$)?

(*f*) Equipped with rail wheels, this transformed vehicle can also be used to inspect railroad track. Find F_R (use $f_v = 0.0025$).

(*g*) At 72 km/h, how far can each of the two designs theoretically coast before coming to a full stop [use $0.5mv^2 = F_R$ (distance traveled)]? Because of wind resistance, the actual distance will be much shorter.

P9-10 A truck carrying a load of 9 kN has a mass of 450 kg (Fig. 9-29). The driving wheels are 1.066 m in diameter. The differential ratio is 5:1. The efficiency of the differential is 0.95. On a level road at 15 km/h, the traction (total

Figure 9-28 Small vehicle. (Problem 9-9)

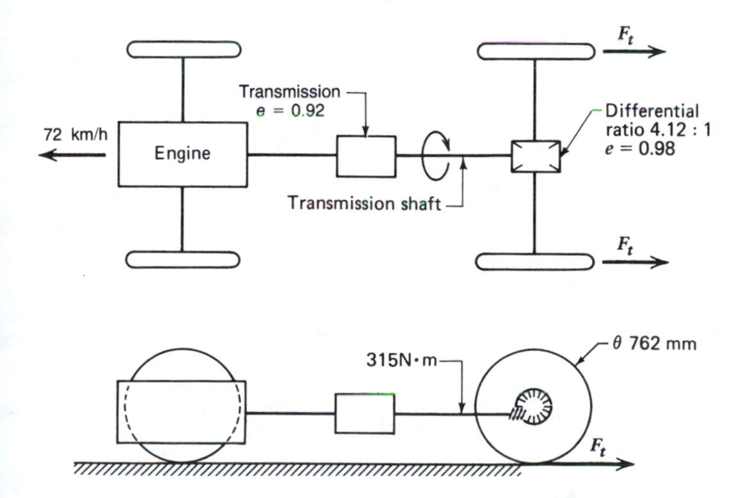

amount of driving push on a given surface) is 3 kN.

(*a*) What is the total weight W ($g = 9.81\ \text{m/s}^2$)?

(*b*) Calculate the wheel speed and the speed of the transmission shaft at 15 km/h ($v = \pi D n$).

(*c*) Calculate the power necessary to move the vehicle at a speed of 15 km/h ($P = Fv/1000$).

(*d*) What is the power requirement at the transmission at a speed of 15 km/h?

(*e*) At high gear, the transmission output torque is 550 N · m. Will it be necessary to shift down to climb a 10% grade ($\tan\,\theta = 0.1$) at the same speed?

Figure 9-29 Small truck. (Problem 9-10)

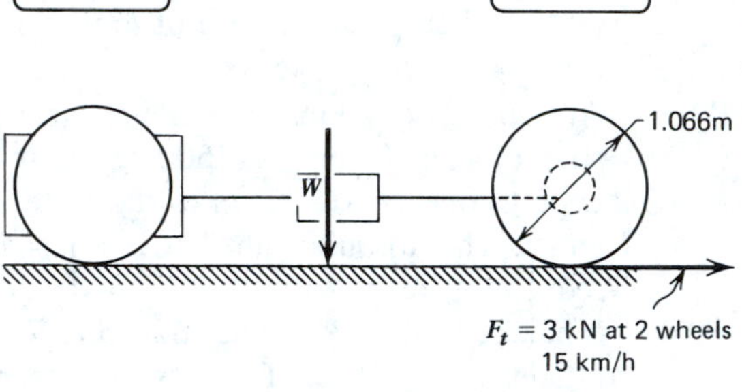

Chapter 10
Flexible Couplings

One machine can do the work of fifty ordinary men.
No machine can do the work of one extraordinary man.

ELBERT HUBBARD

Flexible couplings exist because the world is imperfect. Machines never line up perfectly, never run smoothly, and never stay where put. In today's high-speed machinery, transmission of power from one shaft to another is a potential source of destructive forces. At high speeds, shaft misalignment is the number one source of premature downtime.

10-1 FUNCTION AND DESIGN

A shaft coupling is a *constant* torque device for joining two shafts end to end in such a manner that continuous rotary motion of the drive shaft in one direction causes continuous rotary motion of the driven shaft in the *same* direction with *no* reduction in torque and little if any reduction of speed. No limitation on speed is usually imposed, but careful aligning of shafts is required.

Couplings are fairly permanent shaft connections; that is, they are permanent until servicing or rebuilding becomes necessary. (Less permanent connectors are termed clutches.)

Couplings differ widely in size and appearance. However, nearly all couplings consist of three basic members: two shaft hubs and a connector or connecting element (Fig. 10-1). A coupling usually must accomplish several objectives in addition to transmitting rotary power. These include allowing or compensating for misalignment between the rotating coupled shafts and allowing axial or end movement of the coupled shafts. Some couplings also provide a means of damping vibration and insulating the coupling halves from electrical current transfer. Couplings are also produced that perform added special jobs such as preventing shaft overload, acting as a brake, or allowing the driver and driven parts to be spaced some distance apart. Such compound designs save space and reduce cost.

Couplings are indispensable in machine systems because power units and driven machines are often manufactured separately instead of as a

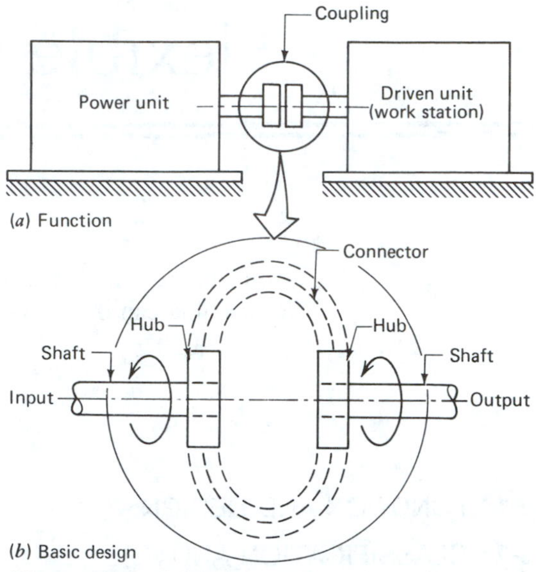

Figure 10-1 Schematic outline of coupling.

unit. Within a machine system couplings may also provide convenient drives for auxiliary equipment. During repair or testing, couplings facilitate temporary disconnection of machine components to permit one member to rotate while the other is stationary. This is often a safety feature.

10-2 CLASSIFICATION AND STANDARDIZATION

Couplings can be classified according to:

1. Relative shaft position (parallel, collinear, intersecting).
2. Basic design (fluid, mechanical, electromechanical).
3. Function (accommodation of misalignment, shaft extension, or special function).
4. Performance (heavy-duty, high-speed, light-duty).

Figure 10-2 shows schematically a classification based on three of these criteria. With the exception of Hooke's coupling, only collinear shaft couplings will be discussed.

The mechanical group is the most widely used

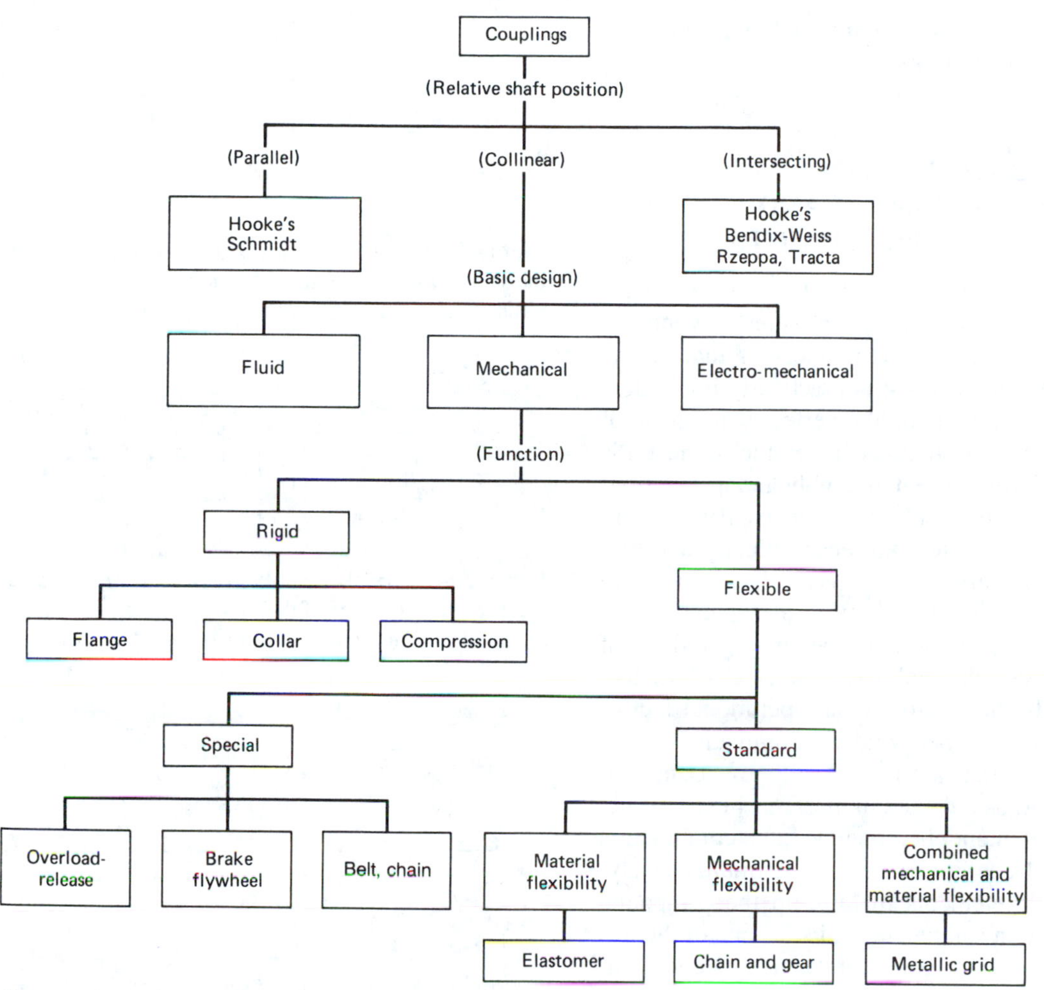

Figure 10-2 Schematic outline of couplings.

because of its predominance in motor applications. This group includes both rigid and flexible couplings. Flexible types are highly standardized items available from manufacturers in many configurations.

Standard as well as special couplings fall into one of three groups that provide flexibility by:

1. Mechanical components.
2. Resilient materials.
3. A combination of both.

The following sections will discuss these types in greater detail.

Flexible couplings are covered by these AGMA standards.

510.02 – 1969 Nomenclature for Flexible Couplings.

511.02 – 1969 Bore and Keyway Sizes for Flexible Couplings.

512.03 – 1974 Keyways for Flexible Couplings.

513.01 – 1969 Taper Bores for Flexible Couplings.

514.02 – 1971 Load Classification and Service Factors for Flexible Couplings.

515.02 – 1976 Balance Classification for Flexible Couplings.

10-3 CONDITIONS OF MISALIGNMENT AND AXIAL DISPLACEMENT

Operating conditions that lead to shaft misalignments are caused by factors often beyond the control of designers or builders of machinery. Experience indicates it is practically impossible to achieve and maintain perfect alignment of coupled rotating shafts, so a flexible connector is needed. During initial assembly and installation, precise alignment of the shaft axes is difficult and, in many cases, not economically feasible. During operation, alignment is even more difficult to maintain. Shaft misalignment caused by flexure of structural members, settling of foundations, thermal expansion, shaft deflection, and other factors is an operational fact of life that cannot be avoided. Compounding this difficulty is the fact that misalignment and displacement, as it occurs in most machinery, is not a single-element phenomenon but a combination of several conditions occurring simultaneously. The single elements making up the compound phenomenon can be classsified as one of the following: angular, axial, torsional, and end-float. These conditions are shown in Figs. 10-3 to 10-6 and represent *momentary* positions of the shaft ends. Since several of these conditions may be present at any instant, the actual path described by the shaft ends may be very intricate and subject to rapid changes. For this reason, misalignment is often referred to as *multidirectional*.

Axial displacement comes in two forms. One is caused by uneven temperature changes during some part of operation (e.g., by starting and stopping). The second, referred to as end-float, is cyclic in nature and will occur when changing axial forces are present together with axial bearing play. End-float can often be observed in

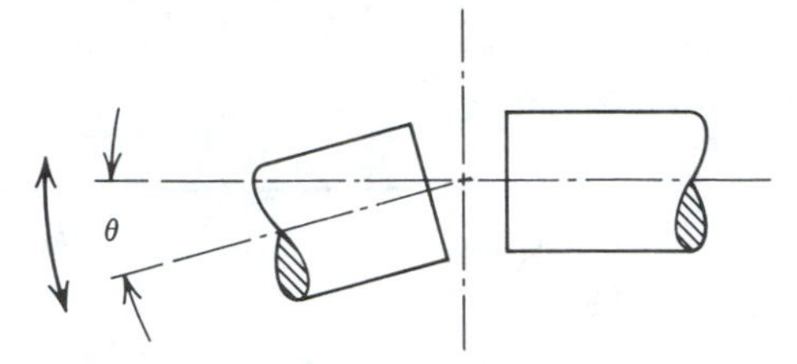

Figure 10-3 Angular misalignment is present when the shafts' axes are inclined one to the other. Its magnitude can be measured at the coupling faces.

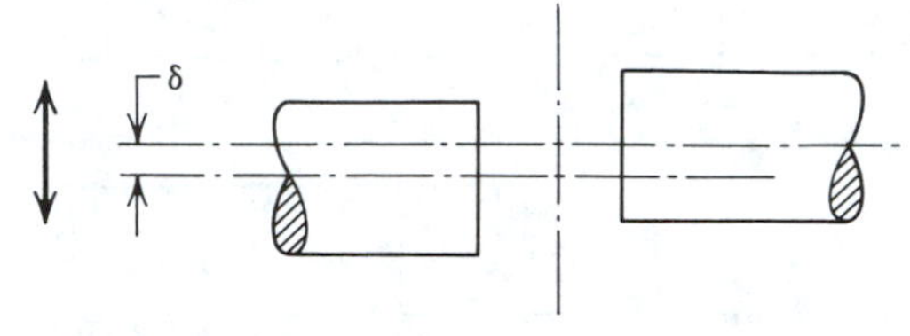

Figure 10-4 Axial or parallel-offset misalignment is present when the axes of the driving and driven shafts are parallel, but laterally displaced.

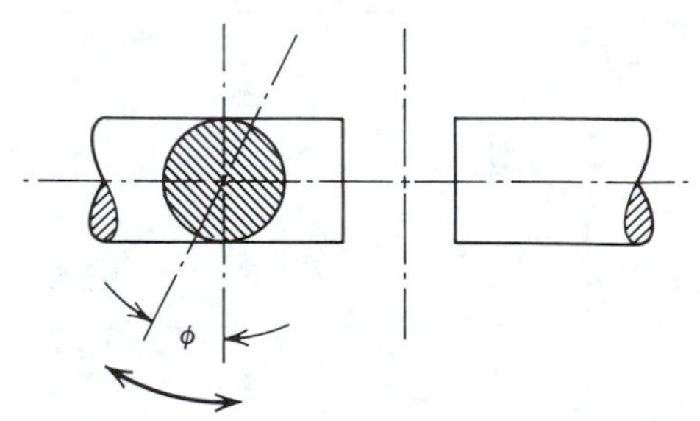

Figure 10-5 Torsional misalignment is present when the two shafts undergo angular displacements during operation.

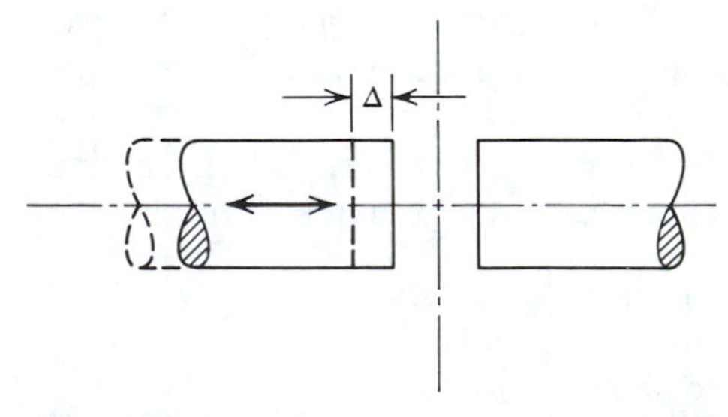

Figure 10-6 End float is present when the two shafts move axially relative to each other.

slow-moving shafts, particularly when extensive wear has taken place.

When misalignment is present in a given design, the introduction of some form of flexibility becomes mandatory. Without some degree of flexibility, shaft stresses due to reaction forces can become very high and bearings overloaded. Under such conditions a fatigue failure becomes a virtual certainty. Since high-speed motors and engines are common today, an accumulation of 10^7 cycles or more occurs in a relatively short time. Designers must exercise every precaution to ensure that normal operational stresses and stresses from possible misalignment do not exceed the fatigue limit. The fatigue limit is the stress that can be applied to a machine member of 6 million cycles without causing failure (additional cycles beyond this number rarely have an additional adverse effect).

EXAMPLE 10-1

A 1200-rpm electric motor drives a pump through a flexible coupling 8 hr a day, 5 days a week. How long does it take to accumulate 10 million cycles?

SOLUTION

$$X = \frac{10^7 \text{ cycles}}{(5 \text{ day/week})(8 \text{ hr/day})(60 \text{ min/hr})(1200 \text{ cycles/min})}$$

$$= \mathbf{3.47\ weeks}$$

EXAMPLE 10-2

What is the misalignment stress of an overhanging steel shaft that is out of line with an adjoining shaft of an electric motor?

Shaft diameter:	$d = 50$ mm
Shaft overhang:	$L = 125$ mm
Displacement:	$\delta = 0.25$ mm

This situation is shown in Fig. 10-7.

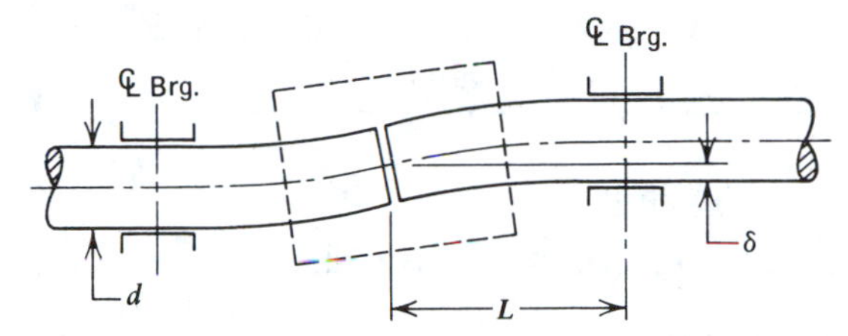

Figure 10-7 Bending stress in overhanging steel shaft due to parallel offset misalignment.

Maximum stress occurs at the bearing support and is due primarily to bending. To calculate this stress accurately, it would be necessary to know the shaft and its loading beyond the bearing. However, a good approximation is obtained by treating the overhanging end as a cantilever beam. Since the bearing support is not rigid, the calculated stresses tend to be too high.

$$\sigma_b \max = \frac{M}{Z} = \frac{PL\,32}{\pi d^3}$$

Only P is unknown; hence we must find it. P is found by means of the formula for maximum deflection, Table F-3 in the Appendix.

$$\delta = \frac{PL^3}{3EI} = \frac{64PL^3}{3E\pi d^4}$$

$$P = \frac{3E\pi d^4\delta}{64L^3}$$

$$\sigma_b \max = \frac{3E\pi d^4\delta}{64L^3}\left(\frac{L\,32}{\pi d^3}\right)$$

$$= \frac{3Ed\delta}{2L^2}$$

For steel, $E = 207\,000$ MPa. By substituting the given values

$$\sigma_b \max = \frac{3(207\,000 \text{ N/mm}^2)(50 \text{ mm})(0.25 \text{ mm})}{2(125 \text{ mm})^2}$$

$$= \mathbf{248\ N/mm^2 = 248\ MPa\ (36{,}000\ psi)}$$

Clearly, stresses of this magnitude are an invitation to material fatigue.

10-4 FLEXIBLE COUPLINGS

A coupling is termed flexible when it is compatible with reactive loads generated by the coupling or transferred to it. Such couplings should always be used whenever a prime mover is directly coupled to a gear unit or machine shaft, or when the gear unit is directly coupled to a machine shaft or short shaft with two bearings (one being fixed). In these circumstances, a rigid coupling should *never* be used. A flexible coupling should not be used to compensate for careless or deliberate misalignment of shafts; it can be used only to absorb initial assembly inaccuracies and possible operational misalignments of a transient nature. Maintenance alignment checks should be scheduled to catch shifts in basic position.

Flexible couplings accommodate a certain amount of misalignment because the reaction forces due to angular, axial, or parallel loadings are absorbed in the linkages or joints or in the pivoting or sliding components. The ratings of flexible couplings are therefore largely determined by the acceptable stress levels in their flexible members. Compensation for end-float, for instance, is achieved by sliding members or flexure of resilient members. Torsional flexibility (the ability to deflect torsionally when subjected to normal cyclic torque variations) is achieved by the provision of springs between the two halves of the coupling or achieved by the provision of a flexible medium such as rubber. Rubber is a marvelous material for couplings but, from a design point of view, an odd one. It is marvelous because it combines resiliency with high damping capacity, roughly sixty times that of steel. It is odd because it deviates greatly from Hooke's law, which confuses designers accustomed to the straight-line characteristics of steel.

Considerable ingenuity is embodied in today's flexible couplings. A discussion of the basic types follows.

1. Material Flexibility

(*a*) *Elastomer-type Coupling*. Members of this group include the most versatile of all coupling systems. The basic design concept—two metal hubs joined by molded rubber—provides a resilient shaft connection. The elastomer will flex in any direction and compensate for all types of misalignment, including end-float. The most important advantage, however, is torsional flexibility coupled with vibrational damping capacity. This helps protect the system against torsional vibrations, a major cause of power transmission systems' failure (Fig. 10-8).

Perhaps the most versatile elastomer coupling is the tirelike design shown in Figs. 10-9 and 10-10. It accommodates angular misalignment, parallel misalignment, and end-float singly or in any combination. It cushions shock loads and damps torsional vibrations. Requiring no lubrication, elastomer couplings are easy to install, and the flexible element can be removed without moving either shaft or the hubs on the shafts. Ample radial space is the only design requirement.

(*b*) *Cushion-type Flexible Coupling*. The spider coupling shown in Fig. 10-11 is typical. It transmits torque through an oil-resistant rubber spider assembled between two pairs of axially overlapping rigid jaws. Flexibility is derived from clearances in mating parts, permitting differential movement during rotation or in deflection of elastomeric members. Reaction forces may result from distortion of parts during angular misalignment, friction in sliding parts, and lockup in parallel misalignment. This inexpensive design is available in many sizes, from fractional ratings up to 75 kW (100 hp).

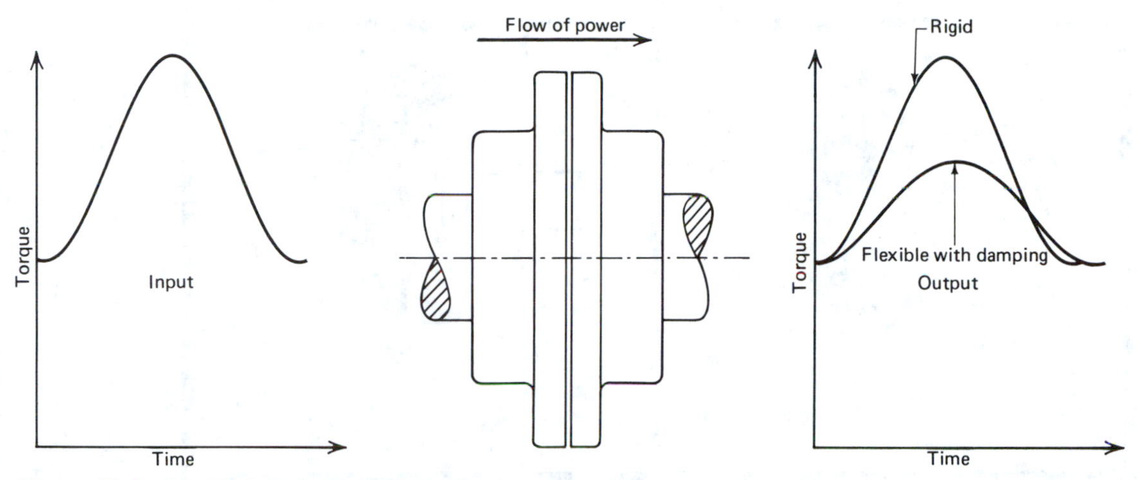

Figure 10-8 Mechanics of couplings.

Figure 10-9 Elastomer-type coupling. (Courtesy Reliance Electric Company.)

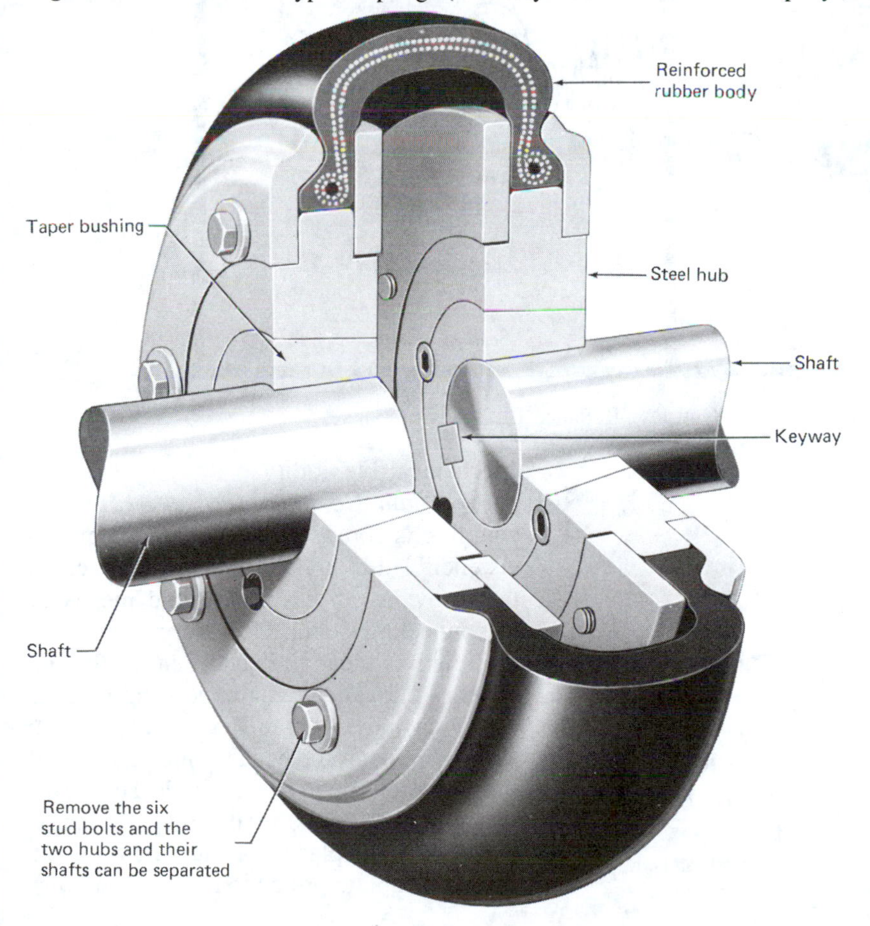

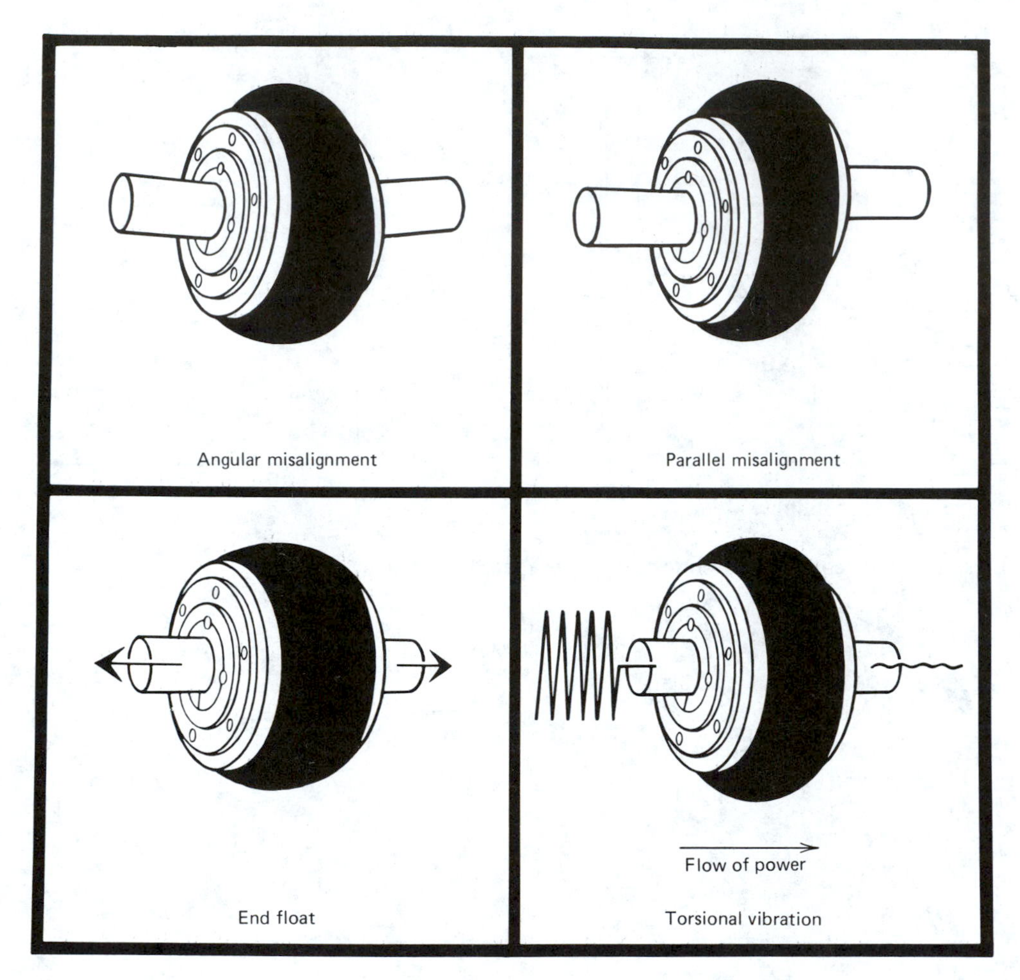

Figure 10-10 Elastomer-type coupling. (Courtesy Reliance Electric Company.)

Figure 10-11 Cushion-type flexible coupling.

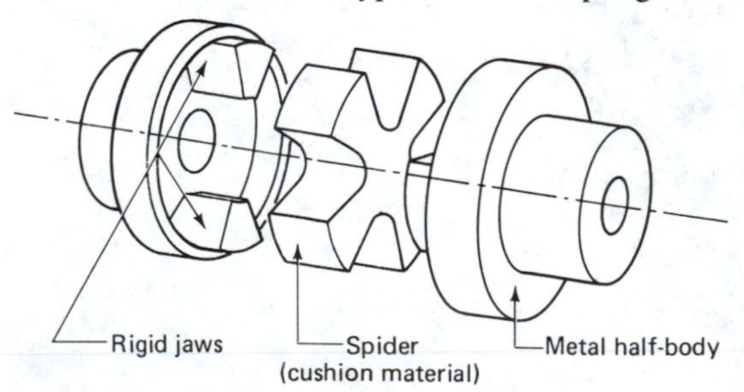

2. Mechanical Flexibility

(*a*) *Chain Coupling (Fig. 10-12).* This coupling is composed of two chain sprockets encircled by a detachable, duplex chain that provides an easy and rapid means of disconnecting shafts. Flexibility is provided by relative clearance between teeth and rollers. The silent chain version reduces operating noise, while the use of a nylon chain eliminates the need for lubrication and thus also eliminates a possible source of contamination in the food and textile industries. Oil-tight plastic or aluminum covers ensure complete protection of the coupling.

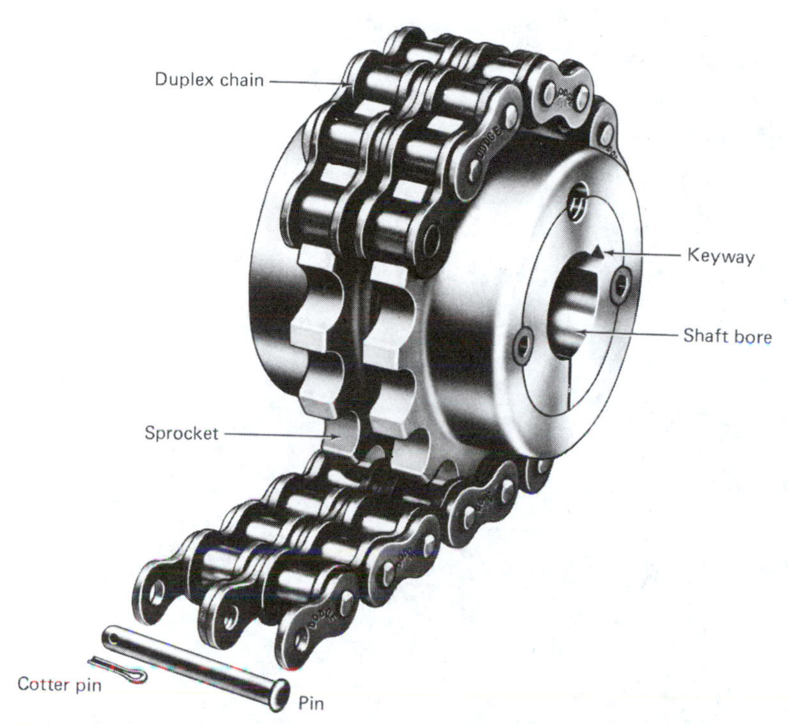

Figure 10-12 Chain coupling. (Courtesy Reliance Electric Company.)

(*b*) *Gear Couplings (Fig. 10-13).* These couplings can be used for practically all types of driving and driven machinery. Gear couplings are designed to compensate for angular and axial misalignment, with the particular advantage of allowing end movement without transmitting thrust loads. The crowned gear tooth design provides for a wide degree of angular misalignment without excessive backlash.

The basic design consists of two identical, externally geared hubs keyed, respectively, to the driving and driven shafts. These hubs are connected and enclosed by covers internally geared to engage with the hub teeth. The covers are bolted together and function as a single rigid unit, transmitting torque from one hub to the other, with seals fitted to limit lubricant seepage and dust infiltration. The floating cover assembly is restricted in axial direction to ensure that the teeth always maintain adequate contact for power transmission.

Gear couplings have a large power capacity but lack damping ability and torsional flexibility. They are widely adapted to general industrial drives, pumping installations, heavy rolling mills, rotary compressors, turbines, and drives operating under the most severe conditions. (Fig. 12-ld)

3. Combined Mechanical and Material Flexibility

(*a*) *Metallic Grid Coupling (Fig. 10-14).* This connector consists of two metal half-bodies with slots cut into the peripheries to receive a serpentlike spring steel alloy grid, which forms the resilient transmission member. The slots into which the springs fit are

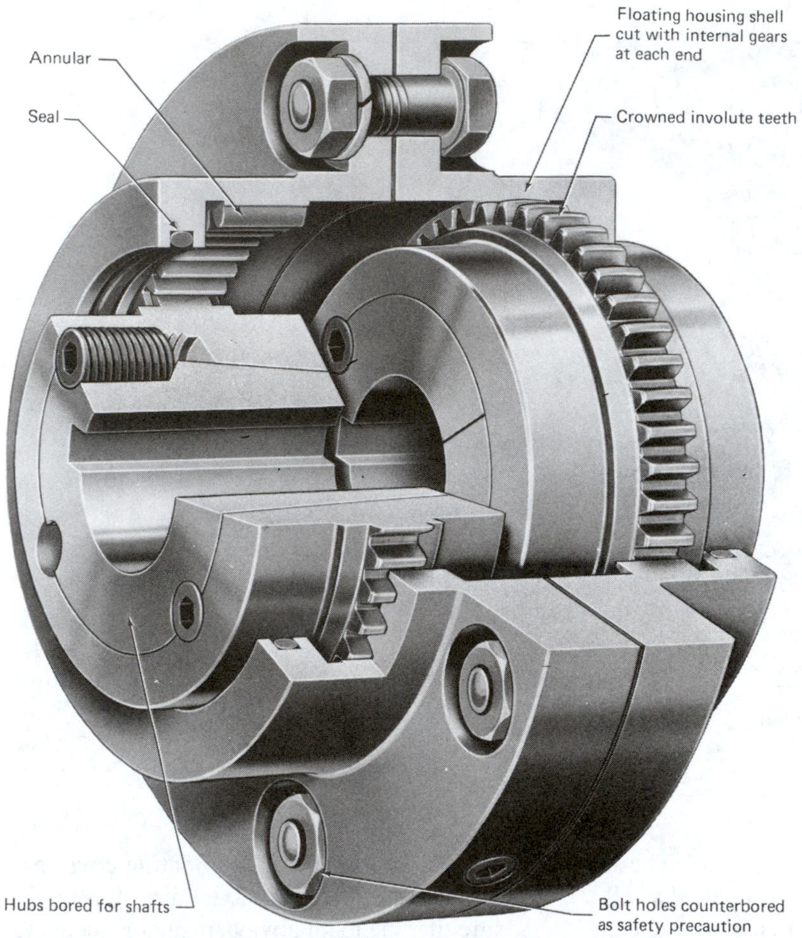

Figure 10-13 Gear coupling. (Courtesy Dodge Manufacturing Company.)

parallel in each half-body and inwardly flared in a parabolic configuration. Under normal load the serpentlike springs have a long, flexible span. As the load increases, the springs deflect, the flexible span is reduced, and greater support is afforded to the springs by the walls of the flared grooves, thus reducing significantly the stresses in the springs. This progressive resistance is most desirable when torsional vibrations are present, reducing shock loading by as much as 30%.

The couplings are totally enclosed in oil-tight, dust-proof covers. The two metal half-bodies may be readily disconnected by removing the steel grid. These units are adaptable to almost any type of equipment and are widely used by the mechanical industries.

(*b*) *Diaphragm Coupling*. This lightweight, high-speed, high-temperature coupling will operate economically in the speed range of 15 000 to 90 000 rpm at temperatures up to 500°C (900°F). For this reason, it is often used with gas turbines.

Diaphragm couplings transmit torque

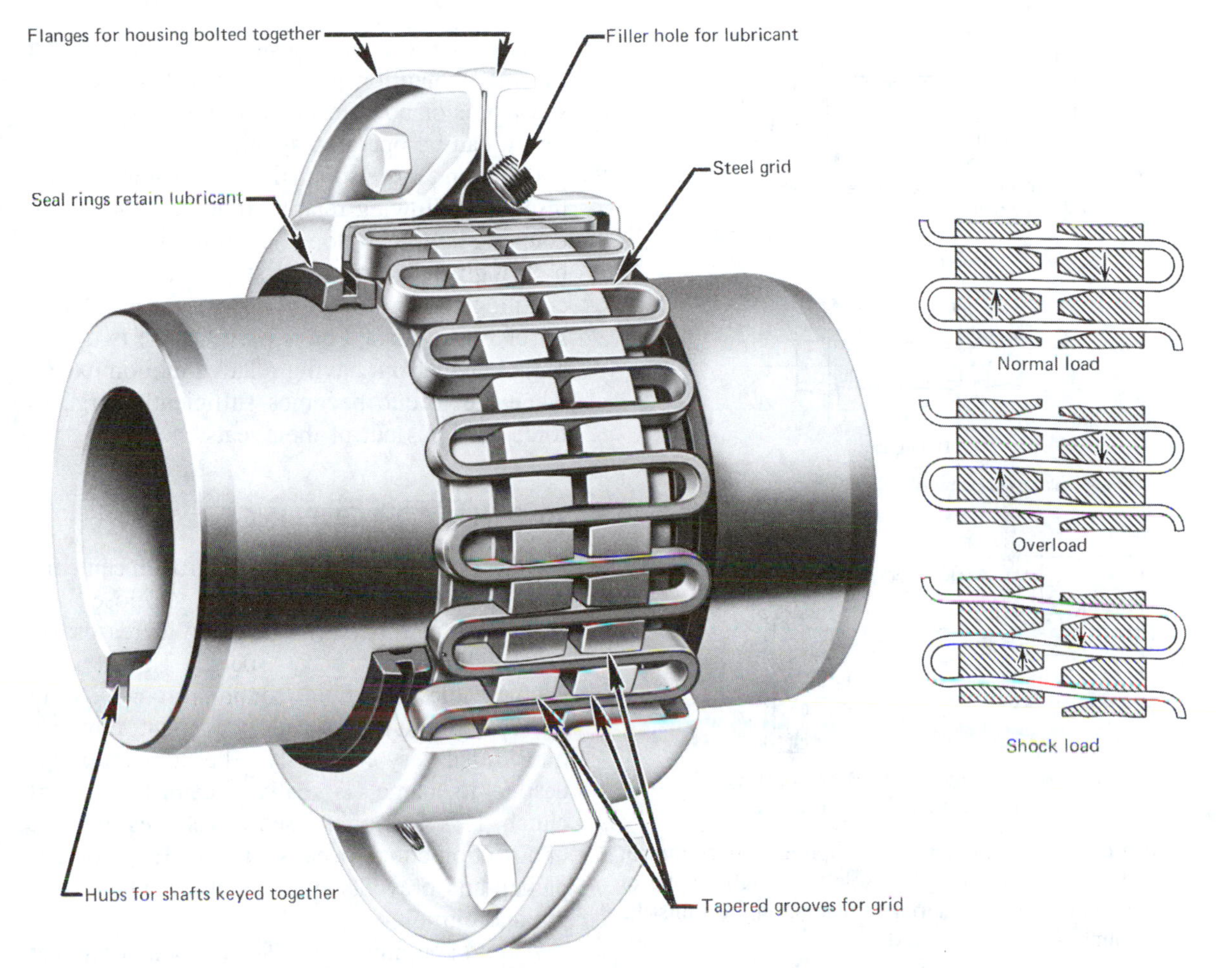

Figure 10-14 STEELFLEX® metallic grid coupling. (Courtesy Falk Corporation.)

through diaphragms bolted directly to the hub (Fig. 10-15). The diaphragms, in turn, are bolted onto the spacer tube. Misalignment and end float are absorbed through flexing of the metal plates. The torsional spring rate can be increased or decreased by changing the spacer tube. Because all members flex, instead of moving relative to each other, there is no wear and no need for lubrication. The original balance is therefore permanent. Limitations on misalignment, end float, and operating temperatures are imposed by the induced plate stresses, which must be held well below the material fatigue limit.

10-5 SPECIAL COUPLINGS

It is often desirable to combine the flexible coupling with a secondary function such as overload protection or braking. The coupling is thus a prime example of how machine parts may be combined for greater service, compactness, and lower cost. The following combinations with other machine members are common

Integrated brake wheel.

Integrated V-belt sheaves or chain sprockets.

Inbuilt flywheel for compressor drive.

Inbuilt counterweight for balancing out inherent imbalance of a machine.

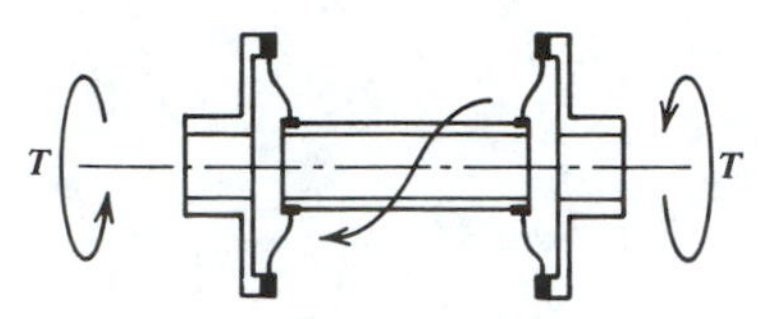

(*a*) No angular misalignment, no end float, only torsional misalignment

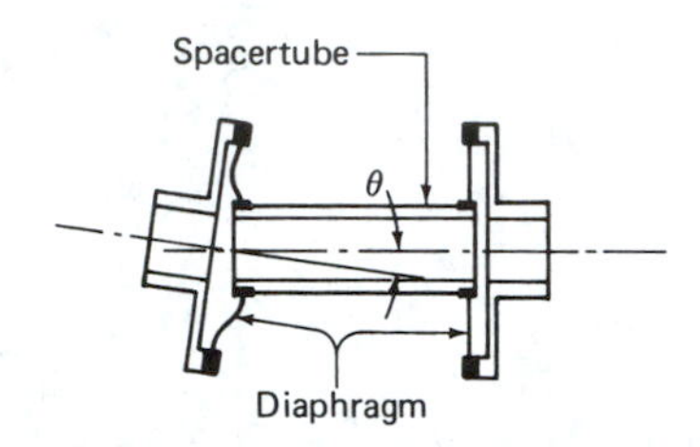

(*b*) Angular misalignment

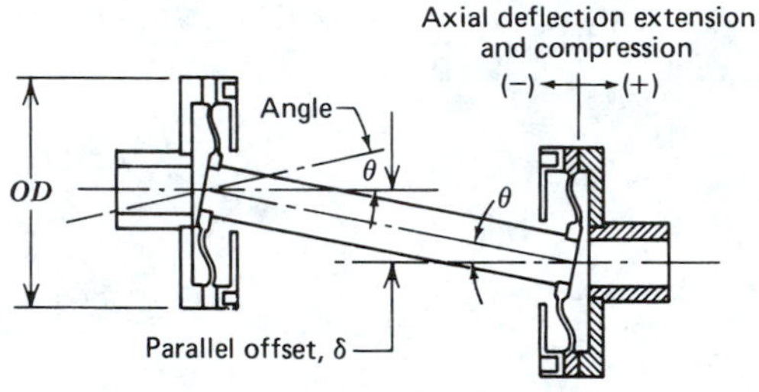

(*c*) Parallel offset misalignment. Generally, $0 \text{ deg} < \theta < \frac{1}{2} \text{ deg}$, $0 < \delta < 0.4$ in., 6 in. $< OD <$ 16 in.

Figure 10-15 Diaphragm coupling. Misalignment and end float are absorbed solely through flexing of the metal plates. Material fatigue strength limits the amount of flexing allowed.

10-6 OVERLOAD-RELEASE COUPLINGS

Flexible couplings with overload-release devices set at a predetermined torque have often saved driven machines from costly breakdowns. Torque limitations can be attained by using different mechanical designs. This overload protection is the mechanical equivalent of an electric fuse.

One of the most popular designs relies on one or more pins to shear at the desired torque value. A second design incorporates notched pins or bolts that have been designed to snap when the transverse forces reach a predetermined value. In either of the foregoing designs, the cause of the overload condition must be corrected and new shear pins or bolts installed in the coupling before it can be operated again.

Other and more sophisticated designs need no resetting following relative motion between the coupling halves. Automatic resetting is achieved by bringing the drive to a stop. The ball detent coupling shown in Fig. 10-16 is typical. Here a set of spring-loaded balls prevents the two coupling halves from having relative motion except when the torque becomes sufficiently large to force the balls out of their seats.

10-7 FLUID COUPLINGS

Fluid couplings transmit power through centrifugally generated fluid action in a closed recirculating path. Centrifugal force increases exponentially with rotary speed. Thus, when engines idle, the fluid coupling transmits no power at all. At high speed, in contrast, power is transmitted very efficiently. The fluid coupling, despite its name, is really a centrifugal fluid clutch. It is thus often used in vehicles in place of a friction clutch because it provides smoother pickup and prevents engine stalling when starting and climbing.

A fluid coupling is a self-contained pump and turbine unit. The pump (impeller) is coupled directly to the engine and the turbine (runner) to the output shaft. A fluid coupling looks like two halves of a bowl, each half with three dozen straight radial blades leading from the hub to the outside edge (Fig. 10-17*a*). The two halves are set closely together but do not touch. Both are enclosed in a tight housing filled with an oil fluid (Fig. 10-17*b*).

At low speed, nothing happens except that the oil begins to move slowly between the two halves. As speed increases, the oil is forced outward with greater momentum. At the same time it is also whirled around in the plane of rotation. Thus, when flung outward, the oil crosses over

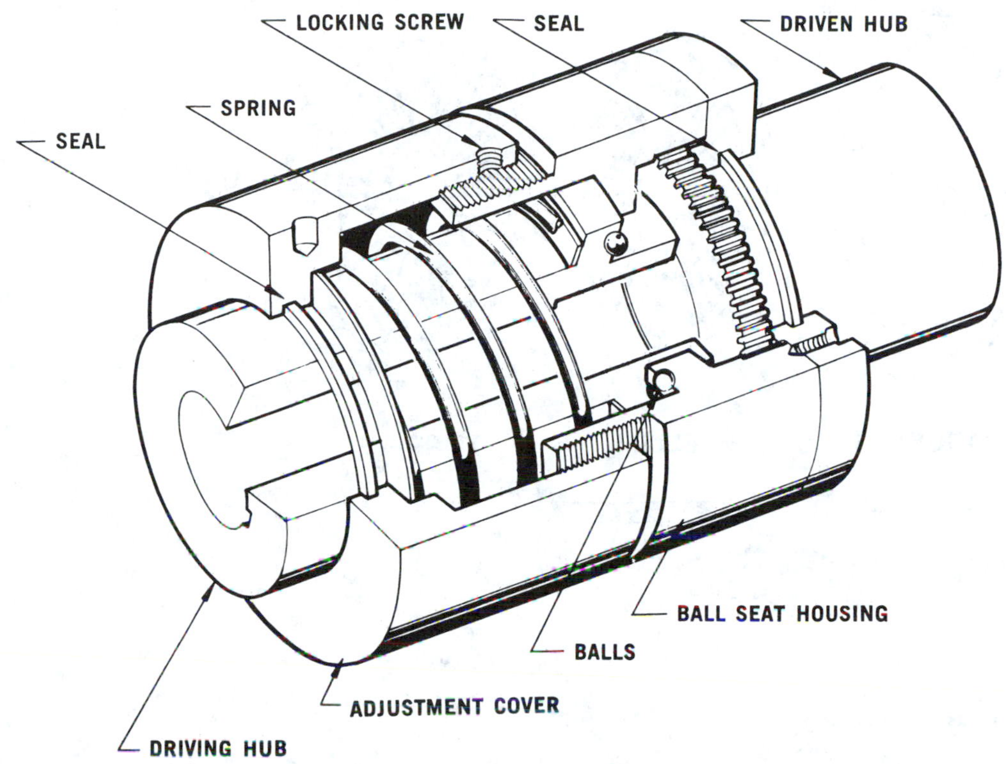

Figure 10-16 Ball detent coupling. (Courtesy Ferguson Machine Co.)

into the driven member and is caught by the vanes of the runner in a tangential driving direction. The fluid is now decelerated as it travels toward the hub of the turbine and back across into the driving member. Thus, the oil is continually circulating, gaining momentum in the impeller and losing momentum in the runner. At the same time it is traveling in a direction at right angles to this path, being pushed by the blades of both members.

At coupling speed the two halves run together and act as a nearly solid connection—a fluid flywheel. At this speed the efficiency may be as high as 98%. Below this speed efficiency decreases rapidly and more so the lower the speed.

Although fluid couplings cost more, weigh more, and occupy more space than equivalent mechanical couplings and clutches, they also outperform them and have a number of distinct advantages. Torque is transmitted unchanged as in mechanical couplings, a major advantage. Fluid couplings also have the advantage of transmitting power but blocking the transfer of torsional vibrations and load shocks from the engine to the transmission. This protects not only the gears and the drive shaft but the entire drive line, whether power is transmitted to a pair of wheels or a ship's propeller. Fluid couplings will also compensate for the slight misalignments inherent in any mechanical drive system.

10-8 MAGNETIC COUPLINGS

The typical magnetic coupling is a permanent magnet, eddy-current slip coupling used to transmit power from a source to a load (Fig. 10-18). This coupling consists of two basic components, a bimetallic rotor and a magnetic rotor. The bi-

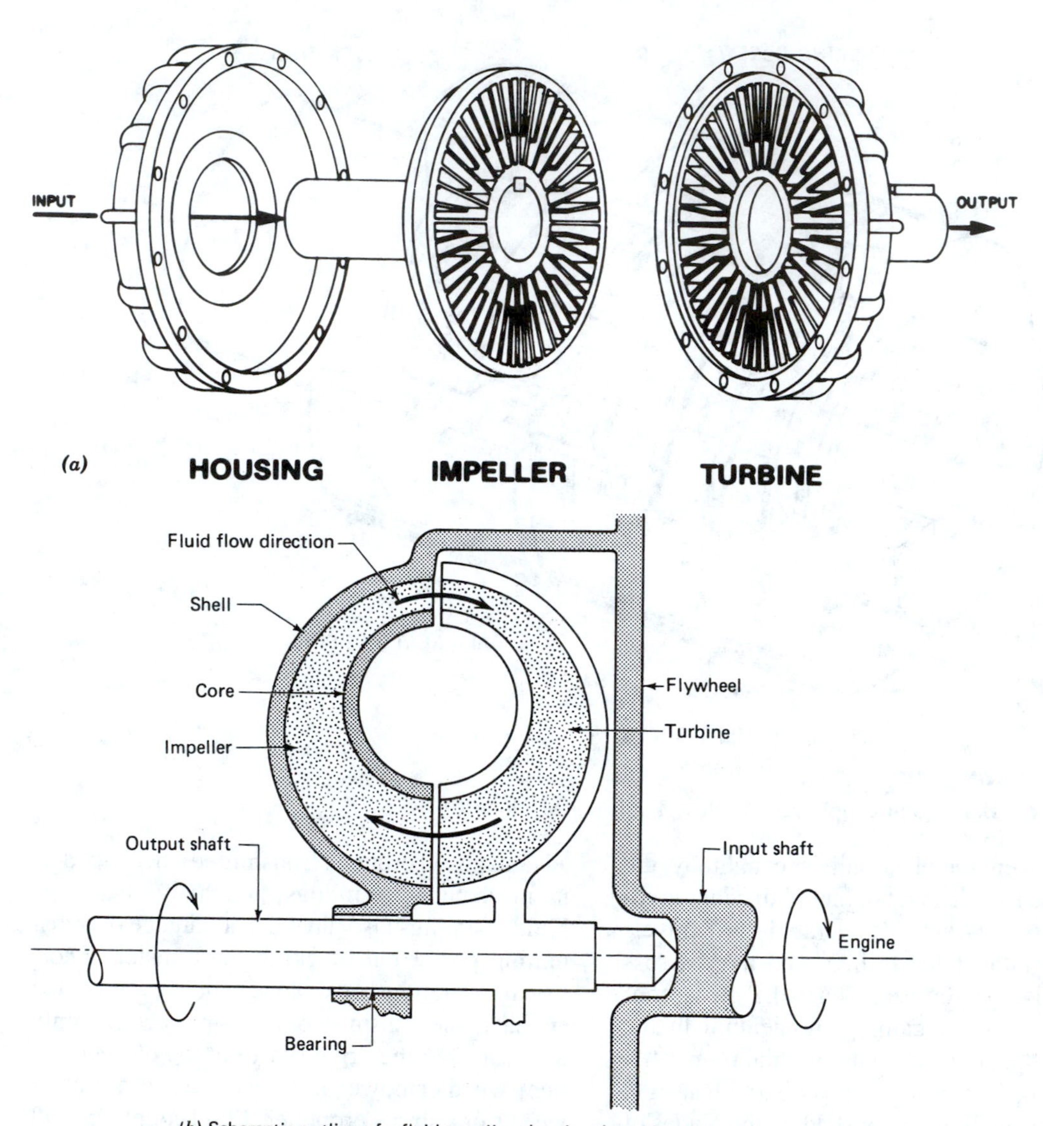

(b) Schematic outline of a fluid coupling showing the main components

Figure 10-17 A fluid coupling is a hydrodynamic drive that transmits power without the ability to change torque. (*a*) A hydraulic coupling has three main parts, housing, impeller and turbine, which are die castings of good-quality aluminum alloy, accurately machined and balanced. Impeller and turbine each have over 50 symmetrical vanes, radially oriented. (*b*) Schematic outline of a fluid coupling showing the main components. (Courtesy Plessey Dynamics Corporation.)

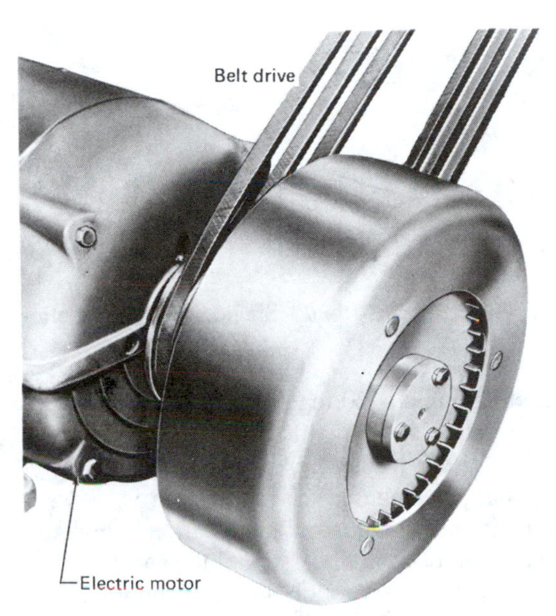

Figure 10-18 Magnetic coupling. (Courtesy Fairchild Industrial Products Company.)

metallic rotor is a cylindrically shaped steel shell with a copper insert lining the inner surface (part A in Fig. 10-19). The magnetic rotor is an aluminum diecasting with permanent magnets embedded in the periphery which face radially outward (part B in Fig. 10-19). The two rotors are mounted together so that the magnetic rotor fits inside the bimetallic rotor. This construction then allows the lines of magnetic force emanating from the permanent magnets to pass through the copper and steel of the bimetallic rotor. During rotation of the bimetallic rotor, driven by the power source, the magnetic lines are cut and magnetic attraction created. Torque is thus transmitted between the two shafts.

The main advantages of magnetic couplings are simplicity of design, easy installation, high efficiency, no wear, and smooth start. The soft starting characteristics permit the use of smaller motors on high-inertia loads or lighter gearing on high-shock loads. Power input on starting is low. Because of the weightless connector or lack of physical contact between the two rotors, neither shocks nor vibrations can be transmitted. In fact, power may be transmitted through a thin wall to a totally enclosed machine. No provision needs to be made for shaft misalignment or axial displacement.

10-9 THE UNIVERSAL COUPLING

This simple, low-cost mechanism derives its name from its ability to carry power between two shafts that intersect at an angle other than 180 deg and continue to do so even when this angle constantly varies as, for instance, in automotive

Figure 10-19 Magnetic coupling. (Courtesy Fairchild Industrial Products Company.)

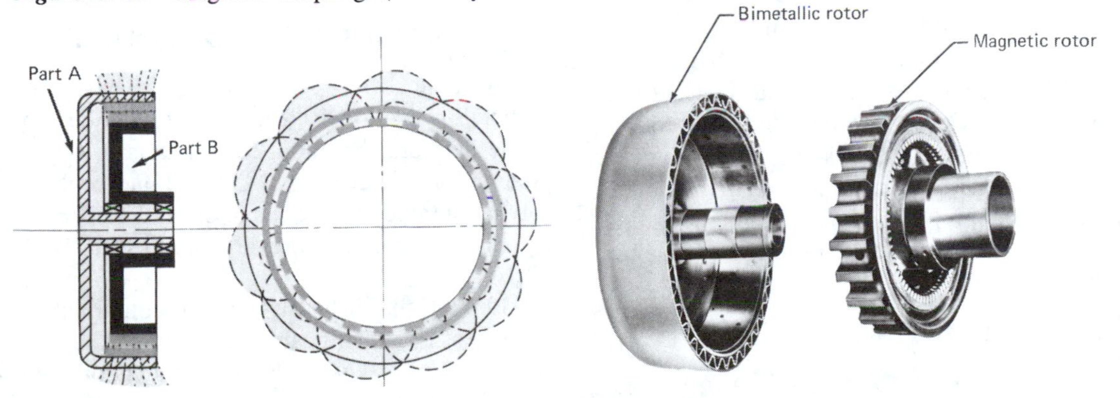

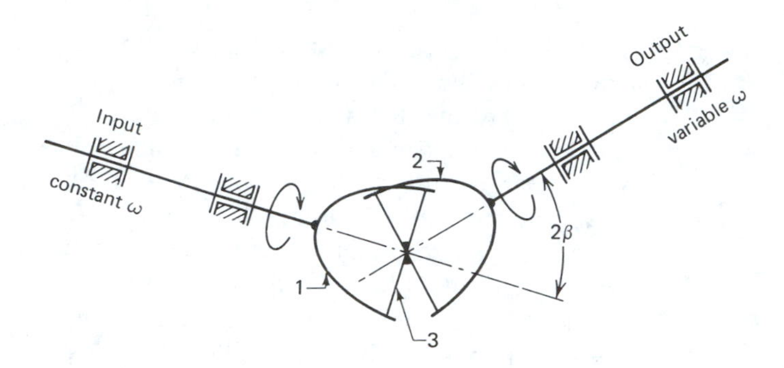

Figure 10-20 Single universal coupling.

drive shafts. Here it compensates for misalignment inherent in most rear wheel drives, plus that superimposed by reaction of the suspension system to bumps and changing loads.

The universal coupling, also known as the *universal joint* or *Hooke's*[1] coupling, is a common linkage, essentially of the form shown in Fig. 10-20. It is made up of two U-shaped pieces (1) and (2) at right angles to each other and fastened together by a cross (spider) having arms of equal length (3). The U-shaped yokes can pivot on the arms of the cross. *Because there are two of these pivots, the two shafts can be at any angle to one another and can still rotate and transmit power.* Although the driver (1) and the follower (2) make complete revolutions during the same time, the velocity ratio is not constant throughout one revolution. When the driver has uniform angular velocity, the ratio of the angular velocities between driver and driven varies between extremes of cos β and 1/cos β (where 2β is the acute angle between the connected shafts). These variations of angular output velocity mean acceleration is present. This, in turn, gives rise to undesirable inertia forces and subsequent noise and vibration. Because the differences in angular motion increase as the shaft angle increases, there is a practical limit to the operating angle that can be tolerated. For high speeds the limiting angle is 20 to 25 deg. For lower speeds the limiting angle is 40 to 45 deg.

Because it leads to excessive wear and material fatigue, angular speed variation is *undesirable* in most industrial applications. For this reason, most universal couplings are used in *pairs*. By using a double joint, as shown in Fig. 10-21*a* and 10-21*b*, the variation of angular motion is entirely avoided. In both compensating arrangements an intermediate shaft is placed between the two main shafts, making the same angle with both. When the two forks on the intermediate shaft lie in the same plane, all speed variation is eliminated. (One might say that the couplings are tuned.) As can be seen, the double joint arrangement can be adapted to parallel as well as to intersecting shafts. The automotive differential drive shaft is a well-known example of a double joint universal coupling.

10-10 SPECIFICATION OF COUPLINGS

To select the correct type and size, the basic data outlined in Section 10-2 must first be collected. These data are then applied to charts in appropriate vendors' catalogs. To facilitate problem solving, a table of service factors and a listing of machinery characteristics are supplied in Table 10-1.

Although the different types of couplings are somewhat interchangeable, each type is also restricted to a fairly well-defined application area.

---

[1]Named after its inventor Robert Hooke (1635-1703), British scientist.

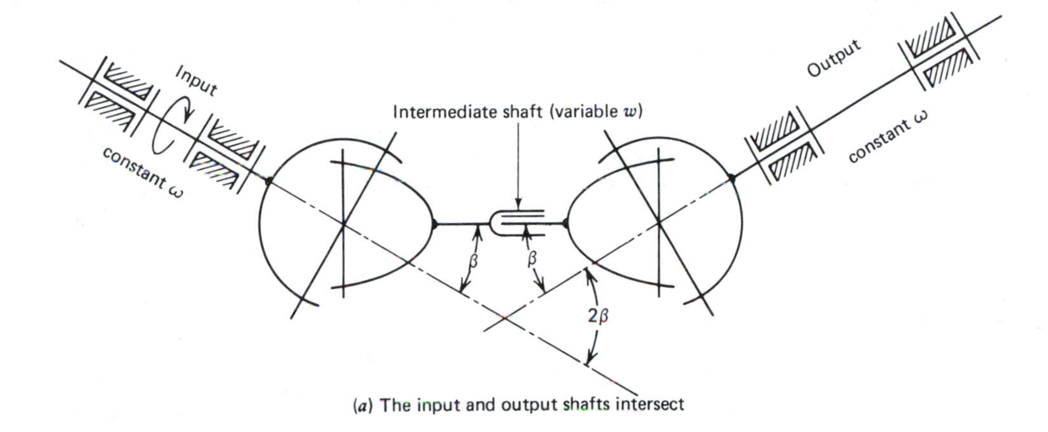

(a) The input and output shafts intersect

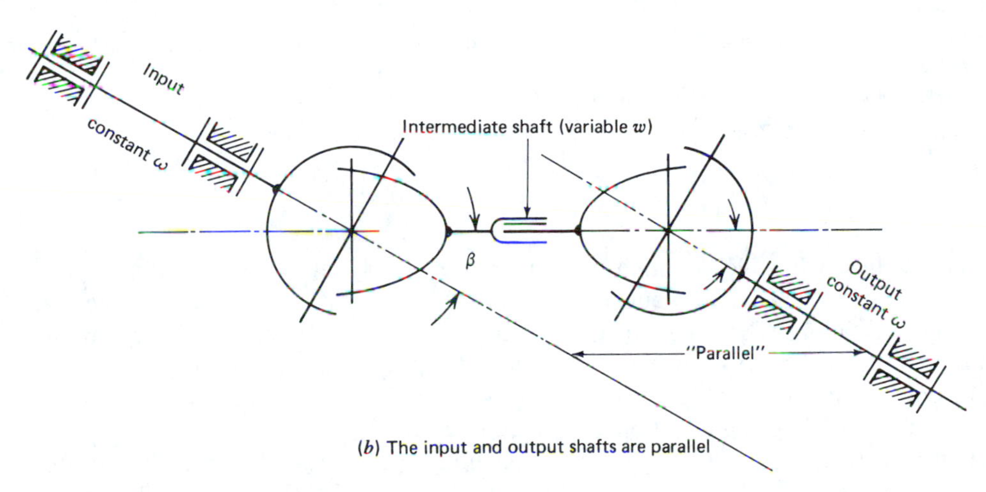

(b) The input and output shafts are parallel

Figure 10-21 Double joint universal coupling. Essentially a double-hinged joint that transmits torque in spite of constantly changing relative angles.

If more than one type will do the job, cost or ease of assembly usually dictates the selection.

EXAMPLE 10-3

Find the service factor for a shear pin coupling of a small outboard motor. The pin protects the propeller from damage due to stoppage when hitting sand or other obstructions.

SOLUTION

Small outboard motors are usually single cylinder with highly impulsive input. The propeller is in the category of "agitators and mixers," with medium impulse characteristics. The service factor is therefore **2.00.**

A coupling is part of a system of masses subject to torsional vibrations. The danger of resonant conditions at certain speeds is always present. Thus one cannot rely entirely on inherent damping to prevent torsional vibrations. The coupling spring rate must fit the mass-elastic system. A flexible coupling can prevent torsional vibration only if selected to fit the particular sit-

TABLE 10-1 Service Factors

Prime Mover (Drive Input)	Machinery Characteristics[a] Steady	Medium Impulse	High Impulse
Electric motor (steady input)	1.00	1.50	2.50
Multicylinder engine (medium impulsive input)	1.50	2.00	3.00
Single cylinder engine (highly impulsive input)	2.00	2.50	3.50

[a]Machinery Characteristics (typical)

Steady: Agitators and mixers (liquid, semi-liquid), alternators and generators (generators-lightning), compressors (centrifugal)

Medium impulse: Compressors (reciprocating—3 or more cylinders), conveyors and elevators (nonuniform feed, screw feeders)

High impulse: Alternators and generators (welding), compressors (reciprocating—1 or 2 cylinders).

Note: The complete list available from Renold Crofts, Inc., contains 58 separate cases of driven machinery.

Source: *Internal Gear Flexible Couplings*. Westfield, New York: Renold Crofts, Inc., 1972.

uation. In fact, an ill-chosen flexible coupling may cause torsional vibrations when none would occur with a more rigid coupling. The danger of torsional vibrations is present primarily when an engine drives a machine of medium or high impulsive characteristics. In such cases, a vibration specialist should be consulted.

SUMMARY

The primary purpose of flexible couplings is transmission of torque between collinear shafts while compensating for misalignment in both systems. Couplings provide a simple, direct, semipermanent connection. What they lack in complexity, they make up for in numbers. Only gears are used in greater numbers. Because power units and driven machines are often manufactured separately instead of as a unit, couplings are indispensable in machine systems.

Couplings differ widely in size and appearance, yet nearly all consist of three basic members: two shaft hubs and a connecting element. Secondary functions of couplings include damping vibration; balancing eccentric loads; and acting as a brake, flywheel, or overload device.

Couplings are classified according to relative shaft position, basic design, and function. In terms of basic design, there are mechanical, fluid, and magnetic couplings. The mechanical group is the largest because of low cost and predominance in motor applications. Despite their flexibility, they provide a one-to-one speed ratio. Fluid and magnetic couplings, in contrast, slip in response to load conditions. They are essentially starter couplings with high torsional flexibility, minimum wear, and built-in overload protection.

The universal coupling transmits power between two shafts that intersect at an angle other than 180 deg and continue to do so even when this angle varies constantly. Because of the inherent angular speed variation in a single joint, most practical applications require a double joint.

Couplings are selected from vendors' catalogs. Although available types are somewhat interchangeable, each type is also restricted to fairly well-defined application areas. If more than one type will do the job, cost or ease of assembly usually dictates the selection.

REFERENCES

10-1 Hagler, P., H. Schwerdin, and R. Esleman, "Effects of Shaft Misalignment," *Design News,* January 22, 1979.

10-2 *Shaft Couplings*. Farmingdale, N.Y.: Renold Crofts, Inc., 1972.

PROBLEMS

P10-1 When is a shaft connector a coupling and when is it a clutch? Why do motors generally use couplings and not clutches? Why do engines require clutches and rarely mechanical couplings? See Chapter 1.

P10-2 (*a*) Why is design of mechanical couplings based on interlocking and not friction?

(*b*) What is a special coupling? What are their advantages and disadvantages?

P10-3 Indicate four different coupling designs based on interlocking. For each design specify the kind of torque (magnitude and variation) and the power sources they are best suited for.

P10-4 How would you design the special couplings referred to in Section 10-5? Which of the standard couplings are best suited for additional functions?

P10-5 A motor drives a reciprocating air compressor. The torsional moment to be transmitted from the engine to the compressor is 180 N · m. What is the design torque for this coupling? Review the various types of couplings presented in this chapter and indicate which would be suitable for this application. Give reasons. Which coupling would you specify?

P10-6 The motor on an overhead crane is connected to its reduction gear by means of a flexible coupling. The motor yields 15 kW at a speed of 900 rpm. The design torque used in selecting the coupling was 600 N · m. Was this selection justified? Select a suitable flexible coupling.

P10-7 A metallic grid coupling (Fig. 10-14) connects two 25-mm steel shafts with an allowable stress of 40 MPa. If the connecting strip is 75 mm in diameter, how many folds of 0.25 by 2.5 mm are needed if the strip has a shear strength of 175 MPa? The connector is to have a torque capacity not to exceed 95% of the shaft. *Hint:* Equate the shaft torque with the shear torque in the grid, $d^3 = 16T/\pi\tau_w$.

P10-8 Figure 10-22 shows a self-adjusting overload coupling. The torque is transferred from the outer ring to the inner section, mounted on the driven shaft, by means of three bell cranks, each carrying a roller and a spring. Maximum torque occurs when contact at point A ceases.

(*a*) Maximum torque is transferred for the position shown. For this position develop an expression for the moment that can be transmitted. Ignore friction and centrifugal forces.

(*b*) If $R = 100$ mm, $d = 30$ mm, $a = 10$ mm, $b = 30$ mm, $\alpha = 40$

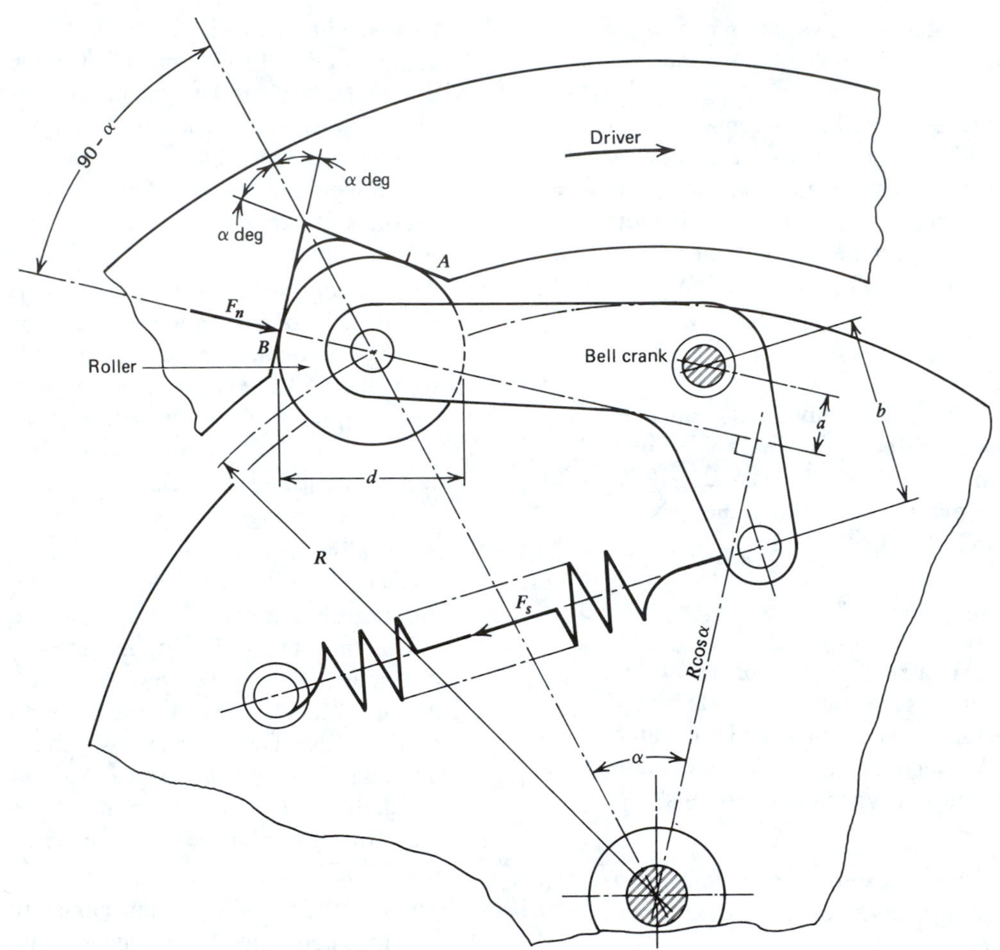

Figure 10-22 Overload coupling. (Problem 10-8)

deg, and $F_s = 200$ N, determine the power that can be transmitted for $\omega = 25$ rad/sec ($P = T\omega$).

(*c*) How would you provide torque adjustment?

P10-9 A shear pin coupling connects two shafts having diameters $D = 50$ mm that rotate at 1800 rpm (Fig. 10-23). The material of the shear pin has an ultimate shearing strength of 335 MPa, and the centerline distance R is 60 mm. If the shearing stress in the shafts is not to exceed 41 MPa, what should be the shear pin neck diameter d? *Hint:* The torsional resistance of the pin connection must equal the torsional strength of the shaft.

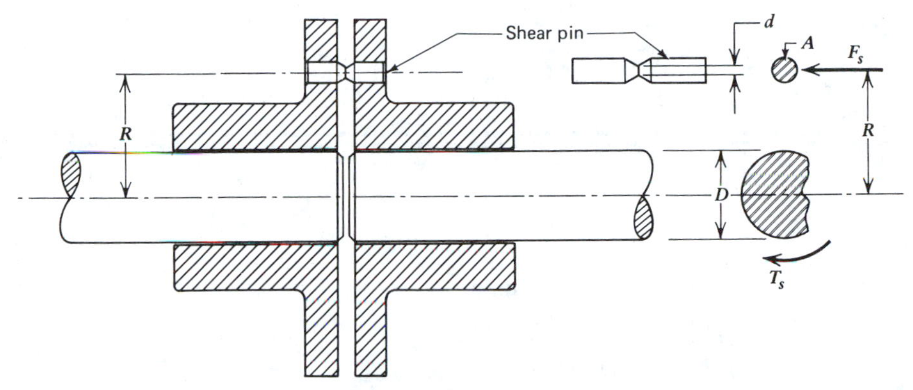

Figure 10-23 Shear pin coupling. (Problem 10-9)

P10-10 Two shafts I and II are connected by a universal joint. The shaft angle is 20 deg. Shaft I has a speed of 1000 rpm. Find the minimum and maximum speeds for Shaft II (Fig. 10-24).

Figure 10-24 Universal joint. (Problem 10-10)

P10-11 Power is transmitted between two parallel shafts by means of a universal joint. The input shaft rotates at 1200 rpm and transmits 10 hp. Calculate the following (Fig. 10-25).

(*a*) The angle β.
(*b*) The speed variation of shaft II.

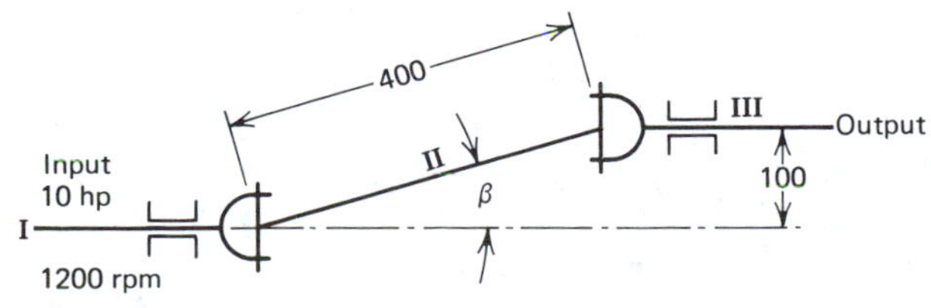

Figure 10-25 Universal joint. (Problem 10-11)

Chapter 11
Belt Drives

Theory is the captain
Practice, the soldiers.

LEONARDO DA VINCI

NOMENCLATURE

C = center distance
D, d = pitch diameters
f' = wedge factor
F_0 = initial tension
F_1 = tight tension
F_2 = loose or slack tension
F_c = centrifugal tension
F_{max} = maximum tension
F_n = normal force
F_t = tensile force
i = speed ratio
L = belt length
S = belt slip
σ_1 = stress due to F_1
σ_c = stress due to F_c
σ_{max} = stress due to F_{max}
ρ = mass density
θ = contact angle
2α = angular deviation from a 180 deg angle of contact

A belt drive is a low-cost means of transmitting power to one or more power-absorption units. Belt drives are smooth running, quiet, resistant to momentary surges or overloads, clean (requiring no lubrication), and inexpensive to maintain. Their disadvantages, compared with chains and gears, are lower strength and durability. Recently, however, improved reinforcing materials such as steel, nylon, polyester, and kevlar have made belts practical where formerly only chains would have been specified.

11-1 DESIGN, USES, TERMINOLOGY, AND FUNCTION

Design Applications

Belts are used when *large* distances between shafts make gears impractical or when the designated speed is too high for chain drives. They can be used only where some variation in speed can be tolerated. Additional design features, such as clutching and a variable-speed ratio, can be combined with belt drives to make them very versatile. Figure 11-1 shows a variety of belt drives.

In its simplest form a belt is a flexible band or loop passing around two wheels that communicates motion and/or power from one to the other. Wheels with a flat profile are called pulleys; when grooved, they are termed sheaves. A sheave or pulley that does not transmit power but is used as a belt tensioner to take up slack or change the direction of rotation is called an *idler* (Fig. 11-1*a*).

The term *belt drive* is used for a combination or assembly of belts and pulleys and would include the means of locking sheaves or pulleys to their respective shafts. Engines, motors, shafts, or bearings, however, are usually not thought of as parts of the drive itself.

Figure 11-2 shows the most common forms of belt drives. Note that drives are not limited to two pulleys or to pulleys mounted on parallel shafts. With nonparallel shafts, however, one condition must be fulfilled; the centerline of the belt as it approaches a pulley should lie in a plane perpendicular to the axis of that pulley; otherwise, the belt could leave the pulley (Fig. 11-3).

Modern belting is virtually all of one-piece or "endless" construction. A machine must, therefore, have some provisions—movable shafts, removable pulleys, or multipiece pulleys—to allow for belt adjustment and replacement.

Terminology

Regardless of type, a few common names and definitions apply to all belts (Fig. 11-4).

The *driver sheave or pulley* is mounted on the drive shaft, frequently that of an electric motor. It is usually the smaller of the two and has the higher rotative speed.

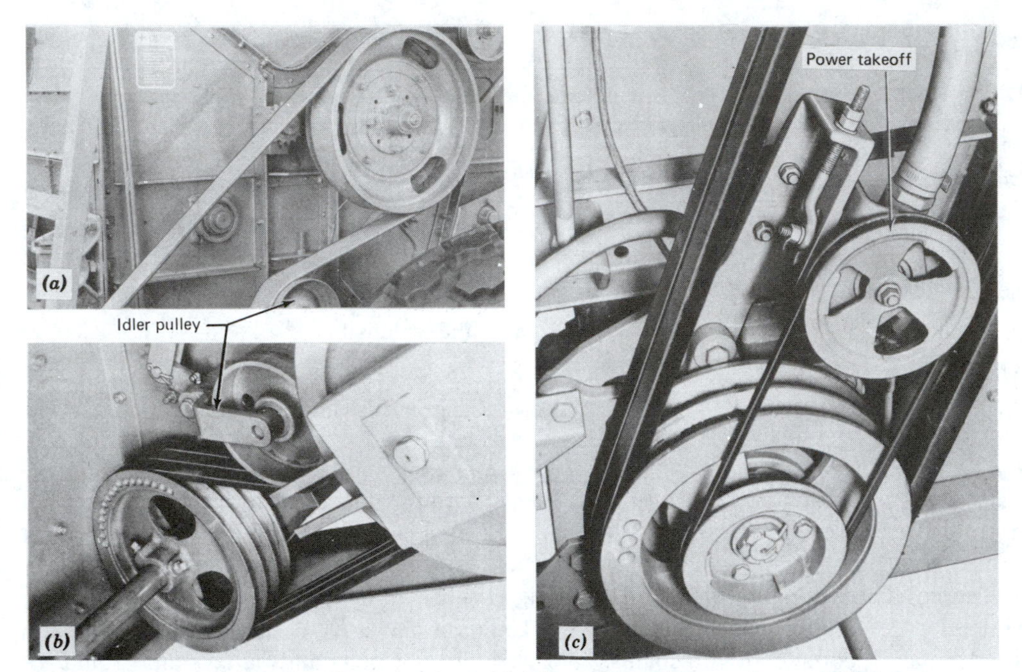

Figure 11-1 Belt drives. (a) Flat-belt drive with idler pulley as belt tensioner. (b) Triple V-belt drive with flat idler pulley as belt tensioner. (c) Main drive with power takeoff (PTO) for auxiliary equipment. (Courtesy Deere and Company Technical Services.)

The *driven sheave or pulley* on the driven shaft frequently is the input shaft of a driven machine. Usually this is the larger of the two sheaves and has the lower rotative speed.

Belt length (L) is dependent on both the pulley diameters and center distances. When the diameters are specified as pitch diameters, the belt length is termed pitch length. Sometimes belt length is specified as outside length, a term that signifies the dimension obtained by wrapping a tape measure around the outside of the belt in the installed position.

Initial tension (F_0), also referred to as installation or static tension, is the strand tension as measured when the drive is stopped or idle.

Sheave diameter (d, D) for a flat belt is the outside diameter of the sheave. For V-belts, it is the sheave diameter at a point where the neutral axis of the belt contacts the sheave, commonly called the pitch diameter. At the pitch diameter, the belt and sheave have identical linear speeds.

Contact angle (θ) is the angle of belt wrap on the smaller sheave or pulley (usually the driver).

Center distance (C) is the distance between centers of driver and driven sheaves.

Speed ratio (i) is obtained by dividing speed of driver shaft by that of driven shaft. It is also the ratio of pitch diameters in the absence of slip.

Open belt is the term describing a belt drive in which the belt proceeds in a direct line from the top of one pulley to the top of another, without crossing (Fig. 11-2). In an open belt drive both pulleys rotate in the same direction.

Crossed belt is the term describing a belt drive whose belt crosses over itself between the pulleys, causing them to have opposite rotative di-

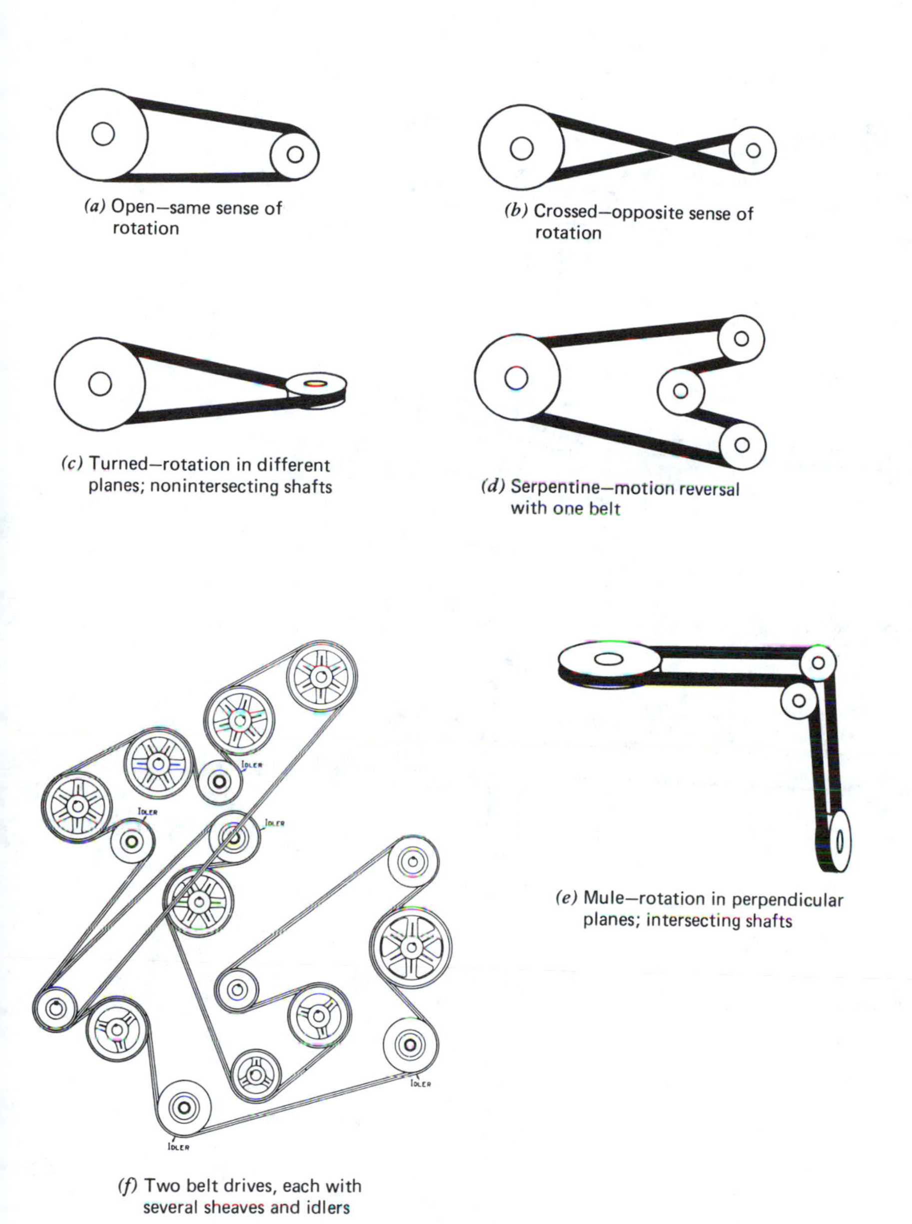

(*a*) Open—same sense of rotation

(*b*) Crossed—opposite sense of rotation

(*c*) Turned—rotation in different planes; nonintersecting shafts

(*d*) Serpentine—motion reversal with one belt

(*e*) Mule—rotation in perpendicular planes; intersecting shafts

(*f*) Two belt drives, each with several sheaves and idlers

Figure 11-2 Forms of belt drives. (Courtesy Deere and Company Technical Services and Browning Manufacturing Division.)

rections. Crossed belt drives should be limited to large center distances to minimize the damage that occurs when a belt rubs against itself.

Function

To function properly, belt drives must maintain certain tension levels. Too much tension shortens belt life (due to fatigue) and the life of other drive components such as shaft bearings. Too little tension allows slip, generating heat and wear that also reduce component life.

Adequate tension can be achieved by adjusting one or more pulleys. A more permanent arrangement is to install a spring-loaded idler pul-

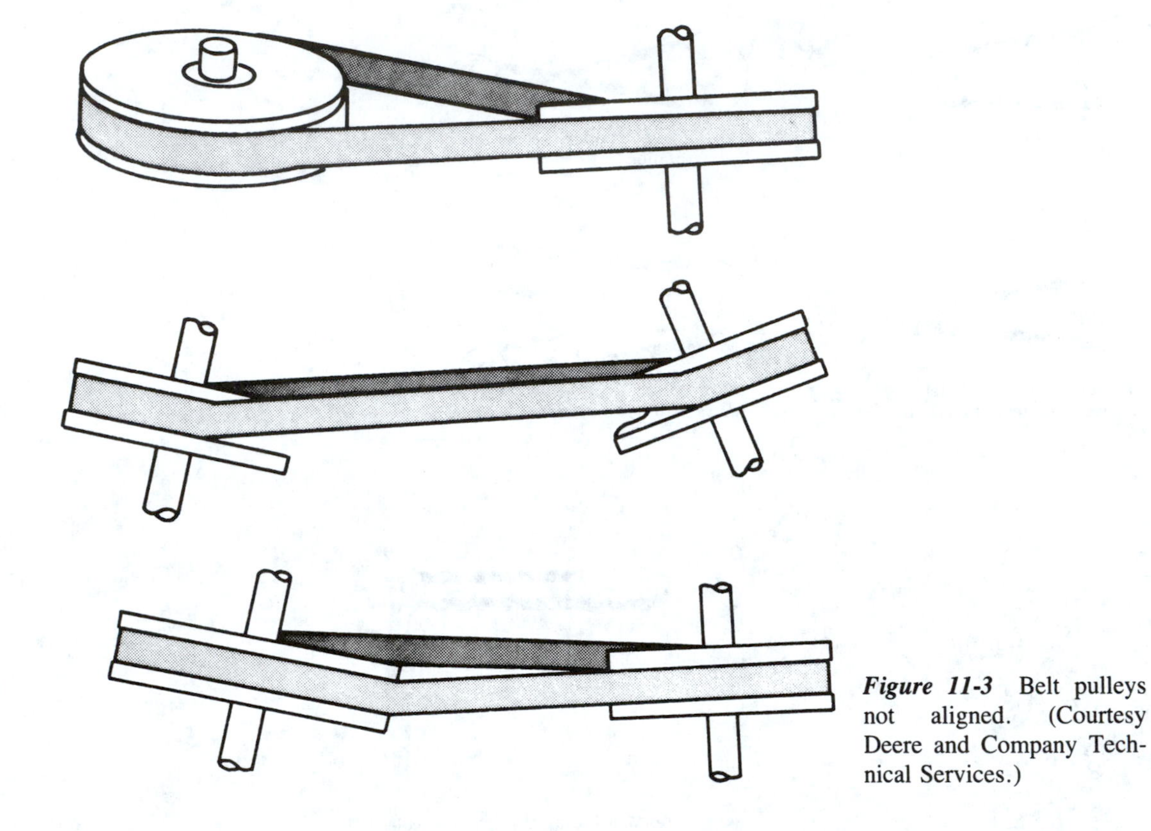

Figure 11-3 Belt pulleys not aligned. (Courtesy Deere and Company Technical Services.)

Figure 11-4 Belt terminology and geometry. For flat belts: D and d are pulley diameters. For V-belts: D and d are pitch diameters.

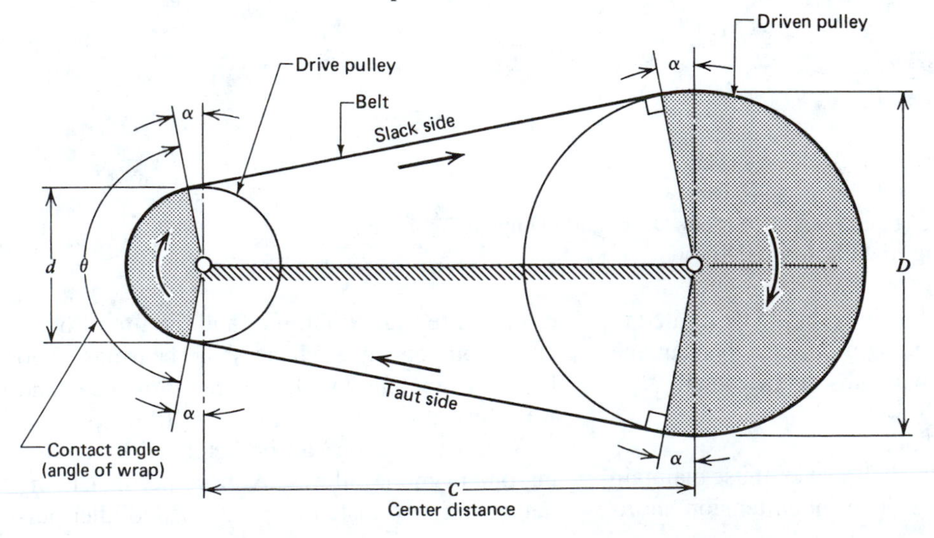

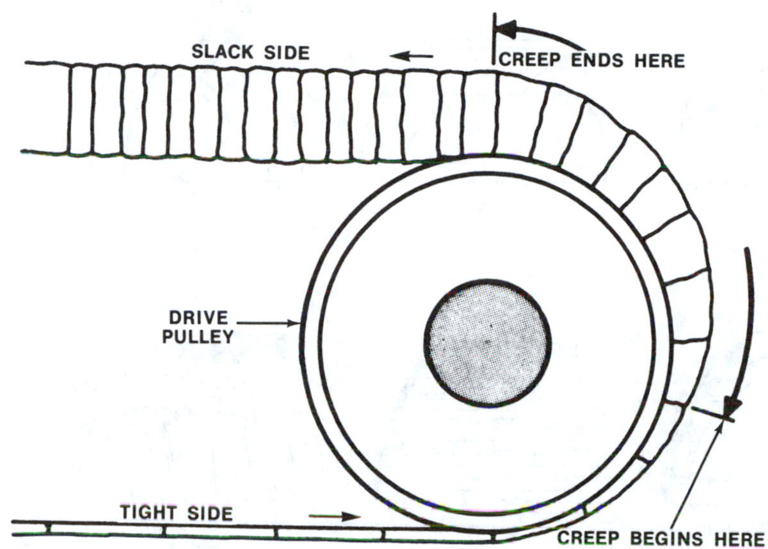

Figure 11-5 Belt creep. (Courtesy Deere and Company Technical Services.)

ley as an automatic belt tensioner (Fig. 11-1). Despite adequate normal tension, a belt may still slip when sudden overloads are encountered. Slip limits efficiency but avoids serious overloading of the driver and driven machine. Adequate tension under normal dry conditions may also prove inadequate when oil, dust, and humidity lower the coefficient of friction.

A phenomenon peculiar to belting is known as *creep*. This is a slight relative motion between belt and pulley that causes a loss of driven speed. Creep, therefore, is the result of minute alternate lengthening and shortening of each portion of the belt as it goes through a complete cycle. Figure 11-5 shows the mechanics of creep greatly exaggerated for clarity. Flat belts are most affected by this phenomenon; V-belts are least affected. A toothed timing belt would not be affected at all.

11-2 CLASSIFICATIONS AND STANDARDS

For cost and efficiency, many sizes of belts are provided for different load levels. Three basic types are *round*, *flat*, and *V*, as shown in Fig. 11-6. This figure illustrates several types of V and timing units. Classification may also be based on material, as for rubber belts, or on use, as for automobiles and agriculture.

The V-belt[1] has been standardized extensively as to nomenclature, size, length, tolerances, power ratings, and sheave-groove dimensions. Ordinarily, size of a V-belt refers to its cross section, and the expression "nominal cross section" is used to define it more specifically. The word "nominal" is intentionally ambiguous and allows for slight variations in size among different manufacturers' products.

V-belts are available in light-duty and heavy-duty construction. Within the heavy-duty classification there are conventional and narrow sections. In the conventional or standard group, belt sizes are currently designated by the letters A, B, C, D, and E, as shown in Fig 11-7. Heavy-duty narrow belts come in three sizes—3V, 5V, and 8V. Light-duty belts are designated 2L, 3L, 4L, and 5L.

---

[1]Also referred to as V-V in contrast to V-flats.

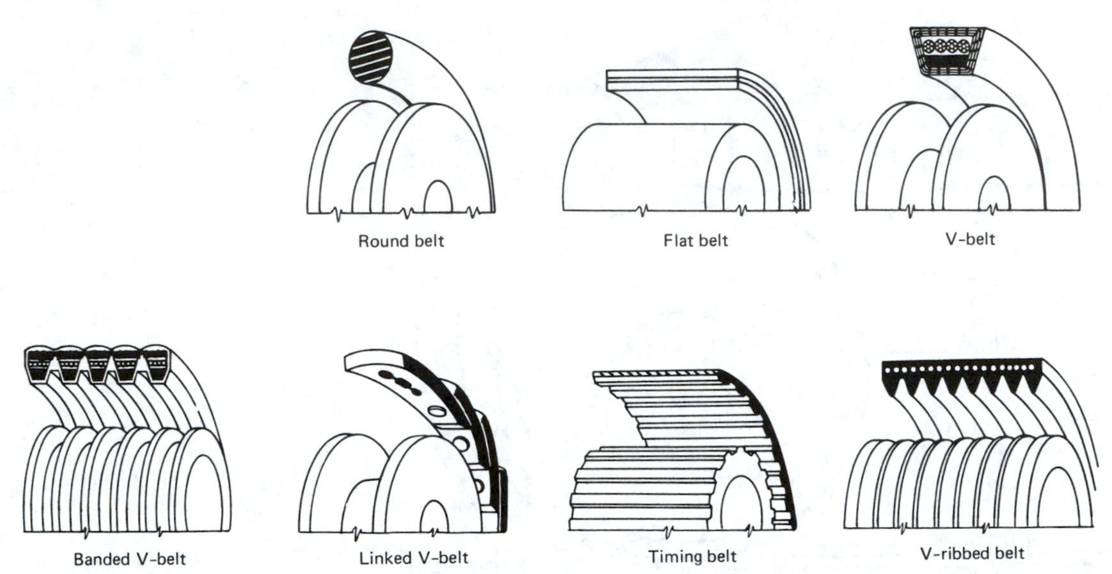

Figure 11-6 Types of belts. (Courtesy Deere and Company Technical Services.)

Figure 11-7 Standard V-belt sections. (Courtesy *Machine Design*.)

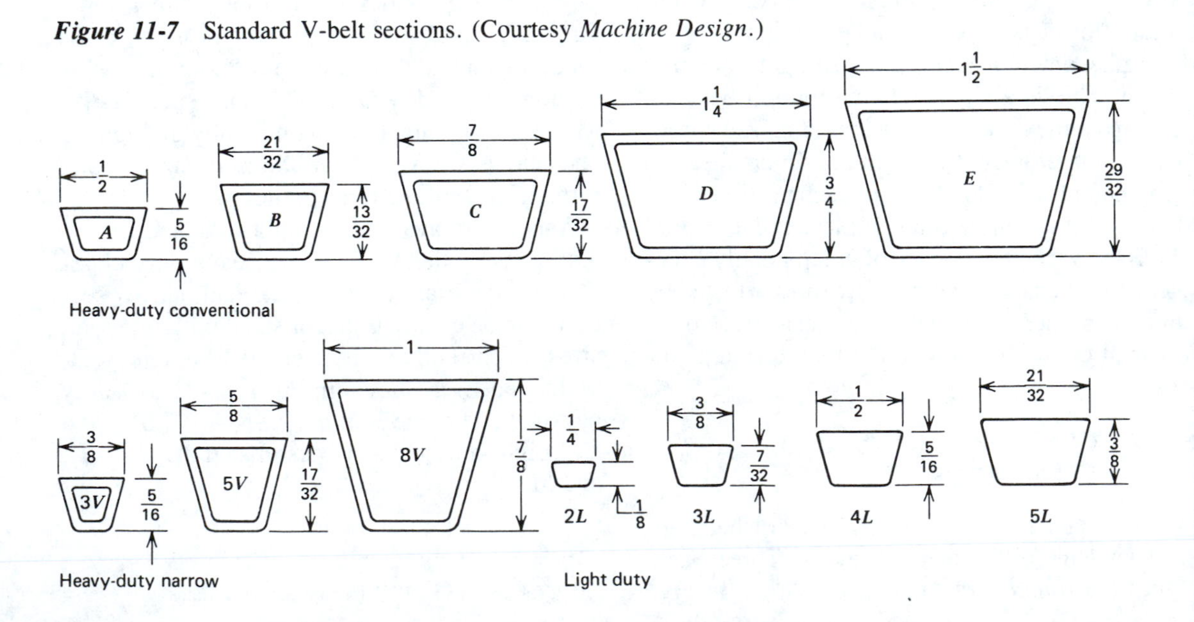

11-3 ROUND-BELT DRIVES

Round belts, modern versions of the now obsolete rope drive (Fig. 11-6), are produced as all-textile, textile-elastomer combinations, and all elastomers.[2] Both endless and spliced variations are available. Round belts can operate satisfactorily over pulleys in several different planes and are thus well suited for serpentine drives, reverse bends, and 90-deg twists, as shown in Fig. 11-2. They are used in some dishwasher drives.

11-4 FLAT-BELT DRIVES

A belt in the form of a thin, flat band, designed to run on cylindrical pulleys, is known as a flat belt. Figures 11-1 and 11-6 show this design. Flat belts are the simplest and least expensive type. They are made from fabric, rubber, plastic, leather, or paper-thin steel bands.

Flat belts are satisfactory at high speed and relatively low power. Here their low cost and flexibility give them advantages over other types. When the primary consideration is high power, however, flat belts become overly large and are no longer competitive with V-belts.

With their thin sections flat belts offer low resistance to bending as they move around pulleys. They can therefore operate on pulley diameters too small for V-flats. For this reason, they are used extensively in business machines. On relatively long center distances, flat belts can be crossed to increase the arc of contact, thereby increasing their capacity to transmit power.

Compared to V-belts, flat belts must maintain higher tensions to transmit a given power level; therefore drives must be designed with rigid shafts and high-capacity bearings. Shaft alignment is also important to utilize the full width of the belt.

The *V-ribbed, poly-V* belt, or *guided* flat belt, is basically a flat belt with a longitudinal ribbed underside (Fig. 11-8). The flat-belt section carries the load, while ribs provide a better grip on the pulley by their wedging action. This type of flat belt is well suited to high-speed drives using small-diameter pulleys.

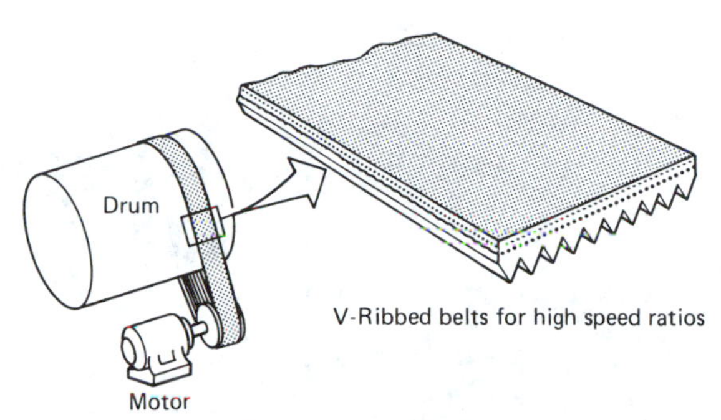

Figure 11-8 V-ribbed belt. V-ribbed belts are typically used in light-duty applications requiring high speed ratios. For instance, in a clothes dryer, a V-ribbed belt is flexible enough to operate around a small-diameter motor pulley to produce a 30:1 speed ratio. The ribs provide enough traction so the belt can ride directly on the drum like a flat belt. (Courtesy *Machine Design*.)

11-5 SYNCHRONOUS BELT DRIVES

Flat belts and V-belts, subject to creep and slip, are not suitable for locked or synchronous[3] motion. Synchronous or timing belts were developed to solve this problem and make belts more competitive with chains. Basically, synchronous belts are flat belts with a series of evenly spaced teeth on the inside of the circumference, designed to make positive engagement with matching sprocket teeth, as shown in Fig. 11-9. The belt ridges enter and leave the mated grooves in a smooth, rolling manner and with low frictional resistance. In short, timing belts combine the high-velocity characteristics of flat belts with power capacity approaching that of chains. Rubber timing belts are replacing steel chains in ap-

[2] An *elastomer* is generally a synthetic, rubberlike material employed in place of natural rubber.

[3] In step or in phase.

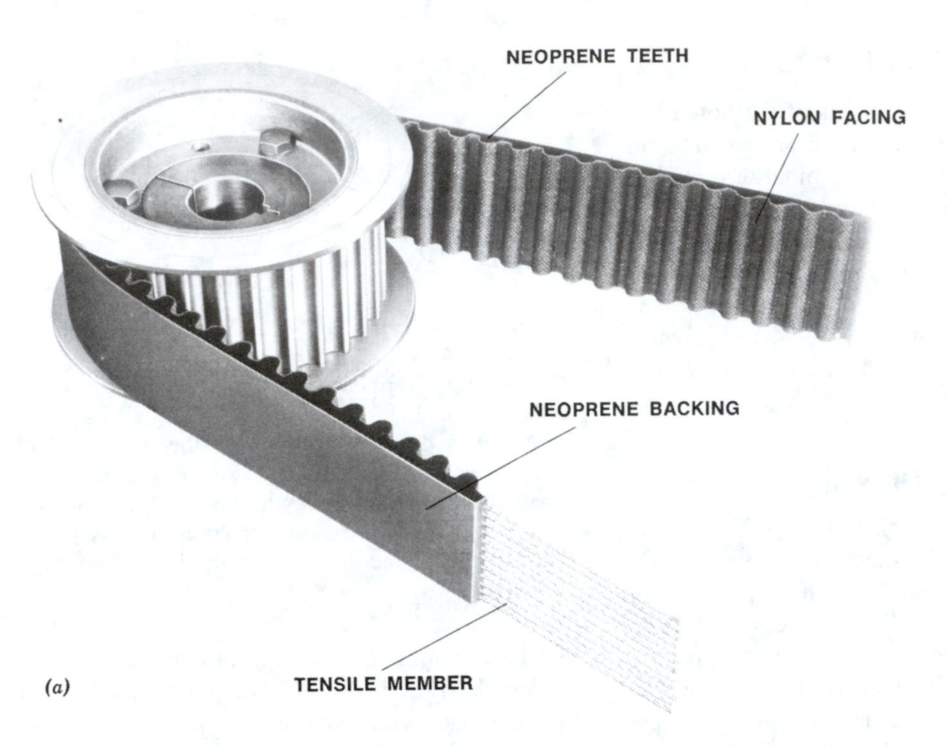

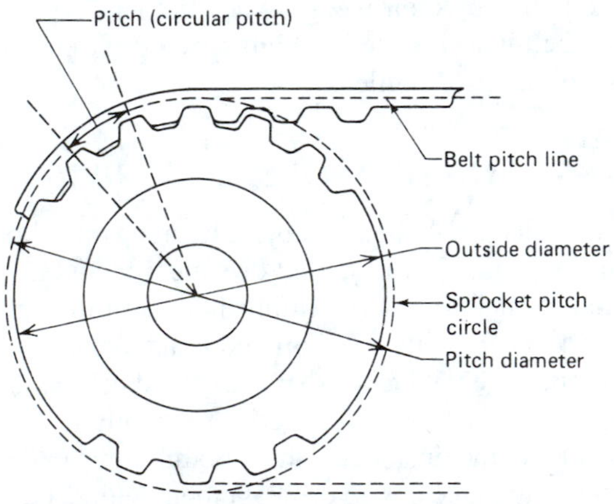

(*b*) Details of belt and sprocket

Figure 11-9 Synchronous (timing) belt. (*a*) Power Grip® timing belt (Courtesy Uniroyal, Inc.). (*b*) Details of belt and sprocket.

plications ranging from miniature, high-speed (above 1500 rpm) computer drives to low-speed, high-torque units up to 200 kW.

11-6 V-BELT DRIVES

A V-belt drive is an assembly of one or more endless flexible belts of trapezoidal cross sections that pass through the grooves of two or more V-grooved pulleys. Under load the belts wedge into the sheave to minimize slippage so that initial belt tension can be low. In terms of low cost and space, V-belts provide the best overall power transmission capability within its range of power transmissibility.

Most V-belts are of the closed-loop type and thus may require disassembly of shaft and bearings for service and maintenance (except for overhanging or cantilever applications). Figure 11-10 shows a typical belt drive with sheaves mounted on overhanging shafts.

Banded V-Belts

This compound design is shown in Fig. 11-11. For most drives, multiple V-belts are dependable and trouble-free. However, on some machines

Figure 11-10 V-belt drive using multiple belts. (Courtesy Deere and Company Technical Services.)

with pulsating or shock loads, V-belts can become unstable, whipping sideways, slapping together, and entering the sheave at an angle, effects that cause them to turn over, wear out quickly, and even jump off the sheaves. To prevent this, the banded V-belt was developed. Essentially, it is a tie-band or flat belt to which several V-belts have been attached. These pull exactly like V-belts, with minimum slip at the same low tensions, while the tie-band gives the belt internal rigidity to keep it from bending sideways and wearing unevenly.

11-7 V-BELT CONSTRUCTION

Regardless of manufacturer, basic V-belts consist of five component sections: (1) load-carrying or tensile, (2) cushion section surrounding the tensile, (3) flexible top section, (4) bottom compression member, and (5) cover or jacket. These sections are combined so that a belt is created to meet specific design needs. Figure 11-12 shows the cross section of a typical V-belt.

Because a belt is fundamentally a tension drive, strength of the tensile members is of prime importance. Presently, rayon, nylon, polyester, glass, and steel, all inexpensive and strong, are preferred.

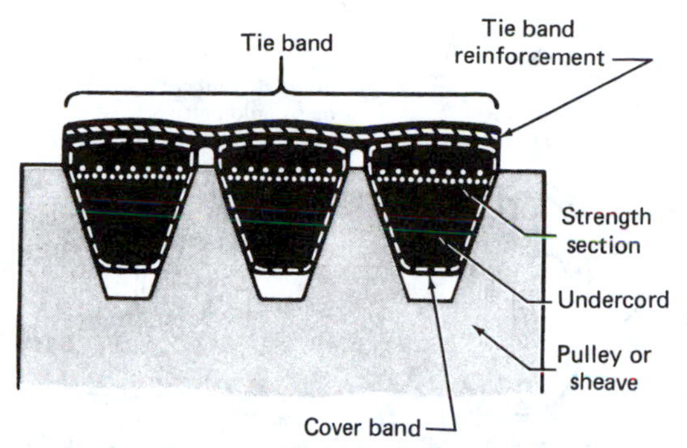

Figure 11-11 Construction of banded V-belts. (Courtesy Deere and Company Technical Services.)

11-8 PULLEYS AND SHEAVES

Flat-belt pulleys are generally made from cast iron, but they are also available in steel. They may be solid or split but, in either case, the hub may be split for clamping onto the shaft. Pulley width should exceed belt width by roughly 10%. Flat belts can be made to center themselves on their pulleys by the use of a crowned contour (Fig. 11-13). With small pulleys, excessive flexing of belts and short center distances result in reduced life. To assume a reasonable life, the

Figure 11-12 V-belt construction.

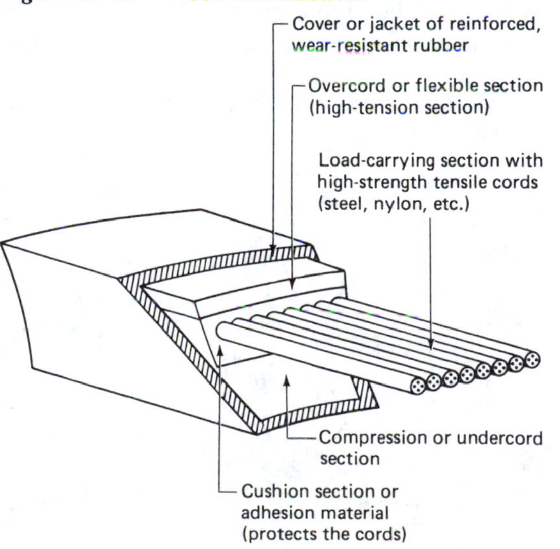

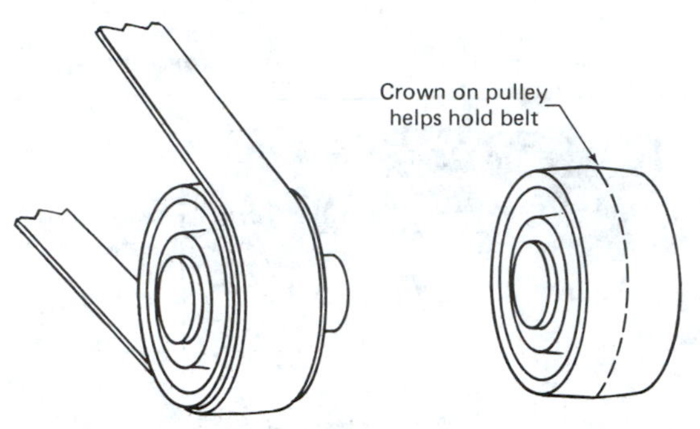

Figure 11-13 Crown helps keep flat belt on pulley. (Courtesy Deere and Company Technical Services.)

smallest pulley diameter should be at least 30 times belt thickness.

For most V-belt drives, sheaves are available either in formed steel or cast iron, as in Fig. 11-14. Formed steel sheaves are used primarily in light-duty applications such as automotive and agricultural services. Cast-iron drives are preferred for fluctuating loads where flywheel effect is important.

11-9 ANGLE OF CONTACT

The skeletal outline of an open-belt drive, shown in Fig. 11-4, applies to both flat belts and V-belts. For flat belts D and d are pulley diameters; for V-belts, they represent pitch diameters. C is the center distance, θ is the contact angle for the smaller pulley, and 2α is the angular deviation from a 180-deg angle of contact.

The geometry of Fig. 11-15 shows that

$$\theta = \pi - 2\alpha = \pi - \frac{D - d}{C} \text{ rad, approximately} \tag{11-1}$$

θ is always less than or equal to 180 deg or π rad. Contact angle is one of several factors determining the power capacity of belts. Its *lower* guideline is roughly *150 deg*. Below this value, increasing tension and slip, with resulting decrease in life, must be expected. The limit on θ also imposes a lower limit on the center distance C. As C decreases, so does θ. Conversely, for a given center distance, there will be a practical limit to the speed ratio attainable.

EXAMPLE 11-1

What is the angle of wrap for $D = 200$ mm, $d = 75$ mm, and $C = 375$ mm?

SOLUTION

$$\theta = \pi - \frac{200 - 75}{375} = \pi - 0.333$$

$$= 2.81 \text{ rad} = \mathbf{161\ deg}$$

This would be a satisfactory belt drive, since

Figure 11-14 Types of sheaves for V-belt drives. (Courtesy Deere and Company Technical Services.)

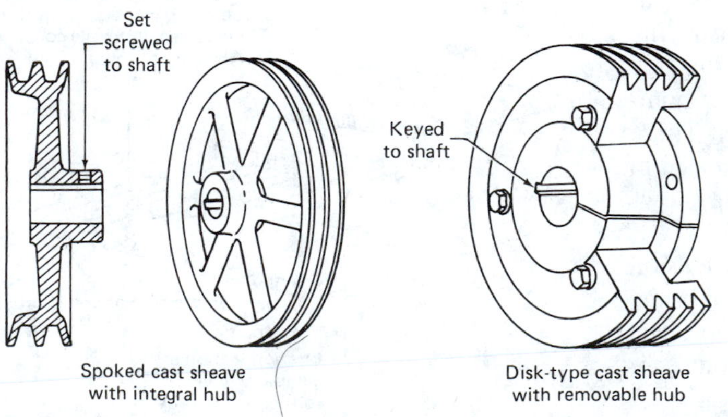

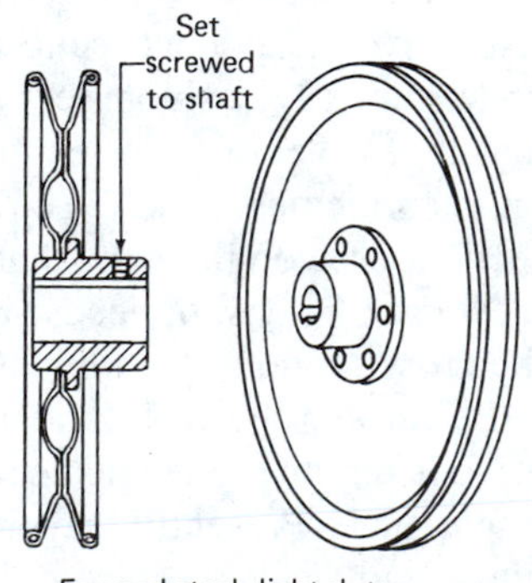

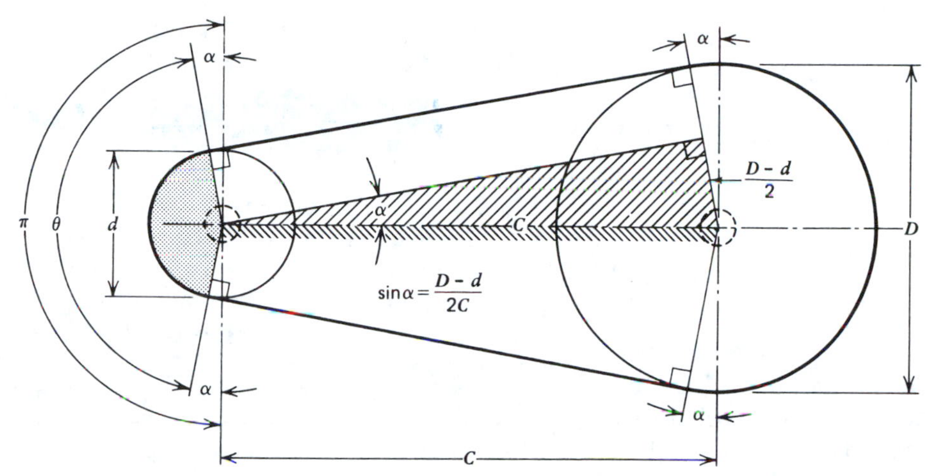

Figure 11-15 Belt geometry.

the angle of wrap exceeds the guideline figure of 150 deg.

11-10 CENTER DISTANCE AND BELT LENGTH

Shorter center distances are preferred for economy of space and stability of operation. Minimum drive centers are limited either by the physical dimensions of the sheave (pulley) diameters or the minimum angle of wrap ($\theta = 150$ deg). Maximum drive centers are limited only by available standard lengths.

When center distances are unknown, they can be estimated for preliminary calculations by the following.

Approximate center distance,

$$C = \begin{cases} D + 1.5d & \text{for ratios} \quad < 3.0 \\ D \cdots & \text{for ratios} \quad > 3.0 \end{cases}$$

$$D = \begin{cases} \text{Pitch diameter of driven sheave} \\ \text{Outer diameter for driven pulley} \end{cases}$$

$$d = \begin{cases} \text{Pitch diameter of drive sheave} \\ \text{Outer diameter of drive pulley} \end{cases}$$

The industrial formula for calculating center distance is

$$C = 0.0625\{b + [b^2 - 32(D - d)^2]^{1/2}\} \qquad (11\text{-}2)$$

where

$$b = 4L - 6.28(D + d)$$

$$L = \begin{cases} \text{V-belt pitch length} \\ \text{flat-belt length} \end{cases}$$

This formula can be rearranged to solve for L.

$$L = 2C + 1.57(D + d) + \frac{(D - d)^2}{4C} \qquad (11\text{-}3)$$

Modern belts are nearly all of the closed-loop type and come in standard lengths. Belt length is calculated to ensure that some standard unit will fit the projected drive or to establish dimensions for building belts to the size requirements of original equipment manufacturers.

For drives with more than two sheaves, belt length is calculated from sheave coordinates and dimensions are determined by layout, either trigonometric or scaled.

EXAMPLE 11-2

Calculate the belt length for the drive given in Example 11-1 by substituting the given values into Eq. 11-3.

SOLUTION

$D = 200$ mm $d = 75$ mm $C = 375$ mm

$$L = (2)(375) + 1.57(200 + 75) + \frac{(200 - 75)^2}{(4)(375)}$$

$$= \mathbf{1192\ mm}$$

11-11 SPEED RATIO

Most prime movers rotate at higher speeds than are desirable for driven machines. Therefore speed reduction is necessary for most belt drives. The speed ratio between the drive shaft and the driven is designated i.

$$i = \frac{\text{speed of drive shaft}}{\text{speed of driven shaft}} \tag{11-4}$$

For example, the speed ratio of a 3600-rpm motor driving a 2000-rpm compressor is $i = 3600/2000 = 1.8$. This can also be expressed in terms of pulley diameters as

$$i = \frac{\text{diameter of larger pulley}}{\text{diameter of smaller pulley}} \tag{11-5}$$

Slip and creep cause an additional, but small, reduction in speed. Slip occurs primarily at the smaller pulley. If we use the nomenclature of Fig. 11-4 and disregard slip, we obtain a belt speed v.

$$v = \pi D n_D = \pi d n_d \tag{11-6}$$

where n_D and n_d are the speeds in revolutions per minute or revolutions per second of the driven and driver sheaves. Therefore

$$n_D = \frac{d}{D}(n_d) = \frac{n_d}{i} \tag{11-7}$$

Slip leads to a reduction in belt speed as well as output speed. Therefore

$$n_D = n_d \frac{1 - S}{i} \tag{11-8}$$

where S is slip in percent divided by 100.

When slip occurs at both sheaves, Eq. 11-8 assumes the form

$$n_D = \frac{n_d}{i}(1 - S_1)(1 - S_2) \tag{11-9}$$

Slip, greater at the drive pulley, varies from 1 to 2%, depending on the type of belt used. Power losses due to bending *hysteresis*[4] and entrance and exit friction are 3 to 5% for most belt drives. Loss in speed can be compensated for by a small change in one pulley diameter.

EXAMPLE 11-3

Using the data from Example 11-1, calculate the following.

(*a*) Speed ratio.

(*b*) Speed of the driven shaft if the input shaft speed is 900 rpm.

(*c*) Speed of the output shaft if there is a slip of 2% at the drive sheave and a slip of 1% at the driven sheave.

(*d*) How would one compensate for the given slip?

SOLUTION

(*a*) $i = \frac{D}{d} = \frac{200}{75} = \mathbf{2.67}$

(*b*) From Eq. 11-7.

$$n_D = \frac{75}{200}(900) = \mathbf{338\ rpm}$$

(*c*) From Eq. 11-9.

$$n_D = (900)\frac{75}{200}(1 - 0.02)(1 - 0.01)$$

$$= \mathbf{327\ rpm}$$

(*d*) Increase d to d_1 (adjustable pitch diameter).

[4]A kind of molecular friction within the material.

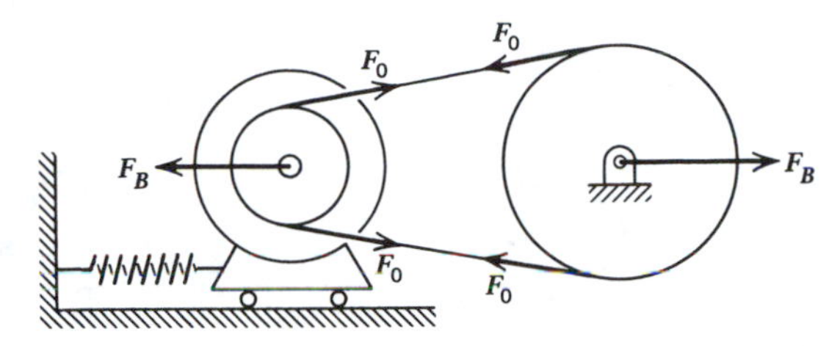

Figure 11-16 Belt drive with adjustable center distance, unloaded but with initial tension.

$$d_1 = (75)\frac{338}{327} = \mathbf{77.5\ mm}$$

or

$$d_1 = \frac{75\text{ mm}}{(1 - 0.02)(1 - 0.01)} = \mathbf{77.3\ mm}$$

11-12 STATIC FORCE ANALYSIS

A force analysis, disregarding centrifugal loads, will be made first. Figure 11-16 shows an open, flat-belt drive with an initial tension F_0 to clamp the belt firmly against the pulley surface. F_B is the reaction of the base.

As shown in Fig. 11-17, the motor torque is being transmitted from the drive to the driven pulley through a tensile belt force $F_t = 2T/d$.

From start to operating speed, the tensile force in the bottom strand increases from F_0 to F_1, called tight tension. In the upper strand, the tensile force drops to F_2, termed slack or loose tension.

The relationship between the contact angle θ of the drive pulley, the coefficient of friction f, and the forces F_1 and F_2 is given by

$$F_1 = F_2 e^{f\theta} \qquad (11\text{-}10)$$

where

e = 2.718, the basis of the natural logarithm
f = coefficient of sliding friction (for rubber against steel or cast iron, use $f = 0.30$)
θ = contact angle; rad

$$e^{f\theta} = \frac{F_1}{F_2} = \text{force ratio}$$

For $f = 0.30$ and $\theta = 3.0$ rad, the ratio is roughly 2.46.

For V-belts the *wedge factor* enters the basic equation for band friction.

$$\text{V-belt:} \quad \frac{F_1}{F_2} = e^{f'\theta} \qquad (11\text{-}11)$$

where

$$f' = f(\sin\beta)^{-1} \qquad (11\text{-}12)$$

2β = included groove angle; deg (Fig. 11-18)
$(\sin\beta)^{-1}$ = wedge factor

Notice that for $2\beta = 180$ deg we obtain the expression for *flat* belts. The flat belt is thus a special case of the V-belt and from now on will be treated as such.

For standard V-belts the included-groove angle varies from 32 to 40 deg. Because the wedge factor is 3.24 at an average value of 36 deg for the included angle, it is obvious why V-belts are preferred to flat belts in spite of a somewhat higher initial cost. The force ratio is now roughly $18.47 \gg 2.46$.

A review of Fig. 11-18 illustrates the difference between flat belts and V-belts. Both show an element of a belt acted on by a radial force F_n. In the case of a flat belt the friction force will, of course, be fF_n. In the case of a V-belt, the reaction will be taking the form of two component forces N perpendicular to the sides of the V-groove. Therefore:

$$\text{Total normal force:} \quad 2N = F_n(\sin\beta)^{-1} \qquad (11\text{-}13)$$

$$\text{Friction force:} \quad F_f = fF_n(\sin\beta)^{-1} = f'F_n \qquad (11\text{-}14)$$

Figure 11-17 Open belt drive showing static forces.

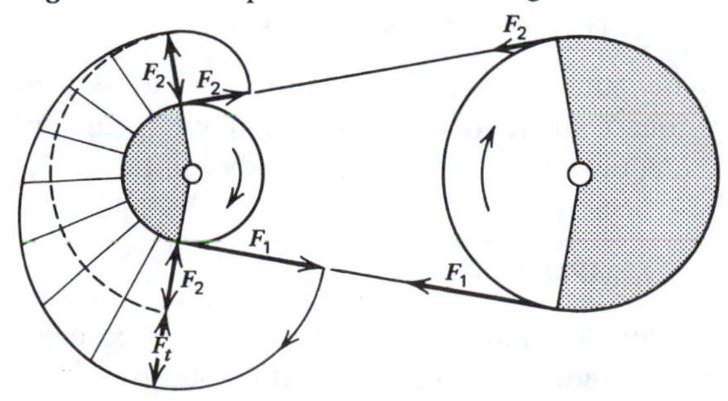

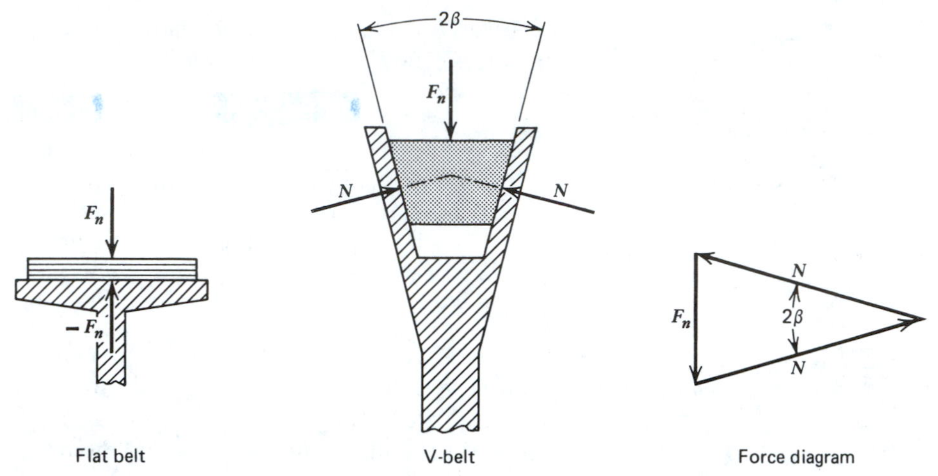

Figure 11-18 Element of flat belt on pulley and V-belt in sheave.

Because of wedging action, the same radial force generates a larger friction force at the V-belt interface, with subsequent increase of tight tension.

EXAMPLE 11-4

A flat belt transmits power between two 700-mm diameter pulleys. The shaft load is not to exceed 1000 N. Find the limiting values of F_1 and F_2. Disregard service factors.

SOLUTION

The shaft load is roughly the sum of tight and loose tensions.

$$F_1 + F_2 = 1000 \text{ N}$$

$$F_1 = F_2 e^{f\theta} = F_2 e^{0.30\pi}$$

$$(\theta = 180 \text{ deg} = \pi \text{ rad})$$

$$F_1 = F_2(2.57)$$

Substituting for F_1 leads to

$$1000 - F_2 = 2.57F_2$$

$$3.57\, F_2 = 1000$$

$$F_2 = \mathbf{280\ N} \qquad F_1 = \mathbf{720\ N}$$

11-13 POWER TRANSMITTED BY STATIC FORCES

The force F_t necessary to transfer given torque can be obtained by taking the moment with respect to the center of the drive pulley.

$$F_t(0.5d) + F_2(0.5d) - F_1(0.5d) = 0$$

$$F_t = F_1 - F_2 = F_1(1 - e^{-f'\theta}) \qquad (11\text{-}15)$$

The power transmitted is

$$P = F_1(1 - e^{-f'\theta})v10^{-3} \text{ kW} \qquad (11\text{-}16)$$

where

F_1 = tight tension; N
v = belt speed; m/s

$$P = \frac{F_1(1 - e^{-f'\theta})v}{550} \text{ hp} \qquad (11\text{-}17)$$

F_1 = tight tension; lb
v = belt speed; fps

When the assignment requires an estimate of torque T and revolutions per minute n, the power equations, Eqs. 1-7 and 1-10, should be used.

EXAMPLE 11-5

Calculate the power transmitted by the drive in Example 11-4 at a speed of 600 rpm.

SOLUTION

$$P = (720 \text{ N} - 280 \text{ N})(v \text{ m/s})(10^{-3})$$

$$v = \pi(0.70 \text{ m})(600 \text{ rpm})(60^{-1} \text{ s/min}) = 22 \text{ m/s}$$

$$P = (440 \text{ N})(22 \text{ m/s})(10^{-3}) = \mathbf{9.68\ kW}$$

11-14 EFFECT OF CENTRIFUGAL FORCE

To make a belt go around a pulley requires initial belt tension separate from that needed to generate friction forces for power transmission. Figure 11-19 shows diagrammatically a belt and a pulley. The pulley is free to rotate, and moment of inertia is ignored. The belt, in which there is an initial tension F_0, approaches the pulley at speed v, travels around part of the circumference, and leaves at the same speed. During its circular passage, any belt element has to be accelerated radially. The belt tension F_0 is the only force available to provide this acceleration. If F_0 is larger than necessary, part of F_0 will be used to press the belt against the pulley surface (and power may be transmitted in addition to motion). When F_0 matches the required centrifugal force F_c, contact between belt and pulley will be lost and the belt will have reached maximum speed.

The magnitude of the centrifugal force F_c for a given belt speed v (in meters per second) can be obtained by considering a small element of mass per unit length (kilograms per meter). To this element we apply Newton's second law: $F = ma$. During its circular passage, the mass is accelerated radially due to the radial component of F_c. Therefore

$$F_c = mv^2$$

The effect of initial tension is to balance the centrifugal force, which will cause the belt to move away from the pulley in a radial direction ($F_0 = F_c$).

If information about the belt had been provided in the form of a cross-sectional area A and a mass density ρ,

$$F_c = \rho(1.0\text{m})Av^2 \tag{11-18}$$

Mean tensile stress in the belt would be

$$\sigma_t = \frac{F_c}{A} \text{ MPa} \tag{11-19}$$

Figure 11-19 Initial tension required to transmit motion.

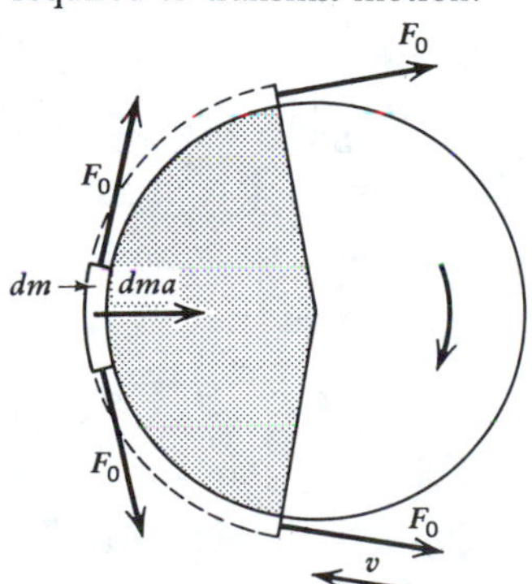

EXAMPLE 11-6

Use the data from Examples 11-4 and 11-5 to calculate centrifugal tension F_c, when the belt has a mass of 0.50 kg/m. Calculate the tensile stress for a cross section of 5mm × 100 mm.

SOLUTION

$$m = 0.50 \text{ kg/m} \quad v = 22 \text{ m/s} \quad A = 500 \text{ mm}^2$$

$$F_c = (0.50 \text{ kg/m})(22 \text{ m/s})^2 = 242 \text{ N}$$

$$\sigma_t = \frac{242 \text{ N}}{500 \text{ mm}^2} = \mathbf{0.48\ MPa}$$

11-15 POWER TRANSMISSION BY BELTS

The net effect of the centrifugal force is increased belt tension (Fig. 11-20). It is a *waste* load, however, because it increases tension without increasing power capacity. When the centrifugal effect is included:

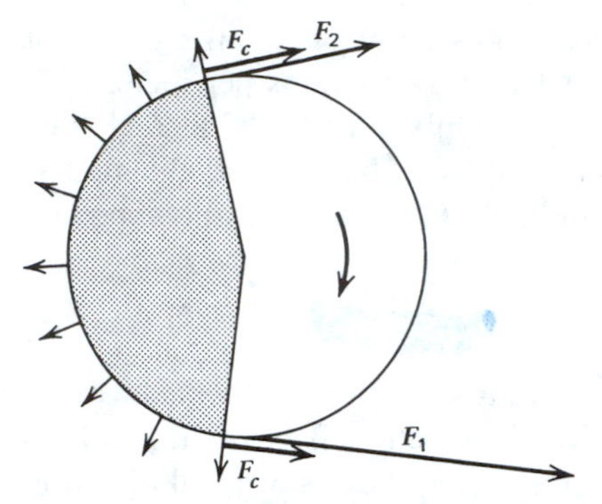

Figure 11-20 Centrifugal belt tension.

Tight tension: $F_{1,c} = F_1 + F_c$

Loose tension: $F_{2,c} = F_2 + F_c$

$$F_t = F_{1,c} - F_{2,c} = F_1 - F_2$$

Naturally, tight tension should never exceed the allowable belt load F_{max}. Therefore

$$F_1 + F_c \leq F_{max}$$

$$F_1 \leq F_{max} - F_c$$

$$\sigma_1 = \sigma_{max} - \sigma_c \qquad (11\text{-}20)$$

Substitution for F_1 in Eqs. 11-16 and 11-17 yields

$$P = (F_{max} - F_c)(1 - e^{-f'\theta})v\,10^{-3} \text{ kW} \qquad (11\text{-}21)$$

$$P = \frac{(F_{max} - F_c)(1 - e^{-f'\theta})v}{550} \text{ hp} \qquad (11\text{-}22)$$

Interpretation of this equation will serve as a guideline for belt construction. High capacity for power is achieved for a large value of F_{max}; hence the belt must have great tensile strength. In practice, this is achieved by tensile members in the load section.

Power capacity seems to be directly proportional to velocity v; thus high velocity is desirable. Modern belts do indeed operate at high speeds of 15 to 30 m/s. However, v also appears in the equation for centrifugal tension. Since v^2 is a large number, the mass m should be minimized. Thus rubber, with its low mass and high friction, is the main material in belts. For small velocities, the centrifugal effect is negligible. As the speed increases, however, F_c grows and eventually a speed is reached where no power is transmitted.

$$F_{max} - mv^2 = 0$$

$$v = \sqrt{\frac{F_{max}}{m}}$$

Large values for f and θ also favor power capacity. Vulcanized rubber is ideal in this respect. θ generally should not be less than 150 deg.

Since power capacity peaks, it is desirable to find the corresponding theoretical speed. This is achieved by differentiating Eq. 11-21 with respect to v and equating it to zero.

EXAMPLE 11-7

For the data used in previous examples, calculate the speed for which power capacity peaks.

$$P = (720 - 0.5v^2)[1 - e^{-f\theta}(\sin^{\beta-1})]v\,10^{-3}$$

SOLUTION

$$\frac{dP}{dv} = \frac{d}{dv}(720 - 0.5v^2)v$$

$$\frac{dP}{dv} = 720 - 1.50v^2 = 0$$

$$v = \sqrt{\frac{720}{1.5}} = \mathbf{22\ m/s}$$

EXAMPLE 11-8

Use the data given in Examples 11-5 to 11-7 to calculate the actual power transmitted.

SOLUTION

$$P = (720 \text{ N} - 242 \text{ N})(1 - e^{-0.30\pi})(22 \text{ m/s})$$
$$(10^{-3} \text{ kW/W})$$
$$= \mathbf{6.41\ kW}$$

Service factors for belts compared with those for couplings, chains, and gears are low, averaging 1.3 and rarely exceeding 1.6. The ability to slip under sudden load, plus the damping effect of rubber, are the inherent characteristics accounting for the low service factor.

11-16 V-FLAT DRIVES

When large velocity ratios and short center distances are needed, the grooving of the *driven* wheel can be omitted without sacrificing tractive power. The large pulley has a greater contact angle that compensates for the loss of wedging action. Termed V-flat, a drive of this makeup consists of a grooved drive sheave, a set of V-belts or a V-ribbed belt, and a flat-faced (cylindrical) driven pulley. Belt contact thus *alternates* between sides and bottom (Fig. 11-21).

Considerable savings can be made by using a flat pulley or a flywheel already on hand. Flat belts replaced by V-flats show increased capacity and operate on shorter center distances.

V-flats are feasible only when the arc of contact of the *driven* pulley exceeds about 210 deg. This corresponds to $(D - d)/C$ greater than 0.5. Best results are obtained for ratios between 0.8 and 0.9. For V-flats, *drive* sheave contact angles may be as low as 130 deg, less than the recommended value of 150 deg for V-belts. A larger belt or more belts may therefore be required than if the contact angle were 150 deg or more.

11-17 VARIABLE-SPEED V-BELT DRIVES

The main advantage of this type of drive is its ability to provide any ratio between an upper and lower limit—and to do so while running under full load. Thus a variable-speed V-belt drive is ideal for vehicles requiring continuous speed variation in response to changing road conditions. Limited capacity (less than 200 hp) is the only major drawback to this "ideal transmission." As a result, use is presently limited to snowmobiles, motorscooters, small automobiles (e.g., the Volvo 243, 70 hp), Reeves 50-hp industrial motors, and 150-hp combines.

Speed variation is achieved by means of two variable pitch sheaves and a belt several times wider than standard. Each sheave consists of two halves that can be moved either away from or toward each other. When the halves are together, the belt rides high in the sheave, providing a large pitch diameter. When the halves move apart, the belt sinks lower, resulting in a smaller effective sheave diameter (Fig. 11-22). Thus the degree of speed increase or decrease from driver to driven shaft can be varied easily while the belt is moving simply by synchronized adjustment of sheave separation. Speed control can be manual

Figure 11-21 V-flat drive.

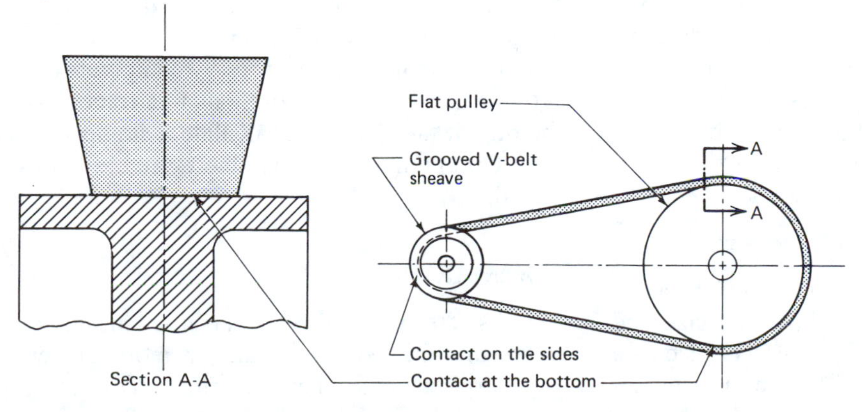

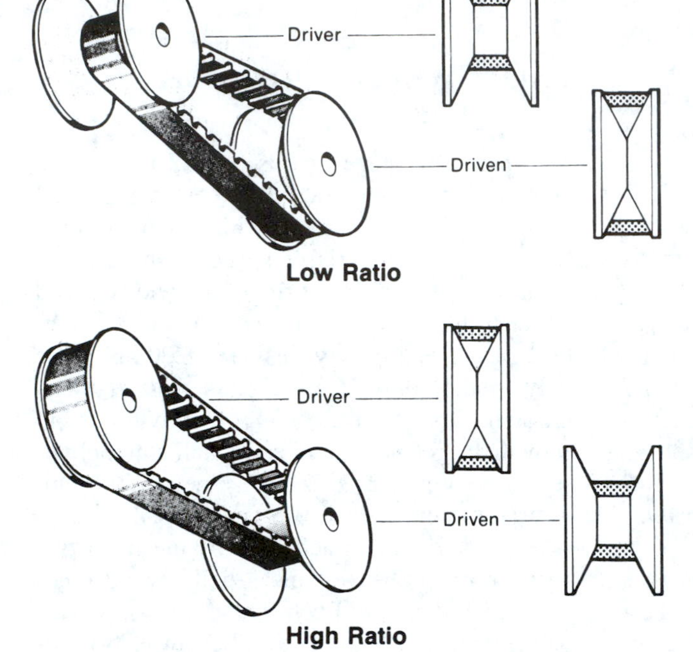

Figure 11-22 Variable-speed drive. The prime attraction of the variable-speed belt drive is its ability to provide smooth, infinite variations of ratio while running. The drive ratio is changed as the sheave faces move either away from or toward each other. (Courtesy *Machine Design*.)

or automatic. Automatic control is more sensitive to speed variations and therefore is generally preferred.

Although less durable and less costly than equivalent gear reducers, belt reducers have a tendency to wander from a set ratio. Operation at low speed is not economical.

Variable-speed V-belt drives have been used for decades, but not until recently have they provided any competition for variable-speed drives based on gearing, chains, and fluid mechanics. The advent of the snowmobile caused this change. The commerical success of this vehicle hinged on development of a durable "belt transmission."

11-18 DESIGN OF BELT DRIVES

The type of belt suitable for a specific job can be found by means of Table 11-1. The size and length are then found by applying the basic data given in Chapter 9, Section 9-2, to a manufacturer's catalog or a computer program.

In designing belt drives, additional considerations may be necessary. For example, designers may be required to use parts presently in stock or take into account special work environments and safety measures.

According to B. F. Goodrich, the design and construction of power transmission V-belts has progressed so much over the past 30 years that many existing drives should be reexamined. For a given cross section, a V-belt rated at 1 hp in 1950 is rated today at 2.2 hp.

Drives can now be designed by computers that provide many different solutions in a fraction of the time it takes to hand calculate a single solution. Optimum solutions are therefore easily obtained.

SUMMARY

Belts are inexpensive, versatile means of transmitting power to one or more power-absorption units. They can efficiently transmit large loads while absorbing shocks. Their

TABLE 11-1 Comparison of Five Belt Types

	Flat Belt	Ordinary V-belt	Linked V-belt	Timing Belt	V-Ribbed Belt
Tension-bearing loads	Highest	Low	Low	Lowest	High
Best operating range; m/s	5 to 50	5 to 30	5 to 25	5 to 50	5 to 30
Performance above 25 m/s	Good	Fair	Not recommended	Good	Fair
Performance below 5 m/s	Fair	Fair	Fair	Good	Fair
Resistance to shock loads	Good	Good	Good	Fair	Good
Mechanical efficiency	Good	Good	Good	Best	Good
Misalignment tolerance	None	Compensates well	Compensates well	None	None
Noisiness	Most	Very little	Little	Little	Little
Synchronizes operations	No	No	No	Yes	No
Creep	Some	Negligible	Some	None	Some
Initial cost	Low	Low	Moderate	Moderate	Moderate
Resistance to weather	Good	Good	Fair	Good	Good
Maintenance needed	Some	Negligible	Some	Negligible	Some

Source. John Deere Service Publications: Belts and Chains. (Courtesy Deere & Company Technical Services.)

ability to slip prevents overloading. For maximum life, however, belts must operate within prescribed tension ranges and in environments compatible with material life. Belts are used when large distances between shafts make gears impractical or when the designated speed is too high for chain drives. Modern belting is virtually all of one-piece or "endless" construction. For this reason, belts are used almost exclusively with pulleys or sheaves mounted on overhanging shafts.

Three basic types of belts are available: flat, round, and V. V-belts dominate. In terms of low cost and space, V-belts provide the best overall power transmission capability for the normal range of power requirements. Increasingly, however, hybrid designs (synchronous, banded, and V-ribbed) combining the best features of flat belts and V-belts are used in place of V-belts and even chains. Although the different types of belt drives are often considered interchangeable, each type is actually restricted to a fairly well-defined application area. The popularity of belts is primarily due to their compatibility with motors, where they simultaneously serve as speed reducers, shock absorbers, and regulators of overloads.

Even though belt drives are primarily constant-speed drives, recent developments have greatly increased their use in applications where speed variation is required, such as in snowmobiles and light vehicles. Belt calculations are summarized in Fig. 11-23.

REFERENCES

11-1 *Dayco Power Transmission Handbook*, Third Edition. Dayton, Ohio: Dayco Corporation, 1975.

11-2 Hyatt Bearing Division. *Engineering Handbook*. Detroit: General Motors Corporation, 1964.

11-3 *Machine Design. Mechanical Drives Reference Issue*.

11-4 Oliver, L. R., C. O. Johnson, and W. F. Breig. *Agricultural V-Belt Drive Design*. Dayton, Ohio: Dayco Corporation, 1977.

11-5 *Rubber Products Application Manual*. Denver: Gates Rubber Company, 1967.

PROBLEMS

P11-1 Which simple machines are found in belt drives? What makes V-belts more powerful than flat belts? Why do cen-

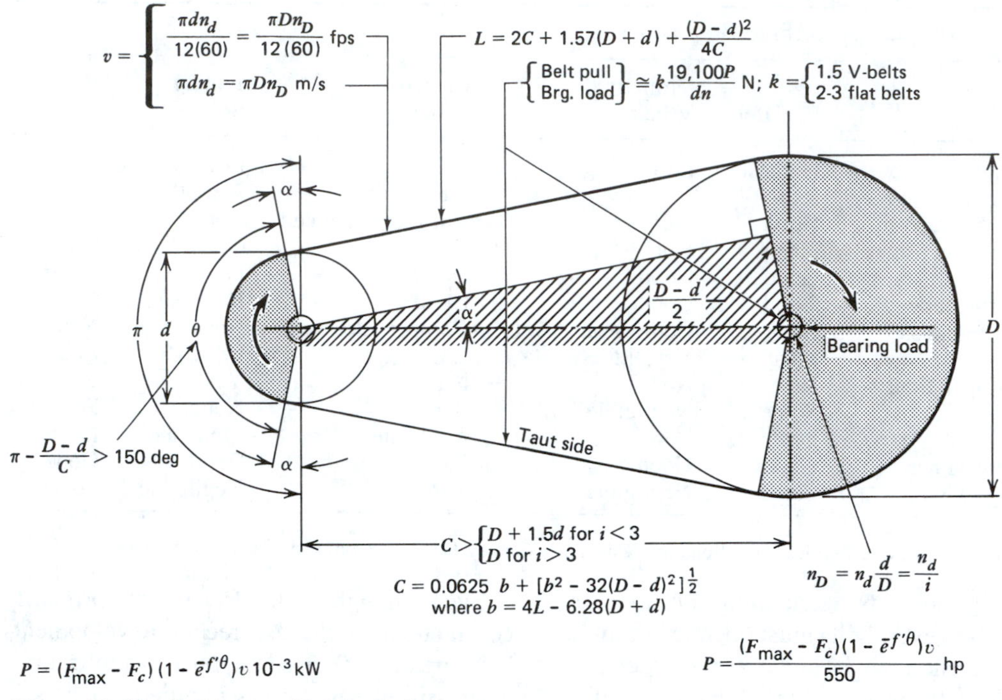

Figure 11-23 Summary of belt calculations.

trifugal forces add to belt tension without increasing traction? Why are flat-belt pulleys crowned and V-flat pulleys "flat"? What would happen if the two were switched? How can a belt idler be used as a clutching device?

P11-2 A 15-hp, 700-rpm motor drives an industrial fan by means of a flat belt. The motor has a 6-in. diameter pulley, and the fan has an 18-in. diameter pulley. Calculate (*a*) the fan speed if the belt slip is 3%, and (*b*) power output if each of the two bearings has an efficiency of 0.99 and the power loss due to slip and creep is 3%.

P11-3 What is the maximum speed ratio of a belt drive when the angle of wrap θ is 150 deg and the center distance C is 400 mm? Confirm your calculations with a suitable layout (scale 2.5:1). (Maximum ratio for $D + d = 2C$.)

P11-4 Calculate the power that can be transmitted by a flat rubber belt when tight tension is 115 lb, belt speed 40 fps, and minimum angle of contact 165 deg. Disregard centrifugal forces.

P11-5 Calculate the power capacity of a rubber V-belt when tight tension is 115 lb, belt speed 40 fps, and minimum contact angle 165 deg. Included sheave angle is 38 deg. Compare this, on a percentage basis, with the power capacity of the equivalent flat-belt drive in P11-4. Ignore centrifugal forces.

P11-6 A 1.5-kW, 1200-rpm motor drives a rotary pump through a V-belt drive. The sheave diameters are d = 100 mm and D = 400 mm. Belt slip is 2% at the drive sheave and 1% at the driven sheave. Assume a bearing efficiency of 0.98 and a sheave angle $2\beta = 38$ deg. For a single V-belt drive, calculate the following.

(*a*) Total drive efficiency.
(*b*) Belt speed v and revolutions per minute of the driven sheave.
(*c*) Belt length L for a center distance of 600 mm.
(*d*) Tight tension F_1 (disregard slip and power loss).
(*e*) Transverse load on the supporting cantilever shafts. (See Section 9-11.)
(*f*) To obtain a more compact drive, reduce the center distance C. How much can C be reduced?

P11-7 A 30-kW motor drives a compressor by means of a flat rubber belt. Pulley diameters are 300 mm for the drive and 1200 mm for the driven sheave. Belt speed is 15 m/s and center distance is 1500 mm.

(*a*) Draw a skeletal outline in two views of the drive.
(*b*) Determine the speeds of the two shafts when there is slip of 2% at the input and 1% at the output ($v = \pi dn$).
(*c*) Determine the power transmitted when bearing efficiency is 0.99 and power loss in the belt due to friction, flexure, and creep is 3%.
(*d*) Calculate the force ratio. Assume $f = 0.30$.
(*e*) Estimate the transverse load.

P11-8 Prove that V-belts are more effective than flat belts by determining the percentage of horsepower gained in changing from a flat belt to a V-belt. Assume that the diameter of the flat pulley equals the pitch diameter of the grooved pulley and that speed, coefficient of friction, material, angle of contact, and cross-sectional area are the same for both types of belt. The V-belt pulley groove angle is 2β. Ignore the effect of centrifugal forces. Calculate the percentage using the figures $f = 0.30$, $2\beta = 38$ deg, and $\theta = 150$ deg. *Hint:* Calculate the ratio of power of flat belts to V-belts.

P11-9 In *Machinery's Handbook* (twenty-first edition), review the material on "Selection of Multiple V-Belts," pp. 1044–1050, with emphasis on the example on p. 1047. The solutions to problems P11-10 to P11-13 may vary because individual judgment enters into them.

P11-10 Use the data and procedures outlined in *Machinery's Handbook* to select the appropriate size and number of regular quality V-belts for the following drive.

Motor: 40 hp, 1200 rpm.

Motor sheave: 9.0-in. diameter.

Compressor (rotary) sheave: 29.0-in. diameter.

Center distance: 54 in. approximately.

P11-11 Use *Machinery's Handbook* to select a V-belt drive for a 10-hp, 1200-rpm electric motor to drive a ventilating fan at approximately 425 rpm. The minimum center distance between sheaves is 41 in.

P11-12 An air compressor is to be driven by a 40-hp, 1200-rpm motor. Use *Machinery's Handbook* to select suitable V-belting. Suggested pitch diameters for sheaves are 9 in. and 30 in. approximately. The shaft center distance should not be less than 50 in.

P11-13 The compressor of a small air-conditioning unit requires a 2-hp, 1200-rpm motor. Compressor speed must be 300 rpm. The center distance should be minimized to provide a compact design. Use data from *Machinery's Handbook* to select a suitable V-belt drive. What should be the minimum center distance allowance for installation and take-up, using a standard V-belt length?

P11-14 Two multistep V-belt sheaves are used with an open belt drive, as shown in Fig 11-24. By moving the belt from one set of pulleys to another, the speed can be changed. The drive pulley has a speed of 900 rpm, which imparts speeds of 900, 600, and 450 rpm to the driven shaft. The shaft separation distance is 500 mm and, for step 1, $d_1 = d_4 = 225$ mm. Calculate d_2, d_3, and d_6. Disregard belt thickness and creep. Assume the belt is taut all the time. *Hint:* The belt length remains unchanged.

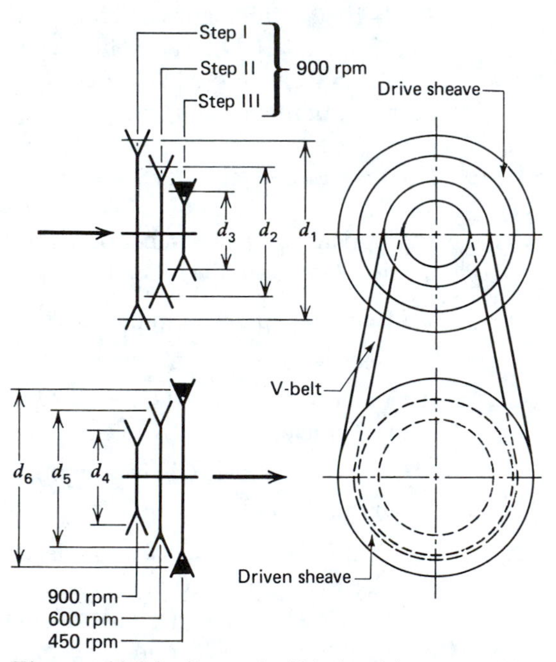

Figure 11-24 Stepped V-belt drive. (Problem 11-14)

P11-15 According to G. Nieman, belts should not be exposed to flexing or bending at a rate exceeding 40 cps. Flexing raises the temperature and precipitates fatigue failure. The frequency ν (Greek nu) is given by the formula

$$\nu = \frac{vz}{L}$$

where

v = belt speed; m/s
z = number of sheaves
L = belt length; m

Calculate the frequency of a belt drive for:

$d = 40$ mm $\qquad D = 400$ mm
$C = 400$ mm $\qquad n_d = 3600$ rpm

Also estimate the limiting speed for the drive sheave when a third driven sheave is added.

P11-16 What causes a flat belt to stay centered on a crowned pulley?

Chapter 12
Chain Drives

I never did anything worth doing by accident
nor did any of my inventions come by accident.
They came by work.

THOMAS EDISON

NOMENCLATURE

a = centrifugal acceleration
D, d = pitch diameters of sprocket
F = total chain tension
F_c = centrifugal force
F_{max} = static tensile strength
F_t = tensile force
L = chain length
N, n = number of teeth in driven and drive sprockets, respectively
p = pitch
2α = angular deviation from a 180-deg angle of contact
β = twice the angle of articulation
θ = angle of contact or angle of wrap

Chain drives combine some of the more advantageous features of belt and gear drives. Chains provide almost any speed ratio for any practical shaft separation distance. Their chief advantage over gears is that chains can be used with arbitrary center distances. Compared with belts, chains offer the advantage of positive (no slip) drive and therefore greater power capacity. An additional advantage is that not only two but many shafts can be driven by a single chain at different speeds, yet all have *synchronized* motions.

Primary applications are in conveyor systems, farm machinery, textile machinery, automotive timing, and motorcycles. The modern power transmission chain is, in part, an American invention growing out of a need for mechanized farm machinery. The roller chain, however, was invented by Renold (England) in 1880. As a result, an inch-based pitch was developed that has been retained.

12-1 BASIC DESIGN

A chain can be defined as a series of links, usually composed of metal, connected and fitted into one another to form what is in effect a flexible rack with a series of integral journal bearings. The center distance from one hinge or joint to the next is known as the *pitch* of the chain and is the primary identifying dimension. Figure 12-1 shows several types of chains and their application.

In its simplest form a chain drive consists of two sprockets of arbitrary size and a chain loop, as depicted in Fig. 12-2. Sprockets are wheels with external teeth shaped so that they can fit into the links of the drive or driven chain. The shape of the teeth varies with the number of teeth. In some recent automotive applications, tooth shape and/or size is modified to reduce noise generation.

The term *chain drive* therefore denotes a combination of chains and sprockets and their means of shaft mounting. It sometimes also includes their enclosure. Engines, motors, shafts, or bearings are usually not thought of as parts of such a drive.

Standardization of chains is under the jurisdiction of the American National Standards Institute (ANSI), a cooperative group supported by manufacturers. Because of the great variety of sizes available in each type of chain, an identification code or system was developed to avoid errors in specifying and ordering.

Chains are available in a range of accuracies extending from *precision* to *nonprecision*. Nonprecision chains are low in cost and intended primarily for noncritical drives of less than 40-kW power ratings. Precision chains, by contrast, are designed for high speeds and high power capacity.

12-2 TERMINOLOGY

Like belt drives, chains also have specific terms describing various components (Fig. 12-2).

Pitch (p), the distance between any point on a link and the same point on the next link in a

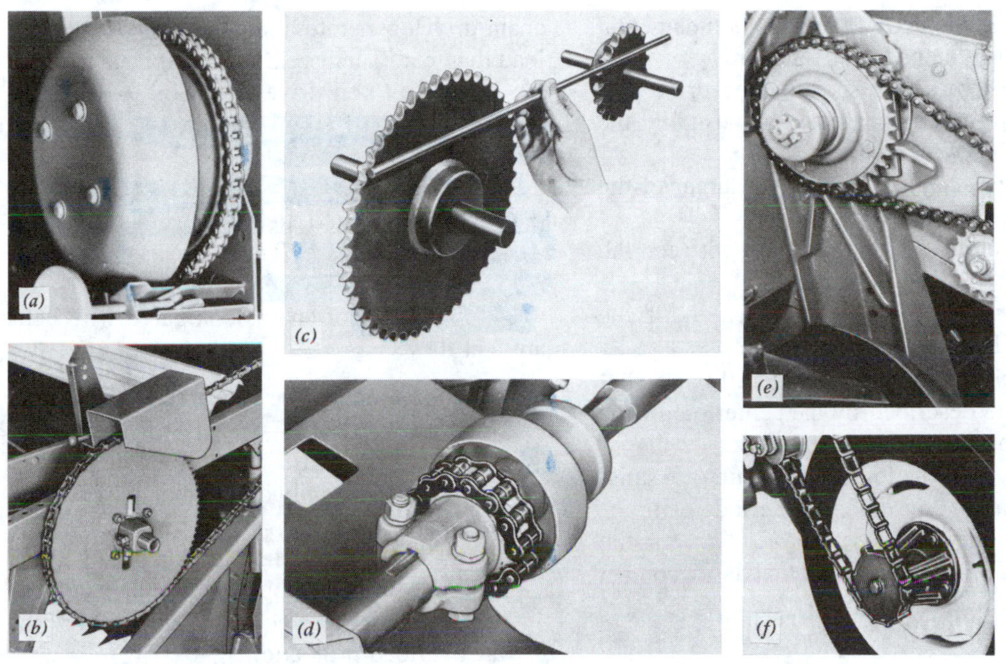

Figure 12-1 Application of chains. (*a*, *b*, *e*) Roller chain drives. (*c*) Alignment of sprockets. (*d*) Chain coupling. (*f*) Detachable link chain drive. (Courtesy Deere and Company Service Publication.)

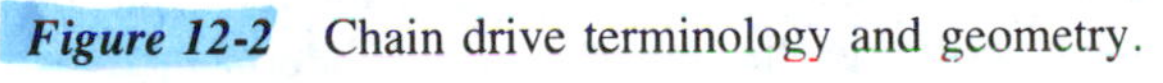

Figure 12-2 Chain drive terminology and geometry.

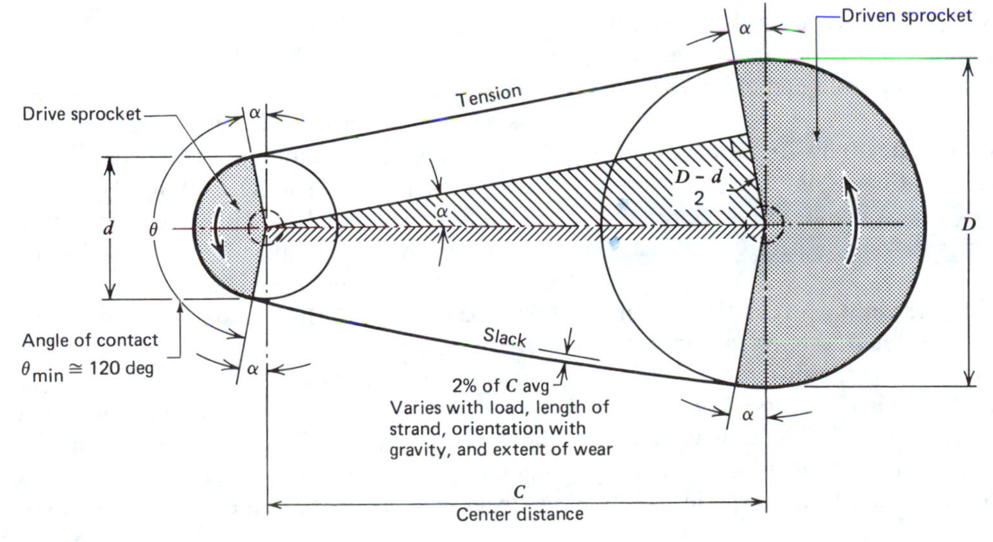

straight (unarticulated) chain, is a fundamental unit of measurement for all standards.

Drive sprockets correspond to the drive pulleys of belt assemblies and are usually the smaller of the two sprockets.

Driven sprockets are usually the larger of the two and have the slower speed.

Center distance (C) is the length between the centers of driver and driven shafts.

Chain length (L) is the total length of the chain, usually measured in pitches.

Pitch diameter (d or D) is generally on a theoretical circle described by the centerline of the chain as it passes over the sprocket.

Chain rating is the load that a chain will satisfactorily handle over extended periods of time.

Angle of wrap (θ) or angle of contact is the angular section of a sprocket that is in contact with the chain.

12-3 FUNCTION AND DESIGN

Chains transmit power through mechanical interlocking of the driver and driven sprockets. The driven sprocket is forced to rotate under the constant tension imparted to the chain from the driving sprocket. A principal advantage of chain drives is their high load capacity, which results from hardened steel links being loaded in simple tension and having many teeth (one-third to one-half) engaged on the sprocket. Chain drives are thus more compact, powerful, and efficient than belt drives.

Chains derive their flexibility from pin-jointed links that articulate at each joint during operation. *Articulation* is the relative movement between adjacent links as a joint enters and leaves a sprocket.

For a drive to transmit power, the driven sprocket must be rotated against a resisting torque. As with belt drives, a difference in tension of the two strands is necessary. In contrast to belt drives, one strand of a chain drive must always be slack. Thus power is transmitted solely by the tension side, which explains why chain drives generate a smaller transverse shaft load than equivalent gear drives and, in particular, equivalent belt drives (Section 9-10). Consequently, chain drives require smaller, less costly bearings and shafts.

Chain drives are less sensitive to dust and humidity than belts and are not adversely affected by sun, oil, or grease. They can also operate at much higher temperatures than belts.

Disadvantages of chains compared with belts are that they:

1. Require frequent or, in some cases, continual lubrication.
2. Accept very little misalignment and so should be used only on parallel and horizontal shafts.
3. Provide no overload protection because they will not slip. (For this reason, chain sprockets are sometimes equipped with shear pins as overload protection.)
4. Cost more for the same application.

12-4 CLASSIFICATION OF CHAINS

Chains are designed for three basic applications.

1. Transmitting power.
2. Conveying or moving parts and materials by utilizing the linear motion inherent in chain drives.
3. Timing or synchronizing motion.

The major use of chains for power transmission has already been discussed in some detail. In conveying motion, chains are used to convey materials by sliding, pulling, or carrying. In timing or synchronizing, chains are used as devices for synchronization of movements such as valve timing in engines, raising or lowering loads on a forklift, or balancing counterweights on machine tools.

Six major types of chains are used in power transmission and conveying, but only roller and silent chains will be discussed (Fig. 12-3). The roller chain is the most widely used type; its nonmetallic counterpart is the V-belt. It will not

Roller chain

Rollerless chain

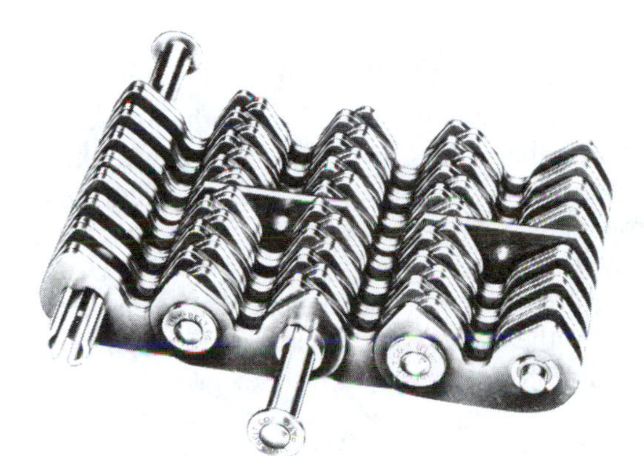

Silent chain

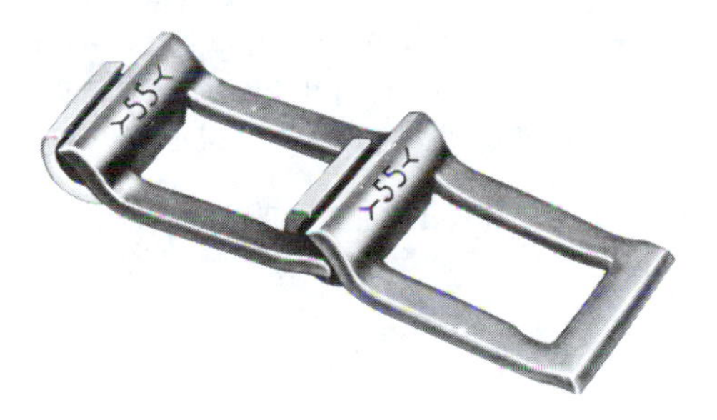

Detachable link chain

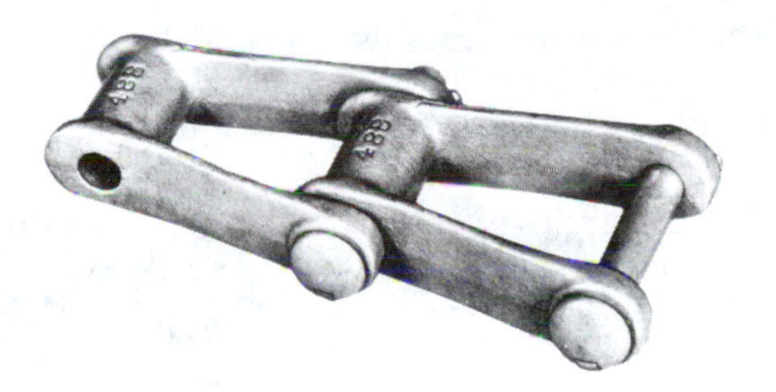

Pintle chain

Block chain

Figure 12-3 Types of chains. (Courtesy Deere and Company Technical Services.)

only be discussed first, but all examples and problems, with one or two exceptions, will relate to roller chains. Timing and synchronizing of heavy loads often require chains of the tension-linkage type, specifically designed for such purposes. Much of the discussion on roller chains applies to the remaining types of chains.

12-5 ROLLER CHAINS

The outstanding advantage of roller chains is the ability of the rollers to rotate when contacting sprocket teeth (rolling versus sliding friction). Consequently, roller chains have a high efficiency (0.97) and can be used for heavy loads at speeds up to 20 m/s. Also, roller chain drives are normally the most economical means of power transmission for speeds of 0 to 15 m/s.

Figure 12-4 shows component parts of the pin link and roller link of a single-strand roller chain. As illustrated, a precision steel roller chain is a series of journal bearings held in precise relationship to each other by the connecting link plates. Each bearing consists of a pin and bushing on which the roller revolves. The pin and bushing are case-hardened to allow articulation under high pressures and to contend with the load-carrying pressures imparted by the chain rollers.

Chains are not endless, and herein lies a major advantage over belts and gears. One link is always detachable, so that the chain can be mounted and dismounted at will. Since each roller link requires a pin link for assembly, a chain normally has an *even* number of links. If an odd number of links is required, it becomes necessary to use an *offset* link—sometimes called a *hunting* link—as shown in Fig. 12-5. An offset link is a pin link and a roller link combined. Offset links wear faster than straight links, however, and should be avoided whenever possible.

Figure 12-6 shows several important dimensions: pitch, roller diameter, and width. On a chain the width is an *internal* dimension because it corresponds to the width of a sprocket tooth. For exact dimensions, consult *Machinery's Handbook,* p. 1055.

Double-pitch chains, which mesh with every other tooth, are intended for lighter speeds and lighter loads than the equivalent single-pitch chain. However, double-pitch chains will not fit

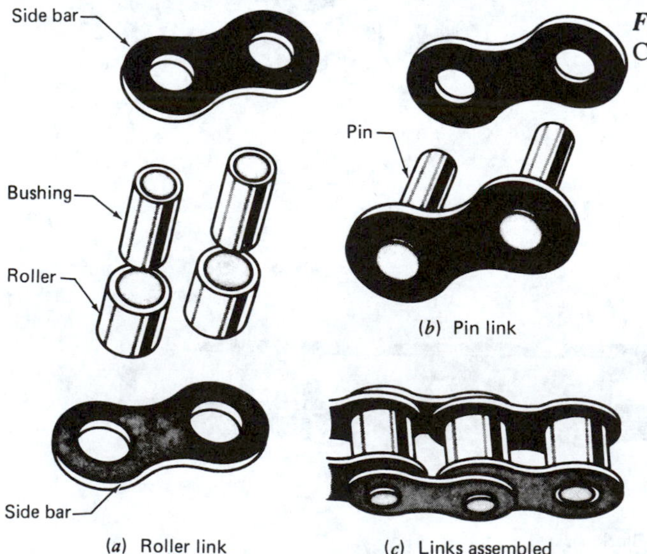

Figure 12-4 Roller chain. (Courtesy Deere and Company Technical Services.)

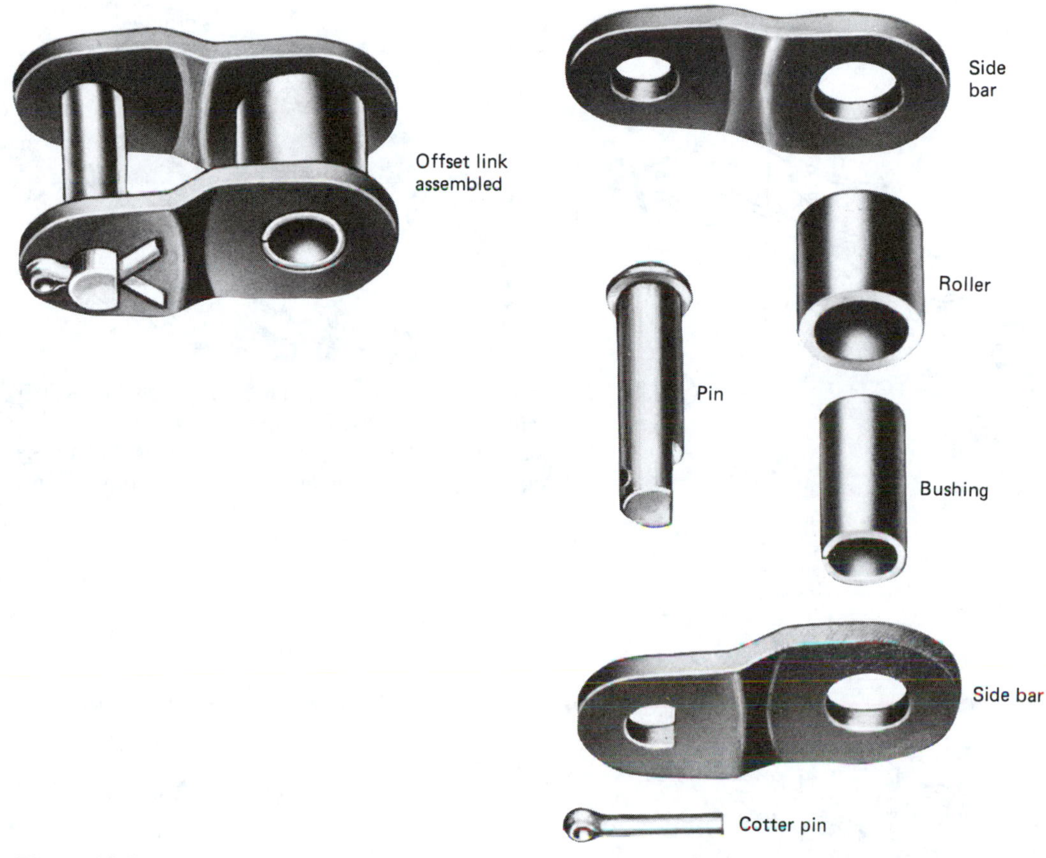

Figure 12-5 Offset link for roller chain. (Courtesy Deere and Company Technical Services.)

sprockets made for standard chains because of the difference in chordal pitch of the sprockets. Double-pitch chains are particularly useful on low-speed equipment at long center distances, where cost and weight must be minimized.

The compactness of roller chains is further enhanced by the use of double-, triple-, and four-strand chains. (Banded and V-ribbed belts are the equivalent in belt drives.) This is achieved by means of common pins that maintain the alignment of the rollers in the different strands. Figure 12-7 shows a chain with double strands.

12-6 STANDARDIZATION OF ROLLER CHAINS

All chains are classified according to the pitch, roller diameter, and width between roller link plates. Collectively, these dimensions are known as the gearing dimensions, since they determine the form and width of sprocket teeth.

Table 12-1 shows major proportions for medium to large roller chains. The right-hand figure in the chain number stands for size. Thus 0 stands for medium to large chains. The numbers

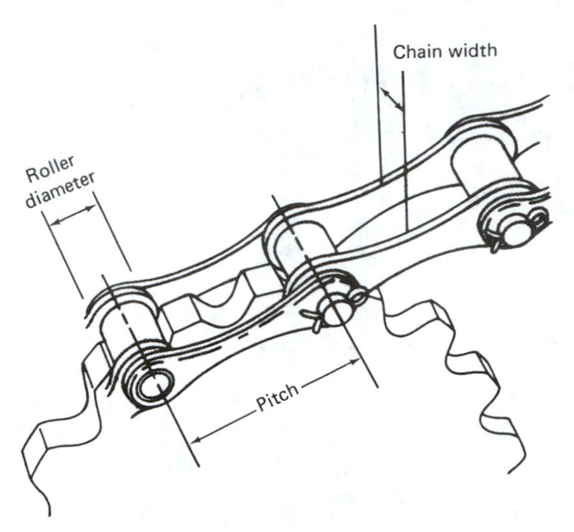

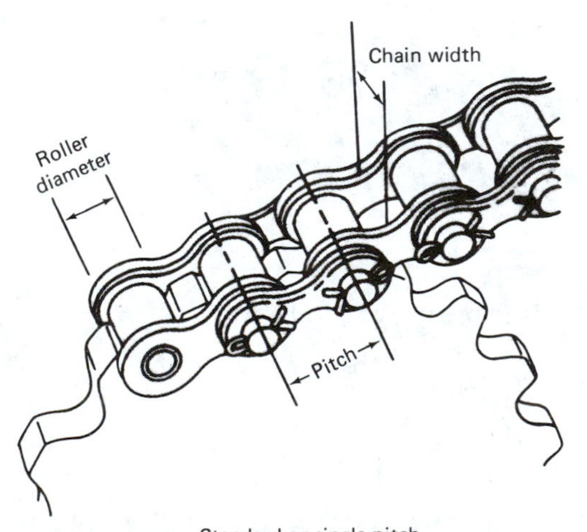

Figure 12-6 Roller chain pitch, width, and roller diameter. (Courtesy Deere and Company Technical Services.)

Figure 12-7 Double-strand roller chain. Roller chain is generally used for applications requiring high horsepower at low to moderate speeds. (Courtesy Morse Chain, Division of Borg Warner Corporation.)

to the left—4, 5, 18, 20—denote the number of $\frac{1}{8}$ in. in the pitch. Thus the pitch is based on multiples of $\frac{1}{8}$ in. The remaining dimensions, of which only two are shown, are based on the pitch. The system is shown next and illustrated by means of RC 80.

$$\text{Pitch} = (\text{number to the left})\ (\tfrac{1}{8}\ \text{in.})$$

$$[\text{RC 80:} \qquad p = 8(\tfrac{1}{8}) = 1.0\ \text{in.}]$$

$$\text{Width} = \text{constant (pitch)} = \tfrac{5}{8}\ \text{(pitch)}$$

$$[\text{RC 80:} \qquad W = \tfrac{5}{8}(1.0) = \tfrac{5}{8}\ \text{in.}]$$

12-7 DRIVE ARRANGEMENT

As observed in Chapter 11, flat belt drives can be used with almost any shaft position. Torsional flexibility plus low weight make this possible. Not so with chain drives. The effect of gravity requires that shafts be horizontal but not necessarily on the same level. Furthermore, the centerline should have a slope preferably not exceeding 60 to 70 deg from horizontal. Also, to avoid interference, the slack side should be the lower strand. Figure 12-8 shows examples of good and bad practices in drive arrangements.

Adequate tension is as important for chain drives as for belts. Chain life will be shortened if chains are run too tight or too loose. A chain too tight may carry an unnecessary additional load. Under certain circumstances, however, static preloading is desirable and does not impose any

TABLE 12-1 Standard Roller Chain Sizes (New Chains)

Chain Number	Pitch in.	Pitch mm	Width in.	Roller Diameter in.
40	$\frac{1}{2}$	12.700	$\frac{5}{16}$	$\frac{5}{16}$
50	$\frac{5}{8}$	15.875	$\frac{3}{8}$	0.400
60	$\frac{3}{4}$	19.050	$\frac{1}{2}$	$\frac{15}{32}$
80	1	25.400	$\frac{5}{8}$	$\frac{5}{8}$
100	$1\frac{1}{4}$	31.750	$\frac{3}{4}$	$\frac{3}{4}$
120	$1\frac{1}{2}$	38.100	1	$\frac{7}{8}$
140	$1\frac{3}{4}$	44.450	1	1
160	2	50.800	$1\frac{1}{4}$	$1\frac{1}{8}$
180	$2\frac{1}{4}$	57.150	$1\frac{13}{32}$	$1\frac{13}{32}$
200	$2\frac{1}{2}$	63.500	$1\frac{1}{2}$	$1\frac{9}{16}$

Source. *Belts and Chains,* John Deere Service Publications. (Courtesy Deere & Company Technical Services.)

Figure 12-8 Drive arrangements. (Courtesy Deere and Company Technical Services.)

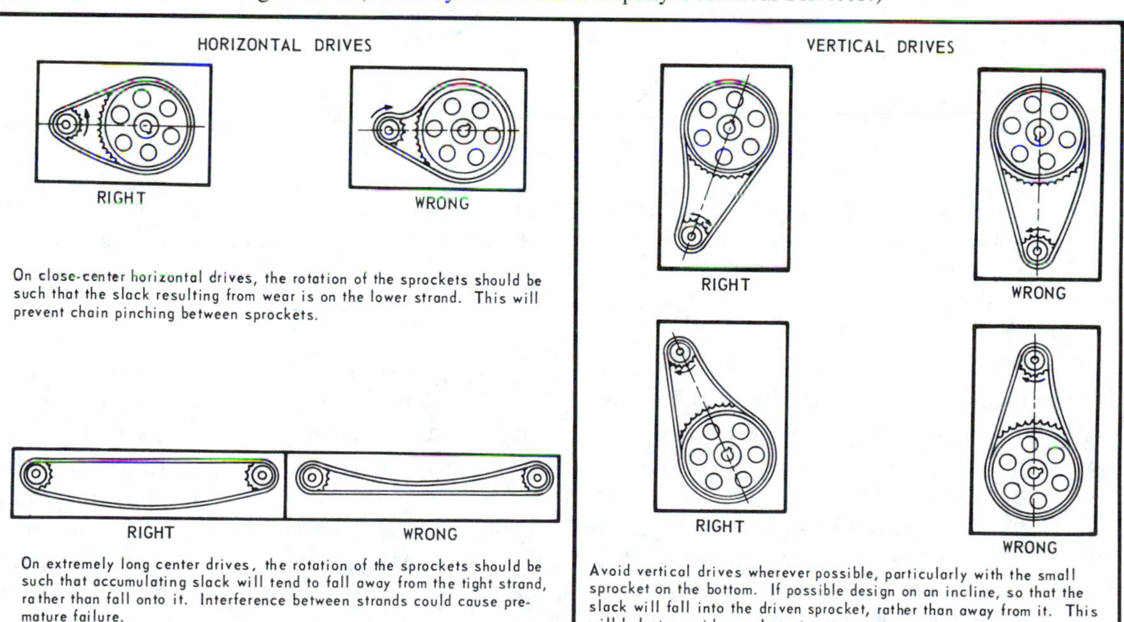

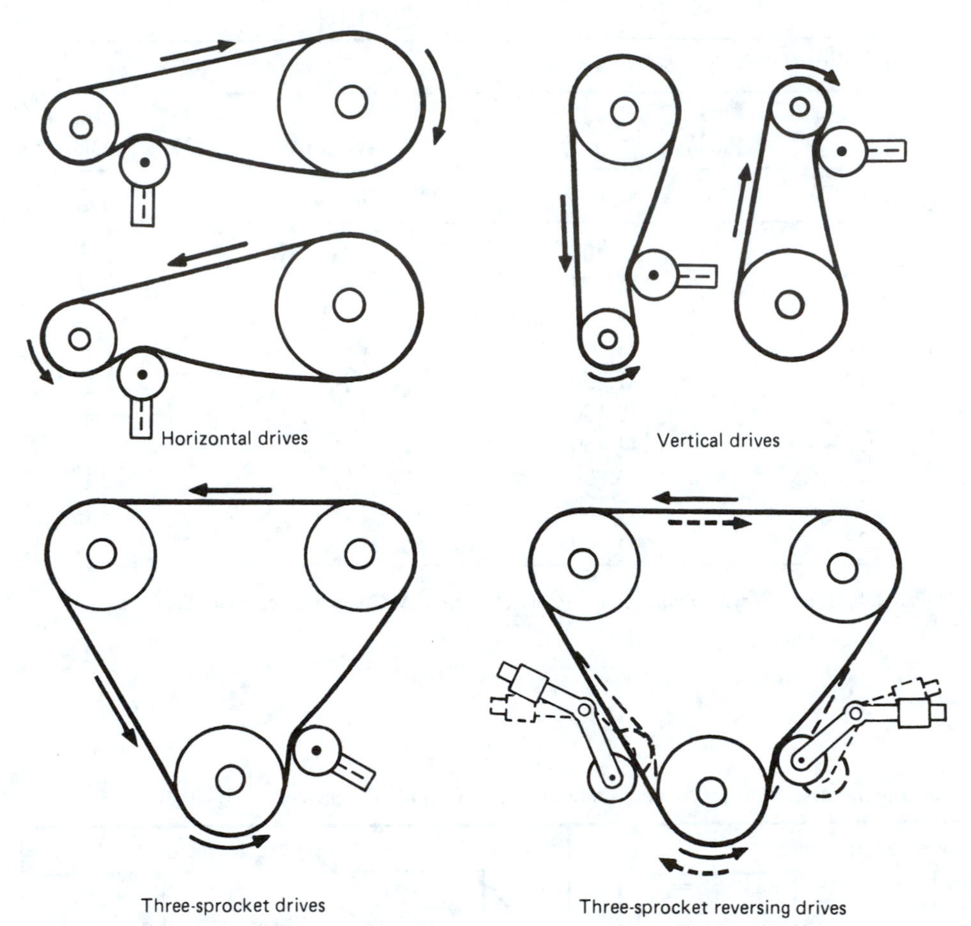

Figure 12-9 Location of chain tighteners. (Courtesy Deere and Company Technical Services.)

additional load on a fully utilized chain. A chain too loose causes whip or surge in the slack strand. The extra load superimposed on all joints by surge and whip will produce wear and material fatigue.

Provision to take up excess slack should always be included in the design. Two methods are available: change of center distance and application of chain tighteners. Adjusting center distance is often a simple matter of moving an electric motor or adjusting the bearings. Chain tighteners become necessary when two or more sprockets move on fixed centers. Proper location of chain tighteners is important, as shown in Fig. 12-9. The tighteners should be located on the inside to save space. Idlers should be no smaller than the minimum recommended drive sprocket.

12-8 DRIVE SPROCKETS

Proportions of sprockets are standardized and are available in *Machinery's Handbook* and manufacturers' catalogs. Sprockets may be designed with arms, lightening holes, or as solid units (Fig. 12-1). For easier mounting, they may come with a split center or removable rim. Sprockets are generally made from cast iron, cast steel, or fabricated steel. Standard sprockets are classified as either semiprecision or precision.

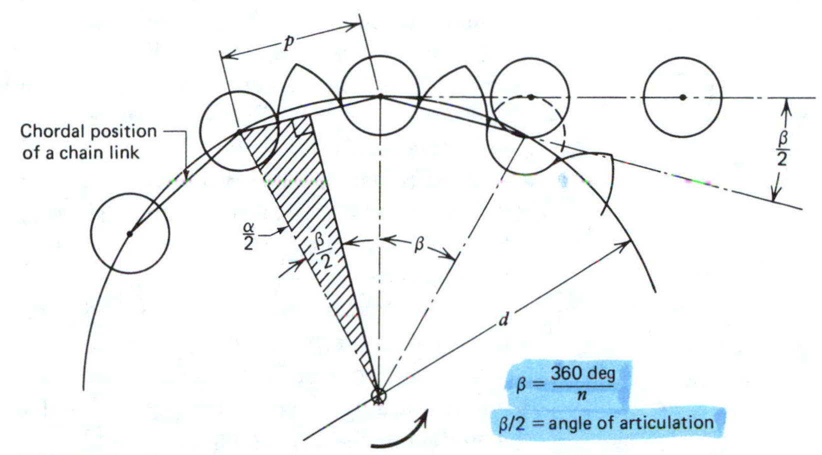

Figure 12-10 Geometry of chain engaging sprocket.

Figure 12-10 shows the action of a chain engaging sprocket teeth. The angle $\beta/2$ is that through which a chain link turns on its pin as it engages the sprocket. From the right triangle (hatched) one obtains:

$$\sin(0.5\beta) = \frac{0.5p}{0.5d} = \frac{p}{d}$$

$$d = \frac{p}{\sin(0.5\beta)} = \frac{p}{\sin(180\ \text{deg}/n)} \qquad (12\text{-}1)$$

where

d = pitch diameter of small sprocket
n = number of teeth in small sprocket
p = pitch (from Table 12-1)

Since p is always in inches, a conversion factor must be applied when d is in SI units. This equation also applies to the large sprocket. It contains three variables and can be solved for any one. Note that the $i = N/n \neq D/d$.

EXAMPLE 12-1

Find the number of teeth in a sprocket for $p = 0.625$ in. and $d = 85$ mm approximately.

SOLUTION

Rearrange Eq. 12-1.

$$\sin\left(\frac{180\ \text{deg}}{n}\right) = 25.4\ (p/d)$$

$$\frac{180\ \text{deg}}{n} = \arcsin\left[\left(\frac{25.4\ \text{mm}}{1.0\ \text{in.}}\right)\left(\frac{0.625\ \text{in.}}{85\ \text{mm}}\right)\right]$$

$$\frac{180\ \text{deg}}{n} = 10.76\ \text{deg}$$

$$\boldsymbol{n = 16.72}$$

Use $n = 17$ teeth. Always use an odd number if there is a choice. The exact value of d is

$$d = \frac{(25.4\ \text{mm/in.})(0.625\ \text{in.})}{\sin(180\ \text{deg}/17)} = 86.39\ \text{mm}$$

Therefore

$$\boldsymbol{n = 17 \qquad d = 86.39\ \text{mm}}$$

12-9 CENTER DISTANCE, CHAIN LENGTH, AND ANGLE OF WRAP

For a simple two-sprocket drive, the following considerations apply.

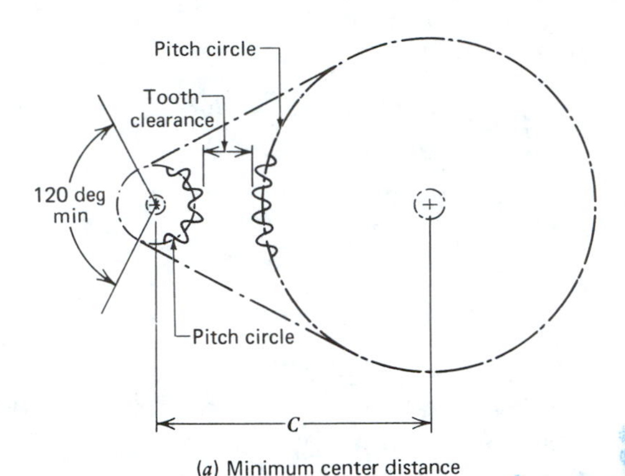

(*a*) Minimum center distance

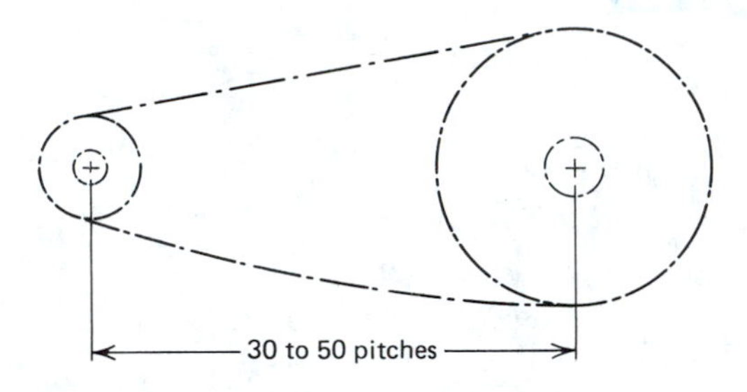

(*b*) Preferred center distance

Figure 12-11 Shaft center distance. (Courtesy American Chain Association.)

1. Minimum center distance is limited to the distance at which the two sprockets contact each other, or slightly more than half the sum of the outer diameters of the sprockets (Fig. 12-11*a*). For exact dimensions, consult *Machinery's Handbook*, p. 1058.
2. Based on experience, a center distance equal to the diameter of the large sprocket plus half the diameter of the small sprocket is a good compromise with regard to angle of wrap, wear, and initial cost (Fig. 12-11*b*).

$$C = D + 0.5d \qquad \text{for} \qquad D \gg d \qquad (12\text{-}2)$$

3. With large ratios, the angle of contact becomes smaller and the number of teeth engaged with the chain decreases. For angles less than 120 deg, θ increasingly becomes a critical factor in the design of chain drives.
4. For maximum life and minimum wear, the center distance should be chosen so as to provide an *even* number of links in the chain. This arrangement, coupled with an *odd* number of teeth in each sprocket, will minimize wear.
5. A short center distance provides a compact design (desirable) and allows for a shorter, less expensive chain. But wear is more rapid on a drive with a short center distance because the chain has fewer links and each joint must therefore articulate more often.
6. When the center distance exceeds 60 pitches, a long chain will be needed and a manufacturer's representative should be consulted.

Angle of Wrap

This angle is the same as for belts and may be calculated from Fig. 12-2.

$$\theta = \pi - 2\alpha = \pi - \frac{D - d}{C} \text{ approximately} \qquad (12\text{-}3)$$

The chain length L is generally measured in *pitches* because it consists of a whole number of links, each being of length p. Chain length is a function of the number of teeth in the two sprockets and the distance between sprocket centers. The following formula applies.

$$L = \frac{N + n}{2} + \frac{2C}{p} + \frac{p(N - n)^2}{39.5C} \text{ pitches} \qquad (12\text{-}4)$$

where

L = chain length, p
N = number of teeth in the driven sprocket
n = number of teeth in the driver sprocket
C = center distance; in.
p = pitch; in.

In such calculations, L often does not come out evenly and should be rounded off to a *whole* number, preferably an *even* one to avoid specification of an offset link.

When several sprockets are used, center distances and chain lengths are best found by means of an accurate layout or a combination of layout and calculations.

EXAMPLE 12-2

The sprocket in Example 12-1 is part of a drive with a speed ratio of roughly 3.5. Use the center distance given in Eq. 12-2. Calculate the chain length. For minimum wear, use an odd number of teeth in the sprocket. Given: $p = 0.625$ in., $d = 86.39$ mm, and $n = 17$. Confirm $i = N/n \neq D/d$.

SOLUTION

$$N = (17)(3.5) = 59.5 \qquad \text{Use } N = 59 \text{ (odd)}$$

From Eq. 12-1,

$$D = \frac{(0.625 \text{ in.})(25.4 \text{ mm/in.})}{\sin(180 \text{ deg}/59)} = 298.28 \text{ mm}$$

From Eq. 12-2, $C = 342$ mm approximately. Use C = 13.46 in. approximately. From Eq. 12-4,

$$L = \frac{59 + 17}{2} + \frac{2(13.46 \text{ in.})}{0.625 \text{ in.}} + \frac{(0.625 \text{ in.})(59 - 17)^2}{39.5(13.46 \text{ in.})}$$

$L = 83.15$ pitches. Use $L =$ **84 pitches** (even)

$$i = \frac{59}{17} = 3.470 \neq 3.452 = \frac{298.28}{86.39}$$

12-10 CHAIN SPEED

The basic formula for belt velocity v also applies to chains.

$$v = \pi d n_s \, 10^{-3} \text{ m/s} \qquad (12\text{-}5)$$

$$v = \pi n_s (d/12) \text{ fps} \qquad (12\text{-}6)$$

where

d = pitch diameter; mm, in.
n_s = shaft speed; rps

Since chain length is measured in pitches, chain speed can obviously be measured by the number of pitches reeled off a sprocket per unit of time or $\pi d \sim np$. Hence

$$v = (pn)n_s \, 10^{-3} \text{ m/s} \qquad (12\text{-}7)$$

$$v = \frac{np \, n_s}{12} \text{ fps} \qquad (12\text{-}8)$$

EXAMPLE 12-3

Calculate the speed of a chain in feet per second using both methods and the following data. The chain is RC 80; the drive sprocket has 15 teeth and rotates at 700 rpm.

SOLUTION

(*a*) $n_s = (700 \text{ rpm})(60 \text{ s/min})^{-1} = 11.67$ rps

From Table 12-1,

RC 80: $\quad p = 8(0.125 \text{ in.}) = 1$ in.

From Eq. 12-1,

$$d = \frac{1 \text{ in.}}{\sin(180 \text{ deg}/15)} = 4.81 \text{ in.}$$

From Eq. 12-6,

$$v = \pi(4.81 \text{ in.})(1/12 \text{ ft/in.})(11.67 \text{ rps})$$
$$= \mathbf{14.70 \text{ fps}}$$

(*b*) For velocity based on pitch (Eq. 12-8),

$$v = p(n/12)n_s$$
$$= (1 \text{ in.})(15)(1/12 \text{ in./ft})(11.67 \text{ rps})$$
$$= \mathbf{14.59 \text{ fps}}$$

12-11 CHORDAL ACTION

The rise and fall of each link (pitch) of chain as it engages a sprocket is termed *chordal action*. It causes repeated chain speed variations (pulsations) and is a serious limiting factor in roller chain performance. As illustrated by Fig.

12-12*a*, chordal action and speed variation decrease as the number of teeth in the small sprocket is increased and, in most applications, become negligible when 21 teeth or more are used.

Figure 12-12*b* shows schematically a roller chain link entering a sprocket. The line of approach is not tangent to the pitch circle of the sprocket. The link makes contact below the tangency line and is then lifted up to the top of the sprocket—a distance of $(R - r)$—termed *chordal rise*. Then it is dropped back down again (Fig. 12-12*c*) as sprocket rotation continues.

That speed variations take place, despite a constant input speed, is obvious from the formulas for maximum and minimum speeds given in Fig. 12-12*a* and 12-12*b*. The average speed occurs at a circular pitch radius, which is greater than r but less than R. The general expression for speed variation, using the nomenclature of Fig. 12-12, is:

$$v = \pi(R/12)(n_s)(1 - \cos\theta) \qquad (12\text{-}9)$$

$$0 < \theta < 180 \text{ deg}/n$$

where

v = speed variation; fpm
R = pitch radius; in.
n_s = speed of sprocket; rpm
θ = angle that varies between the angle of articulation 180 deg$/n$ and zero
n = number of teeth of smallest sprocket

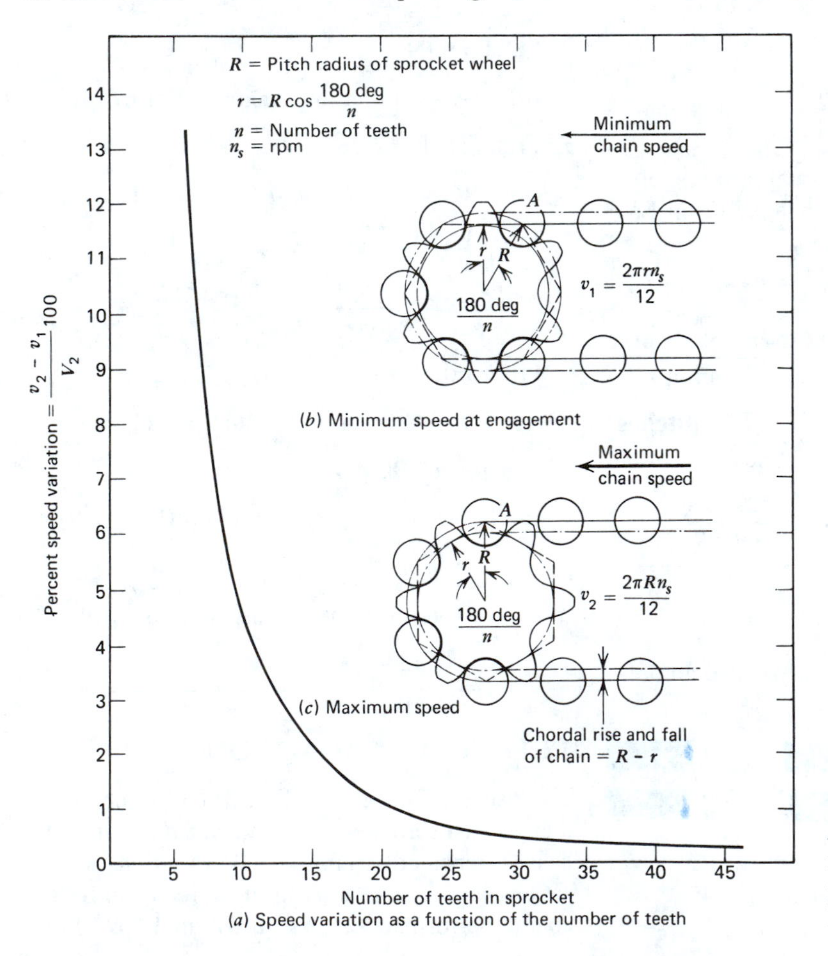

Figure 12-12 Chordal action of roller chain. (Courtesy PT Components, Inc., Link-Belt Chain Division.)

The adverse effects of chordal action are augmented by input speeds that are always subject to some variation. The net effect is waste loads that increase chain tension without improving power transmission.

The rising and falling motion of the chain becomes harmful when resonance occurs. It is then known as *chain whip*.

The chordal rise and fall is estimated as:

$$\Delta r = \left(\frac{R}{12}\right)(1 - \cos\theta) \qquad (12\text{-}10)$$

where Δr is chordal rise.

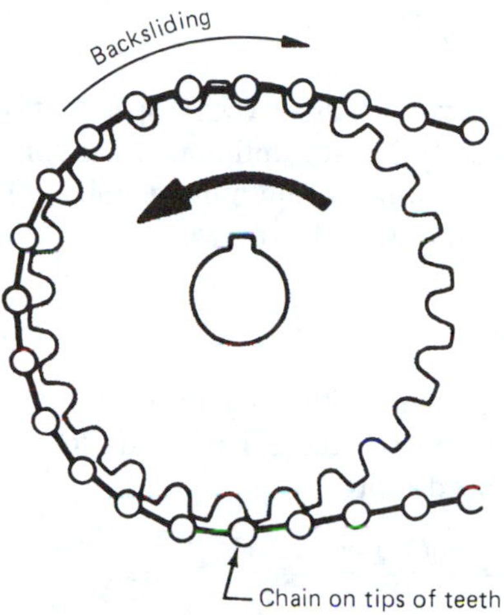

Figure 12-13 Effect of wear on roller chains. The chain compensates for its greater length by moving up and outward on the tooth flanks and eventually out of mesh. (Courtesy Deere and Company Technical Services.)

12-12 WEAR CONSIDERATIONS

Although chains are tensile devices, they are not designed to fail in tension but to malfunction gradually because of combined wear in the chain and on the sprockets (preferred mode of failure). Rupture of the drive chain on a bicycle or motorcycle disrupts power transmission totally, often with catastrophic consequences. By contrast, backsliding of a worn chain across a few worn sprocket teeth impedes but does not disrupt power transmission. The jerking motion and noise generated also are early warnings of greater troubles ahead.

A chain has as many pin joints as it has links. However, each time a link articulates over a sprocket, wear occurs between the chain bushing and the pins. As wear slowly progresses, the pitch of each pin link increases and with that the entire chain length. A corresponding change does not occur on the sprocket. The chain therefore compensates for its greater length by moving up and outward on the tooth flanks and eventually out of mesh (Fig. 12-13). Backsliding thus indicates that the drive should be replaced.

Fatigue strength of a chain is the maximum tensile load that a chain can endure for a specified number of cycles, say 2.5 million. It is the basis for chain ratings published by chain manufacturers. The ratings ensure that fatigue failure does not precede failure due to wear. Most ratings are based on a life of 15,000 to 20,000 hr with correct alignment, lubrication, and maintenance.

To guard against fatigue and wear, tooth numbers should be *odd* or, better yet, *prime* numbers.[1] Since chains usually have an *even* number of links, this choice allows each sprocket tooth to mesh with all links, one after another, instead of meshing with the same link continually. Wear will thus be more evenly distributed and total wear will be lower.

As with many machine members, choice of material is a compromise among cost, strength, and resistance to wear. Low-capacity chains are made from cast iron or carbon steels, while high-capacity chains require high-strength, heat-treated alloy steels. Surface hardness of sprocket teeth ranges from 63 to 59 Rockwell C hardness.

---

[1]A prime number, in arithmetic, is only divisible by unity or by itself.

EXAMPLE 12-4

Select sprockets for a drive with a speed ratio of approximately 3:1 for optimum wear considerations. Assume a minimum number of teeth in the drive sprocket of 15.

SOLUTION

The prime numbers following 15 are 17, 19, 23, 29, 31, and so forth. This leads to the following speed ratios.

$$i_1 = \frac{(3)(17) + 2}{17} = \frac{53}{17} = \mathbf{3.12}$$

$$i_2 = \frac{(3)(19) + 2}{19} = \frac{59}{19} = \mathbf{3.10}$$

$$i_3 = \frac{(3)(23) - 2}{23} = \frac{67}{23} = \mathbf{2.91}$$

If wear takes precedence over an exact speed ratio, any of these ratios should be adequate.

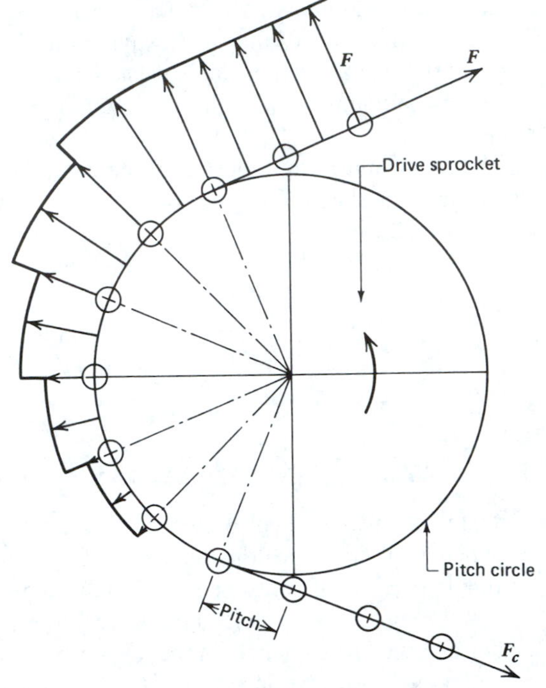

Figure 12-14 Gradual buildup of chain force. For clarity, the forces in the diagram are shown perpendicular to their actual direction, which is tangential to the sprocket pitch circle.

12-13 POWER TRANSMISSION BY CHAINS

Power transmission computation for chains is simpler than for belts because there is no exponential relationship nor initial tension value with which to contend. All power is transmitted by the tight-tension side. In contrast to gears, force transmission takes place simultaneously through or by means of several teeth over angles ranging from 90 to 250 deg. Thus total force is being built up step by step from tooth to tooth (Fig. 12-14). A high torque capacity is therefore characteristic of chain drives as compared with belts and gears.

The expression for chain tension is analogous to that for belts.

$$F = F_t + F_c \tag{12-11}$$

where

F = total chain tension; N, lb
F_t = useful working load; N, lb
F_c = centrifugal pull (waste load); N, lb

From definition of torque,

$$F_t = \frac{2T}{d} \tag{12-12}$$

where

T = torque on drive sprocket; N · m, lb-in.
d = pitch diameter of sprocket; mm, in.

Usually T is obtained from the power equation. From Newton's first law,

$$F = ma \tag{12-13}$$

where

m = mass per unit length; kg, slug
a = centrifugal acceleration; m/s², ft/s²

The basic equations for power transmission are:

$$P = (F - mv^2)(v)(10^{-3}) \text{ kW} \tag{12-14}$$

$$P = (F - mv^2)(v)(550^{-1}) \text{ hp} \tag{12-15}$$

where

m = mass per unit length; kg, slug
v = chain speed; m/s, fps

At low speeds, the centrifugal force is small and thus need not be included. Power then increases in proportion to velocity. As speed increases, however, centrifugal force grows exponentially; thus, theoretically, a speed could be reached at which power capacity becomes zero. Maximum operating speed, however, is limited not by centrifugal force but by the impact that sprocket teeth and chain rollers can endure. The destructive action caused by impact between rollers and teeth is proportional to the mass of a chain link and to the square of the velocity with which the roller strikes the sprocket. The following empirical formula applies to maximum speed of roller chains.

$$n = \left(\frac{2000}{p}\right)\sqrt{\frac{A}{Wp}} \tag{12-16}$$

where

n = maximum speed; rpm
A = projected area of roller; in.^{2}
W = weight per foot of chain; lb

In general, chain speeds should be no greater than 80% of maximum speed.

The operating conditions for chains clearly resemble those for belts, except that moving masses are larger for chains. Consequently, chains have lower operating speeds than belts. In many drives there is a kind of "labor division" between belts and chains. Often a belt drive is used on a high-speed motor as the first step in a power train, while a chain is used farther down the line where the speed is lower and the torque higher. Some examples include large snowmobiles, go-carts, small off-the-road vehicles, and snowblowers.

Centrifugal forces have an adverse effect on lubrication; they force lubricants outward and eventually off the chain if the speed is sufficiently high, thus imposing another upper limit to chain speed. To increase the chain's power capacity, two options are available: (1) a larger pitch, and (2) duplex or multistrand chains. A larger pitch, however, increases undesirable effects of chordal action, including noise.

In V-belt drives, we observed, designers can always double power capacity by doubling the number of individual belts. The same simple relationship does *not* exist in chain drives. When a duplex chain instead of a single strand of the same pitch is specified, capacity increases by only 70%. Triple stranding increases capacity by a mere 150%. Designers should therefore specify single-strand chains except where noise, power, or space considerations require more strands. A multistrand chain needs less radial, but more axial, space than single-strand chain drives do.

Service factors are somewhat higher than for belts partly because of chordal action and potential resonant conditions. Chains, for instance, are rarely coupled directly to engines but work well when a shock-absorbing hydraulic coupling is interposed.

The service factors in Table 12-2 for roller chain drives are recommended by the American Chain Association.

EXAMPLE 12-5

Find the maximum power that can be transmitted by a single-strand RC 40 roller chain. Average ultimate strength is 3700 lb. Average weight per foot is 0.41 lb. The drive sprocket has 15 teeth and rotates at 1200 rpm. What is the factor of safety on this type of chain when *Machinery's Handbook* gives a horsepower rating of 5.64?

TABLE 12-2 Service Factors for Roller Chain Drives

	Type of Input Power		
Type of Driven Load	Internal Combustion Engine with Hydraulic Drive	Electric Motor or Turbine	Internal Combustion Engine with Mechanical Drive
Smooth	1.0	1.0	1.2
Moderate shock	1.2	1.3	1.4
Heavy shock	1.4	1.5	1.7

SOLUTION

From Eq. 12-14,

$$P = (F - mv^2)(v)(550^{-1}) \text{ hp}$$

$$F = 3700 \text{ lb (ultimate chain pull)}$$

$$m = \frac{0.41 \text{ lb}}{32.16 \text{ ft/s}^2} = 0.0127 \text{ lb-ft/s}^2$$

From Table 12-1, $p = 4\ (0.125) = 0.5$ in./ tooth. From Eq. 12-8,

$$v = \frac{npn_s}{12} \text{ fps}$$

$$= \frac{(15 \text{ teeth})(0.5 \text{ in./tooth})(1200 \text{ rpm})}{(12 \text{ in./ft})(60 \text{ s/min})}$$

$$= 12.50 \text{ fps}$$

$$P = [3700 - (0.0127)(12.50^2)](12.50)(550^{-1})$$

$$= \mathbf{84.45\ hp}$$

$$fs = \frac{84.45 \text{ hp}}{5.64 \text{ hp}} \sim \mathbf{14.97}$$

A factor of safety of roughly 15 is what one would expect of a product used on bicycles and motorcycles.

12-14 LUBRICATION OF CHAINS

Because a chain, in effect, is *a series of connected journal bearings forming a flexible rack*, close attention must be paid to lubrication and cleaning. Adequate lubrication is largely a matter of oil penetration to the many vital, hard-to-reach bearing surfaces. Figure 12-15 shows

Figure 12-15 Lubrication points on chains. (Courtesy Deere and Company Technical Services.)

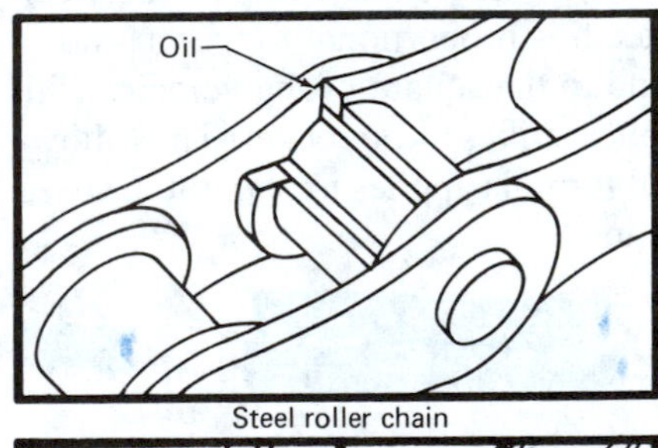

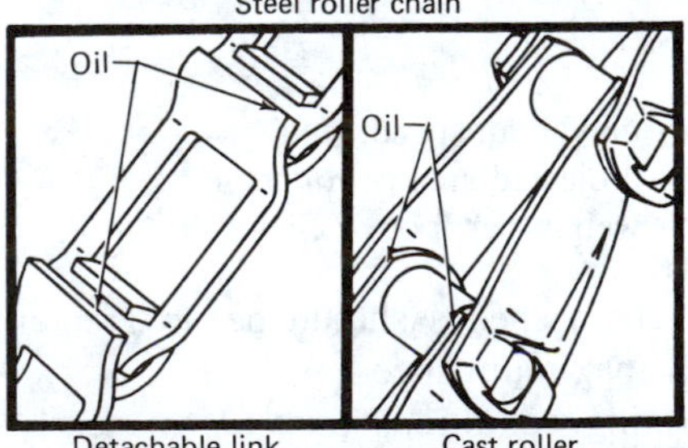

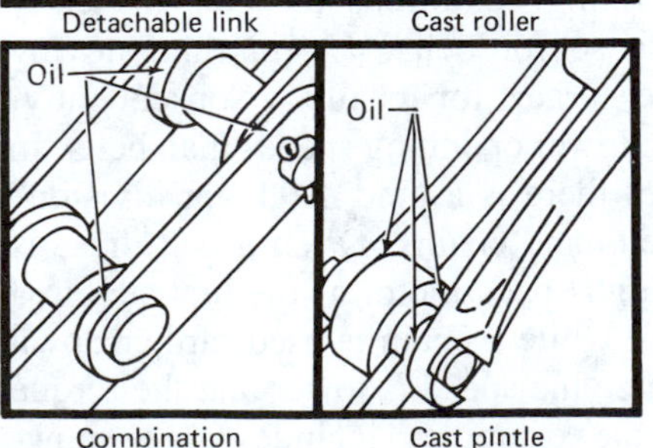

METHOD OF LUBRICATION	MANUAL		SEMI-AUTOMATIC	AUTOMATIC		
APPLICATION OF LUBRICANT	Brush or oil can	Pressure lubrication	Drip Cup	Oil Bath	Oil Disk	Oil Stream
			Contact brush, direct dip	Chain dips into oil	Oil disk throws lube up on chain	Pump sprays oil on chain
KIND OF EQUIPMENT	 Conveyors, Elevators, Simple Drives	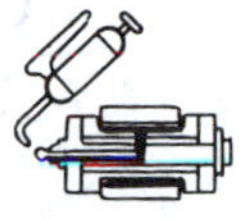 Conveyors	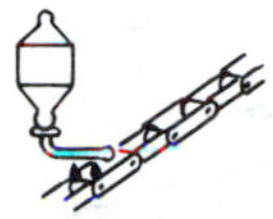 Conveyors and Elevators. Drives in low horsepower and speed range	 Drives–low to moderate horse-power and speed	 Drives–moderate to high horsepower and speed	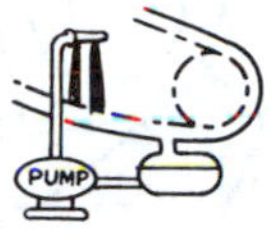 Drives–high horsepower and speed

Figure 12-16 Methods of lubrication. (Courtesy Deere and Company Technical Services.)

important lubrication points on chains. First, the oil must penetrate the gap between the inside and outside plates of the roller chains. Penetration can be helped by oiling the chain while it is warm and running it for a short time. Factory pre-lubrication protects the chain before installation and extends its operating life. Where the chain would otherwise be exposed to considerable dirt and dust, a protective casing should be used. Such an enclosure also protects workers.

Proper lubrication should be maintained because it

1. Reduces wear.
2. Protects against rust, corrosion, and heat.
3. Prevents seizing of pins and bushings.[2]
4. Cushions shock loads.

Adequate lubrication is probably the single most important factor in long service life, particularly in high-speed chains. Once a lubricated chain has run dry, failure may occur in less than 200 additional hours of operation.

[2]*Seizing* is the stiffening or freezing of a joint due to heat expansion (often called galling).

Figure 12-16 presents methods of lubricating chain drives. Note that oil is applied to the slack side for maximum penetration and that its application on the inside of the strand is best because centrifugal force and gravity tend to move oil into the links instead of spinning it off.

12-15 SILENT CHAIN

This powerful chain is so named because it operates with less noise than conventional roller chains. A silent or *inverted* tooth chain is made from a series of flat metal links pinned together as in Fig. 12-17. The U-shaped links are tooth-shaped at each end and, when assembled or stacked, form rows of straight-sided or involute teeth extending across the width of the chain. Since the links are small in proportion to their strength, silent chains can be used on smaller sprockets, thereby reducing radial space.

Power is transmitted through positive engagement of chain teeth with sprocket teeth, combining the smooth operation of belt drives with the compactness, high strength, and long life of gear drives. Because of its lower vibration and

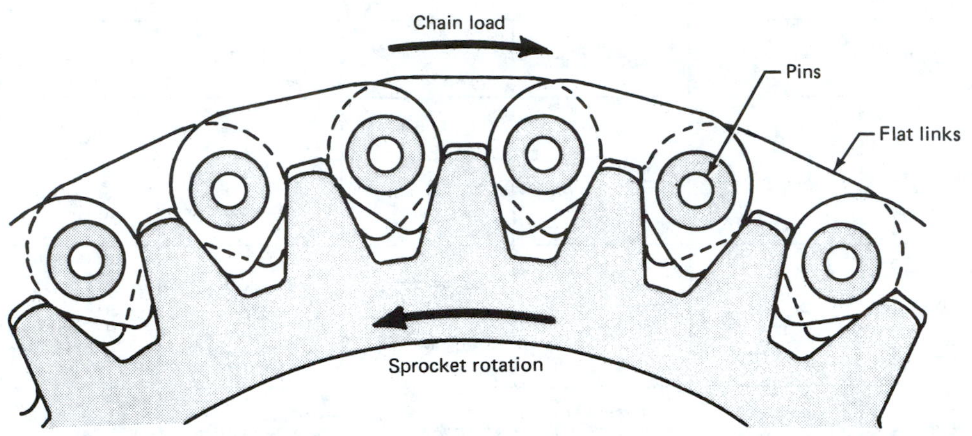

Figure 12-17 Silent chain. Engagement of chain and sprocket. (Courtesy Deere and Company Technical Services.)

frictional power losses, a silent chain operates at higher speeds (30 m/s as opposed to 15 m/s) and higher loads than a roller chain. It can transmit up to 900 kW at efficiencies of 0.97 to 0.99.

The type of joint varies. Some designs have a round pin in a round aperture (Fig. 12-17). Another design uses pins with segmental bushings. A third design uses rocker joints for improved performance. Figure 12-18 shows how chains are held in place on the sprocket.

The superior performance of silent chains is due to (1) great tensile strength, (2) reduced sliding action, and (3) automatic compensation for wear. The chain teeth and wheel teeth are shaped so that as the chain pitch increases through wear at the joints, the chain shifts outward on the teeth, engaging the wheel on a pitch circle of increasing diameter. As a result, the sprocket pitch increases at the same rate as the chain pitch.

On the negative side, silent chains must be protected from abrasives and corrosives and

Figure 12-18 Retention of silent chain. (Courtesy Deere and Company Technical Services.)

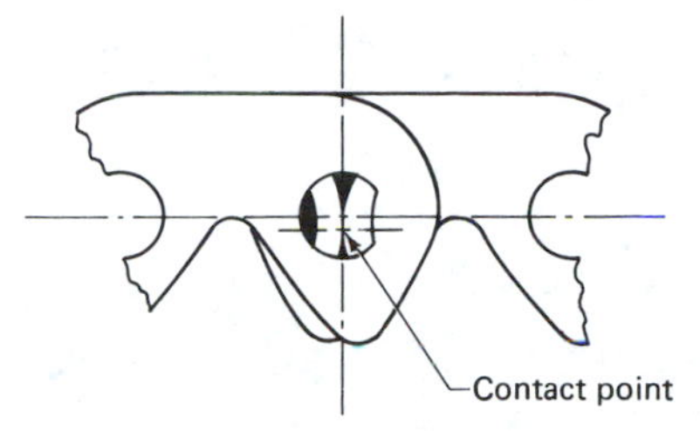

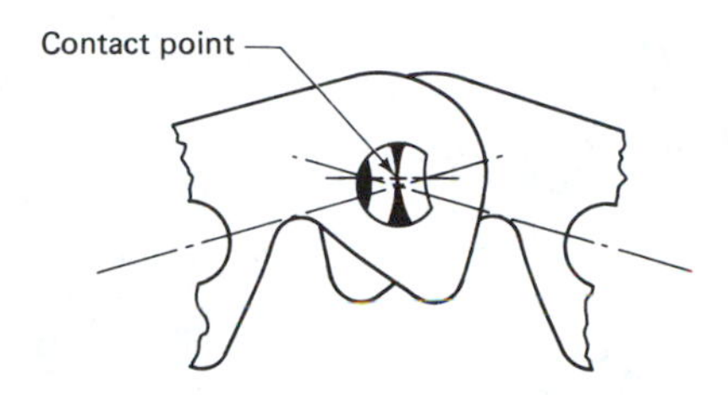

Figure 12-19 Concentric pin-rocker joint for HY-VO chain. As shown on the left, the contact point is below the pitch line of the chain prior to articulation. As the joint articulates, the contact point moves upward and the pitch of the chain elongates an amount equal to that required for the chain to wrap the sprocket along the pitch circle. This is known as "chordal compensation." (Courtesy Morse Chain, Division of Borg Warner Corporation.)

must be lubricated continuously. They are also more expensive than roller chains.

12-16 THE HY-VO CHAIN

A special type of inverted tooth chain is the HY-VO by Morse.[3] This chain is capable of quietly transmitting power in excess of 1500 kW (2000 hp) at speeds in excess of 50 m/s (5000 fpm) without overheating. The high performance is due to minimal sliding friction, resulting in an efficiency of 0.995. This chain is a successful application of the old rule: *For higher speed and efficiency, substitute rolling motion for sliding.* The HY-VO rocker joint is a modified, anti-friction bearing. As shown in Fig. 12-19, a typical rocker joint consists of two pins with convex surfaces placed back to back so that they roll on each other as the chain flexes to pass around the sprocket. The result is much less friction and therefore less wear than in the typical silent chain joint, which consists of a cylindrical pin and bushing that slide on each other under load as the chain flexes.

In addition, the pins function as cams, causing first a slight increase in pitch length and then a slight decrease. This lengthening and shortening moves in phase opposite to the speed changes generated by chordal action and hence alleviates the adverse effects of this undesirable form of vibration. Although the chain is made from rigid links, it seemingly behaves elastically.

The up-and-down movement is also counteracted by the lengthening and shortening of the pitch. A lengthening of the pitch will force a link to ride higher on the teeth—away from the center—when chordal action tends to force it toward the center, and vice versa.

[3]HY-VO is the trade name for a unique chain design manufactured by Morse Chain, Division of Borg Warner Corporation.

12-17 NONSTANDARD DRIVES

The chains discussed so far, except HY-VO, have been standard, general-purpose types. The unique design of roller chains, featuring side plates, pins, and rollers, makes them well suited for dozens of adaptations. Here, by extending the side plate, designers can obtain a saw tooth plus a lug for conveying. A pin extended may trigger a switch. Possibilities are limited only by the designers' imaginations (Fig. 12-20).

SUMMARY

Chain drives, a form of flexible gearing, are rugged, powerful, versatile, and positive. Unlike gears, they can operate at arbitrary center

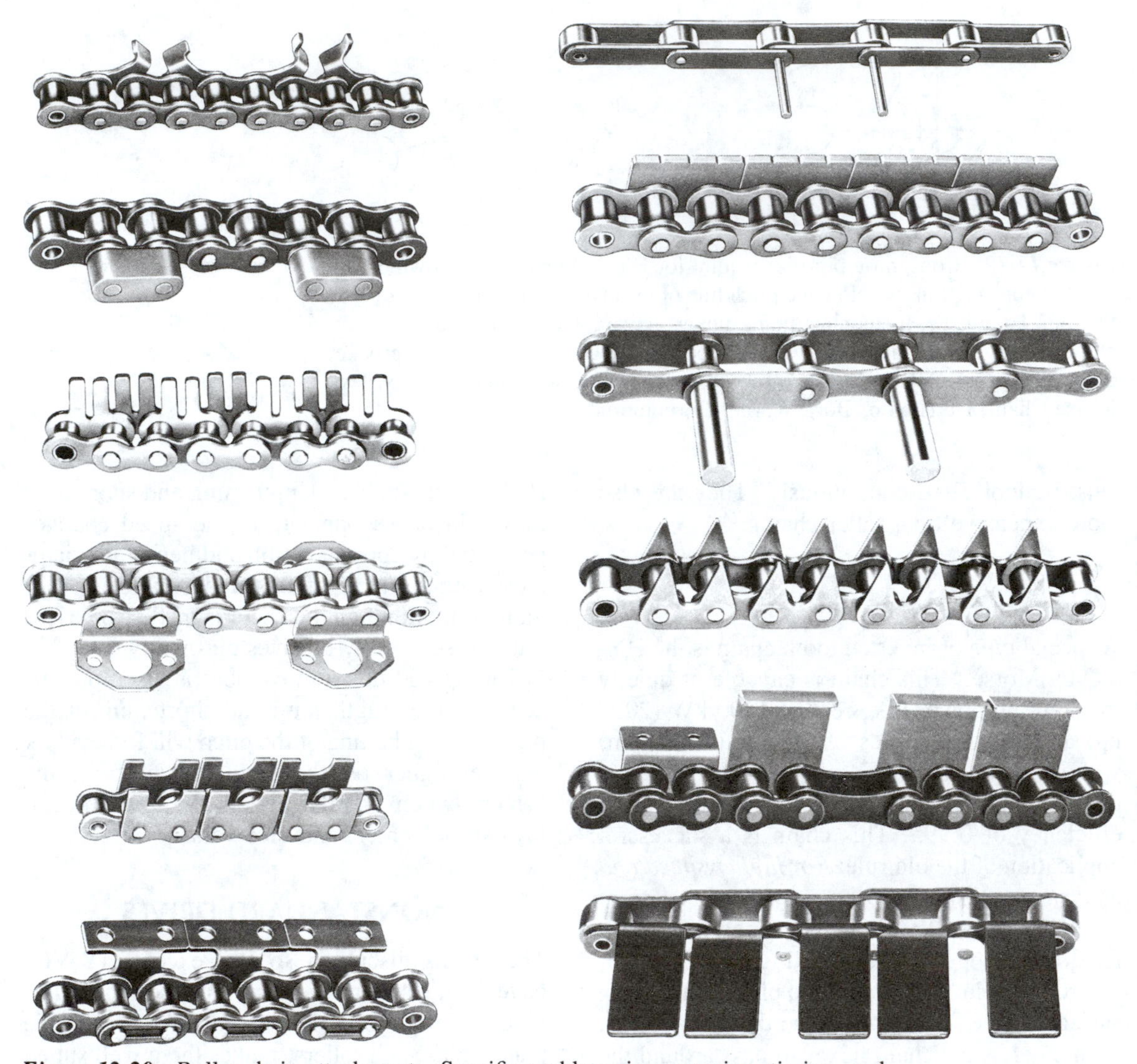

Figure 12-20 Roller chain attachments. Specific problems in conveying, timing, and indexing can often be solved by means of custom attachments, some adapted from standards, others completely original. (Courtesy Acme Chain.)

distances, but they occupy more space. Chain drives are more compact than belt drives and can operate at higher ambient temperatures, lower speeds, greater loads, and shorter center distances than belts. However, chains will wear out quickly if lubrication fails. They are also noisy, accept little misalignment, and present a work hazard when unprotected. Chains generate smaller bearing loads than belts and gears. They can be equipped with numerous attachments for special jobs, a feature unique to chains.

Of the six major types of chain available, roller chain dominates. The outstanding ad-

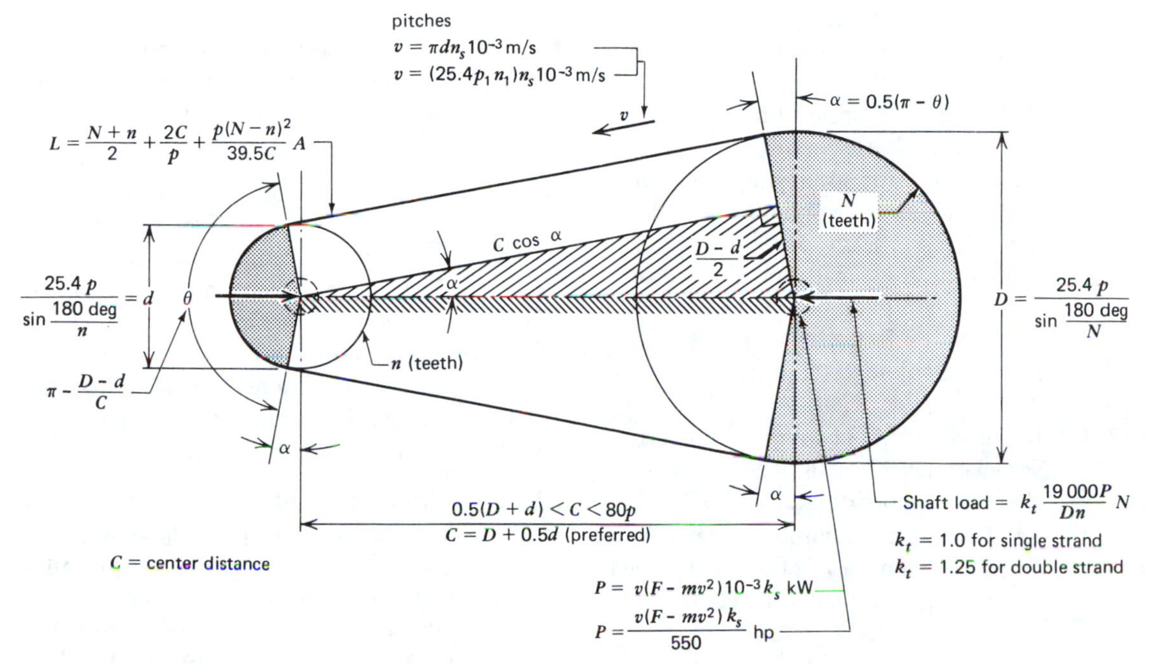

Figure 12-21 Summary of chain calculations.

vantage of roller chains is the ability of the rollers to rotate when contacting sprocket teeth. As a consequence, roller chains have a slightly higher efficiency (0.96) than V-belts. Roller chains are usually the most economical means of power transmission for speeds from 0 to 15 m/s. For speed ranges of 15 to 50 m/s, normally the domain of belt drives, designers may retain the superior load capacity of chains by specifying an inverted tooth chain. The initial cost is much higher, but the operation is silent and the space occupied is minimized.

Roller chains are standardized worldwide, with the pitch as a base for all other dimensions. Nonstandard chains are often adaptations of roller chains with numerous attachments for special jobs. Figure 12-21 summarizes chain calculations.

REFERENCES

12-1 Binder, R. C. *Mechanics of the Roller Chain Drive*. Englewood Cliffs, N.J.: Prentice-Hall, 1956.

12-2 *Chain Transmission*. Farmingdale, N.Y.: Renold Crofts, 1973.

12-3 *Design Manual, Roller and Silent Chain Drives*. St. Petersburg, Fla.: American Chain Association, 1974.

12-4 *Fundamentals of Service: Belts and Chains*. Moline, Ill.: John Deere Service Publications, 1971. (Available to the public along with a slide collection, well suited for teaching.)

12-5 Jones, F. D., and P. B. Schubert. *Engineering Encyclopedia,* Third Edition. New York: Industrial Press, 1963.

12-6 *Morse, Standard Power Transmission Products, SP-80*. Ithaca, N.Y.: Morse Chain, 1980. (Well suited for use in class.)

PROBLEMS

P12-1 Why are roller chains designed to fail not in tension but because of wear? Why do motorcycles use chains in preference to belts? Why are roller chains used in more than half of all chain drives? Why are roller chains often used as tension-linkage chains (forklift trucks)?

P12-2 A good rule for specifying center distance is to have it equal the diameter of the large sprocket plus half the diameter of the small sprocket. Prove that this is a good arrangement for all speed ratios less than or equal to 10. *Hint:* Calculate θ; compare θ to 120 deg.

P12-3 In a given roller chain drive, the drive sprocket has 17 teeth. The speed ratio should be approximately 2.70.

(*a*) Calculate the number of teeth N in the driven sprocket when (1) speed ratio is important, and (2) long life is important.
(*b*) Calculate the pitch diameters d and D based on maximum wear and a pitch $p = 0.50$ in.
(*c*) Calculate center distance $C = D + 0.5d$.
(*d*) Calculate the corresponding chain length L.
(*e*) Calculate the chain speed using both methods given in the text. Assume a speed of 900 rpm for the drive sprocket.
(*f*) Use the tables in *Machinery's Handbook* or another source to estimate the power capacity of this chain.
(*g*) Indicate chain number, width, roller diameter, and outside width (needed for clearance). Use the same source as in part (f) to find these dimensions.
(*h*) How much space does this drive require? Make a sketch in three views on a standard sheet of paper showing the minimum space required. Add ⅛ in. on each side for clearance. (*Suggested scale:* Half size.)

P12-4 A 10-hp, 500-rpm hydraulic motor is to drive a vane-pump at 150 rpm through an RC 60 chain. The drive sprocket has 13 teeth.

(*a*) Find the number of teeth appropriate for the driven sprocket to achieve minimum wear.
(*b*) Calculate pitch diameters d and D.
(*c*) Calculate chain speed in feet per second.
(*d*) Calculate tensile load F of the chain assuming an RC 60 chain has a weight of 1.03 lb/ft.

P12-5 A roller chain driven by an engine with a hydraulic drive operating under moderate shock conditions is to transmit approximately 12 kW from a shaft operating at 900 rpm to one rotating at roughly 200 rpm. Use information given in *Machinery's Handbook* to:

(*a*) Select a duplex chain for this drive (a duplex cuts noise) and indicate type of lubrication needed. (See Roller Chain Ratings, p. 1069.)
(*b*) Calculate the chain length for a center distance of $C = 0.5d + D$ and make the corresponding correction of the center distance.
(*c*) Check the contact angle to meet minimum requirements.
(*d*) Specify maximum hub dimension and sprocket bore for drive sprocket. (See p. 1062.)
(*e*) To reduce the center distance and obtain a more compact drive, it was decided to reduce the angle of wrap to 130 deg. What is the new center distance? Does this cause the sprockets to interfere?

P12-6 Use the roller chain drive ratings in *Machinery's Handbook* to solve the following problem. A roller chain is to transmit 4 hp at 900 rpm to a shaft running at 300 rpm. The power source is an electric motor; operating conditions are moderate shock.

(*a*) Select a single-strand chain for this application.
(*b*) Specify number of teeth and pitch diameters of the sprockets.
(*c*) Check the angle of contact θ for a center distance $C = D + d/2$.
(*d*) Calculate the chain length and readjust the center distance using the formula given in *Machinery's Handbook,* p. 1063.
(*e*) Calculate the chain speed in feet per second.
(*f*) How should this chain be lubricated?

P12-7 A special-purpose vehicle (dune buggy) equipped with a 40-hp Wankel engine uses a roller chain as the final drive to the rear axle. The drive sprocket runs at 300 rpm and the driven sprocket at 100 rpm. The transmission efficiency between the engine and the driving sprocket is 85%.

(*a*) Specify a standard four strand chain and sprocket for this drive. Use moderate shock for the Wankel; it is a rotary engine.
(*b*) Determine the number of teeth in each sprocket and the type of lubrication needed.
(*c*) Would it be practical and economical to replace the chain drive by a V-belt type? Give your reasons.

P12-8 Use *Machinery's Handbook* (Table 4, p. 1058) to specify a roller chain drive to operate under steady-load conditions while transmitting 3.75 kW from a shaft rotating at 575 rpm to one operating at 750 rpm. Low cost and compact design (short center distance) are major considerations.

The following problems, reproduced with permission from *Design Manual, Roller and Silent Chain Drives* by the American Chain Association, can be solved by means of this useful manual. Make a suitable sketch of each drive. For initial guidance, the step-by-step procedure is given in the back of the manual.

P12-9 *Tumbling Barrel for Metal Stampings Driven by a Motor with a Speed Reducer*. A tumbling barrel is to be driven at 26 rpm by a speed reducer powered by a 10-hp electric motor. The reducer output speed is 124 rpm and the output shaft is $2\frac{1}{2}$ in. in diameter. The shaft diameter of the tumbling barrel is $2\frac{15}{16}$ in. The shaft center distance must be approximately 48 in.

P12-10 *Heavy-Duty Apron Conveyor Driven by a Gear Motor*. The head shaft of a heavy-duty apron conveyor, handling rough castings from a shakeout, operates at 66 rpm and is driven by a gear motor whose output is 70 hp at 100 rpm. The head shaft is $4\frac{1}{2}$ in. in diameter and the gear motor output shaft is $3\frac{1}{2}$ in. The shaft center distance should not exceed 48 in.

P12-11 *Gear Pump Drive on Hydraulic Press*. A gear type lubrication pump located in the base of a large hydraulic press is to be driven at 860 rpm from a $1\frac{1}{4}$-in. diameter counter shaft operating at 1000 rpm. The pump is rated at 3 hp and has a $1\frac{3}{8}$-in. diameter shaft. The shaft center distance must not be less than 10 in.

P12-12 *Electric Motor Drive to Line Shaft*. A 200-hp, 480-rpm electric motor is to drive a line shaft, subject to heavy service, at 160 rpm. The motor shaft will be in approximately the same horizontal plane as the line shaft. The diameters of the motor shaft and line shaft are $4\frac{7}{8}$ and 6 in., respectively. A shaft center distance of from 4 ft to 5 ft will be acceptable.

P12-13 *Centrifugal Fan Drive*. A centrifugal fan is to be driven at 2800 rpm by a 10-hp electric motor. The motor speed is 1800 rpm and the shaft is $1\frac{3}{8}$ in. in diameter. The compressor shaft is $1\frac{1}{4}$ in. in diameter. The center distance is to be approximately 20 in. The overall drive must not exceed a 5-in. radius on the motor or a 3-in. radius at the fan.

P12-14 Select a silent chain drive to meet the following conditions.
Source of power, electric motor.
Transmitted horsepower, 38.
Driven equipment, mine fan.
Size of driver shaft, $2\frac{1}{8}$-in. diameter.
Speed of driver shaft, 1160 rpm.
Size of driven shaft, $2\frac{7}{8}$-in. diameter.
Speed of driven shaft, 400 rpm.
Approximate wheel centers, $22\frac{1}{4}$ in.
Service requirement, 24 hr per day operation.
Space limitations, none.

P12-15 Select a silent chain drive to meet the following conditions.
Source of power, six-cylinder gasoline engine.
Driven equipment, coal breaker.
Size of driver shaft, $4\frac{7}{8}$-in. diameter.
Speed of driver shaft, 500 rpm.
Size of driven shaft, 6-in. diameter.
Speed of driven shaft, 150 rpm.
Desired wheel centers, 48 in.
Service requirement, 16 hr per day.
Transmitted horsepower, 325.
Space limitations, none.

Chapter 13
Spur and Helical Gears

Imagination is more important than knowledge,
for knowledge is limited, whereas imagination
embraces the entire world—stimulating progress,
giving birth to evolution. . . .

ALBERT EINSTEIN

The *terminology*, *nomenclature*, and *symbols* used in this chapter are consistent with most gear literature and publications of the American Gear Manufacturers Association (AGMA).

NOMENCLATURE

a = addendum
b = dedendum
B = backlash, linear measure along pitch circle
C = center distance
C_v = velocity factor
D = pitch diameter of gear
d = pitch diameter of pinion
D_b = base diameter of gear
d_b = base diameter of pinion
D_o = outside diameter of gear
d_o = outside diameter of pinion
D_r = root diameter of gear
d_r = root diameter of pinion
F = face width
h = effective height (parabolic tooth)
h_t = whole depth (tooth height)
L = lead (advance of helical gear tooth in 1 revolution)
$L_p(L_G)$ = lead of pinion (gear) in helical gears
M = measurement over pins
m = module
m_f = face contact ratio
m_G = gear or speed ratio ($m_G = N_G/N_p$)
m_n = normal module
m_p = profile contact ratio
m_t = total contact ratio
$N_p(N_G)$ = number of teeth in pinion (gear)
N_c = critical number of teeth for no undercutting
$n_p(n_G)$ = speed of pinion (gear);
p_a = axial pitch
p_b = base pitch (equals normal pitch for helical gears)
p_c = circular pitch
P_d = diametral pitch
P_n = normal diametral pitch of helical gear
p_n = normal circular pitch of helical gear
R = pitch radius gear
r = pitch radius pinion
$r_b(R_b)$ = base radius of pinion (gear)
$r_o(R_o)$ = outside radius of pinion (gear)
S_e = endurance limit
t_c = circular tooth thickness (theoretical)
t = tooth thickness at root
v = pitch line velocity
W = tooth load, total
W_a = axial load
W_r = radial load
W_t = tangential load
y = tooth form factor
Z = length of action
$\emptyset$ = pressure angle
$\emptyset_n$ = pressure angle in normal plane
ψ = helix angle

Gears are toothed wheels used primarily to transmit motion and power between rotating shafts. *Gearing* is an assembly of two or more gears. The most durable of all mechanical drives, gearing can transmit high power at efficiencies approaching 0.99 and with long service life. As precision machine elements gears must be designed, manufactured, and installed with great care if they are to function properly.

Relative shaft position—parallel, intersecting, or skew—accounts for three basic types of gearing, each of which can be studied by observing a single pair. This chapter will discuss the fundamentals, kinematics, and strength of gears in terms of spur gears (parallel shafts), which are the easiest type to comprehend. Spur gears compose the largest group of gears, and many of their fundamental principles apply to the other gear types.

13-1 FUNCTION AND DESIGN

Probably the earliest method of transmitting motion from one revolving shaft to another was by contact from unlubricated friction wheels (Fig. 13-1). Because they allow no control over slippage, friction drives cannot be used successfully where machine parts must maintain contact and constant angular velocity. To transmit power without slippage, a positive drive is required, a condition that can be fulfilled by properly designed teeth. Gears are thus a logical extension of the friction wheel concept (Fig. 13-2).

Gears are spinning levers capable of performing three important functions. They can provide a positive displacement coupling between shafts, increase, decrease, or maintain the speed of rotation with accompanying change in torque, and change the direction of rotation and/or shaft arrangement (orientation).

To function properly, gears assume various shapes to accommodate shaft orientation. If the shafts are parallel, the basic friction wheels and gears developed from them assume the shape of cylinders (Fig. 13-3*a*). When the shafts are intersecting, the wheels become frustrums of cones, and gears developed on these conical surfaces are called bevel gears (Fig. 13-3*b*). When the shafts cross (skew, one above the other), the friction wheels may be cylindrical or of hyperbolic cross section (Fig. 13-3*c*). In addition, a gear is sometimes meshed with a toothed bar called a rack (Fig. 13-3*a*), which produces linear motion. Besides shaft position and tooth form, gears may be classified according to:

1. *System of Measurement*. Pitch (EU) or module (*SI*).
2. *Pitch*. Coarse or fine.
3. *Quality*. Commercial, precision, and ultraprecision or tolerance classifications per AGMA 390.03.

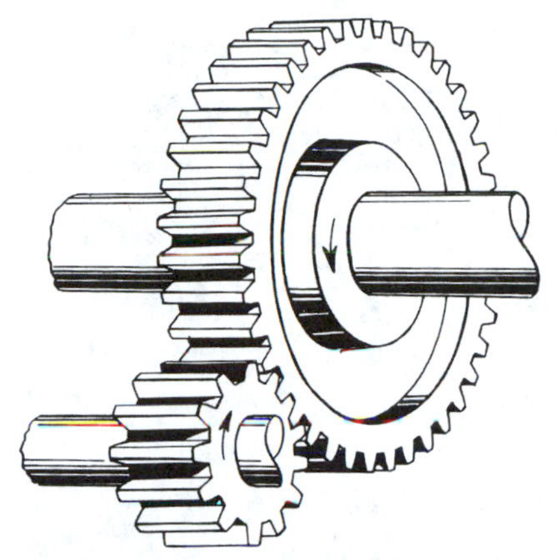

Figure 13-2 Spur gears. The teeth of these gears are developed from blank cylinders. (Courtesy Mobil Oil Corporation.)

Figure 13-1 Friction wheels on parallel shafts. (Courtesy Mobil Oil Corporation.)

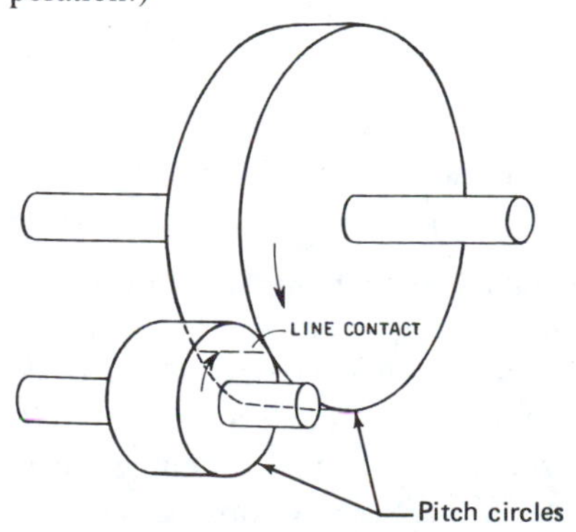

Law of Gearing

Gears are provided with teeth shaped so that motion is transmitted in the manner of smooth curves rolling together without slipping. The rolling curves are called pitch curves because on

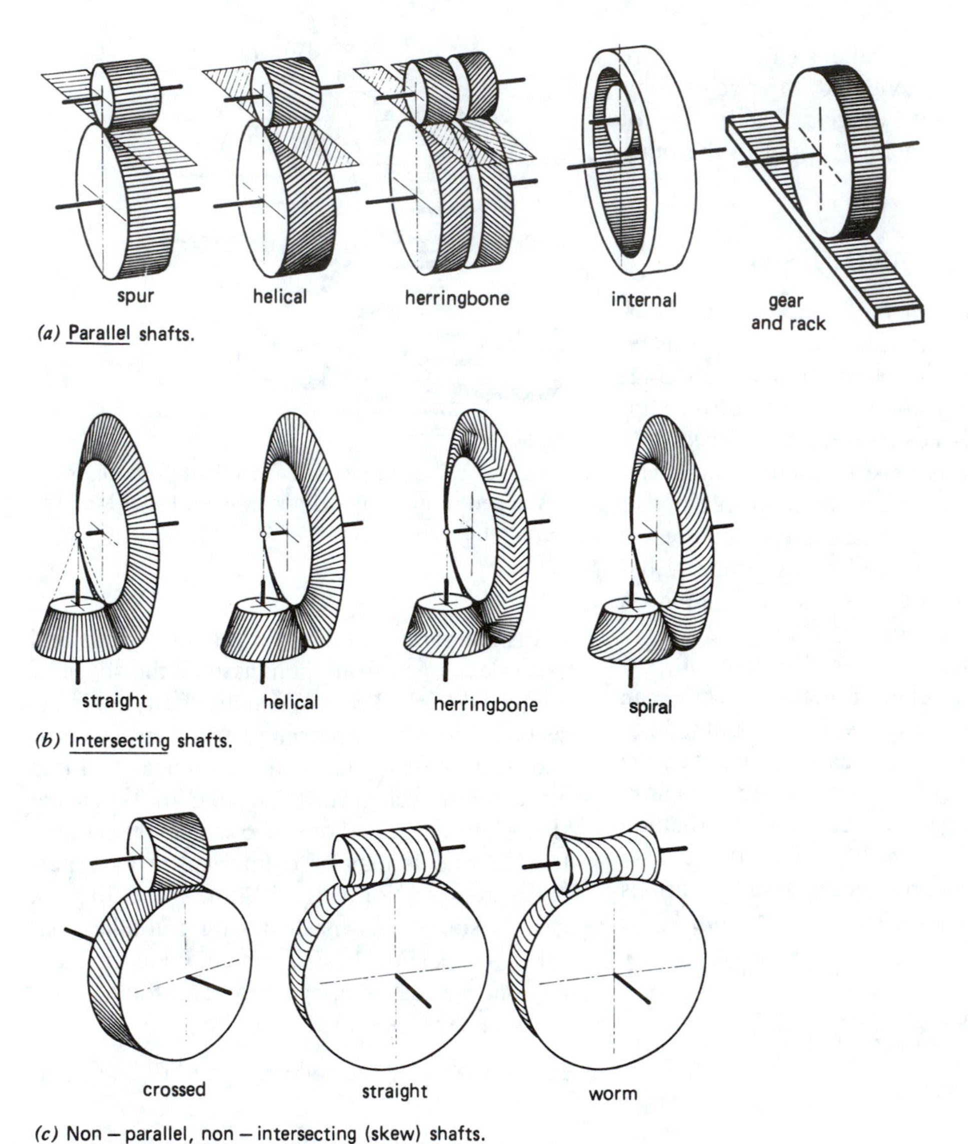

Figure 13-3 Important types of gears.
Toohtermann/Bodenstein, Konstruktions elemente des Maschinenbanes, Springer Verlag Berlin 1979

them the pitch or tooth spacing is the same for both engaging gears. The pitch curves are usually circles or straight lines, and the motion transmitted is either rotation or straight-line translation at a constant velocity.

Mating tooth profiles, as shown in Fig. 13-4, are essentially a pair of cams in contact (back to back). For one cam to drive another cam with a constant angular displacement ratio, the common normal at the point of contact must at all times intersect the line of centers at the *pitch point*. This *fixed* point is the point of tangency of the pitch circles. To ensure continuous contact and the existence of one and only one normal at

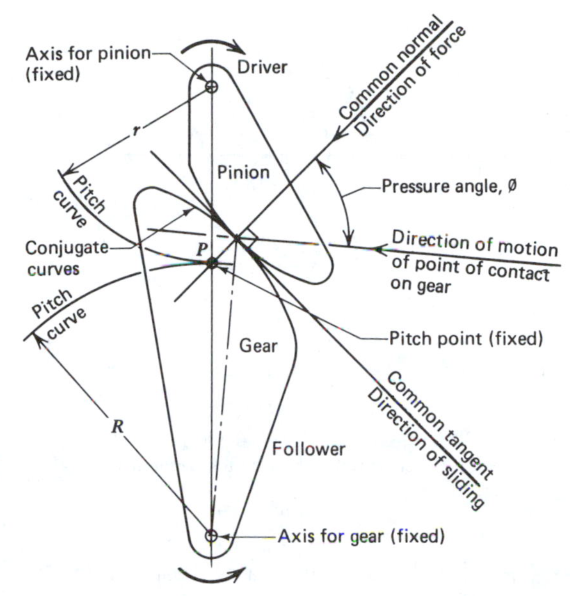

Figure 13-4 Two cams showing the law of gearing. The common normal must, for all useful positions, go through the fixed point *P*.

each point of contact, the camlike tooth profiles must be continuous differentiable curves.

Mating cam profiles that yield a constant angular displacement ratio are termed *conjugate*. Although an infinite number of profile curves will satisfy the law of gearing, only the cycloid and the involute have been standardized. The involute has several advantages; the most important is its ease of manufacture and the fact that the center distance between two involute gears may vary without changing the velocity ratio.

13-2 INVOLUTE GEAR PRINCIPLES[1]

An involute curve is generated by a point moving in a definite relationship to a circle, called the *base* circle. Two principles are used in mechanical involute generation. Figures 13-5*a* and 13-5*b* show the principle of the *fixed base circle*. In this method the base circle and the drawing plane in which the involutes are traced remain fixed. This is the underlying principle of involute compasses and involute dressers (where the motion of a diamond tool gives an involute profile to grinding wheels).

The second principle, that of the *revolving base circle,* is used in generating involute teeth by hobbing, shaping, shaving, and other finishing processes (Fig. 13-5*c*). This method is employed primarily where the generating tool and the gear blank are intended to work with each other like two gears in mesh for purposes of gear manufacturing.

Cord Method

In using this method the involute path is traced by a taut, inextensible cord as it unwinds from the circumference of the fixed base circle (Fig. 13-6). The radius of curvature starts at zero length on the base circle and increases steadily as the cord unwinds. After one revolution, the radius of curvature equals the circumference of the base circle (πD_b). *It is significant that the radius of curvature, to any point, is always tangent to the base circle and normal to one and only one tangent on the involute.*

From Fig. 13-6 it can be seen that the full involute curve is a *spiral* beginning at the base circle and having an infinite number of equidistant coils (distance πD_b). However, only a small part of the innermost cord is used in gearing. The character of the involute near the base reveals the existence of a *cusp* at point 0 and a *second* branch of the involute going in the opposite direction (shown as a dash-dot curve). The second branch serves, as we will see later, to form the back side of the gear tooth after a space is left for a meshing tooth.

Properties of the Involute

The following properties can be seen in Fig. 13-6.

---

[1]The material in Sections 13-2 to 13-4 was in part extracted from a gear manual formerly used at International Harvester, Farmall Works, courtesy Robert Custer and Frederick Brooks.

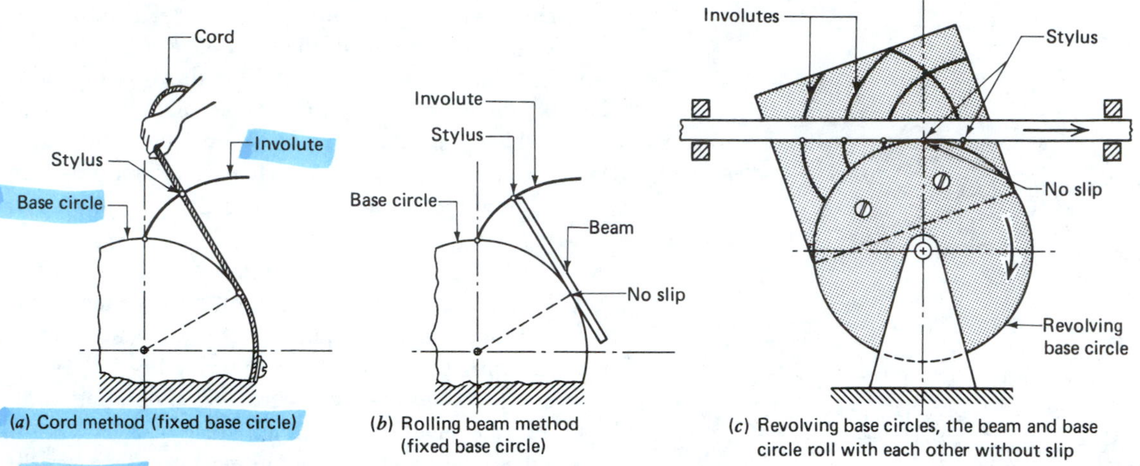

Figure 13-5 Generation of the involute.

1. Any tangent to the base circle is always *normal* to the involute.
2. The length of such a tangent is the radius of curvature of the involute at that point. The center is always located on the base circle.
3. For any involute there is only *one* base circle.
4. For any base circle there is a family of equivalent involutes, infinite in number, each with a different starting point.
5. All involutes to the same base circle are *similar* (congruent) and equidistant; for example, the distance of any two of such involutes in normal direction is constant πD_b (Fig. 13-6).
6. Involutes to *different* base circles are *geometrically similar* (Fig. 13-7). That is, corresponding angles are equal, while corre-

Figure 13-6 Generation of an involute by the cord method.

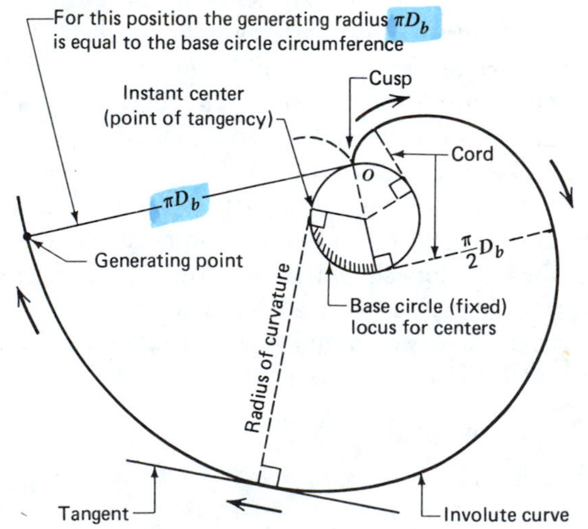

Figure 13-7 Geometric similarity of all involutes explains why the teeth of a large gear can mesh properly with those of a small gear.

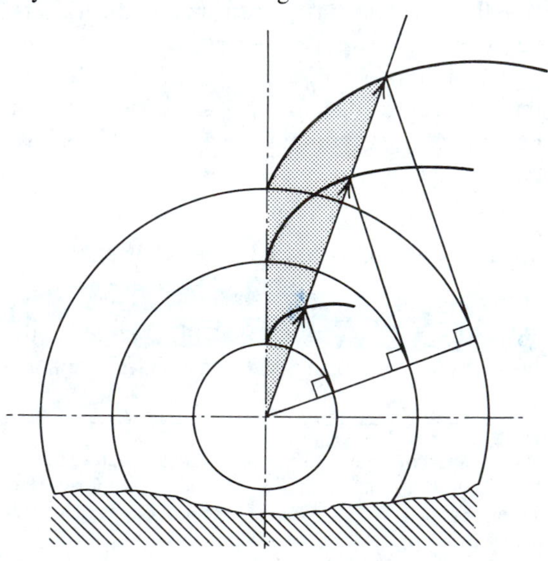

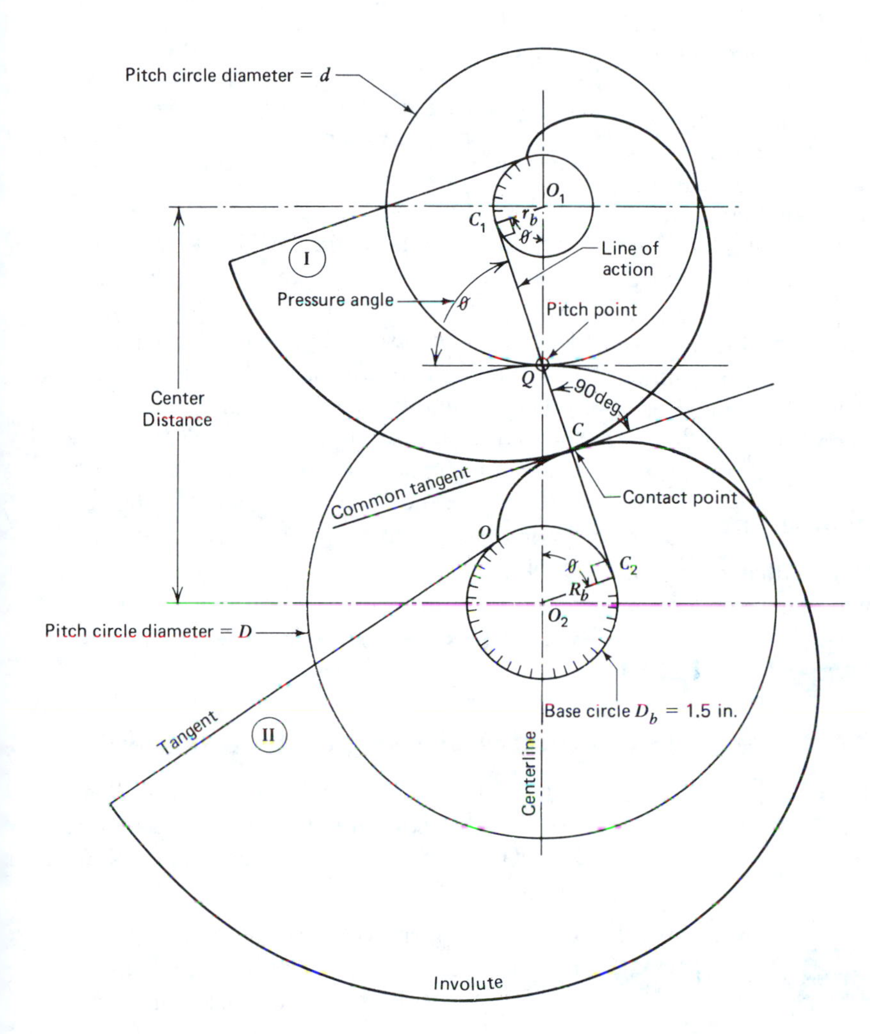

Figure 13-8 Curved involutes in contact.

sponding lines, curves, or circular sections are in the ratio of the base circle radii. Thus, when the radius of the base circle approaches infinity, the involute becomes a straight line. Geometric similarity explains why the teeth of a large gear can mesh properly with those of a small gear.

Involutes in Contact

Mathematically, the involute is a continuous, differentiable curve; that is, it has only one tangent and only one normal at each point. Thus two involutes in contact (back to back) have one common tangent and one common normal (Fig. 13-8). This common normal, furthermore, is a common tangent to the base circles. Since this normal for all positions intersects the centerline at a fixed point, conjugate motion is assured. Thus *conjugate motion* is the term used to describe this important characteristic of involute gear action.

The action portrayed is that of two oversize gear teeth in contact. The oversized teeth are the result of using too large a center distance. This situation is presented only for reasons of clarity.

When two involute gear teeth move in contact, there is a positive drive imparted to the two shafts passing through the base circle centers, thus ensuring shaft speeds proportional to the base circle diameters. This is equivalent to a positive drive imparted by an inextensible connecting cord as it winds onto one base circle and unwinds from the other. It is analogous to a pulley with a crossed belt arrangement. Note that the surfaces of both involutes at the point of contact are moving in the same direction.

A rack is a gear with its center at infinity. It is a simplified gear in which all circles concentric with the base circle and all involutes have become straight lines. A rack therefore has a *baseline* and a *linear* tooth profile.

Relationship of Pitch and Base Circles

Referring to Fig. 13-8, we find that triangles QC_1O_1 and QC_2O_2 are similar. Therefore

$$\cos \phi = \frac{r_b}{r} = \frac{R_b}{R} \tag{13-1}$$

where

r_b, R_b = base circle radius of pinion and gear, respectively; mm, in.
r, R = pitch circle radius of pinion and gear, respectively; mm, in.

Also,

$$\cos \phi = \frac{d_b}{d} = \frac{D_b}{D} \tag{13-2}$$

where

d_b, D_b = base diameters of pinion and gear, respectively; mm, in.
d, D = pitch diameters of pinion and gear, respectively; mm, in.

Figure 13-9 shows the smaller of the two gears in Fig. 13-8 meshing with a rack (obtained by moving the center of the larger gear to infinity). The rack is represented by a single tooth that can move horizontally, as shown. If the involute is turned counterclockwise the "rack" will move to the right because of a horizontal force component. The motion of rack and pinion is conjugate because the pitch point has not changed and the normal to the rack tooth goes through this point.

13-3 THE MECHANICS OF INVOLUTE TEETH

Effect of Changing Center Distance

Figure 13-10 shows the same two involutes as in Fig. 13-8 brought into contact through appropriate rotation on a reduced center distance (2 in.). Consequently:

1. A *new* pitch point was established.
2. The pressure angle was reduced from 70 to 50 deg (still large by normal standards).

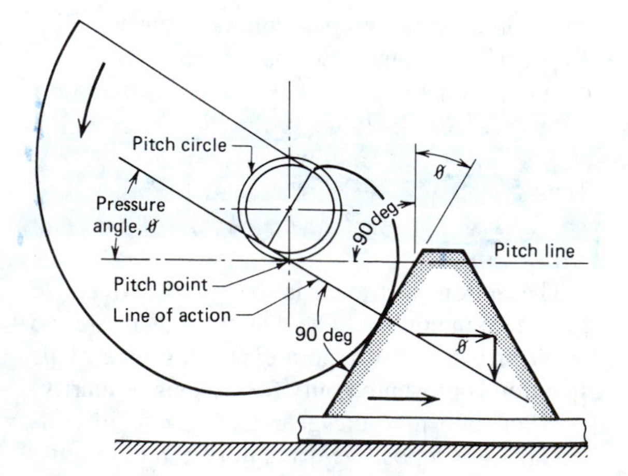

Figure 13-9 Curved involute contacting a flat surface (rack tooth).

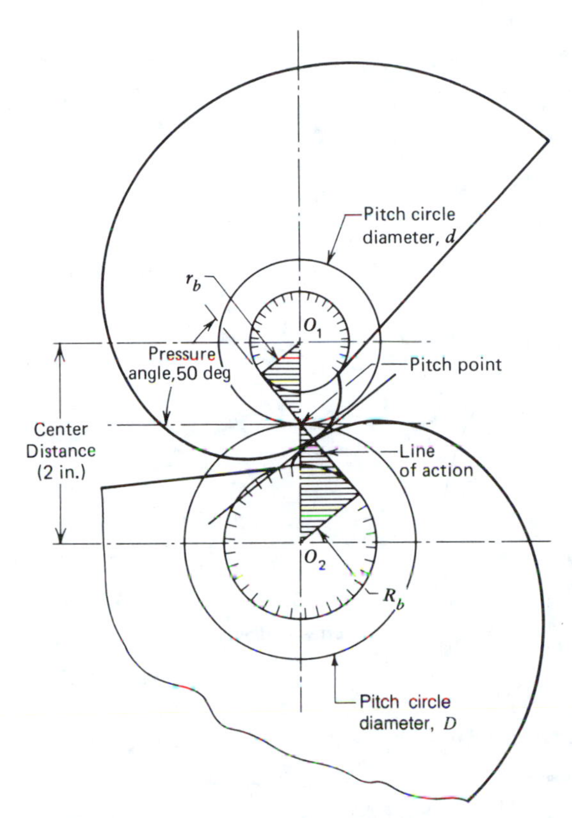

Figure 13-10 Effect of changing center distance.

3. The line of action was shortened.
4. The pitch diameters were reduced (halved), but their ratio remained unchanged (similar triangles).

The speed ratios are not affected by altering center distance because they are functions of base radii only. The two triangles (crosshatched) remain similar, regardless of center distance alterations. Furthermore, two corresponding sides, the base radii, do not change; hence the ratio of pitch radii *cannot* change.

Rolling and Sliding Action Between Contacting Involutes

Figure 13-11*a* returns the two involutes from Fig. 13-8 to their former, larger center distance (4 in.) but in a different relative position—the only position for which corresponding arc lengths are equal (arc 12 = arc 12′). By the belt analogy, an equal length of cord has been exchanged between the smaller base cylinder of the pinion and the larger base cylinder of the gear. The angular but *equal* displacements of the gear are therefore smaller than those of the pinion by a ratio of 1:1.5.

For clarity, the angular increments of the pinion were chosen as 22.5 deg, making those of the gear 15 deg. The length of arc corresponding to each pair of increments will be in contact during rotation. However, since each pair varies in length, the rolling motion of one involute on another inevitably must be accompanied by sliding because the time elapsed to cover corresponding but *unequal* lengths is the same.

For counterclockwise rotation of the pinion, the lengths of arc 11, 10, 9, and so forth, will be in *synchronized* contact with the arc 11′, 10′, 9′, and so on (Fig. 13-11*b*). The former steadily increase in length, but the latter steadily decrease in length, making *sliding* inevitable. Rotation in opposite direction produces the same effect, but this time gear teeth have the greater surface speed. Maximum sliding takes place when the point of contact is close to either base circle. Thus, in gearing, only a small section of any involute is useful if sliding is to be minimized.

13-4 GEAR TERMINOLOGY

Basic Terminology

To make further discussion more meaningful, geometric quantities resulting from involute contact will now be defined, discussed, and assigned nomenclature, as shown in Figs. 13-12 to 13-14.

Pinion. A pinion is usually the *smaller* of two mating gears. The larger is often called the gear.

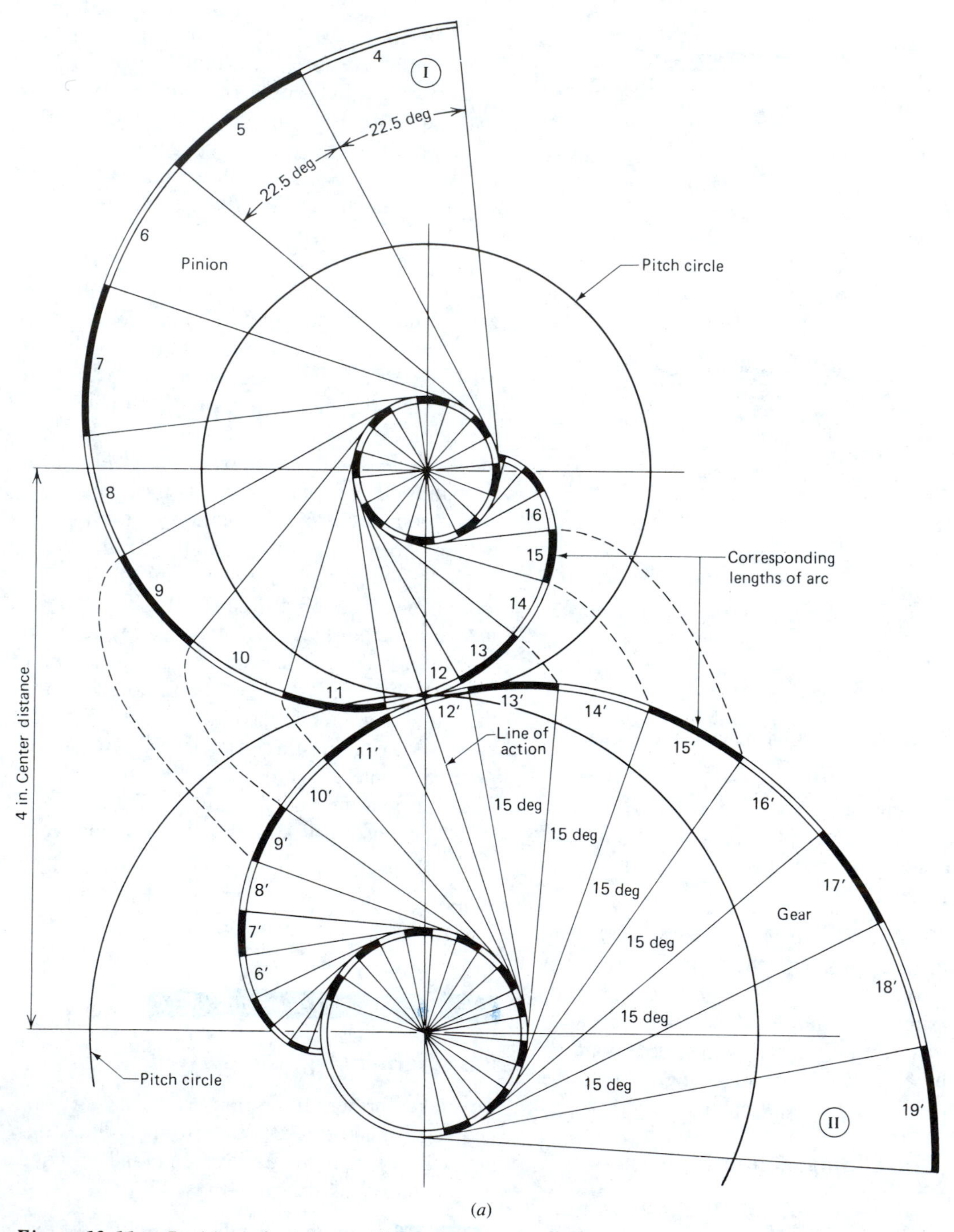

(a)

Figure 13-11a Position of minimum sliding is the one shown here where arc 12 equals arc 12′. For rotation in either direction, sliding action will increase until the contact point reaches the base circle.

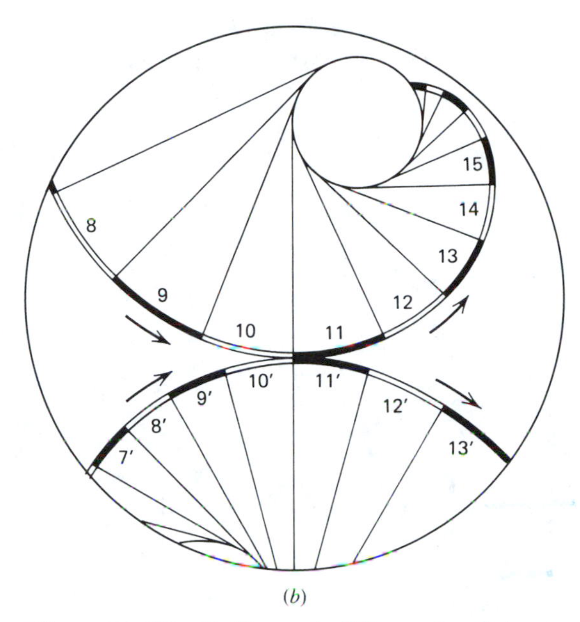

(*b*)

Figure 13-11b Rolling and sliding action between two involutes on fixed centers. Contact between arcs 10 and 10′ involves much sliding, since arc 10 is almost one-third longer than arc 10′.

Center distance (C). The distance between the centers of the pitch or base circles.

$$C = 0.5(D + d) \tag{13-3}$$

where

D = pitch diameter of gear
d = pitch diameter of pinion

Base circle. The circle from which an involute tooth curve is developed.

Base Pitch (p_b). The pitch on the base circle (or along the line of action) corresponding to the circular pitch.

Pitch Circle. Since the pitch point is fixed, only two circles, each concentric with a base circle, can be drawn through the pitch point. These two imaginary circles, tangent to each other, are the *pitch circles*. They are visualized as rolling on each other, without sliding, as the base circles rotate in conjugate motion.

The ratio of pitch diameters is also that of the base diameters (similar triangles). Because the pitch circles are tangent to each other, they are used in preference to the base circles in many of the calculations. Note that pitch circles must respond to any center distance variation, for a meshing gear pair, by enlarging or contracting. In contrast, the base circles never change size.

Circular Pitch (p_c). The identical tooth spacing on each of the two pitch circles.

Pressure Angle ($\emptyset$). The pressure angle lies between the common tangent to the pitch circles and the common tangent to the base circles, shown exaggerated in Fig. 13-8. The pressure angle is also the acute angle between the common normal and the direction of motion, when the contact point is on the centerline. Since a pair of meshing gear teeth is, in essence, a pair of cams in contact, the pressure angle of gearing is identical to the one encountered in cam design. The pressure angle of contacting involutes, as opposed to the one on

Figure 13-12 Gear geometry and terminology.

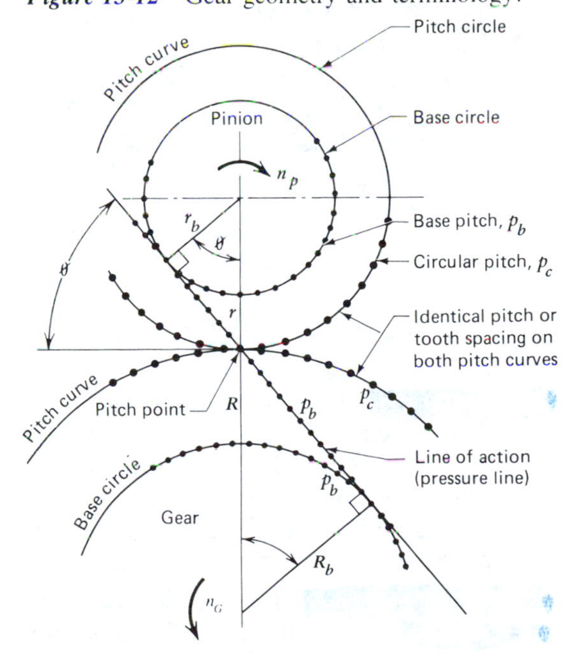

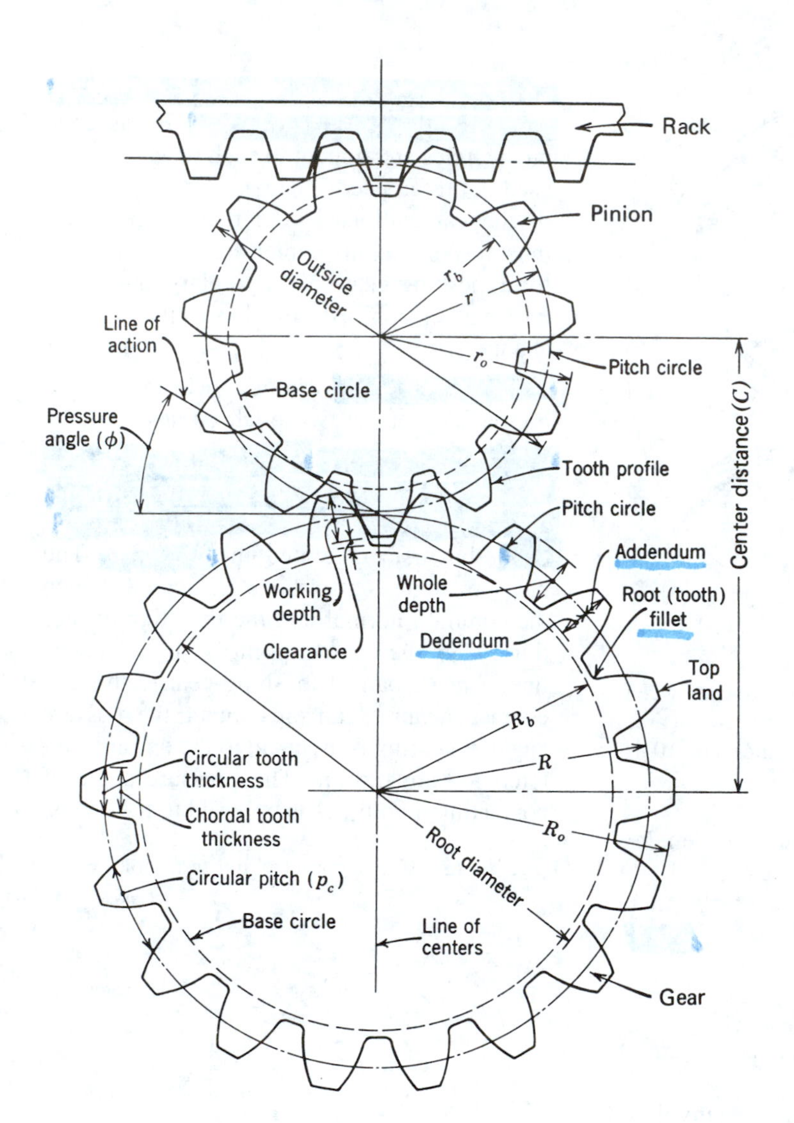

Figure 13-13 Spur gear geometry. (G. W. Michalec, *Precision Gearing,* Wiley, 1966.)

cams, is *constant* throughout its entire cycle, a feature of great practical importance.

Line of Action. This is the common tangent to the base circles. Contact between the involutes must be on this line to give smooth operation. Force is transmitted between tooth surfaces along the line of action. Thus a constant force generates a constant torque.

Velocity Ratio (m_G). This ratio, also called speed ratio, is the angular velocity of the driver divided by the angular velocity of the driven member. Because the line of action cuts the line of centers into the respective pitch radii, the speed ratio becomes the inverse proportion of those distances and related quantities (e.g., base and pitch diameters).

Because most gears are designed for speed reduction, one generally finds the pinion driving the gear; from now on we will assume that this is the case. Therefore

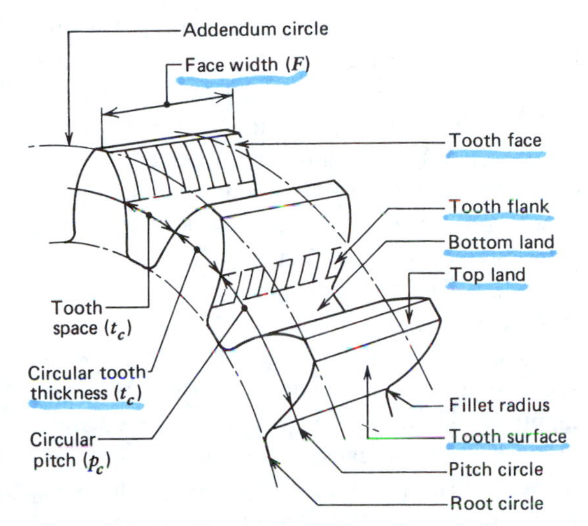

Figure 13-14 Tooth parts of spur gears.

$$m_G = \frac{n_P}{n_G} = \frac{D}{d} = \frac{N_G}{N_P} \tag{13-4}$$

where

m_G = speed ratio (gear ratio)
n_P = speed of pinion; rpm
n_G = speed of gear; rpm
D = pitch diameter of gear; mm, in.
d = pitch diameter of pinion; mm, in.
N_G = number of teeth in the gear
N_P = number of teeth in the pinion

In practice, speed ratios are determined principally from ratios of tooth numbers because they involve *whole* numbers only.

Tooth Parts

The following tooth parts are shown in Figs. 13-13 and 13-14.

Addendum. Height of tooth above pitch circle (Fig. 13-13).

Bottom land. The surface of the gear between the flanks of adjacent teeth (Fig. 13-14).

Dedendum. Depth of tooth below the pitch circle (Fig. 13-13).

Face width (F). Length of tooth in axial direction (Fig. 13-14).

Tooth face. Surface between the pitch line element and the top of tooth (Fig. 13-14).

Tooth fillet. Portion of tooth flank joining it to the bottom land (Fig. 13-13).

Tooth flank. The surface between the pitch line element and the bottom land (Fig. 13-14).

Tooth surface. Tooth face and flank combined (Fig. 13-14).

Top land. The surface of the top of the tooth (Fig. 13-14).

Circular tooth thickness (t_c). This dimension is the arc length on the pitch circle subtending a single tooth. For equal addendum gears the theoretical thickness is half the circular pitch (Fig. 13-14).

Overpins measurements (M). The pitch circle is an imaginary circle; hence pitch diameters cannot be measured directly. However, indirectly the pitch diameter of spur gears can be measured by the pin method. When spur gear sizes are checked by this method, cylindrical pins of known diameter are placed in diametrically opposite tooth spaces; or, if the gear has an odd number of teeth, the pins are located as nearly opposite one another as possible (Fig. 13-15). The measurement M over these pins is then checked by using any sufficiently accurate method of measurement. For further detail, see *Machinery's Handbook,* pp. 945–964.

Figure 13-15 Measurement over pins.

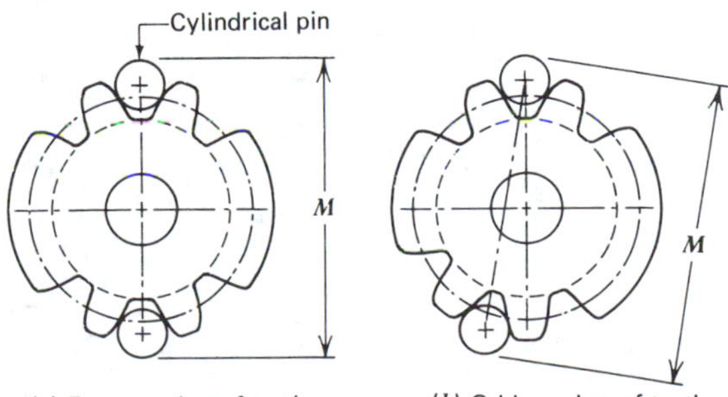

13-5 INVOLUTE GEAR TEETH

So far we have considered only two profiles in contact. However, successive revolutions are merely successive contacts of two profiles. Figure 13-16 shows how a series of symmetrical involute profiles, alternately clockwise and counterclockwise and with a tooth space allowed for meshing, will produce a complete set of pointed teeth. By using only the portion near the base circle, mating gear tooth profiles can be formed with two or more teeth in contact at all times, thus permitting continuous rotation in either direction.

Figure 13-17 shows in greater detail the development of curved and straight teeth. Continuous motion of a rack necessitates a series of short, symmetrical, equally spaced involute teeth on the base circle circumference. The pitch, in this case, is named base pitch (p_b). Point 1 is the point of tangency for the line of action. Thus the distances 1–2, 2–3, and the like, measured along the base circle, are all equal; by definition, this is the base pitch. From points 2 to 7 involutes have been extended until they intersect the line of action, dividing it into distances equal to the base pitch. From the equidistant points 1′ to 7′, lines have been drawn perpendicular to the line of action. The full line portion represents one side of the rack.

As the gear rotates, the gear teeth will contact successive rack teeth in a continuous, overlapping motion. The force will be exerted along the line of action, causing the rack to move horizontally. The pressure angle, as shown, is the acute angle between the directions of force and motion.

When the gear rotates in the direction shown, the surface of an involute gear tooth contacts the flat-surfaced rack tooth. As rotation continues,

Figure 13-16 Involute gear teeth are generated by a series of symmetrical involutes oriented alternately in a clockwise and counterclockwise direction. For the two gears to mesh properly, they must have the same base pitch.

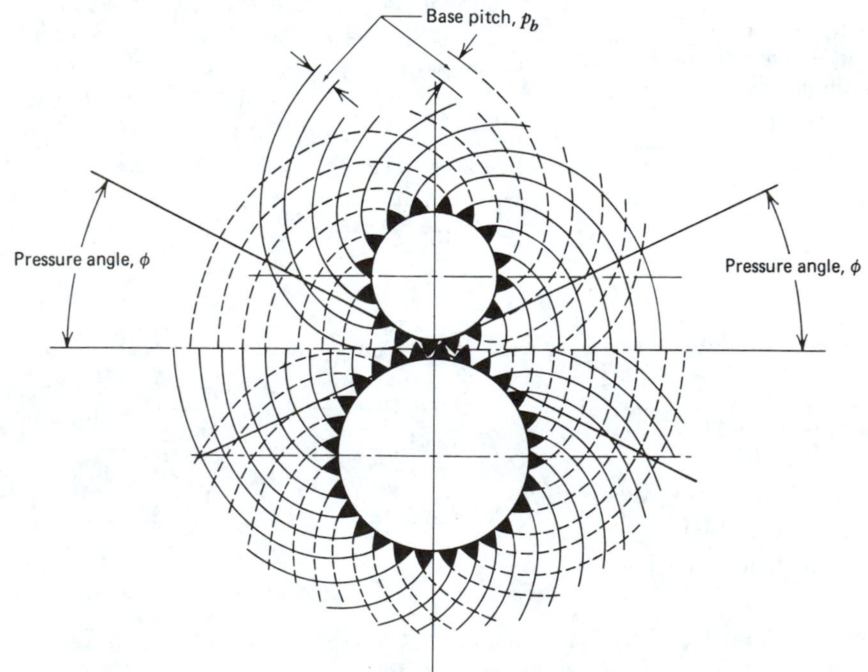

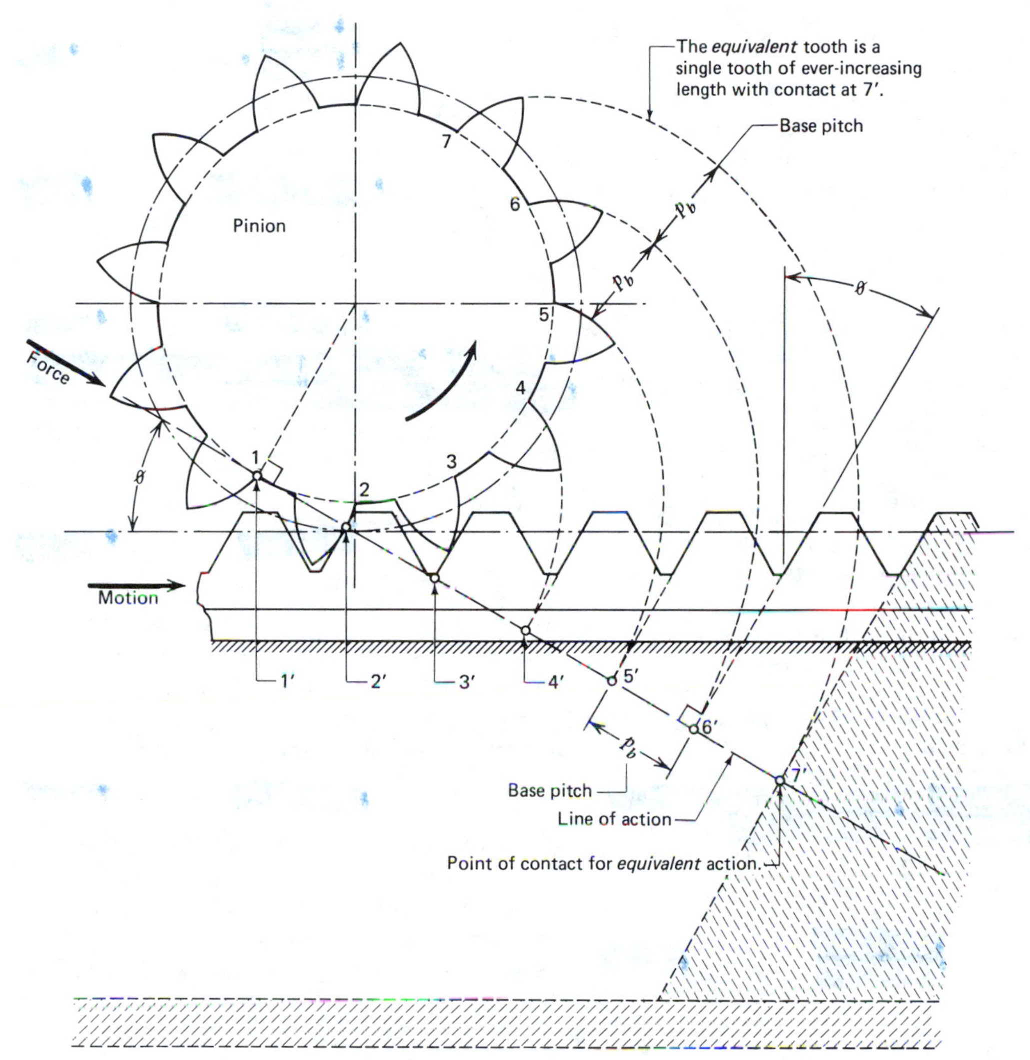

Figure 13-17 A succession of short symmetrical involutes gives continuous motion to a rack in either direction.

the contact points move down the line of action away from the base circle. This continues until the tooth surfaces lose contact at the upper end of the line of action represented by the full line. Before contact is lost, another pair of teeth come into contact, thus providing continuous motion. This tooth action is equivalent to the action of a single tooth of ever-increasing length contacting one ever-increasing flat surface along an ever-increasing line of action. The outward motion of the involute originating at point 7 mirrors the equivalent single tooth action. For the position shown, contact is at point 7′. Rack and pinion, like meshing gears, have two pressure lines and hence permit motion in both directions.

Summary of Involute Gears

The simplicity and ingenuity of involute gearing may be summarized as follows.

1. Involute profiles fulfill the law of gearing at any center distance.
2. All involute gears of a given pitch and pressure angle can be produced from one tool and are completely interchangeable.
3. The basic rack has a straight tooth profile and therefore can be made accurately and simply.

13-6 STANDARD SPUR GEARS

Spur gears can be made with greater precision than other gears because they are the least sophisticated geometrically. All teeth are cut across the faces of the gear blanks parallel to the axis, a procedure that greatly facilitates manufacturing and accounts for the relatively low cost of spur gears compared to other types. Spur gears are therefore the most widely used means of transmitting motion and are found in everything from watches to drawbridges.

Pitches and Modules

The *base pitch* (p_b) is the distance between successive involutes of the same hand, measured along the base circle. It is the base circle circumference divided by the number of teeth.

$$p_b = \frac{\pi D_b}{N} \qquad (13\text{-}5)$$

Mating teeth must have the same base pitch (Figs. 13-16 and 13-17).

The *circular pitch* (p_c) is the distance along the pitch circle between corresponding points of adjacent teeth. Meshing teeth must have the same circular pitch (Fig. 13-13). The pitch circle circumference is thus the circular pitch times the number of teeth.

$$p_c N = \pi D$$

$$p_c = \frac{\pi D}{N} \qquad (13\text{-}6)$$

Module: $$m = \frac{D}{N} \frac{\text{mm}}{\text{tooth}} \text{ (definition)} \qquad (13\text{-}7)$$

Diametral pitch: $$P_d = \frac{N}{D} \frac{\text{teeth}}{\text{in.}} \text{ (definition)} \qquad (13\text{-}8)$$

By substituting $D_b = D \cos \emptyset$ into Eq. 13-2, we obtain

$$p_b = p_c \cos \emptyset \qquad (13\text{-}9)$$

Diametral pitch is related to the module as follows.

$$mP_d = 25.4 \qquad (13\text{-}10)$$

Module, the amount of pitch diameter per tooth, is an index of tooth size. A higher module number denotes a larger tooth, and vice versa. Because module is proportional to circular pitch, meshing gears must have the same module.

$$p_c = \pi m \qquad (13\text{-}11)$$

Diametral pitch, the number of teeth per inch of pitch diameter, is also an index of tooth size. A large diametral pitch number denotes a small tooth, and vice versa. Because diametral pitch is inversely proportional to circular pitch, meshing gears must have the same diametral pitch.

$$P_d p_c = \pi \qquad (13\text{-}12)$$

The diametral pitch is the number of teeth per inch of pitch diameter and is not a pitch. A misnomer, it is easily confused with base and circular pitch. To avoid confusion, the word "pitch," when used alone from now on, refers solely to diametral pitch.

13-7 STANDARD TOOTH PROPORTIONS OF SPUR GEARS

Gears are standardized to serve those who want the convenience of *stock gears* or standard tools for cutting their own gears. To meet these needs, however, gear standards must provide users with sufficient latitude to cover their requirements. Optimum design requires a wide range of pitches and modules, but only a few pressure angles. There should also be an extensive choice in the number of teeth available. A practical range of stock gears is from 16 to 120 teeth with suitable incremental steps. The corresponding ratios vary from 1:1 to 7.5:1.

Pressure Angle

The preferred pressure angle in both systems—module and pitch—is 20 deg, followed by 25, 22.5, and 14.5 deg. The 20-deg angle is a good *compromise* for most power and precision gearing. Increasing the pressure angle, for instance, would improve tooth strength but shorten the duration of contact. Decreasing the pressure angle on standard gears requires more teeth in the pinion to avoid undercutting of the teeth.

Diametral Pitch System

This system applies to most gears made in the United States and is covered by AGMA standards. These standards are outlined in 65 technical publications available from AGMA.[2] For gear systems we have 201.02–1968: Tooth Proportions for Coarse-Pitch Involute Spur Gears.

Despite the rapid transition to SI by the mechanical industries, the change to the module system will probably be slower. The reasons are:

1. AGMA has yet to complete its SI standards.
2. Many existing gear hobs (tools for making gears), for reasons of economy, will be kept in service and not be replaced until worn out.
3. The need for repair of older gears will continue for several decades.

Thus future gear reduction units may be all metric except for the pitch system.

Selection of pitch is related to load and gear size. Optimum design is achieved by varying the pitch, but rarely the pressure angle; hence there is a wide selection of "preferred values" (Table 13-1). Small pitch values yield large teeth; large pitch values yield small teeth. Table 13-2 gives involute tooth dimensions based on pitch.

Module System

Tooth proportions for metric gears are specified by the International Standards Organizations (ISO). They are based on the ISO basic rack (not shown) and the module m. A wide variety of modules is available to cover every tooth size required from instrument gears to gears for steel mills. Table 13-3 shows only the preferred values ranging from 0.2 to 50 mm. Specific tooth dimensions are obtained by multiplying the dimension of the rack by the module (Table 13-4).

Because of the simple relationship between pitch and module ($mP_d = 25.4$), metrication of gearing does not seem overly difficult. However, *the transition from pitch to module rarely yields standard values*. Thus module gears are *not interchangeable* with pitch gears. Herein lies the difficulty of metric conversion in gearing.

EXAMPLE 13-1

The tools for cutting gear teeth according to the diametral pitch system range 10 to 13 are to be replaced by tools based on the module system. Find equivalent modules.

SOLUTION

From Eq. 13-10: $mP_d = 25.4$

$$P_d = 10, \qquad m = \frac{25.4}{10} = 2.54, \qquad \mathbf{m = 2.50}$$

$$P_d = 11, \qquad m = \frac{25.4}{11} = 2.31, \qquad \mathbf{m = 2.25}$$

$$P_d = 12, \qquad m = \frac{25.4}{12} = 2.12, \qquad \mathbf{m = 2.00}$$

$$P_d = 13, \qquad m = \frac{25.4}{13} = 1.95, \qquad \mathbf{m = 2.00}$$

---

[2] A complete index is available from the American Gear Manufacturers Association (AGMA), 19101 Fort Meyer Drive, Arlington, Virginia, 22209.

TABLE 13-1 National Pitch System

Coarse Pitch										
0.5	0.75	1	1.5	2	2.5	3	3.5	4	5	6
7	8	9	10	11	12	13	14	15	16	18
Fine Pitch										
20	22	24	28	30	32	36	40	44	48	
50	64	72	80	96	120	125	150	180	200	

TABLE 13-2 Involute Gear Tooth Dimensions Based on Pitch-Coarse

	Stub	Full Depth	Stub	Full Depth
Pressure angle, ∅ deg	20	20	25	25
Addendum, a	$0.80/P$	$1/P$	$0.80/P$	$1/P$
Dedendum, b	$1.00/P$	$1.25/P$	$1.00/P$	$1.25/P$
Tooth thickness, t_c (theoretical)	$\pi/2P$	$\pi/2P$	$\pi/2P$	$\pi/2P$

TABLE 13-3 Preferred Values for Module m (mm)

0.2	0.6	0.9	1.75	2.75	3.75	5	7	14	24	42
0.3	0.7	1.0	2.00	3	4	5.5	8	16	30	50
0.4	0.75	1.25	2.25	3.25	4.50	6	10	18	36	
0.5	0.80	1.50	2.50	3.50	4.75	6.5	12	20		

TABLE 13-4 Involute Gear Tooth Dimensions Based on Module m (mm)

	Stub	Full Depth	Stub	Full Depth
Pressure angle, ∅ deg	20	20	25	25
Addendum, a	$0.8m$	m	$0.8m$	m
Dedendum, b	$1.25m$	$1.25m$	$1.25m$	$1.25m$
Tooth thickness, t_c (theoretical)	$0.5\pi m$	$0.5\pi m$	$0.5\pi m$	$0.5\pi m$
Circular pitch, p_c	πm	πm	πm	πm

13-8 LIMITATIONS ON SPUR GEARS

Two spur gears will mesh properly, within wide limits, provided they have the same pressure angle and the same diametral pitch or module. Limitations are set by many factors, but two in particular are important: *contact ratio* and *interference*. To obtain the contact ratio, the length of action must first be introduced. The length of action (Z) or length of contact is the distance on an involute line of action through which the point of contact moves during the action of the tooth profiles. It is the part of the line of action located between the two addendum circles or outside diameters (Fig. 13-18).

Contact ratio (m_p)

As two gears rotate, smooth, continuous transfer of motion from one pair of meshing teeth to the following pair is achieved when contact of the first pair continues until the following pair has established initial contact. In fact, considerable overlapping is necessary to compensate for contact delays caused by tooth deflection, errors in tooth spacing, and center distance tolerances.

To assure a smooth transfer of motion, overlapping should not be less than 20%. In power gearing it is often 60 to 70%. Contact ratio, m_p, is another, more common means of expressing overlapping tooth contact. On a time basis, it is the number of pairs of teeth simultaneously engaged. If two pairs of teeth were in contact all the time, the ratio would be 2.0, corresponding to 100% overlapping.

Contact ratio is calculated as length of contact Z divided by the base pitch p_b (Fig. 13-18).

$$m_p = \frac{Z}{p_b} = \frac{\sqrt{R_0^2 - R_b^2} + \sqrt{r_0^2 - r_b^2} - C \sin \emptyset}{p_c \cos \emptyset} \qquad (13\text{-}13)$$

where

p_c = circular pitch; mm, in.
R_0 = outside radius, gear; mm, in.
R_b = base circle radius, gear; mm, in.
r_0 = outside radius, pinion; mm, in.
r_b = base radius, pinion; mm, in.
C = center distance; mm, in.
$\emptyset$ = pressure angle; deg

Figure 13-18 Contact ratio, m_p.

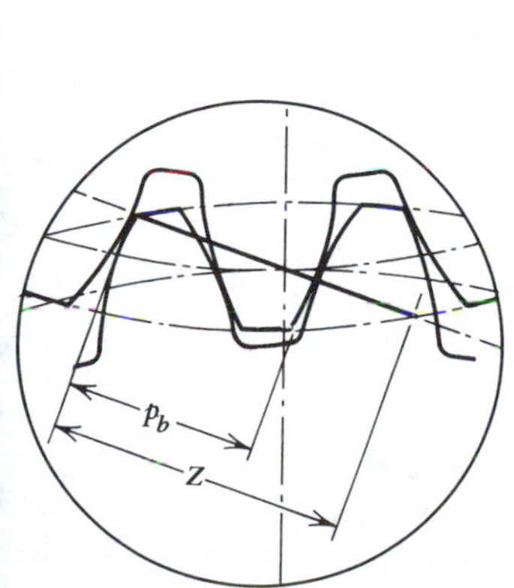

(*a*) Contact length Z and base pitch p_b

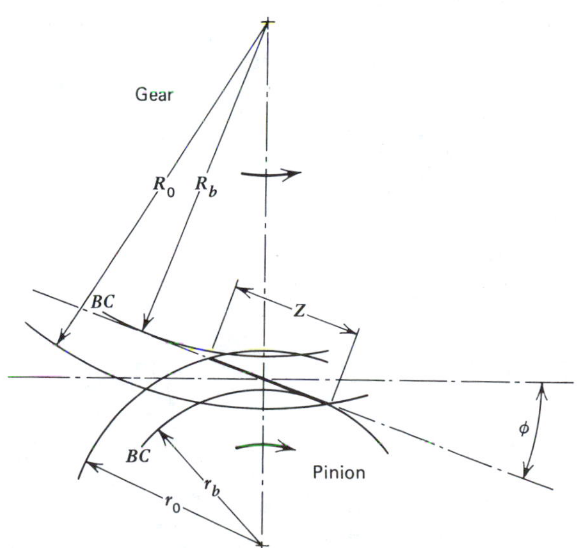

(*b*) Layout showing dimensions needed to calculate m_p

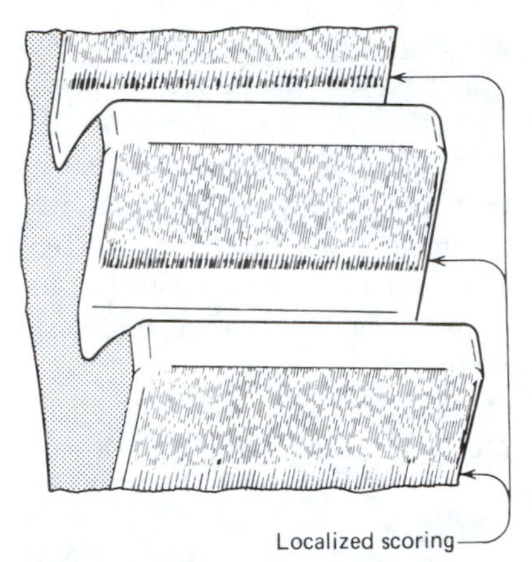

Figure 13-19 Tip and root interference. This gear shows clear evidence that the tip of its mating gear has produced an interference condition in the root section. Localized scoring has taken place, causing rapid removal in the root section. Generally, an interference of this nature causes considerable damage if not corrected. (Extracted from AGMA Standard Nomenclature of Gear Tooth Failure Modes (AGMA 110.04), with permission of the publisher, the American Gear Manufacturers Association.)

Note that base pitch p_b equals the theoretical minimum path of contact because $m_p = 1.0$ for $Z = p_b$.

Contact ratios should always be calculated to avoid intermittent contact. Increasing the number of teeth and decreasing the pressure angle are both beneficial, but each has an adverse side effect such as increasing the probability of interference.

Interference

Under certain conditions, tooth profiles overlap or cut into each other. This situation, termed interference, should be avoided because of excess wear, vibration, or jamming. Generally, it involves contact between involute surfaces of one gear and noninvolute surfaces of the mating gears (Fig. 13-19).

Figure 13-20 shows maximum length of contact being limited to the full length of the common tangent. Any tooth addendum extended beyond the tangent points T and Q, termed interference points, is useless and interferes with the

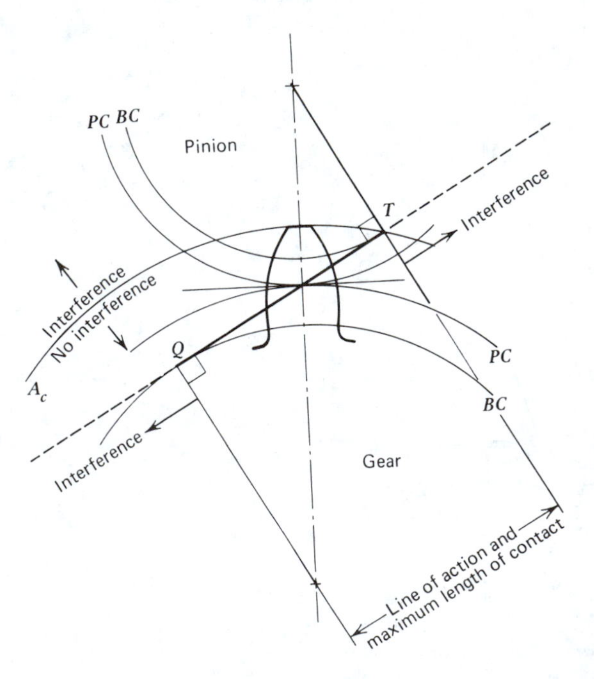

Figure 13-20 Interference sets a geometrical limitation on tooth profiles. For standard tooth forms interference takes place for contact to the right of point T and to the left of point Q.

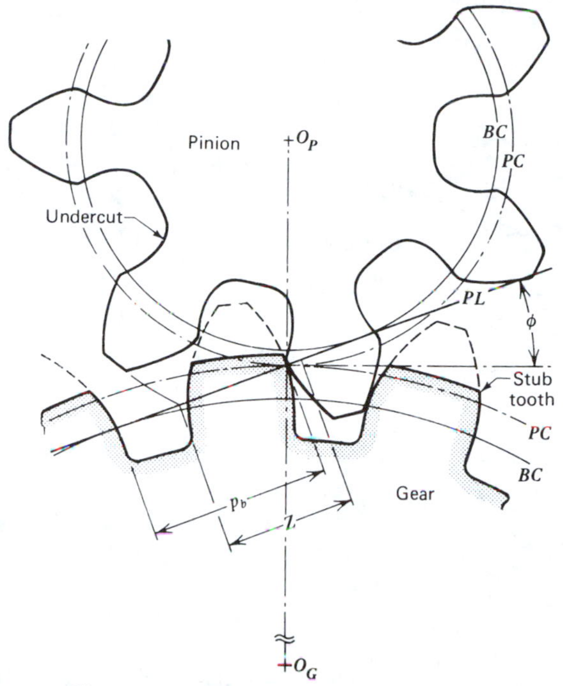

Figure 13-21 To operate without interference, either the pinion must be undercut or the gear must have stub teeth. Although interference is avoided, intermittent contact persists as p_b is greater than Z.

root fillet area of the mating tooth. To operate without profile overlapping would require undercut teeth. But undercutting weakens a tooth (in bending) and may also remove part of the useful involute profile near the base circle (Fig. 13-21).

Interference is first encountered during "approach," when the tip of each gear tooth digs into the root section of its mating pinion tooth. During "recess" this sequence is reversed. Thus we have both tip and root interference as shown in Fig. 13-19. Because addenda are standardized ($a = m$), the interference condition intensifies as the number of teeth on the pinion decreases. The pinion in Fig. 13-21 has 7 teeth. The minimum number of teeth N_c in a pinion meshing with a rack to avoid undercut is given by the expression

$$N_c = \frac{2}{\sin^2 \phi} \tag{13-14}$$

The minimum number of teeth varies inversely with the pressure angle. By increasing the pressure angle from 14.5 to 20 deg, the limiting number drops from 32 to 17. The corresponding increase in the maximum speed ratio potential indicates one of several reasons why the 20 deg pressure angle is preferred in power gearing.

Interference can be avoided if:

$$R_0 \leq \sqrt{R_b^2 + C^2 \sin^2 \phi} \tag{13-15}$$

$$r_0 \leq \sqrt{r_b^2 + C^2 \sin^2 \phi} \tag{13-16}$$

For a given center distance, an increase in pressure angle, with the resulting decrease in base radius, lengthens the involute curve between the base and pitch circles, thereby diminishing interference (Fig. 13-22).

When stock gears to suit a specific ratio are selected, it may not be sufficient to provide gears of the same module, pressure angle, and width. A pair must also have an acceptable contact ratio and mesh without interference.

EXAMPLE 13-2

Two identical 20-deg, full-depth gears, $m = 3$ and $N_G = N_p = 17$, are available from stock for a particular application. Would you recommend them for continuous operation?

SOLUTION

The gears should be checked for contact ratio and possible interference.

$$R = (0.5)(17)(3 \text{ mm}) = 25.5 \text{ mm}$$

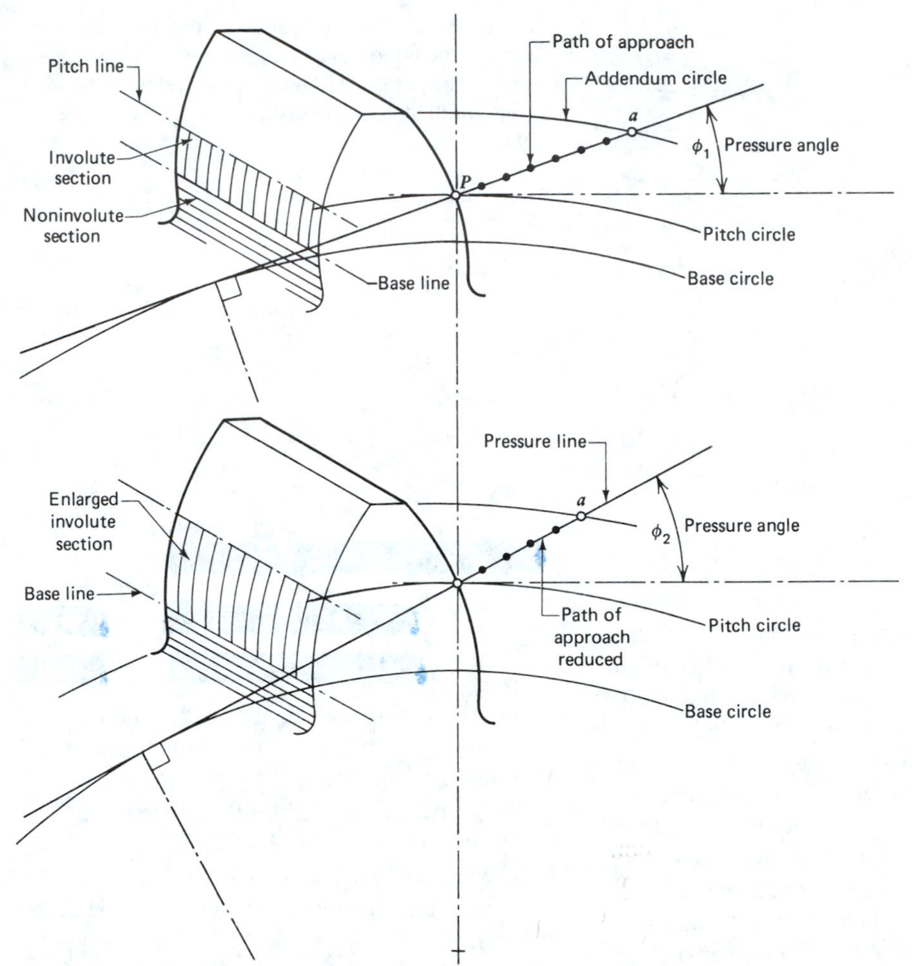

Figure 13-22 Effect of changing the pressure angle. Interference and contact ratio vary inversely with the pressure angle. When the pressure angle increases from ϕ_1 to ϕ_2, the involute section between the pitch line and the base line lengthens, tending to alleviate interference. The path of contact, however, shortens, thereby effectively lowering the contact ratio. (Only the path of approach is shown.)

$$R_0 = (0.5)(17 + 2)(3 \text{ mm}) = 28.5 \text{ mm}$$

$$R_b = (25.5 \text{ mm}) \cos 20 \text{ deg} = 23.96 \text{ mm}$$

$$C = (2)(25.5 \text{ mm}) = 51 \text{ mm}$$

Substitution into Eq. 13-13 yields:

$$m_p = \frac{2\sqrt{28.5^2 - 23.96^2} - 51 \sin 20 \text{ deg}}{3\,\pi \cos 20 \text{ deg}}$$

$m_p = \mathbf{1.51}$ This is acceptable because $m_p \gg 1.20$ (corresponding to the 20% minimal acceptable overlapping). Substitution into Eq. 13-15 or 13-16 yields:

$$R_0 \leq \sqrt{23.96^2 + 51^2 \sin^2 20 \text{ deg}} = 29.63 \text{ mm}$$

$$R_0 = \mathbf{28.5\ mm} < 29.63 \text{ mm}$$

Clearly, 17 teeth is the limit, as stated.

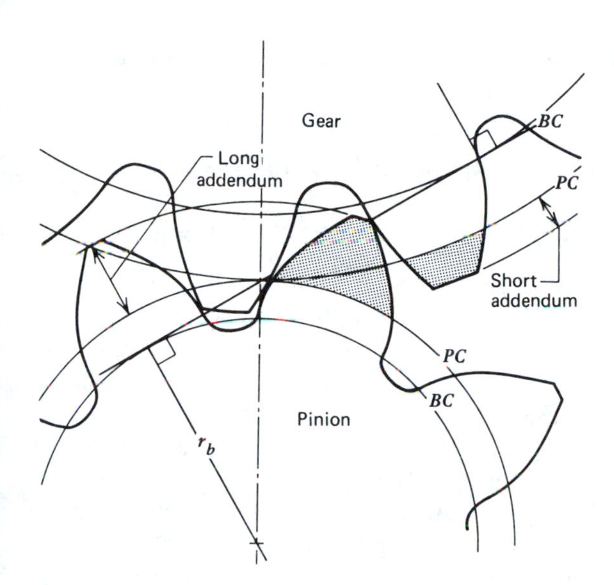

Figure 13-23 Long and short addenda.

Other limitations on spur gears are set by speed and noise level. When standard spur gears mesh, overlapping is less than 100%. The transmitted load is therefore briefly carried by one tooth on each gear. The sudden increase in load causes deflection of both meshing teeth and thus affects gear geometry adversely. The ideal constant velocity is no longer achieved. At low speed, this is not a serious factor but, as speed and load increase, deformation and impact may cause noise and shock beyond acceptable limits. Consequently, spur gears are seldom used for pitch line velocities exceeding 50 m/s (10,000 fpm).

13-9 MODIFICATIONS OF SPUR GEARS (NONSTANDARD)

The teeth of a pinion will always be weaker than those of the gear when standard proportions are used. They are narrower at the root and are loaded more often. If the speed ratio is three, each pinion tooth will be loaded three times as often as any gear tooth. Furthermore, if the number of teeth is less than the theoretical minimum, undercutting—with its resulting loss of strength—cannot be avoided. These adverse conditions can be circumvented by specifying nonstandard addenda and dedenda.

Long and Short Addenda[3]

In order to strengthen the pinion tooth, avoid undercutting, and improve the tooth action, its dedendum may be decreased and the addendum increased correspondingly. In practice, this is done by retracting the gear cutter a predetermined distance from its standard setting prior to cutting. Each pinion tooth becomes thicker and therefore stronger (Fig. 13-23). For such pinions to mesh properly with the driven gear, on the *same* center distance, the addendum of each driven tooth is correspondingly decreased and its dedendum increased. Although the gear teeth have thus become weaker, the net effect has been one of equalizing tooth strengths. The increased outer diameter of the pinion and decreased outer

[3]*Profile shifted gears* is an alternate term.

diameter of the gear have been achieved without changing the pitch diameters.

Extended Center Distance

In this arrangement a modified pinion is meshed with a standard gear. Pinions with decreased dedenda and increased addenda have thicker teeth than equivalent standard gear teeth. They also provide less space for any mating gear tooth. Consequently, proper mesh requires a larger center distance.

Both modifications are widely used because they can be achieved by means of standard cutters. A different setting of the generating tool is all that is required.

Backlash (B) (tooth thinning), in general, is play between mating teeth (Fig. 13-24). It occurs only when gears are in mesh. In order to measure and calculate backlash, it is defined as the amount by which a tooth space exceeds the thickness of an engaging tooth. The general purpose of backlash is to prevent gears from jamming together (making contact on both sides of their teeth simultaneously). Backlash also compensates for machining errors and heat expansion. It is obtained by decreasing the tooth thickness and thereby increasing the tooth space or by increasing the center distance between mating gears.

These modifications will improve primarily the kinematics of spur gears. Modifications aimed at improving the dynamic conditions of spur gears will be discussed in Section 13-14.

Figure 13-24 Backlash.

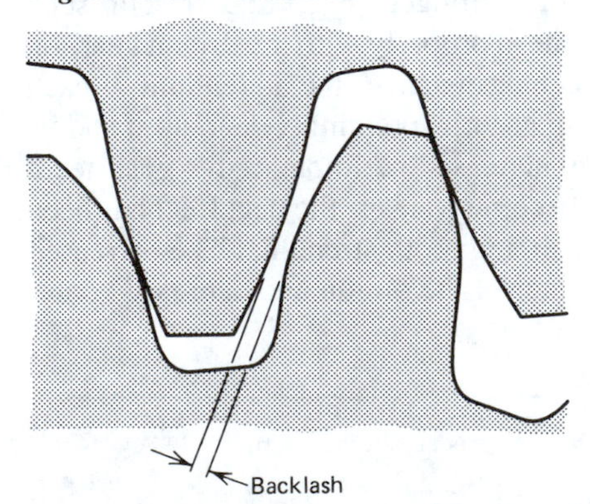

13-10 SPUR GEAR FORCE ANALYSIS

Fundamentals of Gear Calculations

By today's standards, early industrial gears were made of brittle materials; tooth breakage was the dominant mode of failure. This problem was largely overcome in 1893 when Wilfred Lewis presented his pioneering equation for estimating tooth strength in bending. From then on, tooth breakage could be predicted and controlled.

Soon, however, with the advent of automobiles and other machines in which power is transmitted continuously, wear became the dominant mode of failure, and Earl Buckingham developed his famous equation for wear.

Both methods have since undergone numerous refinements and have become the standard methods of AGMA for sizing gears. These methods are covered in *Machinery's Handbook* pp. 810–821, so the material from this source will be used in solving numerous problems at the end of this chapter. The mechanics of tooth breakage and pitting will be explained in the following sections so that the role of each of the many derating factors involved may be understood.

The objective of gear calculations is to provide basic manufacturing data: module or pitch, number of teeth, face width, gear blank diameter, and surface hardness for given or selected materials. Calculations are based on the two primary modes of failure.

1. *Tooth breakage,* a bending fatigue phenomenon caused by the bending action of a load (Fig. 13-25).
2. *Pitting,* a subsurface fatigue phenomenon induced by contact stresses exceeding the material fatigue limit (Fig. 13-26).

Resistance to breakage is called *tooth strength; surface durability* is the term for resistance to pitting.

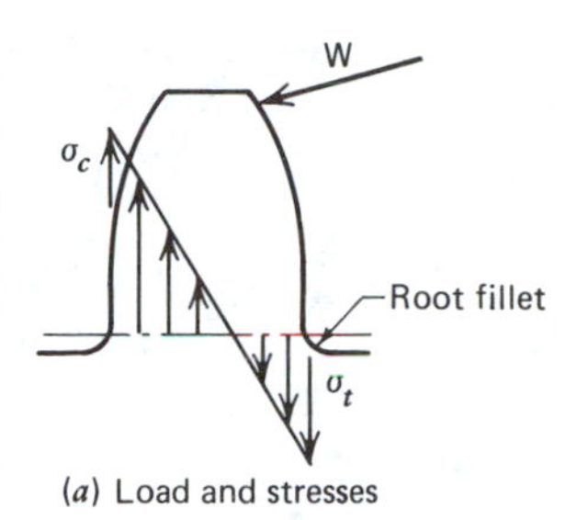

(*a*) Load and stresses

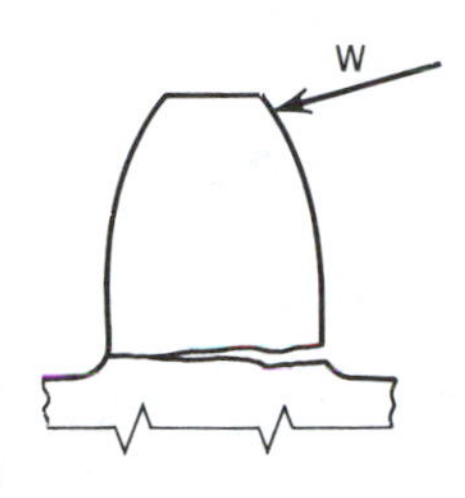

(*b*) Fracture at the root fillet

Figure 13-25 Tooth breakage.

The loads affecting gear teeth are often more severe than those encountered in chain sprockets because the load is not shared by as many teeth and because it acts on a smaller moment arm. In addition to tangential and service loads, both found in chain drives, gear designers must accommodate large dynamic loads caused by higher pitchline speeds, minute tooth errors, and small deviations in center distance.

Gear calculations are typical of design procedures because they involve more unknowns than equations. As with springs, these calculations are a series of approximations following well-established procedures. Preferred values for module or pitch and pressure angle, plus limitations on tooth number and gear width, impose additional restraints.

Static Gear Forces

Load is transmitted through contact *across* the active tooth face, as shown in Fig. 13-27*a*. This fairly uniform load can be replaced by a single force W. In transmitting power the driving force W acts along the line of action so that the tooth on the driven gear experiences a force moving from tip to base, as shown in Fig. 13-27*b*. Although contact with each tooth is brief, the sequence is rapid and the net effect is a constant force W acting along the pressure line and generating a constant shaft torque (Fig. 13-27*c*).

The normal or driving force W can be resolved into two components: (1) a tangential or transmitted force W_t, and (2) a radial or separating force W_r (Fig. 13-27*d*). For simplicity, we assume that both act at the pitch point. The components are:

$$W_t = W \cos \phi \qquad (13\text{-}17)$$

$$W_r = W \sin \phi = W_t \tan \phi \qquad (13\text{-}18)$$

Although a waste load, the separating force W_r is important in that it contributes to shaft bending and to the magnitude of bearing loads; it contributes nothing to transmitted power from the gear pair. The tangential or transmitted load W_t, in contrast, is used in calculating beam strength and dynamic load. Note that for a pressure angle of 20 deg, the transmitted load W_t is only slightly less than the normal force W ($W_t = 0.94W$). The

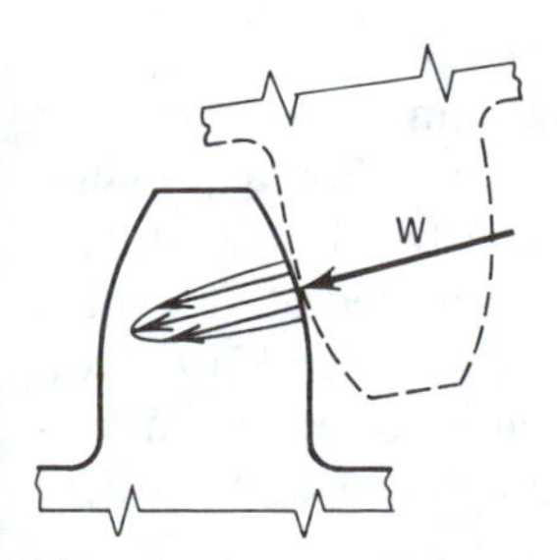

(*a*) Load and compressive stress

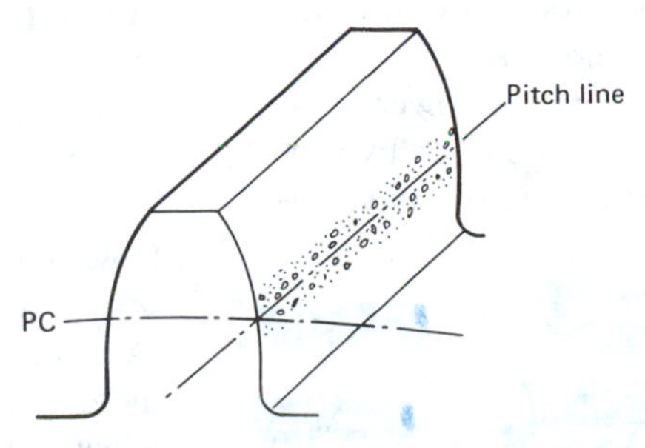

(*b*) Pit-marks start at the pitch line and congregate around it as pitting progresses.

Figure 13-26 Pitting.

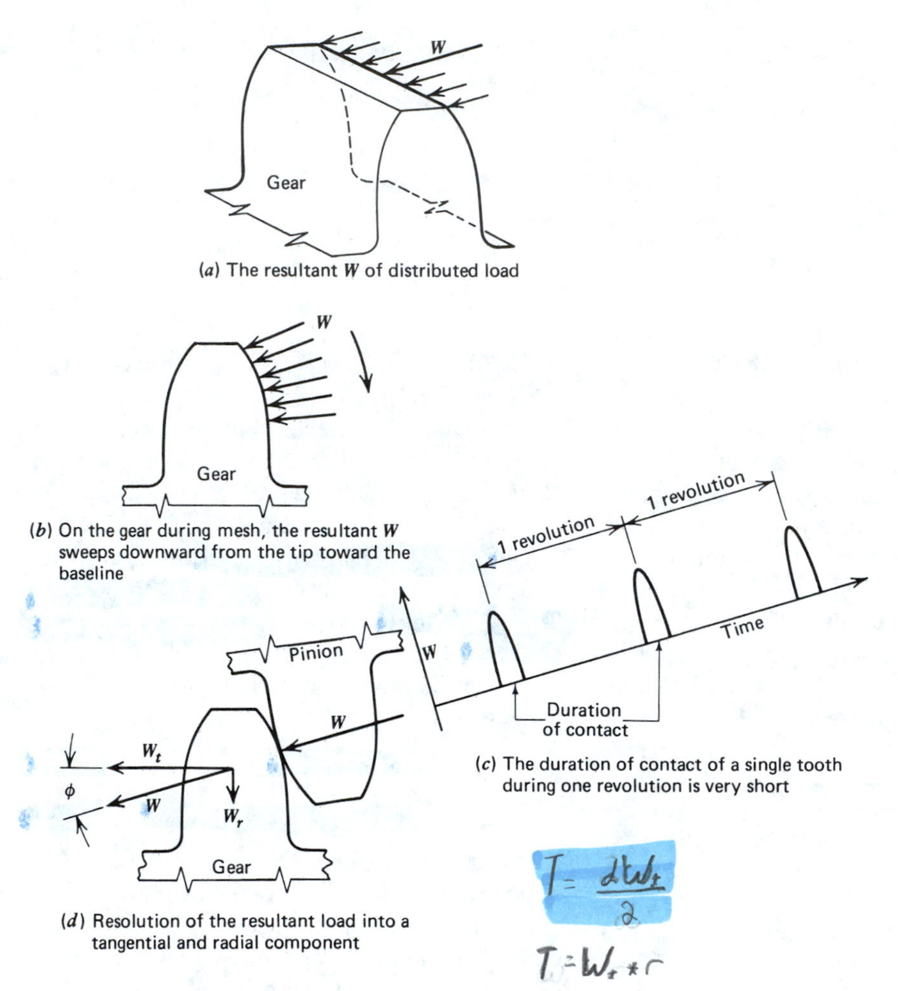

Figure 13-27 Static gear forces.

separating load W_r is only one-third of the normal force ($W_r = 0.34W$). Since W_t does useful work, this is a most fortunate development.

W_t is obtained indirectly from the power equations $P = Tn/9550$ or $P = Tn/63{,}000$ by substituting $T = 0.5d\ W_t$ where d is the pitch diameter. Thus

$$W_t = \frac{19\ 100P}{dn}\ \text{N} \qquad (13\text{-}19)$$

$$W_t = \frac{126{,}000P}{dn}\ \text{lb} \qquad (13\text{-}20)$$

EXAMPLE 13-3

A spur pinion, $d = 100$ mm, has a torque of 225 N·m applied to it. (*a*) For a pressure angle of 20 deg, calculate W_t, W, and W_r. (*b*) How much power is transmitted at 750 rpm? (*c*) What is the transverse shaft load? (*d*) Indicate the percentage change of W when Ø is increased to 25 deg or reduced to 14.5 deg. What is your conclusion?

SOLUTION

(*a*) $W_t = \dfrac{T}{0.5d}$ (from definition of torque)

$$= \frac{225\ \text{N}\cdot\text{m}}{0.5(0.100\ \text{m})} = \mathbf{4500\ N}$$

From Eq. 13-17: $W = \dfrac{4500\ \text{N}}{\cos 20\ \text{deg}}$

$$= \mathbf{4789\ N}$$

From Eq. 13-18:

$W_r = (4789\ \text{N}) \sin 20\ \text{deg.} = \mathbf{1638\ N}$

(*b*) From Eq. 1-6:

$$P = Fv 10^{-3}\ \text{kW}$$
$$= (4500\ \text{N})v 10^{-3}\ \text{kW}$$

$$v = \pi\, dn$$
$$= \pi(0.10\ \text{m})(750\ \text{rpm})(60^{-1}\ \text{min/s})$$
$$= 3.93\ \text{m/s}$$

$$P = (4500\ \text{N})(3.93\ \text{m/s})10^{-3}$$
$$= \mathbf{17.69\ kW}$$

(*c*) The transverse shaft load is $\boldsymbol{W} = \mathbf{4789\ N}$.

(*d*) The change of W depends solely on the change of cos Ø.

$$RC(\cos Ø) = \frac{\cos 20\ \text{deg} - \cos 25\ \text{deg}}{\cos 20\ \text{deg}} 100$$

$$= \mathbf{3.6\%}$$

$$RC(\cos Ø) = \frac{\cos 20\ \text{deg} - \cos 14.5\ \text{deg}}{\cos 20\ \text{deg}} 100$$

$$= \mathbf{3.0\%}$$

A change in pressure angle by its effect on the driving force W has only negligible effect on the power transmission capacity of involute gears. As we will see later, a larger pressure angle yields a stronger tooth and thus greatly increases the power capacity.

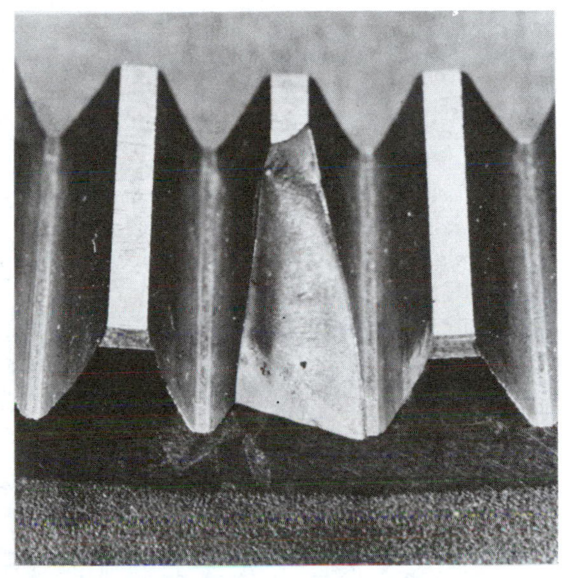

Figure 13-28 Tooth breakage. A typical fatigue failure with its smooth zone. The failure probably resulted from a heavy load or shock load (improper shifting or clutching). (Courtesy Deere and Company Technical Services.)

13-11 SPUR GEAR STRESS ANALYSIS

Tooth Breakage

Tooth fracture, unlike profile pitting, is catastrophic. It usually cannot be alleviated by reducing load or changing lubricant. This mode of failure, in which major parts of a tooth break off, is shown in Fig. 13-28. It may be a static fracture brought about by a sudden overload. More often it is a fatigue fracture—the result of millions of accumulated cycles of bending stress at the root of the tooth that barely exceed the material fatigue limit. Usually failure starts with a microscopic crack in the tooth fillet on the tensile side. In time, the crack slowly opens up across the root sectional area. The tooth, having been greatly weakened, may now fail due to a sudden overload.

Breakage may sometimes be caused by misalignment that places a disproportionate share of the load at one end of the tooth. Misalignment, in turn, can often be traced to loose bearings or excessive shaft deflection.

Beam Strength

By treating gear teeth as stubby cantilever beams, bent in one direction only, Lewis was able to calculate, with fair accuracy, the maximum induced bending stress for any given load and tooth size (Fig. 13-29). A check of the induced stress against the allowable fatigue stress for a given material would disclose the margin of safety. For a given load and material the method would also provide the size of a tooth.

The Lewis equation rests on the following simplifying assumptions.

1. The entire load is carried by one tooth (implies a contact ratio of one).
2. The load is evenly distributed across the full tooth width (disregards misalignment).
3. The uniformly distributed load can be replaced by a single static force acting at the top of the tooth only, the position that pro-

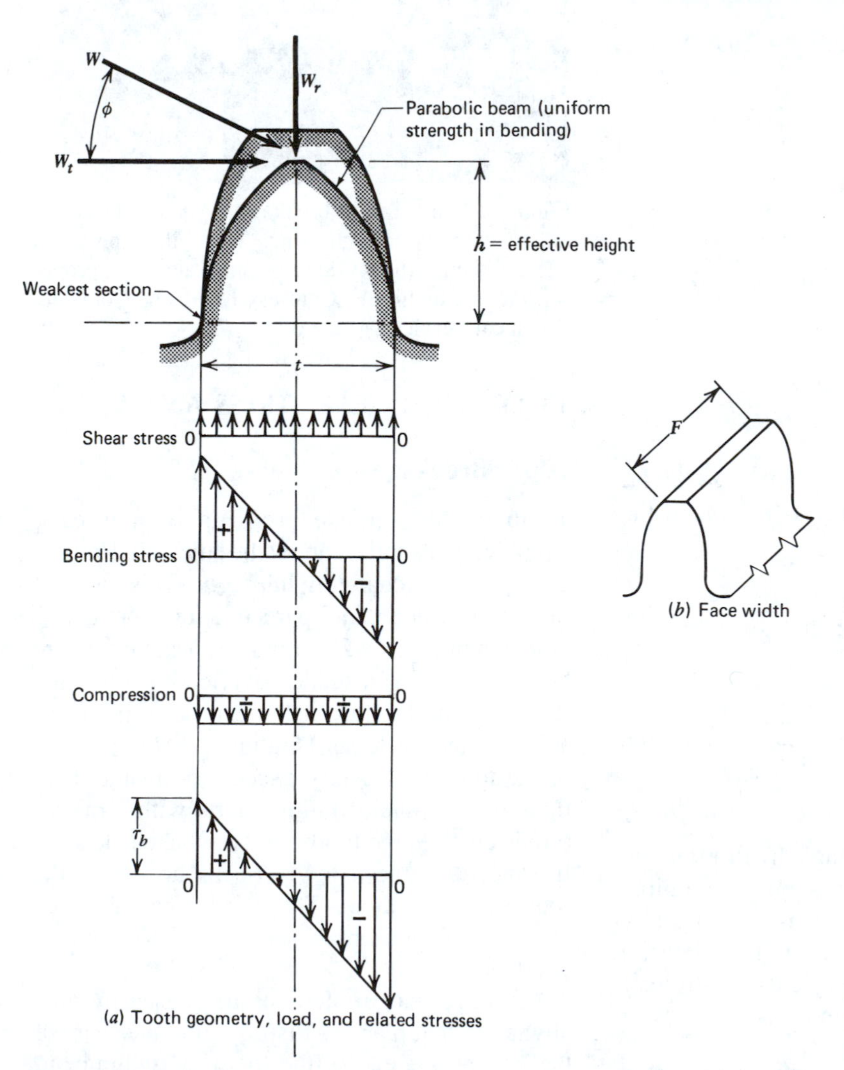

Figure 13-29
Beam strength based on the Lewis Formula.

duces maximum induced stress (disregards combined rolling and sliding of mating teeth).

4. The involute tooth, in calculating the induced stress, can be replaced by a shorter tooth of parabolic shape. This parabola represents the outline of a cantilever beam of uniform stress. Because the involute tooth is wider than the parabolic tooth, its maximum stress will occur at the common points of tangency at the root fillet and equal that of the parabolic beam. (This shortcut disregards stress concentration at the root fillet.)
5. Friction forces can be disregarded because their direction is largely radial.

The Lewis equation is obtained by calculating the induced bending stress at the fillet (Fig. 13-29*a*), thus

$$\sigma_b = \frac{Mc}{I} = \frac{W_t h(0.5t)12}{Ft^3} \qquad (13\text{-}21)$$

where

h = effective height (parabolic tooth); in.

t = root thickness of tooth; in.

F = tooth width (Fig. 13-29*b*); in.

$M = W_t h \doteq$ bending moment; lb-in.

$c = 0.5t$ = distance to outer fiber; in.

$I = Ft^3/12$ = moment of inertia; in.[2]

Since there is *one equation*, containing *five variables*, a succession of simplifying steps will be necessary before one can "size" the pinion by computing the diametral pitch.

First, introduce the circular pitch p_c by multiplying and dividing by p_c. Following rearrangement of terms, single out a new quantity y: the *Lewis form factor*.

$$\sigma_b = \frac{6W_t h p_c}{Ft^2 p_c}$$

$$= \frac{W_t}{Fp_c}\frac{6hp_c}{t^2} = \frac{W_t}{Fp_c y} \qquad (13\text{-}22)$$

TABLE 13-5 The Lewis Form Factor y as a Function of the Number of Teeth N for a 20-deg Full-Depth Standard Tooth

N	15	16	17	18	19	20
y	0.092	0.094	0.096	0.098	0.100	0.102
N	22	24	26	28	30	34
y	0.105	0.107	0.110	0.112	0.114	0.118
N	38	43	50	60	75	100
y	0.122	0.126	0.130	0.134	0.138	0.142
N	150	300	Rack			
y	0.146	0.150	0.154			

where

$$y = \frac{t^2}{6p_c h} \qquad \text{(Lewis form factor)}$$

Thus, two parameters t and h have been replaced by a single parameter y.[4]

The form factor is a *dimensionless* quantity that depends only on tooth number, tooth shape, and pressure angle. Herein lies its importance. It simplifies calculations because it can be tabulated (Table 13-5) and applied regardless of gear size.

Further simplifications are obtained by eliminating F, the gear width, from the Lewis equation. Empirically, the width should always be less than the pitch diameter, generally less than $4p_c$ but greater than $2p_c$. If $F = kp_c$

$$\sigma_b = \frac{W_t}{kp_c^2 y} \qquad (13\text{-}23)$$

where k is a parameter, 3 to 4 for ordinary use. In preliminary calculations, use $k = 3.5$, thereby reducing the number of unknowns to three.

Material strength, always a limiting factor in design, must now replace the induced bending

[4]A parameter is a quantity that is constant under one set of conditions but assumes different values under other conditions.

stress. Since the load is repeated (Fig. 3-21), fatigue strength is the limit. For reasons of simplicity, only one derating factor C_v is applied. Termed velocity factor, it is essentially an impact derating factor that compensates for dynamic effects caused by speed and manufacturing inaccuracies. Thus

$$\sigma_w = S_e C_v \tag{13-24}$$

$$C_v = \frac{600}{600 + v} \tag{13-25}$$

where

σ_w = working (allowable) fatigue strength; psi
S_e = endurance limits; psi
C_v = velocity factor
v = pitch line velocity; fpm

When v is unknown, use $C_v = 0.34$. Values for S_e are found in *Machinery's Handbook*, Table 6, p. 815. Thus

$$S_e C_v = \frac{W_t}{k p_c^2 y}$$

Since the diametral pitch rather than the circular pitch is the preferred measure of tooth size, a more practical form of the Lewis equation is obtained by substituting P_d for p_c $(P_d p_c = \pi)$. Thus

$$S_e C_v = \frac{W_t P_d^2}{k \pi^2 y}$$

We can solve for P_d and obtain one or perhaps two suitable values leading to an acceptable tooth size. Since the pinion is the weaker, the Lewis equation should be applied to the pinion. Thus

$$P_d^2 = \frac{S_e C_v k \pi^2 y_p}{W_t} \tag{13-26}$$

This equation is preferred only when the center distance is known.

In most designs, the center distance is not specified, and for good reasons. Any downward limitations on center distance would prevent optimum utilization of available space. Furthermore, the torque rather than the tangential force is usually given. Equation 13-26 must therefore be further transformed by substituting torque for tangential force and introducing the number of teeth in the pinion.

$$W_t = \frac{2T_p}{d} = \frac{2T_p P_d}{N_p}$$

where

d = pitch diameter of pinion; in.
T_p = torque on the pinion; lb-in.
N_p = number of teeth in the pinion

Equation 13-26 then becomes

$$P_d^3 = \frac{S_e C_v k \pi^2 N_p y_p}{2T_p} \tag{13-27}$$

This is the most practical form of the Lewis equation and the one that is closest to the AGMA formula for fillet stress (Fig. 13-33). The form factor, traditionally called y, is now the tooth form factor J as shown in Table 5 in *Machinery's Handbook*, p. 813.

EXAMPLE 13-4

Determine the size of a pair of gears capable of transmitting 10 hp from a shaft running at 1200 rpm to one with a speed of 400 rpm. The starting torque exceeds the running torque by 25 percent. Use a standard 20 deg, full-depth tooth system with 24 teeth in the pinion.

SOLUTION

Since the center distance is unknown, we must use Eq. 13-27. Thus

$$\text{Pinion:} \qquad P_d^3 = \frac{S_e C_v k \pi^2 N_p y_p}{2T_s}$$

T_s = starting torque; lb-in.

From Table 6, p. 815 in *Machinery's Handbook*, we select a low-cost material steel 200 for which S_e is 27,000 psi.

Assume: $C_v = 0.34 \quad k = 3.5 \quad k_r = 1.25$

From Table 13-5 we find for $N = 24$ that $y = 0.107$.

$$T_s = k_r T_r = \frac{k_r P\,63{,}000}{n}$$

$$= \frac{1.25(10 \text{ hp})63{,}000}{1200 \text{ rpm}} = 656 \text{ lb-in.}$$

$$P_d^3 = \frac{(27{,}000 \text{ psi})(0.34)(3.5)\pi^2(24)(0.107)}{2(656 \text{ lb-in.})}$$

$$P_d^3 = 621; \qquad P_d = \mathbf{8.53}$$

According to Table 13-1, the closest standard values for P_d are 8 and 9. Which pitch to use depends on factors such as cost and available space. We will assume $P_d = 8$, which yields a less expensive gear because of a larger tooth size.

$$P_d = 8 \qquad d = \frac{24 \text{ teeth}}{8 \text{ teeth/in.}} = \mathbf{3 \text{ in.}}$$

$$N_G = m_G N_p = \frac{1200 \text{ rpm}}{400 \text{ rpm}} 24 = \mathbf{72}$$

$$D = \frac{N_G}{P_d} = \frac{72 \text{ teeth}}{8 \text{ teeth/in.}} = \mathbf{9 \text{ in.}}$$

Dynamic Tooth Load

Gears in mesh never operate under a smooth, continuous load. Factors such as manufacturing errors in spacing and tooth profile, tooth deflection under load, and imbalance all interact to create a dynamic load on gear teeth. The resulting action is similar to a dynamic load W_d superimposed on the transmitted tangential load W_t (Fig. 13-30). Thus

$$W_T = W_t + W_d \qquad (13\text{-}28)$$

where

W_T = total load; N, lb
W_t = transmitted tangential load; N, lb
W_d = dynamic tangential load; N, lb

The dynamic load is largely a function of pitch line velocity, width of gear (mass), and accuracy. Essentially a waste load, it places an upper limit on gear speed. Earl Buckingham devised an empirical formula for W_d that, with modifications, has become part of the AGMA standard procedure for calculating gear parameters.

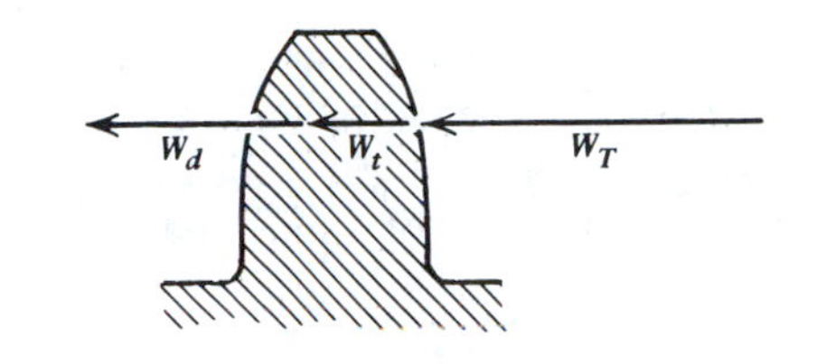

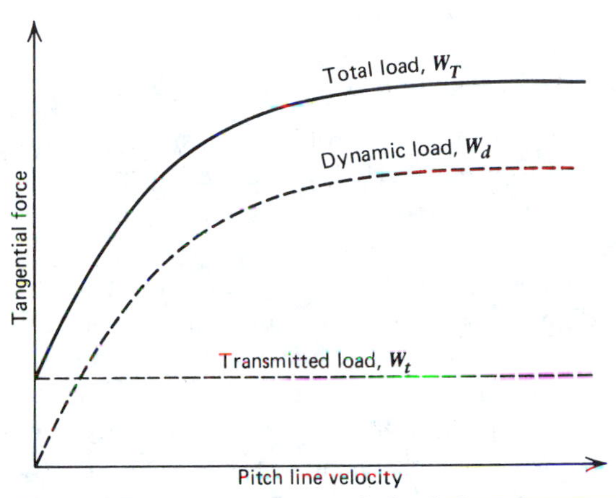

Figure 13-30 Dynamic tooth load for spur gear.

Dynamic Strength

The Lewis formula, using the transmitted load W_t but simulating dynamic conditions by means of a velocity factor C_v, leads to a complete set of gear dimensions. These, in turn, provide a basis for estimating the dynamic tangential load W_d. A second major approximation makes use of W_T to check tooth beam strength.

The capacity of a gear to support its total load W_T is termed dynamic strength. It is checked by applying W_T to the Lewis equation and omitting the velocity factor C_v. By comparing loads and not stresses, we arrive at a simple inequality.

$$W_e = \frac{k\pi^2 y S_e}{P_d^2} \geq W_T \qquad (13\text{-}29)$$

where W_e is the allowable endurance load.

From this we specify minimum hardness required to maintain adequate strength in bending by changing S_e. Satisfying this inequality completes our investigation of tooth breakage.

13-12 DESIGN FOR SURFACE DURABILITY

Surface Destruction

As gears mesh, both rolling and sliding occur, causing pitting and wear; pitting is primarily a consequence of rolling contact and wear is the result of sliding contact.[5] Pitting is initiated when contact stresses are high; it is hastened by sliding action. Although sliding causes wear, it also creates hydrodynamic action that counteracts wear. This relative motion, minute as it may be, is sufficient to provide conditions necessary for hydrodynamic action: a narrow, wedge-shaped opening between surfaces that have relative motion. Each time a gear goes through mesh with its accompanying rolling and sliding, surface and subsurface material is subjected to compressive, shear, and tensile stresses. The result is slow surface destruction in the form of pitlike pockmarks and metal erosion. These phenomena, for clarity, are often described as occurring separately, but this is not the case in practice. Two or more may occur simultaneously. In fact, one type may promote the other. The result is roughening of tooth surface, alteration of tooth profile, and loss of conjugate motion. Failure has definitely occurred when performance is unsatisfactory because of noise, vibration, or overheating (Fig. 13-31).

Some forms of surface wear and deterioration are actually corrective in nature and will cease after an initial period of running under favorable conditions. Following this, the teeth may polish up, work-harden, and begin a long period of normal, trouble-free service (see Section 4-9, Fig. 4-8).

Pitting first occurs at the pitch line, where the absence of sliding favors early breakdown of protective oil films. Excessive contact stress causing surface fatigue is the real cause of pitting. This may be due to (1) a narrow gear width, (2) radii of the involute surfaces being too small, and (3) frequent overloads. The smaller the radius of either surface, the narrower is the contact band and the greater the unit stress. After what is usually a very large number of stress repetitions, surface failure may occur. Minute cracks form in and below the surface, then grow and join. Eventually, small bits of metal are separated and forced out, leaving pits (Figs. 13-26*b* and 13-31).

To control pitting, contact stresses should not exceed allowable established values. A pair of spur gear teeth in action can be approximated by a pair of contacting cylinders on parallel axes (Fig. 13-32).

The contact stresses generated are termed surface durability stress by AGMA and can be calculated by means of formula (1) on p. 810 of *Machinery's Handbook*. Figure 13-33 shows the AGMA criteria for sizing spur gears according to the basic formulas (1) and (3) on pp. 810 and 811 of *Machinery's Handbook*.

Gear Lubrication[6]

Lubrication serves several purposes, but its most important function is to protect the rolling and sliding tooth surfaces from direct *metal-to-metal* contact. As a lubricant, oil predominates. Grease and solid lubricants are adequate only for very low speeds and/or intermittent operations.

To ensure penetration and maintain surface separation, viscosity should decrease as pitch line velocity increases. Generally, the lubricant

---

[5]According to F. L. Heine, a gear consultant, more than 50% of all gear failures are due to pitting.

[6]See Reference 2-5, pp. 292–293, "Typical Methods of Providing Lubrication for Gear Systems."

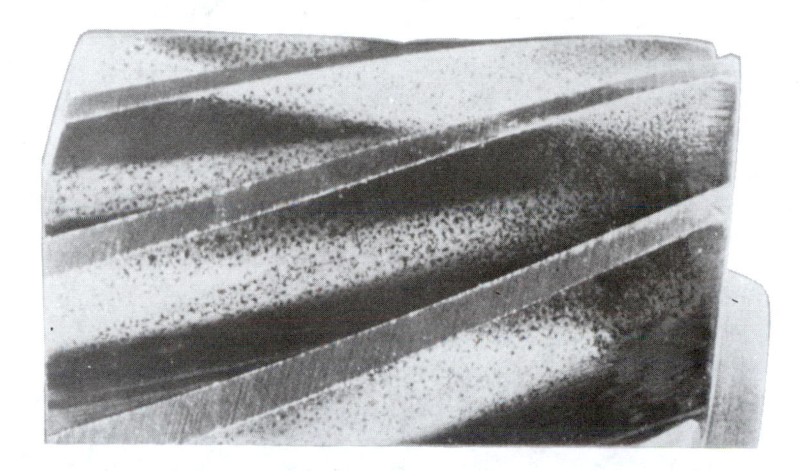

(*a*) Corrosive pitting

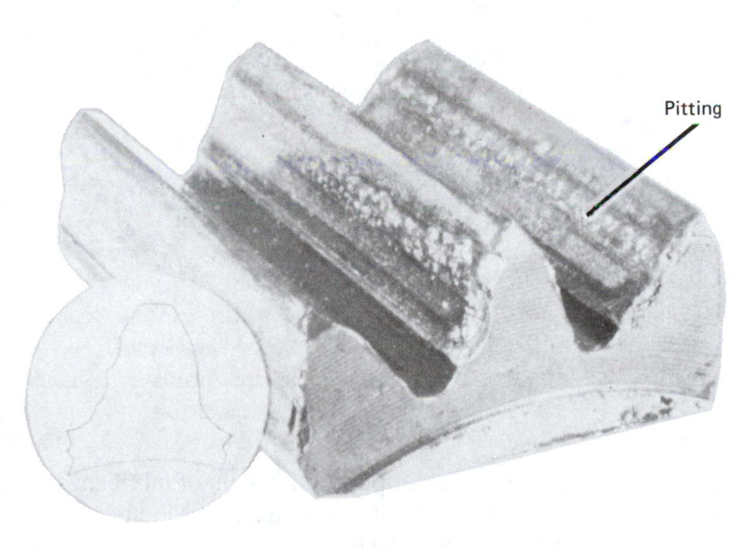

Abrasive wear

(*b*) Pitting and abrasive wear. The insert, on the left, shows advanced destruction of the involute profile

Figure 13-31 Pitting and wear of gear teeth. (Courtesy Caterpillar Tractor Company.)

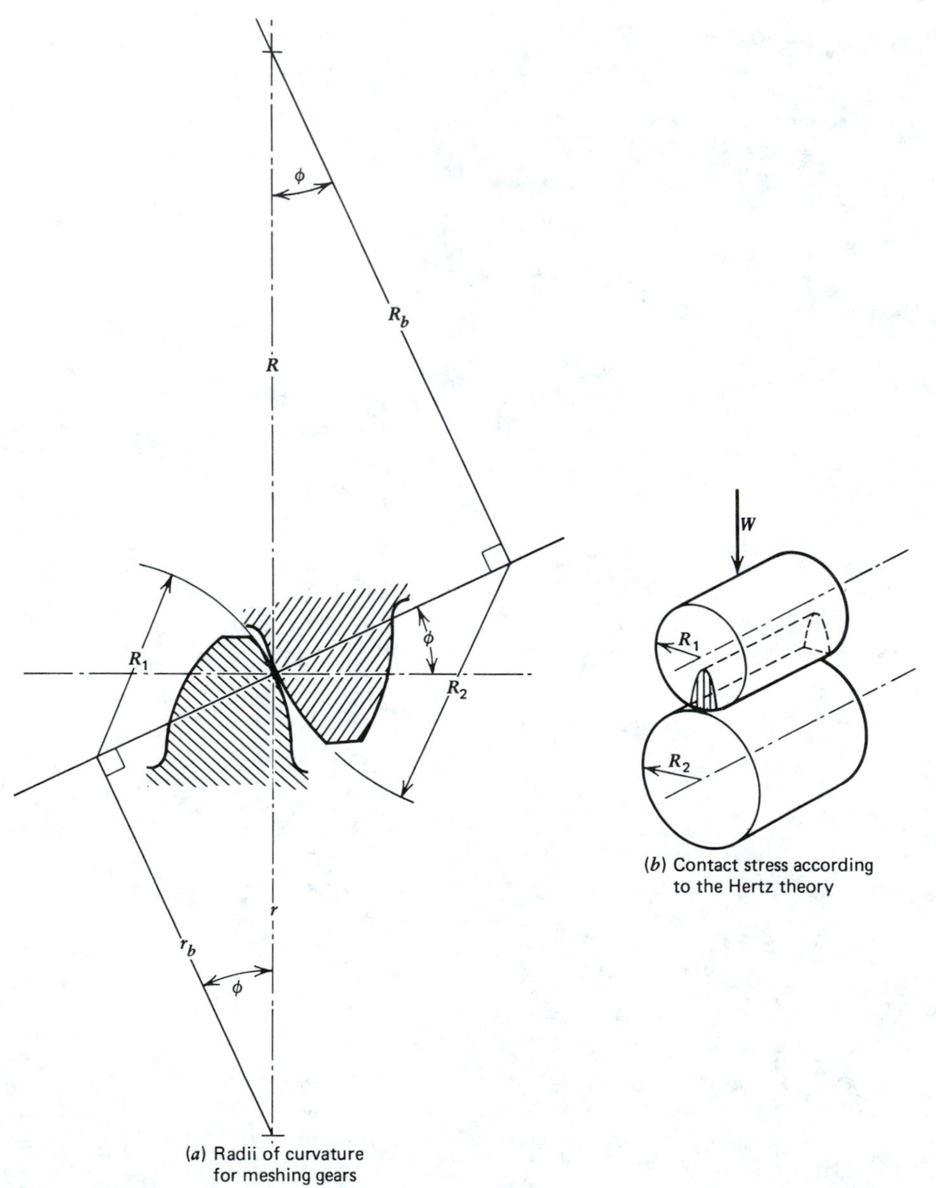

(a) Radii of curvature for meshing gears

(b) Contact stress according to the Hertz theory

Figure 13-32 Surface durability stress.

should prevent temperatures from exceeding 71°C during continuous runs and 93°C during shorter runs.

Methods of lubrication depend on pitch line velocity and on whether gearing is *open* or *closed*. Enclosed gearing predominates. In this arrangement the gears are in a sealed enclosure that provides ample containment of lubricant and protection against contamination. Methods of lubrication are essentially the same as those for chain drives: manual, drip, splash, forced-feed, and spray. Gear lubricants also serve as bearing lubricants and remove the heat generated by friction.

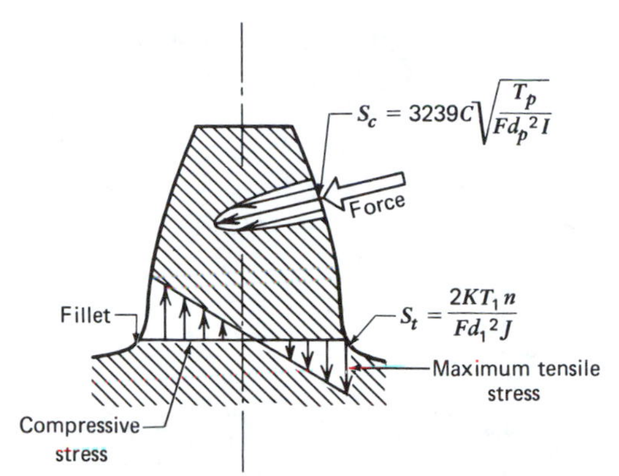

Figure 13-33 AGMA standards for spur gear sizing of standard tooth form is based on the *surface durability stress* S_c and the *fillet tensile stress* S_t. Because maximum tensile stress occurs at the root fillet, it is of the utmost importance that hardening penetrates the material along the entire length of this fillet.

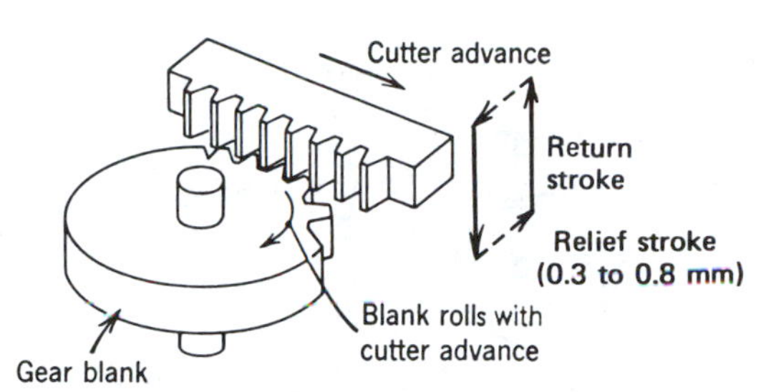

Figure 13-34 Rack-cutter generation. (Courtesy George Michalec.)

Heat of Operation

When gears operate under conditions of high speed or high pressure, the heat developed from the frictional losses may become appreciable. If a greater flow of oil is required for the purposes of cooling than is necessary for lubrication, the additional oil should be directed onto the gear *blanks* and not into the mesh of the teeth. Excessive oil at the tooth mesh will create further heating because of its rapid expulsion from mesh.

13-13 GEAR MANUFACTURE AND MATERIAL

Formation of Gear Teeth

Gear teeth are formed by a variety of methods including casting, shaping, molding, hobbing, cold forming, cutting, and grinding. Of these, shaping, hobbing, and grinding are the most important.

The involute rack as a cutting tool is indispensable to the manufacture of precision gears. Its geometric simplicity, together with conjugate motion of rack and pinion, makes it an ideal precision cutter. Most machine tools for cutting gears are based on the principle of the revolving base circle plus addenda (Fig. 13-5*c*). The circular disk (gear blank) in which the teeth are to be cut takes the place of the revolving base circle (Fig. 13-34). The rack is the substitute beam. The linear movement of the beam becomes the feed movement of the cutter. The rack cutter moves the length of one circular pitch during cutting and then returns to its starting position. Since the beam rolls on the base circle without slipping, the linear feed movement of the cutter must be synchronized with the rotation of the gear blank.

The rotary equivalent of the rack is the gear hob. Since rotary motion is easier to generate than oscillating motion, most external gears (70%) are cut by hobbing (Fig. 13-35). The hob is essentially a worm that has been interrupted by a series of transverse gashes to form teeth and cutting faces. Gear hobbing is a continuous, indexing cutting process. The hob and the gear blank rotate in synchronized motion while the hob is fed across the blank. Because the hob cuts to full depth, a single pass through the edge of the blank will finish the gear.

Design of Gear Blanks

Finished gears are obtained by cutting teeth in gear *blanks*. The term "blank" is commonly applied to castings, disks, forgings, and weldments

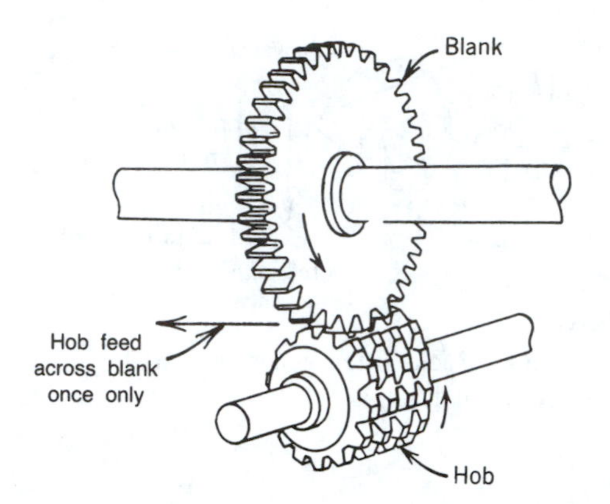

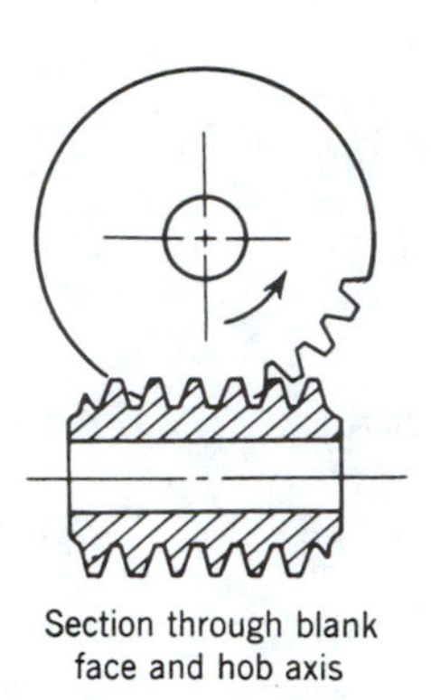

Figure 13-35 Gear hobbing. (Courtesy George Michalec.)

that are in a preliminary, unfinished form. The material of the blank depends in part on available manufacturing facilities. Manufacturing considerations therefore enter gear calculations early with the choice of material and gear size.

For simplicity and low cost, pinions are often forged integrally with the shaft (Fig. 13-36*a*). A small gear may also be obtained from a forged and rough-machined blank (Figs. 13-36*b* and 13-36*c*). Blanks for larger gears ($OD = 3 \times$ shaft diameter) often have a forged or machined web (Fig. 13-36*d*).

Welded blanks combine simplicity with low cost.[7] Inexpensive materials (carbon steel) can be used for hub and web, while alloy steel goes into the rim (Figs. 13-36*e* and 13-36*f*). Another

[7]For further information, see Section 5.4, "How to Design Large Steel Gears," *Design of Weldments,* Cleveland, Ohio: The James F. Lincoln Arc Welding Foundation.

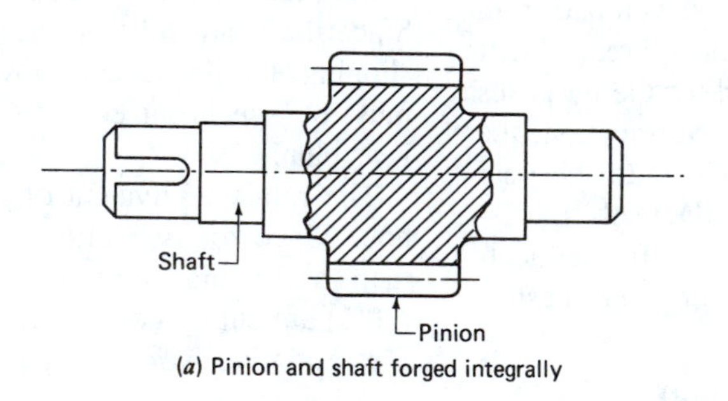

(*a*) Pinion and shaft forged integrally

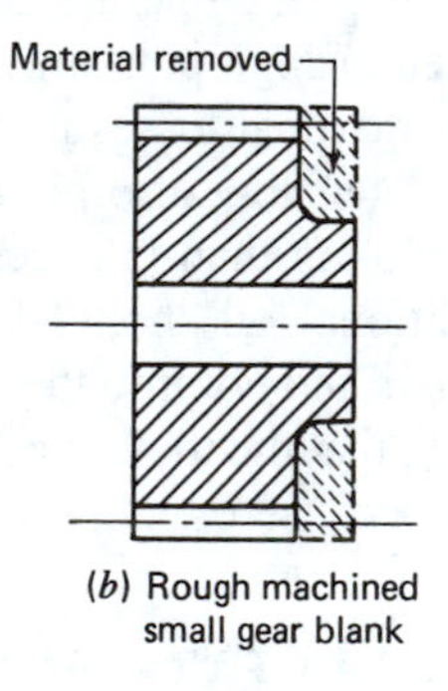

(*b*) Rough machined small gear blank

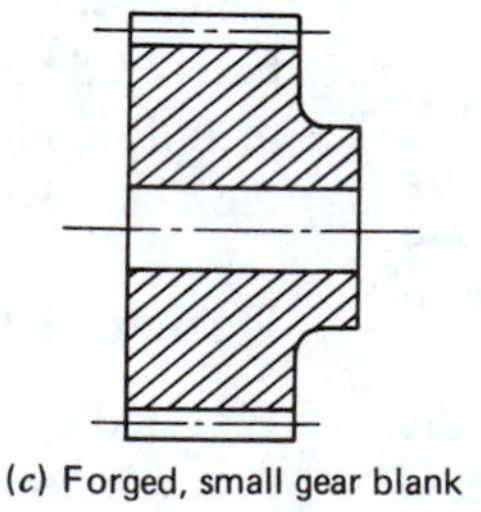

(*c*) Forged, small gear blank

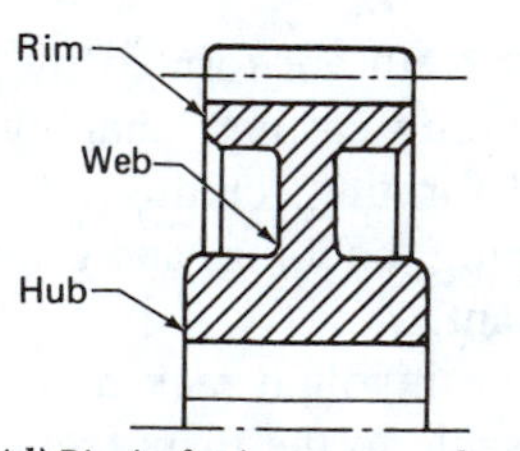

(*d*) Blanks for larger gears often have a forged or machined web

design combining high- and low-cost materials has the shaft welded onto the pinion (Fig. 13-36*g*).

A cast gear is shown in Fig. 13-36*h*. Large ring gears, 3 m (10 ft) in diameter, are often made in two or three cast sections and bolted together. When the diameter of a gear is more than double the shaft diameter, it is usually keyed to its shaft. Because of stress concentration in the keyway, the blank must accommodate a radial distance of at least four modules from the top of the keyseat to the outside of the gear (Fig. 13-36*i*).

Note that accepted gear blank designs call for removal of excess material (web, lightening holes, etc.). The aim is not merely to save material but also to minimize the flywheel effect of individual gears in order to minimize the total inertia of a train of gears. Too much inertia can be the cause of sudden overloads which, if unrecognized, lead to premature failure and, if recognized, require larger design factors and higher initial cost.

Gear Materials

Metallic Gearing. Common gear materials include ferrous and nonferrous metals, plastics, and phenolic resins. Material selection centers on the pinion which, because of its size and possible undercutting, is usually the weaker. The pinion also tends to wear faster due to more

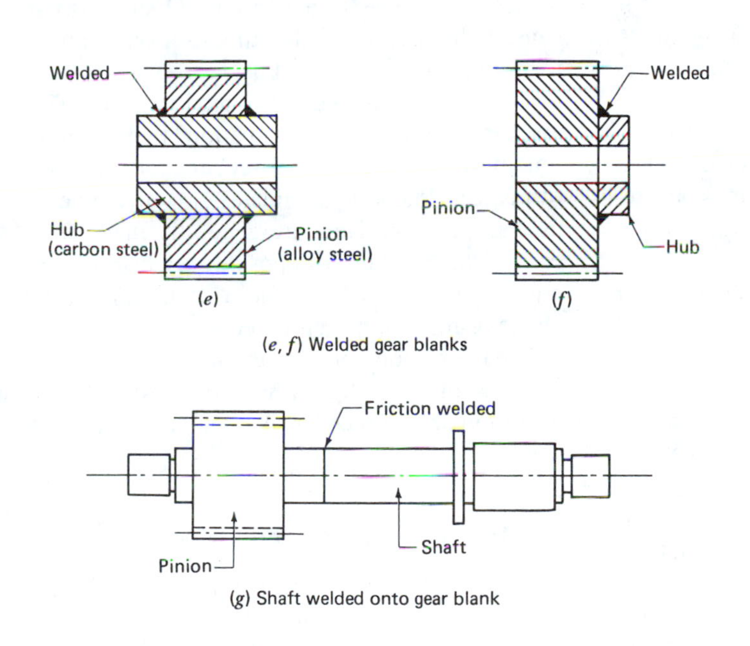

(*e, f*) Welded gear blanks

(*g*) Shaft welded onto gear blank

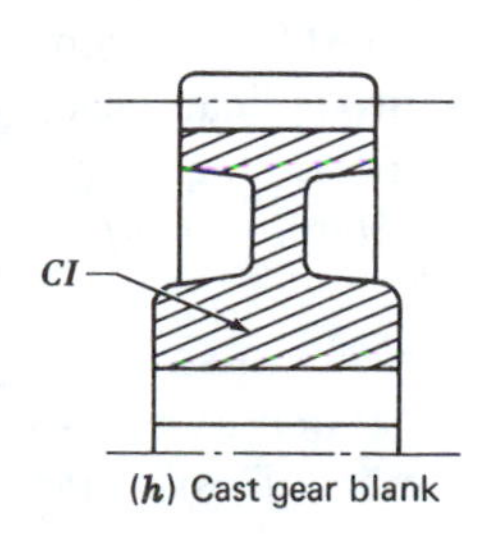

(*h*) Cast gear blank

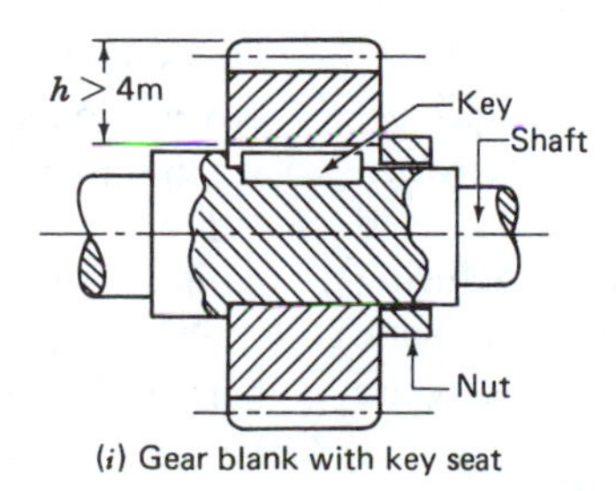

(*i*) Gear blank with key seat

Figure 13-36 Design of gear blanks.

frequent contacts. For the gear, load and wear considerations are less severe; therefore less expensive materials may often be used. One rule, however, should be observed. *Wear is generally less severe in sliding contact when dissimilar materials are used.*

Other considerations relate to available space, mass, speed, power, and operating conditions. Often these considerations are incompatible, so compromise is necessary.

Steel combines superior characteristics of strength per unit of volume and low cost per unit of mass. Both plain steel and alloy steel are used in gear design. The *carbon* steels offer low initial cost, great surface hardness, and low manufacturing cost, but they do not resist corrosion. *Alloy* steels cover a wide range, from low grade types to special alloys offering great strength. A wide range of heat-treatment properties makes this group extremely versatile. Industrial power gears need high wear resistance at the surface and hardness in sufficient depth to prevent crushing of the tooth surface, yet they must be resilient enough to resist tooth breakage. Proper heat treatment provides these qualities.

Hardened gears have many advantages over unhardened gears, since the elevated mechanical properties permit higher stress levels. Surface durability of gear teeth is roughly proportional to the *square* of the surface hardness. Thus a gear tooth with 600 Bhn hardness on the surface may be able to carry as much as nine times the power of a gear with only a surface hardness of 200 Bhn. Additionally, hardness increases tooth fillet strength and resistance to wear. A hardened gear also offers advantages where limitations of space and mass are prime considerations, since the drive will be smaller and lighter.

Cast iron is used for large units or intricate shapes. Gears cast from nodular iron are tough and resilient. Powdered metal gears are also used in many applications. Lower manufacturing costs often result from the use of powdered metal gears when they are of small sizes and many are produced.

Other commonly used materials are bronze, aluminum alloys, and zinc die-cast alloys. Brass is an inexpensive material for clock gears. Bronze offers good sliding properties when mated with steel. These qualities are utilized in worm gearing, where sliding is far more extensive than in spur gears.

Nonmetallic Gearing. These gears offer quiet operation, wear resistance, damping, and little or no need for lubrication. However, lack of strength and surface durability limits their application to light loads. Pitch line velocities for nylon gears may be as high as 10 m/s (1800 fpm), but the allowable tensile stress for nylon is only 40 MPa (6000 psi). Nonmetallic gearing, especially nylon gears, often serve as overload devices in high-cost gear trains. Overloads can cause nylon gears to fail without serious damage to more expensive parts. For best results, plastic gears should be mated with metallic gears. This arrangement aids heat dissipation, thereby minimizing distortion and subsequent reduced performance. Phenolic laminates are also widely used in gearing. They are generally made from canvas, linen, glass fiber, or other substrates impregnated with 30% or more of thermosetting phenolic resin and then cured.

The use of nonmetallic gearing is increasing in the areas of technology where low cost and quiet operation take precedence over space requirements.

AGMA Gear Classes and Tolerances

To aid designers in the proper selection of gears for various applications, the American Gear Manufacturers Association has prepared a Gear Classification Manual, AGMA 390.03, which covers coarse and fine pitch spur, helical, herringbone, bevel, and hypoid gears. Data extracted from this manual are available in *Machinery's Handbook,* pp. 766–782.

Selections are based on AGMA class numbers, which combine a dimensional quality number and a material treatment number. Designers

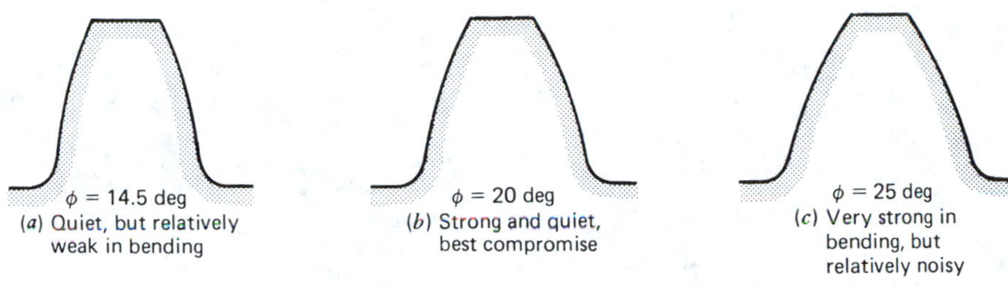

Figure 13-37 Tooth form as a function of the pressure angle, Ø.

should work closely with manufacturing engineers in this area.

13-14 MODIFICATIONS FOR STRENGTH AND NOISE ABATEMENT

Strength and surface durability can generally be increased by the following means.

1. *Enlarging the pressure angle* generates a broader tooth base and yields greater resistance to bending (Fig. 13-37). Gears with a 25-deg pressure angle have higher load capacities and better wear characteristics than gears with lower pressure angles. However, these gears do not run as smoothly or quietly because they are much more sensitive to center distance variations.
2. *Crowning; root and tip relief* generates slightly barrel-shaped gear teeth. This modification prevents excessive loading at the ends due to misalignment, tooth inaccuracies, deflection, or heat treatment. As the load increases, a smooth "spreading" of the contact occurs until the entire flank is loaded. Crowning thus increases bending strength and wear resistance (Fig. 13-38). In gear couplings, crowning greatly augments the ability of the coupling to adjust itself to angular shaft misalignment.

Gears are inherently noisy because of metal-to-metal contact, load variations, changing inertia forces, and fluctuating friction forces. Gear teeth, resembling the short prongs of a tuning fork, will vibrate in much the same manner when struck by mating teeth. The following measures, some already discussed, will lead to quiet gear design.

1. Modification of standard profiles: crowning, root, and tip relief.
2. Hunting or nonintegral tooth ratios.[8]
3. Small modules and/or large tooth numbers.
4. Improved surface finish.
5. Helical instead of straight teeth (for $m_p > 2$).
6. Pairing of metallic and nonmetallic gears.
7. Increased rigidity of housing, bearings, and shafts.
8. Minimized bearing play.

13-15 INTERNAL GEARS

Internal gears are used when rotational direction for both shafts must be the same or when a much shorter center distance is advantageous. An internal or annular gear is a ring with teeth on the inner circumference (Fig. 13-39). Pertinent definitions are the same as those for external gears, except that a new element—the internal diameter—has replaced the outside diameter.

Internal gears can mesh only with external gears. Contact stresses are thus lower than for equivalent external gears, a condition that helps increase load capacity or life. For the same tooth proportions, internal gears have greater length of

[8] A nonintegral tooth ratio refers to a ratio containing one or two prime numbers.

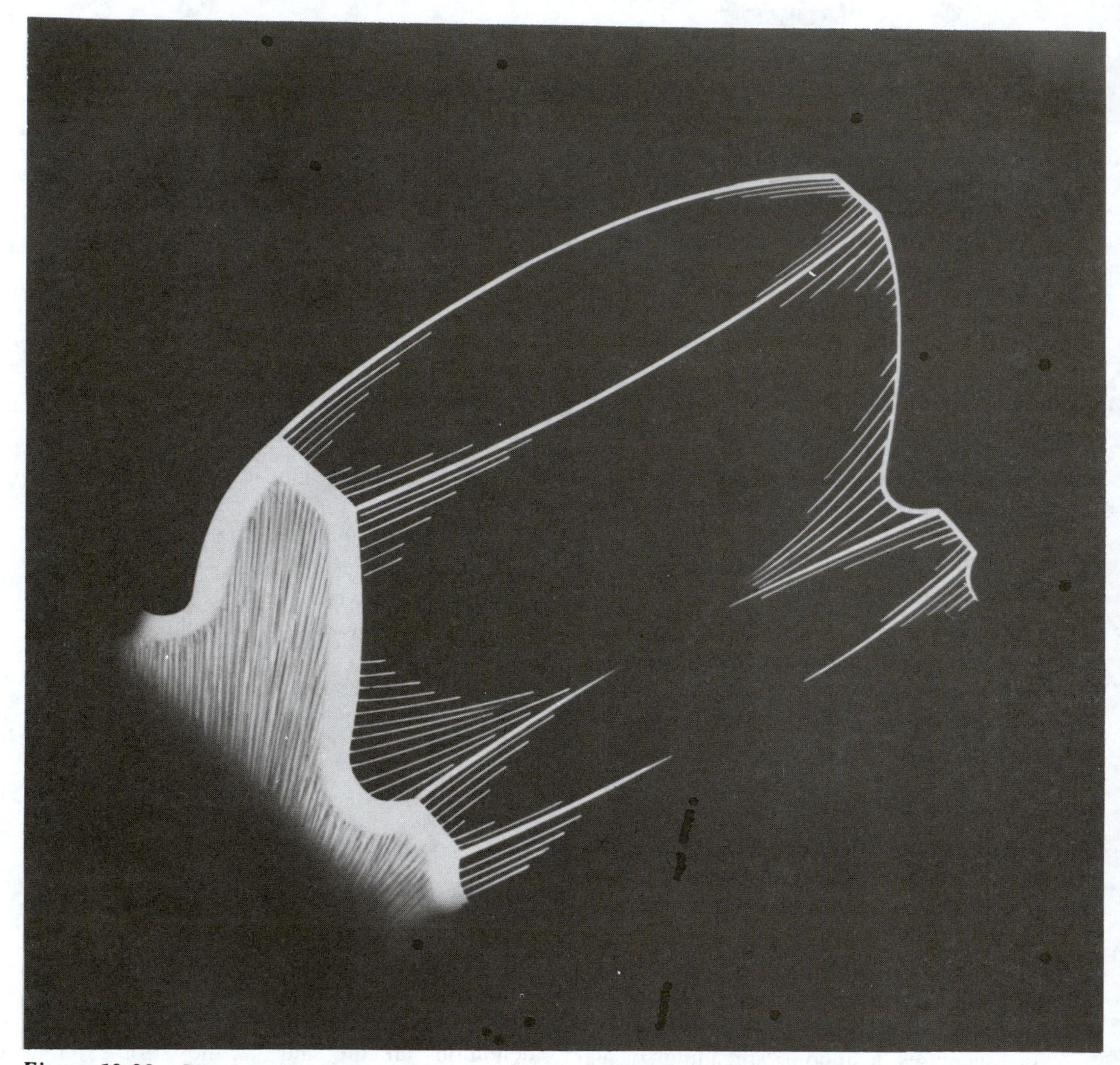

Figure 13-38 Crowned gear tooth for shaft coupling. (The crowning is greatly in excess of what is used on gears for power transmission.) (Courtesy Xtek Inc., TSP Coupling Division.)

contact, tooth strength, and pitting resistance, plus lower relative sliding. The result is greater torque capacity and reduced wear. Compact designs are achieved by using the outer "cylindrical" surface as a pulley, sheave, gear, worm, or brake surface.

However, internal gears have limitations. For instance, the speed ratio for a single pair is much more restricted than for two external gears. Also, tooth action is such that interference may occur. Interference can be alleviated by tooth modifications to the internal gear. Mounting of

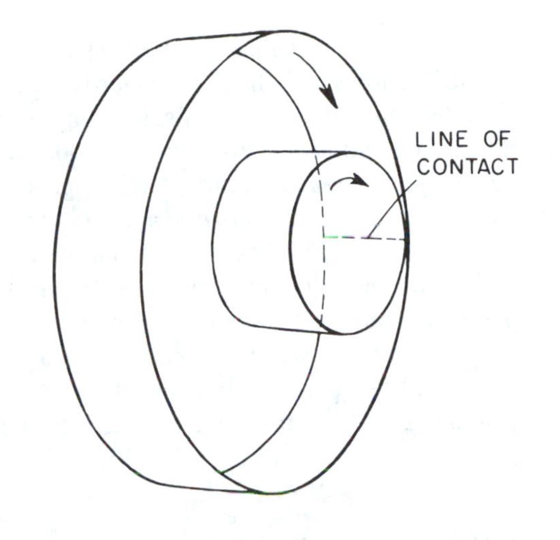

Figure 13-39 Internal spur gears. The pitch surfaces on friction wheels contact along a line (upper view) to transmit motion between parallel shafts. The teeth are developed from internal and external blank cylinders, as shown in the lower view. (Courtesy Mobil Oil Corporation.)

the internal gear can also be more difficult. Because the hub is offset, more attachment rigidity is required for single mesh arrangements.

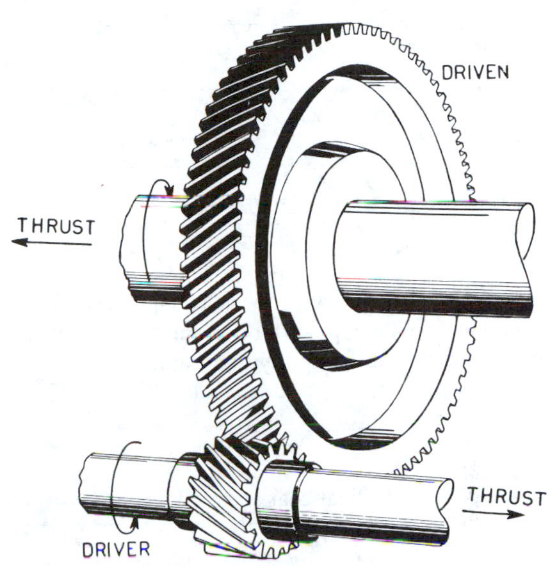

Figure 13-40 Helical gear and pinion. (Courtesy Mobil Oil Corporation.)

13-16 HELICAL GEARS FOR PARALLEL SHAFTS

Function and Design

Helical gears were "invented" to overcome the limitations of speed inherent in spur gears.[9] Because helical gears have teeth cut at an angle to the axis of rotation, tooth engagement becomes gradual and smooth, and much higher operating speeds are made possible (Fig. 13-40). The versatility of the helical gear is shown in Fig. 13-41, where two helical gears have replaced a single spur gear. By varying the slant of the teeth, many helical gears can replace a single spur gear, thus lending great flexibility to design of gear trains by providing any number of workable pitch diameters.

Helical gearing gets its name from the tooth

[9]DeLaval of Sweden in 1883 designed a helical gear reduction unit for his newly developed steam turbine. The unusually high speeds inherent in steam turbines required gears with slanted teeth.

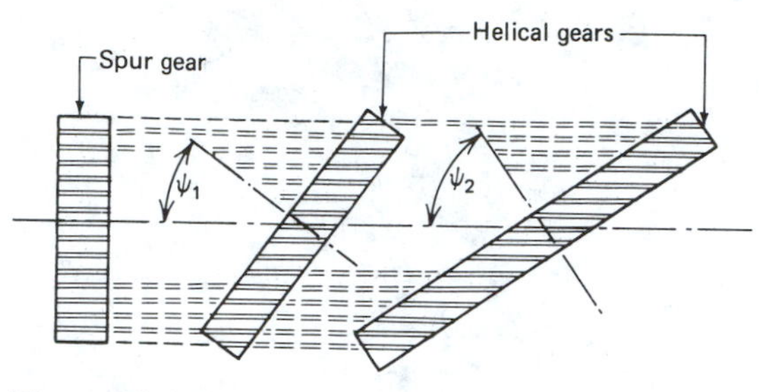

Figure 13-41 By varying the helix angle ψ, a single spur gear can be replaced by a number of helical gears, thus adding flexibility to design.

curvature, which, like a short, steep screw thread, follows a helical path across the base cylinder (Fig. 13-42*a*). This seemingly insignificant deviation from a spur gear tooth makes helical gears *dual* in nature. Helical gears have two significant planes, *normal* and *front* or *transverse* (Fig. 13-42*b*). The normal plane is perpendicular to the teeth; the transverse plane is the plane of rotation. There are two pitches (normal and transverse), two contact ratios (profile and face), and two parasitic loads (radial and axial). A dual driving action occurs between helical

Figure 13-42 Helical gear and rack features. (Courtesy General Motors Corporation.)

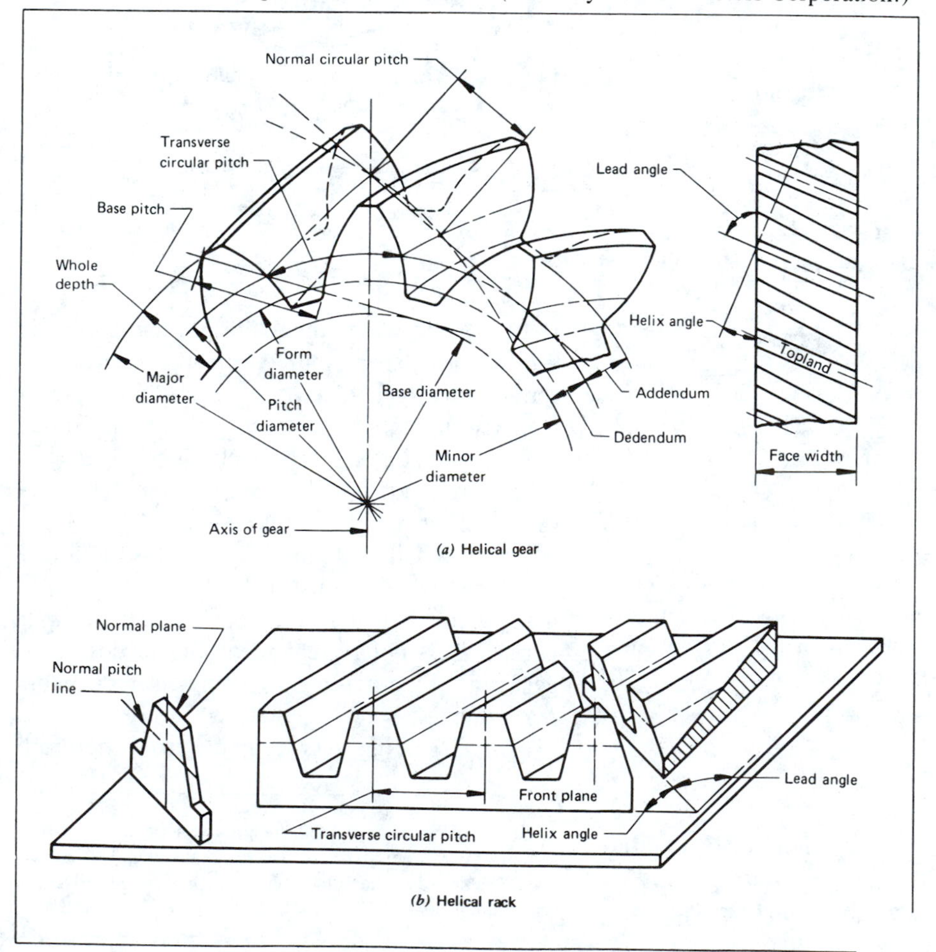

gears: a conjugate involute cam action in the plane of rotation, plus a traveling action between the mating helices of meshing tooth pairs.

Helical gears have the following advantages over spur gears.

1. A much higher total contact ratio, profile plus face (2 to 5 as against 1.2 to 1.80).
2. Much higher permissible speeds, with pitch-line velocities exceeding 200 m/s (40,000 fpm) as opposed to 50 m/s for spur gears.
3. Greater power capacity for the same space occupied.
4. Reduced noise and wear.

These advantages are obtained against higher initial tooling cost and sizable axial force (not found in spur gears) that requires bearings capable of sustaining thrust loads.

Stepped Gears

A helical gear is essentially a stepped gear with an *infinite* number of teeth. To simplify matters, a stepped gear with a *finite* number of teeth will be discussed. In this construction an assembly of two or more narrow gears is substituted for each single, wide gear. These narrow gears are mounted on the shaft in such a way that the teeth are staggered (Fig. 13-43). The gear shown meshes with one having teeth staggered in the opposite direction.

Stepped gears were an early attempt to overcome the limitations on speed inherent in spur gears. Since only a few teeth of a pair of spur gears are in contact at the same time, meshing of these teeth may be accompanied by a slight impact as the load shifts from tooth to tooth. In stepped gears the sudden transfer of load from one tooth to another is minimized as each adjacent staggered tooth absorbs part of the load before the preceding tooth leaves mesh. This increases the number of teeth in contact at any one time, and a substantially higher contact ratio is achieved. In general, stepped gears are seldom used for ordinary transmission of power because of the difficulty of equalizing the load among the various tooth faces and manufacture.

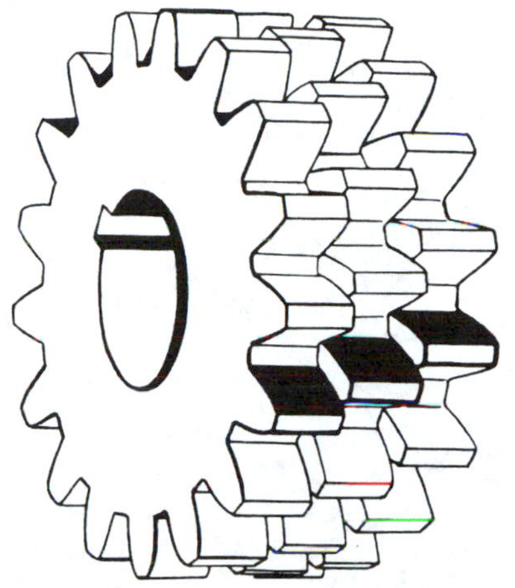

Figure 13-43 Stepped gear. (Courtesy Mobil Oil Corporation.)

Generation of the Helical Tooth

The helical tooth form is involute in the plane of rotation or transverse sheet and can be developed in a manner similar to the spur gear. Figure 13-44*a* shows a base cylinder from which a taut sheet in the shape of a trapezoid is unwrapped. This is analogous to the unwinding of a taut string used in generating spur gear teeth. The angular straight edge or line *AB* on the sheet was generated when a section of a helix was unwrapped from the base cylinder.

Recall that a helix is a curve generated by a point moving about a cylindrical surface at a constant rate in the same direction as the cylinder's axis. The *lead* (*L*) of a helix is the distance it advances in an axial direction in one complete turn about the cylindrical surface.

As the taut sheet is unwrapped, any point on line *AB* will trace an involute from the base cylinder. Together these involutes, infinite in number, will generate a tooth surface. But, unlike spur gearing, the involutes are *phase displaced* along the helix. The surface obtained is called an *involute helicoid*. Figure 13-44*b* shows the base

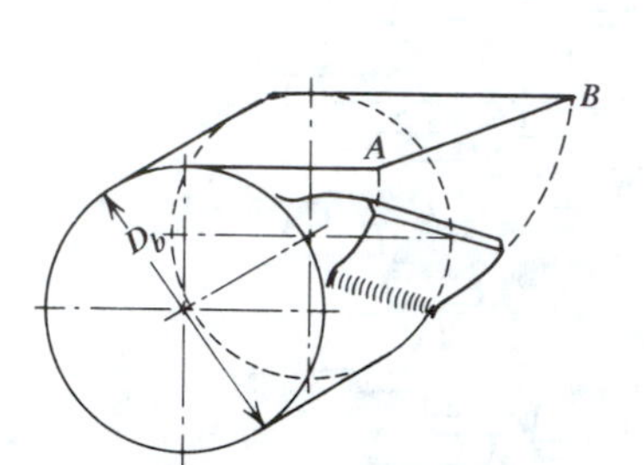

(a) Pictorial view of base cylinder and taut plane generating a helicoidal tooth

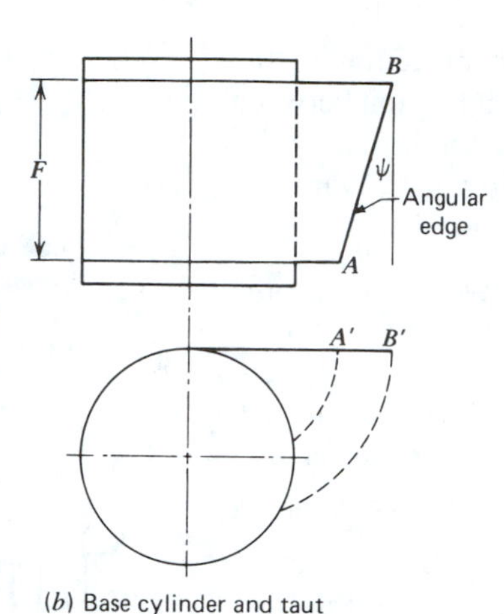

(b) Base cylinder and taut plane shown in two views

Figure 13-44 Generation of a helical tooth profile.

cylinder and the plane in two views. The general direction of each tooth is fixed by means of the angle between the tangent to the helicoidal tooth profile and an element of the pitch cylinder (Fig. 13-45). This, by definition, is the *helix* angle.

To obtain conjugate helical involutes, it is necessary to visualize the taut sheet being unwrapped from the base cylinder and wound onto another base cylinder, not necessarily of the same size but rotating in the opposite direction. If a reverse direction of rotation is assumed and a second tangent plane is arranged so it crosses the first, a complete involute helicoidal tooth shape is formed.

Figure 13-45 Definition of helix angle. (Courtesy Sterling Instrument.)

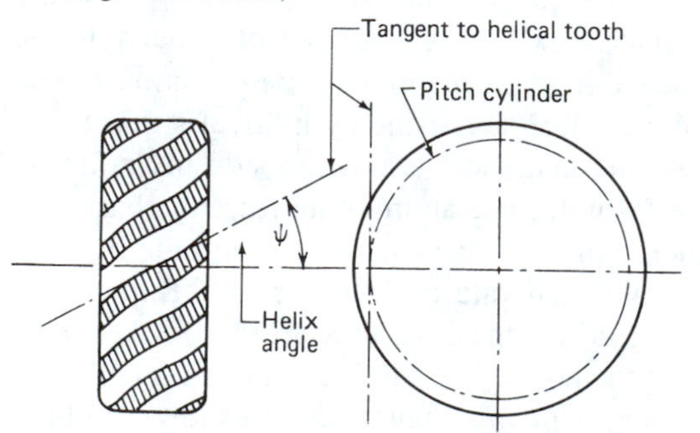

Spur gears have a single major identifying parameter—the base circle diameter—and helical gears have *two* such parameters—the diameter of the base cylinder and the helix angle.

Helical Gear Relationship

For helical gears there are *two* measurements of the usual pitches: one in the transverse plane and the other in a plane normal to the tooth. In addition, there is a transverse (axial) pitch. These are defined and related as follows. As indictated in Fig. 13-46, triangle ACD, the two circular pitches are equated as

$$p_n = p_c \cos \psi \tag{13-30}$$

where p_n is the circular pitch.

The relationship between the diametral pitches P_d and P_n is obtained by substituting p_c and p_n in Eq. 13-30 as follows $p_n = \pi/P_n$ and $p_c = \pi/P_d$

$$\frac{\pi}{P_n} = \frac{\pi}{P_d} \cos \psi$$
$$P_n = \frac{P_d}{\cos \psi} \tag{13-31}$$

where P_n is the normal diametral pitch.

Figure 13-46 shows the *axial* pitch of a gear as the distance between corresponding points of adjacent teeth measured parallel to the gear's axis.

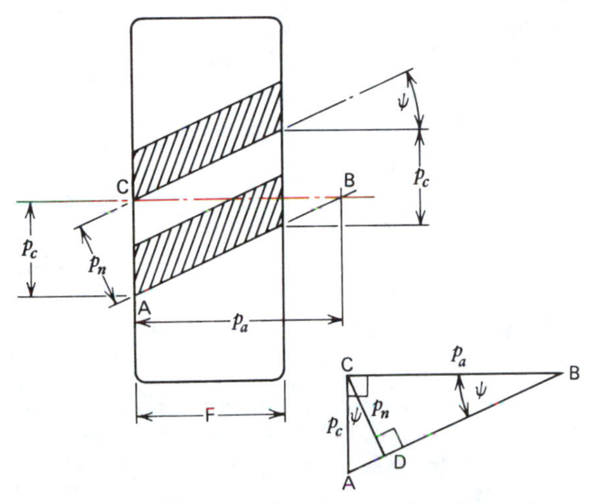

Figure 13-46 Geometric relationship of circular, normal, and axial pitches.

From triangle ABC $\quad p_c = p_a \tan \psi$

From triangle BCD $\quad p_n = p_a \sin \psi$

or

$$p_a = \frac{p_c}{\tan \psi} = \frac{p_n}{\sin \psi} \tag{13-32}$$

where p_a is the axial pitch.

For parallel helical gears mounted on parallel shafts to mesh properly, they must have:

1. Equal helix angles.
2. Equal pitches.
3. Helices of opposite hands; that is, one helix must be left-handed, the other right-handed.

The velocity ratio is the same as for spur gears,

$$m_G = \frac{n_p}{n_G} = \frac{D}{d} = \frac{N_G}{N_p} \tag{13-33}$$

The center distance is

$$C = \frac{N_G + N_p}{2P_d} \tag{13-34}$$

The lead is

$$L = \frac{\pi D}{\tan \psi} \tag{13-35}$$

The pressure angle located in the transverse plane is

$$\tan \emptyset = \frac{\tan \emptyset_n}{\cos \psi} \tag{13-36}$$

where $\emptyset_n$ is pressure angle in normal plane.

EXAMPLE 13-5

What is the axial and normal pitch of a parallel helical gear in which the circular pitch is 26.594 mm and the helix angle 30 deg? What is the module?

SOLUTION

From Eq. 13-32: $p_a = \dfrac{p_c}{\tan \psi} = \dfrac{26.594 \text{ mm}}{\tan 30 \text{ deg}} = \mathbf{46.062\ mm}$

From Eq. 13-30: $p_n = p_c \cos 30 \text{ deg} = (26.594 \text{ mm}) \cos 30 \text{ deg} = \mathbf{23.031\ mm}$

$m = p_c \pi = (26.594 \text{ mm})\pi = \mathbf{83.548\ mm}$

13-17 HELICAL GEAR ANALYSIS

Force Analysis

When helical gears mesh, engagement is gradual and the load is propagated diagonally across the tooth face (Fig. 13-47). With helical gears, the average tooth lever arm is roughly 75% of the tooth height,[10] whereas spur gear teeth have a maximum lever arm nearly equal to full tooth height. Helical gear teeth are thus capable of transmitting greater forces than equivalent spur gears.

---

[10]This depends on the helix angle.

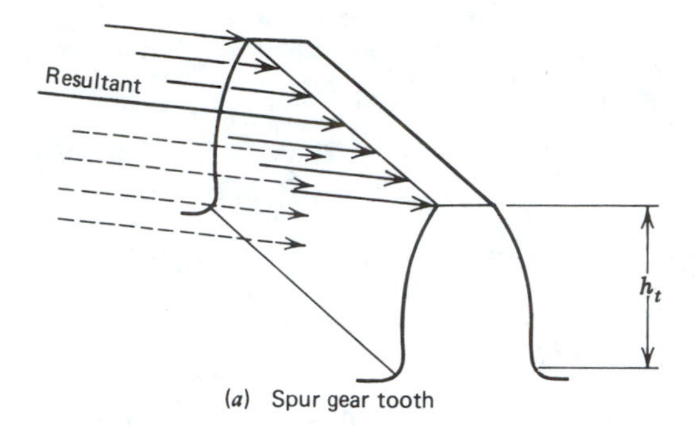

(a) Spur gear tooth

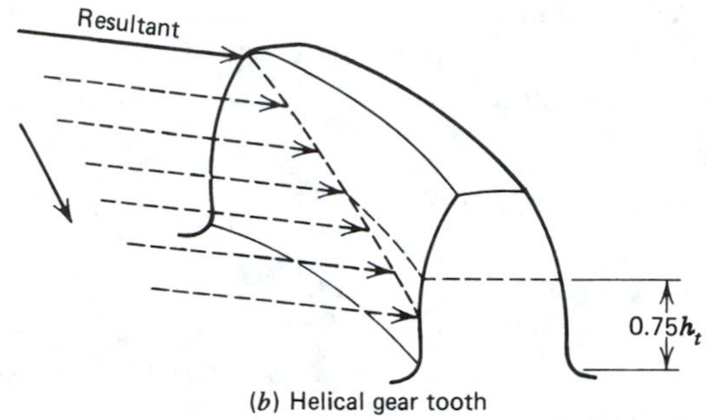

(b) Helical gear tooth

Figure 13-47 Maximum tooth loading of spur and helical gears.

Figure 13-48 Force analysis of helical gear tooth.

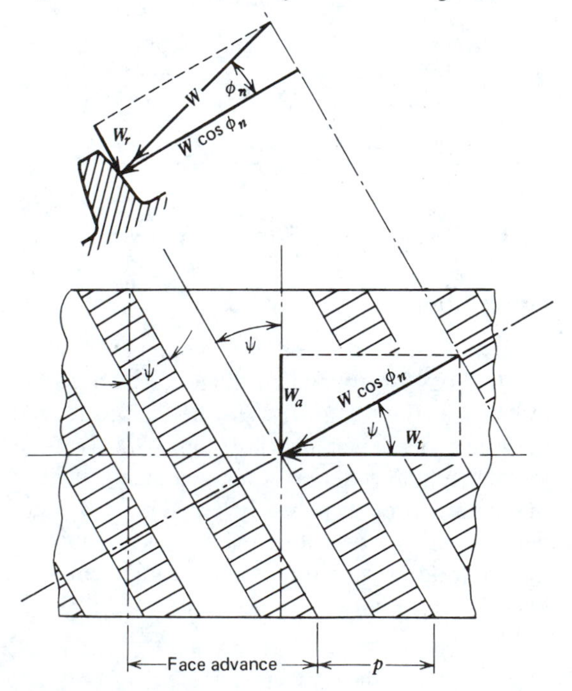

Figure 13-48 shows the total load W, which can be resolved into a radial load W_r and a component W_t tangent to the pitch cylinder. Thus we have

$$W_r = W \sin \phi_n \tag{13-37}$$

$$W_t = W \cos \phi_n \cos \psi \tag{13-38}$$

$$W_a = W_t \tan \psi \tag{13-39}$$

where

W = total load; N, lb
W_t = tangential component; N, lb
W_r = radial load; N, lb
W_a = axial load; N, lb

W_a and W_r are *waste* loads, generating bearing loads but contributing nothing to power transmission. W_t, also referred to as the *transmitted* load, appears in the power equations. Usually W_t is the given force from which the others are found. Thus

$$W = \frac{W_t}{\cos \phi_n \cos \psi} \tag{13-40}$$

$$W_r = W_t \frac{\tan \phi_n}{\cos \psi} \tag{13-41}$$

EXAMPLE 13-6

A parallel helical gear, 250 mm in diameter with a helix angle of 30 deg, carries a torque of 340 N · m. Determine the force components and the total tooth load.

SOLUTION

Transmitted load:

$$W_t = \frac{T}{0.5D} \quad \text{(from definition of torque)}$$

$$= \frac{2(340 \text{ N} \cdot \text{m})}{0.250 \text{ m}} = \mathbf{2720\ N}$$

Total tooth load:

$$W = \frac{W_t}{\cos \phi_n \cos \psi} \quad \text{find } \phi_n$$

$$\tan \phi_n = \tan \phi \cos \psi$$
$$= \tan 20 \text{ deg} \cos 30 \text{ deg}$$
$$\phi_n = 17.495 \text{ deg}$$
$$W = \frac{2720\ N}{\cos 17.495 \text{ deg} \cos 30 \text{ deg}} = \mathbf{3293\ N}$$
$$W_r = W \sin \phi_n = (3209 \text{ N}) \sin 17.495 \text{ deg}$$
$$= \mathbf{965\ N}$$
$$W_a = W_t \cos \phi_n \sin \psi$$
$$= (2720\ N)(\cos 17.495 \text{ deg})(\sin 30 \text{ deg})$$
$$= \mathbf{1297\ N}$$

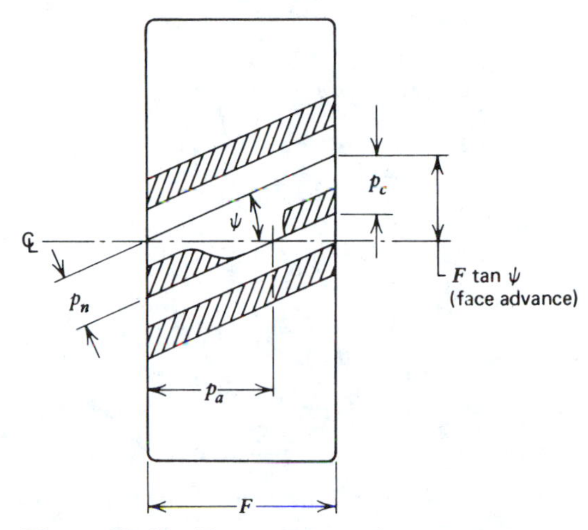

Figure 13-49 Face contact ratio.

Helix Angle and Contact Ratio

The advantage of helical gears over spur gears lies first of all in their much higher contact ratio, often triple that of spur gears. The higher contact ratio arises from the *helical effect* superimposed on the spur gear effect, which is always retained. Thus

$$m_t = m_p + m_f \qquad (13\text{-}42)$$

where

m_t = *total* contact ratio
m_p = *profile* contact ratio (spur gear ratio)
m_f = *face* contact ratio (overlap ratio)

The profile contact ratio is due to the inherent spur gear characteristics and may be calculated by means of Eq. 13-13. The face contact ratio is a function of face width and helix angle. It is found from

$$m_f = \frac{F \tan \psi}{p_c} \qquad (13\text{-}43)$$

where F is the face width of gear in millimeters or inches. Figure 13-49 shows $F \tan \psi$ as the *face advance per tooth*. For smooth operation AGMA recommends that the face advance exceed the circular pitch by at least 15%, or

$$F \tan \psi \geq 1.15 p_c$$
$$\frac{F \tan \psi}{p_c} \geq 1.15 \qquad (13\text{-}44)$$

or

$$m_f \geq 1.15$$

Thus the face contact ratio should always exceed 1.15. Rearranging terms, we obtain

$$F \geq \frac{1.15 p_c}{\tan \psi} \qquad (13\text{-}45)$$

In general, $2p_c < F < 4p_c$.

A certain amount of face width is thus required to obtain the full benefit of helical gear action. This benefit is increasingly offset, however, by the thrust force, which increases in proportion to $\tan \psi$. The 10 to 20 deg helix angles are preferred because they give relatively low thrust. Helix angles of 35 to 45 deg are used in double-helical gears only. In double-helical gears (or herringbone) there is no external thrust-bearing load because the thrust of one helix is opposed by an equal and opposite thrust from the other helix (Fig. 13-50).

EXAMPLE 13-7

Calculate the contact ratio m_t of a pair of identical 20-deg, full-depth, parallel helical gears, $m = 3$, $N_p = N_G = 17$, $F = 40$ mm, and

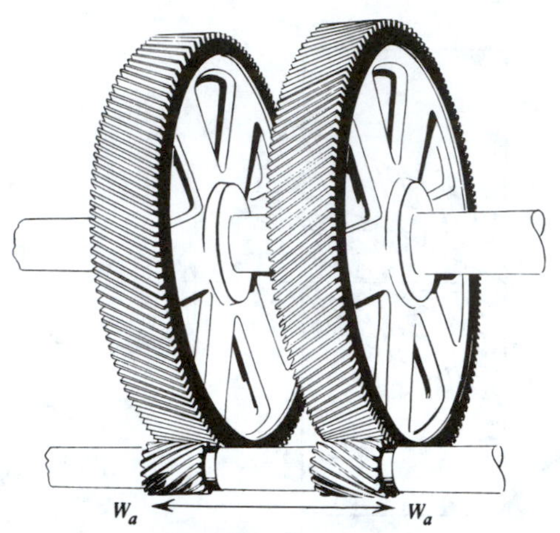

Figure 13-50 Double-helical or herringbone gear and pinion. (Courtesy Mobil Oil Corporation.)

SOLUTION

The equivalent spur gears were discussed in Example 13-2; m_p, the profile contact ratio (spur gear ratio), was 1.51. Thus, according to Eq. 13-42,

$$m_t = 1.51 + m_f$$

From Eq. 13-43:

$$m_f = \frac{F \tan \psi}{p_c} = \frac{F \tan \psi}{m\pi}$$

$$= \frac{(40 \text{ mm}) \tan 15 \text{ deg}}{3\pi \text{ mm}}$$

$$= 1.14$$

$$m_t = 1.51 + 1.14 = \mathbf{2.65}$$

Replacement of a pair of spur gears with an equivalent (same m_G) pair of helical gears, in this instance, increased the contact ratio by 75%.

Normal Plane Geometry

The normal plane contains the geometry needed for cutting teeth, and herein lies its importance. By means of this reference plane, helical gear teeth can be cut by the same tools as spur gears. The involute features, however, are contained in the transverse plane. They will differ from the standard normal values and, hence, there is a real need for relating parameters in the two reference planes.

Figure 13-51 shows the transverse or plane of rotation and a side view of a single helical gear. The normal plane, which is a reference plane, is shown on the left, along with an auxiliary view. The real and the reference gears have certain features in common, but they differ in other respects. Common features are number of teeth and radial dimensions (addendum, whole depth, and tooth thickness). The two gears differ with respect to diametral pitch; consequently, all diameters vary. The ratio is cosine of the helix angle, making the reference gear the *smaller* of the two.

The gear teeth are cut by moving or feeding the tool perpendicularly to the normal plane as indicated by the arrow in Fig. 13-51. Calculations thus start with the reference gear, which must first have a whole number of teeth. Standard formulas are then used to calculate addendum, whole depth, and tooth thickness. These are transferred without any changes to the real helical gear. For all other dimensions, the factor of transformation contains $\cos \psi$. The diametral pitch is usually given for the reference gear and should be a preferred value. The helix angle is usually less than 20 deg to avoid excessive end thrust. The helix angle, like all angles, cannot be measured with great accuracy unless measured indirectly. By measuring the lead, designers can calculate and check the helix angle using Eq. 13-35:

$$\tan \psi = \frac{\pi D}{L}$$

D = pitch diameter
L = lead

EXAMPLE 13-8

A pair of existing tractor gears are to be recalculated based on these data.

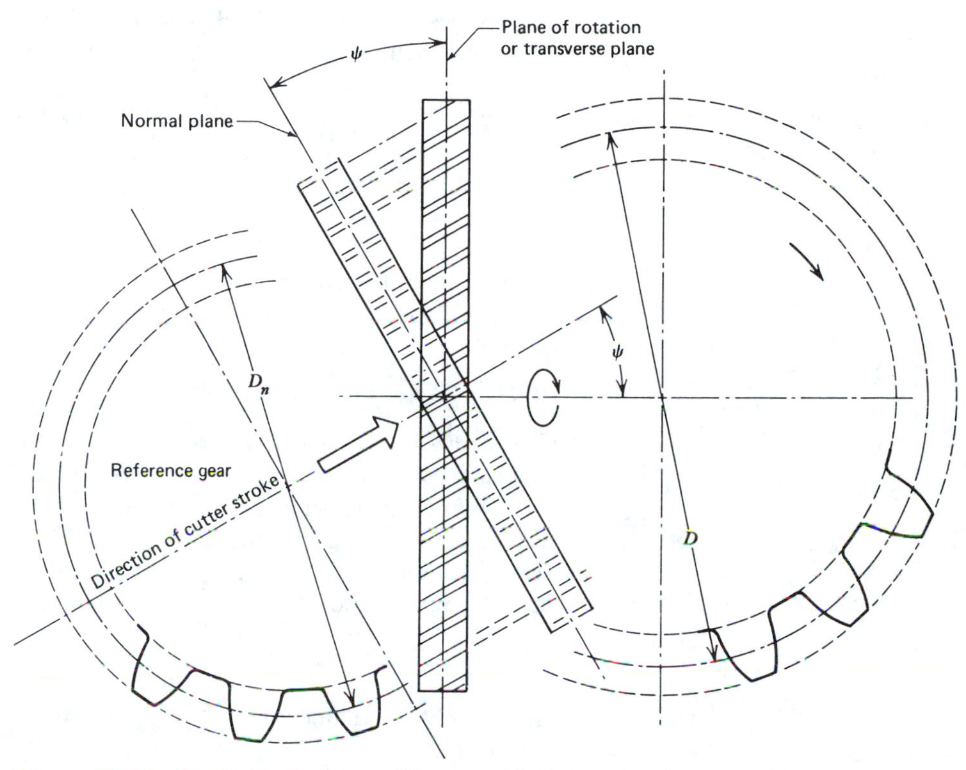

Figure 13-51 Parallel helical gear. The arrow indicates the direction of cutter stroke across the blank. Manufacturing considerations are the primary reason for calculating the reference gear.

$m_G = 1.5$ $\quad C = 5$ in. $\quad P_n = 7$

$\psi = 31$ deg $\quad \emptyset_n = 20$ deg

SOLUTION

In the *transverse* plane

$$D = m_G d = 1.5d$$

$D + d = 2C$, substituting D and C

$$1.5d + d = 2(5\text{ in.}) \Rightarrow \boldsymbol{d = 4}\textbf{ in.}$$

$$D = 1.5(4\text{ in.}) = 6\ in. \Rightarrow \boldsymbol{D = 6}\textbf{ in.}$$

Equivalent pitch diameters are found in the *normal* plane.

$$d_n = d \cos \psi = (4\text{ in.}) \cos 31\text{ deg} = 3.429\text{ in.}$$

$$D_n = D \cos \psi = (6\text{ in.}) \cos 31\text{ deg} = 5.143\text{ in.}$$

Both gears have the same number of teeth, but calculations relate to the normal plane since the normal diametral pitch is given.

$$N_p = P_n d_n = 7(3.429) = \mathbf{24}$$

$$N_G = P_n D_n = 7(5.143) = \mathbf{36}$$

The standard tooth proportions apply to both gears.

$$a = \frac{1}{P_n} = \frac{1}{7} = \textbf{0.143 in.}$$

$$b = \frac{1.25}{P_n} = \frac{1.25}{7} = \textbf{0.179 in.}$$

$$t_{c_n} = \frac{\pi}{2P_n} = \frac{\pi}{14} = \textbf{0.224 in.}$$

In the transverse plane

Pinion: $\quad d_0 = d + 2a = 4 + 2(0.143\text{ in.})$

$$= \textbf{4.286 in.}$$

Gear: $D_0 = D + 2a = 6 + 2(0.143 \text{ in.})$

$$= \mathbf{6.286 \text{ in.}}$$

The transverse pressure angle ϕ corresponding to a 20 deg normal pressure angle is

$$\tan \phi = \frac{\tan \phi_n}{\cos \psi} = \frac{\tan 20 \text{ deg}}{\cos 31 \text{ deg}}$$

$$= 0.425 \qquad \phi = \mathbf{23.01 \text{ deg}}$$

$$d_b = d \cos \phi = (4 \text{ in.}) \cos 23.01 \text{ deg}$$

$$= 3.682 \text{ in.}$$

$$D_b = D \cos \phi = (6 \text{ in.}) \cos 23.01 \text{ deg}$$

$$= 5.523 \text{ in.}$$

In the normal plane

$$p_n = \frac{\pi D_b}{N_G} = \frac{\pi 5.523}{36} = \mathbf{0.482 \text{ in.}}$$

To mesh properly, helical gears must have the same helix angle, but different leads.

Pinion: $$L_p = \frac{\pi d}{\tan \psi} = \frac{\pi(4 \text{ in.})}{\tan 31 \text{ deg}} = \mathbf{20.914 \text{ in.}}$$

Gear: $$L_G = \frac{\pi D}{\tan \psi} = \frac{\pi(6 \text{ in.})}{\tan 31 \text{ deg}} = \mathbf{31.371 \text{ in.}}$$

For further discussion of helical gears, see *Machinery's Handbook*, pp. 905–933.

13-18 GEAR DRAWINGS

The purpose of gear calculations is primarily to provide gear tooth data for production drawings. Figure 13-52 shows a gear-specification drawing for a spur gear as suggested by AGMA. Gears of simple design may thus be shown satisfactorily by a section through the axis with or

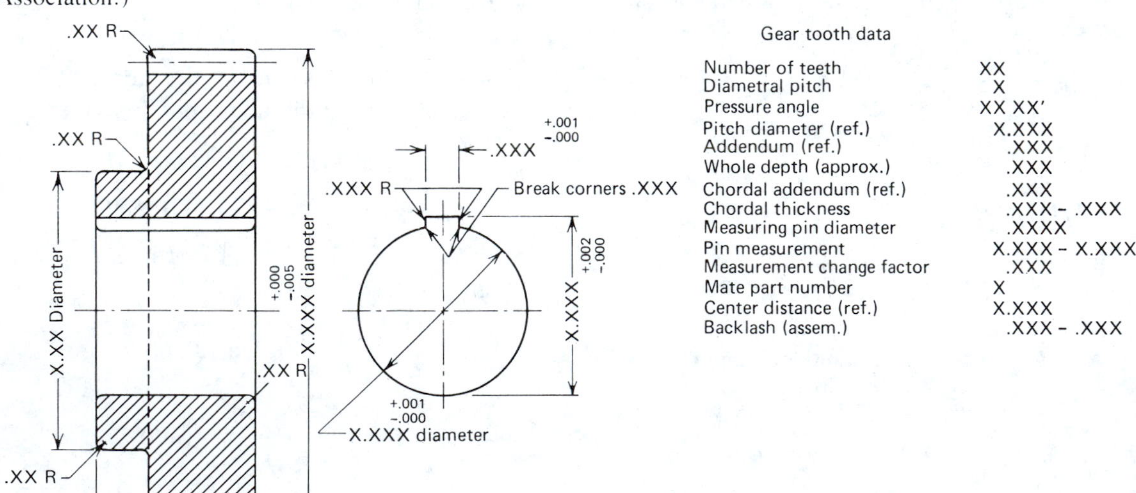

Gear tooth data

Number of teeth	XX
Diametral pitch	X
Pressure angle	XX XX′
Pitch diameter (ref.)	X.XXX
Addendum (ref.)	.XXX
Whole depth (approx.)	.XXX
Chordal addendum (ref.)	.XXX
Chordal thickness	.XXX – .XXX
Measuring pin diameter	.XXXX
Pin measurement	X.XXX – X.XXX
Measurement change factor	.XXX
Mate part number	X
Center distance (ref.)	X.XXX
Backlash (assem.)	.XXX – .XXX

Figure 13-52 Suggested standard dimensioning. External spur gear blank. (Extracted from AGMA Reference Information, Gear-Specification Drawings (AGMA 113.01 - 1952, R 1970), with permission of the Publisher, the American Gear Manufacturers Association.)

without a view in the direction of the axis. For more complicated gears, additional views may be needed.

A gear-specification drawing should give basic data on the gear itself, such as number of teeth, diametral pitch, and pressure angle. Additionally, it should furnish information that establishes the relations of the gear to its mating gear and to other parts, such as center distance, backlash, and the part number of the mating gear.

For further detailed information, consult AGMA 113.01-1952 (R 1970), Reference Information—Gear-Specification Drawings.

13-19 GEAR TRAINS FOR POWER TRANSMISSION

Any combination or series of interconnected gears is a gear train. The most common function of gear trains is to reduce the speed of power sources to match that required for a particular job. As can be seen from the power equation ($P = Tn/1000$), this will have the effect of increasing torque, thus providing a very desirable design feature. Trains of gears are needed because of practical limitations on speed ratios when only a pair of gears is used. Higher efficiency relative to ratio makes gear trains preferable to hydrostatic and worm gear drives in most continuous operations.

Gear trains designed to increase speed are also available, and most commercial speed reducers will function with the power flow in reverse (back-drive). Two common applications of speed-increasing drives are turbochargers and centrifugal refrigeration compressors operating at speeds of 9000 to 10 000 rpm. In the automotive industry a step-up gear ratio is known as an "overdrive."

Classification

Gear trains are classified according to shaft position, speed ratio, speed and mounting (Fig. 13-53). Tooth form—helical, bevel, or worm—is used to identify further the classification. An in-line helical speed reducer is an example of the terminology that might be used.

Speed reducers usually have a gear train that provides a single, fixed, output-input ratio, although special features such as an infinitely variable speed ratio are also available. Speed reducers are commercially produced as packaged units to which other drive components can be attached.

The term *gear box* should not be confused with speed reducer. Both have the same function, but gear boxes contain sets of gears *integrally* assembled within a piece of equipment, such as machine tools.

For compactness of design, motors and gear trains are often available as a package in what is known as a *gear motor*. In this integral design the motor shaft itself becomes the input or high-speed shaft.

A *transmission* is a device for transmitting power at a multiplicity of interdependent speed and torque ratios.

Speed reducers, like electric motors, are available in a wide variety of sizes to cover many design situations. Transmissions, in contrast, are usually built to satisfy a specific need. Both are rated by torque or power at a specified speed.

Basic Reducer Styles

The four basic types of speed reducers are spur or helical, bevel, worm, and planetary. Spur and helical are the most common gear types for ratios from 1:1 to 7:1 per stage. When total power exceeds 7000 to 8000 kW (10,000 hp), they are the only reducers available. Ratios up to 350:1 can be achieved with three sets of gears in a compound arrangement. Parallel shafts add to simplicity and accuracy, while helical teeth reduce noise.

SUMMARY

Involute gears are the most common, most compact, most versatile, most efficient, and most durable means of mechanical power

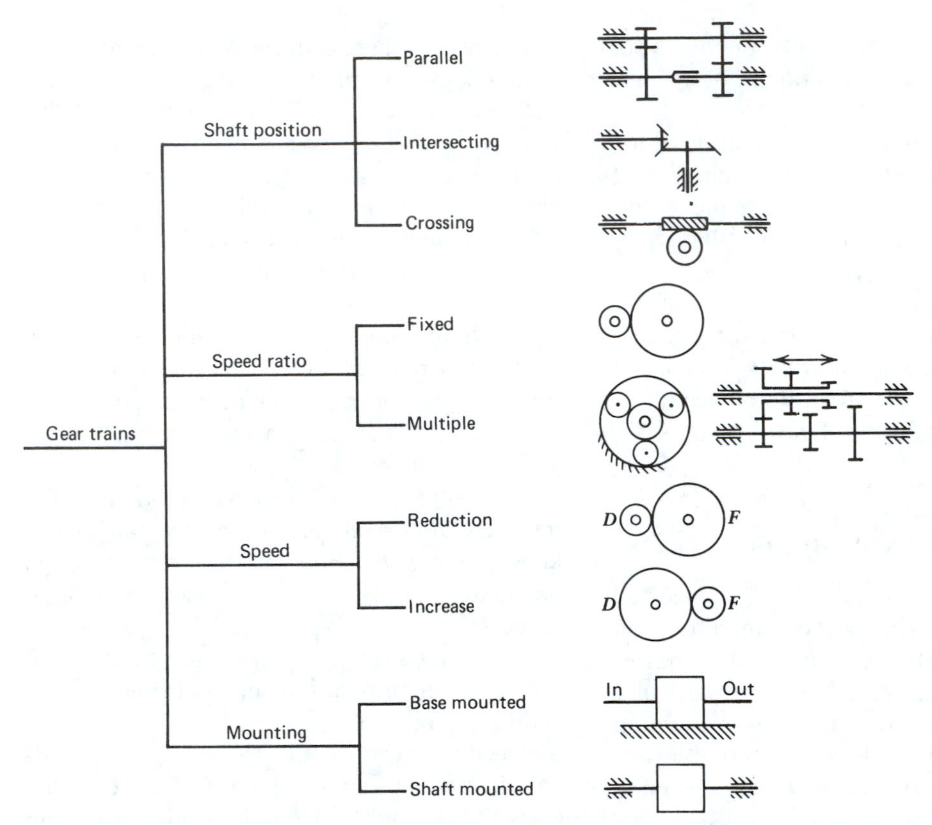

Figure 13-53 Classification of gear trains.

transmission. These advantages, however, are obtained against higher initial cost and noisy operation. Gears are classified according to shaft position, tooth form, tooth system, and quality. Extensive national standards facilitate design and manufacturing. Proportioning is based on two criteria: surface durability and fillet strength. For each design criterion, AGMA has specified exact formulas and procedures.

The spur gear is the most widely used type because of its simplicity, low tooling cost, and the great precision with which it can be made. The higher cost of helical gears is justified in terms of greater tooth strength, much higher speeds, and greater power capacity. Internal gears are advantageous where space is at a premium.

REFERENCES

13-1 Buckingham, E. *Analytical Mechanics of Gears*. New York: McGraw-Hill, 1949. (The "Bible" of gearing.)

13-2 Dudley, D. W., Ed. *Gear Handbook: The Design, Manufacture and Application of Gears*. New York: McGraw-Hill, 1962. (Covers in 21 chapters all major aspects of gearing.)

13-3 *Handbook of Stock Gears. Catalog 75: Inch & Metric Sizes*. New Hyde Park, N.Y.: Stock Drive Products, 1975.

13-4 Michalec, G. W. *Precision Gearing*. New York: Wiley, 1966. (An outstanding book on instrument, computing, and servogears. Combines analytical engineering with the practical art of making precision gears.)

13-5 Mobil Oil Corporation. *Gears and Their Lubrication*. Copyright © 1971. (This 50-page booklet is well illustrated, informative, and easy to read. Contains no math.)

13-6 *Precision Electro-Mechanical Components*. New Hyde Park, N.Y.: Sterling Instrument, 1977. (Contains 40 pages on gear design written by Dr. George Michalec.)

13-7 Tucker, A. I. "Gear Design: Dynamic Load." *Mechanical Engineering*, October 1971.

13-8 Tucker, A. I. "The Gear Design Process." *ASME Publication 80-C2/Det-13*. (A very lucid discussion of the step-by-step procedure required to design a new gear unit starting from a blank piece of paper.)

SUMMARY OF FORMULAS

Name and Symbol		SI	EU	SI and EU
Pitch diameter	D	mN	N/P_d	Np_c/π
Circular pitch	p_c	$m\pi$	π/P_d	$\pi D/N$
Module	m	$25.4/P_d$, D/N		
Diametral pitch	P_d		$25.4/m$, N/D	N/D
No. of teeth	N	D/m	DP_d	$\pi D/p_c$
Addendum	a	m	$1/P_d$	
Dedendum	b	$1.25m$	$1.25/P_d$	
Outside diameter	D_0	$D + 2m$ $m(N + 2)$	$D + 2/P_d$ $(N + 2)/P_d$	
Root diameter	D_r d_r	$D_0 - 4.50m$ $d_0 - 4.50m$	$D_0 - 4.50/P_d$ $d_0 - 4.50/P_d$	
Base circle diameter	D_b d_b			$D \cos \phi$ $d \cos \phi$
Base pitch	p_b	$m\pi \cos \phi$	$(\pi/P_d) \cos \phi$	
Center distance	C	$0.5m(N_p + N_G)$	$(N_p + N_G)/2P_d$	$0.5(D + d)$
Speed ratio	m_G	$\dfrac{n_p}{n_G} = \dfrac{D_G}{D_p} = \dfrac{N_G}{N_p}$		
Contact ratio	m_p	$\dfrac{(r_0^2 - r_b^2)^{0.5} + (R_0^2 - R_b^2)^{0.5} - C \sin \phi}{p_c \cos \phi}$		
Minimum no. of teeth for no undercutting	N_c		$2/\sin^2 \phi$	
To avoid interference			$R_0 \leq (R_b^2 + C^2 \sin^2 \phi)^{0.5}$ $r_0 \leq (r_b^2 + C^2 \sin^2 \phi)^{0.5}$	

PROBLEMS

Unless stated otherwise, the following problems relate to 20-deg, full-depth, involute spur gears based on preferred pitches and modules. Spur gear sizing is done primarily by means of the AGMA standard procedures outlined in *Machinery's Handbook,* 21st Edition, pp. 810–821. As an additional assignment, instructors may ask students to provide a gear specification drawing.

P13-1 A machine shop has tools for gear cutting, ranging in pitch size from 5 to 10. Metric "replacements" are to be ordered. Specify appropriate replacements. Are these gears interchangeable?

P13-2 Previously when a lathe was used in cutting threads, in both the English and metric units, the conversion from inches to centimeters (1 in. = 2.54 cm) was done by means of a single pair of lathe change gears connecting the lathe spindle with the lead-screw. Find such a pair. For a center distance not to exceed 90 mm, find a suitable preferred module.

P13-3 A pinion has an outside diameter of 84 mm and 32, 25-deg stub teeth. Determine module, circular pitch, base pitch, tooth height, and theoretical tooth thickness.

P13-4 A pinion turns at 500 rpm, driving a gear at 200 rpm. Center distance is fixed at 126 mm, and the module is 4 mm. Find the number of teeth required for gear rotation in (*a*) opposite directions, and (*b*) the same direction. Make a simple sketch of both arrangements. *Hint:* Use Eqs. 13-3, 13-4, and 13-7.

P13-5 For a pair of spur gears the following data are given.

$$m = 4 \text{ mm} \qquad C = 162 \text{ mm}$$

$$n_p = 950 \text{ rpm} \qquad m_G = 3.5$$

Find N_p, N_G, n_G, tooth height h_t, and thickness, t_c.

P13-6 A pair of meshing spur gears has 18 and 28 teeth. Outside diameter of the pinion is 80 mm. Determine (*a*) module, (*b*) pitch diameters, (*c*) center distance, and (*d*) base circle radii.

P13-7 For the gears in P13-6, find the contact ratio and check for interference.

P13-8 In a pair of gears the pinion has 18 teeth and the gear has 63 teeth. The pitch is 9 and the gear rotates at 270 rpm. Find the center distance and pinion speed for internal and external contact. Calculate the reduction of the center distance in percent.

P13-9 A pair of standard spur gears has a module of 6 mm. The pinion and gear have 20 and 24 teeth, respectively. Make a double-scale geometric layout. Measure the length of contact and calculate the contact ratio. Check for interference graphically. What is the advantage of a graphical layout compared with computations?

P13-10 A 20-deg, 20-tooth stub-tooth pinion has an addendum of 16 mm. Calculate module, dedendum, theoretical tooth width, circular pitch, and base pitch.

P13-11 A pair of spur gears with a pitch of 6, a speed ratio of 4, and a center distance of 10 in. is to be replaced by a pair of module-based gears. Specify such gears when (*a*) exact center distance is important, and (*b*) exact speed ratio is important.

P13-12 Two gears $P_d = 8$, $N_p = 15$, and $N_G = 17$ are available from stock and fit a particular design as far as ratio, center distance, and strength are concerned. Would you recommend them for continuous run, particularly if noise is a factor?

P13-13 Consider two gears $N_G = N_p = 63$ and $P_d = 9$, in mesh. Calculate the contact ratio m_p as a function of the pressure angle $\emptyset$ for the following values: 14.5 deg; 17.5 deg; 20 deg; 22.5 deg; 25 deg; 27.5 deg; and 30 deg. Plot m_p as a function of $\emptyset$. What is your conclusion?

P13-14 Calculate and plot N_c (the minimum number of teeth required to avoid undercutting of standard involute teeth), as a function of the pressure angle, for the following values of $\emptyset$: 30 deg to 14 deg in increments of 2 deg, plus the preferred values 25 deg, 22.5 deg, and 14.5 deg. Use a scale in which one degree corresponds to 5 mm on the abscissa. For the ordinate let 5 mm correspond to one tooth. Discuss the significance of this curve.

P13-15 Figure 13-54 depicts the countershaft in a two-stage reduction gear that is to transmit 5.25 kW. The countershaft, containing input gear and output pinion, rotates at 250 rpm. The pressure angle is 20 deg and the gears are spur gears. The output pinion has a pitch diameter d = 57.15 mm and the input gear a pitch diameter D = 146.05 mm. Find the bearing forces.

P13-16 Use the Lewis equation to determine the size and center distance of two gears to transmit 7.5 hp. The pinion rotates at 1200 rpm, the gear at 200 rpm. Assume steel 400 Bhn S_e = 42,000 psi. *Hint:* Use 24 teeth in the preliminary calculations of the pinion.

P13-17 Two parallel shafts, 15 in. apart, are connected by two gears. The drive shaft running at 270 rpm will transmit 40 hp to the driven shaft with a speed reduction of 3:1.

(*a*) Calculate the pitch diameters.
(*b*) Use the Lewis equation to find a suitable diametral pitch. Assume steel 200 (S_e = 27,000 psi). A preliminary value for y may be obtained by assuming N_p = 24.
(*c*) Find the actual number of teeth in both gear and pinion.
(*d*) Find k by means of the Lewis equation and estimate the gear width.

P13-18 Use the Lewis equation to estimate the torque capacity of a 24-tooth steel pinion, S_e = 32,000 psi, P_d = 6, and F = 2 in. Draw a curve for T, for

Figure 13-54 The countershaft in a two-stage reduction gear. Forces are shown on the countershaft as a free body (Problem 13-15).

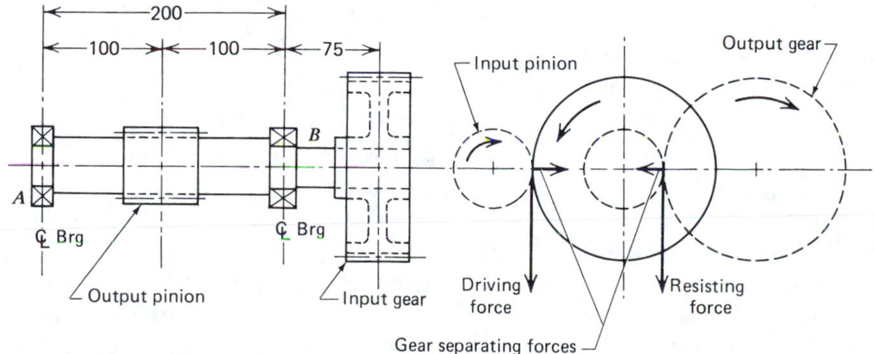

speeds in increments of 100 rpm up to 1200 rpm. *Hint:* Use Eq. 13-27, solve for T, find k and C_v.

P13-19 Prove that the teeth of a pinion will always be weaker than those of a mating gear when standard proportions are used. Use the Lewis equation to prove this for beam strength applied to a 20-tooth pinion and a 60-tooth mating gear. *Hint:* Compute the ratio of beam strength using Eq. 13-22.

P13-20 Reference to P13-17. How much can the center distance be reduced if carburized steel, S_e = 55,000 psi, is used? Use C_v = 0.53 obtained in problem 13-17. For maximum reduction in size, use k = 4 and a minimum number of teeth as specified by Eq. 13-14.

P13-21 How much power can the gears in P13-8 transmit at a speed of 900 rpm if the fatigue strength in bending is S_e = 27,000 psi? Specify the gear width. *Hint:* Use the Lewis equation.

P13-22 How much power can the gears in P13-11 transmit at a speed of 1200 rpm if the material has a bending fatigue strength of 45,000 psi? Specify the gear width and center distance. Use the Lewis equation.

P13-23 Use the Lewis equation to calculate the torque capacity of the gears in P13-12. Specify width and center distance for S_e = 36,000 psi.

P13-24 A manufacturer of tractors decided to increase the torque capacity of the tractor transmission by increasing the pressure angle of all gears from 20 deg to 25 deg. Estimate the approximate torque increase of a 20-tooth pinion driving a 60-tooth gear, assuming operating conditions remain the same and that the limiting value for tensile strength has not decreased. Does this require a change in surface durability? If so, state your reasons. *Hint:* Calculate the ratio of beam strength using the J values, based on ded. = 1.25 from Table 5 in *Machinery's Handbook,* p. 813.

P13-25 Two parallel shafts, 15 in. apart, are connected by two spur gears. The drive shaft running at 270 rpm will transmit 40 hp to the driven shaft with a speed reduction of 3 : 1. Use the AGMA standards in *Machinery's Handbook* to "size" these gears. Only preferred values for P_d should be used. *Hint:* Find the pitch diameters and select a suitable pitch, which must be a whole number. Select a material that will provide adequate "fillet tensile stress." Check the gears for surface durability using AGMA standards.

P13-26 Determine size and material for two spur gears to transmit 7.5 hp using AGMA standards. The pinion rotates at 1200 rpm; the gear rotates at 240 rpm. Assume through-hardened steel gears, 400 Bhn. Check the gears for surface durability using AGMA standards.

P13-27 Specify a pair of ferrous commercial gears to transmit 10 hp at 1200 rpm with a reduction ratio of 4:1. The pinion shaft is driven by V-belts. The output shaft has cams mounted on it. A high reliability is required of the gear set. Check the gears for surface durability. Use AGMA standards.

P13-28 Estimate the torque capacity of a 25-tooth steel pinion with a surface hardness of 200 Bhn, meshing with a suitable steel gear having 25 teeth. The gear width is 7.0 in. and the pitch is 6. Assume precision gearing. What are your conclusions? Use AGMA standards.

P13-29 A reciprocating compressor is driven by a 900 rpm motor through a pair of commercial spur gears. The recommended operating speed is roughly 210 rpm. The corresponding torque is 2500 lb-in. For a starting overload of 30%, select a pair of suitable ferrous gears. Check the gears for surface durability. Use AGMA standards.

P13-30 For helical gears in general, the face width most commonly used lies somewhere between two and four times the circular pitch. Find the corresponding limits imposed on the helix angle.

P13-31 What is the pitch diameter and circular pitch of a helical gear having 30 teeth and a diametral pitch of 3?

P13-32 The helix angle of a gear is 25.84 deg. What is the normal diametral pitch if the diametral pitch is 3.6?

P13-33 What is the normal and axial pitch of a parallel helical gear in which the circular pitch is 26.594 mm and the helix angle 30 deg? What is the module?

P13-34 A helical gear has a normal diametral pitch of 8, 40 teeth, and a helix angle of 35 deg. What is the outside diameter?

P13-35 A helical gear has 40 teeth, a circular pitch of 8.179 mm, and a helix angle of 35 deg. What is the lead?

P13-36 A 20-tooth helical gear has a pitch diameter of 260.720 mm. The helix angle is 23 deg and the pressure angle is 20 deg. Find the normal module, the normal circular pitch, and the normal pressure angle.

P13-37 Calculate the total contact ratio for the helical gear discussed in Example 13-8.

P13-38 A parallel helical gear, 125 mm in diameter, with a helix angle of 15 deg, carries a torque of 150 N · m. Determine the force components and the total tooth load.

P13-39 For the gear in Example 13-7, calculate the total contact ratio, m_t for ψ, assuming the values 5 deg, 10 deg, 15 deg, 20 deg, 25 deg, 30 deg, and 35 deg. Plot m_t as a function of ψ.

P13-40 For the gear in Example 13-7, calculate the limits of the face width. Calculate and plot m_t when F varies in increments of 0.25 mm between its minimum and maximum value.

Chapter 14
Gears for Nonparallel Shafts

There is no great concurrence between learning and wisdom.

FRANCIS BACON

14-1 BEVEL GEARS

14-2 HYPOID GEARS

14-3 HELICAL GEARING

14-4 WORM GEARING

14-5 WORM GEAR TERMINOLOGY AND KINEMATICS

14-6 THERMAL RATINGS

14-7 EFFICIENCY OF WORM GEARING

14-8 OPTIMUM DESIGN

14-9 APPLICATIONS

14-10 DESIGN DETAIL OF WORM GEARING

NOMENCLATURE

C = center distance
d = pitch diameter of bevel pinion and worm
D = pitch diameter of bevel gear or worm gear
D_G = pitch diameter of worm gear
D_w = pitch diameter of worm
e = efficiency
L = lead of worm
m = module
m_G = speed ratio
n_G, n_w = speed of gear and worm, respectively
N_p = number of teeth in pinion
N_G = number of teeth in gear
N_w = number of teeth in worm
p_a = axial pitch
p_c = circular pitch
P_d = diametral pitch
TR = thermal rating
λ(lambda) = lead angle
Γ(gamma) = pitch angle
Σ(sigma) = shaft angle
ψ(psi) = helix angle
θ(theta) = friction angle

Transmission of power between nonparallel shafts is inherently more difficult than transmission between parallel shafts, but is justified when it saves space and results in more compact, more balanced designs. Where *axial* space is limited compared to *radial* space, *angular* drives are preferred despite their higher initial cost. For this reason, angular gear motors and worm gear drives are used extensively in preference to parallel shaft drives, particularly where couplings, brakes, and adjustable mountings add to the axial space problem of parallel shaft speed reducers.

In angular drives, the gears not only rotate in different planes, but contact is frequently diagonal across the face of mating teeth (Fig. 14-1). Such gears are generally more difficult to design, manufacture, and install and thus cost more than equivalent spur and parallel-axis helical gears. They are also more sensitive to mounting and manufacturing errors. In addition, mounting on overhanging shafts makes some types sensitive to shaft deflections. As with spur and helical gearing, optimum design becomes a compromise. Two general classifications cover the right-angle gears: *coplanar* types, which have intersecting axes, and *offset* types, which have nonintersecting or skew axes (their axes do not lie in a common plane).

14-1 BEVEL GEARS

Bevel gears provide the most *efficient* means of transmitting power between intersecting shafts. The related friction wheels are frustrums of cones, and the gears developed on these conical surfaces are called bevel gears. If teeth are cut straight across the faces of conical blanks, the gears are called *straight* bevel; when the teeth are twisted along a curved path, the gears are termed *spiral* bevel (Figs. 14-1 and 14-2). The involute tooth form is used.

Customarily, tooth dimensions are determined in a transverse plane (perpendicular to the common element of the pitch cones) at the *large* end of the teeth (Fig. 14-3*a*). Intersection of tooth surfaces with this plane gives a tooth profile as shown in Fig. 14-3*b*. Figure 14-3*a* also shows that the shaft angle equals the sum of the pitch angles. Thus

$$\sum = \Gamma_p + \Gamma_G \qquad (14\text{-}1)$$

where

Σ(sigma) = shaft angle; deg
Γ_p(gamma) = pitch angle of pinion; deg
Γ_G(gamma) = pitch angle of gear; deg

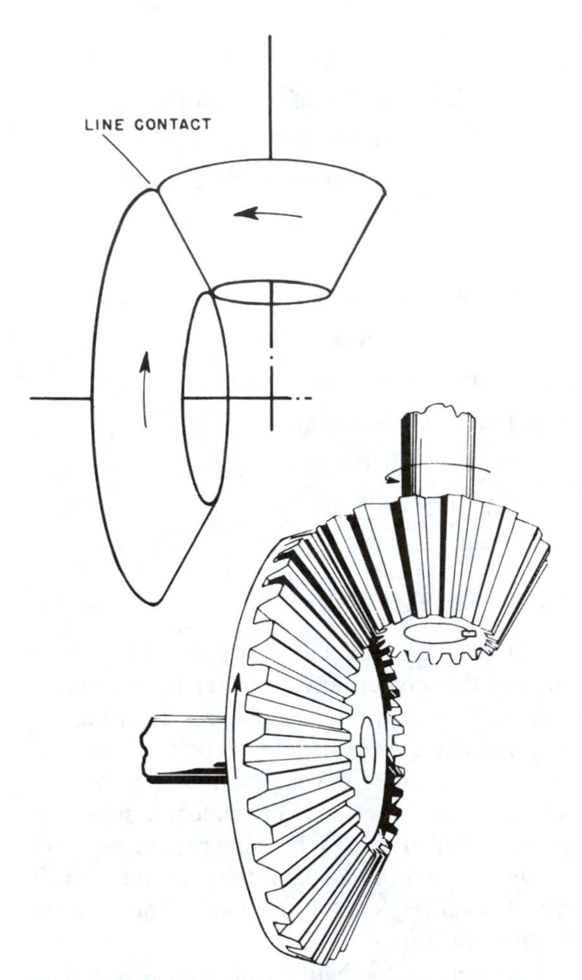

Figure 14-1 Bevel gears. These shafts intersect at a right angle, although bevel gears may also be used between shafts that intersect at larger or smaller angles. (Courtesy Mobil Oil Corporation.)

Figure 14-3*c* indicates that the following relationship exists for $\Sigma = 90$ deg.

$$\tan \Gamma_p = \frac{d}{D} = \frac{N_p}{N_G} \tag{14-2}$$

$$\tan \Gamma_G = \frac{D}{d} = \frac{N_G}{N_p} \tag{14-3}$$

where

d = pitch diameter of pinion; mm, in.
D = pitch diameter of gear; mm, in.

Figure 14-2 Spiral bevel gears. (Courtesy Mobil Oil Corporation.)

N_p = number of teeth in pinion
N_G = number of teeth in gear

Nomenclature and symbols commonly used for straight bevel gears are shown in Fig. 14-4. The similarity to spur gearing should be noted. With few exceptions, the nomenclature also applies to spiral bevel gears.

Equivalent spur gears is a term commonly used in connection with bevel gears. Two bevel gears roll on each other in the same manner as a pair of spur gears with pitch diameters equal to those of the bevel gears. In Fig. 14-4 the equivalent spur gear would have a pitch diameter D_G.

The difference between spiral and straight bevel gears is that spiral teeth have a *gradual* pitch line contact and a larger number of teeth in contact. Their teeth, instead of engaging in a full line contact at once, engage with one another gradually. This continuous contact makes it possible to obtain smoother action than is possible with straight bevels.

Arrangement

Bevel gears are widely used where a right-angle change in direction of shafting is required, although the shafts occasionally may intersect at

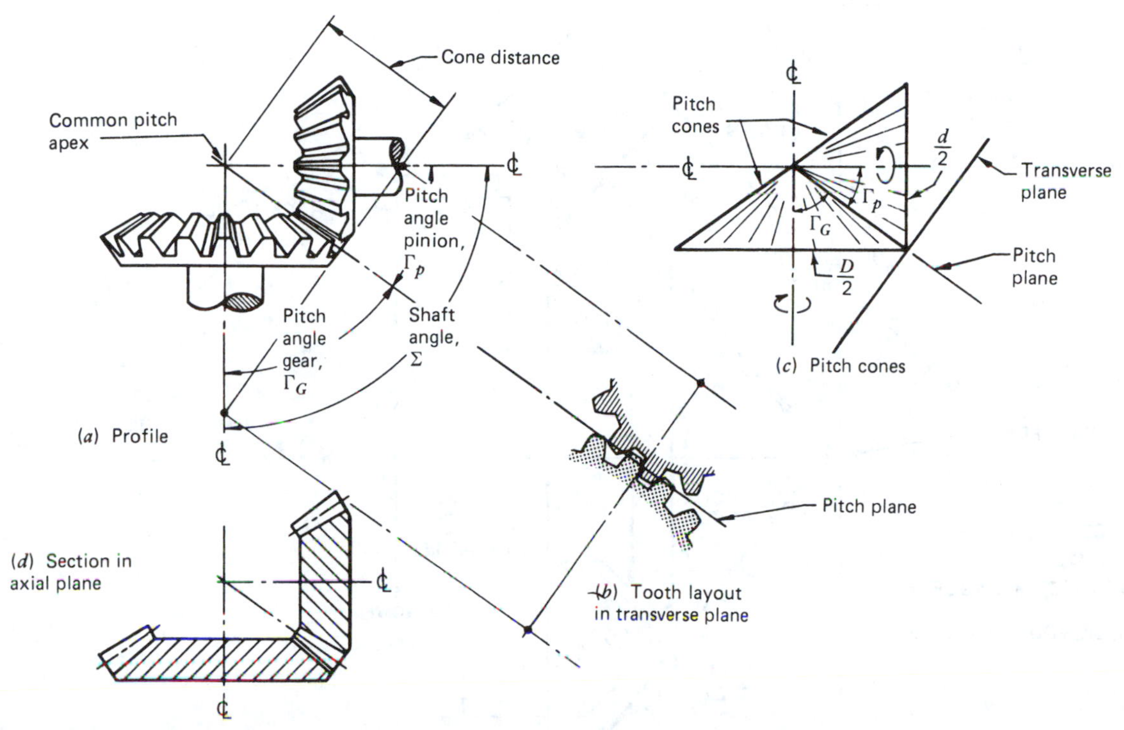

Figure 14-3 Basic bevel gear sections. (Courtesy General Motors Corporation.)

acute or obtuse angles (Fig. 14-5). When of equal size and mounted on shafts at right angles, they are referred to as *miter* gears (Fig. 14-5*b*). A bevel gear with a right (90 deg) pitch angle is a *crown* gear (Fig. 14-5*e*). Internal bevel gearing, like internal spur or helical gearing, is sometimes used in planetary or internal gear arrangements (Fig. 14-5*f*).

Application

Straight bevel gears, like spur gears, are well suited for manual and low-speed operations, such as in small hoists, valves, gates, or doors. When greater speed and more power are required, spiral bevel gears are preferable.

Mounting

Bevel gears require larger shaft diameters and heavier bearings because they impose high reaction loads on bearings.

Speed Ratio

As in spur gears, the speed ratio is the ratio of tooth numbers. For bevel gears, it is also the ratio of corresponding pitch radii or diameters of the pitch cones (Fig. 14-4).

Design of Bevel Gears

Bevel gears are simple modifications of spur gears. Beam strength and surface durability determine size and surface hardness. The AGMA formulas used are therefore simple modifications of the formulas used for calculating spur gears. These formulas, along with basic information on bevel gears, can be found in *Machinery's Handbook,* pp. 837–874.

EXAMPLE 14-1

A pair of straight-toothed bevel gears are to be designed for a shaft angle of 90 deg and a

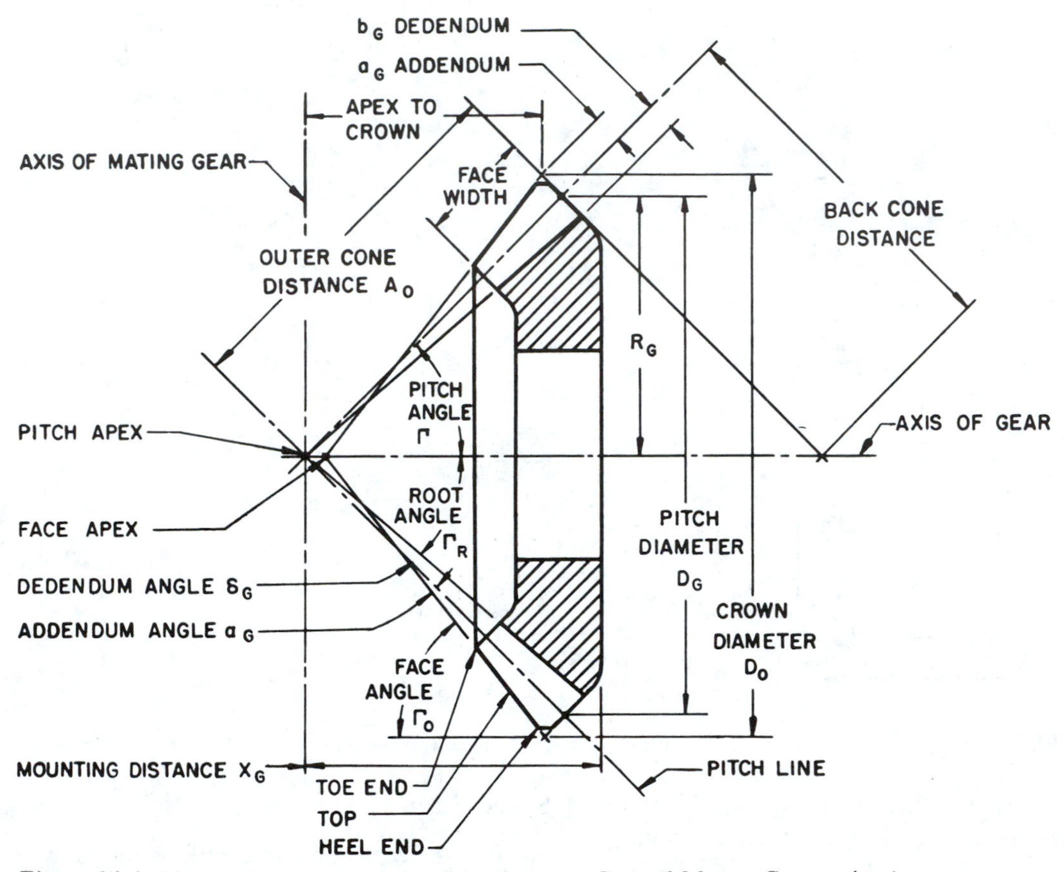

Figure 14-4 Nomenclature for bevel gears. (Courtesy General Motors Corporation.)

reduction ratio of roughly 3:1. If the pinion is to have a minimum of 17 teeth, find pitch angles and the number of teeth in the gear.

SOLUTION

$N = (3)(17) + 1 = \mathbf{52}$ leads to a hunting ratio equal to 3.06:1. Assuming a module m, we obtain for the equivalent spur gears:

$$d = m\ (17\ \text{mm}) \qquad D = m\ (52\ \text{mm})$$

$$\tan \Gamma_p = \frac{N_p}{N_G} = \frac{17}{52} \longrightarrow \Gamma_p = \mathbf{18.1\ deg}$$

$$\tan \Gamma_G = \frac{N_G}{N_p} = \frac{52}{17} \longrightarrow \Gamma_G = \mathbf{71.9\ deg}$$

Check:

$$\sum = \Gamma_p + \Gamma_G = 18.1\ \text{deg} + 71.9\ \text{deg} = 90\ \text{deg}$$

14-2 HYPOID GEARS

Hypoid gears closely resemble spiral bevel gears except that the pinion does not meet the ring gear at its center. It meets it at a *lower* point (Fig. 14-6). The pitch surfaces are hyperboloids from which the term "hypoid" was derived. Curved teeth contribute to smooth, *noiseless* operation even at high speed.

Hypoid gears grew out of a need for *silent* automotive differentials that would also allow

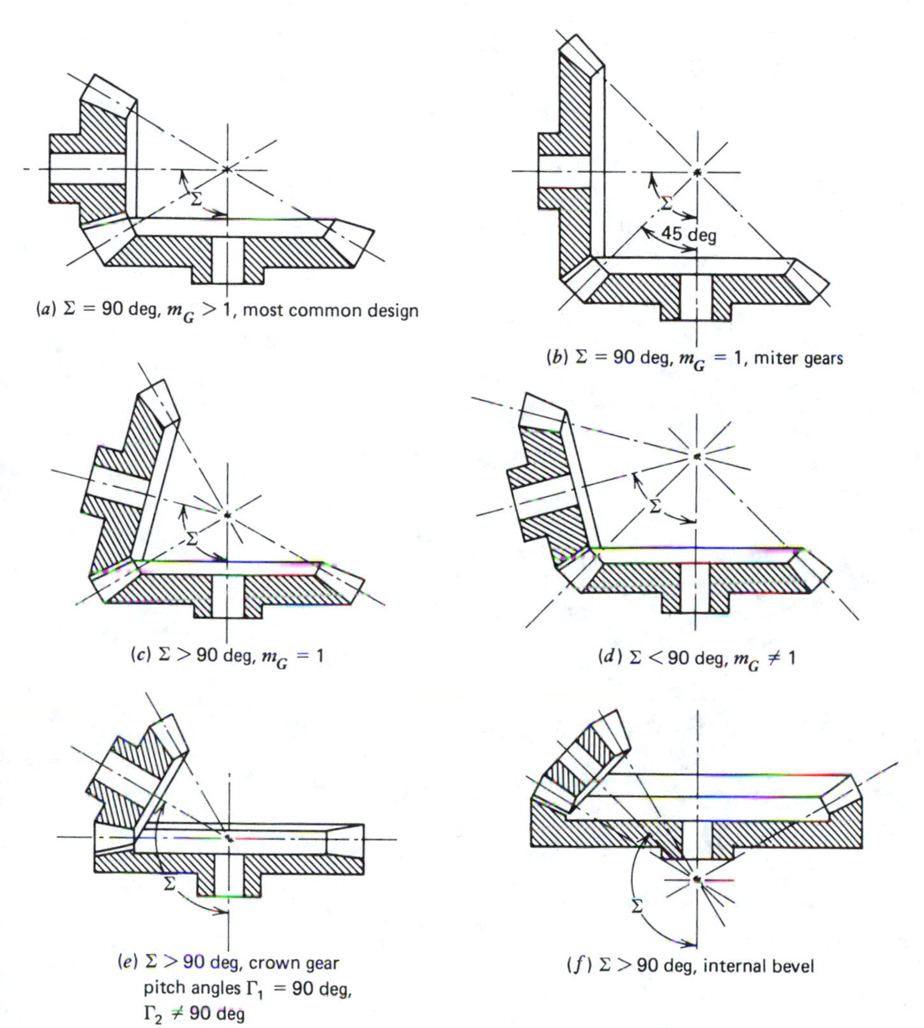

Figure 14-5 Bevel-gear arrangements. (Courtesy General Motors Corporation.)

the drive shaft to be placed well below the centerline of the rear axle, thus contributing to a lower body design. Because the two shafts do not intersect, (1) two rear axles, instead of one, can be successively driven from the same transmission shaft, and (2) bearings can be mounted on both sides of the pinion. Although the altered tooth shape results in more sliding, lower efficiency, and the need for special lubricants, it provides a perfectly smooth drive and solves a major automotive problem—noise.

14-3 HELICAL GEARING

Helical gearing is a term applied to all types of gears whose teeth are of helical form. Helical gears work equally well to connect parallel shafts (discussed in Chapter 13) and nonparallel, nonintersecting shafts. In the latter case, however, a distinction must be made between a gear and a worm, even though both have helical teeth. As seen in Fig. 14-7, the teeth on this gear make only a *fraction* of a turn on the base cylin-

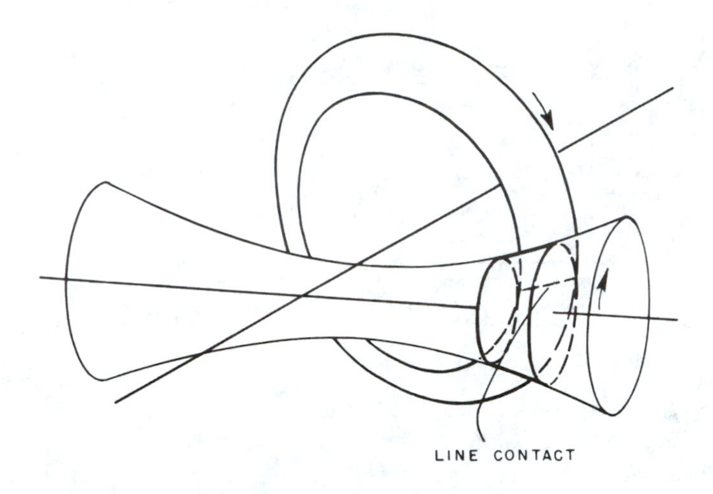

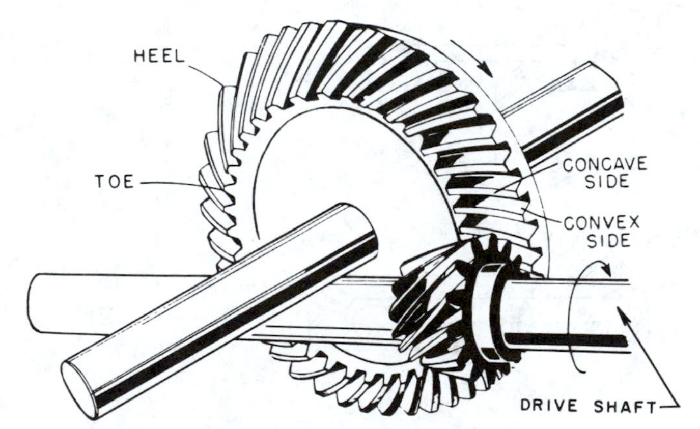

Figure 14-6 Hypoid gear and pinion. These gears transmit motion between nonintersecting shafts crossing at a right angle. The pitch surfaces are hyperbolic in form. (Courtesy Mobil Oil Corporation.)

der. That the tooth curvature is helical is not even obvious. What the teeth lack in length, however, they make up for in number, which always exceeds 10.

Crossed Helical Gearing

Figure 14-7 shows this form of gearing. For what they can do kinematically, crossed helical gears are the acme of simplicity. Tooth contact, however, is only a point, which greatly limits their power transmission capability.

14-4 WORM GEARING

Worm gear drives are used on right-angle applications with nonintersecting shafts. They provide smooth, quiet action and maximum reduction ratios for a given center distance. These favorable characteristics are obtained by using a worm gear combination. As seen in Fig. 14-8, the gear has teeth inclined at the same angle as the threads in the worm. For speed reduction or torque amplification, the worm is the driver.

In a *worm,* the number of teeth *rarely* exceeds

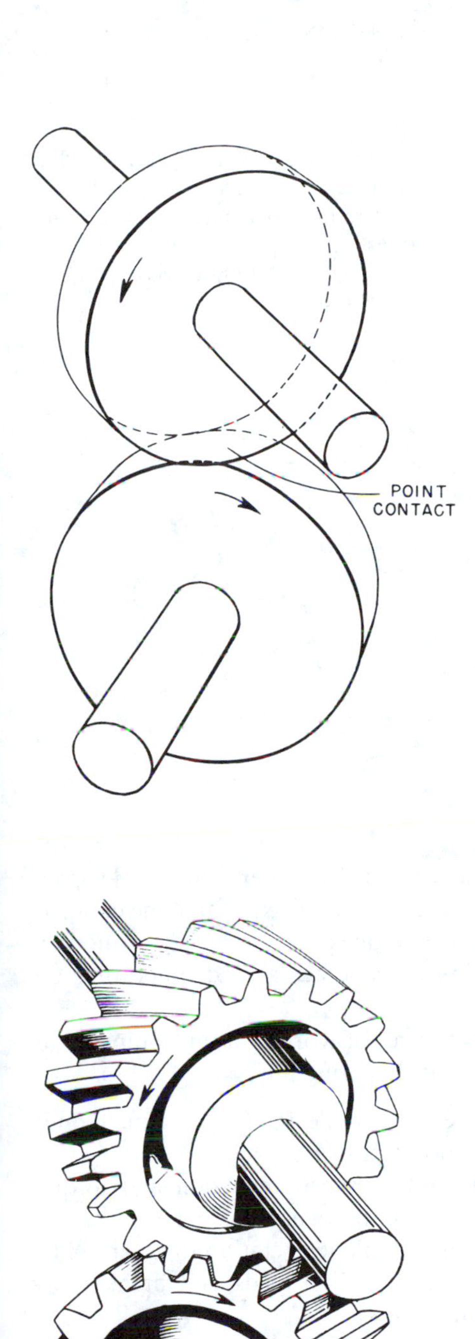

10 (one to four teeth is common, as shown in Figs. 14-8 and 14-9). Each tooth, however, makes at least one revolution on the base cylinder. If only one tooth is used, it winds around the base cylinder several times, like a screw thread.

Worm gearing derives its characteristics from the two simple machines of which it is composed: the screw and the lever. From the screw it obtains a *large* mechanical advantage but a somewhat *lower* efficiency because of larger friction forces. The conjugate tooth action is identical to that of a large spur gear and rack (Fig. 14-8). As the worm revolves, the thread form advances along its axis and the worm gear rotates a corresponding amount. The presence of a screw instead of a rack ensures a vastly greater output torque on the gear shaft. The net effect is a torque converter of superior capacity but, because of inherent sliding action, one of reduced efficiency.

Worm gear terms are shown in Fig. 14-9. In this particular case the *lead,* the distance advanced during one revolution, is three times greater than the pitch. Triple-threaded worms are more efficient than single-threaded worms (due to less friction), but their reduction ratio is only one-third that of the single-threaded worm.

Figure 14-10 shows the basic difference between single- and double-thread worms. The slope or helix angle of the double-thread worm is *twice* that of the single-thread worm, as is the lead. Thus, for one revolution the double-thread worm will advance or turn its mating gear an angle *twice* that of the single-thread worm.

Tooth breakage from bending action is not prevalent in worm gear sets. With relatively high sliding velocities, the design criteria are usually based on scoring and pitting. *Scoring* is wear

Figure 14-7 Crossed helical gears. These gears transmit motion between nonintersecting shafts crossing at an acute angle. The teeth are developed on cylindrical pitch surfaces. Since only point contact exists between the gears, this arrangement is rarely used to transmit loads of any magnitude. (Courtesy Mobil Oil Corporation.)

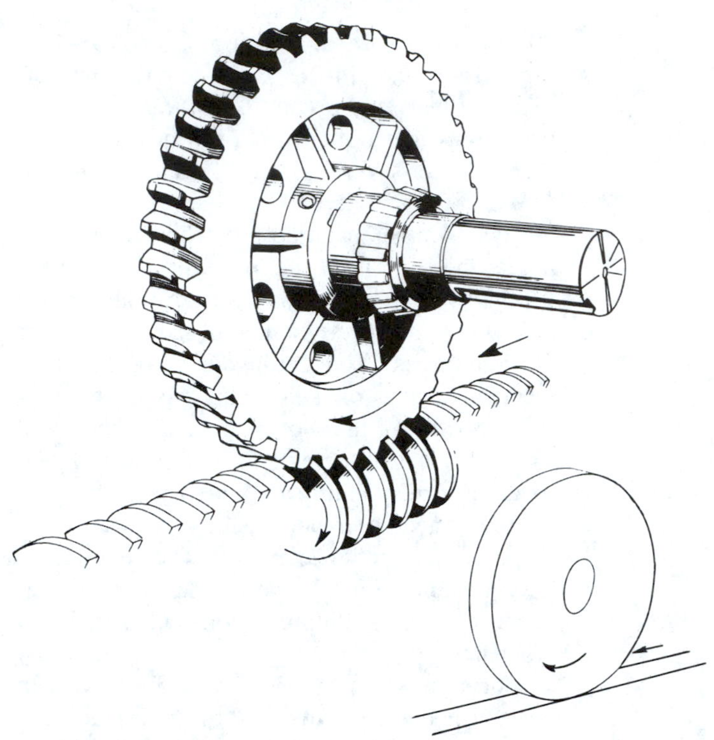

Figure 14-8 Worm gear. In this drawing the worm is represented as an endless rack. The resulting pitch surfaces are a plane and a cylinder. (Courtesy Mobil Oil Corporation.)

resulting from failure of the lubricant film due to localized overheating of the mesh, permitting metal-to-metal contact. Because more teeth are in contact simultaneously, worm gearing provides smoother operation than involute gearing.

Figure 14-9 Worm gear terms. (Courtesy Mobil Oil Corporation.)

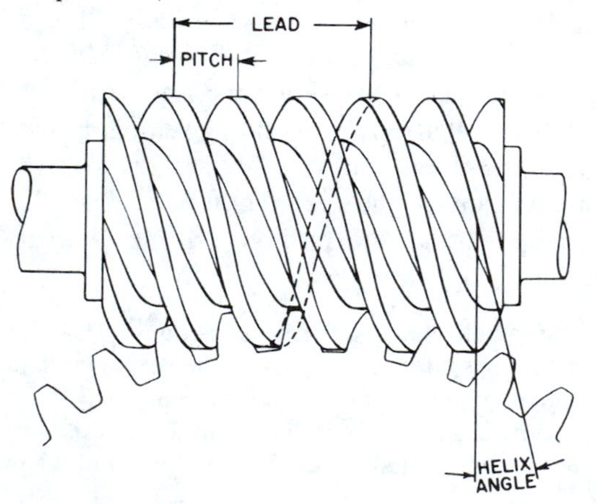

The contact area is also larger. Thus load capacity is high despite sliding action and line contact. Entering side wedges are produced on modern worm gears to generate a load-supporting oil film.

The advantages of worm gearing compared to spur and helical gearing are:

1. A more compact design for the same reduction ratio or power capacity.
2. Much greater speed reduction and torque amplification in a single step.
3. Smooth and silent operation that can withstand higher shock loads and higher momentary loads.

The disadvantages compared with ordinary gearing are:

1. Lower and varying efficiency.
2. Greater axial forces requiring costlier bearings.
3. Overheating that limits duration and capacity for power transmission.

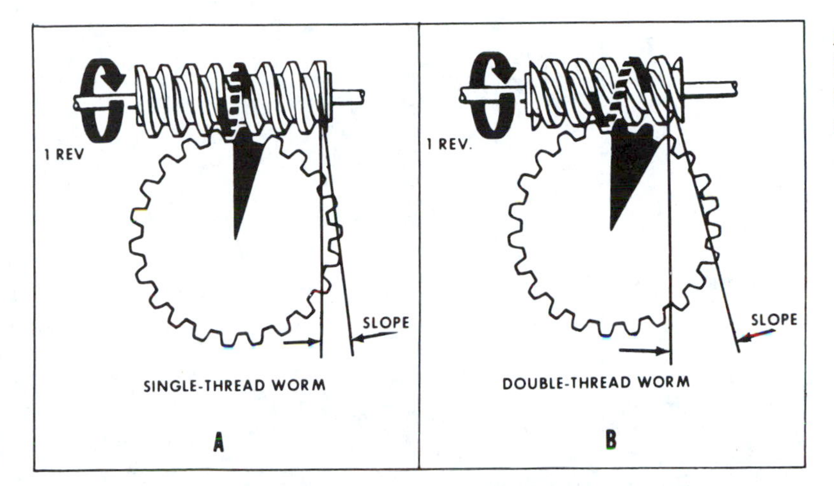

Figure 14-10 The difference between a single- and a double-thread worm. (Courtesy Bureau of Naval Personnel.)

14-5 WORM GEAR TERMINOLOGY AND KINEMATICS

Figure 14-11 shows worm gear terminology and development of a worm thread. For reasons of clarity, a triple-threaded worm was used. A worm can be single, double, triple, quadruple, or multithreaded, plus left or righthanded. Threads in excess of 10 are rarely advantageous.

Axial pitch (p_a) of a worm (Fig. 14-11) is the distance measured axially from a point on one thread to the corresponding point on the next thread. For proper mesh, p_a must equal the circular pitch p_c of the gear.

Lead (L) is the distance L that a thread advances in one turn of the worm. Thus

$$L = N_w p_a \qquad (14\text{-}4)$$

where N is the number of threads on the worm; e.g., $N_w = 2$ for a double-threaded worm.

Lead angle (λ_w) is the angle between a tangent to the thread at the pitch diameter and a plane normal to the worm axis.

Velocity ratio (m_G) is the ratio of pitch circumference of gear to lead of worm, which equals tooth ratio. It is also the ratio of worm speed to gear speed. Thus

$$m_G = \frac{p_a N_G}{L} = \frac{N_G}{N_w} = \frac{n_w}{n_G} = \frac{D_G}{D_w \tan \lambda_w} \qquad (14\text{-}5)$$

For worm and gear to mesh properly, the lead angle of the worm must equal the helix angle of the gear ($\lambda_w = \psi_G$), and axial pitch of the worm must equal circular pitch of the gear ($p_a = p_c$). Since the circumference of the pitch circle of the gear can be expressed as πD_G or $p_c N_G$, this leads to

$$p_c = \frac{\pi D_G}{N_G} = p_a \qquad (14\text{-}6)$$

Substitution into Eq. 14-5 gives

$$n_G = \frac{L}{\pi D_G} n_w \text{ rpm} \qquad (14\text{-}7)$$

as another expression of the speed relationship.

Development of one turn of the worm leads to a triangle, as shown in Fig. 14-11. This triangle shows that

$$\tan \lambda_w = \frac{L}{\pi D_w} \qquad (14\text{-}8)$$

Center distance is 0.5 ($D_w + D_G$), which can also be expressed as

$$C = \frac{L}{2\pi}(m_G + \cot \lambda_w) \qquad (14\text{-}9)$$

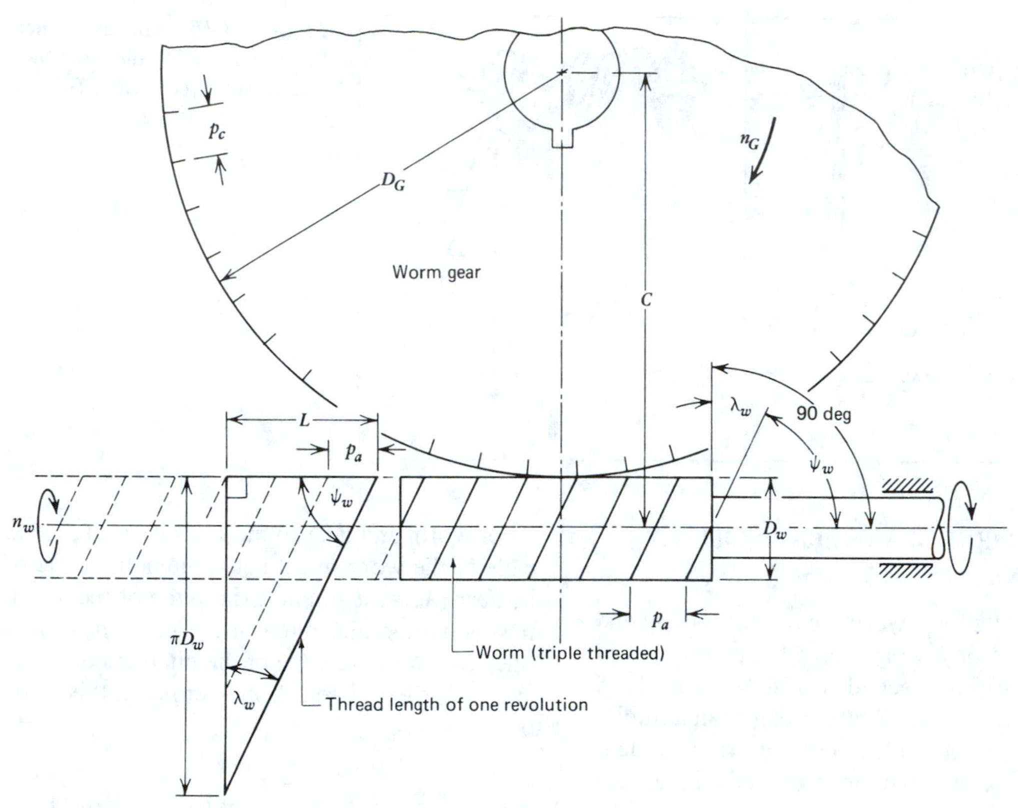

Figure 14-11 Worm gear terminology and development of a worm thread. (For clarity, a multithreaded worm was used.)

14-6 THERMAL RATINGS

When worm gear teeth slide across the surfaces of mating worm threads, far more heat is generated than when the same load is carried by any other type of gearing. These thermal conditions raise the operating temperature of the oil, thereby reducing its load-carrying capacity. To keep oil temperatures below critical levels, manufacturers publish thermal ratings for each unit that indicate the maximum power input that produces a safe rise in oil temperature. Because common gear lubricants deteriorate rapidly at temperatures above 90°C (195°F), operating levels are usually kept at or below 75°C (170°F). Clearly, design of worm gearing must include temperature effects, as well as strength and wear, as a limiting factor. Of the three, overheating is the controlling parameter. For example, a worm gear may have a thermal rating of 5 kW and a mechanical rating of 7 kW for the higher speed ranges. This means that the gearing is capable of transmitting more than 5 kW as far as strength and wear are concerned—but not without overheating. Recently, however, the use of computers has improved worm geometry, thereby narrowing the gap between thermal and mechanical capacity.

14-7 EFFICIENCY OF WORM GEARING

While spur and helical gearing exhibit very high and virtually constant efficiencies (0.98 to 0.99), those of worm gearing may range from a low of 0.50 to a high of around 0.98. Generally, efficiency varies inversely with speed ratio provided the coefficient of friction does not change. The various parameters determining efficiency do not have a linear relationship (Fig. 14-12). Instead, efficiency e increases with decreasing coefficients of friction. Efficiency is greatly influenced by the lead angle for small values, but less and less as λ increases. A maximum is reached for $\lambda = 45$ deg.

For a well-designed, well-lubricated unit, the following is a fair approximation to efficiency.

$$e = \frac{\tan \lambda}{\tan(\lambda + \theta)} \qquad (14\text{-}10)$$

where

$$\theta = \arctan f = \text{friction angle}$$

Back-driving is the term used when the gear drives the worm. In this reverse action speed is increased at the expense of force. Back-driving can thus be used to advantage in speedup drives

Figure 14-12 Efficiency as a function of lead angle and coefficient of friction.

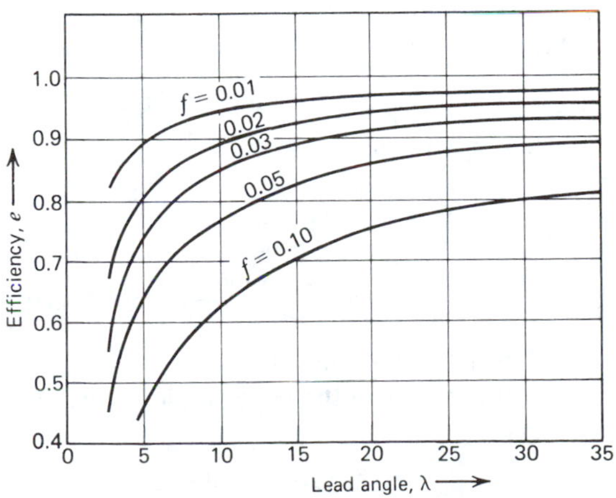

for centrifuges and turbochargers. The corresponding expression for efficiency is

$$e = \frac{\tan(\lambda - \theta)}{\tan \lambda} \qquad (14\text{-}11)$$

Theoretically, back-driving is possible only for $\lambda > \theta$, that is, when the lead angle is greater than the friction angle. In reality, this reverse action occurs for higher values of θ. Vibration, present in most mechanical equipment, effectively lowers friction, thereby reducing "self-locking" to "a mechanical fringe benefit." Only a brake can effectively prevent back-driving.

14-8 OPTIMUM DESIGN

Even though worm gear drives are among the oldest mechanisms, they are one of the least understood. Because of their inherent complexities, precise analytical methods have not evolved, so design relies heavily on trial-and-error testing. Worm gears have, however, reached a high degree of perfection. Thus, by analyzing the expression for efficiency, we may single out major design parameters and compare them with the industrial end product.

Low coefficients of friction are obtained by (1) using dissimilar metals for worm and gear, (2) providing smooth tooth surfaces, and (3) ensuring adequate lubrication. For instance, hardened and ground steel worms are used with gears of phosphor bronze or cast iron. Special lubricants are available for worm gearing.

Large lead angles are also desirable but are obtained at the expense of lower speed ratios. Consider the following equation.

$$\tan \lambda = \frac{L}{\pi D_w} = \frac{p_a N_w}{\pi D_w} \qquad (14\text{-}12)$$

A large efficiency requires a large lead; therefore the worm should be multithreaded. Consequently, when worm gearing is designed primarily for transmitting power, it should be multithreaded, as is common practice.

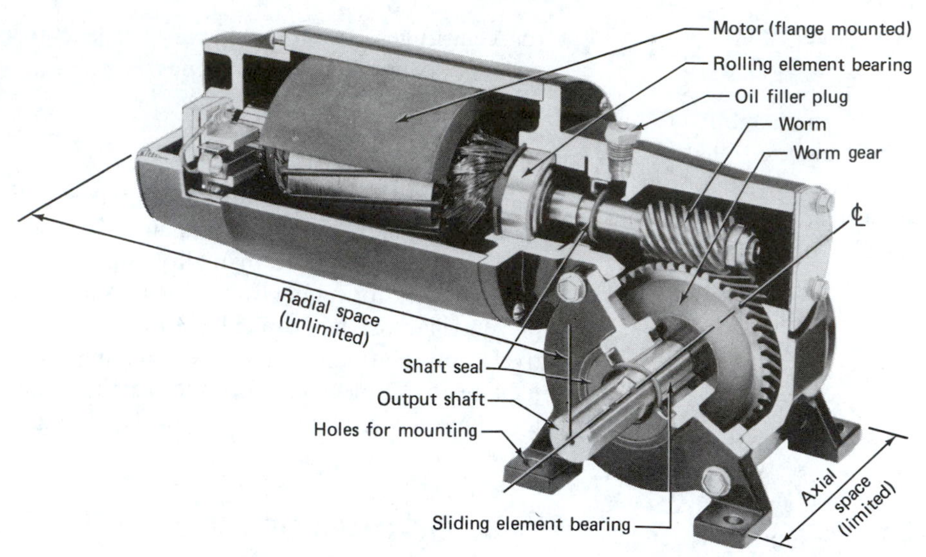

Figure 14-13 Typical worm gear reduction unit. (Courtesy Bodine Electric Company.)

To obtain a given ratio, some number of worm wheel teeth divided by some number of worm threads must equal the ratio. Thus, if the ratio is 6:

$$m_G = \frac{6}{1} = \frac{12}{2} = \frac{18}{3} = \frac{24}{4} = \frac{30}{5} = \frac{36}{6} = \frac{42}{7}$$

Any of these combinations may be used. The numerators represent the number of worm wheel teeth, and the denominators are the number of worm threads. As the total number of teeth increases, so does efficiency, but only at higher initial cost.

The expression for λ also indicates that a small worm diameter D_w is most desirable because it lowers rubbing velocity. As can be seen in Fig. 14-13, the worm diameter is small relative to the thread height.

EXAMPLE 14-2

A right-angle speed reducer has a triple-threaded worm and a 41-tooth gear. Find the speed ratio, lead, lead angle, helix angle, pitch diameter of gear, center distance, efficiency, power output, and transmitted force for the following data.

$p_c = 32$ mm $\qquad n = 900$ rpm $\qquad D_w = 44$ mm

$f = 0.05 \qquad P = 0.75$ kW (input)

Will the unit back-drive?

SOLUTION

$$m_G = \frac{N_G}{N_p} = \frac{41}{3} = \mathbf{13.67}$$

From Eq. 14-4:

$$L = N_w p_a = 3(32 \text{ mm}) = \mathbf{96 \text{ mm}}$$

From Eq. 14-8:

$$\lambda_w = \arctan \frac{L}{\pi D_w}$$

$$= \arctan \frac{96 \text{ mm}}{\pi(44 \text{ mm})} = \mathbf{34.78 \text{ deg}}$$

For proper mesh, the lead angle of the worm must equal the helix angle of the gear. Thus

$$\psi_G = \lambda_w = \mathbf{34.78 \text{ deg}}$$

As can be seen in Fig. 14-12

$$\psi_w = 90 \text{ deg} - \lambda_w = \mathbf{55.22 \text{ deg}}$$

From Eq. 14-6:

$$D_G = \frac{p_a N_G}{\pi}$$

$$= \frac{(32 \text{ mm})41}{\pi} = \mathbf{417.62 \text{ mm}}$$

From Eq. 14-9:

$$C = \frac{L}{2\pi}(m_G + \cot \lambda_w)$$

$$= \frac{96 \text{ mm}}{2\pi}(13.67 + \cot 34.78 \text{ deg})$$

$$= \mathbf{230.86 \text{ mm}}$$

From Eq. 14-10:

$$e = \frac{\tan \lambda_w}{\tan(\lambda_w + \theta)}$$

$$= \frac{\tan 34.78 \text{ deg}}{\tan(34.78 \text{ deg} + 2.86 \text{ deg})} = \mathbf{0.90}$$

Since $\arctan f = \arctan 0.05 = 2.86$ deg

Output: $P = 0.90(0.75 \text{ kW}) = \mathbf{0.675 \text{ kW}}$

The unit will back-drive because the lead angle of the worm ($\lambda_w = 34.78$ deg) is much greater than the friction angle ($\theta = 2.86$ deg). From Eq. 9-2:

$$W_t = \frac{9550\, P}{(0.5\, D_w)n}$$

$$= \frac{9550(0.675 \text{ kW})}{0.5(0.044\text{m})(900 \text{ rpm})} = \mathbf{326 \text{ N}}$$

14-9 APPLICATIONS

As indicated in Fig. 14-12, worm gear efficiency varies widely from 0.5 to 0.98. It also varies inversely with speed ratio or mechanical advantage. Thus, single-threaded worm gear drives yield large reduction ratios, but at the expense of efficiency. In contrast, a multithread speed reducer will have high efficiency, but a somewhat lower reduction ratio. This fact leads to three major areas of application for worm gearing.

1. *Intermittent, infrequent* operations where a small, low-cost motor moves a heavy load, as in small hoists. Efficiency, of minor importance, is thus sacrificed to obtain a large mechanical advantage. Typical applications are standby pumps, large valves, and gates.
2. *Intermittent, manual* operations requiring a *large* mechanical advantage, such as in steering mechanisms and opening and closing of valves and gates by means of handwheels (Fig. 14-14).
3. *Motorized, nearly continuous* operations where worm gearing competes with gear reduction units. When space is at a premium, as in machine tools, packaged, motor-driven worm reduction units are used in preference to gear reducers (Fig. 14-13). Depending on size and application, the unit may be self-contained or built integrally with an electric motor. Because of silent operation, such units are preferred in machine tools and also in elevators. These units all require multithreaded worms and ratios not exceeding 1:18. Larger ratios are achieved by connecting two units in series.

14-10 DESIGN DETAIL OF WORM GEARING

The unit shown in Fig. 14-15 is a typical, medium-size worm gear speed reducer. Smaller units of this type usually have housings of cast aluminum alloys for maximum thermal rating. For larger units the preferred material is cast iron. The worm is case-hardened and ground alloy steel of integral shaft design. The gear is cast bronze with generated teeth and keyed to the output shaft. Larger worm gears are often composed of a ring of bronze mounted on a center or hub of less expensive material. A common de-

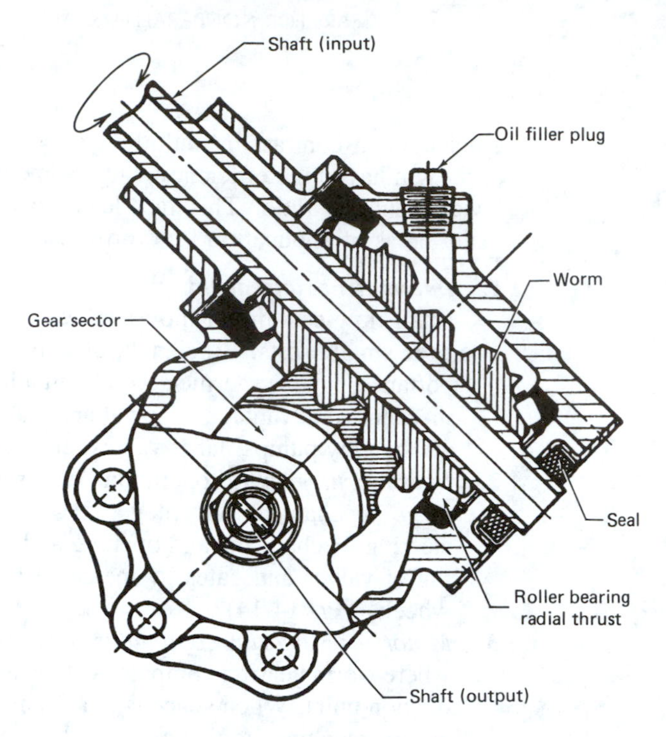

Figure 14-14 Steering mechanism. Note that only a gear sector is needed. (Courtesy The Timken Company.)

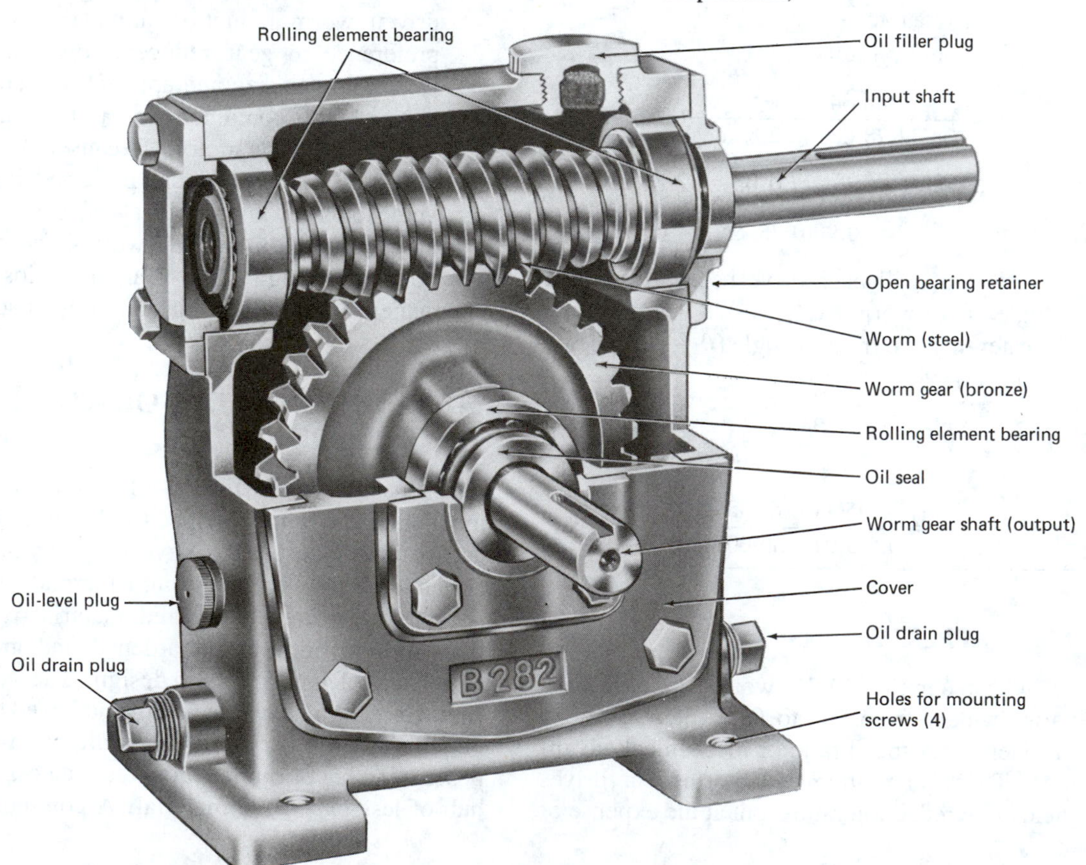

Figure 14-15 Typical medium-size worm gear speed reducer. (Courtesy Morse Chain, Division of Borg Warner Corporation.)

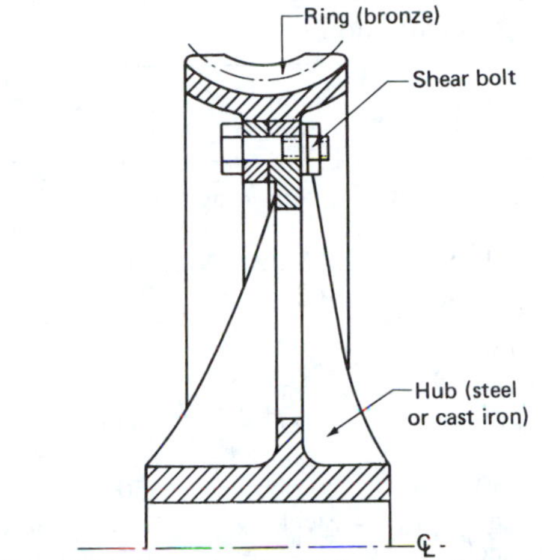

(*a*) Flanged ring mounted on the hub by means of shear bolts

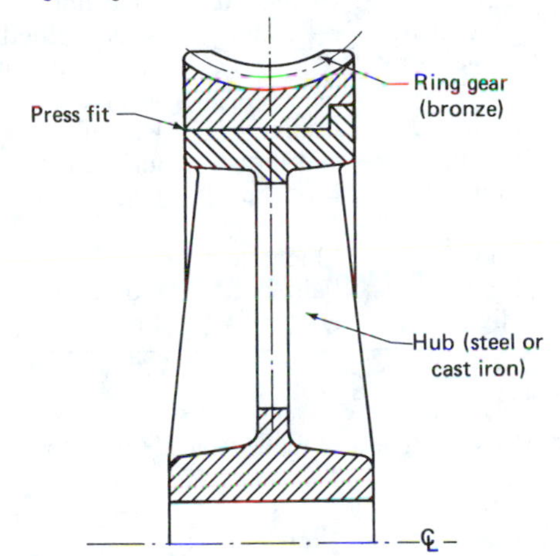

(*b*) Bronze ring mounted on hub by means of a press fit

Figure 14-16 Design of a large worm gear.

sign utilizes a flanged rim mounted on the hub by means of shear bolts (Fig. 14-16*a*). Equally common is mounting by means of a press fit (Fig. 14-16*b*) assisted by a pin connection. The output shaft is high-quality, medium-carbon steel, ground to close tolerances. The worms and output shafts are frequently mounted on roller bearings. All shaft extensions are equipped with lip style, synthetic oil seals.

Lubrication

Generally, oil is contained within the housing and directed by splash to the bearings and to the zone of tooth and thread contact. Natural splash may be augmented by flingers, scrapers, and cups attached to the gear. Channels or ribs may be furnished inside the housing to help direct oil to the bearings.

SUMMARY

Despite higher initial cost, gears for non-parallel shafts are justified because they often save space and lead to a better design. Kinematically, these gears all perform the very difficult task of changing the plane of rotation. With the exception of crossed helical gears, all have reached a high degree of perfection and a long, useful life of transmitting power. Hypoid gears for automotive differentials, for instance, rarely fail during the life of a car. The versatility of worm gearing is due to the inverse relationship of efficiency to torque and reduction ratio. Table 14-1 summarizes comparative characteristics of speed reducer gear families.

REFERENCES

In addition to the references in Chapter 13, the following were used:

14-1 Buckingham, E. K. "Taking Guesswork Out of Worm-Gear Design." *Machine Design,* March 20, 1975.

14-2 MacFarland, W. C. "Straight Talk About Speed Reducers." *Machine Design,* September 18, 1975–October 2, 1975.

14-3 Will, R. J. "Selecting Speed Reducers." *Machine Design,* September 8, 1977.

PROBLEMS

P14-1 A pair of straight bevel gears has a speed ratio of roughly 1.5 and a shaft angle of 90 deg. The pinion has 17 teeth. Find the pitch angles.

P14-2 A pair of straight bevel gears have 19 and 29 teeth, respectively. The module is 12 mm and the shaft angle is 90 deg.

TABLE 14-1 Comparative Characteristics of Speed Reducer Gear Families[a]

	Spur	Helical	Herringbone	Straight bevel	Spiral bevel	Worm gear
Shafts	Parallel	Parallel	Parallel	Intersecting	Intersecting	Non-parallel Non-intersecting
Ratios[b]	1:1-10:1	1:1-15:1	1:1-10:1	1:1-10:1	1:1-10:1	5:1-70:1
Max diameter (in.)	360	300	300	102	102	88
Max face width (in.)	64	64	64	12	12	9[c]
Pitch line velocity (ft/min)(max)	2000	30,000	5000	2000	5000	6000
Input hp (max)	4500	50,000	2000	1500	20,000	1500
Materials of construction	Cast iron, steel	Steel	Steel	Cast iron, steel	Steel	Bronze gear, steel worm
Common manufacturing methods	Cast hobbed	Hobbed	Hobbed	Cast generated	Generated	Gear hobbed, worm milled, hardened and ground
Bearings	Radial	Radial and thrust	Radial	Radial and thrust	Radial and thrust	Radial and thrust
Uniformity of motion transfer	Low	Moderate	Moderate	Low	Moderate	High
Shock resistance[d]	Poor	Moderate	Moderate	Poor	Moderate	Good
Noise[e]	Poor	Fair	Good	Poor	Fair	Good
Initial cost[f]	Very low	Moderate	High	High	High	Low

Source. Courtesy Transamericana Delaval Delroyd Worm Gear Division.

[a]Characteristics depend on quality. Table assumes commercial or AGMA Quality 8 quality, as typically used for power transmission applications.
[b]For single reduction.
[c]Worm gear.
[d]Gear pitch is a major factor in shock resistance. Coarser pitches are more resistant.
[e]For all type gears, precision is the primary determinant of noise, and fine-pitch gears are quieter than coarse-pitch gears. The inherent quietness of worm gearing derives from the wearing in and conformity of the bronze gear to the hardened steel worm, and to bronze's acoustic properties.
[f]Per unit hp or torque.

For the equivalent spur gears, find the pitch diameters, addenda, and dedenda using standard tooth proportions. Assume a face width $F = 9$ m. Sketch the gear blanks needed.

P14-3 Why is it justified in designing a worm gear housing, to use a more expensive material—for instance, aluminum alloy—in a small worm gear unit and a less expensive material—such as cast iron—in larger units?

P14-4 A right-angle worm gear speed reducer has a quadruple-threaded worm and an 87-tooth gear. The worm has an axial pitch of 10 mm and a pitch diameter of 12 mm. The coefficient of friction is 0.032. Calculate (*a*) pitch diameter of gear, (*b*) lead angle of worm, (*c*) center distance, (*d*) speed ratio, (*e*) efficiency, (*f*) output power if input is 20 kW.

P14-5 A 3-tooth worm drives a gear having 43 teeth. The axial pitch is 1.25 in., and the pitch diameter of the worm is 1.75 in. Calculate the lead angle of the worm, the helix angle and pitch diameter of the gear, the center distance, and the speed ratio.

P14-6 The worm gear drive in P14-5 has a power input of 10 hp at 1200 rpm. Calculate the power output and the transmitted force. Assume a coefficient of friction of 0.05. Will the unit back-drive?

P14-7 A worm drives a gear with 43 teeth. The axial pitch is 1.25 in., and the pitch diameter of the worm is 1.75 in. The coefficient of friction is 0.10. Calculate and plot the lead angle and the efficiency as the number of teeth in the worm varies from 1 to 6.

P14-8 A five-threaded worm drives a 31-tooth gear with a shaft angle of 90 deg. The lead angle is 20 deg, and the shaft center distance is 2.75 in. Calculate all basic dimensions, angles, and data.

PART 4

Machine Elements for Carrying and Transmitting Rotary Power

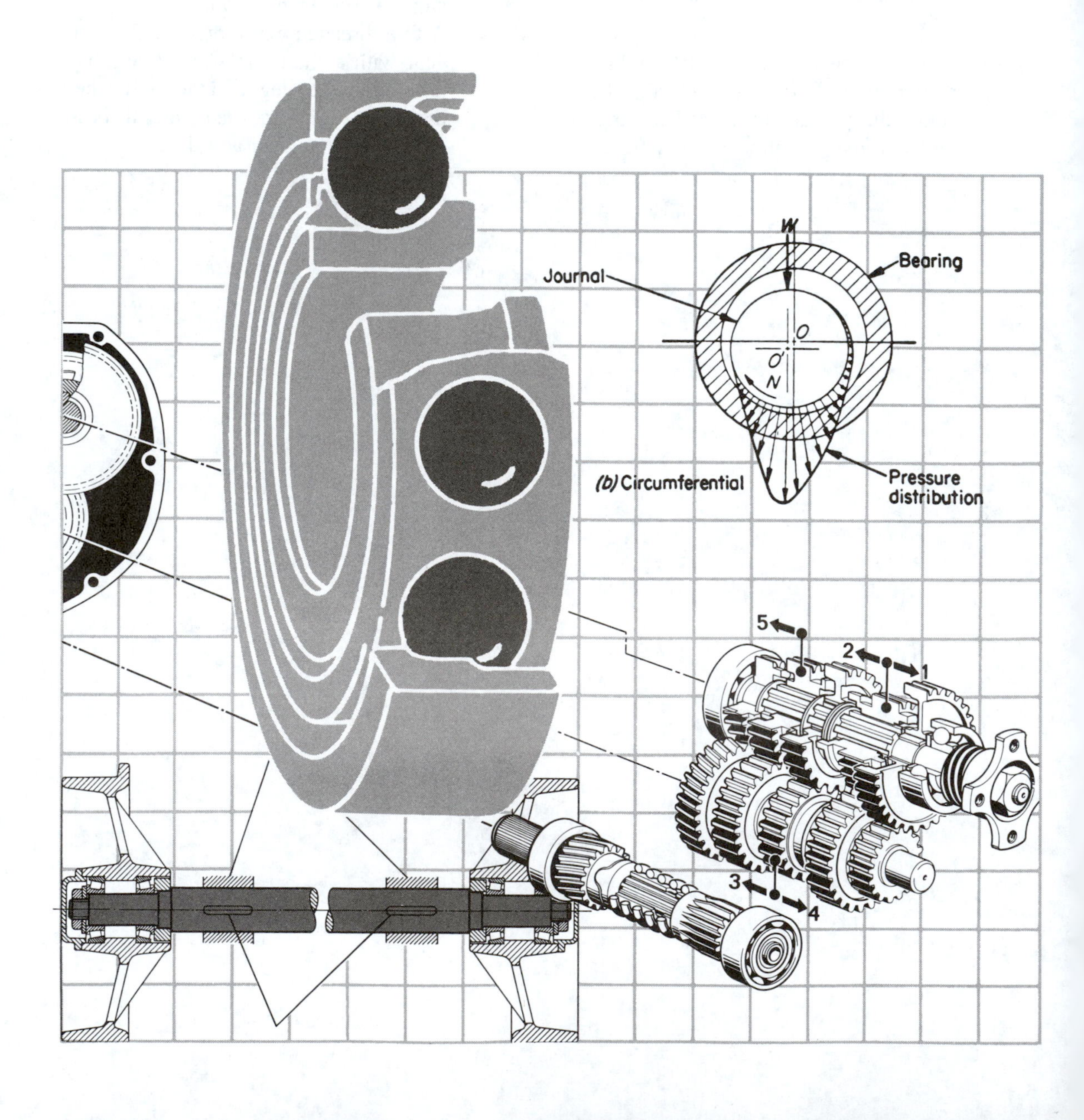

Chapter 15
Axles and Shafts

The great aim of education is not knowledge, but action.

HERBERT SPENCER

NOMENCLATURE

C = ratio of inside to outside diameters
F_a = axial force
F_1, F_2 = band tensions
g = acceleration of gravity; 9.81 m/s^2, 32.17 ft/s^2
k_a = fatigue strength derating factor for surface finish
k_b = fatigue strength derating factor for size
k_c = fatigue strength derating factor for reliability
k_d = fatigue strength derating factor for temperature
k_e = fatigue strength derating factor for stress concentration
k_f = fatigue strength derating factor for shock loading
K = torsion constant for noncircular cross section; mm^4, in.^{4}
K_t = theoretical stress concentration factor
K_f = fatigue stress concentration factor
K_{fs} = fatigue stress concentration factor for shear
l = shaft length
M_a = amplitude of a cyclic bending moment; N · mm, lb-in.
M_m = mean value of a fluctuating bending moment; N · mm, lb-in.
q = notch sensitivity factor
r = average radius of wall for hollow circular cross section
S_e = derated endurance strength
S'_e = endurance strength of laboratory specimen
S_{se} = derated endurance strength in shear
T_a = amplitude of a cyclic torque
T_m = mean value of a fluctuating torque
W = applied load
W = work done by a force
Wt = weight
α = column-action factor
θ = slope, in./in., rad; angular deflection, rad
σ_a = amplitude of a cyclic normal stress
σ_m = mean value of a fluctuating normal stress
σ_i = equivalent static normal stress
σ'_i = von Mises equivalent normal stress
τ_a = amplitude of a cyclic shear stress
τ_m = mean value of a fluctuating shear stress
τ_i = equivalent static shear stress

Axles and shafts are a further development of the wheel—humanity's greatest invention. The wheel, a totally new concept on this planet, greatly added to our total mobility and ability to move heavy objects. The wheel-and-axle machine enabled us to lift heavy burdens with greater ease. From the wheel-and-axle machine evolved the shaft and axle-supported belt, chain, gear, and wheel-drives—the foundation of modern machinery.

15-1 TYPES OF AXLES AND SHAFTS

An *axle* connects a vehicle's wheel to a housing or body. The earliest axles, used in carts, were nonrotating. They supported rotating wheels on simple bearings lubricated with animal fats. Many axles on today's vehicles are also nonrotating, for example, those on towed vehicles and machines. The front axle of the truck shown in Fig. 15-1 is nonrotating, as are the front wheel

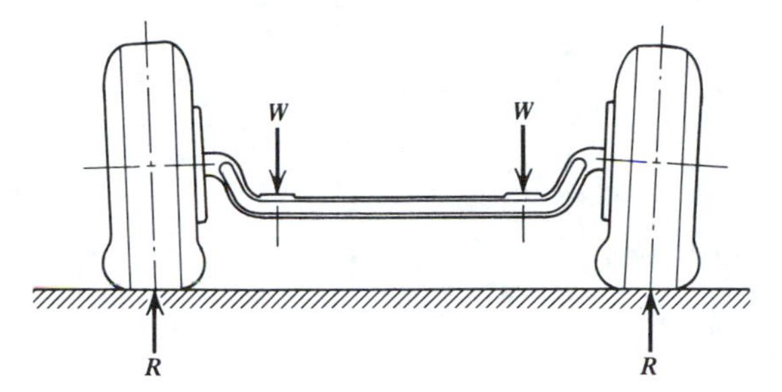

Figure 15-1 Sketch of front axle for truck.

spindles of rear wheel-drive automobiles. Another nonrotating axle is shown in Fig. 15-2*a*.

A *dead axle* may rotate with the wheels but does not transmit power. The railway car axle in Fig. 15-2*b* is fastened to the wheels by interference fits. The wheel and axle assembly turns as a unit. The railway car's weight is supported on bearings at the outer ends of these axles. No driving torque is transmitted.

Live axles transmit torque to provide mobility for self-propelled vehicles. The rear axles on a rear wheel-drive automobile transmit power from the differential to the rear wheels. These axles are often called *axle shafts* to emphasize their ability to carry power.

A *shaft* is a device that supports pulleys, sprockets, gears, cranks, levers, and other attachments and often transmits power between them. The shaft itself is mounted on bearings so that it can turn freely in the housing. It may be subjected to transverse loads, to torque or, more likely, to both. A solid shaft used in a gear transmission is illustrated in Fig. 15-2*c*. The machine shaft shown in Fig. 15-2*d* is hollow to save on weight. The 90 in. long turbine shaft shown in Fig. 15-3 is complex; besides providing support for the turbine rotors, it also allows for bearing supports, seals, and couplings. It is a modern marvel of strength and precision.

In addition to these general types of axles and shafts, designers should also be familiar with the following special-purpose axles and shafts.

A *countershaft* is an intermediate shaft placed between a driving and a driven shaft. In gear transmissions the countershaft makes possible variations in speed ratio.

A *flexible shaft* permits direct transmission of power between two axles at an angle to each other.

A *spindle* is a short shaft. The drive shaft in the headstock of a lathe is an example of a spindle.

Stub axles are short dead axles capable of limited angular motion about the pivots. The front wheels of the typical rear wheel-drive passenger car are supported by stub axles and bearings.

15-2 THE DESIGNER'S PROBLEMS

Successful design of an axle or a shaft demands attention to several matters.

Design for Rigidity

Very often rigidity is the principal criterion in sizing an axle or a shaft. For example, if a shaft supports a gear, excessive bending of the shaft will cause misalignment of gear teeth and uneven distribution of tooth load. In many gear reduction units the shafts must be so heavy to achieve sufficient stiffness that strength ceases to be much of a problem. Designers therefore must be aware of stiffness, both in bending and in torsion, and must be prepared to do the necessary analyses of bending stiffness and torsional stiffness.

Design for Strength

In other cases (e.g., an engine's crankshaft) strength is the determining factor. In any event, strength must be checked, and usually at several critical points on the shaft. The presence of cross-sectional changes, or keyseats, or other stress raisers, complicates the matter. It is seldom obvious in advance where the maximum stress occurs. Designers must be able to calculate the stresses, often caused by a combination

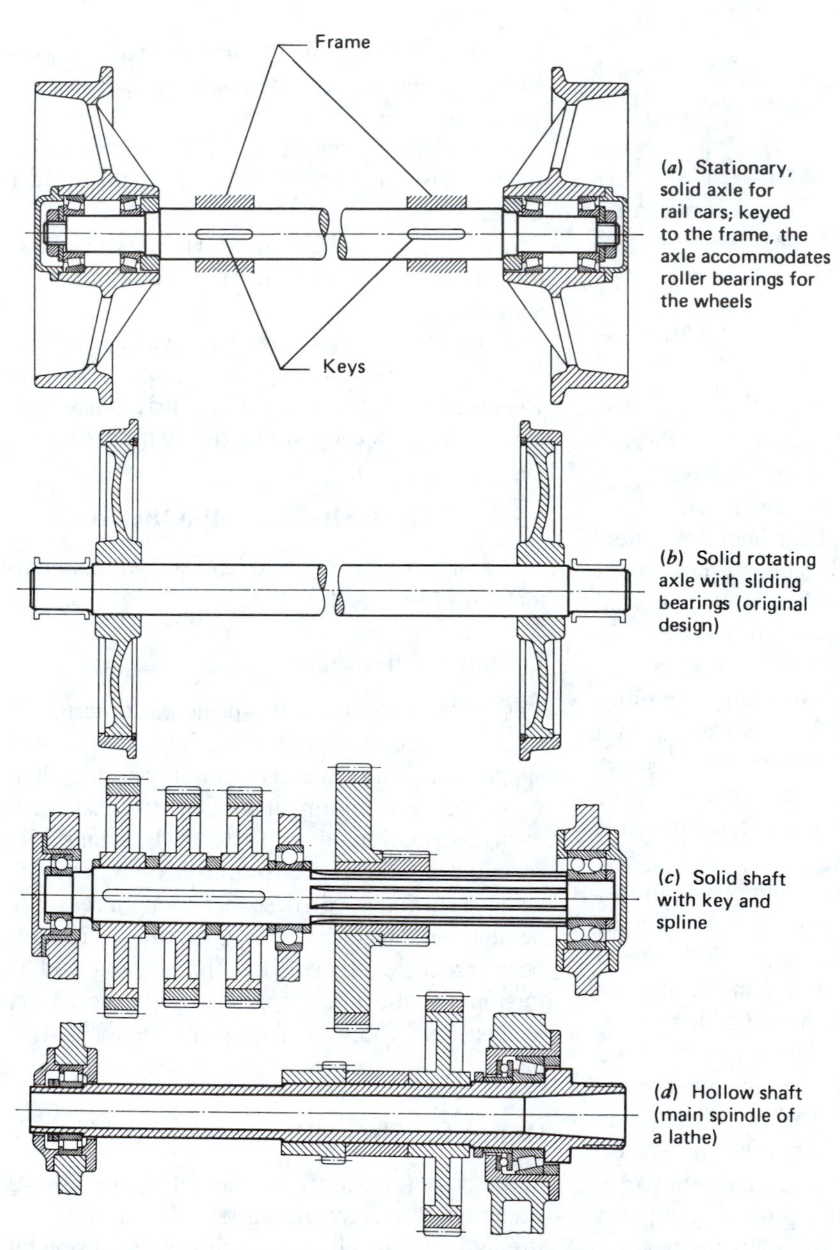

Figure 15-2 Examples of axles and shafts. (Reprinted from *Maschinenelemente, Gestaltung, und Berechnung,* by Karl-Heinz Decker, courtesy of Carl Hanser Verlag, Munich.)

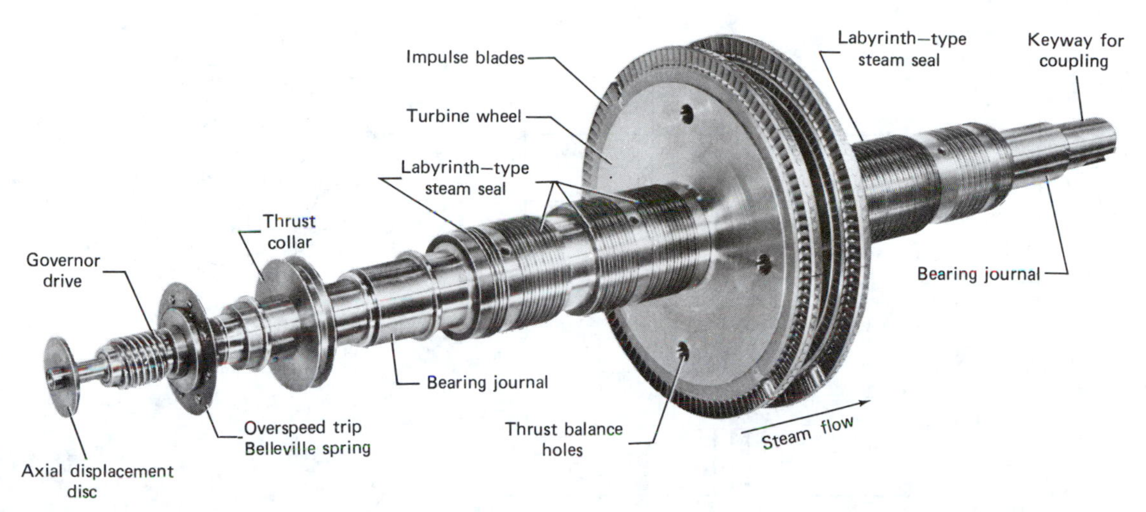

Figure 15-3 The shaft for a 3800-hp, 12,500-rpm steam turbine. The two turbine wheels are shrunk onto the 90-in. long steel shaft. (Courtesy of Elliott, A United Technologies Company.)

of loads, at various points on the axle or shaft, and be able to appraise these stresses by means of a suitable theory of failure. Here the methods of Chapter 3 are essential.

Simplifying Assumptions

The methods of strength of materials provide the single most important set of tools for designers of shafts and axles. However, their application is not always obvious; a great deal of judgment is needed. Consider just the matter of bending stiffness in gear shafts. It seems reasonable to regard the shaft as a beam. But beam deflection theory is full of talk of "knife-edge supports" or "built-in ends." And that does not sound very much like what is encountered in real life. Designers thus have to reconcile theory and reality. For the novice, successful past practice is a starting point.

Designers usually regard a narrow rolling element bearing as a knife-edge support. In contrast, a journal bearing, unless very short, is taken to be a built-in end. How, then, should one classify a double-row ball bearing? Is it a built-in end or a knife-edge support? There is no easy answer. Shrewd designers will consider both alternatives and select the more conservative conclusion.

An analogous problem arises in the statement of applied loads. Figure 15-4*a* shows a shaft supporting a flat belt pulley. The shaft is in turn supported by two single-row ball bearings. The belt's width is small compared to the total length of the shaft, but both the belt load and the bearing reactions are distributed, as Fig. 15-4*b* shows. These forces can be modeled, however, as concentrated forces (Fig. 15-4*c*). This greatly simplifies the designers' tasks and still produces sufficiently accurate results.

Because the weight of a shaft or an axle is usually small compared to the applied loads, it is common practice to ignore it. Sometimes even the weights of the attachments themselves—the gears, pulleys, and so on—are neglected, although it is often easy to include them.

Effects of Fillets, Keyseats, Grooves, and the Like

These common features of a shaft or axle design strongly affect stress distributions. Any strength analysis must therefore take them into account. But they also affect rigidity, both in bending and

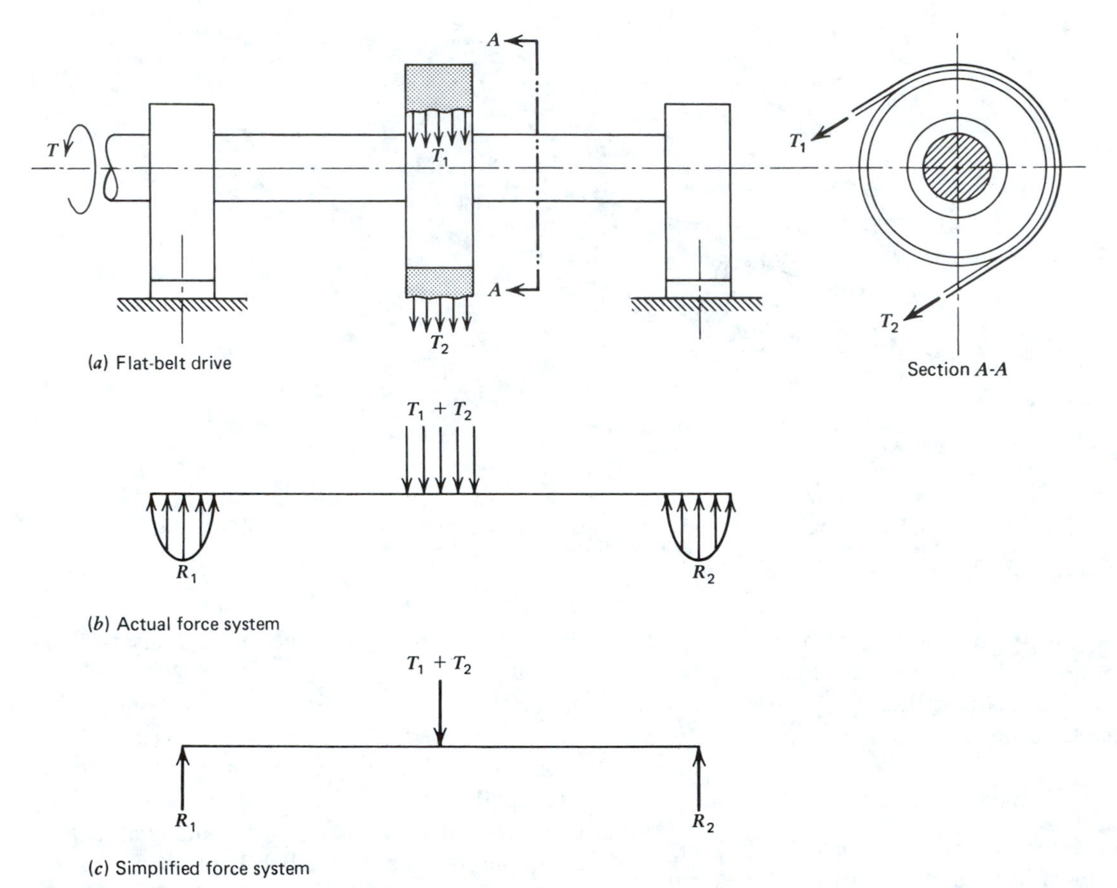

Figure 15-4 Actual and simplified shaft loading.

in torsion, because they make the shaft more flexible than it would be otherwise. They should not be ignored in stiffness analysis.

Nonuniform Cross Sections

The basic analyses one learns in strength-of-materials courses deal with beams or torsion bars of uniform cross section. Thus some change in approach is necessary to cope with designs in which the shaft's cross section varies. Methods of *stress analysis* are not badly compromised. One simply repeats the stress analysis at each cross section—especially at shoulder fillets, keyseats, and retaining ring grooves—where the stress level is likely to be high. The character of the local cross section affects the analysis there.

Appraisal of rigidity or *stiffness* is not quite as straightforward. Hopkins[1] has shown how to handle the *torsional deflection* calculation easily, but the *bending deflection* analysis is more troublesome. The elementary beam deflection equations cannot be used, since they assume constant cross section. Semigraphical or numerical methods are needed. Hopkins' book also deals with this problem.

---

[1] R. Bruce Hopkins, *Design Analysis of Shafts and Beams* (New York: McGraw-Hill, 1970).

Critical Speeds

Imagine a mass suspended from a spring. If the mass were pulled down and then released, the mass and spring would vibrate. The frequency of the vibration is the natural frequency of the mass and spring combination. A shaft is also a spring of sorts, and a rotor mounted on it is a mass. If one were to push hard near the center of the shaft and then let go, the shaft would vibrate transversely, and at its fundamental natural frequency. Books on vibration theory show that if the shaft is operated at a speed corresponding to its natural torsional frequency, large and destructive vibrations can develop. This is the shaft's *critical speed*. Thus designers must ensure that a shaft's operating speeds do not approach its critical speed. Books on mechanical vibrations present methods for calculating critical speed.

Material and Fabrication

This is another design concern. If strength is the critical issue, a heat-treated alloy steel may be the proper material. However, if stiffness is of principal concern and the stresses are not very high, an untreated plain carbon steel will do, since heat treatment has little effect on modulus of elasticity. Considerations of corrosion resistance may indicate the necessity to reduce weight in yet another direction. In short, the material selection process is important. An honest appraisal of competing materials is part of the careful designer's decision making.

15-3 MATERIALS FOR AXLES AND SHAFTS

One or two of the following eight factors are likely to predominate in choosing a shaft or axle material.

1. Rigidity or stiffness (the ability to resist change of form).
2. Strength (the ability to resist loads).
3. Wear resistance.
4. Corrosion resistance.
5. Weight.
6. Cost (the expense of production).
7. Size and availability.
8. Machinability.

Rigidity

In a surprising number of cases (surprising because it is natural to focus on strength), rigidity governs the design. To avoid gear tooth misalignment or misalignment of shaft and journal bearing, for example, the bending deflection of the shaft must be limited.

For a shaft of given length and loading, the bending deflection is inversely proportional to the product EI, where E is the modulus of elasticity and I is the moment of inertia of the cross section. If steel is to be used and shaft rigidity is the principal design requirement, heat-treated steel is unnecessary. Since the heat treatment has almost no effect on E, a low-strength steel will do.

Where rigidity is important but weight must be minimized, aluminum seems a reasonable choice because its density is only about 35% that of steel. Unfortunately, its modulus of elasticity is also lower, about 34% that of steel; so the weight reduction is not quite as great as one would hope. For example, for a solid steel shaft 0.75 in. in diameter,

$$EI = \frac{\pi d^4}{64} E = \frac{(\pi)(0.75 \text{ in.})^4}{64} (30)(10^6 \text{ psi})$$

$$= (4.66)(10^5 \text{ lb-in.}^2)$$

If aluminum is used $[E = (10.3)(10^6 \text{ psi})]$, the shaft diameter must be 0.98 in. to get the same value for EI. The shaft's weight, however, is reduced by 41%. The aluminum shaft will thus be bulkier and more costly, but substantially lighter.

Strength

If operating stresses are light, a low-carbon steel will be satisfactory. Also, because cold-finished bars are made to low tolerances, shafts are sometimes made of cold-finished steel with little or no machining. Chapter 3 pointed out that notch sensitivity is lower for low-strength steels. Thus, low-strength steels are more "forgiving" of careless design and machine shop errors that result in sharp notches.

Where greater toughness, shock resistance, and strength are needed, heat-treated alloy steels (e.g., AISI 1347, 3140, 4140, 4150, 4340, 5145) are sensible choices. AISI 1137 steel is both free-cutting and hardenable.

Shafts over about 3.5 in. in diameter can be machined from hot-rolled carbon steel (e.g., AISI 1040 or 1045). Hot-rolled steel must be machined to remove the soft surface layer and obtain more precise dimensions.

For forged shafts, AISI 1045 steel is a good choice.

Wear Resistance

Wear resistance (e.g., of a shaft in a journal bearing) is obtained by *hardening* the surface. Low-carbon steels (AISI 1020, 1117, 2315, 4320, 4820, 8620) can be given a carburizing treatment that adds carbon to the surface layer, which then is heat treatable to a high hardness.

Nitriding is another way to harden the surface layer, in this case by absorbing nitrogen during extended heating in the presence of ammonia gas. The cyaniding process likewise provides a very hard surface by briefly immersing the part in a molten sodium cyanide bath.

If the steel has sufficient carbon content, localized surface heating by flame or an induction coil, followed by quenching, will produce a hard surface layer with increased wear resistance.

Detailed knowledge of these heat treatments is the province of metallurgists, but designers must be aware of the various choices.

Corrosion Resistance

In some cases a shaft may be exposed to corrosive substances. For example, the shaft joining an inboard engine to the propeller on a boat will be exposed to seawater. Bronze or monel metal might be used. In many industrial applications stainless steels are used for their corrosion resistance.

Weight

As mentioned, substantial weight can be saved by using aluminum instead of steel in designing shafts of equal bending stiffness. One might wonder if the weight savings are comparable for shafts of equal strength. No authoritative answer can be given here because design problems vary too much, but some intriguing hints can be offered.

Consider a very simple case: a solid circular bar supported at its ends by single-row ball bearings and with a bending load at midspan. Whatever material is used, the length and loading are the same. For bars having the same bending stiffness, a wrought aluminum alloy bar will weigh only about 59% as much as one made of steel. But if the design is done on the basis of strength and the same factor of safety based on yield strength is used, a wrought aluminum bar will weigh only 35% as much as one of hot-rolled plain-carbon steel. A bar of heat-treated AISI 4340 alloy steel will weigh only 49% as much as the plain carbon bar.

For equal stiffness, considerable weight can be saved with either magnesium alloys or aluminum alloys. The use of a heat-treated alloy steel is pointless. But for an equal factor of safety with respect to yield strength, the results are strikingly different. If weight reduction is the overriding concern, an aluminum alloy is indicated; if some compromise is possible to keep down cost, the heat-treated alloy steel is an attractive option.

Weight Saving with Composite Materials

The aerospace and automotive industries are investing heavily in the developing technology of composite materials whose ratios of strength to weight promise great savings in vehicle weight.

High-strength, lightweight shafts and axles

TABLE 15-1 Preferred Shaft Sizes

Metric Shaft Sizes

4	5	6	7	8	9	10	12	15	17	20
25	30	35	40	45	50	60	65	70	75	80
85	90	95	100	105	110	120	130	140	150	160
170	180	190	200	220	240	260	280			

The English System

Increments

1/16	1/8	3/16	1/4	5/16	3/8	7/16	1/2
9/16	5/8	11/16	3/4	7/8	11/16		

Whole Numbers

1	2	3	4	5	6	7	8	9	10
11	12	13	14	15					

Preferred shaft sizes are obtained through combinations of the two sets of figures; e.g., 5 11/16″

can be made from fiber-reinforced composite materials such as graphite/epoxy and fiberglass/epoxy. Strength, stiffness, and vibration can be affected by controlling fiber *orientation*, relative *amount* of fiber, and *type* of resin. These materials, which are not weldable, are bonded adhesively or bolted. Where surface abrasion is a problem (e.g., a shaft in a sleeve bearing), a metal sleeve or insert may be needed.

Cost

The cost of an axle or a shaft is so intimately *related to* the costs of the other members of the machine that one would seldom set out to minimize the cost of the shaft itself. Yet cost is closely tied to the material selected. Thus it is feasible to select one of the costlier materials only where weight saving, corrosion resistance, or some other gain justifies it.

Size and Availability

Sometimes a shaft can be made of a cold-finished metal with little or no machining. This is one way of reducing piece cost. Table 15-1 lists the preferred sizes for cold-finished steel bars manufactured in the United States.

Machinability

Machinability is the difficulty or ease with which a material can be machined. The machinability index is a numerical value that designates the degree of difficulty or ease with which a particular material can be machined. This index is available from handbooks.

15-4 DESIGN FOR STRENGTH—THE BASICS

The common axle or shaft designers deal with:

1. Rotates.
2. Is supported by two bearings.
3. Carries a steady torque or no torque at all.
4. Is subjected to steady bending loads (or none at all).
5. May be subjected to a steady axial load (e.g., because of a helical gear).

6. Is circular in cross section, either solid or hollow.
7. Is made of a ductile metal, usually steel.

Admittedly, this description does not cover all of the axles and shafts encountered in machine design practice. It does describe, however, a large class of shaft and axle problems and thus provides a good place to begin.

The design methods illustrated in Sections 15-5 and 15-6 are based on three fundamentals: the designer's rule for handling stress concentrations, the Soderberg relation for fluctuating stresses, and the concept of the von Mises equivalent normal stress. From these fundamentals, design relationships can be developed for torsional loading only, for bending and axial loading only, or for combinations of torsional, bending, and axial loadings.

Stress Concentration

In this chapter and in the appendix, there are stress concentration charts for features such as keyseats, shoulder fillets, locating pin holes, and the like. Here it is essential to learn how machine designers make use of these factors.

If a static stress is applied to a part where there is stress concentration and the nominal stress is below the material's yield strength, localized yielding at points of high stress can redistribute the stress in a way that protects the part from failure. However, if a completely reversing stress is applied, stress concentration accelerates the growth of the crack network that can lead ultimately to fracture.

The general case of a fluctuating stress (Section 3-19) is one involving a cyclic stress of amplitude σ_a superimposed on a steady or static stress σ_m. The designer needs to know how to apply fatigue stress concentration factors to this general case. The rule is to *apply the fatigue stress concentration factor only to the amplitude stress*.

In practice, this is done in either of two ways.

1. Use the fatigue stress concentration factor K_f to calculate the fatigue strength derating factor k_e (see Section 3-15). Do *not* similarly derate the static yield strength.
2. Multiply the amplitude stress σ_a by the fatigue stress concentration factor K_f. Set the fatigue strength derating factor k_e equal to 1.

In this chapter the first method is used because it is the more convenient. In the rare cases where a machine member is subjected to fluctuating normal stresses and fluctuating shear stresses, the second method is preferred because the stress concentration factors in shear and bending are likely to be different. Each stress amplitude can then be multiplied by its own stress concentration factor. The problem of trying to derate fatigue strength in the face of two different stress concentration factors therefore does not arise.

Soderberg Relation

The Soderberg relationship for fluctuating stresses is shown in Fig. 15-5 (review Section 3-18 if needed). The upper line connects the endurance strength on the vertical (amplitude stress) axis to the yield strength on the horizontal (mean stress) axis. It is a conservative estimate of combinations of mean stress and stress amplitude expected to survive to at least 10^6 cycles.

The lower line is parallel to the upper one and

Figure 15-5 The Soderberg lines.

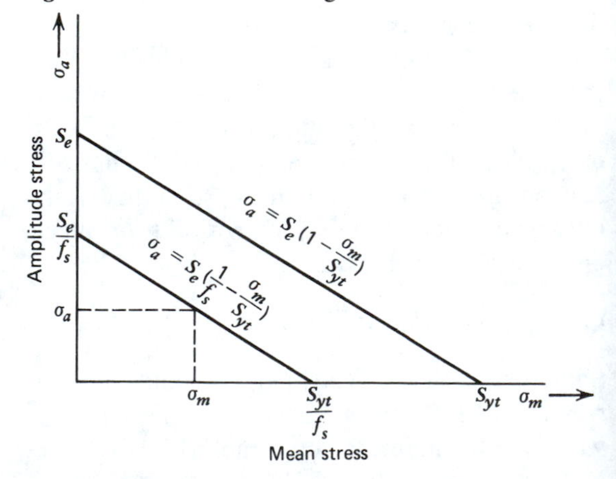

represents combinations of mean stress and stress amplitude corresponding to a factor of safety *fs*. The equation for this lower line is

$$fs = \frac{S_{yt}}{\sigma_m + (S_{yt}/S_e)\,\sigma_a} \qquad (15\text{-}1)$$

The *denominator* in this equation can be regarded as an *equivalent static stress*.

A similar relationship holds for a *fluctuating shear stress*.

$$fs = \frac{S_{sy}}{\tau_m + (S_{sy}/S_{se})\,\tau_a} \qquad (15\text{-}2)$$

The *denominator* is an *equivalent static shear stress*.

To design shafts or axles loaded in bending only or subjected to bending and axial loading, Eq. 15-1 can be used. If only torque is carried, Eq. 15-2 is used where stress concentration is present. In the absence of stress concentration, pure torsional mean stress has no apparent effect on durability and may be ignored (see Section 3-18).

For the cases where combined loadings occur, the concept of the von Mises equivalent stress (or distortion energy) (see Section 3-5) will be used. This approach will be developed in Section 15-6.

15-5 DESIGN FOR STRENGTH—SIMPLE LOADING

Here the methods of Chapter 3 are applied to design. Typically, it is the shaft diameter itself, not a stress or a factor of safety, that is unknown.

Steady Torque

For a solid circular member in steady torsion, the equations of Table 3-2 are applied. Since

$$\tau_m = \frac{Tr}{J} = \frac{16T}{\pi d^3}$$

and

$$fs = \frac{S_{sy}}{\tau_m} = \frac{0.57 S_{yt}}{\tau_m}$$

the design equation for diameter is

$$d = \left(\frac{8.935 T fs}{S_{yt}}\right)^{1/3} \qquad (15\text{-}3)$$

EXAMPLE 15-1

A 28-in.-long piece of solid shaft with a uniform diameter is needed to transmit 45 hp at 3600 rpm from an ac motor to a dc generator. There are flexible couplings keyed to the shaft at each end. Sled runner keyseats are used. If a plain carbon steel with a tensile yield strength of S_{yt} = 47,200 psi is used, what should be its diameter for a factor of safety of 2?

SOLUTION

(*a*) The applied torque can be found by use of the basic power formula in English units. Thus,

$$T = \frac{63{,}000\ P}{n}$$

$$= \frac{(63{,}000)(45\ \text{hp})}{3600\ \text{rpm}} = \mathbf{788\ lb\text{-}in.}$$

(*b*) The required diameter is found by use of Eq. 15-3.

$$d = \left(\frac{8.935 T fs}{S_{yt}}\right)^{1/3}$$

$$= \left[\frac{(8.935)(788\ \text{lb-in.})(2)}{47{,}200\ \text{psi}}\right]^{1/3}$$

$$= \mathbf{0.67\ in.}$$

This, of course, is the minimum diameter. In practice one might round up to the next available size of cold-finished shafting. Note that since the torsional stress is steady, no stress concentration factor is used.

Fluctuating Torque

The Soderberg relation, as expressed in Eq. 15-2, is the basis for this case. For a solid circular shaft, Eq. 15-2 can be solved for shaft diameter d. The very useful result is

$$d = \left\{\frac{8.935[T_m + (S_{sy}/S_{se})T_a]fs}{S_{yt}}\right\}^{1/3} \qquad (15\text{-}4)$$

Earlier it was suggested that the denominator of Eq. 15-1 can be viewed as an equivalent static stress. This applies as well to Eq. 15-2. Compare Eqs. 15-3 and 15-4. The equivalence concept has reappeared. In Eq. 15-4 an equivalent static torque appears in brackets in the numerator.

EXAMPLE 15-2

Suppose that for the shaft described in Example 15-1 there is a cyclic torsional stress of amplitude 240 lb-in. superimposed on the steady torque of 788 lb-in. Recalculate the diameter. Assume that a cold-finished bar is used. Design for a factor of safety of 2 based on average values of strengths. The fatigue stress concentration factor at the keyseats is $K_{fs} = 1.3$. The properties of the steel are $S_{ut} = 85{,}600$ psi and $S_{yt} = 47{,}200$ psi.

SOLUTION

(*a*) The first step is to estimate the fatigue strength by use of the derating factors (see Section 3-10).

$$S_{se} = k_a k_b k_c k_d k_e k_f S'_{se}$$

The values to substitute are

$k_a = 0.76$ (cold-finished surface)
$k_b = 0.85$ (assuming $0.3 \leq d \leq 2$)
$k_c = 1$ (for average value)
$k_d = 1$ (temperature not high)

$$k_e = \frac{1}{K_{fs}} = \frac{1}{1.3} = 0.77$$

$k_f = 1$ (no impact)
$S'_{se} = 0.57S'_e$

$$S'_e = 0.5S_{ut} = (0.5)(85{,}600 \text{ psi}) = 42{,}800 \text{ psi}$$
$$S'_{se} = (0.57)(42{,}800 \text{ psi}) = 24{,}400 \text{ psi}$$
$$S_{se} = (0.76)(0.85)(1.0)(1.0)(0.77)(1)(24{,}400 \text{ psi}) = \mathbf{12{,}140\ psi}$$

(*b*) The shear yield strength is

$$S_{sy} = 0.57S_{yt} = (0.57)(47{,}200 \text{ psi}) = \mathbf{26{,}900\ psi}$$

(*c*) So, by use of Eq. 15-4, the shaft diameter is

$$d = \left\{\frac{8.935[T_m + (S_{sy}/S_{se})T_a]fs}{S_{yt}}\right\}^{1/3}$$

$$d = \left\{\frac{8.935}{47{,}200 \text{ psi}}\left[788 \text{ lb-in.} + \frac{26{,}900 \text{ psi}}{12{,}140 \text{ psi}}(240 \text{ lb-in.})\right]2\right\}^{1/3}$$

$$= \mathbf{0.794\ in.}$$

Compare d and Z_p with those of Example 15-1 and state your conclusion.

Steady Bending Loads

For a solid circular axle or shaft, the equations in Table 3-2 can be applied. The design equation for diameter is

$$d = \left(\frac{10.186M_a fs}{S_e}\right)^{1/3} \qquad (15\text{-}5)$$

Since the shaft or axle is rotating, the stress due to bending is cyclic. There is no static bending stress.

EXAMPLE 15-3

A railroad freight car has a total weight of 1231 500 N when fully loaded. It is supported by four wheel sets like the one in Fig. 15-2*b*. The distance between the load points is $l_W = 1981$ mm, and the mean distance between wheels is $l_R = 1435$ mm. The axle material is hot-rolled AISI 1040 steel. The yield and ultimate strengths are $S_{yt} = 400$ MPa and $S_{ut} = 627$ MPa.

The wheels are assembled to the axle by means of a shrink fit. The fatigue stress concentration factor due to the shrink fit is $K_f = 1.9$. The axle surface at the shrink fit is ground.

What should be the minimum shaft diameter at the wheels? Use a factor of safety of 3 based on average material properties.

SOLUTION

(*a*) The load W on each of the eight wheels is

$$W = \frac{1231\ 500\ \text{N}}{8} = 153\ 940\ \text{N}$$

The bending moment at the wheel is

$$M = \frac{W(l_W - l_R)}{2}$$

$$= \frac{(153\ 940\ \text{N})(1981\ \text{mm} - 1435\ \text{mm})}{2}$$

$$= \mathbf{(4.203)(10^7)\ N \cdot mm}$$

Since the axle is rotating, this is the moment amplitude M_a.

(*b*) The fatigue strength must be found by use of the derating factors. The uncorrected fatigue strength is

$$S'_e = 0.5\ S_{ut}$$

$$= (0.5)(627\ \text{N/mm}^2)$$

$$= 313.5\ \text{N/mm}^2$$

The derating factors are

$k_a = 0.895$ (ground surface)
$k_b = 0.75$ (assume $d > 50$ mm)
$k_c = 1.0$ (for average value)
$k_d = 1$ (no elevated temperature)

$$k_e = \frac{1}{K_f} = \frac{1}{1.9} = 0.526$$

$k_f = 0.7$ (Table 3-4 for medium shock)

The estimated fatigue strength is

$$S_e = k_a k_b k_c k_d k_e k_f S'_e$$
$$= (0.895)(0.75)(1.0)(1.0)(0.526)(0.7)(313.5\ \text{N/mm}^2)$$
$$= \mathbf{77.5\ N/mm^2}$$

(*c*) Now the axle diameter can be found by using Eq. 15-5.

$$d = \left(\frac{10.186 M_a fs}{S_e}\right)^{1/3}$$

$$= \left[\frac{(10.186)(4.203)(10^7\ \text{N}\cdot\text{mm})(3)}{77.5\ \text{N/mm}^2}\right]^{1/3}$$

$$= \mathbf{255\ mm}$$

EXAMPLE 15-4

Figure 15-6 illustrates a method of fatigue testing tractor axle shafts in bending. The axle is mounted on bearings in a housing. The bearings, seals, and housing are not shown in the figure, but the points at which the axle is supported are designated by R_1 and R_2. The shaft is loaded by a weight suspended from a bearing so that the shaft can rotate while the weight hangs downward. The axle shaft is driven by a motor through a chain drive. Because energy is required only to overcome friction, the shear stress is negligible compared to the bending stress.

The axle is made of an alloy steel heat treated to Rockwell C 45 to 50 and has ground surfaces at the bearings. If fatigue failures are expected to occur at the fillet next to R_2, approximately what load W is required to cause failures at the fatigue limit of the material?

SOLUTION

(*a*) In this problem the design is complete; it is load W that is the unknown. The method of analysis involves equating the bending stress at the fillet due to W to the fatigue strength of the steel.

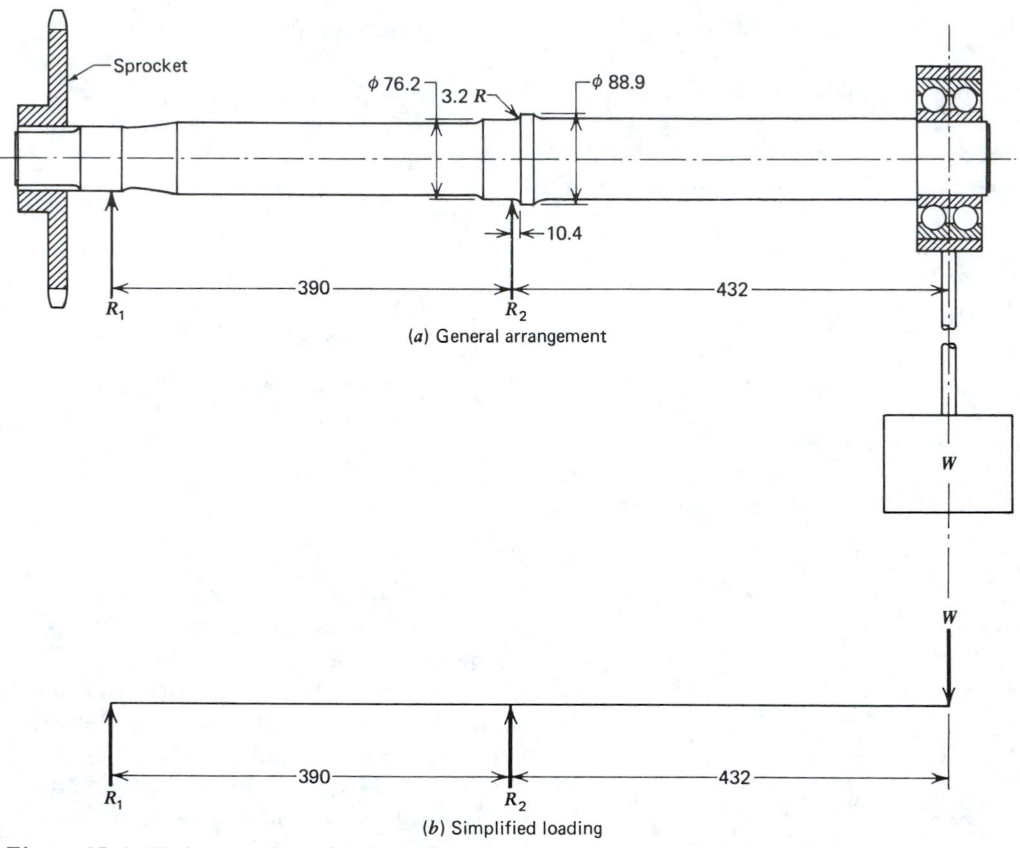

Figure 15-6 Fatigue testing of axle shafts.

$$\sigma_a = \frac{M}{Z} = \frac{32M}{\pi d^3} = S_e$$

(*b*) The bending moment M at the fillet is

$$M = (432 - 10.4)W \text{ N} \cdot \text{mm}$$

$$= \mathbf{421.6W \text{ N} \cdot \text{mm}}$$

(*c*) Next, the fatigue strength must be evaluated. The quantity $k_a S_e'$ can be found directly using Fig. 3-11, which shows the fatigue strength derated only for surface strength. For Rockwell C 45 to 50, an average value is 615 N/mm^2. The other derating factors are

$k_b = 0.75$ (for $d > 50$ mm)
$k_c = 1.0$ (for average strength)
$k_d = 1.0$ (no elevated temperature)

$$k_e = \frac{1}{K_f}$$

where

$$K_f = 1 + q(K_t - 1)$$

A stress concentration chart is given in the appendix for a shoulder fillet in bending. For $D/d = 88.9/76.2 = 1.17$ and $r/d = 3.2/76.2 = 0.042$, the geometrical stress concentration factor is $K_t = 2.06$. The notch sensitivity factor is approximately $q = 0.85$. Thus

$$K_f = 1 + (0.85)(2.06 - 1) = 1.9$$

$$k_e = \frac{1}{1.9} = 0.53$$

$$S_e = (k_a S'_e) k_b k_c k_d k_e k_f$$

$$= (615\ \text{N/mm}^2)(0.75)(1)(1)(0.53)(1)$$

$$S_e = \mathbf{243\ N/mm^2}$$

(*d*) Now W can be found by substitution in

$$\frac{32M}{\pi d^3} = S_e$$

$$\frac{32}{\pi}\frac{(421.6W\ \text{N}\cdot\text{mm})}{(76.2\ \text{mm})^3} = 243\ \text{N/mm}^2$$

$$W = \mathbf{25\ 040\ N}$$

(*e*) The mass m necessary at the end of the axle is

$$m = \frac{W}{g} = \frac{25\ 040\ \text{N}}{9.807\ \text{m/s}^2} = \mathbf{2550\ kg}$$

15-6 DESIGN FOR STRENGTH—COMBINED LOADING

This is a continuation of Section 15-5. More complex loadings demand a more advanced approach based on the concept of the von Mises equivalent normal stress.

Basic Design Equation

The von Mises equivalent stress approach (a consequence of the distortion energy theory of failure) implies that when the equivalent normal stress

$$\sigma'_i = \sqrt{\sigma_i^2 + 3\tau_i^2} \qquad (15\text{-}6)$$

reaches the tensile yield strength of the material, failure occurs. Thus the factor of safety is

$$fs = \frac{S_{yt}}{(\sigma_i^2 + 3\tau_i^2)^{1/2}} \qquad (15\text{-}7)$$

For fluctuating stresses, the Soderberg relationship defines equivalent static stresses. From Eqs. 15-1 and 15-2,

$$\sigma_i = \sigma_m + \frac{S_{yt}}{S_e}\sigma_a \qquad (15\text{-}8)$$

$$\tau_i = \tau_m + \frac{S_{sy}}{S_{se}}\tau_a \qquad (15\text{-}9)$$

In the basic design problem first described in Section 15-4, the applied torque T and the applied bending moment M are steady. Since the member is rotating, the bending stress is cyclic. Thus, in Eqs. 15-8 and 15-9, for this class of design problems, $\sigma_m = 0$ and $\tau_a = 0$. The equation for shaft diameter is

$$d = \left[\frac{10.186\, fs}{S_{yt}}\sqrt{\left(\frac{S_{yt}}{S_e}M\right)^2 + \frac{3}{4}T^2}\right]^{1/3} \qquad (15\text{-}10)$$

Torsion and Bending in One Plane

Here the design equation (Eq. 15-10) can be used directly. Example 15-5 shows this method.

EXAMPLE 15-5

The shaft shown in Fig. 15-7 is driven by an electric motor (not shown) on the left by means of a flexible coupling. A woodworking machine is driven by means of a belt and pulley drive. The driving pulley is on the shaft shown and weighs 29 lb. The belt tensions are as shown. The shaft is supported by two ball bearings.

The shaft is to be made of cold-drawn AISI 1020 steel with the properties $S_{ut} = 78{,}000$ psi and $S_{yt} = 66{,}000$ psi. The shaft surface will not be machined, but sled-runner keyseats will be milled for keys at the flexible coupling and the pulley. The fatigue stress concentration factor is $K_f = 1.3$.

Determine the minimum shaft diameter for a factor of safety of 2.5 based on average material properties.

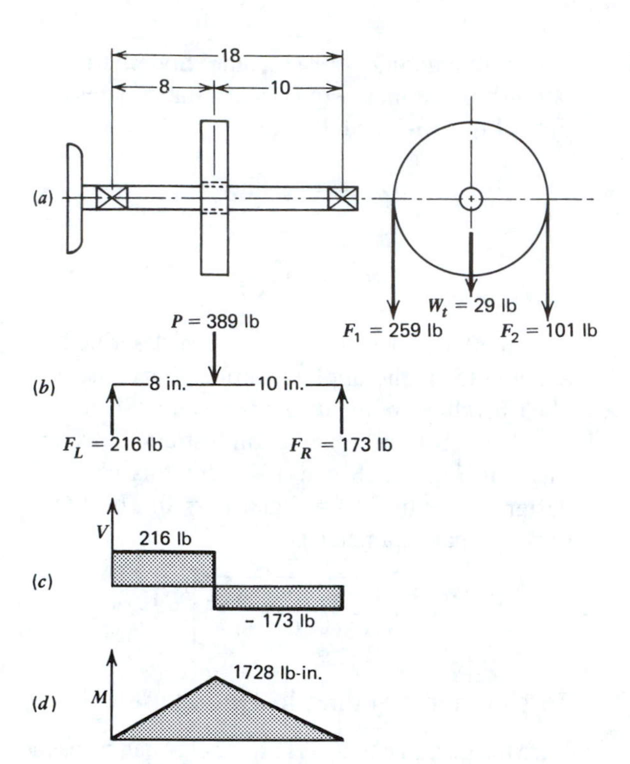

Figure 15-7 Shaft for Example 15-5.

SOLUTION

(*a*) As shown in Fig. 15-7*a*, the bending load *P* at the pulley is

$$W = F_1 + F_2 + Wt = 389 \text{ lb}$$

A loading diagram is shown in Fig. 15-7*b*.

(*b*) The torque *T* on the shaft is

$$T = \frac{(F_1 - F_2)D}{2}$$

$$= \frac{(259 \text{ lb} - 101 \text{ lb})(12 \text{ in.})}{2}$$

$$= 948 \text{ lb-in.}$$

(*c*) The shear and bending moment diagrams have been drawn as Fig. 15-7*c* and 15-7*d*. The maximum bending moment occurs at the pulley and is M = **1728 lb-in.**

(*d*) Here are the calculations for the fatigue strength.

$$S'_e = 0.5S_{ut} = (0.5)(78{,}000 \text{ psi}) = 39{,}000 \text{ psi}$$

$k_a = 0.78$ (for cold-drawn surface)
$k_b = 0.85$ (assume $0.3 < d \leq 2.0$)
$k_c = 1.0$ (use average values)
$k_d = 1.0$ (no elevated temperature)

$$k_e = \frac{1}{K_f} = \frac{1}{1.3} = 0.77$$

$k_f = 0.9$ (light shock)

$$S_e = k_a k_b k_c k_d k_e k_f S'_e = (0.78)(0.85)(1.0)(1.0)(0.77)(0.9)(39{,}000 \text{ psi}) = 17{,}920 \text{ psi}$$

(*e*) Now the diameter can be calculated using Eq. 15-10.

$$d = \left[\frac{10.186\text{fs}}{S_{yt}}\sqrt{\left(\frac{S_{yt}}{S_e}M\right)^2 + \frac{3}{4}T^2}\right]^{1/3}$$

$$= \left\{ \frac{(10.186)(2.5)}{66{,}000 \text{ psi}} \sqrt{\left[\frac{(66{,}000 \text{ psi})(1728 \text{ lb-in.})}{17{,}920 \text{ psi}}\right]^2 + \frac{3}{4}(948 \text{ lb-in.})^2} \right\}^{1/3}$$

$$= \mathbf{1.35 \text{ in.}}$$

Torsion and Bending in Two Planes

If the bending forces are not all in one plane, find their components in mutually perpendicular planes. Plot a bending moment diagram for each plane. Then, at any cross section, the resultant bending moment is the vector sum (square root of the sum of the squares) of the bending moments in the two planes at that cross section. Once the resultant bending moment is known, the method is as before.

15-7 HOLLOW SHAFTS

Suppose a circular shaft has an outside diameter d_o and an inside diameter d_i. Then the section modulus in bending is

$$Z = \frac{I}{r} = \frac{\pi}{64}(d_o^4 - d_i^4)\frac{2}{d_o}$$

$$= \frac{\pi}{32d_o}(d_o^4 - d_i^4) \qquad (15\text{-}11)$$

and the torsional section modulus is $Z_p = J/r$, or

$$Z_p = \frac{\pi}{16d_o}(d_o^4 - d_i^4) \qquad (15\text{-}12)$$

These section moduli can be written in forms that make them apply to both solid and hollow cross sections. Define

$$C = \frac{d_i}{d_o} \qquad (15\text{-}13)$$

Then Eqs. 15-11 and 15-12 can be written as

$$Z = \frac{\pi d_o^3}{32}(1 - C^4) \qquad (15\text{-}14)$$

$$Z_p = \frac{\pi d_o^3}{16}(1 - C^4) \qquad (15\text{-}15)$$

All of the design equations in sections 15-5 and 15-6 can be revised to make them applicable to hollow circular shafts simply by replacing d by d_o and fs by $fs/(1 - C^4)$. For example, Eq. 15-10 becomes

$$d_o = \left[\frac{10.186fs}{S_{yt}(1 - C^4)}\sqrt{\left(\frac{S_{yt}}{S_e}M\right)^2 + \frac{3}{4}T^2}\right]^{1/3} \qquad (15\text{-}16)$$

15-8 AXIAL LOADS

Occasionally, there will be an axial load on a shaft. The axial component of the tooth force between helical gears is an example. The basic equations (Eqs. 15-7, 15-8, and 15-9) still apply because they are very general. But the design methods developed in Section 15-6 assumed that the normal stress σ_i was due solely to bending and was, therefore, cyclic in nature. An axial load, however, causes a steady axial stress that is the mean component σ_m of a fluctuating normal stress.

For the general case of a hollow circular shaft, the mean component of the fluctuating normal stress σ_i is

$$\sigma_m = \frac{4\alpha F_a}{\pi d_o^2(1 - C^2)} \qquad (15\text{-}17)$$

where α is the *column-action factor*. The column-action factor α is unity if the axial load places the shaft in tension. Otherwise, it is

$$\alpha = \frac{1}{1 - 0.0044(L/k)} \quad \text{for} \quad L/k < 115 \qquad (15\text{-}18)$$

$$\alpha = \frac{S_{yc}}{\pi^2 nE}\left(\frac{L}{k}\right)^2 \quad \text{for} \quad L/k > 115 \qquad (15\text{-}19)$$

where

L = length of shaft between bearings; mm, in.
k = radius of gyration of shaft cross section; mm, in.

S_{yc} = yield strength in compression; MPa, psi
E = modulus of elasticity; MPa, psi
n = 2.25 for fixed ends (wide journal bearings)
n = 1.6 for ends partly restrained (as in bearings)
n = 1 for hinged ends (as in self-aligning bearings)

For hollow circular shafts, the radius of gyration is

$$k = \sqrt{\frac{I}{A}} = \frac{d_o}{4}\sqrt{1 + C^2} \qquad (15\text{-}20)$$

In using Eq. 15-19 to calculate the column-action formula, there is some doubt as to the proper value to use for the end factor n. Note that the lowest values for n are the most conservative.

For a rotating circular shaft subjected to steady torque, steady transverse loads, and steady axial loading, the equations for the normal stress and shear stress to use in Eq. 15-7 are

$$\sigma_i = \frac{4F_a\alpha}{\pi d_o^2(1 - C^2)} + \frac{S_{yt}}{S_e}\frac{32M}{\pi d_o^3(1 - C^4)} \qquad (15\text{-}21)$$

$$\tau_i = \frac{16T}{\pi d_o^3(1 - C^4)} \qquad (15\text{-}22)$$

15-9 DESIGN FOR BENDING RIGIDITY

The previous sections have dealt with design for strength. This section discusses rigidity in bending, while the next section deals with torsional rigidity.

Two questions arise regarding rigidity in bending. (1) How does one estimate the bending deflections of a shaft subjected to a number of transverse loads? (2) What is a reasonable upper limit for bending deflections? In this section the first question is answered only for shafts or axles of uniform cross section. A later section provides a simple graphical technique for estimating the bending deflections of nonuniform shafts or axles.

If a shaft supports gears or is itself supported by journal bearings, bending deflections must be controlled to limit misalignments. As usual, until designers gain experience and develop views on the subject appropriate to their own design practice, successful past practice is the proper starting point. Table 15-2 records the pertinent past practices. Note that where gears or journal bearings are involved, the design rules are written in a way that limits the *slope* of the shaft.

TABLE 15-2 Limits for Bending Deflections in Shafts

Shaft Type	Limiting Deflection; mm, in.
General machine shaft	$0.002l$
Shaft supported by journal bearings	$0.0015/a$
Shaft supporting gears	$0.005/f$
Shaft supporting precision gears	$0.001/f$

Notation. l = length of shaft between bearings; mm, in.
a = distance between load point and most remote bearing; mm, in.
f = face width of gear; mm, in.

Now back to the first question. We will make use of the *principle of superposition,* which states that the bending deflection at any shaft cross section is the sum of the deflections at that section due to each of the loads acting separately. An analogous statement holds for stresses. Use this principle for members with only transverse loads or where axial loads are very small because substantial axial loading can cause column bending. So shafts subjected to transverse and axial loads must be handled differently.

To make use of the principle of superposition, simply use the tables for beam deflections listed in *Machinery's Handbook* and in the appendix. At any shaft cross section of interest, find the deflection due to each of the loads, then add the results vectorially. This method is best shown by an example.

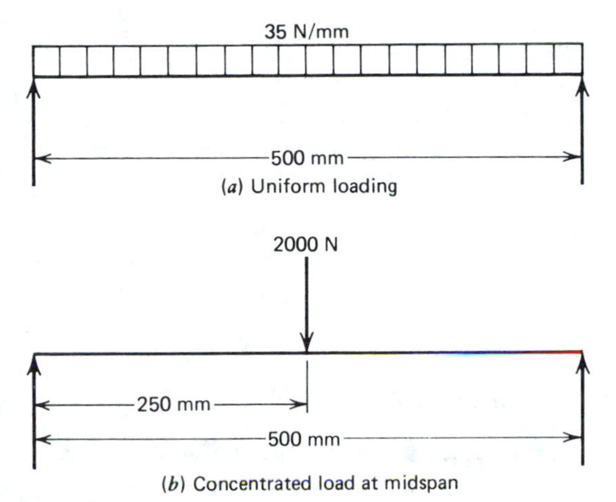

Figure 15-8 Sketch of the two beam loadings for the shaft of Example 15-6.

EXAMPLE 15-6

A steel shaft has a uniform diameter of 50 mm. It is supported by two single-row ball bearings 500 mm apart (center to center). There is a uniform load of 35 N/mm for the 500 mm between the supporting bearings and a concentrated load of 2000 N midway between the bearings and in the same direction as the uniform load. What is the maximum bending deflection?

SOLUTION

(*a*) The two loading patterns are shown separately in Fig. 15-8. The uniform loading is listed as Case 9 in Table F-3 in the appendix. For it, the maximum deflection occurs at the center and is

$$Y_{max} = -\frac{5}{384}\frac{Wl^3}{EI}$$

where W is the total load on the beam.

For the concentrated load at the center, Case 5 in Table F-3, the maximum deflection is at the center and is

$$Y_{max} = -\frac{Wl^3}{48EI}$$

(*b*) A typical value for the modulus of elasticity for steel is $E = 207\ 000$ MPa.

(*c*) The moment of inertia for the solid circular cross section is

$$I = \frac{\pi d^4}{64} = \frac{\pi(50 \text{ mm})^4}{64}$$

$$= 306\ 800 \text{ mm}^4$$

(*d*) The total load W due to the uniform loading of 35 N/mm is

$$W = (35 \text{ N/mm})(500\text{mm})$$

$$= 17\ 500 \text{ N}$$

so the deflection at the shaft's center due to the uniform load is

$$Y_1 = -\frac{5}{384}\frac{Wl^3}{EI}$$

$$= -\frac{(5)(17\ 500 \text{ N})(500 \text{ mm})^3}{(384)(207\ 000 \text{ N/mm}^2)(306\ 800 \text{ mm}^4)}$$

$$= -0.45 \text{ mm}$$

(*e*) The deflection due to the concentrated load of 200 N is

$$Y_2 = -\frac{Wl^3}{48EI}$$

$$= -\frac{(2000 \text{ N})(500 \text{ mm})^3}{(48)(207\ 000 \text{ N/mm}^2)(306\ 800 \text{ mm}^4)}$$

$$= -0.08 \text{ mm}$$

(*f*) The total deflection is

$$Y_t = Y_1 + Y_2$$

$$= (-0.45 \text{ mm}) + (-0.08 \text{ mm})$$

$$= \mathbf{-0.53 \text{ mm}}$$

The deflection is downward, of course, since the loads are downward.

(g) For a general-purpose machine shaft the

maximum bending deflection should not exceed the limits given in Table 15-2.

$$Y_{max} = 0.002l$$
$$= (0.002)(500 \text{ mm})$$
$$= 1 \text{ mm}$$

The shaft is rigid enough.

15-10 DESIGN FOR TORSIONAL STIFFNESS

The general equation for angular deflection of a uniform bar is:

$$\theta = \frac{Tl}{GJ} \tag{15-23}$$

For a shaft with a uniform circular cross section, solid or hollow, the polar moment of inertia of the cross section is

$$J = \frac{\pi}{32}(d_o^4 - d_i^4)$$

$$J = \frac{\pi}{32}(1 - C^4)d_o^4 \tag{15-24}$$

where C is the ratio of inside to outside diameters.

The limits on torsional deflection are customarily stated in terms of *twist per unit length,* that is, θ/l. Thus a useful equation for checking an existing design is

$$\frac{\theta}{l} = \frac{32T}{\pi(1 - C^4)d_o^4 G} \tag{15-25}$$

The corresponding design equation is

$$d_o = \left[\frac{32T}{\pi(1 - C^4)(\theta/l)G}\right]^{1/4} \tag{15-26}$$

The limits for torsional deflection listed in Table 15-3 include the vast majority of applications. For machine tools a rigorous criterion is necessary. For overhead line shafts, on the other hand, deflections about 12.5 times those for machine tool shafts can be tolerated.

TABLE 15-3 Limits for Torsional Deflection of Shafts

	Maximum torsional deflection	
Application	rad/mm	rad/in.
Machine tools	$(4.6)(10^{-6})$	$(1.16)(10^{-4})$
Line shafting	$(57)(10^{-6})$	$(14.5)(10^{-4})$

EXAMPLE 15-7

In Example 15-1 a minimum diameter of 0.67 in. was calculated for shaft strength. Is this diameter large enough to limit the torsional deflection to $(8)(10^{-4})$ rad/in.?

SOLUTION

(*a*) Since the shaft is solid, $C = 0$. The torque applied to this shaft was found to be $T = 788$ lb-in. To check for stiffness, Eq. 15-25 is appropriate. A typical value of G for plain carbon steel (see Table F-4 in the Appendix) is $(11.4)(10^6 \text{ psi})$. Thus

$$\frac{\theta}{l} = \frac{32T}{\pi(1 - C^4)d_o^4 G}$$

$$= \frac{(32)(788 \text{ lb-in.})}{(\pi)(1)(0.67 \text{ in.})^4(11.4)(10^6 \text{ psi})}$$

$$\mathbf{(35)(10^{-4}) \text{ rad/in.} > (8)(10^{-4}) \text{ rad/in.}}$$

This is too large.

(*b*) As is often the case, a diameter adequate for strength is not large enough for stiffness. Equation 15-26 will now be used to calculate the necessary diameter.

$$d = \left[\frac{32T}{\pi(1 - C^4)(\theta/l)G}\right]^{1/4}$$

$$= \left[\frac{(32)(788 \text{ lb-in.})}{(\pi)(8)(10^{-4} \text{ rad/in.})(11.4)(10^6 \text{ psi})}\right]^{1/4}$$

$$= \mathbf{0.97 \text{ in.}}$$

In this instance the diameter needed for

TABLE 15-4 Fatigue Stress Concentration Factors for Keyseats[a]

Material	Profile Keyseat		Sled-runner Keyseat	
	Bending	Torsion	Bending	Torsion
Annealed steel	1.6	1.3	1.3	1.3
Quenched and tempered steel	2.0	1.6	1.6	1.6

[a]Note that these are K_f, not K_t, so no correction for notch sensitivity is needed.

adequate stiffness is about 45% larger than the one needed for strength.

15-11 EFFECTS OF KEYSEATS

Keyseats, grooves, and similar features have two effects. *They raise the stress level and increase flexibility*.

Stress Concentration

Machinery's Handbook has charts of the stress concentration factors for most of the features found on shafts. Fatigue stress concentration factors suggested by Lipson and Juvinall[2] are listed in Table 15-4 for profile and sled-runner keyseats (described in Chapter 16).

Increased Flexibility

Hopkins has dealt with the problem of calculating increased flexibility caused by a number of features common to shafts: keyseats, shoulders, splines, and tapers. For keyseats he recommends, as a conservative procedure, that one calculate the angular twist for the shaft length containing the keyseat by using an *effective* diameter. For the effective diameter he uses simply the shaft diameter less the keyseat's depth. With standard-size square keys, the keyseat width is $d/4$ and its depth is $d/8$. For these proportions the angular twist for the length of the keyseat is 71% greater than for no keyseat.

[2]Charles Lipson and Robert C. Juvinall, *Handbook of Stress and Strength* (New York: Macmillan, 1963).

15-12 BENDING OF NONUNIFORM SHAFTS AND AXLES

The standard formulas for bending of beams and torsion of bars work well for evaluating the stiffness of axles or shafts of uniform cross section. But if the moments of inertia I and J vary along the length, the problem is more complex. In a book and two articles, Hopkins has presented numerical methods for determining torsional deflections and bending deflections for nonuniform axles and shafts. His methods are well suited to a programmable pocket calculator or a digital computer. However, for determining bending deflections, there is an elegantly simple and easy-to-learn graphical technique that produces tolerably accurate results with reasonable care.

That method is based on graphical integration of the basic beam-bending equation

$$\frac{d^2y}{dx^2} = \frac{M}{EI} \tag{15-27}$$

By integrating once, the slope $\theta = dy/dx$ is obtained. Then, by integrating a second time, the beam deflection y is found.

Graphical Integration

The essentials of the method are illustrated in Fig. 15-9. A varying force F has been plotted against displacement x in Fig. 15-9a. The graph-

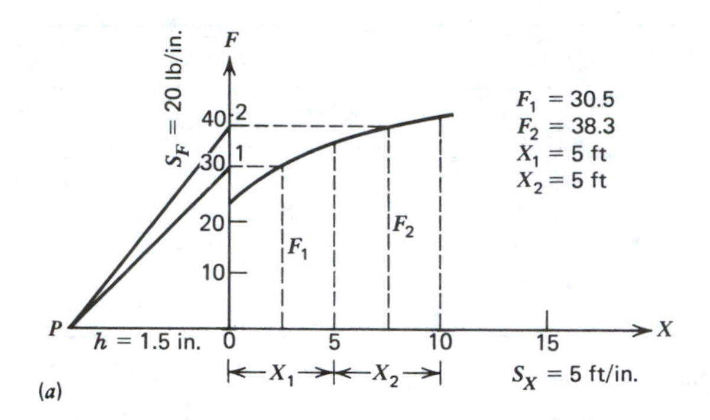

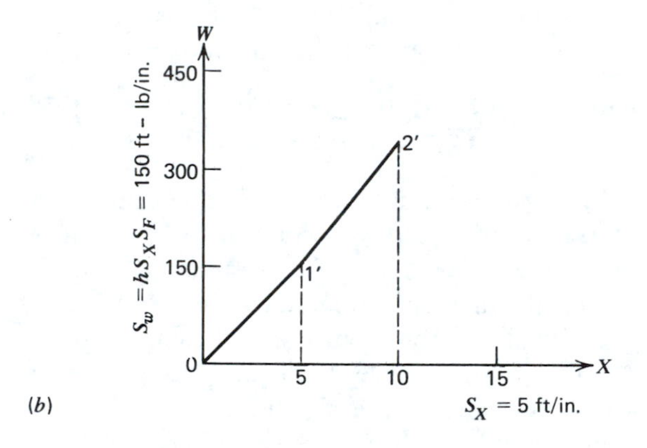

(c) 0 1 in.

Figure 15-9 Graphical integration.

ical scales used (scale = ratio of quantity represented to corresponding drawing distance) are $S_x = 5$ ft/in. and $S_F = 20$ lb/in. The designer must estimate the area under the curve (i.e., the integral) from $x = 0$ to $x = 10$ ft. That area represents the work done by this varying force.

The area has been divided into two strips of width x_1 and x_2. F_1 and F_2 are the values of F at the midpoints of these two strips. The areas of the two strips are approximately

$$W_1 = F_1 x_1$$

$$W_2 = F_2 x_2$$

This method of estimating areas is called the *rectangular rule* and becomes more accurate as more and narrower strips are used.

The plot of the area (work done, W) under the original force-displacement curve is shown in Fig. 15-9*b*. Here is how it was drawn.

1. On the original graph the heights F_1 and F_2 were projected to the vertical axis as points 1 and 2.
2. At a convenient distance h to the left on the horizontal axis, a pole point P was selected. Lines $P - 1$ and $P - 2$ were drawn.
3. For the new graph (Fig. 15-9*b*) ordinates were erected at the same spacing as in the old graph.
4. The line $O - 1'$ was drawn parallel to $P - 1$ and $1' - 2'$ parallel to $P - 2$.

The new plot can only be interpreted, however, when the vertical axis scale S_W is known. Now, since

$$\text{Scale} = \frac{\text{quantity represented}}{\text{drawing distance}}$$

it follows that

$$\text{Drawing distance} = \frac{\text{quantity represented}}{\text{scale}}$$

Lines $P - 1$ and $O - 1'$ were made parallel on the drawing. Hence the relationship of *drawing distances* is

$$\text{slope} = \frac{F_1/S_F}{h} = \frac{W_1/S_W}{x_1/S_x}$$

However, since it is approximately true that $W_1 = F_1 x_1$, this simplifies to

$$S_W = hS_xS_F \tag{15-28}$$

The vertical (work) scale for this example is

$$S_W = (1.5 \text{ in.})(5 \text{ ft/in.})(20 \text{ lb/in.})$$
$$= 150 \text{ ft-lb/in.}$$

The vertical scale in Fig. 15-9*b* has been labeled accordingly.

Bending Deflection

Here is an example of the method.

EXAMPLE 15-8

The steel shaft in Fig. 15-10*a* is supported by narrow bearings at R_1 and R_2. Two concentrated loads are shown. Find the maximum bending deflection.

SOLUTION

(*a*) In Fig. 15-10*b* and 15-10*c* the shear and bending moment diagrams have been drawn using conventional methods.

(*b*) The graphical solution is shown in Fig. 15-11. First, the M/EI diagram was plotted by dividing the bending moment at any cross section by the value of EI there. Note that where there is an abrupt change in shaft cross section there is a correspondingly abrupt change in M/EI. A convenient trick was also used. For E the value 30 instead of 30 million was used. This has the effect of stating the shaft deflection in microinches instead of inches.

(*c*) The M/EI diagram was divided into convenient widths, using points of reaction, load application, or change in cross section as the boundaries. In each of the six strips, the value of M/EI at the midpoint was located and projected to the vertical axis.

(*d*) A pole point P was located $h = 2$ in. to the left, and lines $P - 1$, $P - 2$, . . . , $P - 6$ were drawn.

(*e*) Then the slope ($\theta = dy/dx$) diagram was constructed. From any point O on the vertical axis, line O $-$ 1 in. was drawn parallel to $P - 1$, $1 - 2$ in. parallel to $P - 2$, and so on.

(*f*) The x-axis is drawn *after* the curve is constructed by estimating where the point of zero slope is. In this case, we consulted our intuitions and decided it occurs at $x = 2$ in. The x-axis was drawn accordingly. The guess typically will be wrong but, as it turns out, no great harm is done.

(*g*) The scale for the slope is

$$S_\theta = hS_xS_{M/EI}$$
$$= (2 \text{ in.})(1 \text{ in./in.})(100\ \mu\text{in.}^{-1}/\text{in.})$$
$$= 200\ \mu\text{rad/in.}$$

where radian is used in place of inches per inch.

(*h*) The method of graphical integration is applied again, and again the curve is started at any point O′ on the vertical axis. This time it is known that the deflection is zero both at $x = 0$ and $x = 6$. The x-axis is drawn through the points where the $x = 0$ and $x = 6$ ordinates cross the deflection curve. As it happened, the estimate that the slope is zero at $x = 2$ was nearly right, but not entirely correct. Thus the x-axis

Figure 15-10 Bending deflection problem for Example 15-8.

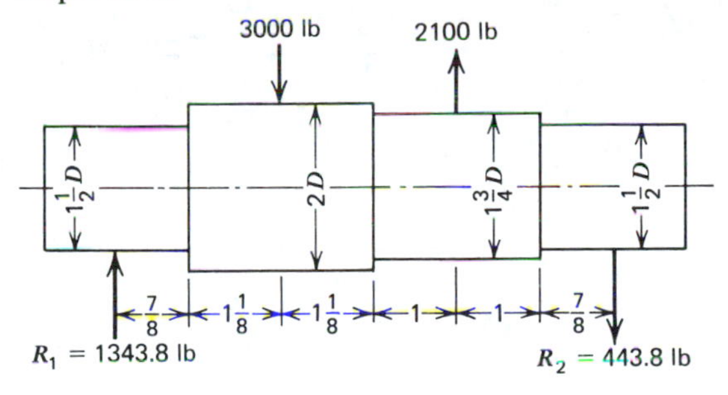

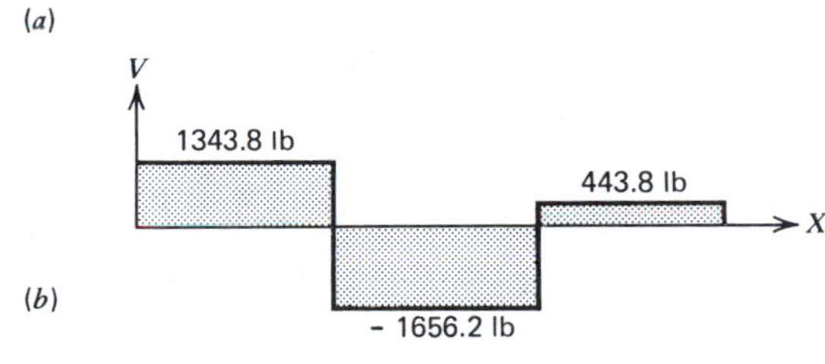

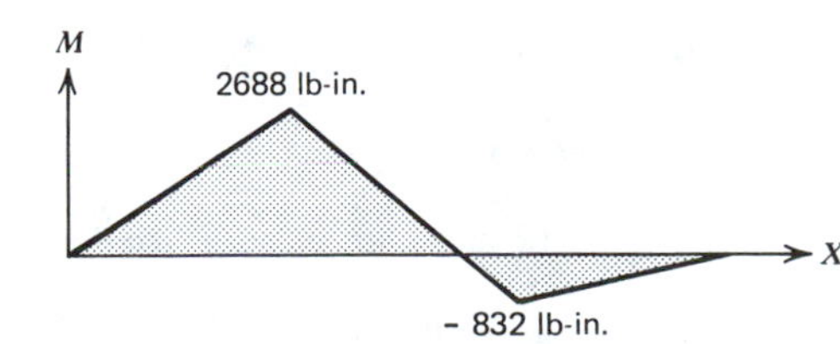

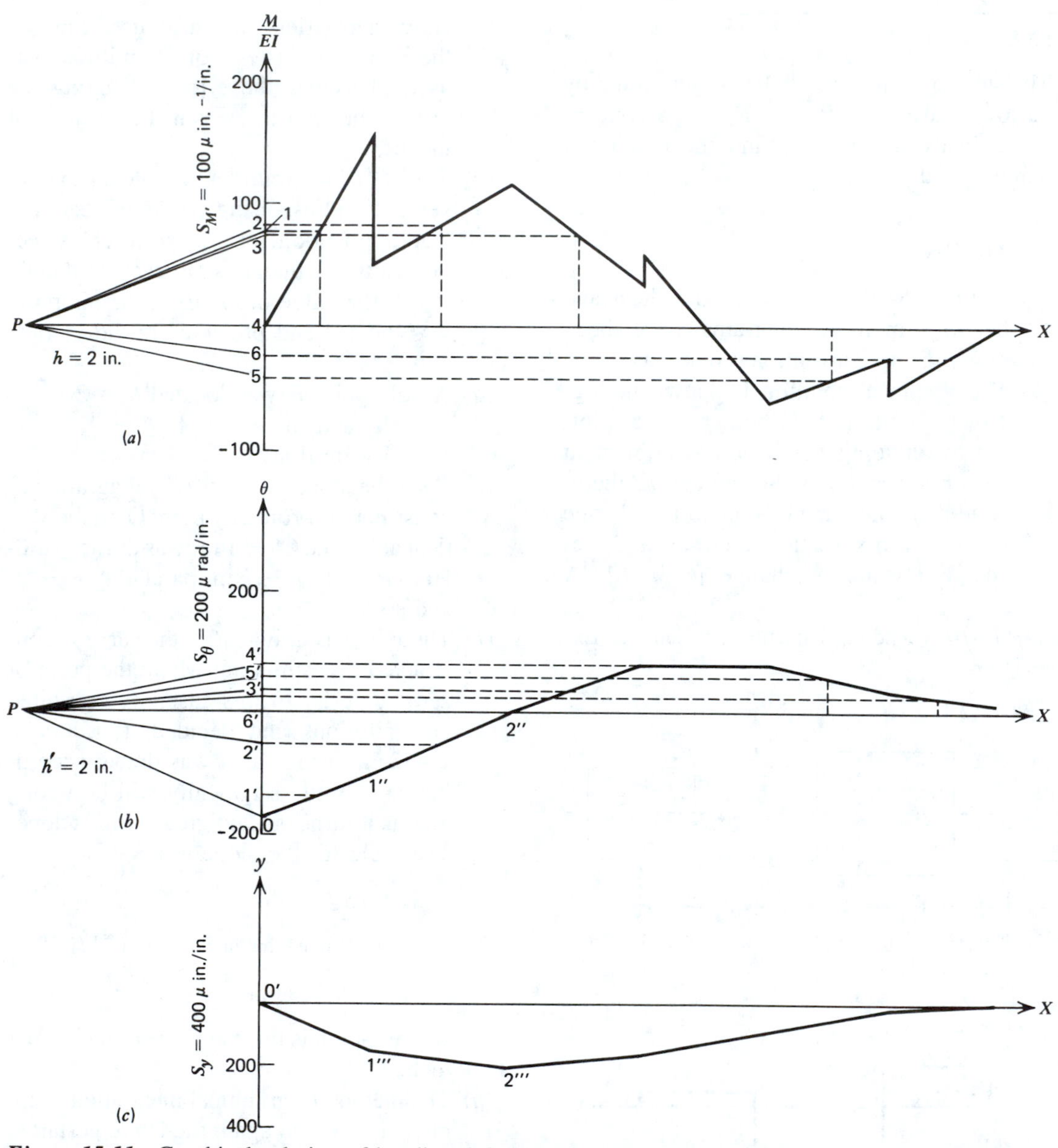

Figure 15-11 Graphical solution of bending deflection problem.

slopes slightly down and to the right. If the estimate had been even worse, the slope would be steeper. No matter. Simply measure deflections from the x-axis, but *vertically,* not perpendicular to the x-axis.

(i) The scale for deflection is

$$
\begin{aligned}
S_y &= h'S_xS_\theta \\
&= (2\text{ in.})(1\text{ in./in.})(200\ \mu\text{rad/in.}) \\
&= 400\ \mu\text{in./in.}
\end{aligned}
$$

(*j*) In this problem the maximum deflection due to bending is approximately **200 μin. at $x = 2$ in.**

15-13 TORSION IN NONCIRCULAR SECTIONS

For calculations involving stresses or deflections in a beam, the basic equations apply to any cross-sectional shape. With torsion, however, there is a problem. Sections other than closed circular sections warp in torsion. This affects the shear stress and the angular deflection.

In the appendix, Table F-2 tabulates the properties of various cross sections in bending and in torsion. Information is provided to calculate the magnitude and position of the maximum shear stress.

For angular deflection, the basic equation is

$$\theta = \frac{Tl}{GK} \tag{15-29}$$

This resembles Eq. 15-23, although instead of the polar moment of inertia J, a factor K appears. In general, K is less than J for noncircular cross sections. For some sections it is much less. Equations for calculating K for a number of cross sections are included in Table F-2. Additional cross sections are tabulated in References 2 and 6.

The ends of a member with a noncircular cross section are usually fastened securely enough to prevent warping near the ends. This reduces the effective length l in Eq. 15-29 and reduces the angular deflection. In short torsion members, end restraint also affects the stress level. Figure 15-1 provides an example of end restraint in which the I-section changes to a rectangular section at each end of the axle.

EXAMPLE 15-9

Drive shafts for automobiles are usually constructed of a welded circular tube connecting the yokes of the universal joints. Someone has proposed cutting costs by substituting an open tube without the longitudinal weld. What effect would this have on stress level and on angular deflection for a long steel tube with an outside diameter of 102 mm and a wall thickness of 3.5 mm?

SOLUTION

(*a*) The inside diameter is

$$\begin{aligned} d_i &= d_o - 2t \\ &= 102\text{ mm} - 2(3.5\text{ mm}) \\ &= 95\text{ mm} \end{aligned}$$

The ratio of diameters is

$$\begin{aligned} C &= d_i/d_o \\ &= 95\text{ mm}/102\text{ mm} \\ &= 0.9314 \end{aligned}$$

(*b*) The polar moment of inertia and the polar section modulus are, by Eqs. 15-24 and 15-15,

$$\begin{aligned} J &= \frac{\pi}{32}(1 - C^4)d_o^4 \\ &= \frac{\pi}{32}(1\text{-}0.9314^4)(102\text{ mm})^4 \\ &= (2.629)(10^6\text{ mm}^4) \end{aligned}$$

$$\begin{aligned} Z_P &= \frac{\pi}{16}(1 - C^4)d_o^3 \\ &= \frac{\pi}{16}(1 - 0.9314^4)(102\text{ mm})^3 \\ &= 51\,557\text{ mm}^3 \end{aligned}$$

(*c*) For the closed circular tube, the maximum shear stress is

$$\tau_{\text{closed}} = \frac{T}{Z_P} = \frac{T}{51\,557} = (1.94)(10^{-5})T\text{ N/mm}^2$$

(*d*) The maximum shear stress for an open tube is

$$\tau_{open} = \frac{T(6\pi r + 1.8t)}{4(\pi rt)^2}$$

where

r = average radius; mm, in.
t = wall thickness; mm, in.
T = applied torque; N · mm, lb · in.

The average radius r is half the average diameter.

$$r = \frac{1}{2}\left(\frac{d_i + d_o}{2}\right)$$

$$= \frac{1}{4}(95\text{ mm} + 102\text{ mm})$$

$$= 49.25\text{ mm}$$

So the maximum shear stress is

$$\tau_{open} = \frac{T[(6\pi)(49.25\text{ mm}) + (1.8)(3.5\text{ mm})]}{4[(\pi)(49.25\text{ mm})(3.5\text{ mm})]^2}$$

$$= \mathbf{(7.97)(10^{-4})}\boldsymbol{T}\ \mathbf{N/mm^2}$$

(*e*) The ratio of shear stresses is

$$\frac{\tau_{open}}{\tau_{closed}} = \frac{(7.97)(10^{-4})T\text{ N/mm}^2}{(1.94)(10^{-5})T\text{ N/mm}^2}$$

$$= \mathbf{41.1}$$

Clearly, omitting the weld is not a trivial matter.

(*f*) The ratio of the angular deflections is

$$\frac{\theta_{open}}{\theta_{closed}} = \frac{(Tl/GK)}{(Tl/GJ)} = \frac{J}{K}$$

The equation for K is

$$K = \frac{2\pi rt^3}{3}$$

Thus

$$K = \frac{(2\pi)(49.25\text{ mm})(3.5\text{ mm})^3}{3}$$

$$= 4422.5\text{ mm}^4$$

The ratio of angular deflections is

$$\frac{\theta_{open}}{\theta_{closed}} = \frac{J}{K} = \frac{(2.629)(10^6)\text{ mm}^4}{4422.5\text{ mm}^4}$$

$$= \mathbf{595}$$

Because of end restraints where the tube is welded to the yokes of the universal joints, the increase in angular deflection will be slightly less than this—not much comfort in this instance.

SUMMARY

Axles support wheels and shafts support gears, pulleys, sprockets, and other rotating elements.

To design axles or shafts, one must be able to calculate stresses and deflections. The effects of keyseats and other features must be accounted for in estimating strength and stiffness.

Steel is the most common material for shafts and axles, but other materials such as aluminum or composites are occasionally advantageous. Even where steel is used, the question of the proper type of heat treatment (if any) remains. In addition to strength and rigidity, wear resistance, corrosion resistance, weight, cost, size, and availability influence the choice of material. Weight can be saved by using hollow shafts.

In calculating the strength of axles or shafts, the Soderberg line is used to define a static stress equivalent to a fluctuating stress. The von Mises equivalent normal stress combines the effects of applied normal stresses and torsional shear stresses. Where stress concentration occurs, designers apply the stress concentration factor, as corrected for notch sensitivity, to the cyclic component of a fluctuating stress.

The most common shaft or axle design problem involves a rotating member subjected to one or more of these loads: steady torque, steady transverse forces, steady axial force. By the elementary formulas for bending and shear stresses, the Soderberg relationship and the von Mises equivalent normal stress, a number of design equations have been developed for determining minimum diameter.

The stresses caused by axial loads are handled by using the standard column formulas, which depend on length and cross section.

For axles or shafts of uniform cross section, bending deflections can be calculated by using standard beam-bending formulas along with the principle of superposition. Torsional deflections can be calculated by the basic equation for twisting of a circular bar (correcting for keyseats, if present).

To estimate the bending deflections on shafts or axles of varying cross section, a graphical integration method can be used.

If axles or shafts are made with noncircular cross sections, warping of the section produces a higher shear stress and a higher torsional deflection than for circular sections. Shear stress and torsional deflection must be calculated using special equations.

REFERENCES

15-1 Hopkins, R. Bruce. "Angular Deflection of Stepped Shafts." *Machine Design*, Vol. 34, No. 16, pp. 149-152, 1962.

15-2 Hopkins, R. Bruce. "Calculating Deflections in Stepped Shafts and Nonuniform Beams." *Machine Design*, Vol. 33, No. 14, pp. 159-164, 1961.

15-3 Hopkins, R. Bruce. *Design Analysis of Shafts and Beams*. New York: McGraw-Hill, 1970.

15-4 Lipson, Charles, and Robert C. Juvinall. *Handbook of Stress and Strength*. New York: Macmillan, 1963.

15-5 Peterson, R. E. *Stress Concentration Factors*. New York: Wiley, 1974.

15-6 Roark, R. J., and D. C. Young. *Formulas for Stress and Strain*, Fifth Edition. New York: McGraw-Hill, 1975.

PROBLEMS

P15-1 A 5-hp, 1725-rpm electric motor has a 0.875-in.-diameter shaft extension with a sled-runner keyseat for a 0.1875-in.-square key. The shaft is made of cold-drawn AISI 1040 steel with S_{ut} = 100,000 psi and S_{yt} = 88,000 psi. The motor is used to drive a machine that frequently requires a torque that is steady, but about 120% of that corresponding to the rated horsepower. Based on average material properties, what is the factor of safety? Why is the factor of safety so high?

P15-2 A shaft transmits 28 hp at 845 rpm. What diameter should be used for a factor of safety of 2.5? Use AISI 4340 steel, quenched and tempered. S_{ut} = 142,000 psi and S_{yt} = 130,000 psi.

P15-3 A short shaft made of cold-drawn AISI 1040 steel is subjected to a fluctuating torque varying between 180 and 236 N · m. Its diameter is 22 mm. The steel's properties are S_{ut} = 690 MPa and S_{yt} = 607 MPa. What is the factor of safety based on average properties? Assume there is a profile keyseat.

P15-4 A solid steel axle 2.5 in. in diameter is subjected to a fluctuating bending moment that varies between 22,400 and 31,900 lb-in. The material is AISI steel with a tensile yield strength of 56,000 psi and a derated fatigue strength of 12,900 psi. What is the factor of safety?

P15-5 A shaft nominally carries 18 kW at 1380 rpm. However, the torque varies rapidly from 80 to 120% of the nominal value. The material is cold-drawn AISI 1030 steel (S_{ut} = 579 MPa and S_{yt} = 524 MPa). There is a sled-runner keyseat at each end (K_f = 1.3 and K_{fs} = 1.3). For a factor of safety of 1.75, what should be the diameter? Assume the surface remains cold-drawn.

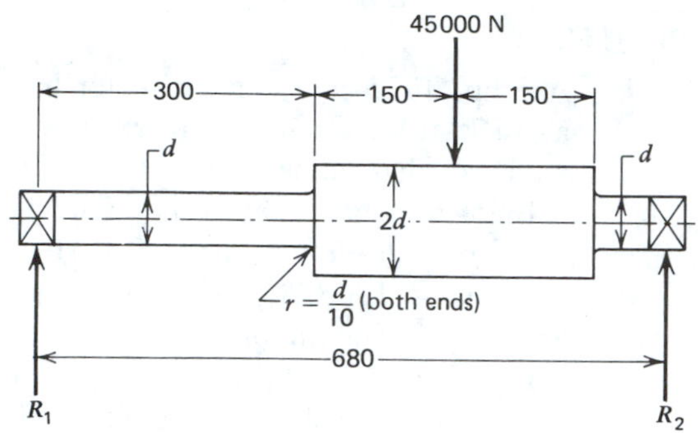

Figure 15-12 Illustration for Problem 15-6.

P15-6 The rotating machined axle shown in Fig. 15-12 has bearing reactions at R_1 and R_2. It is subjected to a transverse load of 45 000 N. The proportions are as shown. A ductile steel with the properties $S_{ut} = 820$ MPa and $S_{yt} = 620$ MPa is to be used. Find the diameter d for a factor of safety of 2.50.

P15-7 Two designs for a shaft supporting a cement kiln are in Fig. 15-13. In the first design the kiln is mounted on a fixed shaft. In the second design, the kiln has been fastened to the shaft and the shaft rotates. In the second design the shaft was the same diameter as in the first and failure occurred. Explain why this happened. Show how to calculate the shaft diameter for the second design so that the same material can be used. The factor of safety will be the same.

Figure 15-13 Illustration for Problem 15-7.

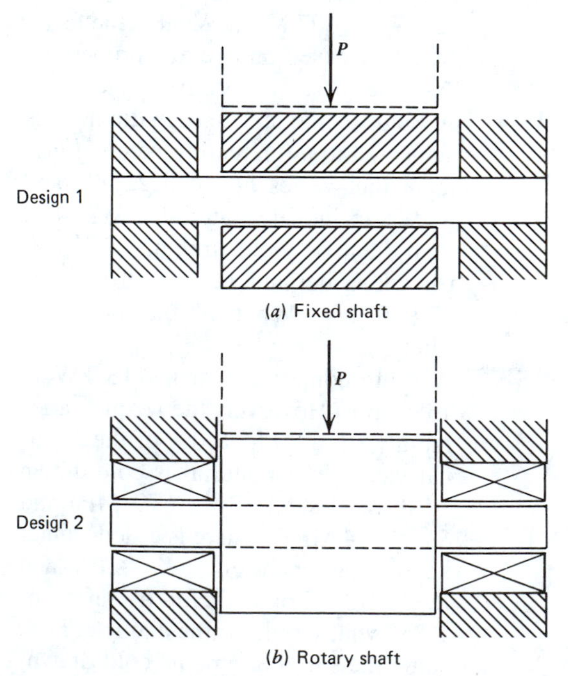

P15-8 A round camshaft is 30 in. from bearing to bearing. The cam lobe is at the midpoint. During one revolution, the transverse force goes from 0 to 1400 lb and back to 0 lb in a sinusoidal manner. The shaft is ground and made from hot-rolled SAE 1040 steel ($S_{yt} = 42{,}000$ psi and $S_{ut} = 76{,}000$ psi). What should the shaft diameter be? State your assumptions.

P15-9 A machine shaft is supported by two single-row ball bearings 1800 mm apart. It is to transmit 150 kW at 200 rpm. There is a transverse load of 4500 N at a location 600 mm inboard of the left-hand bearing. What should be the shaft diameter for a factor of safety of 2? Use a steel with the properties $S_{yt} = 620$ MPa and S_e (derated) $= 185$ MPa.

P15-10 A machine shaft supported on narrow bearings 90 in. apart is to transmit 200 hp at 200 rpm while subjected to a bending load of 1000 lb located 24 in. inboard of the left-hand bearing. The shaft is made by machining hot-rolled AISI 1030 steel. Determine the shaft diameter for a factor of safety of 3.

P15-11 Figure 15-14 shows a shaft supported by two bearings at L and R. Power entering at the pulley is distributed by the two spur gears. Thus the net torque due to the belts is in the direction the shaft is turning while the two tooth forces oppose the shaft's motion. What

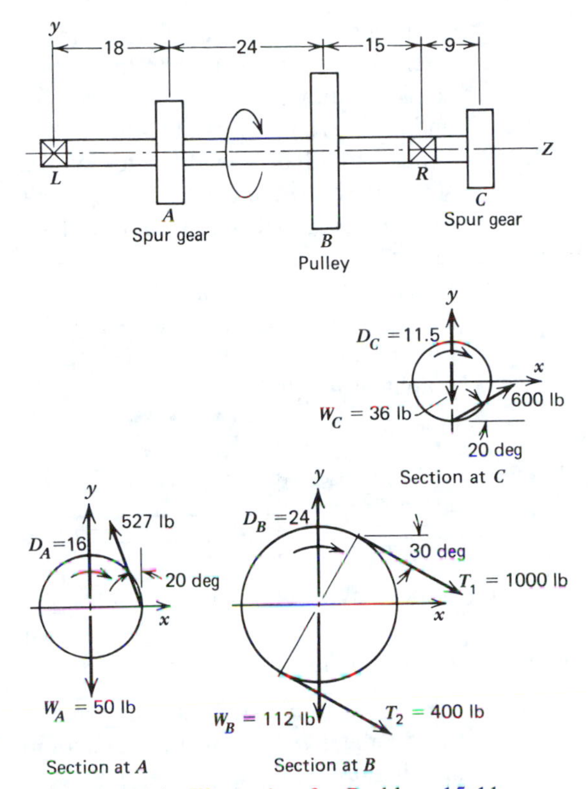

Figure 15-14 Illustration for Problem 15-11.

is the maximum bending moment in the shaft?

P15-12 As shown in Fig. 15-15, a steel shaft is supported by two narrow bearings 300 mm apart. Two spur gears are keyed to it by means of square keys fitted into sled-runner keyseats in the shaft. The tooth loads at the gears are as shown. The pinion (on the right) has a diameter of 125 mm and weighs 37 N. The gear (on the left) has a 400-mm diameter and a weight of 83 N. The shaft is made of AISI 1020 steel with the properties S_{ut} = 448 MPa and S_{yt} = 296 MPa. The shaft is machined. The fatigue stress concentration factors for the keyseats are K_f = 1.3 and K_{fs} = 1.3. Find the shaft diameter for a factor of safety of 2.

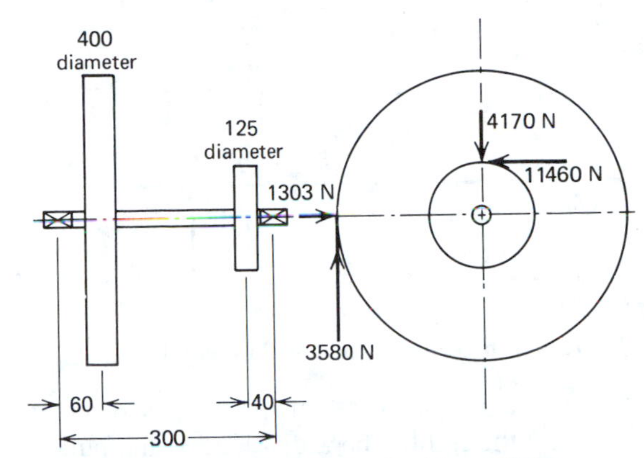

Figure 15-15 Illustration for Problem 15-12.

P15-13 Compare a solid, circular steel shaft, d = 100 mm, with a hollow, circular shaft of the same outside dameter.

(*a*) For d_0 = 100 mm and d_i = 50 mm, calculate (1) the percentage change in mass (proportional to cross-sectional area A); (2) the percentage change in section modulus Z; and (3) the ratio of Z/A.

(*b*) Repeat calculations for inside diameter ranging 90 to 10 mm. Express A, Z, and Z/A in terms of C = d_0/d_i and plot the corresponding curves. Suggested scales: abscissa 100 c = 100 mm (total length); ordinate: 1 percent = 2 mm.

(*c*) How do you interpret these data?

P15-14 Do P15-2 assuming a hollow circular shaft is to be used for which C = 0.80.

P15-15 What weight saving is achieved if the shaft in P15-5 is made hollow with C = 0.82?

P15-16 Do P15-10 but with a hollow shaft for which C = 0.78.

P15-17 A loading diagram for a speed reducer shaft is shown in Fig. 15-16. A torque of T = 411 N · m is transmitted from a worm on the left to a pulley on the right. The worm subjects the shaft to radial and axial loads, as shown. The

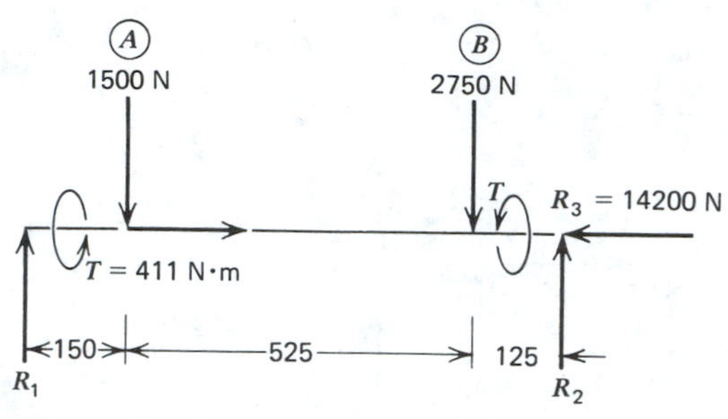

Figure 15-16 Illustration for Problem 15-17.

axial load is taken up by the bearing on the right. The radial load at the pulley is the net effect of the belt tensions. The shaft is made of heat-treated plain carbon steel, and all surfaces are ground. Strength data are S_{ut} = 760 MPa and S_{yt} = 560 MPa. The shaft is solid and 46 mm in diameter. At worm and pulley there are profile keyseats (K_f = 2 and K_{fs} = 1.6).

(*a*) What is the mean normal stress due to the axial load?

(*b*) Calculate the equivalent normal static stress by the Soderberg rule.

(*c*) Calculate the von Mises stress at section *B*.

(*d*) What is the factor of safety?

P15-18 A steel axle is supported by two single-row ball bearings 40 in. apart. Two transverse loads of 250 lb each are applied 8 in. inboard of each bearing. The axle diameter is 1-5/8 in. Find the maximum bending deflection.

P15-19 A solid steel shaft is supported by two journal bearings (built-in ends) 360 mm apart. There is a concentrated load of 544 N at a position 75 mm to the right of the left-hand bearing. What diameter should be used for the shaft to have adequate rigidity?

P15-20 How much weight is saved by using a hollow shaft with C = 0.75 for the design in P15-19?

P15-21 What is the angular deflection of a hollow steel shaft 21 in. long that is subjected to a torque of 1950 lb-in. over its entire length? Its cross-sectional dimensions are d_o = 2.5 in. and d_i = 2.1 in. G = (11.4)(10^6 psi).

P15-22 What should be the diameter of the shaft described in P15-12 to limit the torsional deflection to (18)(10^{-6} rad/mm)?

P15-23 Determine the bending deflection at each load for the steel shaft shown in Fig. 15-17. The shaft is solid. Use the graphical integration method.

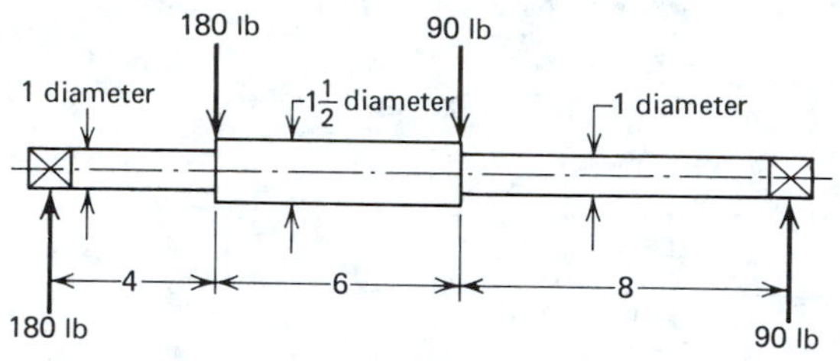

Figure 15-17 Illustration for Problem 15-23.

P15-24 For the shaft in Fig. 15-18, determine the diameter d to limit the deflection at either 400-N load to 0.035 mm. Use the graphical integration method.

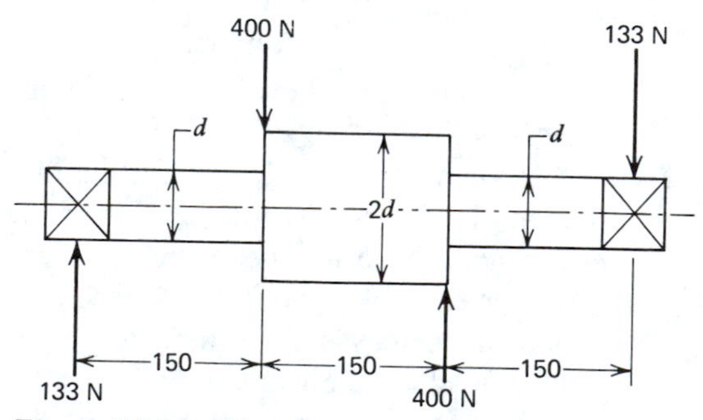

Figure 15-18 Illustration for Problem 15-24.

P15-25 What would be the size of a square cross-sectional shaft to replace the round shaft of P15-3?

P15-26 What should be the size of a square cross-sectional torsion bar to limit the torsional deflection per unit length to the same level as a solid circular bar of 30-mm diameter? How much heavier will the square bar be?

Chapter 16
Detachable Fastenings for Shaft and Hub

An inventor is an engineer who doesn't take his education too seriously.

CHARLES F. KETTERING

NOMENCLATURE

A_c = area of compression
A_s = shear area
b = width of key
D = shaft or major spline diameter
D_b = bolt circle diameter
D_h = outer diameter of collar or hub
D_m = mean diameter
d = pin diameter
F_a = axial force on shaft cone
F_c = force of compression
F_n = normal force
F_s = shear force
F_t = tangential force at shaft outside diameter
L = length of key or spline
t = height of key
α = taper angle of cone, deg
θ = friction angle, rad

The main function of these semipermanent fastenings is to transmit unidirectional torque between the shaft and the machine parts assembled on it. Although such parts—wheels, sheaves, sprockets, and gears—vary widely, they all have similar hubs. The hub is the central, strengthened, inner portion or mounting from which the web or sprockets radiate and that is bored to receive a shaft or axle.

Design of detachable fastenings for shaft and hub is based on either the principle of interlocking (pins, keys, splines) or the principle of friction (setscrews, split hub, tapered shaft ends). Some also serve as locking devices (pins), while others allow axial movement (splines). As with screw fasteners, it is "the in-place cost that counts." For this reason, easy-to-mount shaft connectors (Ringfeder, snap rings, etc.) increasingly replace costly splines, keyways, pins, and tapered shafts.

For safety and economy, shaft and hub should be designed with a greater margin of safety than any detachable fastener. The nominal shear stress in a shaft is a means to this end and is calculated as

$$\tau = \frac{16T}{\pi D^3} \qquad (16\text{-}1)$$

To transfer a given torque T from shaft to hub or vice versa, the fastening device must resist the tangential force F_t acting at the shaft surface. This force is found by means of the power equations, Eqs. 1-7 and 1-10, by substituting $T = F_t D/2$

$$F_t = \frac{9550P}{0.5Dn}\ \text{N} \qquad (16\text{-}2)$$

$$F_t = \frac{63{,}000P}{0.5Dn}\ \text{lb} \qquad (16\text{-}3)$$

To be seated properly, fasteners for shaft and hub require holes, grooves, flats, or similar features. Unfortunately, these features raise stress levels and increase shaft deflection. Hence, their location and shape should be chosen to minimize stress and deflection.

16-1 SETSCREWS

Function and Design

One of the simplest devices used for shaft and hub connections is the setscrew. It is used in light service only. As shown in Fig. 16-1, a hole is drilled radially through the hub and sometimes a short distance into the shaft. The hub section is threaded, and a setscrew is inserted and turned until it is firmly seated against the shaft.

Setscrews are available in various head and point configurations to serve a multitude of design situations (Fig. 16-2). Setscrews may be classified as those with and those without a head. The headless type is more popular because it can be mounted flush with the surface and thus

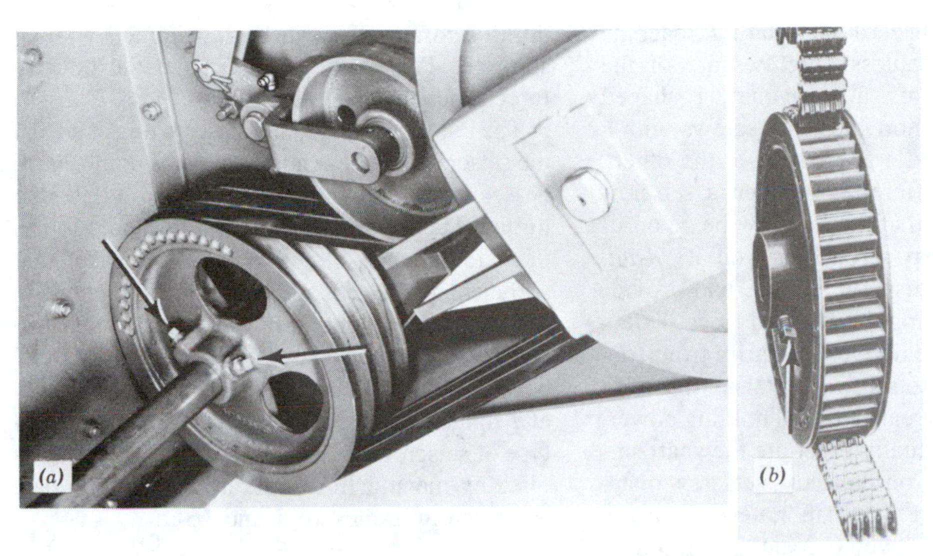

Figure 16-1 Setscrews used to effect a shaft-to-hub connection. (*a*) Two square head setscrews 90 deg apart, used on a V-belt sheave. (*b*) Single square head, setscrew in a hub of a silent chain sprocket. (Courtesy Deere and Company Technical Services.)

Figure 16-2 Setscrew head and point styles (R. P. Hoelscher, C. H. Springer, and J. S. Dobrovolny, *Graphics for Engineers,* New York: Wiley, 1968.)

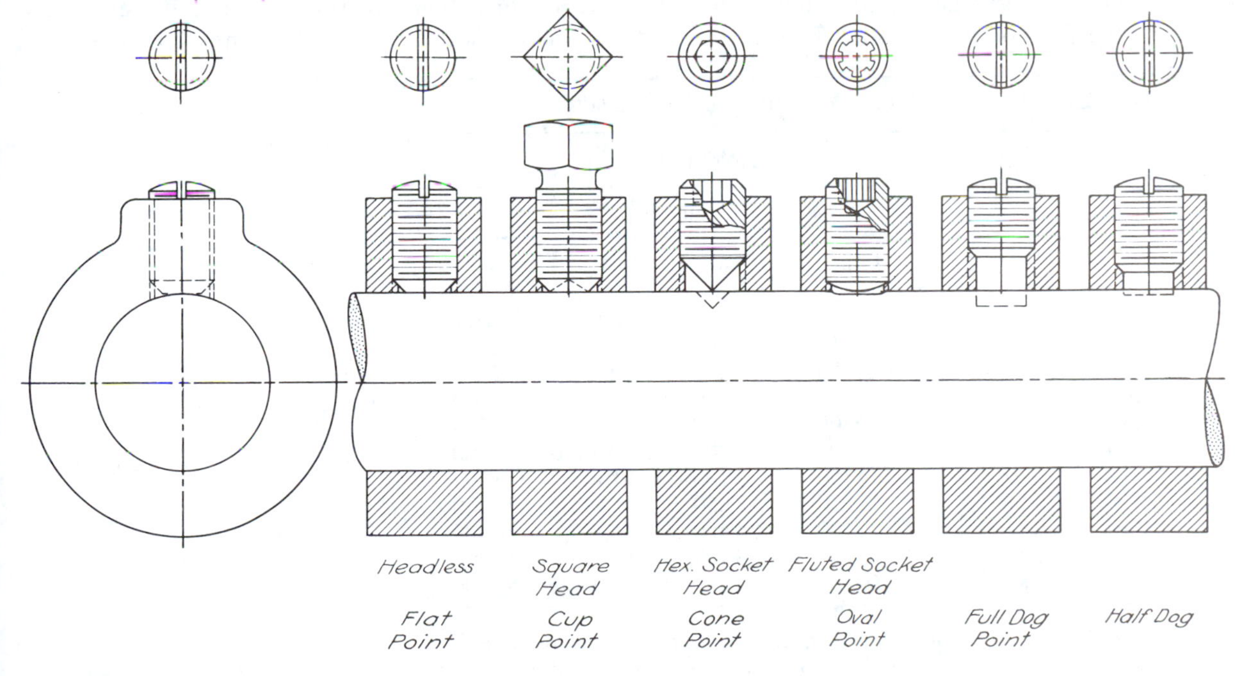

present no work hazard on rotating machine members. Most headless setscrews are of the Allen type turned by a hexagonal, L-shaped wrench. They are short, headless screws with a point at one end and a sockethead at the other.

The principal difference between a setscrew and a cap screw is that the setscrew bears on its *point* while the cap screw bears on its *head*. Threads hold the setscrew in place and provide the means for exerting the required force. Effectiveness is therefore derived primarily from friction between the setscrew point and the shaft surface. In order to realize its full holding power, the point should actually penetrate the shaft surface on assembly. Consequently, setscrew material should be harder than shaft material by about 10 points on the Rockwell C Scale. The full dog, half dog, and cone points require matching holes in the shaft prior to assembly and attain additional holding power due to this interlocking feature.

Holding Capacity

The holding capacity of a setscrew is equal to the tangential force generated at the shaft surface. Often the holding capacity is expressed as the limiting torque that can be transmitted without slippage. This torque is the tangential holding force times the shaft radius.

The holding capacity increases with the size of the setscrew and depends on type of point, hardnesses of setscrew and shaft, and tightening torque. Setscrew diameter is usually one-fourth of shaft diameter. Length need only be such that sufficient thread engagement occurs to prevent stripping when tightening torque is applied. To ensure safety, the setscrew should not extend beyond the surface of the hub. Holding capacity and tightening torque for a particular style and type of setscrew are best obtained from data supplied by manufacturers. Examples of such data are given in Tables 16-1 and 16-2.

Arrangement of Setscrews

Multiple setscrews may be used to increase the holding capacity of a connection, but the increase is not always proportional to the number of setscrews. When two setscrews are used, holding power is approximately doubled when the second screw is installed on an axial line with the first (Fig. 16-3). But holding power is only about 30% greater when the screws are di-

TABLE 16-1 Theoretical Holding Power of Setscrews

Screw Size	Tightening Torque, N · m	Holding Force, N Point Type Cup	Oval	Flat	Cone
M3	1.04	868	780	798	929
M4	2.43	1515	1360	1390	1620
M5	4.90	2450	2210	2250	2620
M6	9.35	3900	3510	3590	4170
M8	20.20	6300	5670	5800	6740
M10	40	10 000	9000	9200	10 700
M12	70	14 550	13 100	13 400	15 600
M14	111	19 850	17 850	18 300	21 250
M16	173	27 100	24 400	24 900	29 000
M20	338	42 300	38 100	38 900	45 300
M24	585	60 900	54 800	56 000	65 000

Torsional holding power in N · m is equal to one-half of the shaft diameter in meters (m) times the axial holding power in newtons (N).

TABLE 16-2 Cup Point Setscrew Holding Power

Nominal Screw Size	Seating Torque, lb-in.	Axial Holding Power, lb
No. 0	0.5	50
No. 1	1.5	65
No. 2	1.5	85
No. 3	5	120
No. 4	5	160
No. 5	9	200
No. 6	9	250
No. 8	20	385
No. 10	33	540
$\frac{1}{4}$ in.	87	1000
$\frac{5}{16}$ in.	165	1500
$\frac{3}{8}$ in.	290	2000
$\frac{7}{16}$ in.	430	2500
$\frac{1}{2}$ in.	620	3000
$\frac{9}{16}$ in.	620	3500
$\frac{5}{8}$ in.	1225	4000
$\frac{3}{4}$ in.	2125	5000
$\frac{7}{8}$ in.	5000	6000
1 in.	7000	7000

Source. Data by F. R. Kull, "Fasteners Book Issue," *Machine Design,* March 11, 1965.

Torsional holding power in inch-pounds is equal to one-half of the shaft diameter in inches times the axial holding power.

Cone points will develop a slightly greater holding power than cup points; flat, dog, and oval points, slightly less.

ametrically opposed (Fig. 16-4). Where it is necessary to locate two setscrews on the same circumferential line, a spacing of 45 to 120 deg will result in a holding capacity about 1.5 to 1.75 times greater than that of a single setscrew (Fig. 16-5). Use of more than two setscrews is seldom advantageous. Where strain is greater than normal or the setscrew is hard to get at, the threaded hole may be deep enough for *two* setscrews, the upper one locking the lower one (Fig. 16-6).

Setscrews may be used to advantage where loads are light and frequent disassembly is not required, since marring of the shaft surface due to point penetration often makes disassembly

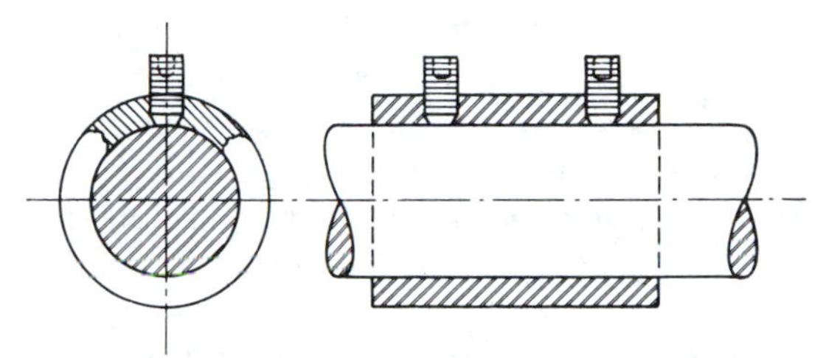

Figure 16-3 Two setscrews in line have twice the holding power of one.

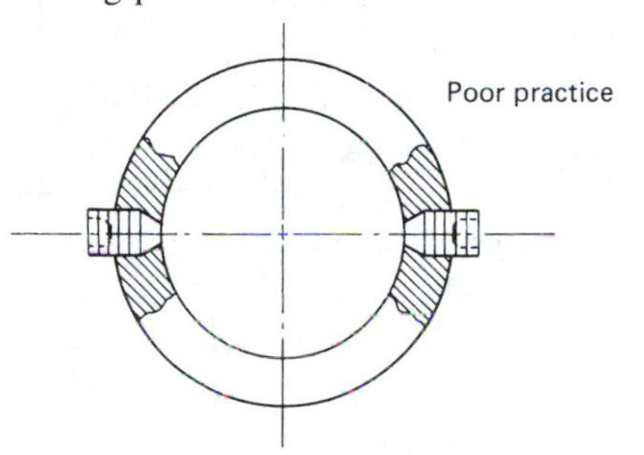

Figure 16-4 The holding power is only 30% greater when the screws are diametrically opposed.

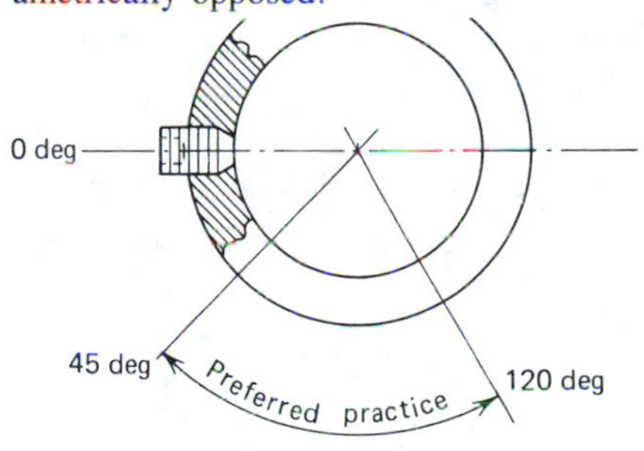

Figure 16-5 A spacing of 45 to 120 deg will result in a holding capacity 1.5 to 1.75 times greater than that of a single setscrew.

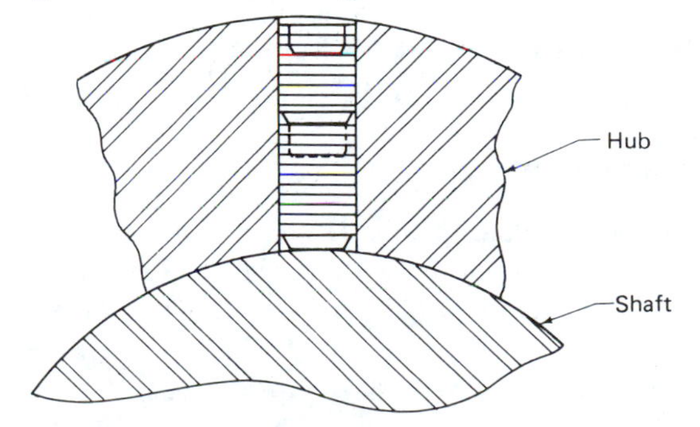

Figure 16-6 Where space permits, two setscrews may be used, one locking the other.

difficult. Setscrews are also used as a means of obtaining slight adjustments, either to change the location of a tool or other part or to eliminate unnecessary play by means of gibs.[1] Setscrew connections are not suited to impact loading, nor should they be used where accuracy of location is required. Setscrews are often used in conjunction with keys, as will be discussed later.

EXAMPLE 16-1

A spur gear is secured to a 40-mm shaft with an M10 cup point setscrew. What power can be transmitted through this connection at 900 rpm with a factor of safety of 2 against slipping?

SOLUTION

From Table 16-1, tangential holding force of an M10 cup point setscrew is 10 000 N. Incorporating the factor of safety, design force becomes 5000 N. Torque to be transmitted is

$$T = (0.5D)F_t$$

$$= (0.5)(40 \text{ mm})(5000 \text{ N}) = 100 \text{ N} \cdot \text{m}$$

Applying the power equation $P = Tn/9550$ gives

$$P = \frac{(100 \text{ N} \cdot \text{m})(900 \text{ rpm})}{9550} = \mathbf{9.42 \text{ kW}}$$

which can be transmitted.

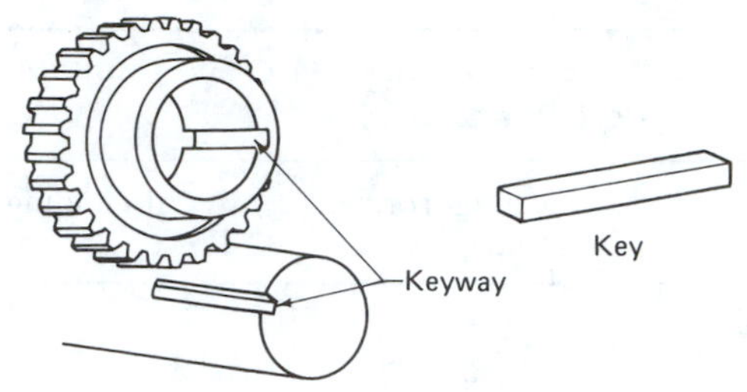

(*a*) A key and the keyways in a gear and shaft

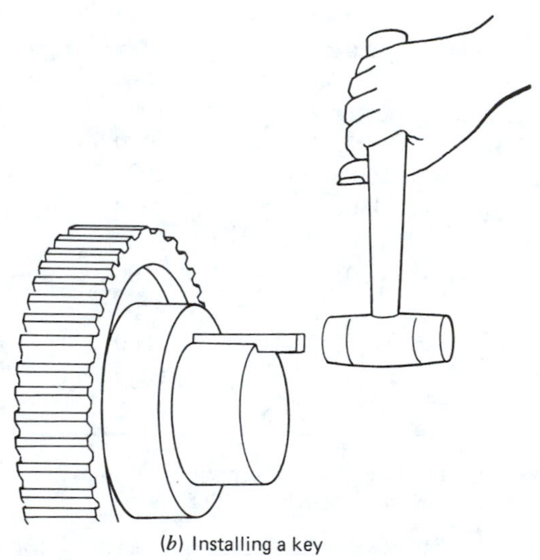

(*b*) Installing a key

Figure 16-7 An interlocking shaft-to-hub connection with key and keyway. (Courtesy Deere and Company Technical Services.)

16-2 KEYS: FUNCTION, DESIGN, CLASSIFICATION

Function and Design

The main function of a key is to transmit unidirectional torque between a shaft and common machine members such as sheaves, sprockets, gears, and couplings. Keys are axially oriented metal inserts designed so that one-half is in the shaft while the other half is in the hub (Fig. 16-7). The matching grooves are called keyways. Thus relative rotary motion of shaft and hub is prevented due to resistance offered in both shear and compression of the key. For maximum torque capacity a key must be long relative to its width, depth, or diameter and must be mounted axially. Resistance to relative axial motion is inherent in most keys because of friction forces

[1]A gib is a wedge or adjusting shoe that ensures a proper sliding fit between two machine parts.

on two or all four sides. When axial motion is required, as in multistep gear transmissions, multiple keys (splines) with a sliding fit are provided. Therefore, the advantages of keys are simplicity, low cost, and ease of assembly. Their disadvantage lies in a weakening of the shaft. When such weakening could become critical, one or two *flat* keys should be used.

Classification

Keys may be distinguished as straight or tapered; rectangular, round, or disk shaped; and radial or tangential. They may also be distinguished according to their use as light duty or heavy duty. However, these features so overlap that a single all-embracing classification is impossible. Keys are covered by AGMA 620.02 and ANSI B 17.1, B 17.2 standards. For further detail, see *Machinery's Handbook*, pp. 967–977.

16-3 SQUARE AND RECTANGULAR KEYS

Square keys, due to their geometric simplicity, are preferred for light- and medium-duty service. Since the square key is a special case of the rectangular key, calculations will be based on the latter. A shaft transmitting torque to a hub is shown in Fig. 16-8, which illustrates the mutual forces exerted by hub and key to achieve torque transfer. Force F, induced by torque T, results from a force distribution along half of one side of the key. The force can be assumed to act at the shaft's outside diameter without significant error. Torque is then related to force exerted on the key by $T = 0.5\ DF$. This induced force on the key and keyways leads to four possible modes of failure: (1) the key may fail in direct shear, (2) the side of the key may be crushed, (3) the side of the keyway in the shaft may be crushed, or (4) the side of the keyway in the hub may be crushed. The latter two modes of failure cannot be tolerated and are avoided by using higher-strength materials in both shaft and hub. Calculations are therefore based on failure of the key in shear or compression (Fig. 16-8*a*). Shearing stress in the key is

$$\tau = \frac{F}{bL} \tag{16-4}$$

By introducing torque through the expression $F = 2T/D$, a more useful expression for τ is obtained:

$$\tau = \frac{2T}{bLD} \tag{16-5}$$

where

τ = shear stress; MPa, psi
T = applied torque; N · mm, lb-in.
b = width of key; mm, in.
L = length of key; mm, in.
D = shaft diameter; mm, in.

The compressive stress σ_c on the sides of the key (and on the keyway walls) is:

$$\sigma_c = \frac{F}{(0.5t)L} = \frac{4T}{tLD} \tag{16-6}$$

where

σ_c = compressive stress; MPa, psi
t = height of key; mm, in.

Experience indicates that $b = 0.25D$. Standard sizes are thus based on this relationship, as shown in Tables 16-3 and 16-4. With key size fixed by shaft diameter, stress equations may be used to calculate the key length. A given hub length, however, is a limiting factor; the key length cannot, of course, exceed that of the hub. Experience indicates that the length of a single key should not, in general, exceed $1.5D$. This limitation is caused by the twisting of the key within the hub. The torsional displacement on a shaft section of length $1.5D$ is small, indeed, even when the material is stressed to its allowable limit. Yet this minute displacement is sufficient to cause a pronounced nonuniform distribution of the tangential force F_t, as shown by the exponential curve in Fig. 16-9. Thus, when calculations indicate a key length in excess of

TABLE 16-3 Standard Metric Sizes of Square Keys

Shaft Size D, mm	Key Size b, mm	Shaft Size D, mm	Key Size b, mm
12–15	3	50–60	14
15–20	4	60–70	16
20–30	6	70–80	18
30–40	8	80–90	20
40–50	10	90–100	24

TABLE 16-4 Standard Dimensions of Square Keys

Shaft Size D, in.	Key Size b, in.	Shaft Size D, in.	Key Size b, in.
$\frac{1}{2}-\frac{9}{16}$	$\frac{1}{8}$	$2\frac{5}{16}-2\frac{3}{4}$	$\frac{5}{8}$
$\frac{5}{8}-\frac{7}{8}$	$\frac{3}{16}$	$2\frac{13}{16}-3\frac{1}{4}$	$\frac{3}{4}$
$\frac{15}{16}-1\frac{1}{4}$	$\frac{1}{4}$	$3\frac{5}{16}-3\frac{3}{4}$	$\frac{7}{8}$
$1\frac{5}{16}-1\frac{3}{8}$	$\frac{5}{16}$	$3\frac{13}{16}-4\frac{1}{2}$	1
$1\frac{7}{16}-1\frac{3}{4}$	$\frac{3}{8}$	$4\frac{9}{16}-5\frac{1}{2}$	$1\frac{1}{4}$
$1\frac{13}{16}-2\frac{1}{4}$	$\frac{1}{2}$	$5\frac{9}{16}-6$	$1\frac{1}{2}$

Standard dimensions for rectangular keys can be found in *Machinery's Handbook*, p. 969.

Figure 16-8 Forces on key due to transmitted torque.

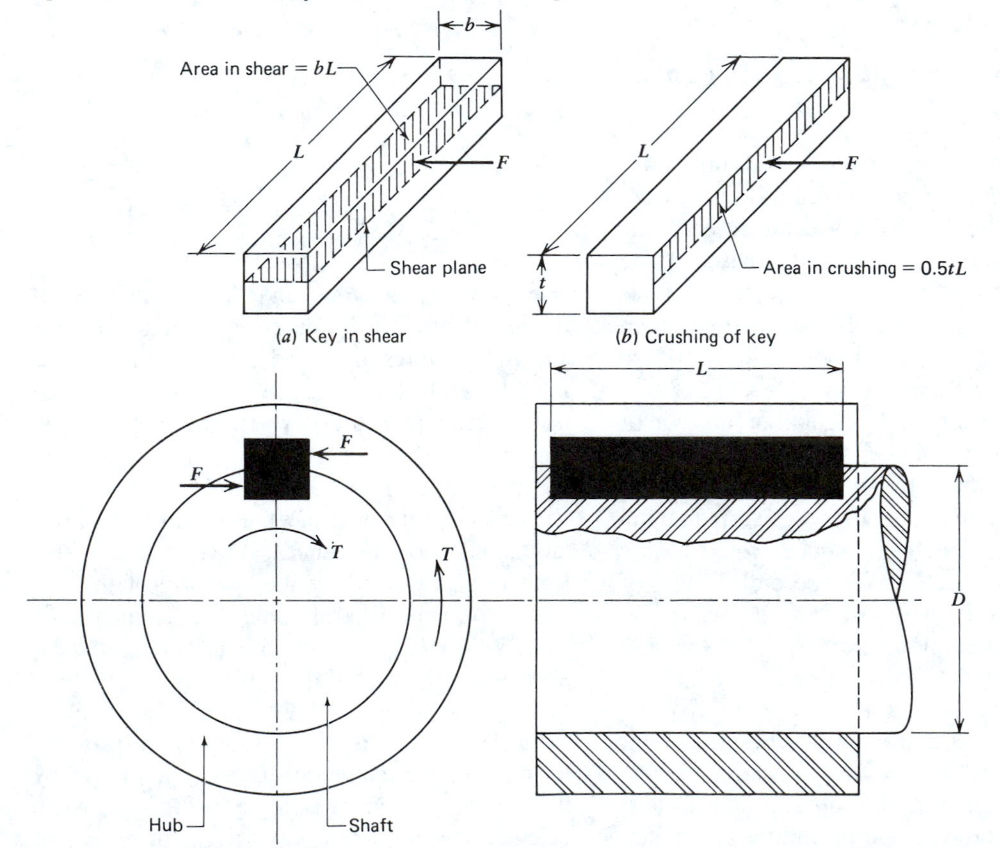

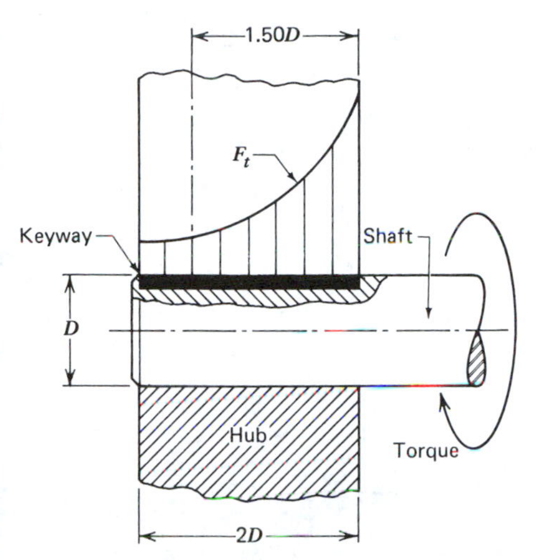

Figure 16-9 Load diagram for a rectangular key. The curve shows the approximate distribution of the tangential load, F_t, along the length of a rectangular key.

1.5*D*, designers should specify two keys. When two keys become necessary, they should be spaced 90 deg apart. In all preliminary calculations, $L = 1.2D$ is a safe value.

As usual, the choice of material is a compromise. A key should be strong enough to transfer the design torque yet *weaker* than the shaft. Mild steel and low-strength alloy steels are preferred, as indicated by Table 16-5.

A key should fit tightly in its keyways in order to prevent tipping of the key under load. A tight fit will also help prevent axial motion of the key. However, friction forces resulting from the fit between key and keyway should not be expected to support an axial load arising, for example, from a helical gear. For relatively light axial loading, a setscrew is often used in conjunction with a key. The setscrew should be located so as to bear on *top* of the key. This will hold the key in place and prevent marring of the shaft surface.

EXAMPLE 16-2

Design a key connection for a steel shaft and a cast iron hub. The working strengths are as follows.

	Shaft, MPa	Key, MPa	Hub, MPa
τ_w	200	165	—
σ_w (compression)	400	330	550

The shaft transmits 125 kW at 150 rpm. Calculations may be based on pure torsion and a factor of safety of 2.

Analysis

Find the shaft diameter first, then select the corresponding standard key. Next, obtain the key length by examining the key for strength in both shear and crushing. Because the key material is weaker in compression (by 18%) than the material of the shaft and hub, there is no need to base the key length on crushing of the shaft and hub. Thus calculations are reduced by 50%.

TABLE 16-5 Allowable Key Stresses

Key Material	Brinell Hardness, Bhn	Allowable Stresses, psi	
		Shear	Compression
AISI 1018	None specified	7,500	15,000
AISI 1045	225–265	10,000	20,000
	255–300	15,000	30,000
AISI 4140	310–360	20,000	40,000

Source. Extracted from AGMA Standard-Design of Components-Enclosed Gear Drives-Bearings, Bolting, Keys and Shafting (AGMA 260.02), with the permission of the publisher, the American Gear Manufacturers Association.

Note. The allowable stresses provide for (1) effective stress concentrations arising from key joints, shoulders, grooves, splines, etc., not exceeding a value of 3.0; (2) a service factor of 1.0.

SI. For conversion use 1 MPa = 145 psi (round off to the nearest whole number).

SOLUTION

From the power equation (Eq. 1-7)

$$T = \frac{9550P}{n} = \frac{9550(125 \text{ kW})}{150 \text{ rpm}} = 7958 \text{ N} \cdot \text{m}$$

Shaft in torsion: $\tau_{max} = \frac{16T}{\pi D^3} \leq \frac{\tau_w}{fs}$

$$D = \sqrt[3]{\frac{16(7958 \text{ N} \cdot \text{m})10^3}{\pi(0.5)(200 \text{ N/mm}^2)}} = 74 \text{ mm}$$

Use a standard size: $\mathbf{D = 75 \text{ mm}}$

The corresponding key, 18 mm × 18 mm, is obtained from Table 16-3.

Shearing:

$$L_1 = \frac{2Tfs}{\tau_w bD} \quad \text{(See Eq. 16-5)}$$

$$= \frac{2(7958\,000 \text{ N} \cdot \text{mm})2}{(165 \text{ N/mm}^2)(18 \text{ mm})(75 \text{ mm})} = 143 \text{ mm}$$

Crushing:

$$L_2 = \frac{4Tfs}{\sigma_w tD} \quad \text{(See Eq. 16-6)}$$

$$= \frac{4(7958\,000 \text{ N} \cdot \text{mm})2}{(330 \text{ N/mm}^2)(18 \text{ mm})(75 \text{ mm})} = 143 \text{ mm}$$

Conclusion:

$$\mathbf{t \times b \times L = 18 \text{ mm} \times 18 \text{ mm} \times 75 \text{ mm}}$$

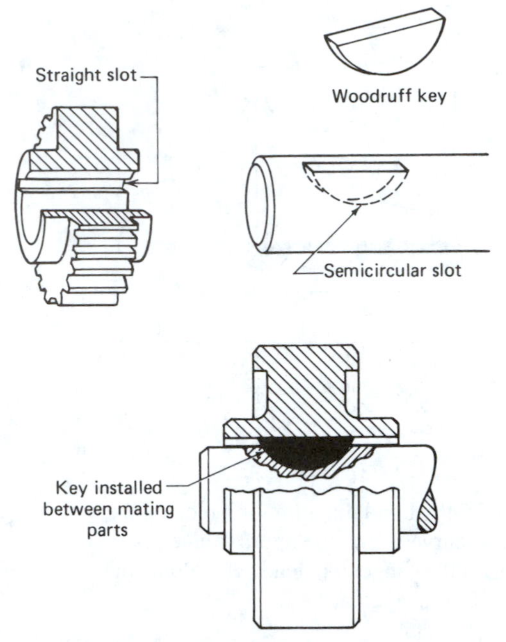

Figure 16-10 Woodruff key. (Courtesy Deere and Company Technical Services.)

16-4 SPECIAL-PURPOSE KEYS

Woodruff keys are self-aligning and resistant to rolling. In the Woodruff key system semicircular steel disks are used as keys. The half-circular side of the key rests in a curved slot in the keyway; the straight part projects above the shaft and enters into a straight hub keyway (Fig. 16-10). Using a deep, semicircular keyway in the shaft has three major effects on the system. It allows the key to rotate in its own plane and align itself with the hub keyway. For this reason, the Woodruff key is well suited for tapered shafts. The depth of the keyway relative to shaft diameter also provides much greater resistance to rolling than prismatic keys. Unfortunately, however, it also weakens the shaft, thereby limiting its use to small and medium-size shafts (up to 65 mm) in light service. Standard dimensions of Woodruff keys are given in *Machinery's Handbook,* pp. 974–977.

Tapered keys are essentially press-fit devices that function on the basis of elastic deformation of one or more of the components, ensuring a sound self-locking action in the assembly. Tapered keys are tapered on one side (1 : 100) and fit a matching tapered keyway in the hub (Fig. 16-11*a* and 16-11*b*). When such a key is driven into place, the tapered fit draws shaft and hub together on the side opposite the key. Thus the key bears on both top and bottom and initially power is transmitted due to friction (Fig. 16-11*b*). The key also secures the hub axially, thereby increasing overall reliability. Tapered keys are therefore well suited for heavy-duty service. However, because of the slight eccentricity created by taper keys, they are best suited for *large* diameter parts, which negate the adverse effect of eccentricity (Fig. 16-11*c* and 16-11*d*).

Normally tapered keys are driven out at disassembly by pounding at the "small" end. When lack of access hinders this method of removal, designers should use a *gib head taper key* (Fig. 16-12). This design has a hooklike head used to *pull* the key out. (Unless covered, this protruding head may cause accidents.)

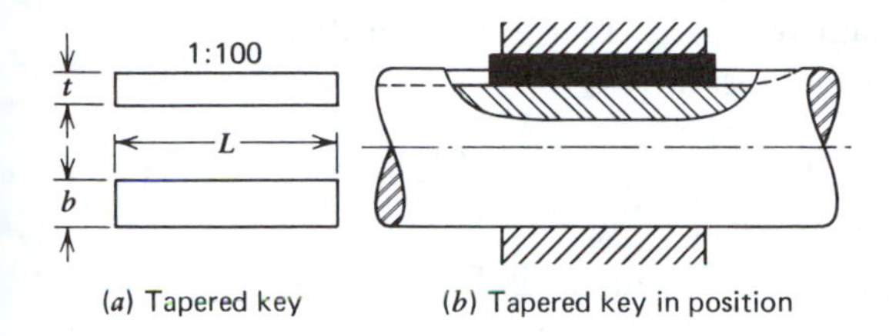

(*a*) Tapered key (*b*) Tapered key in position

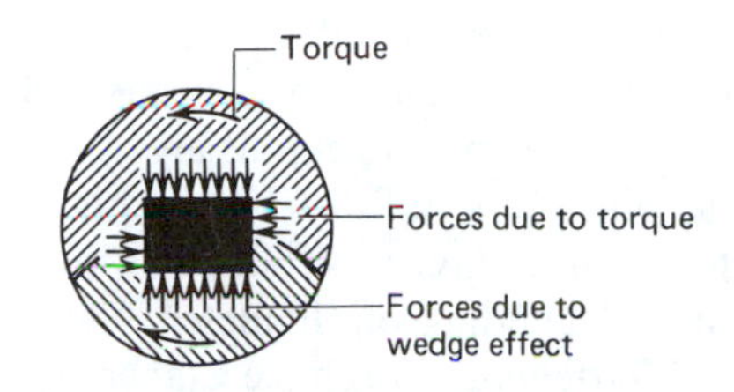

(*b*) Forces on key due to torque and wedge effect

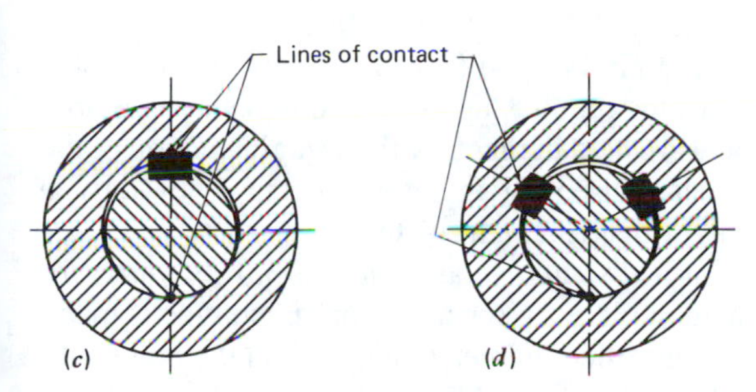

(*c*, *d*) Eccentricity caused by tapered keys (exaggerated for clarity)

Figure 16-11 Tapered keys.

Figure 16-12 Gib head tapered key.

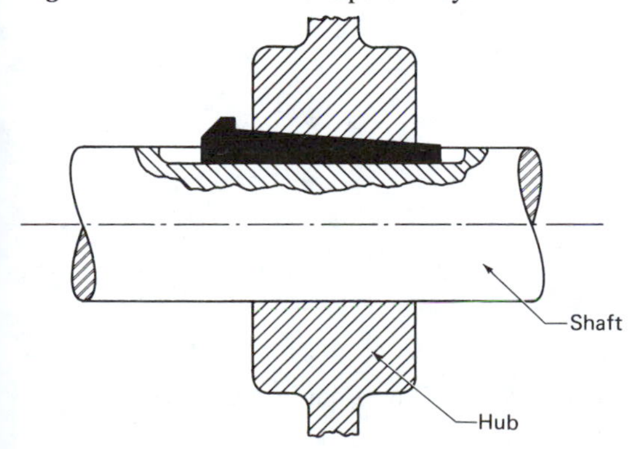

16-5 PINS

An effective and inexpensive shaft-to-hub connection can be made with pins. Commonly used pins are shown in Fig. 16-13. Connections made with these pins are classified as semipermanent and are not intended for rapid or frequent disassembly. Each is held in place by interference or elastic reaction and must be driven or pressed out for removal. Pins are hardened and ground for precision. Commonly used materials include ISO 5.8, AISI 1070, and AISI 1095 carbon steels and types 302, 410, and 420 stainless steels.

Figure 16-13 Pins are effective, inexpensive means of joining shaft and hub. (Courtesy Deere and Company Technical Services.)

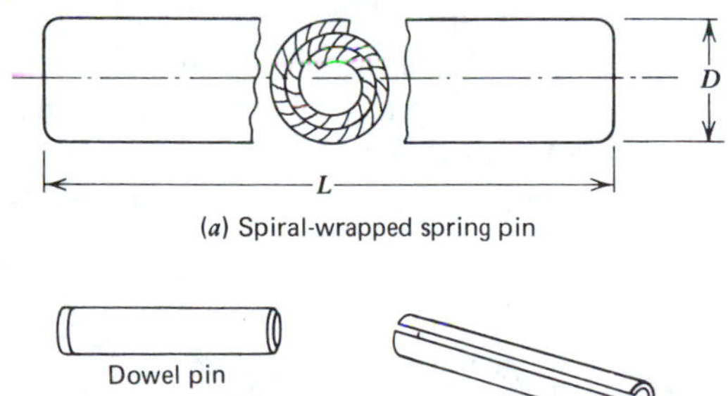

(*a*) Spiral-wrapped spring pin

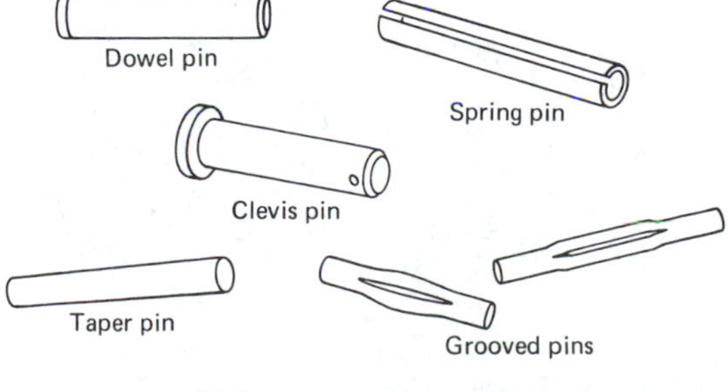

(*b*) Some commonly used pins

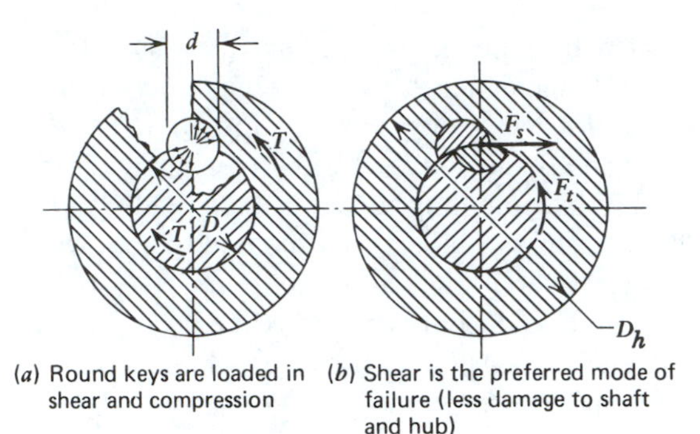

(*a*) Round keys are loaded in shear and compression (*b*) Shear is the preferred mode of failure (less damage to shaft and hub)

Figure 16-14 Dowel pin used as a round key.

Straight, solid pins are called *dowel pins* and can be used to make two types of shaft-to-hub connections. When used as in Fig. 16-14*a*, the pin is called a *round key* and transmits torque through shear along an axial plane through the pin center. Shear stress in the pin is

$$\tau = \frac{2T}{dLD} \tag{16-7}$$

where

T = transmitted torque, N · m, lb-in.
D = shaft diameter, mm, in.
d = pin diameter, mm, in.
L = the shorter hub or pin length, mm, in.

The failure mode of a round key connection is shown in Fig. 16-14*b*.

A second type of connection is shown in Fig. 16-15*a*. Here a dowel pin is inserted in matching *transverse* holes drilled in the shaft and hub. Torque is transmitted through *double* shear in the diametral plane of the pin. In this connection, shear stress is

$$\tau = \frac{4T}{\pi D d^2} \tag{16-8}$$

and failure occurs as in Fig. 16-15*b*. Transverse pin connections offer the advantage of precision positioning of hub relative to shaft but have the disadvantages of reduced area and introduction of stress concentration because of the drilled hole. Standard dowel pins are available in nominal diameters of 0.8 to 20 mm and lengths of 3 to 140 mm. Oversize ranges from 0.008 to 0.02 mm, depending on diameter.

Metric *taper pins* have a taper of 2% measured on the diameter. Nominal size is the diameter at the large end. In shaft-to-hub connections, taper pins are used in the same manner as dowel pins (Figs. 16-14 and 16-15). Shear stress formulas need only be modified to account for taper. Taper pins require taper-reamed holes at assembly and are held in place by the wedging action of the taper. Disengagement can occur with only a slight displacement of the pin.

Spiral-wrapped and *slotted spring pins* are held in place by spring action. Each type is made in controlled diameters larger than the holes into which they are inserted. When pressed into a hole, such pins are compressed and exert elastic spring forces against the inside surface. Spring pins are used as *transverse* shaft-to-hub connectors and transmit torque in double shear at the shaft surface. Load *ratings* for spring pins are

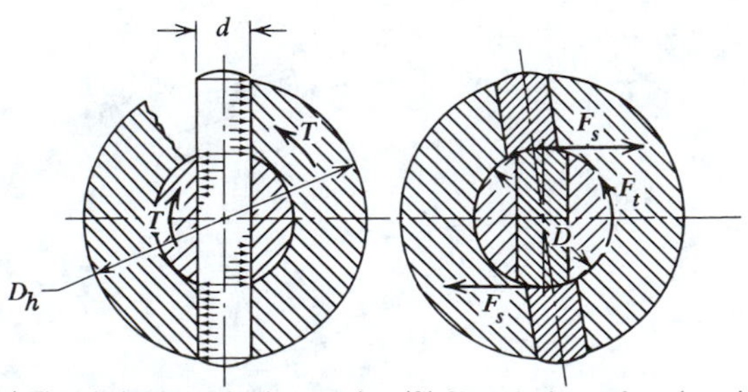

(*a*) The pin is loaded in shear and compression; the hub is loaded in compression only (*b*) Shear is the preferred mode of failure (less damage to shaft and hub)

Figure 16-15 Dowel pin in transverse alignment.

TABLE 16-6 Ultimate Double Shear Values of Spiral Wrapped Pins

	Standard Duty		Heavy Duty	
Nominal Diameter, mm	AISI 302, kN	SAE 1070 AISI 1420, kN	AISI 302, kN	SAE 1070 AISI 1420, kN
1	0.43	0.54		
1.5	1.08	1.32	1.57	1.96
2.0	1.67	2.11	2.45	3.14
2.5	2.45	3.14	3.53	4.41
3.0	3.33	4.22	4.90	6.28
4.0	6.86	8.53	10.79	13.24
5.0	9.81	12.26	15.69	19.61
6.0	13.73	17.16	20.6	25.5
8.0	27.5	34.3	40.2	51.0
10.0	39.2	49.0	62.8	78.5
12.0	53.9	68.6	78.5	98.1

Source. CEM Company, Inc. With permission.

stated in terms of the *minimum ultimate double shear load* that can be supported by the pin. This is another example of preferred mode of failure, since shear pins can also fail in compression. Typical values for spiral-wrapped pins are given in Table 16-6.

EXAMPLE 16-3

The propeller of a small outboard motor is secured to its shaft by a $\frac{3}{16}$-in. shear pin (Fig. 16-16). The shaft is $\frac{5}{8}$ in. in diameter and the hub is $1\frac{1}{4}$ in. in diameter. How much torque can safely be transmitted if the allowable shear stress in the pin is 6000 psi? How much power can be transmitted at 600-rpm propeller speed?

SOLUTION

The pin is sheared in two places located on the same diameter, but 180 deg apart. The shear areas are circular. The two unknown, parallel shear forces, oppositely directed, form a couple. This couple is the unknown torsional resistance. Thus

$$T_s = F_s D$$
$$= F_s(0.625 \text{ in.})$$
$$F_s = \tau_w A_s = \tau_w(0.25\pi d^2)$$
$$= (6000 \text{ lb/in.}^2)(0.25\pi)(0.1875 \text{ in.})^2$$
$$= 166 \text{ lb}$$
$$T_s = (0.625 \text{ in.})(166 \text{ lb}) = \mathbf{104\ lb\text{-}in.}$$
$$P = \frac{T_s n}{63{,}000} = \frac{(104 \text{ lb-in.})(600 \text{ rpm})}{63{,}000} = \mathbf{0.99\ hp}$$

Figure 16-16 Illustration for Example 16-3.

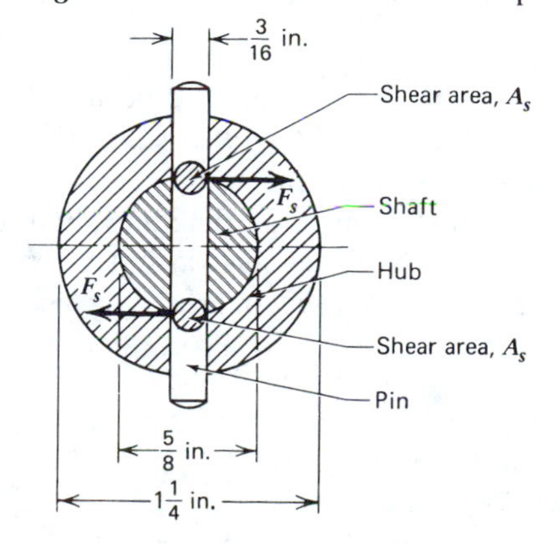

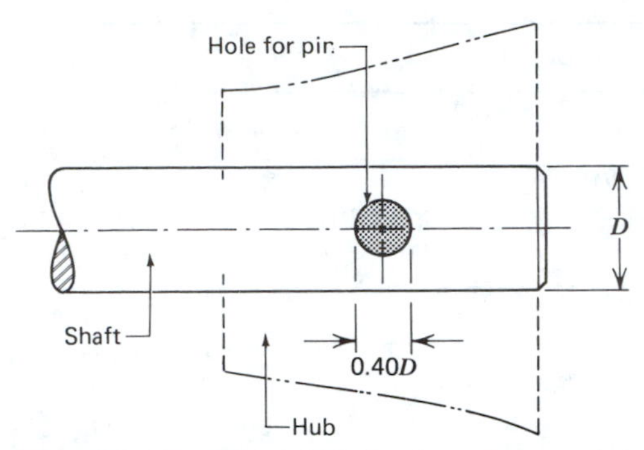

Figure 16-17 Pin and shaft of equal strength in shear.

Conclusion: The shear pin will transmit roughly 1 hp.

Pin and Shaft of Equal Strength

Figure 16-17 shows a round shaft drilled transversely to receive a round driving pin. The question often arises as to the relative size of the two mating parts, if equal strength in shear is desired. When the two parts are made from the same material, equal strength occurs for $d = 0.40D$.[2]

16-6 SPLINES

Machine tool and automotive designs frequently require a positive and very accurate connection between shafts and their related members. Splines provide such a connection. Splines are *multiple keys* that are machined *integrally* with the shaft and that mate with grooves cut in a hub (Fig. 16-18). They are produced most economically in large quantities and can be made with great accuracy. Both *sliding* and *fixed* connections are therefore easily made and used extensively. In a sliding connection, hub and shaft are free to have relative axial motion but *must* rotate together because of the interlocking action. This principle is utilized in transmissions to

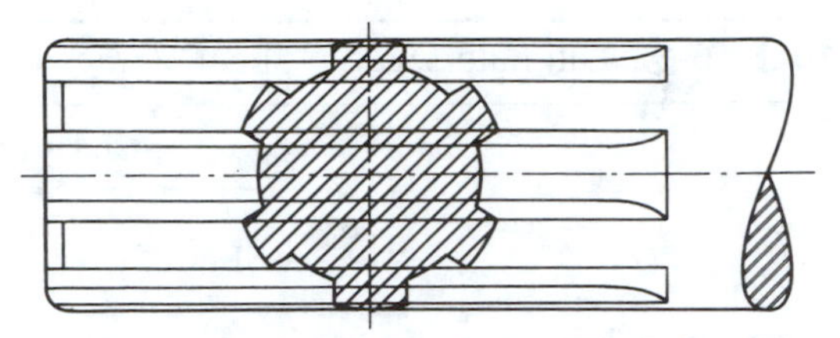

Figure 16-18 Six-tooth, parallel-sided spline.

achieve multiple speed ratios (Fig. 16-19). In a fixed connection axial motion is prevented by clamping or restraining against shoulders. Splined connections are especially useful for coupling shafts end to end when heavy torques are to be transmitted without slippage.

Parallel splines have sides that are straight and parallel (Fig. 16-18). Standard dimensions for parallel 4-, 6-, 10-, and 16-spline fittings have been established by the Society of Automotive Engineers as given in *Machinery's Handbook*, pp. 1016–1019. A look at the tables and the accompanying illustrations shows that four dimensions are required to specify a spline. The nominal diameter is the shaft outside diameter. The torque capacity T is given in pound-inches per inch of bearing length. Note that torque capacity, for the same nominal diameter, increases with the number of splines.

EXAMPLE 16-4

Specify a spline to slide under load while transmitting a torque of roughly 1100 lb-in. The hub is 2 in. long.

SOLUTION

The torque capacity per inch must be roughly 550 lb-in. Surveying the table, we find two suitable splines.

$$\text{6-splines:} \quad T = 608 \text{ lb-in.;}$$

$$\text{Nominal Diameter} = \mathbf{1\tfrac{1}{2}\ in.}$$

$$\text{10-splines:} \quad T = 672 \text{ lb-in.;}$$

$$\text{Nominal Diameter} = \mathbf{1\tfrac{1}{4}\ in.}$$

[2]D. C. Greenwood, *Engineering Data for Product Design*, p. 220. See Ref. 16-1.

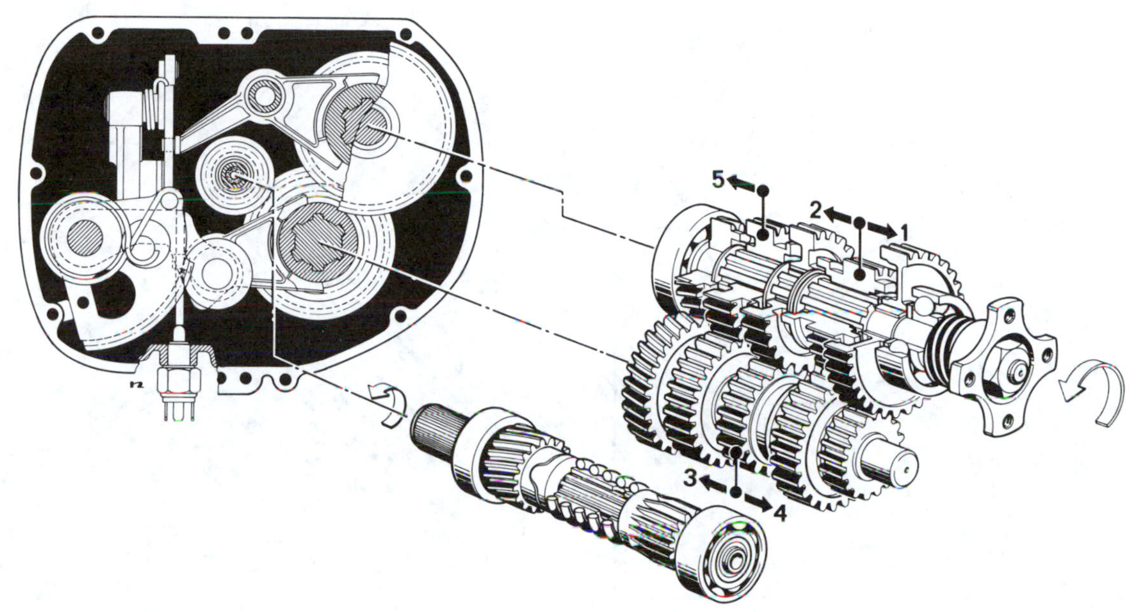

Figure 16-19 Application of splines to a triple-shaft five-speed transmission in a BMW motorcycle. (Courtesy Bavarian Motor Works.)

Which spline to choose depends on considerations such as initial cost, space, tooling, and shaft size.

Involute splines are similar in form to internal and external involute stub gear teeth having a pressure angle of 30 deg or more. Basically, a splined connection is a pair of internal gears in mesh in which the pinion has been enlarged so that each tooth meshes permanently with only one tooth in the annular. A splined connection is therefore subject to static or light dynamic loads only. Hence strength in bending is the main criterion for tooth design. A broad-based tooth of superior bending strength is obtained by specifying one of the three available standard pressure angles: 30, 37.5, or 45 deg (Fig. 16-20).

Involute splines are more popular than straight splines due to greater strength relative to size, particularly where many teeth are required. Involute splines are self-centering and tend to equalize stresses among themselves. The general practice is to form the external splines by hobbing, rolling, or on a gear shaper and to form the internal splines by broaching or on a gear shaper. The internal spline is held to basic dimensions, and the external spline is varied to control it.

Rolled splines, available up to 90 mm (3.5 in.), are preferable to cut splines in high-torque applications. Cold-forming has the effect of increasing fatigue strength by reducing stress concentration and improving grain flow.

Involute splines are discussed in great detail in *Machinery's Handbook,* pp. 991–1015.

16-7 TAPERED SHAFTS

When highly accurate concentricity is required and the hub is to be mounted at the end of a shaft, a connection formed by matching tapered surfaces is often used (Fig. 16-21). The shaft end is threaded for a retaining nut and washer that hold

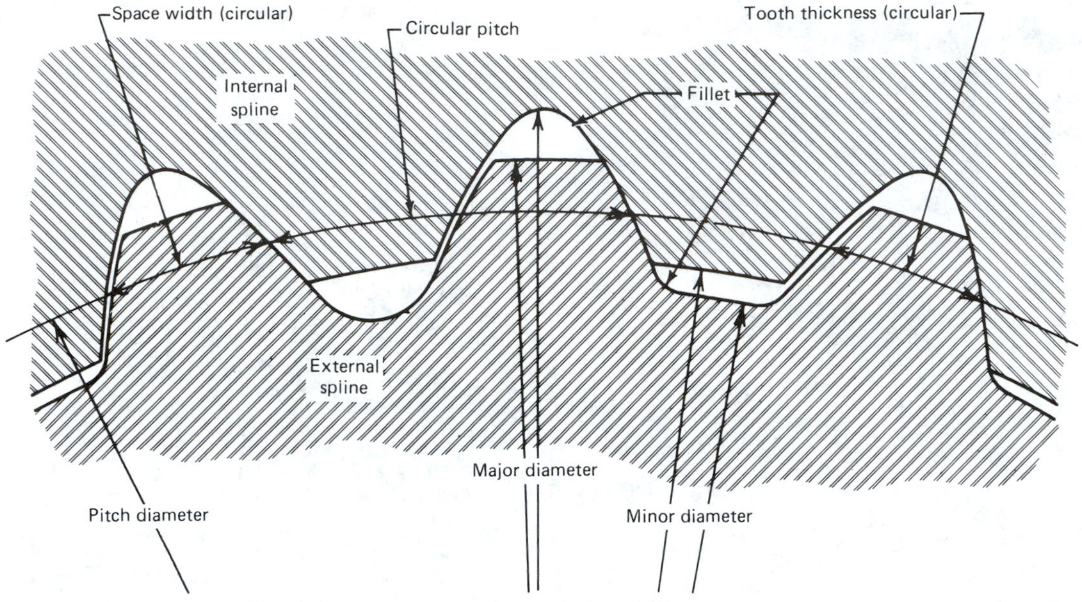

Figure 16-20 Involute spline. Note the broad base of all teeth and the large fillet radii. (Courtesy General Motors.)

the hub in place and provide assembly force when tightened. Such a connection is commonly used for mounting impellers in centrifugal pumps and blowers. The disadvantage of tapered connections is that a minute outward axial displacement eliminates surface contact on the taper.

With key and keyway incorporated, a rigid, heavy-duty connection based on interlocking results. Torque capacity is determined by key strength, and tapered surfaces provide accuracy of location.

Without a key, the tapered connection derives its holding power from friction in a manner similar to an interference fit. Tightening of the retaining nut wedges the tapered surfaces together and produces large normal forces. The maximum torque that can be transmitted without slip is

$$T = 0.5fF_nD_m \qquad (16\text{-}9)$$

where

f = coefficient of friction
F_n = normal force at the mating surfaces; N, lb
$D_m = 0.5(d_1 + d_2)$ = mean diameter; mm, in.

Figure 16-21 Tapered shaft end for highly accurate concentricity of mounting.

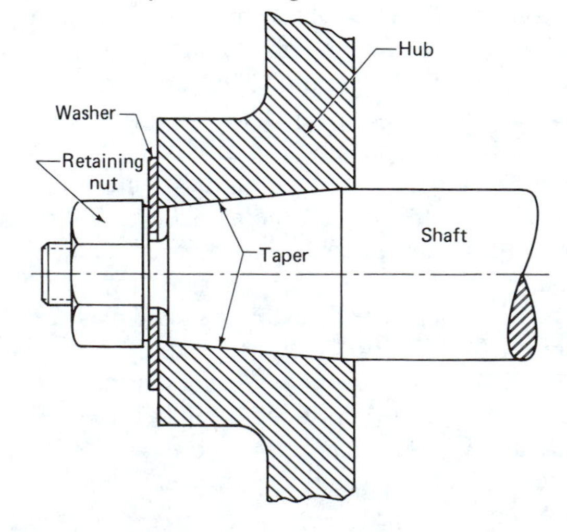

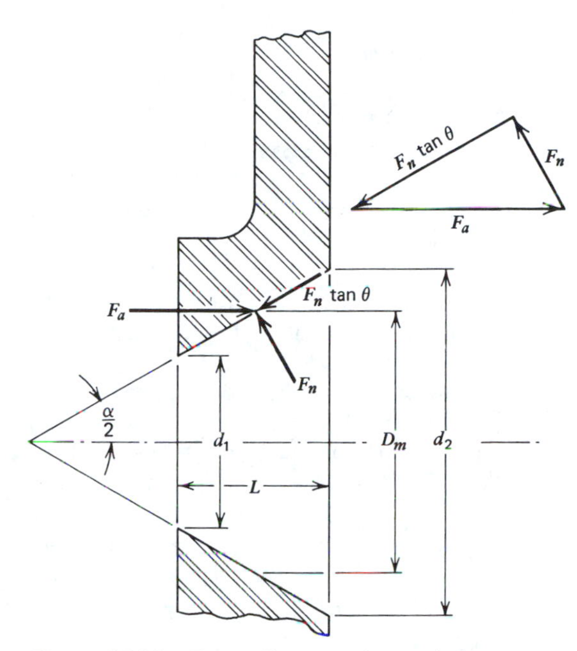

Figure 16-22 Force diagram of taper hub.

Applying the equations of static equilibrium to the hub free-body diagram (Fig. 16-22) gives, for *horizontal* summation,

$$F_a - F_n \tan\theta \cos(0.5\alpha)$$

$$- F_n \cos(90 \text{ deg} - 0.5\alpha) = 0$$

$$F_a \cos\theta = F_n[\sin\theta \cos(0.5\alpha) + \cos\theta \sin(0.5\alpha)]$$

$$F_a \cos\theta = F_n \sin(\theta + 0.5\alpha)$$

$$F_n = \frac{F_a \cos\theta}{\sin(\theta + 0.5\alpha)} \qquad (16\text{-}10)$$

where

F_n = normal force; N, lb
F_a = axial force produced by the nut; N, lb
α = included angle; deg
θ = friction angle; deg

In terms of axial force, the maximum torque capability of the connection is

$$T = \frac{fD_mF_a \cos\theta}{2\sin(0.5\alpha + \theta)} = \frac{D_mF_a \sin\theta}{2\sin(0.5\alpha + \theta)} \qquad (16\text{-}11)$$

The magnitude of applied axial force depends on tightening of the retaining nut and can generally be calculated by the torque-tension relationship for screw fasteners.

The stress state here is analogous to that existing in interference fits, as discussed in Section 6-13. Wedging of the tapered surfaces results in a surface compressive stress given by:

$$\sigma_c = \frac{F_a \cos\theta}{D_mL \sin(0.5 + \theta)} = \frac{2T}{D_m^2 fL} \qquad (16\text{-}12)$$

where

L = length of contact measured axially; mm, in.

Also, because of wedging action, tangential stresses exist in the shaft and hub. If the hub is cast iron or a similarly brittle material, particular care must be exercised to ensure that tensile stress does not exceed tensile strength of the material.

In addition to transmitting torque, tapered shaft connections can also support external axial loads. In fact, if the axial thrust load acts so as to increase wedging (Fig. 16-23*a*), torque capability increases because of larger normal forces.

Figure 16-23 Taper hub connection with external thrust.

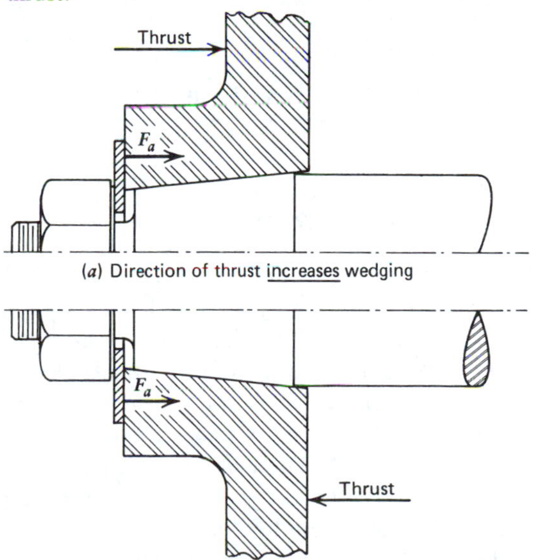

Conversely, if wedging is decreased (Fig. 16-23*b*), the torque at which slipping occurs decreases and the external axial load is supported by the retaining nut. The equations of this section are valid in any case if F_a is taken as *net* axial load.

EXAMPLE 16-5

What is the torque rating of a taper shaft end with a prevailing coefficient of friction of 0.1? The shaft tapers from 60 mm at the small end to 75 mm at the large end, in 100 mm axial length. The axial load that can be applied to clamp the tapered surfaces together is limited to a contact stress of 340 MPa.

SOLUTION

From Eq. (16-12):

$$T = 0.5\sigma_c D_m^2 fL$$

$$\sigma_c = 340 \text{ N/mm}^2$$

$$D_m = 0.5(60 \text{ mm} + 75 \text{ mm}) = 67.5 \text{ mm}$$

$$T = 0.5(340 \text{ N/mm}^2)(67.5 \text{ mm})^2(0.1)(100 \text{ mm})$$

$$= 7746 \text{ N} \cdot \text{m}$$

Torque rating ~ **7750 N · m**

16-8 FRICTIONAL SHAFT CONNECTORS

Machine components can be mounted, adjusted, and removed quickly by means of these unique connectors based on friction. They commonly consist of one or more sets of matching opposite-tapered concentric rings that expand radially when compressed axially, locking everything in place (Fig. 16-24). The mating conical surfaces are on the prefabricated rings, not on the shaft or in the hub, and herein lies their main advantage.

Because stress concentration due to splines, keys, or pins has been eliminated, *smaller* shafting of equal strength can be used. Shaft sizes may be reduced by as much as 12% since most keyways have a depth of one-eighth of the shaft diameter. This in turn leads to smaller bearing sizes and further cost reduction. Unaffected by vibration and torque reversals, these simple precision connectors can transmit torque up to 20 000 N · m. For further detail, see Reference 16-4.

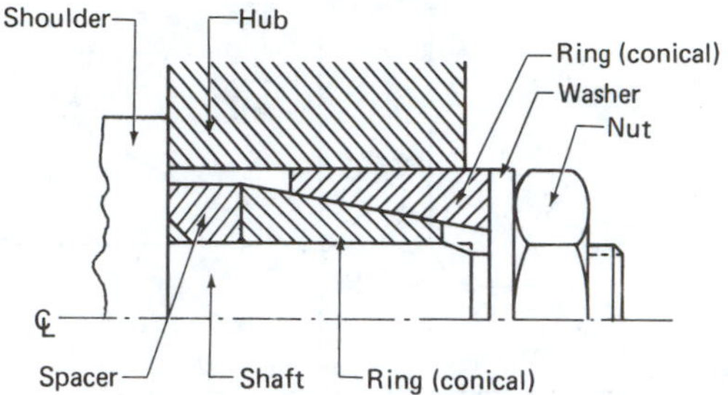

Figure 16-24 Frictional shaft connector. Note the external (shaft) and internal (hub) cylindrical surfaces, which reduce cost and facilitate mounting. The absence of keys and splines accommodates smaller shafts of equal strength and smaller bearings.

16-9 RIGID COUPLINGS

Rigid couplings are permanent couplings that, by virtue of their construction, have essentially no degree of angular, axial, or rotational flexibility. Essentially, they are means of extending a rigid shaft or uniting two rigid shafts.

Rigid couplings are suitable for applications only where accurate shaft alignment can be maintained. For this reason, they are seldom used between motors and driven equipment. The selection of rigid couplings is determined by the diameter of the shafts to be connected; their torque ratings are directly related to the torque capacity of the shafts to which they are fitted.

The three principal types of rigid couplings are the *collar, compression,* and *flange*. The collar and compression types are used when a large flange is objectionable. All three types may be used to connect shafts of different diameters.

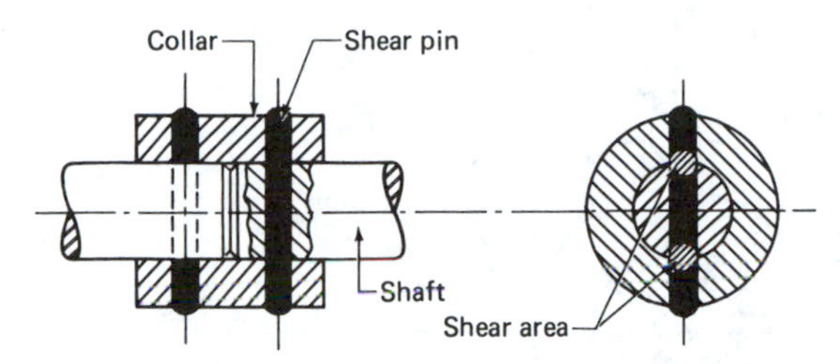

Figure 16-25 Collar or sleeve coupling.

Although the use of rigid couplings has greatly declined in recent decades, they will be discussed in part for instructional purposes. Analyzing the design of rigid couplings includes applying the fundamentals of friction and strength of materials.

Collar or Sleeve Coupling

This rigid type, shown in Fig. 16-25, is simply a hollow cylinder fitted over the ends of the two shafts to be connected and then pinned or keyed to provide the necessary torque capacity. Its advantages are simplicity, low cost, and safety, the last because of the absence of projecting parts. The sleeve coupling, however, is difficult to remove and requires an extra space of half its length on the shaft so that it may be slipped back.

Compression Coupling

This rigid type, illustrated in Fig. 16-26, is a modification or improvement of the sleeve coupling. It is simply the latter split in halves. The halves are recessed and equipped with holes for through bolts, which are used to clamp the two halves together. The coupling can be designed with or without a key, depending on the amount of torque to be transmitted. Its construction in two halves permits ready assembly and disassembly of shafts already in position. The compression coupling may be used for the transmission of large torques.

Figure 16-26 Compression coupling.

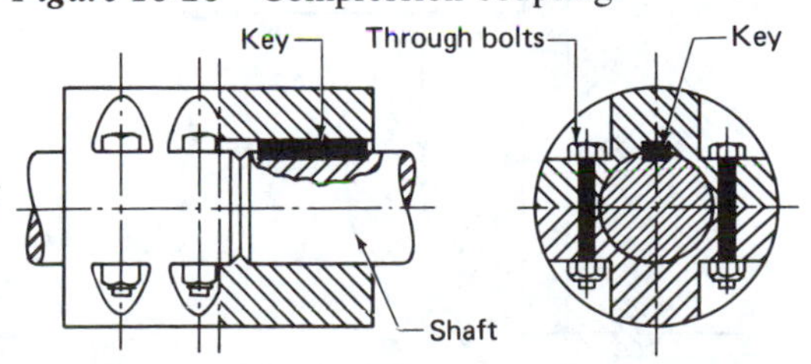

Flange Coupling

This coupling, shown in Fig. 16-27, is composed of two rigid hubs keyed to the shafts. The hubs extend into flanges whose faces are brought and held together by a series of bolts arranged concentrically about the shaft so that their axes are parallel to the collinear axes of the shafts and coupling. Alignment of the shafts is provided either by a cylindrical projection on one flange that fits into a corresponding recess in the other or by extending one shaft through its flange into the bore of the other.

If torque is to be transmitted by shear in the bolts, the bolts must be fitted to the bolt holes in the two flanges. This arrangement is necessary so that the bolts can share equally the loads imposed on them when the coupling transmits torque. If, on the other hand, torque is to be transmitted by friction forces at the flange interface, the bolt holes should be larger than the bolt shank for ease of assembly.

Flange couplings, as opposed to collar and compression couplings, are best suited for *large* shafts (25 mm and over). The large circumferential flange (extended axially in both directions from the bolt flange) avoids the danger of exposed rotating nut and bolt heads. Standard dimensions for safety flange couplings are given in *Machinery's Handbook,* p. 683.

EXAMPLE 16-6

The collar coupling in Fig. 16-25 is mounted on a 15-mm diameter shaft that is to transfer 0.75 kW at 600 rpm. Find the pin diameter d for working shear stress of 40 MPa. Calculate the minimum collar outer diameter D_h when the material has an allowable tensile stress of 40 MPa.

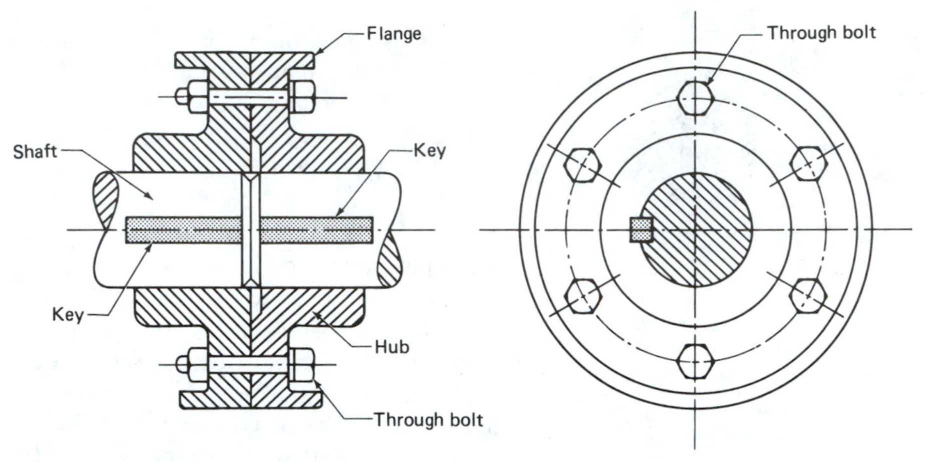

Figure 16-27 Flange coupling.

SOLUTION

The pin is sheared in two places as shown in Fig. 16-28. The two parallel shear planes are 15 mm apart and of equal size. The resistance to shear is therefore a couple T_s equal and opposite to the applied torque T.

$$T_s = F_s(0.015 \text{ m})$$

$$= \tau_w A_s(0.015 \text{ m})$$

$$= (40 \text{ N/mm}^2)(0.25\pi d^2)(0.015 \text{ m})$$

$$= 0.471 d^2 \text{ N} \cdot \text{m}$$

$$T = \frac{9550P}{n} \text{ (See Eq. 1-7)}$$

$$= \frac{9550(0.75 \text{ kW})}{600 \text{ rpm}} = 11.94 \text{ N} \cdot \text{m}$$

Figure 16-28 Steel pin in shear.

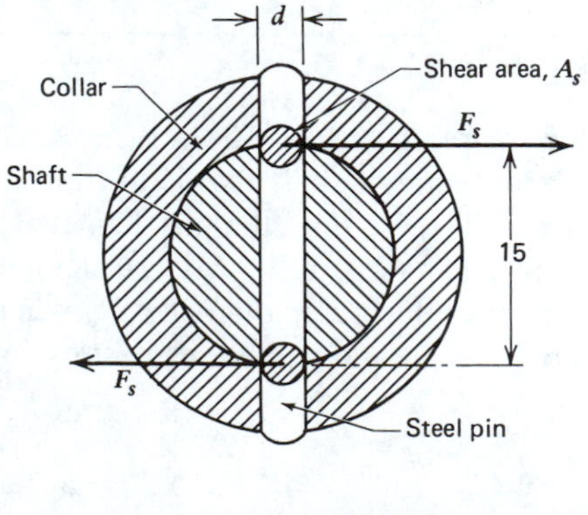

Because $T_s = T$

$$0.471 d^2 \text{ N} \cdot \text{m} = 11.94 \text{ N} \cdot \text{m}$$

$$d^2 = 25.35 \text{ mm}^2 \text{ and } d = 5.03 \text{ mm}$$

Use $\boldsymbol{d = 5}$ **mm**

The minimum outer collar diameter D_h is determined by the compressive strength of the collar material. A force F_c represents the compressive strength of each of the two areas. The applied torque T is balanced by a couple T_c generated by the two forces F_c. Figure 16-29 shows the details.

$$T_c = F_c 0.5(15 + D_h) \text{N} \cdot \text{mm}$$

The area in compression is a rectangle, A_c.

Figure 16-29 Steel pin in compression.

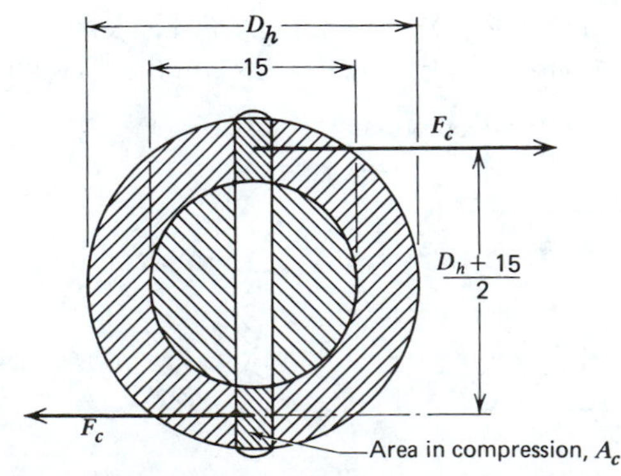

$$A_c = 0.5(D_h - 15\text{ mm})d$$

$$F_c = A_c S_c$$

$$= 0.5(D_h - 15\text{ mm})dS_c$$

$$T_c = F_c 0.5(D_h + 15\text{ mm}) \quad \text{(See Fig. 16-29)}$$

$$= 0.5(D_h - 15\text{ mm})(5\text{ mm})(40\text{ N/mm}^2)$$

$$(0.5)(D_h + 15\text{ mm})$$

$$= 50(D_h - 15)(D_h + 15)\text{N}\cdot\text{mm}$$

Since $T_c = T(= 11.94\text{ N}\cdot\text{m})$

$$50(D_h{}^2 - 225) = 11\,940\text{ N}\cdot\text{mm}$$

$$D_h{}^2 = 463.8\text{ mm}^2 \text{ and } D_h = 21.54\text{ mm}$$

Since D_h is the minimum outer collar diameter, D_h should be **22 mm** or larger.

EXAMPLE 16-7

Find the torque capacity for a safety flange coupling of the type given in *Machinery's Handbook*, p. 683, to fit a 1-in. diameter shaft. Assume the use of shear bolts with a shear strength of 8000 psi.

SOLUTION

Five $\frac{3}{8}$-in. diameter bolts are required. The diameter of the bolt circle is not given. Assuming it is placed in the midst of the flange, one obtains for the bolt circle diameter:

$$D_b = 0.5[(D - 2k) + C]$$

$$= 0.5[(4 - 2(0.25) + 2.25)] = 2.875\text{ in.}$$

Shear area A_s of one $\frac{3}{8}$-in. diameter shear bolt is:

$$A_s = (0.25)(\pi)(0.375\text{ in.})^2 = 0.110\text{ in.}^2$$

Shear force F_s of one bolt is:

$$F_s = (0.110\text{ in.}^2)(8000\text{ psi}) = 880\text{ lb}$$

$$\text{Total shear force} = (880\text{ lb})(5) = 4400\text{ lb}$$

$$\text{Shear torque} = (0.5)(2.875\text{ in.})(4400\text{ lb})$$

$$\text{Torque capacity} = \mathbf{6325\text{ lb-in.}}$$

16-10 RETAINING RINGS (SNAP RINGS)

Generally, "shoulders" seem to be an economical way of positioning parts on a shaft. However, they usually require larger diameter shafts than necessary, and it is a costly procedure to generate shoulders. Material removed is wasted, as shown by the dotted lines in Fig. 16-30*a*. With retaining rings, also called snap rings, designers can allow for a smaller shaft diameter and reduce machining to cutting a narrow circular groove (Fig. 16-30*b*). Retaining rings, however, work only when a small amount of axial play is permissible.

Retaining rings provide a removable shoulder to locate, retain, or lock accurately components on shafts or in bores and housings. They are usually made of spring steel and can be sprung into grooves or recesses and remain seated there in a deformed position. Retaining rings can replace cotter pin and washer, setscrew and collar, or nut and shouldered shafts. Thus they can effectively lower the cost of fastening and assembly.

Snap rings require grooves for seating. But grooves are stress raisers. Hence snap rings should be avoided in areas of high stress, for instance, in the central section of a simple supported shaft carrying a heavy load in bending.

SUMMARY

The main function of detachable fastenings for shaft and hub is to transmit unidirectional torque between a shaft and any machine part assembled on it. Their design is based either on interlocking of matching parts or on friction between close-fitting parts. Some also serve as locking devices, while others allow axial movement. For safety and economy, any detachable fastener should be designed to fail before the hub and shaft.

Setscrews are short, headless cap screws with a point at one end and a square or socket at the other. Available in various head and point configurations, they serve in many de-

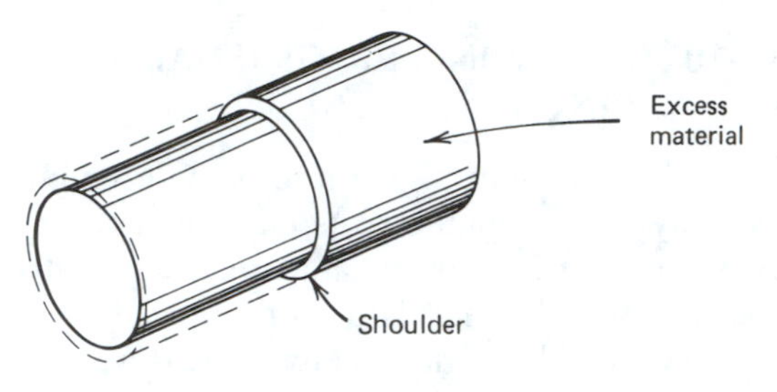

(a) A machined shoulder for locating machine members

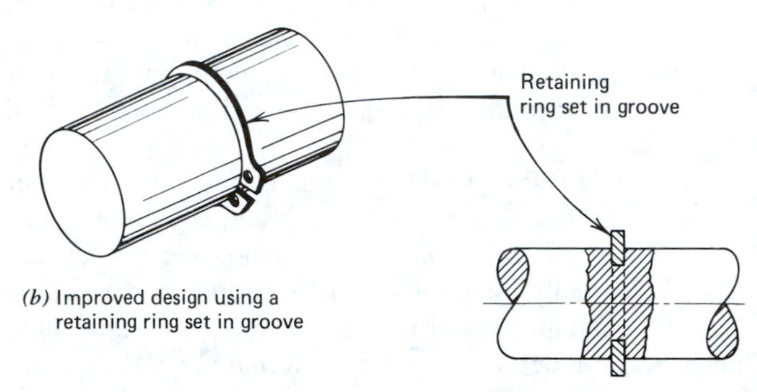

(b) Improved design using a retaining ring set in groove

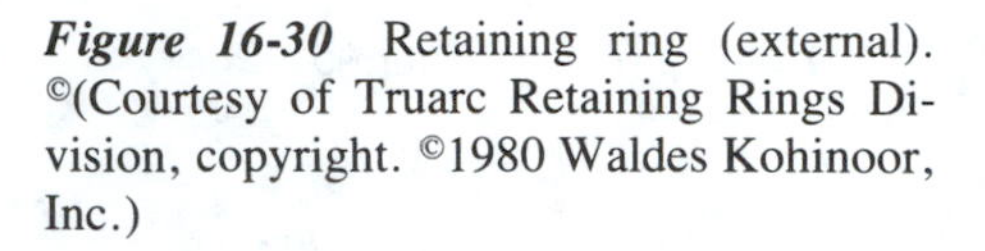

Figure 16-30 Retaining ring (external). ©(Courtesy of Truarc Retaining Rings Division, copyright. ©1980 Waldes Kohinoor, Inc.)

sign situations requiring light service. Their holding power is a function of *relative* position.

Keys are metal inserts designed so that one-half is in the shaft while the other half is in the hub. For maximum torque capacity, a key must be long relative to its width, depth, or diameter and must be mounted axially. The advantages of keys are simplicity, low cost, and ease of assembly. Their disadvantage lies in a weakening of the shaft.

Taper and Woodruff keys are special-purpose keys. The Woodruff key, a semi-circular steel disk, is self-aligning and thus well suited for tapered shafts. Taper keys are used for heavy-duty service of large machine members unaffected by the slight misalignment caused by wedging action. Keys are covered by AGMA standards.

Pins provide inexpensive means of connecting shaft and hub. Some are cylindrical; others are tapered. Spiral-wrapped pins are held in place by spring action; cylindrical and tapered pins rely on friction. Pins can be mounted radially and axially.

Splines are multiple keys that are machined integrally with the shaft and mate with grooves cut in a hub. They provide a positive, very accurate connection between shaft and hub for transfer of heavy torque. Because of ease of assembly and low unit cost, splines are preferred in mass-produced shafts. Tapered shafts are preferred, however, when great accuracy of mounting is required. Such a connection is commonly used for mounting impellers in centrifugal pumps and blowers. For heavy-duty service, a Woodruff key may be incorporated.

Frictional shaft connectors based on the wedging action of mating, tapered concentric rings are becoming more popular because they can be mounted on cylindrical instead of tapered shaft ends. They have little adverse effect on shaft strength.

Rigid couplings are simple means of extending a rigid shaft or uniting two shafts not necessarily of the same diameter. The torque capacity of a rigid coupling should always be less than that of the weakest of the two shafts, thus providing a margin of safety. The three principal types are the collar, compression, and flange.

Retaining rings provide a removable shoulder to locate, retain, or lock accurately components on shafts or in bores and housings. Such rings are usually made of spring steel and can be sprung into grooves or recesses where they remain seated in a deformed position. Retaining rings can replace cotter pin and washer, setscrew and collar, or nut and shouldered shafts. Thus they can effectively lower the cost of fastening and assembly.

REFERENCES

16-1 Greenwood, D. C. *Engineering Data for Product Design*. New York: McGraw-Hill, 1961. (See pp. 336–337, 14 ways to fasten hubs to shafts.)

16-2 Ortwein, W. C. "A New Key and Keyway Design." *ASME*, 78-WA/DE-7. (Describes a new and improved design for keys and keyseats.)

16-3 Parmley, R. O. *Standard Handbook of Fastening and Joining*. New York: McGraw-Hill, 1977. (A comprehensive, up-to-date book on fastenings—excellent.)

16-4 Prause, J. J. "Frictional Shaft Connectors," *Machine Design*, February 21, 1974.

16-5 *Value Engineering with Retaining Rings*. New York: Truarc Retaining Rings Division, Waldes Kohinoor, Inc., 1980.

PROBLEMS

P16-1 Which has the greater torque capability: an M14 setscrew on a 50-mm shaft or an M16 setscrew on a 35-mm shaft?

P16-2 A V-belt pulley is used with a 375-W motor operating at 1800 rpm. What size setscrew should be used if the shaft diameter is 12 mm? Use a factor of safety equal to 3 to allow for load fluctuations. *Hint:* Use Eq. 16-2.

P16-3 A belt pulley is attached to a shaft transmitting 5 hp at 900 rpm by means of two standard setscrews spaced 120 deg apart. Select standard setscrews for this 0.75-in. diameter shaft. The motor drives an industrial fan that operates intermittently 16 hrs each day, five days a week. Select a suitable service factor from Fig. 1-6. What is the factor of safety? *Hint:* Use Eq. 16-3.

P16-4 Although the preferred spacing of setscrews is from 45 to 120 deg, the 45 deg spacing yields a resulting force almost twice that of the 120 deg mounting angle. Prove this by means of suitable layouts of the force vectors involved in each of the two mounting angles.

P16-5 Find a suitable square key to replace the two setscrews for the drive in P16-3.

P16-6 Prove that a square key is the correct shape for materials having the relationship $\sigma_c = 2\tau$. *Hint:* Use Eqs. 16-5 and 16-6.

P16-7 Design a steel key for a solid, cylindrical, 3-in. outside diameter, steel shaft that will transmit 140 hp at 200 rpm. The permissible shear stress in the key is 8000 psi; the permissible compressive stress is 14,000 psi. (Assume $b = 0.25D$.) *Hint:* Use two square keys or an oversized key. Calculate for both.

P16-8 Determine the size and length of a standard square key for a 35-mm shaft that is to transmit 15 kW at 1200 rpm. Use AISI 1018 steel and a factor of safety of 2. *Hint:* Use Table 16-5.

P16-9 A shaft 35 mm in diameter transmits a torque of 900 N·m by means of a chain drive. The chain sprocket is fastened to the shaft by means of a standard square key. The hub length is 50 mm.

(*a*) Determine shear and bearing stress on the key, and select a standard material.
(*b*) How much power can be transmitted at 500 rpm?
(*c*) Determine the shear stress in the shaft and compare it with that of the key. What do you conclude from the answers?

P16-10 A key, the text indicated, often has the size relationship with the shaft of $b = 0.25D$. Assuming a key and shaft of the same material, show that the length relationship is $L = 1.2D$. Assume that the weakening effect of the keyway increases the shearing stress of the shaft by 30%.

P16-11 A gear transmitting 20 hp at 125 rpm is fastened to a 3-in. steel shaft by means of a steel key 0.75 in. wide and 0.5 in. high. What length of key is required if the permissible shear and bearing stresses are 8000 and 15,000 psi, respectively? Make a suitable sketch in two views and to full scale of the upper half of the shaft containing the key.

P16-12 Select two suitable round keys for the connection described in P16-7. Use the material specified in Table 16-5.

P16-13 What are the actual shear and compressive stresses of the key in Example 16-2? Comment on the answer.

P16-14 What is the bearing pressure p_c of the SAE standard 6-spline fitting selected in Example 16-4? *Hint:* Use the nomenclature and figure shown in *Machinery's Handbook*.

P16-15 A 12-mm taper pin in transverse alignment connects a 50-mm shaft to a hub, 100 mm in diameter. What is the shear and bearing or crushing stress for transmitting a torque of 1000 N · m? Show a free body diagram of the pin.

P16-16 Can a SAE standard splined fitting replace the key in Example 16-2? A permanent fit is required.

P16-17 Can a SAE standard splined fitting (permanent fit) replace the key in P16-11? If not, will a longer hub do?

P16-18 What should be the size and duty rating for a spiral wrapped pin to transmit 75 N · m torque with a 20 mm shaft? Use a factor of safety of 2. What is the real factor of safety?

P16-19 What torque can an 8-mm spiral-wrapped pin develop in a shaft whose diameter is 25 mm? Use a factor of safety of 2 and a standard duty SAE 1070 pin. What would be the stress on a solid pin of the same diameter?

P16-20 What is the torque rating of a 35-mm (large end) tapered shaft connection having a 7-deg included angle? Find hub length L rounded off to an even number. The coefficient of friction is 0.12. The axial clamping force is 15 000 N. Contact stress is limited to 200 MPa. *Hint:* Use Eqs. 16-11 and 16-12. Find L by means of Fig. 16-22.

P16-21 Compare the torque rating and contact stress of the following tapered shaft connections. Assume D_m, L, and F_a are constants: (*a*) $\alpha = 6$ deg, $\theta = 6$ deg; (*b*) $\alpha = 5$ deg, $\theta = 6$ deg; (*c*) $\alpha = 6$ deg, $\theta = 7$ deg. *Hint:* Find simplified expressions for T and σ_c.

P16-22 Refer to Example 16-7. Calculate the torque capacity of this coupling for $f = 0.20$ and an allowable tensile bolt stress of 27,000 psi. Compare the torque capacity of the designs. Which is the least expensive to make? *Hint:* Locate the tensile stress area of the bolt, calculate the normal force, etc.

Chapter 17
Sliding Bearings

Bunched in mutual glee
The bearings glint—O murmurless and shined
In oilrinsed circles of blind ecstasy!

HART CRANE

NOMENCLATURE

A = area
C = radial clearance
D = shaft diameter
e = shaft eccentricity
f = coefficient of friction; frequency of oscillation; cps
F = force
h = thickness of oil film
h_o = minimum oil film thickness
H_f = rate of heat generation or dissipation; W, Btu/sec
k_g = flow factor
L = bearing length
m = clearance factor (1000 C/R)
n = shaft speed; rps
P = load per unit of projected bearing area ($P = W/LD$)
Q = flow rate of bearing's lubricant; mm^3/s, cm^3/s, $in.^3/sec$
Q' = minimum lubricant flow rate to support full film; mm^3/s, drops/s, $in.^3/sec$
Q_s = flow rate out of both ends of active bearing arc; mm^3/s, cm^3/s, $in.^3/sec$
R = shaft radius; mm, in.
R' = rate of shearing strain; s^{-1}
S = Sommerfeld number
t_1 = inlet temperature of lubricant
t_2 = outlet temperature of lubricant
t_a = surrounding air (ambient) temperature
t_b = average wall temperature of bearing
V = rubbing velocity
W = bearing load
β = bearing arc
ϵ = eccentricity ratio ($\epsilon = e/C$)
θ = angular position measured from line of centers
μ = dynamic viscosity; mPa · s, microreyns
τ = shear stress
ϕ = angular position of minimum oil film thickness from load line

It has been estimated that one-third to one-half of the energy produced in the world is consumed in overcoming friction. This is a startling and sobering estimate, yet it is not hard to see its validity. Consider the automobile. Even before the energy extracted from the fuel is delivered to the transmission, several elements—the bearings, valve train, timing belt, and piston rings—deduct their tolls. And all through the drive train additional losses occur as the energy flows to the driving wheels, where still more energy is lost in overcoming the rolling resistance of the tires and the air resistance on the body. Designers of machinery, by the very demands of their craft, are compelled to take a strong interest in the problems of minimizing friction and wear.

17-1 FUNCTION OF BEARINGS

A bearing permits relative and controlled motion between two machine members while minimizing frictional resistance. It usually has three elements. Besides the inner and outer members which have relative motion, there is almost always a third element that separates the first two. In *sliding bearings* it is a film of lubricant; in *rolling element bearings* balls or rollers separate the inner and outer members.

Lubrication is an ancient art, but our understanding of the process of forming a load-carrying lubricant film is recent, dating back to the experiments of Beauchamp Tower in England in the 1880s. Today the study of lubrication and

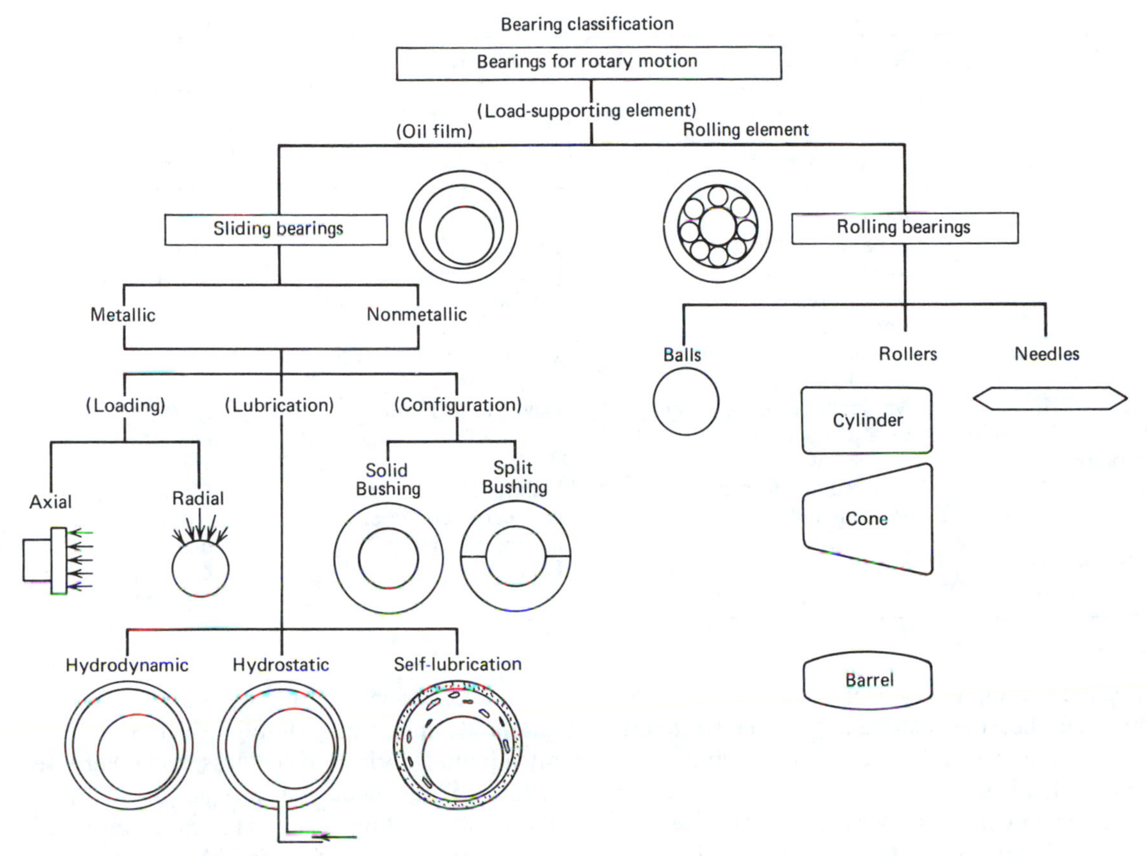

Figure 17-1 Bearing classification.

bearings is an important part of the broader multidisciplinary study called *tribology*, the study of interacting surfaces in relative motion.

17-2 CHOOSING A BEARING TYPE

The classification scheme shown in Fig. 17-1 uses as its basic distinction the type of separating element: a lubricant film for sliding bearings and a form of rolling element for rolling-element bearings. This chapter focuses on sliding bearings, especially *journal bearings* where a shaft rotates inside a sleeve. Chapter 18 will deal with rolling element bearings.

As Fig. 17-1 shows, sliding bearings can be further classified by materials, load direction, lubrication method, and configuration. Method of lubrication is an especially important basis for classifying sliding bearings. In the hydrodynamic bearing, the oil supplied to the bearing is forced by the rotating shaft into a wedge-shaped, high-pressure oil film region. In the hydrostatic bearing, on the other hand, an externally pressurized lubricant is introduced to the bearing and supports the load. This method provides load support and separation of surfaces in cases where low shaft speed precludes formation of a hydrodynamic film.

Self-lubrication is used very successfully in lightly loaded bearings. Nylon or Teflon bearings with no added lubricant work well in 8- or 16-mm movie projectors, for example. In other

TABLE 17-1 Characteristics of Sliding and Rolling Contact Bearings

Bearing Characteristics	Sliding Bearings		Rolling Bearings
	Hydrodynamic	Hydrostatic	
Load carried by	Oil film generated by rotation journal and suitable oil grooves	Oil film generated by a pump placed outside the bearing	Bodies capable of rolling under load
Friction at start-up	Large, due to direct contact between journal and sleeve	Zero; journal and sleeve are separated by an oil film	Low
Friction during operation	Moderate; bearing acts as a pump with low efficiency	Low (fluid friction)	Low (rolling friction)
Life	Limited due to wear at start-up and stop	Unlimited; no contact between metallic parts during operation	Limited (material fatigue in the races)
Relative cost of lubrication	Low	High	Low

cases the bearing provides its own lubricant. Porous metal bearings can be impregnated with oil under pressure to produce adequate lubrication for light loads.

There is no universally suitable bearing. Each type has its special advantages and limitations with regard to load rating, speed, lubrication method, applicability, and cost. Novice designers need some guidance in selecting the type of bearing for a given application. Table 17-1 compares the virtues of rolling element bearings, hydrodynamically lubricated sliding bearings, and hydrostatically lubricated sliding bearings.

A few rough generalizations are possible. Sliding bearings are better suited to the large loads encountered, for example, in a steam turbine. Although sliding bearings may have less running friction than rolling bearings, their starting friction will be much higher. Rolling bearings are also easier to lubricate. However, it may cost less to install and run a sliding bearing, especially at the higher loads and speeds. Sliding bearings are well suited to low-speed applications where shock and vibration occur, as in punch presses and steam hammers. And for many applications—farm machinery, hoists, household appliances—sliding bearings needing only minimal lubrication cost less than otherwise suitable rolling element bearings. In addition, split bushing sliding bearings have one very special advantage: ease of assembly.

Even when designers decide to use a sliding bearing, there are additional decisions to be made. *The Tribology Handbook* offers a set of useful selection tables.

17-3 BEARING LOADS

Most bearings are loaded radially. Sources of this loading are:

1. The weight of the machine parts themselves, such as pulleys, gears, or flywheels, along with the weight of the shaft that supports them.
2. Tension induced by the action of belt or chain drives.
3. The force between meshing gear teeth.
4. Centrifugal forces resulting from unbalanced masses.

TABLE 17-2 Comparative Merits of Lubricants

Characteristic	Petroleum Oils	Synthetic Oils	Grease	Dry Lubricants
Minimize friction	Fair to good	Fair	Fair	Fair
Remain in bearing	Poor	Poor	Good	Very good
Be compatible with bearing materials	Fair	Fair	Fair	Excellent
Cost	Very low	High	Fairly high	High
Life limited by	Deterioration and contamination	Deterioration and contamination	Deterioration	Wear
Temperature range	Good	Fair to excellent	Very good	Excellent
Applicability to boundary lubrication	Fair	Poor to excellent	Good to excellent	Good to excellent

5. Inertia forces accompanying the rapid acceleration and deceleration of machine members.

The load's magnitude will be affected by these factors.

1. Direction of the applied load on the shaft.
2. Point of application of the load on the shaft.
3. Distance between bearing centers.

In addition to radial loads, thrust loads occur from time to time. A shaft supporting a helical gear, for example, will be subjected to the axial component of the tooth force. The weight of a vertical hydraulic turbine and generator unit must be supported by a thrust bearing of impressive size. The bulky, 1 million-lb structure of the 200-in. Hale telescope at Mt. Palomar is supported on six 28-in. square hydrostatic thrust pads.

17-4 LUBRICANTS[1]

It has long been recognized that if a pair of sliding bodies are separated by a fluid or fluidlike film, the friction between them is greatly diminished. The principle of supporting a sliding load on a friction-reducing film is known as *lubrication*. The substance of which the film is composed is a *lubricant,* and to apply it is to *lubricate*. These are not new concepts nor, in their essence, particularly involved ones. But modern machinery has become many times more complicated since the days of the oxcart lubricated by animal fats, and the demands placed on the lubricant have become proportionally more exacting. Even though the basic aim still prevails—the prevention of metal-to-metal contact by means of an intervening layer of fluid or fluidlike material—modern lubrication has become a complex study.

[1]Portions of the text are taken from "Principles of Lubrication" with permission of EXXON, U.S.A.

Types of Lubricants

All liquids will provide lubrication of a sort, but some do it a great deal better than others. Mercury, for example, lacks the ability to wet a metal surface. Alcohol, on the other hand, readily wets a metal surface but lacks the viscosity necessary to form a load-carrying film.

Gases, particularly air, offer potential lubrication where loads are low. Dental drills, gyroscopes, dynamometers, and blowers have been successfully lubricated with gases.

Solid lubricants, either alone or mixed with a liquid lubricant, can be used under conditions of very high pressure, high temperature, and chemically reactive environments.

Table 17-2 compares the merits of liquid,

TABLE 17-3 Physical Units for Dynamic Viscosity

System	Basic Unit	Name	Practical Unit
English	lb-sec/in.^{2}	reyn	microreyn (μreyn)
cgs SI	dyne-sec/cm^2 Pa · s	poise	centipoise (cp) mPa · s

grease, and dry lubricants with respect to severe practical design considerations.

Viscosity of Lubricating Oils

The concept of viscosity is illustrated by Fig. 17-2. A film of lubricant adheres to the stationary plate and supports a moving plate. In order to move the upper plate to the right at a constant velocity V, it is necessary to exert some constant force F. Thus a shear stress is applied at the wetted surface of the moving plate. This shear stress τ is equal to

$$\tau = \frac{F}{A} \tag{17-1}$$

where A is the area of the plate's surface in contact with the lubricant. The rate of shearing strain R' is defined as the ratio of the velocity V to the thickness h of the lubricant film. It is

$$R' = \frac{V}{h} \tag{17-2}$$

The ratio of shearing stress to rate of shearing strain is called the *dynamic viscosity* μ.

$$\mu = \frac{\tau}{R'} = \frac{Fh}{AV} \tag{17-3}$$

Lubricants for which a simple proportionality exists between shearing stress and rate of shearing strain are called *Newtonian*. Petroleum oils are ordinarily Newtonian.

Figure 17-2 Concept of viscous lubrication.

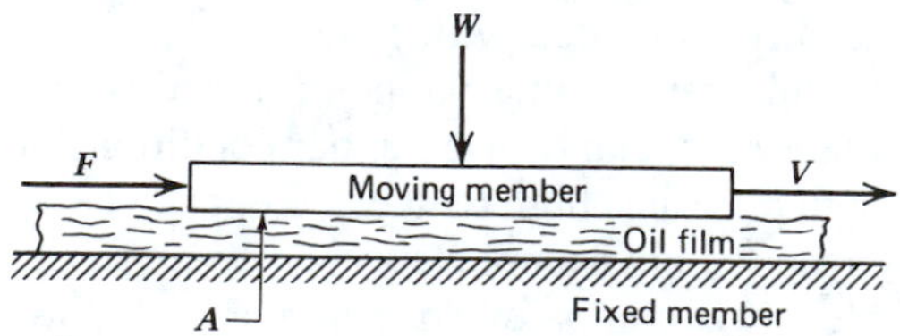

Dynamic viscosity (usually referred to simply as viscosity) is generally dependent on temperature and pressure. For lubricating oils the change in viscosity with pressure is so slight that it need be considered for only the most unusual design cases. There is a substantial change in viscosity with temperature, however, so bearing designers must be very conscious of temperature.

Various physical units have been used to describe dynamic viscosities. A summary of these units is presented in Table 17-3. Here are shown the English units, the old cgs (centimeter-gram-second) version of the metric system, and the current SI system of metric units that has been adopted by international agreement. Happily, the two practical units of the old and the new metric systems are the same; the old centipoise and the new millipascal · second are equivalent. Factors for converting from one system to another are listed in Table 17-4.

The viscosity of water at room temperature is about 1 mPa · s. That is worth remembering; it provides a direct feeling for the viscosities of other substances by comparison with water. The

TABLE 17-4 Conversion Factors for Dynamic Viscosity

To Convert from	Multiply by	To Obtain
microreyns	6.875	mPa · s
mPa · s	0.14545	μreyn
mPa · s	1	cp

TABLE 17-5 Viscosities of Some Common Substances at Room Temperature (21°C)

Substance	Viscosity	
	mPa · s	μreyn
Air	0.018	0.0026
Water	1	0.15
SAE10 oil	70	10
Olive oil	100	15
SAE30 oil[a]	300	44
Honey	1500	218

[a]About 12 mPa · s at an operating temperature of 85°C.

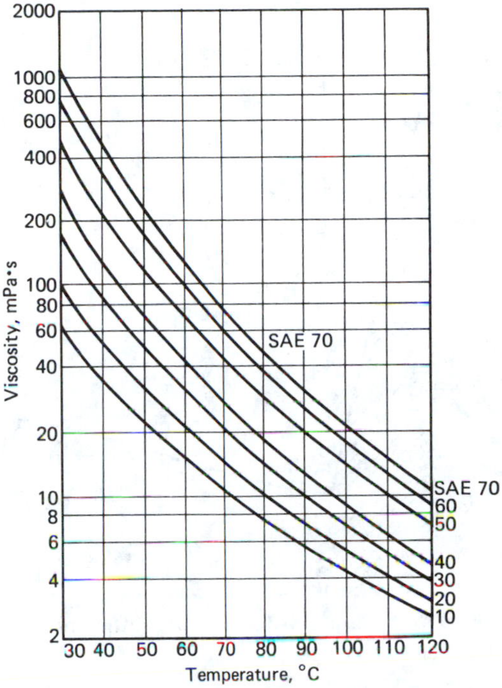

viscosities of several substances at room temperature are listed in Table 17-5.

The viscosities of several SAE (Society of Automotive Engineers) motor oils have been plotted against operating temperature in Figs. 17-3 (SI units) and 17-4 (English units).

17-5 LUBRICATION REGIMES

Three types, or regimes, of lubrication occur in practice. They differ in the degree to which a full-load-carrying film is developed.

Full-film lubrication physically separates the shaft and bearing surfaces by a relatively thick lubricant film, 9 to 25 μm.[2] This prevents any metal-to-metal contact at the operating conditions. The coefficient of friction will be low, usually not above about 0.005. Full-film operation implies minimum power losses and maximum life expectancy of the parts.

Complete boundary lubrication means that the bearing and shaft surfaces are being rubbed together with only a very thin lubricant film (perhaps only a few molecules thick) adhering to each surface and preventing direct contact. The coefficient of friction is high, 0.08 to 0.14 range.

Mixed-film lubrication means that there is

[2]$1\ \mu m = 10^{-6}\ m = 0.001\ mm$.

Figure 17-3 Viscosity-temperature chart for SAE numbered oils (SI units).

Figure 17-4 Viscosity-temperature chart for SAE numbered oils (English units).

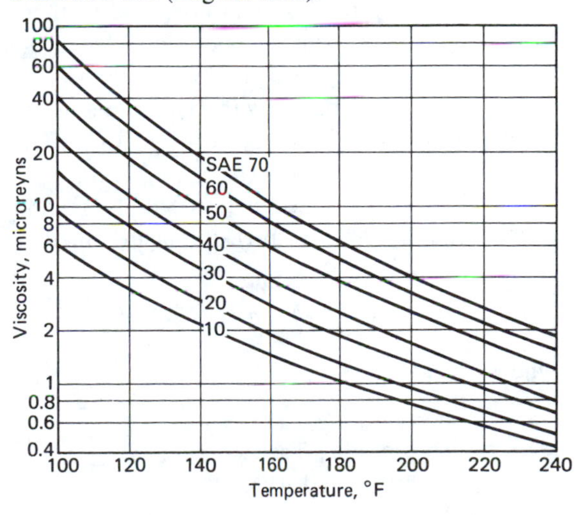

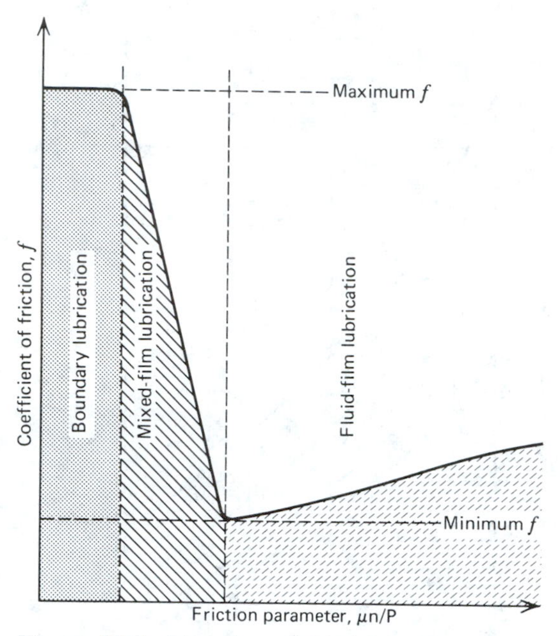

Figure 17-5 Variation of friction coefficient with the μn/P factor.

both boundary and hydrodynamic lubrication. Part of the load is carried by small pools of self-pressurized lubricant. Other areas of the surfaces are rubbing with only a thin film of lubricant separating the peaks. A typical range for the coefficient of friction in this regime is 0.02 to 0.08.

Figure 17-5 provides vivid portrayal of these three regimes of bearing operation. Here the coefficient of friction is plotted against a bearing characteristic number $\mu n/P$. The three operating variables in this bearing characteristic number are the lubricant's viscosity μ (mPa · s or μreyn), the shaft speed n (rps), and the unit bearing load P (MPa or psi), the last defined as

$$P = \frac{W}{LD} \qquad (17\text{-}4)$$

where

W = bearing load; N, lb
D = bearing diameter; mm, in.
L = bearing length; mm, in.

This bearing characteristic number is very convenient. Whatever implies difficulty in forming a complete lubricant film—low viscosity, low shaft speed, or high unit bearing load—implies a low value for $\mu n/P$. Conversely, the higher $\mu n/P$ is, the easier it is to establish a full-load-supporting film.

For the highest values of $\mu n/P$, there is full-fluid-film, or *hydrodynamic,* lubrication. In this regime of operation the coefficient of friction attains a minimum of about 0.001. Wise designers will have the bearing operate at a higher $\mu n/P$ value to be sure of an adequately thick film, paying for this margin of safety with a somewhat higher power loss.

The lowest values of $\mu n/P$ correspond to the regime of complete boundary lubrication. The friction coefficient remains constant throughout this regime; its actual value depends on the character of the surfaces and the lubricant.

The midregime is that of mixed-film lubrication. In this regime a decrease in $\mu n/P$ is accompanied by a sharp increase in friction coefficient.

17-6 SELECTING BEARINGS FOR LIGHT SERVICE

Most of the bearings that the average mechanical designer specifies operate in the mixed-film or boundary lubrication regimes. All around us—in typewriters, office machinery, movie projectors, household appliances, latching mechanisms—we find bearings used with little or even no lubrication and without the proper operating conditions to develop a full-lubricant film. Yet they survive and they provide a low-cost solution to the problem of supporting and controlling machine members in relative motion.

The PV Factor

Manufacturers of bearings for light service usually base bearing selection on the *PV* factor, the product of unit bearing load *P* and rubbing velocity *V*. This factor indicates what bearing temperature will be reached and what rate of wear can be expected. To limit temperature rise and keep the wear rate within reasonable bounds, one limits the PV factor.

The unit bearing load P, already defined by Eq. 17-4, is the ratio of the bearing's load to its projected area. The rubbing velocity V must be calculated differently for oscillating shaft motion than for continuous rotation. For continuous rotation the rubbing speed is, in metric units,

$$V = \frac{\pi D n}{1000} \qquad (17\text{-}5)$$

where

V = rubbing velocity; m/s
D = bearing diameter; mm
n = shaft speed; rps

However, if the shaft is oscillating relative to the bearing, the design value for V is based on the average rubbing speed.

$$V = \frac{\pi D \theta f}{360\,000} \qquad (17\text{-}6)$$

where

θ = total angle traveled per cycle, deg (e.g., for an oscillation of ± 17 deg, the value of θ is (4)(17 deg) = 68 deg)
f = frequency of oscillation; cps

and V and D are defined as in Eq 17-5.

The analogs of Eqs. 17-5 and 17-6 in English units are needed. For continuous rotation,

$$V = 5\pi D n \qquad (17\text{-}7)$$

where

V = rubbing velocity; fpm
D = bearing diameter; in.
n = shaft speed; rps

And, for oscillating shaft motion,

$$V = \frac{\pi D \theta f}{72} \qquad (17\text{-}8)$$

where

θ = total angle traveled per cycle; deg
f = frequency of oscillation; cps

and V and D are defined as in Eq. 17-7.

For convenience, the PV factors in both systems of units are specified in mixed units (MPa · m/s or psi · fpm). No attempt is made to reconcile the units of unit bearing load with those of rubbing velocity.

EXAMPLE 17-1

A nylon bearing has a bore diameter of 0.378 in. and a length of 0.875 in. It supports a load of 1800 lb. The shaft oscillates ± 16 deg at 76 cpm. What is the PV factor?

SOLUTION

(*a*) The unit load is found by use of Eq. 17-4.

$$P = \frac{W}{LD} = \frac{1800 \text{ lb}}{(0.875 \text{ in.})(0.378 \text{ in.})} = 5440 \text{ psi}$$

(*b*) The average rubbing speed is found by Eq. 17-8.

$$V = \frac{\pi D \theta f}{72} = \frac{\pi(0.378 \text{ in.})(64 \text{ deg})(76 \text{ cps}/60)}{72} = 1.337 \text{ fpm}$$

(*c*) So the PV factor is

$$PV = (5440 \text{ psi})(1.337 \text{ fpm}) = \mathbf{7270\ psi \cdot fpm}$$

The Limited *PV* Curve

Later in this section there are tables that list limits on the PV factors for a number of bearing materials. Since it is the *product* of unit bearing load and rubbing speed that must be limited, it is clear that one can, in a sense, exchange load for speed, and vice versa. One can load a bearing heavily if the rubbing speed is low, or light loads can be used at very high rubbing speeds. But there are limits! Too high a load can destroy the bearing material, no matter how low the rubbing speed is. Excessive rubbing speeds can destroy the bearing material due to localized overheating at even very light loads.

Thus designers must select the bearing dimensions so as not to exceed any one of three factors: the maximum unit bearing load, the maximum rubbing speed, or the maximum *PV* factor. Figure 17-6 shows this graphically. Any point between the curve and the origin represents a safe design. Any point beyond the curve indicates a design in violation of at least one of the three design criteria.

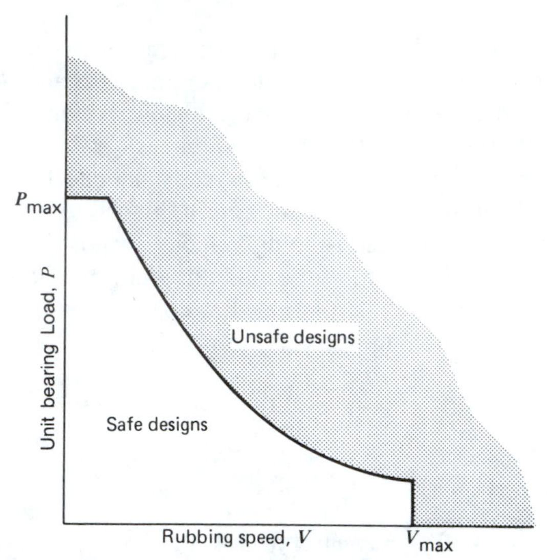

Figure 17-6 Typical limited *PV* curve.

Application Data

For the most current and accurate data, designers must consult the manufacturer of the bearing material. Data tabulated in this chapter are representative. They constitute an adequate basis for example problems and for homework problems but, in professional practice, they should be used only for preliminary design estimates.

Data for a number of porous metal bearing types are presented in Table 17-6. Metal alloy powder is pressed into a mold to shape the bearing, then heated under pressure ("sintered") to form a porous solid. About 10 to 35% of the volume is void. This volume is then filled under pressure with a lubricant. These bearings are said to be *self-lubricating;* if the bearing is running dry, the lubricant flows from the interconnected pores to the surface to provide a lubricant film. When the shaft is stopped and the bearing cools, the lubricant flows back into the pores. Porous metal bearings can be mass-produced inexpensively.

Design data for nonmetallic bearings are presented in Table 17-7. Plastic bearing materials have a number of advantages. They can usually operate dry or with a very limited supply of

TABLE 17-6 Application Data for Porous Metal Bearings

	PV_{max}		P_{max}		V_{max}	
Material	MPa · m/s	psi · fpm	MPa	psi	m/s	fpm
Bronze	1.75	50,000	13.8	2000	6.1	1200
Lead-bronze	2.10	60,000	5.5	800	7.6	1500
Copper-iron	1.23	35,000	27.6	4000	1.1	220
Malleable copper-iron	2.63	75,000	55.2	8000	0.18	35
Iron	1.05	30,000	20.7	3000	2.0	400
Bronze-iron	1.23	35,000	17.2	2500	4.1	800
Lead-iron	1.75	50,000	6.9	1000	4.1	800
Aluminum	1.75	50,000	13.8	2000	6.1	1200

TABLE 17-7 Application Data for Nonmetallic Bearings

	PV_{max}		P_{max}		V_{max}	
Material	MPa · m/s	psi · fpm	MPa	psi	m/s	fpm
Carbon graphite	0.53	15,000	4.14	600	12.7	2500
Rubber	0.53	15,000	0.35	50	7.6	1500
Wood	0.42	12,000	13.8	2,000	10.2	2000
Nylon	0.11	3,100	6.9	1,000	5.1	1000
PTFE (Teflon)	0.035	1,000	3.45	500	0.51	100
Reinforced PTFE	0.35	10,000	17.2	2,500	5.1	1000
PTFE fabric	0.88	25,000	4.4	60,000	0.25	50
Laminated phenolics	0.53	15,000	41.4	6,000	12.7	2500
Polycarbonate	0.11	3,100	6.9	1,000	5.1	1000
Acetal resin	0.11	3,100	6.9	1,000	5.1	1000

lubricant. They are lightweight, easy to machine, and inexpensive. On the other hand, their poor thermal conductivity limits their use to applications where relatively little frictional heat has to be dissipated.

Despite the switch to plastics for many applications, wood bearings—especially rock maple vacuum impregnated with a mixture of oil, wax, and additives—are still used widely in conveyors and in paper and textile machines. They have some of the advantages of porous metal bearings and are quiet running as well. When the shaft rotates up to speed and the bearing is heated, oil flows from the pores to supply an adequate lubricant film for low-load, medium-speed operations. When the shaft stops and the bearing cools, up to 99% of the lubricant is reabsorbed. Wood bearings have also been used in home appliances, office machines, and agricultural machines.

EXAMPLE 17-2

What should be the length of a nylon bearing to support a load of 80 N at a shaft speed of 10 rps? The shaft diameter is 20 mm.

SOLUTION

From Table 17-7 it is learned that one should design for $P_{max} = 6.9$ MPa, $V_{max} = 5.1$ m/s, and $PV_{max} = 0.11$ MPa · m/s.

(*a*) The rubbing speed is, from Eq. 17-5,

$$V = \frac{\pi D n}{1000}$$

$$= \frac{\pi(20 \text{ mm})(10 \text{ rps})}{1000} = 0.628 \text{ m/s}$$

This is well within limits.

(*b*) From the PV_{max} factor, the safe limit for unit bearing load P is

$$P = \frac{PV_{max}}{V}$$

$$= \frac{0.11 \text{ MPa} \cdot \text{m/s}}{0.628 \text{ m/s}} = 0.175 \text{ MPa}$$

Since this is well under P_{max}, it may be used as the basis for determining bearing length.

(*c*) From Eq. 17-4,

$$L = \frac{W}{PD} = \frac{80 \text{ N}}{(0.175 \text{ MPa})(20 \text{ mm})} = \mathbf{23\ mm}$$

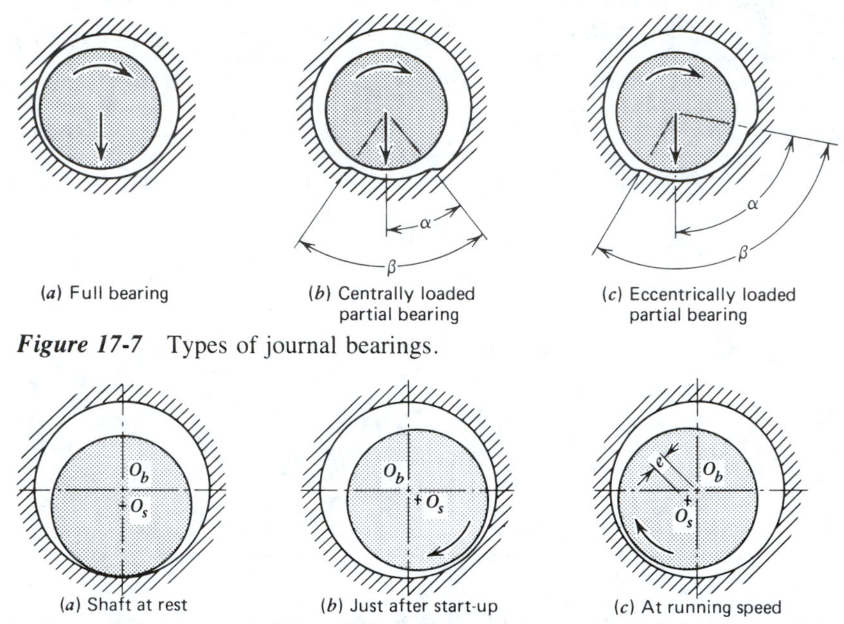

Figure 17-7 Types of journal bearings.

Figure 17-8 Formation of a fluid film in a journal bearing.

In an article by Donald Carswell, a more precise design method for plastic bearings is outlined that takes into account both temperature and duty cycle.[3]

17-7 LUBRICATION OF JOURNAL BEARINGS FOR SEVERE SERVICE

For the rest of this chapter, the objective is to learn how to design journal bearings to produce a thick lubricant film that completely separates the shaft and bearing surfaces. Such lubrication is essential for the more severe service conditions designers occasionally confront.

A journal bearing is said to be a *full bearing* if the active bearing arc extends a full 360 deg. Otherwise, it is a *partial bearing*. A full bearing is illustrated in Fig. 17-7*a*. Two partial bearings are shown in Figs. 17-7*b* and 17-7*c*. If the load line bisects the active bearing arc, the partial bearing is *centrally loaded*. Otherwise, it is *eccentrically loaded*. In the following section design charts are given for full bearings. Similar charts for partial bearings can be found in Reference 4.

While an external pump may be used to supply a lubricant under pressure to the bearing's feed hole, within the bearing itself it is the shaft that acts as a pump and pumps the oil adhering to it into the wedge-shaped oil film that supports the load. Examine Fig. 17-8. With the shaft stationary, the shaft simply rests on the bottom of the bearing. But at start-up the shaft begins to roll up the bearing wall. As it climbs, it also begins to pump oil between itself and the bearing. As this oil is pumped, the shaft lifts off the bearing surface and moves to the left. At operating speed, the shaft has developed a wedge-shaped film between itself and the bearing that supports the shaft and its load. The radial displacement of the shaft's center from the bearing's center is the eccentricity, e.

A typical pressure distribution in the oil film is

---

[3]Donald D. Carswell, "PV Ratings for Plastic Bearings," *Machine Design*, January 23, 1975.

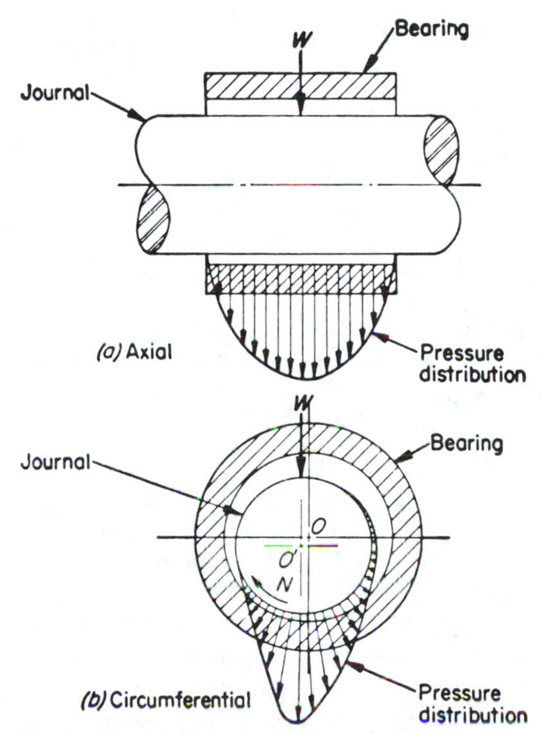

Figure 17-9 Typical pressure distribution in a journal bearing. (Courtesy of Cast Bronze Bearing Institute.)

shown in Fig. 17-9. Here both the axial and the circumferential distributions are shown. The pressure distribution actually achieved depends on factors such as shaft speed, load, lubricant viscosity, bearing clearance, and length-to-diameter ratio.

To describe bearing design and operation, special notation is needed. Many of the more important quantities are defined graphically in Fig. 17-10.

The thickness h of the oil film at any position θ is given by

$$h = C + e \cos \theta \tag{17-9}$$

Since θ is measured from the line of centers, the minimum oil film thickness occurs at $\theta = 180$ deg. It is

$$h_o = C - e \tag{17-10}$$

Equation 17-10 can be written in a different form by using the dimensionless eccentricity ratio $\epsilon = e/C$.

$$h_o = C - e = C - C\epsilon = C(1 - \epsilon) \tag{17-11}$$

Achieving a large enough value of h_o, the minimum thickness of the oil film, is one of the principal design objectives. The film must be thick enough to separate completely the shaft and bearing surfaces and avoid any metal-to-metal contact. To predict oil film thickness for a bearing of given design and for given operating conditions, one of the Raimondi and Boyd design charts is used. These charts are introduced in the next section.

17-8 JOURNAL BEARING DESIGN CHARTS

Our understanding of the behavior of oil films in journal bearings began with the experiments of Beauchamp Tower in England in the 1880s. Tower had been asked to study railroad journal

Figure 17-10 Nomenclature for a journal bearing.

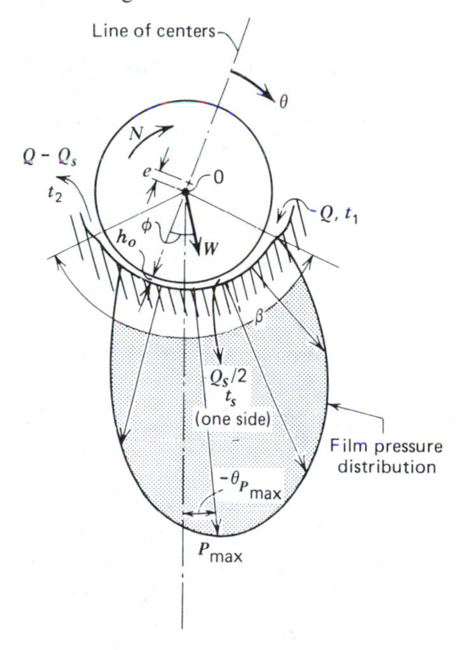

bearings and to determine how best to lubricate them. He studied a bath-lubricated, centrally loaded partial bearing (β = 157 deg) with a 4-in. diameter and a length of 6 in.

Tower was intrigued by the very low coefficient of friction. Apparently, unaware of the substantial pressure developed in a lubricant film, he drilled a $\frac{1}{2}$-in. diameter "lubricator hole" in the top! Of course, when the shaft was started up, oil flowed rapidly out of the hole. He put in a cork to stop the flow, but that popped out immediately, so he drove a wooden peg firmly into the hole with a mallet. That, too, was pushed out!

It became clear to Tower that he had made a discovery. He attached a pressure gauge to the hole and recorded a pressure more than twice the unit bearing load P. Finally, he investigated the bearing film pressure in detail.

Osborne Reynolds, the great English physicist, learned of Tower's experimental results and set about to find a mathematical explanation of this phenomenon. He derived an equation based on several simplifying but reasonable assumptions.

1. The lubricant is Newtonian; that is, the resistance to shearing is simply proportional to the rate of shearing strain.
2. Forces causing any acceleration of the lubricant are negligible compared to other forces involved.
3. The lubricant is incompressible.
4. The viscosity is constant throughout the oil film. (This assumption is obviously at variance with the facts, but we get around it in design by using a mean value of the oil film temperature.)
5. The pressure is uniform in the axial direction.

Reynolds' equation, as it is now called, has resisted all attempts to find an explicit solution by means of classical mathematics. In this chapter the numerical solution achieved by Raimondi and Boyd at the Westinghouse Research Laboratories is used. Their solution has been organized in easy-to-use design charts by plotting the results in terms of dimensionless groups of variables.

Early in the twentieth century, a German physicist, Sommerfeld, solved mathematically a simpler version of Reynolds' equation, one in which flow out the ends was neglected. Mathematically, this was equivalent to assuming an infinitely long bearing. He organized his solution in terms of a dimensionless bearing characteristic number.

$$S = \left(\frac{R}{C}\right)^2 \frac{\mu n}{P} \tag{17-12}$$

This quantity is now usually called the *Sommerfeld number*. Most of the basic design parameters for journal bearings are incorporated into this dimensionless quantity. Two additional quantities needed are the extent of bearing arc β and the length-to-diameter ratio L/D.

Although Eq. 17-12 defines the Sommerfeld number S, it is not the most practical form to use, since it assumes units of viscosity that are not very convenient. Here is an alternate method for calculating the Sommerfeld number. First, calculate the dimensionless clearance factor m.

$$m = 1000 \frac{C}{R} \tag{17-13}$$

where

C = radial clearance; mm, in.
R = bearing radius; mm, in.

Then the equation for S is, in metric units,

$$S = \frac{\mu n}{1000 m^2 P} \tag{17-14}$$

where

μ = viscosity; mPa · s
n = shaft speed; rps
P = unit bearing load; MPa

The comparable equation in English units is

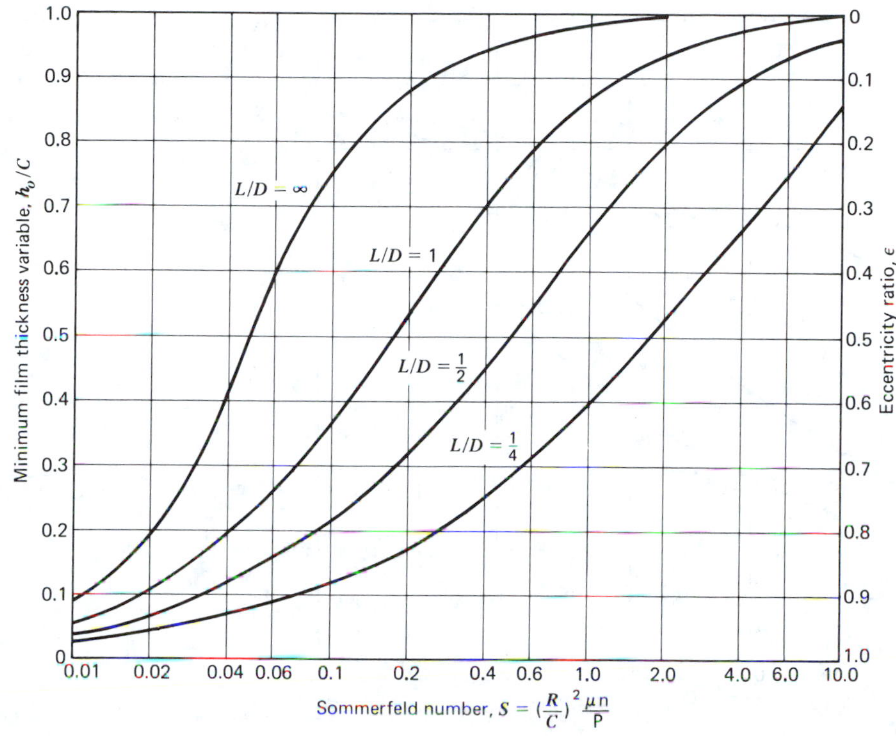

Figure 17-11 Minimum oil film thickness in a full journal bearing.

$$S = \frac{\mu n}{m^2 P} \tag{17-15}$$

where

μ = viscosity; μreyn
n = shaft speed; rps
P = unit bearing load; psi

Raimondi and Boyd have plotted their computer results in design-oriented graphs that enable us to determine things such as minimum oil film thickness, location of that minimum thickness relative to the load line, the coefficient of friction, the total flow rate of lubricant through the film, and the portion of lubricant flow that leaks out the ends of the bearing. They have organized their charts by making use of the Sommerfeld number. A set of design charts for 360-deg full bearings is given here in Figs. 17-11 to 17-15. In each instance curves are shown for a number of L/D ratios.

To find the minimum oil film thickness, Fig. 17-11 is used. The designer simply multiplies the dimensionless film thickness h_o/C found there by the actual radial clearance C to find the actual film thickness h_o. The location of the thinnest part of the oil film, in degrees from the load line and in the direction of shaft motion, is found in Fig. 17-12.

To estimate the rate at which heat is generated by fluid friction, the product of frictional force and surface velocity can be used.

$$\begin{aligned} H_f &= fWV \\ &= fW(2\pi Rn) \\ &= 2\pi fRWn \end{aligned}$$

Since the coefficient of friction given in Fig. 17-13 is in the *dimensionless* form Rf/C, it is convenient to reorganize the equation for H_f to use the dimensionless friction factor directly. The result is, in metric units,

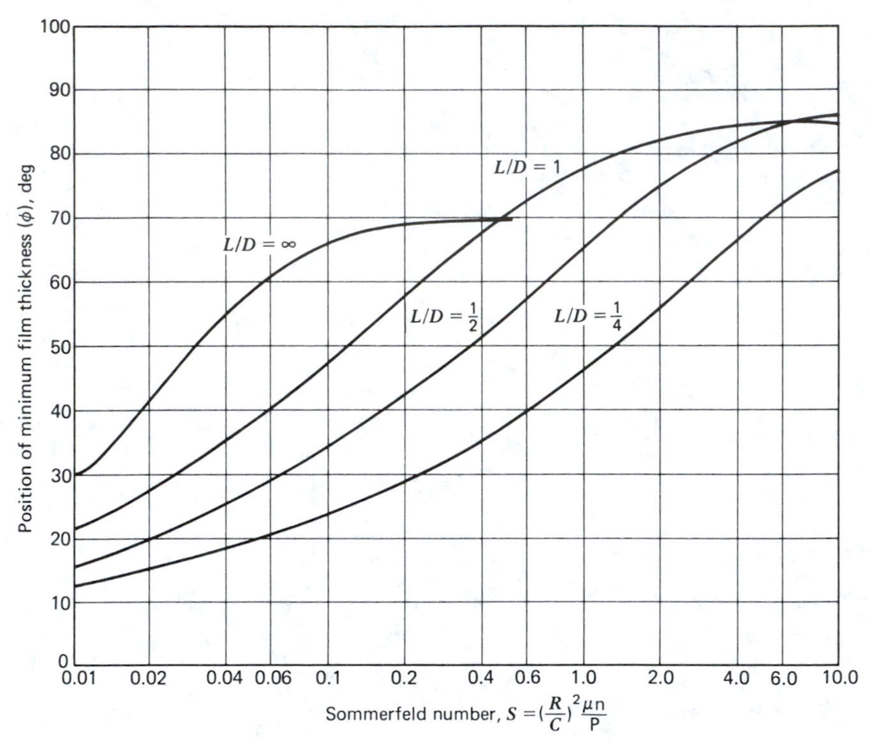

Figure 17-12 Position of the minimum oil film thickness in a full journal bearing.

$$H_f = \frac{WnC}{159.15}\left(\frac{R}{C}f\right) \tag{17-16}$$

where

H_f = rate of heat generation; W
W = bearing load; N
R = bearing radius; mm
C = radial clearance; mm
n = shaft speed; rps

In English units the corresponding equation is

$$H_f = \frac{WnC}{1485.9}\left(\frac{R}{C}f\right) \tag{17-17}$$

where

H_f = rate of heat generation; Btu/s
W = bearing load; lb
R = bearing radius; in.
C = radial clearance; in.
n = shaft speed; rps

The rate at which lubricant is pumped by the shaft is found in Fig. 17-14. Some of this lubricant flows out the *ends* of the bearing and has to be replaced by supply oil. The proportion that leaks out the ends is found in Fig. 17-15.

It is too soon to use these charts to design bearings. First we will have to grapple with the problems of lubricant flow rate and heat balance, problems taken up in the next two sections. And some guidance also will be needed in selecting basic design parameters. This subject is also discussed in a future section. So what are these design charts good for now? They spell out the behavior of a bearing of some *assumed* design. The design process requires us first to make some educated guesses, to use the charts to determine the bearing's performance and then, by a *systematic trial-and-error process*, to revise the bearing design until it is completely satisfactory. Meanwhile, the use of the charts is now demonstrated by a problem in which the behavior of a bearing of assumed design is investigated.

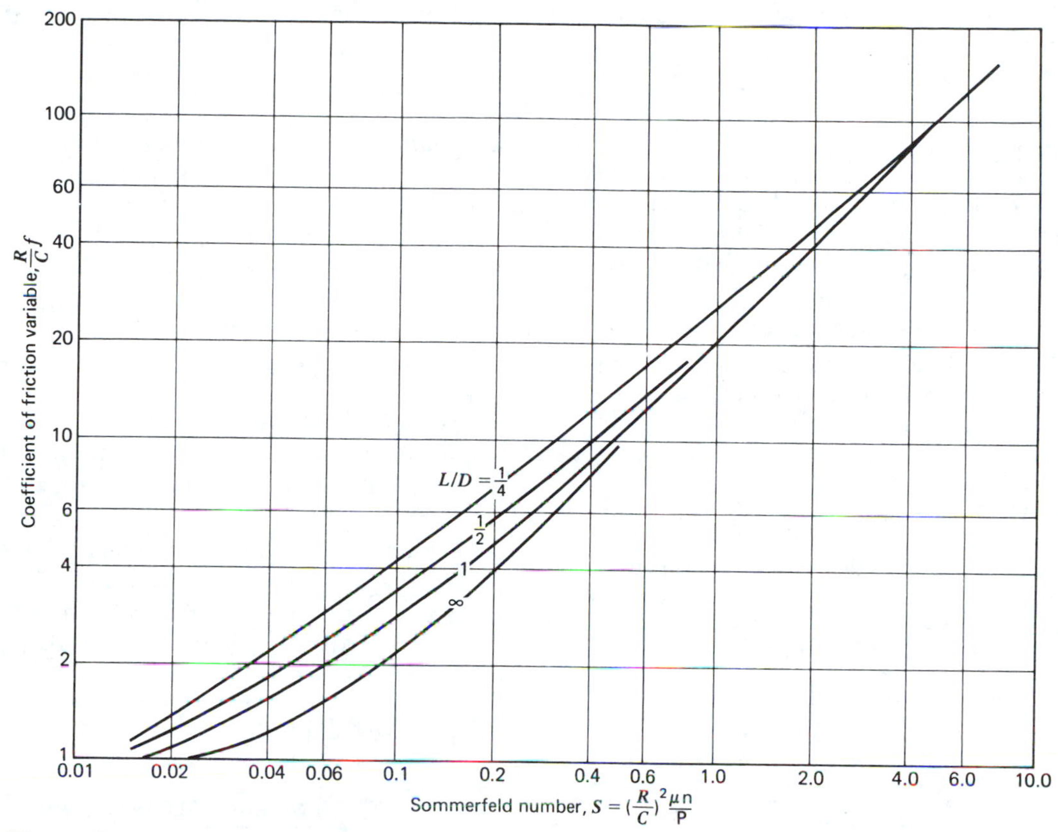

Figure 17-13 Coefficient of friction for a full journal bearing.

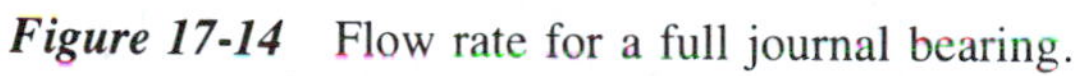

Figure 17-14 Flow rate for a full journal bearing.

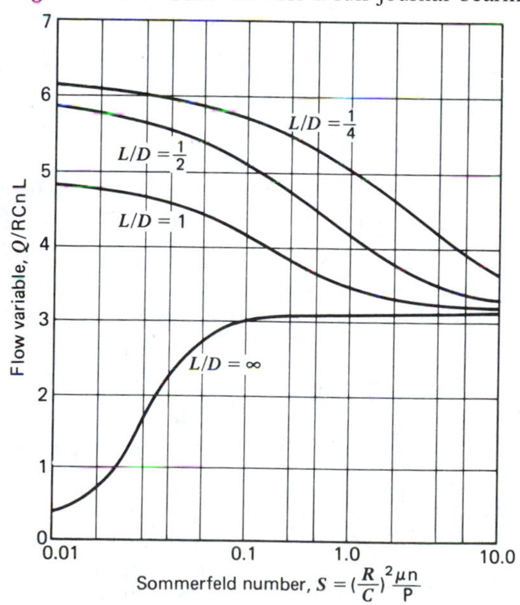

Figure 17-15 Side flow in a full journal bearing.

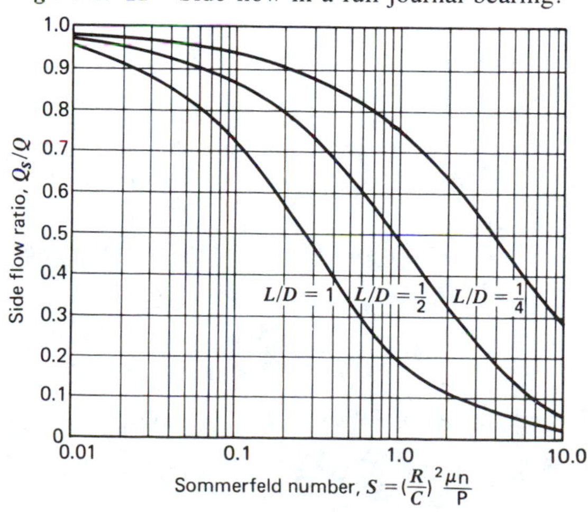

EXAMPLE 17-3

A full bearing supports a shaft turning at 1800 rpm and carrying a load of 18 570 Newtons. The shaft diameter is 127 mm, and the bearing length is 63.5 mm. SAE 20 oil is used at an average film temperature of 90°C. For a radial clearance of 0.0635 mm, determine the minimum oil film thickness, the power loss, and the required lubricant flow rate.

SOLUTION

(*a*) *Sommerfeld number:* The shaft speed is

$$n = \frac{1800 \text{ rpm}}{60 \text{ s/min}} = 30 \text{ rps}$$

The unit bearing load is

$$P = \frac{W}{LD} = \frac{18\ 570 \text{ N}}{(127 \text{ mm})(63.5 \text{ mm})}$$

$$= 2.303 \text{ N/mm}^2$$

The clearance factor is

$$m = \frac{1000C}{R} = \frac{(1000)(0.0635 \text{ mm})}{(0.5)(127 \text{ mm})} = 1.00$$

From Fig. 17-3, the oil's viscosity is 7.3 mPa · s. The Sommerfeld number can now be calculated by use of Eq. 17-14.

$$S = \frac{\mu n}{1000 m^2 P}$$

$$= \frac{(7.3 \text{ MPa} \cdot \text{s})(30 \text{ rps})}{(1000)(1)^2(2.303 \text{ mPa} \cdot \text{s})}$$

$$= 0.0951$$

(*b*) *Length-to-diameter ratio:* From the given data,

$$\frac{L}{D} = \frac{63.5 \text{ mm}}{127 \text{ mm}} = 0.5$$

When using the design charts for this bearing, data will be located for $S = 0.0951$ and $L/D = 0.5$.

(*c*) *Minimum oil film thickness:* From Fig. 17-11, the minimum oil film thickness variable, h_o/C, is 0.21. Therefore the minimum oil film thickness is

$$h_o = \left(\frac{h_o}{C}\right) C = (0.21)(0.0635 \text{ mm})$$

$$= \mathbf{0.0133 \text{ mm}}$$

The angular position of the minimum film thickness is about 33 deg from the load line, in the direction of motion, according to Fig. 17-12.

(*d*) *Power loss:* For $S = 0.0951$ and $L/D = 0.5$, the friction variable from Fig. 17-13 is 3.4. Using Eq. 17-16, the power loss due to lost heat generated by fluid friction is

$$H_f = \frac{WnC}{159.15}\left(\frac{R}{C}f\right)$$

$$= \frac{(18\ 570 \text{ N})(30 \text{ rps})(0.0635 \text{ mm})(3.4)}{159.15}$$

$$= \mathbf{755.8 \text{ W}}$$

The coefficient of friction is

$$f = \left(\frac{R}{C}f\right)\left(\frac{C}{R}\right)$$

$$= \frac{(3.4)(0.0635 \text{ mm})}{(0.5)(127 \text{ mm})} = \mathbf{0.0034}$$

(*e*) *Oil flow rate:* The dimensionless flow variable $Q/RCnL$ is found in Fig. 17-14. For this problem the value is found to be 5.4. Hence, the total oil pumped by the shaft into the pressurized lubricant film is

$$Q = \left(\frac{Q}{RCnL}\right)(RCnL)$$

$$= (5.4)[(63.5 \text{ mm})(0.0635 \text{ mm})(30 \text{ rps})(127 \text{ mm})]$$

$$= 82\ 960 \text{ mm}^3/\text{s} \quad \text{or} \quad 82.96 \text{ cm}^3/\text{s}$$

Some of this oil is recirculated by the

shaft back into the lubricant film, but the rest of it leaks out the ends. This end flow Q_s can be estimated by using Fig. 17-15. For $S = 0.0951$ and $L/D = 0.5$, the value of the dimensionless end-flow variable Q_s/Q is 0.87. Hence the end flow is

$$Q_s = \left(\frac{Q_s}{Q}\right)Q$$

$$= (0.87)(82.96\ \text{cm}^3/\text{s}) = \mathbf{72.18\ cm^3/s}$$

Assuming, as is usually the case, that there is no drainage groove, the quantity of oil that must be supplied to the bearing is **72.18 cm³/s.** The other 13% of the oil flowing in the bearing is recirculated.

This example was a useful *exercise* to introduce the design charts and their use. Why only an exercise? The average oil film temperature was given, but that is exactly what is seldom known in advance. The designer might indeed control the inlet temperature of the oil by means of a suitable oil cooler, but the *average* oil temperature is another matter. It depends on the frictional heat developed in the oil film and on the rate of lubricant flow through the film. These in turn depend on the average viscosity of the oil, which itself depends on the average temperature. The actual operating condition of a bearing must be established by trial and error. First, however, it is necessary to learn how to estimate the rate of oil flow in a bearing and the temperature rise of the oil in the bearing. The next two sections deal with these problems.

17-9 LUBRICANT FLOW IN PRESSURE-FED JOURNAL BEARINGS

When the designer has selected the bearing's basic configuration, one further problem is that of getting an adequate flow of lubricant through it. There are a variety of attractively simple techniques for introducing lubricant to the bearing without having to resort to pumping. In a gear reduction unit, for example, the bearings can often be adequately lubricated by bath oiling or by contriving to have the gears splash oil into channels in the housing that direct the lubricant to the bearings. In small to medium turbine-generator or motor-generator sets, one can often achieve a self-contained bearing design where neither pump nor oil cooler is needed by hanging a ring on the shaft. The bottom of the ring dips into an air-cooled reservoir of oil and picks up the oil for delivery to the top of the rotating shaft. There it is scraped off and directed into the clearance space. In other applications, where only a small flow of oil is needed, bottle oilers, drop-feed oilers, or even waste or pad oilers may suffice.

For high-load, high-speed applications in particular, the designer will often find it essential to supply lubricant under pressure to the bearing to achieve two principal objectives: (1) to maintain a thick enough lubricant film to separate completely the shaft and bearing surfaces; and (2) to maintain temperatures low enough to avoid oxidizing the lubricant or softening the bearing surface.

For a full journal bearing, Figs. 17-14 and 17-15 are used in flow calculations. For a typical full journal bearing, some of the lubricant recirculates so that the amount that must be introduced into the bearing is equal to the amount Q_s that leaks out both ends. The design charts make clear what the bearing needs. The designer must then ensure that the lubricant supply system will, in fact, meet the bearing's need.

For a bearing whose pressurized lubricant is introduced by means of a single feed hole, a single axial groove connected to a feed hole, or a circumferential groove, the flow rate is expressed by

$$Q_s = \frac{k_g m^3 D^3 (1 + 1.5\epsilon^2) p_s}{\mu} 10^{-6} \qquad (17\text{-}18)$$

where

Q_s = flow rate; cm³/s, in.³/s

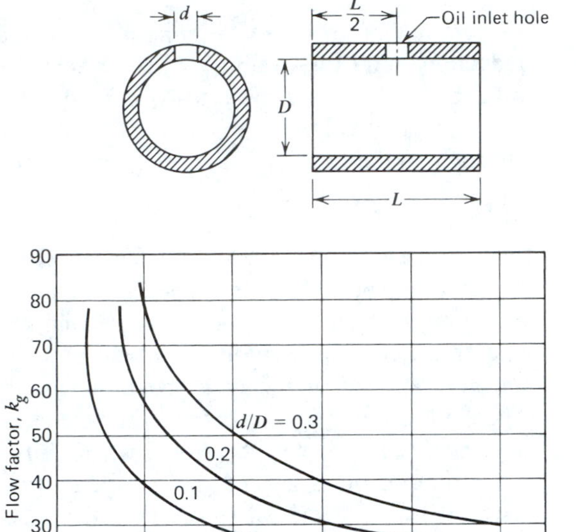

Figure 17-16 Flow factors for a full journal bearing with pressure feed through a single oil hole.

k_g = flow factor
m = clearance factor ($1000C/R$)
D = shaft diameter; mm, in.
ϵ = eccentricity ratio
p_s = supply pressure; MPa, psi
μ = viscosity; mPa · s, microreyns

The flow factor for a full journal bearing with pressure feed to a single inlet hole is plotted against the L/D ratio in Fig. 17-16. The size of the hole has a substantial effect on the flow rate. Curves for three relative hole sizes are shown. In practice, the designer knows the desired flow rate. Figure 17-16 and Eq. 17-18 are used to establish the proper hole size and inlet pressure.

For pressure feed to a single axial groove connected to a feed hole, Fig. 17-17 should be used. As the curves show, the flow factor is dependent on the L/D ratio and on the relative length, l/L, of the groove. Again, the real design problem is likely to be to estimate the inlet pressure needed to achieve a given flow rate.

Consider a main crankshaft bearing in a diesel engine. It is heavily loaded, and lack of space means it is also relatively short. To compound the problem, the load is changing in direction as the crankshaft turns. That is not an easy design problem. One key to success in designing this bearing is to get an ample supply of lubricant flowing through it; however, since the load is changing direction, there does not seem to be any very good place to locate an axial groove. No matter where it is placed, it will be interrupting the formation of a load-carrying lubricant film at least part of the time. A circumferential groove, as in Fig. 17-18, is a good solution. At first glance, it may seem like a poor solution, since it divides the already short bearing into two shorter ones. A glance at the design charts, especially at Fig. 17-11, shows that the effect of a lower L/D ratio is to reduce the oil film's thickness for a given load or to reduce the load capacity for a given film thickness. But if the designer does not provide an adequate flow of oil through the bearing, the lubricant will run

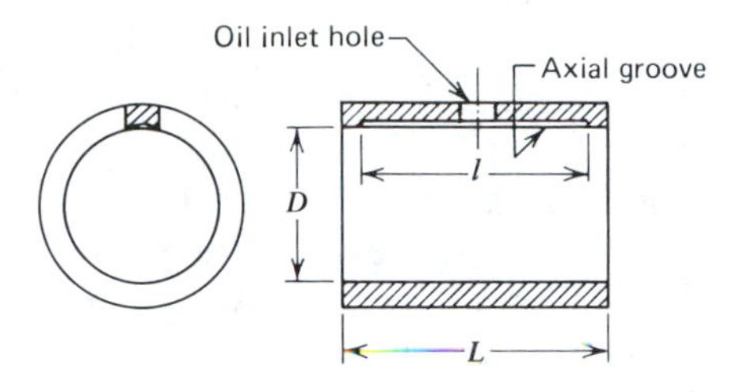

Figure 17-17 Flow factors for a full journal bearing with pressure feed through an axial groove.

at a high temperature. This reduces viscosity and lowers the load-carrying capacity. Experience with engine bearings over the years testifies eloquently to the need for the circumferential groove to obtain sufficient lubricant flow.

For a single inlet oil hole or for a single axial groove, Eq. 17-18, together with the appropriate flow factor chart, indicates the total lubricant flow out both ends of the bearing. In the case of the circumferential groove, Eq. 17-18 yields only the flow from one end. The flows out of the two ends must be summed to get the total flow. The flow factor for one end of a bearing having a circumferential groove is

$$k_g = \frac{32.725}{L'/D} \qquad (17\text{-}19)$$

where L' is the length of that side of the bearing.

EXAMPLE 17-4

An engine bearing has a diameter of 6 in. and a total length of 3.25 in. A 0.25-in. wide circumferential groove is centrally located. The radial clearance is 0.0036 in. SAE 20 oil is used at an average film temperature of 202°F. The inlet pressure of the oil is 60 psi. Find the flow rate for an eccentricity ratio of 0.8.

SOLUTION

Since the groove is centrally located, the length of each side of the bearing is

$$L' = \frac{(3.25 \text{ in.} - 0.25 \text{ in.})}{2} = 1.5 \text{ in.}$$

and the length-to-diameter ratio is

Figure 17-18 Pressurized full journal bearing with circumferential groove.

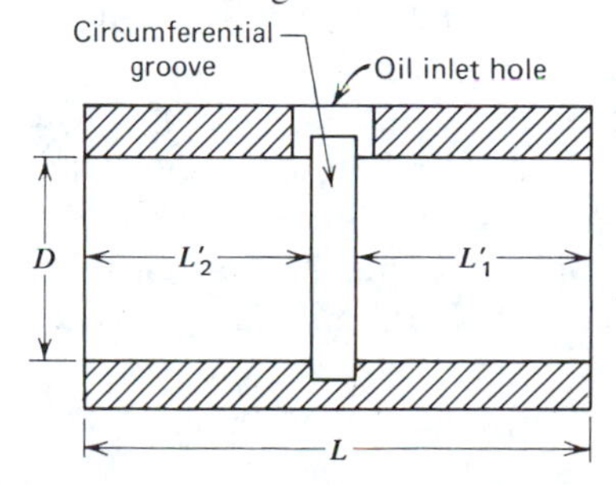

$$\frac{L'}{D} = \frac{1.5 \text{ in.}}{6.0 \text{ in.}} = 0.25$$

From Eq. 17-19, the flow factor is

$$k_g = \frac{32.725}{0.25} = 130.9$$

The clearance factor is

$$m = \frac{1000C}{R} = \frac{1000(0.0036 \text{ in.})}{3 \text{ in.}} = 1.2$$

The oil's viscosity (Fig. 17-4) is 0.76 microreyns, so the flow rate is

$$Q = \frac{(130.9)(1.20)^3(6)^3[1 + 1.5(0.8)^2](60)10^{-6}}{0.76}$$

$$= 7.56 \text{ in.}^3/\text{s}$$

The total flow rate is about **15.1 in.^{3}/s.**

17-10 TEMPERATURE RISE AND HEAT BALANCE FOR PRESSURE-FED JOURNAL BEARINGS

In dealing with the calculations of lubricant flow in the last section, we had to assume a mean temperature for the lubricant flowing through the bearing in order to look up the lubricant's viscosity. In practice, one can establish the mean lubricant temperature only by trial and error. In pressure-fed bearings designers can arrange to cool the lubricant and thus control its inlet temperature, but it is the mean temperature that must be known to establish the lubricant's mean viscosity. We could get an average temperature readily enough if we knew the temperature rise. To calculate that temperature rise, however, requires information that can be found only in terms of some assumed mean viscosity. Thus designers confront the familiar logical circle they enter by making a guess—in this instance a guess as to mean film temperature—and then complete by successively refining that guess iteratively.

The strategy used to discover the mean temperature at which the oil flows through the bearing is simple. Set the rate at which energy is carried away by the oil equal to the rate at which heat is generated by fluid friction within the bearing. This is, admittedly, a conservative approach; in reality, one can expect the shaft and the bearing housing to carry away some of the heat. To estimate how much heat the shaft and bearing housing carry away is difficult, although Section F7 in *The Tribology Handbook* offers some useful counsel. Assuming that the lubricant has the full responsibility for carrying away the heat generated greatly simplifies the analysis and produces realistic results.

In this design it is assumed that no drainage groove is to be used so some of the lubricant recirculates; the rest leaks out the ends. If the total temperature rise of the oil in the bearing is Δt, it is clear that the average temperature rise of the oil leaving the bearing at its ends is less than Δt. Thus, for design purposes, many designers assume that the average temperature rise of the oil flowing out the ends is $\Delta t/2$. This is the approach taken here.

The rate at which the bearing dissipates heat, then, is the rate at which heat is absorbed by the oil flowing out the ends of the bearing. This is the product of flow rate, specific heat, density, and effective temperature rise. That is,

$$H_d = \frac{c\rho Q_s \Delta t}{2} \tag{17-20}$$

where

H_d = rate of dissipation; W, Btu/s
c = specific heat of lubricant; J/g · °C, Btu/lb-°F
ρ = density of lubricant; g/cm^3; lb/in.^{3}
Δt = total temperature rise of lubricant; °C, °F
Q_s = volumetric flow rate of lubricant; cm^3/s, in.^{3}/s

For design purposes it is satisfactory to use average values for density and specific heat of lubricating oils. An SAE motor oil has a density of about 0.85 g/cm^3 or 0.0307 lb/in.^{3}, and its specific heat is about 1.842 J/g · °C or 0.44 Btu/lb-°F.

The rate at which heat is generated in the bearing by fluid friction is given by Eq. 17-16 or 17-17.

EXAMPLE 17-5

Consider the full journal bearing previously dealt with in Ex. 17-3 (W = 18 570 N, D = 127 mm, L = 63.5 mm, C = 0.0635 mm, m = 1.00, n = 30, L/D = 0.50, and P = 2.303 MPa). It is further assumed that oil under pressure is provided to the bearing by means of a single axial groove (l/L = 0.9) and that the net oil flow through the bearing is simply whatever flows out the ends. The rest is recirculated. What will be the maximum oil film temperature if SAE 20 oil is supplied at an inlet pressure of 0.35 MPa and an inlet temperature of t_1 = 90°C?

SOLUTION

The solution is shown in Table 17-8. Here are some brief comments to guide your study of the table.

Line 1. Initially, a guess of 20°C was made for the total temperature rise. As a second guess, a value near the average of the initial guess and the calculated value (line 9) was used.

Line 2. The average temperature of the oil leaking out of the ends is assumed to be equal to the inlet temperature plus half the maximum temperature rise.

Line 3. The oil's viscosity is found for the average temperature calculated in line 2.

Line 4. The Sommerfeld number is calculated using the form in Eq. 17-14.

Line 5. For the Sommerfeld number just calculated and for L/D = 0.5, the dimensionless friction factor is found in Fig. 17-13.

Line 6. Now the rate at which heat is generated by fluid friction can be calculated by Eq. 17-16.

Line 7. For the Sommerfeld number calculated in line 4 and for L/D = 0.5, the eccentricity ratio can be found in Fig. 17-11.

Line 8. The rate of oil flow through the bearing Q_s depends on several design parameters, but also on two operational variables: viscosity and eccentricity ratio ϵ. These are determined in lines 3 and 7. The value of the flow coefficient k_g for L/D = 0.5 and l/L = 0.9 is found to be 115.5 in

TABLE 17-8 Solution of Example 17-5

Line	Quantity	Units	Method	Iterations	
				1	2
1	Δt	°C	Guess	20	23
2	t_{ave}	°C	$t_s + \Delta t/2$	100	101
3	μ	mPa · s	Fig. 17-3	5.4	5.3
4	S		Eq. 17-14	0.0703	0.0690
5	$(R/C)f$		Fig. 17-13	2.65	2.62
6	H_f	W	Eq. 17-16	589	582
7	ϵ		Fig. 17-11	0.825	0.83
8	Q_s	cm^3/s	Eq. 17-18	31.0	31.8
9	Δt	°C	$2H_f/\rho c Q_s$	24.3	23.4
10	t_{max}	°C	$t_s + \Delta t$	114	113

Fig. 17-17. Equation 17-18 is used to calculate the flow rate once eccentricity and viscosity have been determined.

Line 9. To find the temperature rise, a heat balance is made by setting equal the rate of heat generation H_f (Eq. 17-16) and the rate of heat dissipation (Eq. 17-20) and solving for the maximum temperature rise Δt.

Line 10. Finally, the maximum oil film temperature is found by adding the maximum temperature rise to the inlet temperature.

For the first iteration a total temperature rise of 20°C was assumed. The calculated value is 24°C. For the next iteration, a value of 23°C was used as the initial guess. As luck would have it, this turned out to be the calculated value as well, so no further iterations were necessary. Two or three iterations frequently suffice for these problems. The maximum temperature of the oil film is estimated to be **113°C.**

17-11 PRACTICAL CHOICES OF DESIGN PARAMETERS

To some degree, the analyses explained and illustrated in the previous sections tell designers whether the choices of design parameters made for a journal bearing are practical. Still, there are some things these analyses do not reveal. For example, for an assumed bearing design one can calculate the *minimum* oil film thickness, but how thick *should* it be? Based on design experience, experimental investigations, and analytical studies, there are some guidelines designers can use in making preliminary choices of basic bearing dimensions and in selecting limiting conditions for the design.

Length-to-Diameter Ratio

Typically, the shaft diameter is determined by considerations other than bearing capacity. In a gear reduction unit, for example, the shaft diameters are usually set by the requirement of getting sufficient shaft stiffness to keep the misalignment of the gear teeth within strict bounds. An engine's crankshaft, on the other hand, is more likely to have its sizes set by strength requirements. It is the bearing length that designers get to choose, even though space limitations may place severe upper limits on that length.

One might suppose that the longer the bearing the better. After all, the longer the length, the lower the unit bearing load ($P = W/LD$) will be. On the other hand, the longer the bearing, the more likely it is that shaft deflection will lead to metal-to-metal contact at the ends. Also, the longer the bearing, the more difficult it is to get sufficient lubricant flow through it. In the past three decades the trend has been to L/D ratios of 1 or less. A value of L/D greater than 1 is feasible where shaft alignment and cooling the bearing do not present great problems. However, values of L/D less than 1 are needed where shaft deflection and rate of lubricant flow are primary considerations or where there are severe space limitations.

One useful criterion that places a lower limit on bearing length is concerned with start-up. The unit bearing load during start-up should be no greater than 2.1 N/mm^2 or 300 psi. Start-up load is ordinarily that caused solely by static loads such as the shaft's weight.

Maximum Oil Film Temperature

It has already been mentioned that the oil will oxidize excessively if it reaches a maximum operating temperature above about 121°C. If the bearing surface is a soft metal such as a babbitt alloy (tin with antimony, copper, and lead added), it is also necessary to keep the maximum oil film temperature under the softening temperature for the babbitt at the given film pressure. For the babbitts most commonly used, the temperature of softening ranges from 191°C for bearing loads about 1.38 N/mm^2 or less (375°F at 200 psi) to 127°C for steady loads of about 6.9

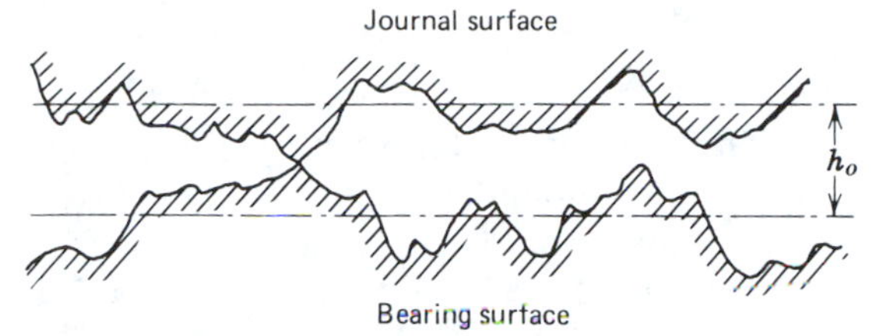

Figure 17-19 Minimum oil film thickness for actual surfaces.

N/mm^2 (260°F at 1000 psi). And, of course, some margin of safety is wanted, so a widely used rule of thumb is to keep the maximum oil film temperature under 93°C (200°F) for babbitt bearings.

Minimum Oil Film Thickness

The surfaces of journal and bearing as they might appear under a microscope are shown in Fig. 17-19. For each surface a centerline is indicated. The oil film thickness required to avoid metal-to-metal contact of all but the highest peaks is h_o. For design purposes, the minimum film thickness should be larger than this to provide some margin of safety. If we can estimate the heights, above the centerline, of the largest peaks on each surface, we have a very good clue as to the magnitude of the oil film thickness required to avoid any metal-to-metal contact. The oil film should also be thick enough to pass particles that escape the oil filter.

Table 17-9 shows some typical root-mean-square (RMS) values of roughness for several of the finer finishing methods. (The root-mean-square value is found by taking the average of the *squares* of a number of height measurements and then taking the square root of this average.) Also listed are the heights of the highest peaks. As a rule, estimate the peak roughness for shaft and for bearing, add them, and double this sum to allow a margin of safety.

EXAMPLE 17-6

What minimum oil film thickness should be maintained in designing a journal bearing if the shaft is precision ground and the bearing surface is lapped?

SOLUTION

Use

$$h_o \geq 2[(4.5)(16) + (6.5)(16)] = \mathbf{352\ \mu in.}$$

Radial Clearance

The normal range for the clearance ratio C/R is 0.001 to 0.002. The larger values are needed sometimes to obtain a greater oil flow through the bearing for cooling. A good initial guess is 0.001 to 0.0012 for the ratio C/R. Of course,

TABLE 17-9 RMS and Peak Roughness of Finely Finished Surfaces

Finishing Method	RMS Roughness (μm)	RMS Roughness (μin.)	Peak Roughness
Commercial grinding	0.4–1.6	16–63	4.5 × RMS
Precision grinding	0.1–0.4	4–16	4.5 × RMS
Honing and lapping	0.1–0.4	4–16	6.5 × RMS
Polishing and buffing	0.05–0.4	2–16	7.0 × RMS
Super finishing	0.025–0.2	1–8	7.0 × RMS

since both shaft and bearing will have to be manufactured to some tolerances, there is a range of clearances. Designers must design the bearing so that it will operate satisfactorily at both the minimum possible clearance and the maximum possible clearance.

17-12 A JOURNAL BEARING DESIGN EXAMPLE

A full journal bearing operates under a load of 752 N at a speed of 133.33 rps (8000 rpm). The shaft diameter is 38 mm. The bearing length cannot exceed 25 mm due to severe space limitations. Inlet oil can be supplied at a temperature of 51°C and a pressure of up to 0.59 MPa. The static load is 365 N. It has been decided to use a bronze bearing with a centrally located circumferential groove 6 mm wide. Thus each side of the bearing supports 376 N. The shaft is precision ground and the bearing is lapped.

Bearing Lengths

Since a groove width of 6 mm is to be used, there will be effectively two bearings sharing the load, each $L' = 9.5$ mm long at the maximum. This corresponds to an L'/D ratio of 0.25. Let us use the maximum length.

Unit Load

During start-up, the unit load will be the static load. For each of the two bearing sections

$$P_{st} = \frac{W'_{st}}{L'D}$$

$$= \frac{(365/2)}{(9.5)(38)} = 0.51 \text{ MPa}$$

This is satisfactory. At full speed, the unit load will be

$$P = \frac{W'}{L'D} = \frac{(752/2)}{(9.5)(38)} = \mathbf{1.04\ MPa}$$

Radial Clearance

Initially, a large radial clearance is assumed. The high shaft speed promises to cause cooling problems unless steps are taken to ensure a copious supply of lubricant. Let us start with $C/R = 0.002$. This implies $C = 0.002(38/2) = 0.038$ mm and $m = 1000C/R = 1000(0.038)/19 = 2$.

Minimum Oil Film Thickness and Eccentricity Ratio

Since the shaft is precision ground and the bearing is lapped, the minimum oil film thickness, using data from Table 17-9, should be

$$h_o = 2[(4.5)(0.4) + (6.5)(0.4)]$$

$$= 8.8\ \mu\text{m} \quad \text{or} \quad 0.0088 \text{ mm}$$

Since

$$h_o = C(1 - \epsilon)$$

the maximum permissible eccentricity ratio is

$$\epsilon_{max} = 1 - \frac{h_{o,\min}}{C}$$

$$= 1 - \frac{0.0088}{0.038} = \mathbf{0.768}$$

According to Fig. 17-11 the corresponding Sommerfeld number is $S = 0.34$. The bearing must operate at a Sommerfeld number no less than this or the oil film will be too thin.

Oil Selection

The Sommerfeld number S is to be no less than 0.34. So, by Eq. 17-14, the minimum acceptable viscosity at the average temperature is

$$\mu = \frac{1000\ Sm^2P}{n}$$

$$= \frac{(1000)(0.34)(2)^2(1.04)}{133.33} = \mathbf{10.61\ mPa \cdot s}$$

Because the flow rate of oil through the bearing can be greatly controlled by regulating the inlet pressure, it is possible to assume a reasonable temperature rise for the oil and then calculate a corresponding inlet pressure. If this pressure is unreasonably high, however, a design revision will have to be made. For this problem a maximum oil temperature of 99°C is assumed. This corresponds to a temperature rise of 48°C and an average film temperature of 75°C. According to Fig. 17-3, at an average film temperature of 75°C, SAE 20 oil is viscous enough. Its viscosity is 11.5 mPa · s. The corresponding value of the Sommerfeld number is 0.369, higher than the 0.34 minimum. For $S = 0.369$ and $L'/D = 0.25$, the dimensionless friction factor is, according to Fig. 17-13, equal to 11.7. Equation 17-16 is used to determine the heat generation rate.

$$H_f = \frac{WnC}{159.15}\left(\frac{R}{C}f\right)$$

$$= \frac{(376)(133.33)(0.038)(11.7)}{159.15} = \mathbf{140\ W}$$

Required Oil Flow

It is assumed (conservatively) that the oil must carry away all of the heat generated. Consequently, the necessary flow rate Q_s for each of the two sides of the bearing is determined from Eq. 17-20 by setting the rate of heat dissipation H_d equal to the rate of heat generation H_f. Thus

$$Q_s = \frac{2H_f}{c\rho\,\Delta t}$$

$$= \frac{2(140)}{(0.85)(1.842)(62)} = \mathbf{2.88\ cm^3/s}$$

Minimum Inlet Pressure

It remains only to find the inlet pressure to supply sufficient lubricant to each of the two sides of the bearing. Here Eq. 17-18 is employed. Since a circumferential groove is being used, the flow factor k_g is evaluated by Eq. 17-19.

$$k_g = \frac{32.725}{L'/D} = \frac{32.725}{0.25} = 130.9$$

The required inlet pressure p_s is, then,

$$p_s = \frac{10^6 Q_s}{k_g m^3 D^3(1 + 1.5\epsilon^2)}$$

$$= \frac{(10^6)(2.88)(11.5)}{(130.9)(2)^3(38)^3[1 + 1.5(0.76)^2]}$$

$$= \mathbf{0.309\ MPa}$$

Since this is well below the maximum available inlet pressure, the design seems to be satisfactory. A higher inlet pressure could be used if desired to force even more lubricant through the bearing and bring down the oil film temperature. Because each side of the bearing requires at least 2.88 cm³/s, the oil pump for the bearing must have a capacity of at least **5.76 cm³/s.**

Effect of Shaft and Bearing Tolerances

The preceding calculations produce a satisfactory result for a nominal radial clearance of 0.038 mm. However, on consulting *Machinery's Handbook,* designers learn that for this given shaft size and for the given methods of finishing, commercially feasible tolerances for the diameters of shaft and bearing are 0.0102 and 0.0076 mm, respectively. This implies a total tolerance of 0.009 mm for the radial clearance. It is essential that the calculations for the inlet pressure be repeated for the two extremes of possible clearance.

This example has shown the principal steps to be taken in designing a pressure-fed journal bearing. There are, however, some additional details. The Cast Bronze Bearing Institute has compiled a group of publications that greatly assist the bearing designer. Reference 5 is a valuable manual for the design of bronze journal bearings.

SUMMARY

For light service, journal bearings can be operated with very little lubrication. Their selection is then based principally on the *PV* factor, the product of unit bearing load P and rubbing velocity V. But there are also individual limits on unit bearing load and rubbing velocity that must be respected. When the shaft is oscillating instead of rotating continuously, the average rubbing velocity during the motion cycle is used for design purposes.

Pressure-fed journal bearings make possible a complete separation of shaft and bearing by a relatively thick film of lubricant. This type of design is necessary for severe conditions of load and speed. The behavior of the bearing is predicted by use of a set of design charts based on a numerical solution of the Reynolds' equation. At the root of the method is the assumption that the heat generated by friction in the bearing is carried away by the oil flowing through it. By trial and error, the average oil film temperature is determined, as well as the power loss, the minimum oil film thickness, and the oil flow rate. The design is then revised as necessary to achieve both a sufficiently thick lubricant film and a safe lubricant temperature.

REFERENCES

17-1 Bierlein, John C. "The Journal Bearing." *Scientific American,* July 1975.

17-2 Carswell, Donald D. "PV Ratings for Plastic Bearings." *Machine Design,* January 23, 1975.

17-3 Neale, M. J., editor. *Tribology Handbook,* New York: Halstead Press, 1973.

17-4 Raimondi, A. A., and John Boyd. "A Solution for the Finite Journal Bearing and Its Application to Analysis and Design" (in three parts). *ASLE Transactions,* 1 (1958), pp. 159–209.

17-5 Rippel, Harry C. *Cast Bronze Bearing Design Manual,* Second Edition. Cleveland: Cast Bronze Bearing Institute, 1971.

PROBLEMS

P17-1 At an average temperature of 96°C, what SAE oil should be used to get a viscosity of about 10 mPa · s?

P17-2 For the oil in P17-1, what are the average film temperature and the viscosity in English units?

P17-3 What is the viscosity of SAE 50 oil at 208°F?

P17-4 What should be the length of a journal bearing to support a load of 3800 N at a unit load of no more than 0.8 MPa? The shaft diameter is 80 mm.

P17-5 A journal bearing 0.8 in. in diameter and 1.5 in. long supports a load of 131 lb. What is the PV factor at 1755 rpm? Which porous metal bearings are sturdy enough for this application?

P17-6 A gear shaft supported by two journal bearings is shown in Fig. 17-20. For each bearing the coefficient of friction is estimated to be $f = 0.025$. (*a*) What is the unit bearing load for each bearing? (*b*) What is the total power loss for the two bearings?

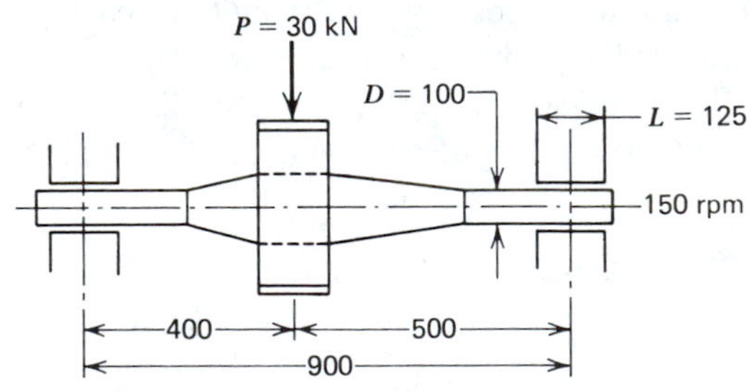

Figure 17-20 Illustration for Problem 17-6.

P17-7 In a journal bearing the shaft is operating at an eccentricity of 0.03 mm. The radial clearance is 0.05 mm. (*a*) What is the minimum oil film thickness? (*b*) What is the eccentricity ratio? (*c*) What is the oil film thickness at $\theta = 135$ deg?

P17-8 In a pressure-fed bearing the average oil film thickness is 90°C. The 45-mm diameter shaft turns at 900 rpm and supports a load of 3500 N. The bearing length is 45 mm and its radial clearance is 0.06 mm. If SAE 10 oil is used: (*a*) What is the minimum oil film thickness? (*b*) What is the angular position of the minimum oil film thickness? (*c*) What is the rate at which frictional heat is generated?

P17-9 For the bearing in P17-8, how much oil would be pumped through the bearing if an axial groove with the relative length $l/L = 0.8$ were used with an inlet pressure of 0.30 MPa?

P17-10 What minimum oil film thickness should be used for design purposes if a shaft is expected to have a surface roughness of 22 μin. root mean square and the bearing a surface roughness of 14 μin. root mean square? Assume the peak roughness is 4.5 times the root-mean-square roughness in each case.

P17-11 What should be the minimum length of a journal bearing if the 76-mm diameter shaft supports a static load of 7600 N? Use P_s = 2.1 N/mm².

P17-12 In a food machine a bearing is to support a load of 300 N. The shaft has a diameter of 20 mm and rotates at 64 rps. It will be only lightly lubricated. Select a suitable porous metal bearing and calculate the bearing length.

P17-13 A water-lubricated wood bearing in a flour mill's small hydraulic turbine must support a load of 231 lb at 580 rpm. There is room for a bearing up to 3 in. in diameter and up to 5.75 in. long. Select the diameter and the length for light service.

P17-14 A center main bearing in a diesel engine operates under a load of 8664 N at a shaft speed of 2100 rpm. The shaft diameter is 114 mm and the bearing length is 122 mm total. Oil can be supplied at 0.34 MPa and 80°C inlet conditions. Use a centrally located circumferential groove. Complete the design by selecting the SAE lubricant, the groove width, and the radial clearance. Evaluate your design by calculating the minimum oil film thickness, the maximum oil film temperature, and the power loss. Try to get a minimum oil film thickness of at least 0.023 mm.

P17-15 Design a pressure-fed journal bearing to support a load of 4134 lb. The shaft is 6 in. in diameter and rotates at 500 rpm. Oil can be supplied at 160°F and 60 psi gauge. A minimum oil film thickness of 0.0020 in. is to be maintained. Use an axial groove to supply the oil.

Chapter 18
Rolling-Element Bearings

The laws of nature are fixed and immutable.
They are also continually enforced.

ANONYMOUS

NOMENCLATURE

C = basic capacity corresponding to base revolutions
F_{rel} = factor of reliability
F_{th} = thrust factor
K (constant) = number of revolutions chosen as a basis
L_d = desired life
L_e = equivalent life
L_{rev} = total life in revolutions
L_w = weighted life
$L_1, \ldots, L_n$ = individual lives in each loading mode
p (dimensionless) = life exponent
R = radial load
RE = equivalent radial load
T = thrust load

Rolling bearings derive their load-carrying ability and high efficiency from the use of revolving hardened elements in rolling contact with hardened, circular tracks. They are preferred whenever there is a need to keep shafts in rigid placement with a minimum of trouble and attention (Fig. 18-1). In all heavy duty equipment they are preferred because of their high-load capacity and because of their low resistance to starting even under extreme loads (Fig. 18-2*a* and 18-2*b*). Finally, rolling bearings are relatively inexpensive. It is not uncommon to find the success of a multi-million dollar piece of equipment dependent upon a bearing costing only a few dollars (Fig. 18-3).

The principle of rolling element bearings is very simple, so simple, in fact, that nobody knows when or by whom it was first conceived. The transition from sliding to rolling was achieved by introducing rolling elements be-

Figure 18-1 Rolling bearings are used when there is a need to keep shafts in rigid placement with a minimum of trouble and attention, as in the worm gear reduction unit shown here. Joint lubrication of worm, gear, and the two roller bearings simplifies maintenance. Axially, each bearing is held in place by a shaft shoulder and a bolted cover. Radially, the bearings are restricted by the housing. (Courtesy The Timken Company.)

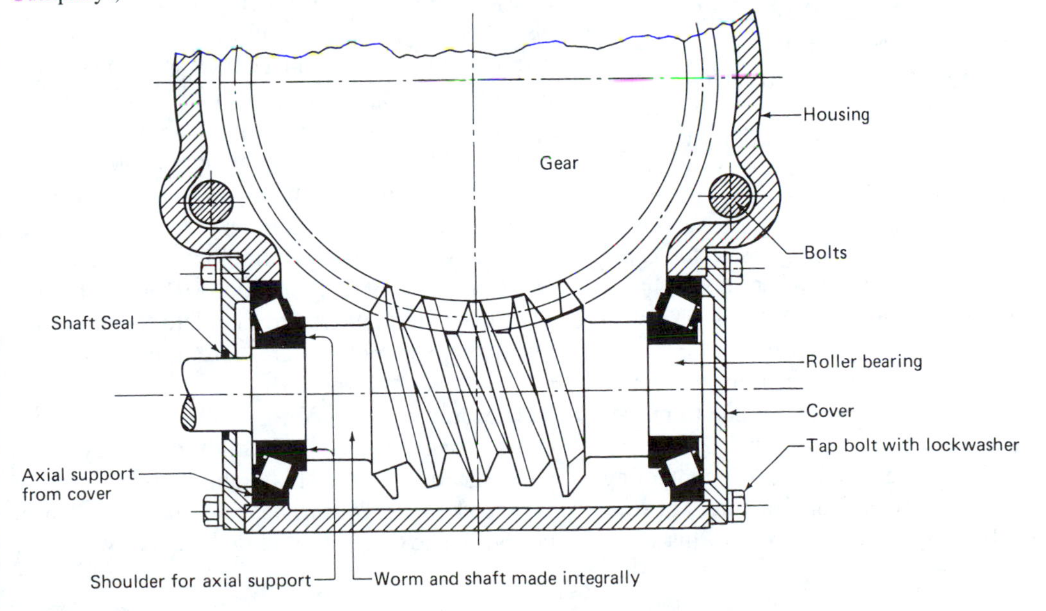

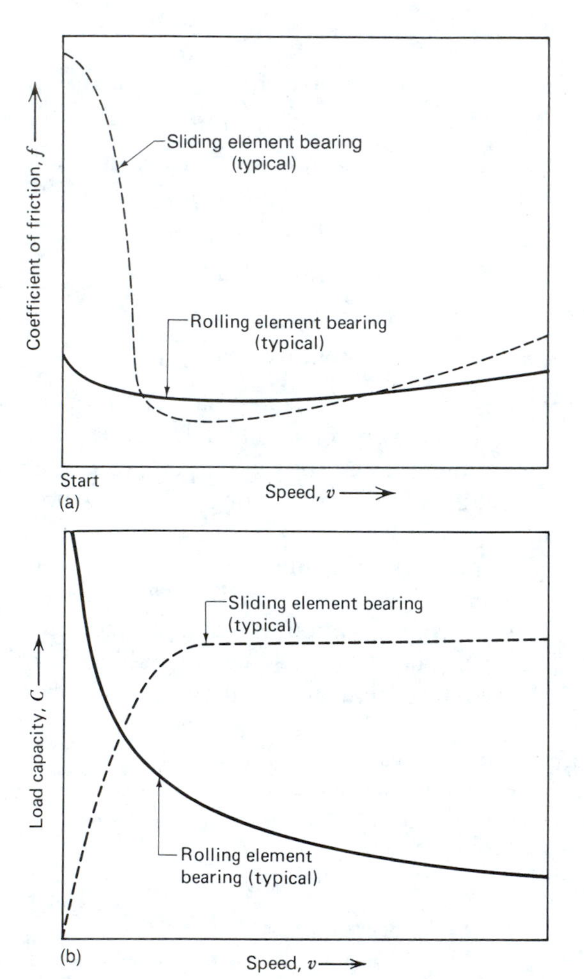

Figure 18-2 Resistance to rotation and load capacity as functions of speed. (*a*) Coefficient of friction versus speed. A low resistance to starting under high load makes rolling bearings preferable in vehicles, trains, airplanes, and mobile equipment in general. (*b*) Load capacity versus speed. High load capacity at low speeds makes rolling bearings ideal for crane hooks, turntables, and similar mechanisms requiring both high load capacity and rotational accuracy.

tween sliding surfaces. The resulting action is characterized by a very low resistance to movement that can be continually reduced by constantly improving the accuracy of contacting parts. As a matter of fact, manufacturers of rolling element bearings, capitalizing on this idea,

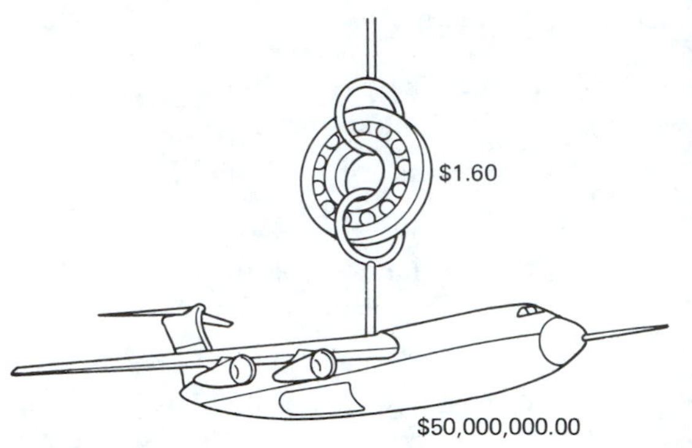

Figure 18-3 Aircraft dependence on bearings. (Courtesy American Society of Mechanical Engineers.)

often refer to their product as *antifriction bearings*. Strictly speaking, the low resistance to motion in a rolling element bearing is due to the low coefficient of rolling friction, although there is some rubbing friction present also. The coefficient of rolling friction is simply the ratio of force to sustain lateral motion (at constant velocity) divided by the gravity force of the object.

18-1 DESIGN AND CHARACTERISTICS

Rolling bearings are finished machine members ready for installation and immediate use (Fig. 18-4). Their simplicity of design permits precision manufacture and economical mass production. These advantages have made rolling element bearings very popular with manufacturers of appliances and many industrial products. To make this simple mechanism work in today's high-speed machinery requires an advanced manufacturing technology that can maintain tolerances of 2.5 μm (0.0001 in.) or better. The finest steel alloys are mandatory, and close metallurgical controls are essential to provide hardened contact surfaces.

The built-in precision of rolling bearings accounts for their incredibly high efficiency, usually exceeding 0.99. This high precision leads to another benefit: rolling bearings need *no* run-in period, since there are no high spots to wear off.

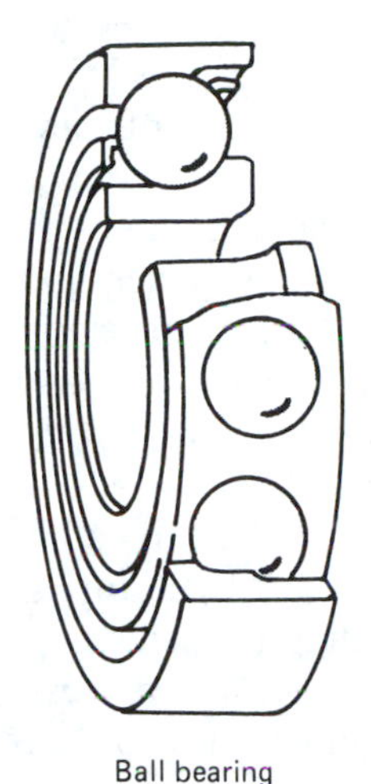

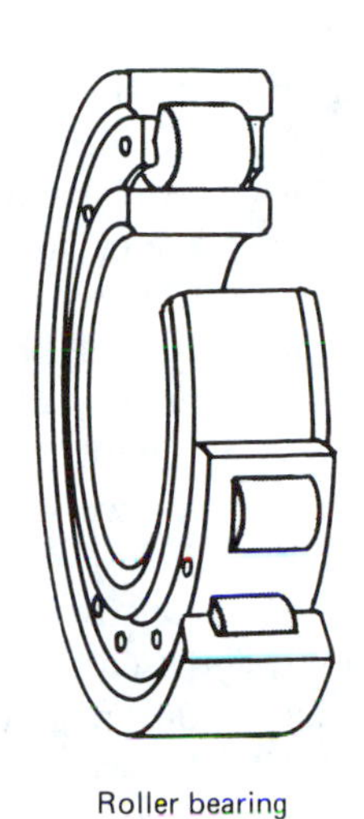

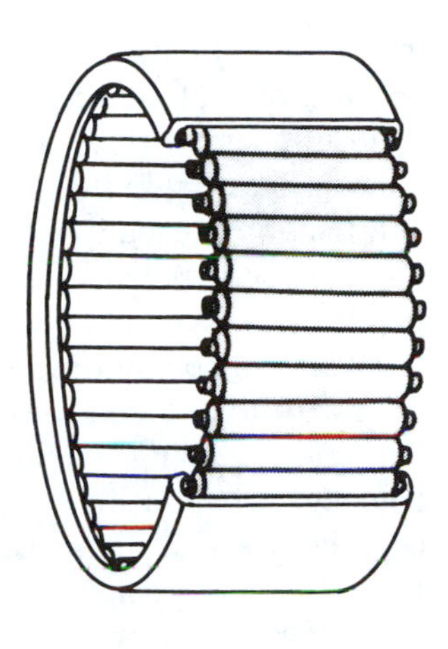

Figure 18-4 Three basic types of rolling element bearings. Note that the needle bearing does not show an inner race. The elements roll on the shaft itself. (Courtesy Deere and Company Technical Services.)

As seen in Fig. 18-5, most rolling bearings *consist of two hardened steel rings called races*. The races form a narrow, dual, circular channel or track within which the rolling elements are confined during operation. For minimum wear and maximum load capacity, the rolling elements are likewise made of hardened steel. To maintain alignment and ensure balance, the rolling elements are equally spaced around the track by means of a *separator* or *cage*.

The shaft and housing dimensions for mounting rolling bearings are very critical. Both housing bore and shaft diameter must be carefully selected and diligently maintained to ensure the proper assembly fit. If the fit is too loose, the bearing may abrade the shaft and/or housing. If

Figure 18-5 Basic parts of antifriction bearings. (Courtesy Deere and Company Technical Services.)

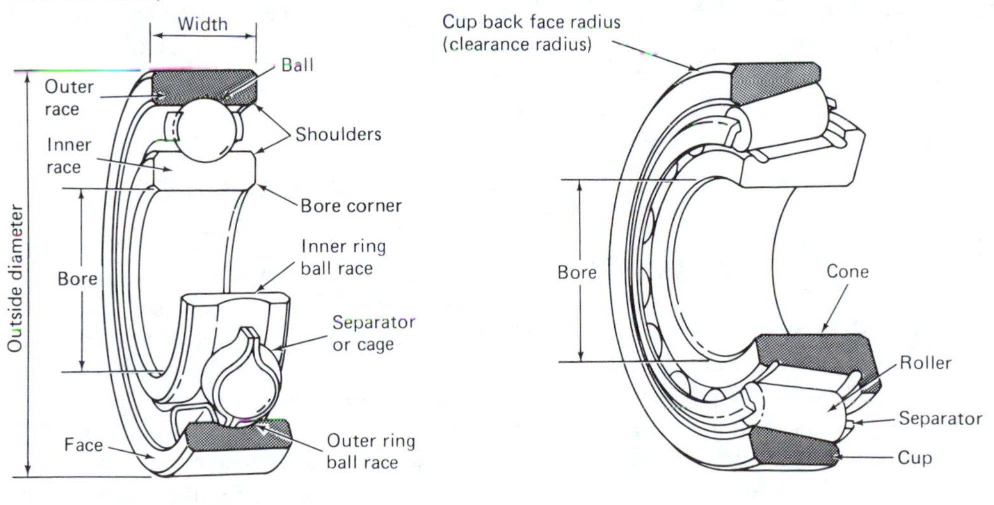

the fit is too tight, the bearing may be damaged because of heat generated while operating with an insufficient internal clearance. This condition is called bearing preload.

Sometimes (usually when a relatively low rotational speed prevails) a prescribed amount of preload may be beneficial. This type of preload is often obtained by rigidly spacing the bearings on a shaft to a dimension slightly different than the corresponding mounting distance in the housing. The housing walls must deflect to accept the oversize, thus resulting in an axial preload on both bearings. In such a case, the axial and radial clearance is zero, providing an ideal arrangement for a spindle or work head on a machine tool.

Rolling bearings require a minimum of lubrication because friction is only a small part of the internal action. Some sliding does occur between the rolling elements and their retainer, and some friction is found at the rubbing surfaces of the lubricant seal. Although a rolling bearing does not generate very much heat, it does transfer considerable amounts, since it is the *intermediary* between a shaft and housing. Meshing gears, for instance, generate heat because of their inherent sliding action. This heat travels through the gear, shaft, and bearing as it finds its way into the housing. For this reason, a gear case should have an adequate supply of lubricant to aid in heat transfer. The case itself should also have a large area exposed to the atmosphere to maintain the required temperature equilibrium.

18-2 TERMINOLOGY AND CLASSIFICATION

Rolling element bearings are often referred to as rolling *contact* bearings and *antifriction* bearings. But the simplest name—*rolling* bearings—adequately describes the product. In this chapter the term *bearing* will be used for rolling element assemblies.

Most manufacturers endorse the standards set up by the AFBMA (Antifriction Bearing Manufacturers Association). These standards are available through cooperating manufacturers and are of great assistance in learning current terminology. Figure 18-5 identifies the components of typical bearings. The three basic dimensions common to all rolling bearings are shown: *bore* (interface with the shaft), *outside diameter* (interface with the housing), and *width*.

Rolling element bearings can be classified by function and form. Classification by form is based on the shape of the rolling elements and differentiates *ball, roller,* and *needle* bearings. Classification by function distinguishes *radial, thrust,* and *combination radial and thrust* groups.

In the *radial* bearing one race rotates relative to the other and the load application is at right angles to the axis of rotation. Examples are radial ball and cylindrical roller bearings (Figs. 18-6 and 18-11).

In the *thrust* bearing (Figs. 18-7 and 18-12) one race rotates with respect to the other and the load application is parallel to the axis of rotation in one or both directions.

Combination radial and thrust bearings combine features of both radial and thrust types, resisting a load applied in an angular direction or the resultant of a combination of radial and thrust loads. Examples are angular contact ball, angular contact spherical, and tapered roller bearings (Figs. 18-6 and 18-11).

18-3 STANDARDIZATION OF ROLLING BEARINGS

Rolling bearings are standardized to a greater extent than perhaps any other basic machine member. Furthermore, ball bearing standards are international in scope and, as such, are predominantly based on the metric system. This was a natural consequence of their development in Germany and Sweden. Tapered roller bearings, in contrast, were developed in the United States and were at first inch based but are now being

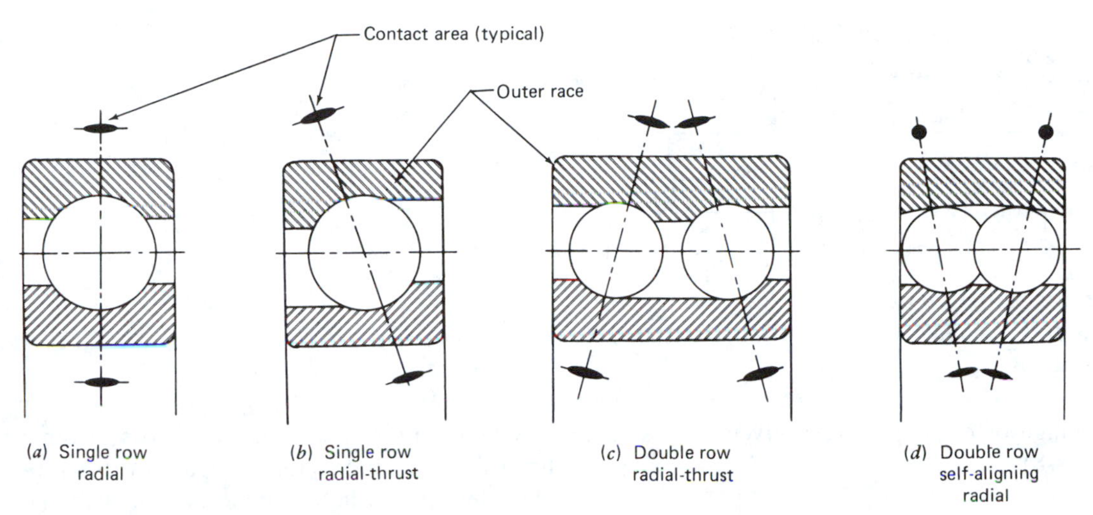

Figure 18-6 Types of ball bearings. The dash-dot lines indicate the axis of the contact between balls and races. (Courtesy Deere and Company Technical Services.)

metricated. Each type of bearing has its own standard but is principally based on *bore, outside diameter*, and *width*. The standards incorporate a comprehensive identification code adopted by the AFBMA. For further detail, designers should consult manufacturers' catalogs or AFBMA/ANSI standards.

As a result of this extensive standardization, a particular size bearing may be available from several manufacturers and still have identical basic dimensions (inside diameter, outside diameter, width, and corner radii).

Ball and roller bearings have been divided into several standard series to meet the requirements most often encountered in design work. Each series is dimensionally different and provides a

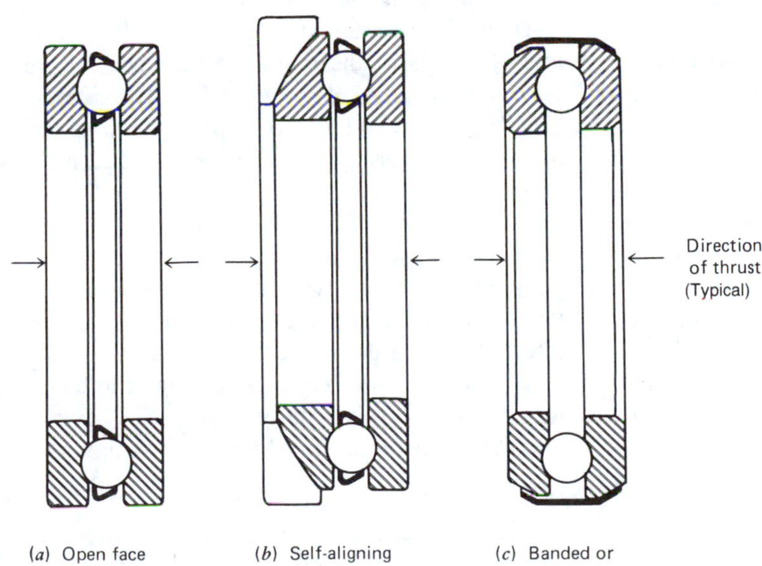

Figure 18-7 Thrust-load ball bearings. (Courtesy Deere and Company Technical Services.)

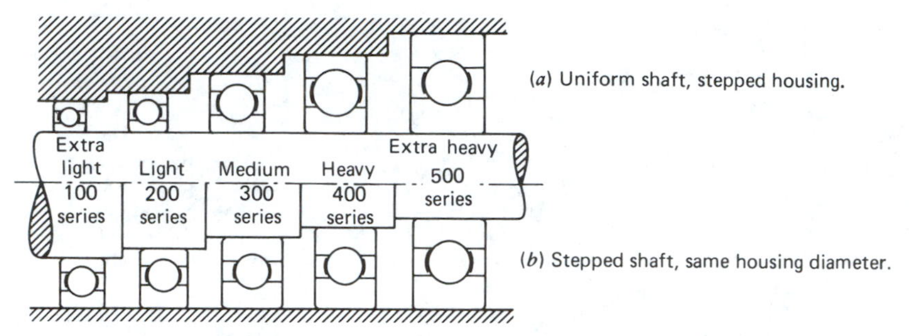

Figure 18-8 Bearing series available. The 100, 200, and 300 series are most commonly used.

wide range of capacities for a given shaft size (Fig. 18-8).

18-4 BALL BEARINGS

Ball bearings can be defined as those whose rolling elements are balls that ride in circumferential grooves. Ball bearings generally operate with less friction than roller bearings and can be used at higher speeds (Fig. 18-18 shows this clearly). Sizes range from minute instrument bearings to huge turntable bearings. Various shapes are available to suit almost any practical application. They come in five basic designs, as shown in Fig. 18-8.

Among ball bearings, the deep-groove (Conrad[1]) types predominate numerically, on a ratio of about 2 : 1 as for all other types combined. Conrad types are available either as single or double row. Some have shields to prevent foreign matter from entering the bearing; others have seals to retain the lubricant and exclude contaminants (Fig. 18-9). Some may have a built-in snap ring in the outer race to locate the bearing in the shaft housing assembly. Conrad bearings are generally lowest in cost of any ball type. Because they can carry thrust loads (loads directed along the axis of the shaft) in both directions, a deep-groove ball bearing is often used for axial location of shafts and rotors.

Within the confines of the Conrad's external dimensions, additional radial capacity can be obtained by suitably notching the two races so that additional balls can be introduced. Radial capacity is thus increased, but thrust capacity is severely reduced because of the interrupted shoulder that the balls must work against. This bearing is referred to as the *Max type* or *loading-groove* type (Fig. 18-10). Since the two bearings, notched and unnotched, have the same external dimensions, they must be carefully distinguished to avoid misapplication. This is usually done by assigning different alphameric[2] designations as well as using their respective names.

To enhance the thrust capabilities of ball bearings, the axis of the balls can be placed in an *off-radial* position, as shown in Fig. 18-6. These bearings are called *angular contact* or *radial and thrust load* bearings. As single-row bearings, they will support both radial and unidirectional thrust loads. A double-row configuration is required if thrust loads will occur in both directions. Proper assembly of the single-row bearing is mandatory, since thrust loads must always be directed toward the higher shoulders.

Alignable radial load ball bearings can be either internally or externally alignable. They can tolerate either *initial* misalignment (by having a

---

[1]Named after a German bearing pioneer.

[2]Same as alphanumeric. All characters including numerals, letters, punctuation marks, and the like.

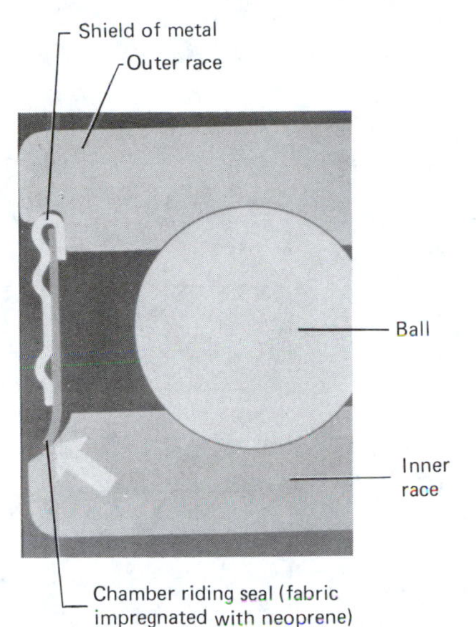

Figure 18-9 Typical seal for ball bearing. (Courtesy SKF Industries, Inc.)

spherical outer race) or *running* misalignment (by having a spherical outer ring ball path).

Thrust load ball bearings are usually single row and may have a shield or seal. These bearings will accept full thrust loads and some light radial loads. They are often used in conjunction with radial load bearings. Figure 18-7 shows three available varieties.

Figure 18-10 Max type ball bearing showing notch for assembly of balls. (Courtesy Deere and Company Technical Services.)

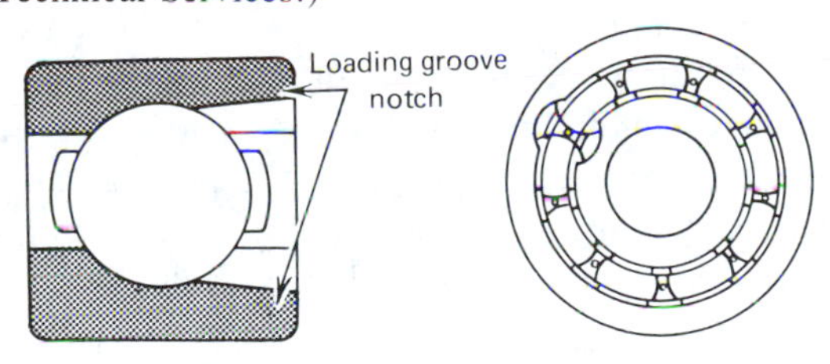

(*a*) Loading grooves in the races

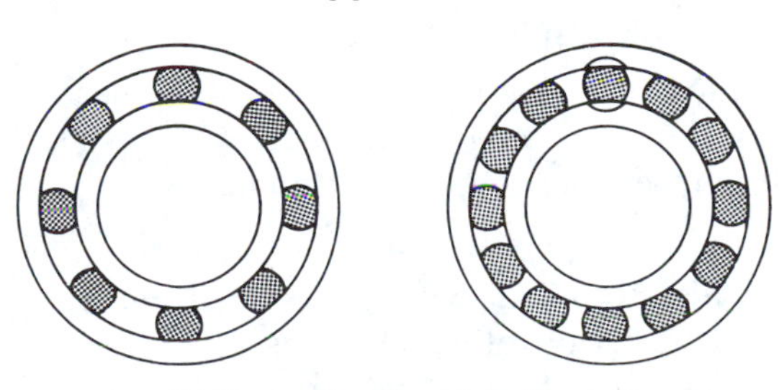

(*b*) More rolling elements in Max type than in comparable Conrad type

18-5 ROLLER BEARINGS

In roller bearings the rolling elements are hardened steel cylinders or truncated cones that revolve in hardened steel races. Roller bearings are preferred when large loads are present. Their load capacity for a given space greatly exceeds that of ball bearings because the basic rolling contact is along a line, not a point. The load is thus spread over a larger area and produces less unit stress and deformation for a given load. Figure 18-11 shows the basic types of roller bearings.

Radial load, straight roller bearings have high radial capacity but cannot accept much misalignment or thrust loads (Fig. 18-11*a*).

Radial and thrust load, tapered roller bearings will accept pure radial loads, pure thrust loads, and combinations of the two (Fig. 18-11*b*). When the taper angle is 45 deg, the thrust and radial capacities are equal. A steeper angle provides more thrust than radial capacity, while a shallower angle offers more radial than thrust capacity. The taper design gives rise to an induced thrust force even though no external thrust is present. For this reason, bearings are mounted in opposition (back to back or front to front) (Figs. 18-21 and 18-22). Thrust forces are thus counteracting, and the shaft is constrained in its assembled position.

Self-aligning radial and thrust load roller

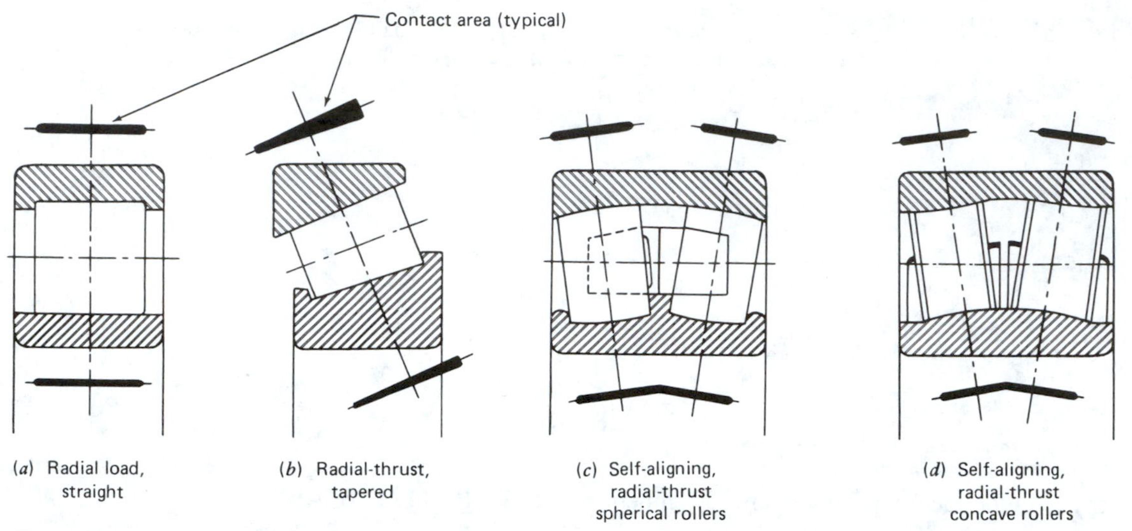

Figure 18-11 Basic types of roller bearings showing roller contact pattern under load. (Courtesy Deere and Company Technical Services.)

bearings have great load-carrying capacity as a result of the two roller paths. They can be obtained with either spherical or concave rollers and will accept operating misalignment resulting from shaft deflections (Fig. 18-11*c* and 18-11*d*).

Thrust load roller bearings (Fig. 18-12) can carry heavy thrust loads but cannot withstand any radial loads. Normally, thrust load roller bearings are paired with radial load bearings to provide stability and support in both directions.

18-6 NEEDLE BEARINGS

Needle bearings, sometimes referred to as quill bearings, have rollers of very small diameter and a relatively high load-carrying capacity. They are effectively used in limited radial space and at low operating speeds. Needle bearings are ideally suited, however, for machine members carrying heavy loads with *oscillatory* motions, such as piston pin bearings in engines or in universal joints (Fig. 18-13). Needle bearings often replace plain bushings, since the small external diameter allows replacement with little or no change in design. Figure 18-14 shows the basic construction of radial and thrust load needle bearings.

Needle bearings are often used without an inner race, running directly on a hardened shaft (Figs. 18-13 and 18-14). In fact, they are sometimes used without either race for maximum space utilization. In addition, the cages may be omitted for maximum load capacity. This situation requires the use of a suitable grease compound to retain the loose needles to ease the assembly problem. The thrust type can also operate without races, depending on the design of adjoining parts, but will always retain the cage element.

18-7 BEARING CAPACITIES

The load-carrying capacity of a rolling bearing depends on the total contact area, which is itself determined largely by four parameters.

1. Series: light, medium, or heavy.
2. Size: inside diameter, outside diameter.
3. Number of rolling elements.
4. Number of pathways for rolling elements: single or double row.

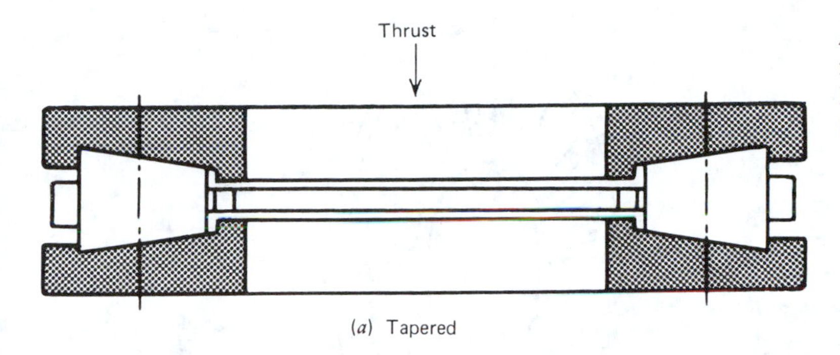

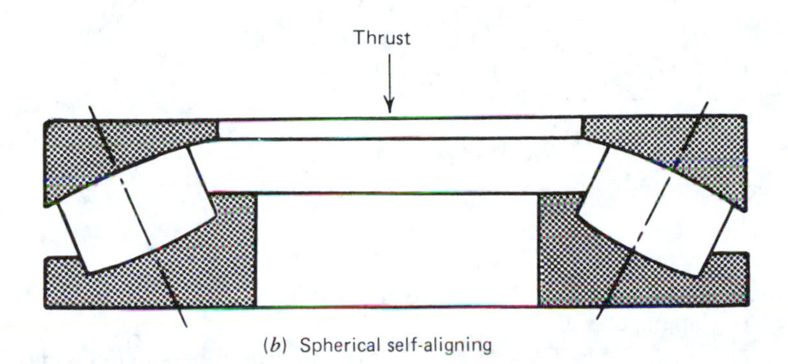

Figure 18-12 Thrust-load roller bearings. (Courtesy Deere and Company Technical Services.)

Contact area, when multiplied by the unit contact stress, gives the total load the bearing can sustain, that is, its capacity.

The stresses on the contact surfaces are usually referred to as *Hertz* stresses, relating to the work of H. Hertz of Germany. Hertz determined the contact area of various geometric shapes whose nominal contact is only a point or a line. In ball bearings, for instance, the initial contact is a point but, under increasing loads, it spreads to an elliptical pattern. The spread is determined by the magnitude of the applied load, the modu-

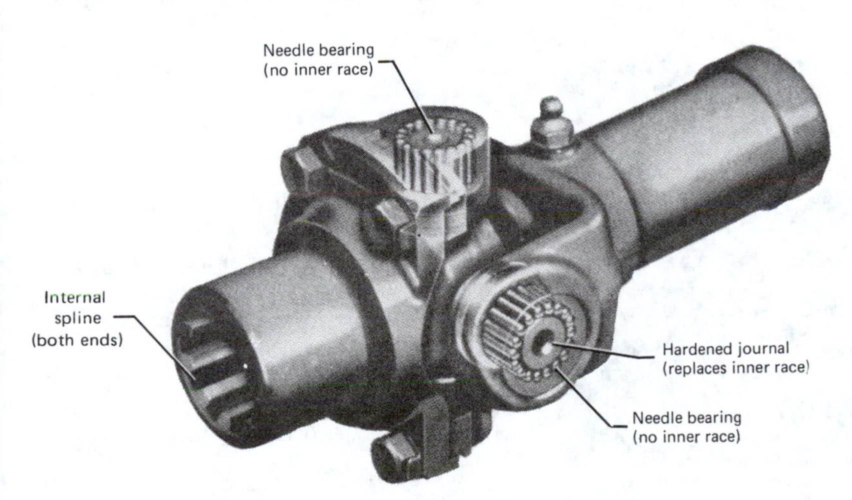

Figure 18-13 Hooke-type universal joint with a needle bearing mounted yoke. (Courtesy Mechanics Universal Joint Division, Borg-Warner Corp.)

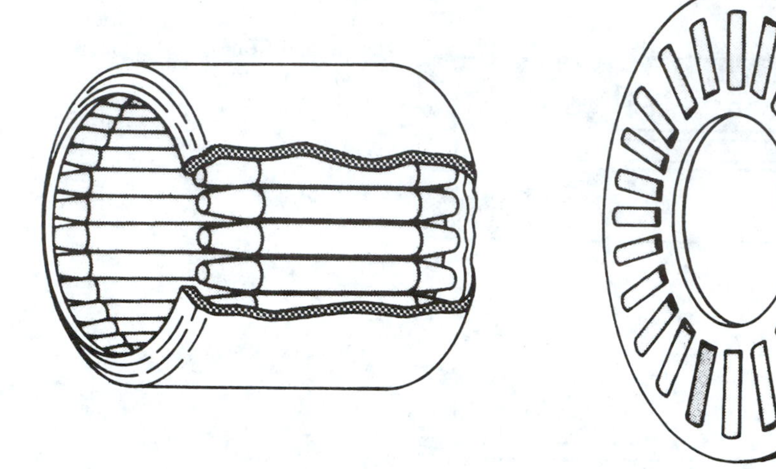

Figure 18-14 Basic types of needle bearings. (Courtesy Deere and Company Technical Services.)

lus of elasticity of the materials in contact, and the radii of the contacting bodies. The contact stress needed to produce the published capacity figures is actually very large; values of 2400 MPa (350,000 psi) are typical for the high-strength steels used. Because the mechanism of bearing failures is complex, manufacturers have found that they must physically test many samples of their product to establish basic capacity ratings.

18-8 BEARING LIFE

If a rolling bearing is correctly mounted, adequately lubricated, and otherwise handled properly, all causes of failure have been eliminated except one: *material fatigue*. This characteristic is a direct consequence of rolling bearings operating at very high stress levels. Each time a rolling element passes through the load zone (Fig. 18-15), a fatigue cycle is created—extreme pres-

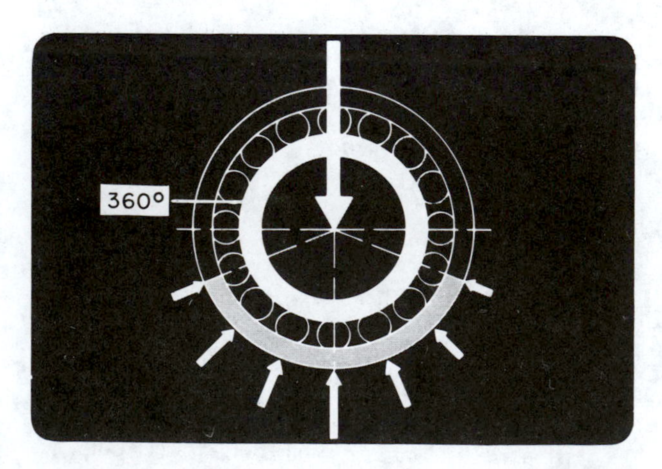

Figure 18-15 An applied load of *constant* direction and magnitude is distributed *unevenly* among the rolling elements of the bearing. The large arrow indicates the applied load, and the smaller arrows show each rolling element's share of the load. The *outer* ring has a load zone of roughly 130 deg (65 deg on each side of the applied load). The *inner* ring, in contrast, has a 360 deg load zone. (Courtesy SKF Industries, Inc.)

Figure 18-16 Fatigue failure (spall marks). (Courtesy Deere and Company Technical Services.)

sure as the element traverses the critical area and then a recovery period before the element again begins its contact. The number of load cycles accumulates and eventually exceeds the fatigue limit of the material. Failure then ensues.

The terms *life, normal life,* and *useful life* are defined as the number of revolutions that a bearing can carry a given load before the first visible evidences of material fatigue develop. These signs may appear either on the balls, rollers, outer race, or inner race; they consist of minute spalls or pit marks (Fig. 18-16). Another way of expressing bearing life is to convert the number of revolutions to hours of operation at a given load and speed. This is the most common way of referring to bearing life. Automobile manufacturers use still another expression for bearing life by stating the number of kilometers (miles) the vehicle can be driven before failure is expected.

Bearings are usually calculated on the basis of *finite* rather than infinite life. That is, some bearings are expected to fail after operating at design load for some preselected length of time. The idea of designing for finite life may at first seem strange to some designers, but it is a valid and necessary concept. Consider a home appliance used 2 h/wk and assume a bearing is mounted on a shaft turning at 2000 rpm. It has been determined that a percentage of bearings will fail after 10^8 revolutions when operating at the design load. The 10^8 revolutions are reached in:

$$\frac{10^8 \text{ revolutions}}{(2000 \text{ rpm})(60 \text{ min/h})(2 \text{ h/wk})(52 \text{ wk/yr})} = 8.01 \text{ yr}$$

This length of time is well within the expected and planned obsolescence of such equipment.

Consider, too, a road-building contractor who uses machines all day, 4 h of which are at peak loads, for 200 days a year. Again, bearings operating at 2000 rpm will reach 10^8 revolutions of peak load in:

$$\frac{10^8 \text{ revolutions}}{(2000 \text{ rpm})(60 \text{ min/h})(4 \text{ h/day})(200 \text{ days/yr})}$$
$$= 1.04 \text{ yr}$$

This contractor will normally expect to rebuild the machines periodically because of such rough, heavy usage and should plan to replace bearings during routine maintenance programs.

The most conclusive evidence in favor of designing to finite life, however, is a consideration of the *alternative*. How would a machine with infinite-life components compare to the finite-life machine? Assuming equal power, the infinite-life unit would be much larger, heavier, and far more expensive because all parts would be proportionally oversized. Furthermore, in performance the infinite-life machine would be no match for the finite-life machine because of added bulk and size. In short, the contractor can make more money using a finite-life machine, even allowing for costs of downtime to replace worn parts.

There is *no* way to predict when any given bearing will fail out of a group of bearings operating under identical conditions. *Statistical* analysis, however, will provide data as to the average life of a group of identically loaded bearings. Or, by statistical correlation, one can determine what percentage of bearings will exceed a certain life. Figure 18-17 is a typical representation of

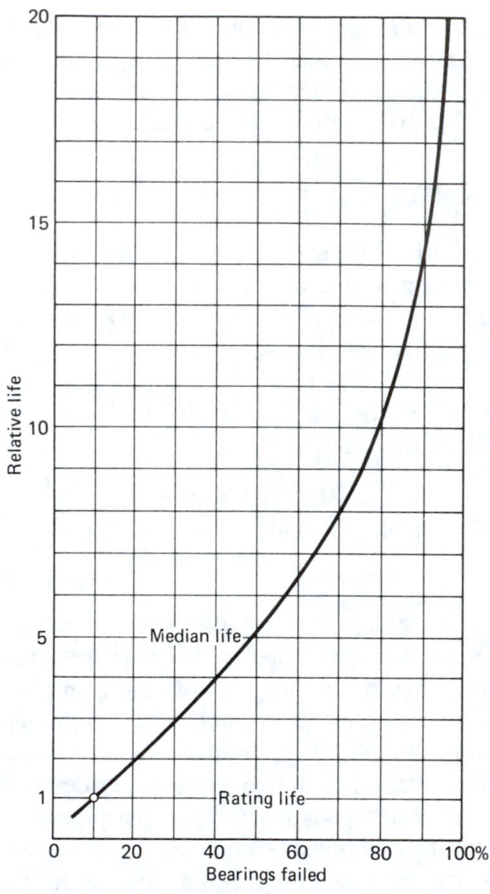

Figure 18-17 Typical life comparisons of a group of bearings tested to fatigue failure under equal loading conditions. (Courtesy SKF Industries, Inc.)

bearing failure rate. Note that there is a spread in life of over 20 to 1. Average life implies a 50% failure rate; for every bearing operating in excess of this life, another will have failed. Of course, a failure rate of this magnitude during the useful life of any product would never be tolerated by manufacturers or customers. For this reason, the suppliers of rolling bearings have standardized using a life-rating based on only 10% failure. Thus the *load rating* of a group of apparently identical bearings is defined as the load that can be sustained by 90% of the bearings for a specified number of revolutions without failure. In technical literature this is referred to as the *basic load rating* or *capacity* of a bearing. Published figures of basic loads in vendor catalogs are obtained by a statistical correlation of actual loads causing failure under laboratory conditions.

The definition of what constitutes a failure is arbitrary. The one adopted more or less universally by bearing manufacturers is that a bearing has failed when a single spall mark attains an area of 6.5 mm^2 (0.01 $in.^2$). This implies that a bearing can run considerably longer before precipitating a total failure and provides for some margin of error in mathematical predictions. Remember that the presence of a spall mark is associated with nonmetallic inclusions in the steel and their proximity to the surface. As the cleanliness of steel improves through metallurgical research, it is not unreasonable to expect higher fatigue lives from the same-sized bearings.

As noted, manufacturers continually test their bearings to verify basic load ratings stated in their catalogs. A test program may consist of many bearings operating under different loads and speeds to allow statistical reduction of the data generated. A study of typical test results reveals that life measured in bearing revolutions varies exponentially with the load applied. On logarithmic graphs a plot of load versus life becomes a straight line that can be defined as follows.

$$L_{rev} = K\left(\frac{C}{R}\right)^p \qquad (18\text{-}1)$$

where

L_{rev} = life of bearing in total accumulated revolutions
K (constant) = the number of revolutions chosen as a basis
R = radial load on bearing
C = the basic capacity of bearing that corresponds to base revolutions
p (dimensionless) = life exponent

For ball bearings $p = 3$ and for roller bearings $p = 10/3$. K is 10^6 revolutions for ball and needle bearings and $(90)(10^6)$ for straight and tapered roller bearings. Individual values for each parameter will be found in respective manufacturers' catalogs.

As mentioned earlier, the most common method of denoting bearing life is on the basis of hours of operation at normal or average speed. To make the transition from life in total revolutions to life in hours, we must introduce the relative speed of the bearing. Thus:

$$\text{Total revolutions} = (\text{hours operation})(60\ \text{min/h, rpm})$$

$$\text{Hours operation} = \frac{\text{total revolutions}}{60\ \text{min/h, rpm}} = \frac{\text{total revolutions}}{60n}$$

where

$$n = \text{rpm}$$

Making this substitution in Eq. 18-1 for each of the bearing types considered, we have:

Ball bearings:

$$L_{(\text{hours})} = \frac{10^6}{60n}\left(\frac{\text{catalog rating}}{\text{applied load}}\right)^3 \tag{18-2}$$

Straight roller bearings:

$$L_{(\text{hours})} = \frac{(90)10^6}{60n}\left(\frac{\text{catalog rating}}{\text{applied load}}\right)^{10/3} \tag{18-3}$$

Needle bearings:

$$L_{(\text{hours})} = \frac{10^6}{60n}\left(\frac{\text{catalog rating}}{\text{applied load}}\right)^{10/3} \tag{18-4}$$

Taper roller bearings:

$$L_{(\text{hours})} = \frac{(90)10^6}{60n}\left(\frac{\text{catalog rating}}{\text{applied load}}\right)^{10/3} \tag{18-5}$$

Notice that 1 million revolutions is equivalent to 500 h of operation at $33\frac{1}{3}$ rpm, 100 h at $166\frac{2}{3}$ rpm, $33\frac{1}{3}$ h at 500 rpm, and so on. These hours of life seem very minimal at what would be considered representative speeds. This is not to say that bearings operating at 1000 to 2000 rpm and higher speeds are doomed to early failure. It simply means that designers must make their bearing choices from those that have catalog ratings several times the loading expected for their particular use.

EXAMPLE 18-1

(*a*) What life can be expected for a ball bearing application at 1500 rpm when the catalog rating is 2.6 times the applied load? (*b*) What ratio of catalog rating to applied load must be chosen to obtain 2000 h life?

SOLUTION

$$L_{(\text{hours})} = \frac{10^6}{60\ n}\left(\frac{\text{catalog rating}}{\text{applied load}}\right)^3$$

(*a*) $$\text{Life} = \frac{10^6\ \text{revolutions}}{(60\ \text{min/h})(1500\ \text{rpm})}2.6^3 = \mathbf{195.3\ h}$$

(*b*) $$2000\ \text{h} = \frac{10^6\ \text{revolutions}}{(60\ \text{min/h})(1500\ \text{rpm})}\left[\frac{\text{catalog rating}}{\text{applied load}}\right]^3$$

$$\frac{\text{catalog rating}}{\text{applied load}} = \left[\frac{2000(60)1500}{10^6}\right]^{1/3} = \mathbf{5.65}$$

18-9 BALL BEARING SELECTION FROM VENDOR CATALOGS

Designers constantly face the task of assigning proper sizes to various components. In the same way that gears, pulleys, shafts, and keys are all sized according to the amount of power transmitted through them, supporting bearings must be discretely sized for the task they will perform. All bearing manufacturers have catalogs of their products showing basic sizes, tolerances, and capacity ratings. This information enables designers to specify the bearing that most efficiently and economically fits a particular design.

At some point in the design sequence, tentative bearing sizes are chosen. Because bearing loads are dependent on axial location of gears, sprockets, wheels, and pulleys and the overall span length of shafts, final bearing selection must wait until these parameters are set. By this time, the shaft size should also be fixed, and this dictates the approximate bearing inside diameter. The question to settle, then, is what bearing outside diameter and width will meet the desired life-hour requirement.

For this discussion, assume that the designer has already determined the bearing type by means of Fig. 18-18 and has chosen ball bearings. Next, the ball bearing type best suited to the design, both by its availability and by its price, must be determined. Again, let us assume that the designer has chosen single-row over double-row or angular-contact types.

Within the single-row classification there is a further breakdown according to series. The three most popular series are extra light, light, and medium. For a given bore size, outside diameter and width vary from series to series, thus accommodating varying ball diameters. Furthermore, the light and medium series are also available as the Conrad or Max type in most sizes. The final size selection is from these five categories: extra light series (Conrad type only), light series (Conrad and Max type), and medium series (Conrad and Max type).

The choice will be based on size compatibility with the design to achieve required life objectives. To aid in this selection, it is convenient to compare outside dimensions and capacity ratings against a common bore size. Such a compilation is shown in Table 18-1, which illustrates the type of data to be found in a bearing manufacturer's catalog. Once the required capacity is determined, it is easy to select the bearing size best suited to the design.

Figure 18-18 Guide to selection of ball or roller bearing. This figure is based on a rated life of 30,000 h. (Note that recent developments may have pushed both speed and load limits upward.) (From *Marks's Standard Handbook for Mechanical Engineers* by T. Banmeister. Copyright ©1978. Used with the permission of McGraw-Hill Book Company.)

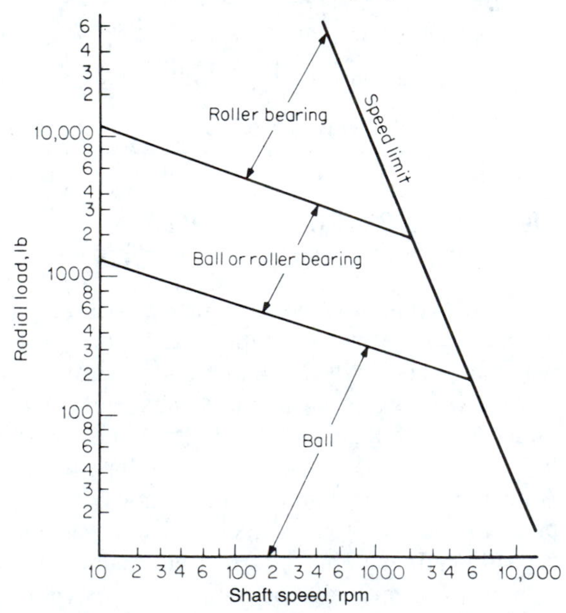

EXAMPLE 18-2

A designer needs a ball bearing with a capacity rating of 30 kN to achieve a desired life criterion. The shaft diameter is 45 mm. What bearing should be selected? What are the basic dimensions of the bearing?

SOLUTION

In the far left column of Table 18-1, the designer locates 45 mm bore, for a shaft diameter of 45 mm. Each corresponding row lists capacity ratings. The light series, Max type, is the first to provide a capacity rating of 32.2 kN > 30 kN. Dimensions of this bearing are: bore = **45 mm**; outside diameter = **85 mm**; and width = **19 mm.**

TABLE 18-1 Capacity Ratings for Single-Row Ball Bearings

	Extra Light Series			Light Series				Medium Series			
						Capacity Rating				Capacity Rating	
Bore, mm	Outside Diameter, mm	Width, mm	Capacity Rating, kN	Outside Diameter, mm	Width, mm	Conrad, kN	Max type, kN	Outside Diameter, mm	Width, mm	Conrad, kN	Max type, kN
10	26	8	3.51	30	9	4.58		35	11	6.22	
12	28	8	3.91	32	10	5.78		37	12	7.46	
15	32	9	4.28	35	11	5.98		42	13	8.77	
17	35	10	4.62	40	12	8.07		47	14	9.07	
20	42	12	7.20	47	14	9.87		52	15	13.9	16.3
25	47	12	7.67	52	15	10.9		62	17	16.3	20.7
30	55	13	10.2	62	16	14.9	18.1	72	19	20.4	25.3
35	62	14	12.4	72	17	19.7	23.9	80	21	25.6	31.8
40	68	15	12.9	80	18	22.3	28.6	90	23	31.3	38.9
45	75	16	16.0	85	19	25.1	32.2	100	25	40.7	50.2
50	80	16	16.6	90	20	27.0	33.8	110	27	47.6	58.9
55	90	18	21.7	100	21	33.2	41.8	120	29	55.0	68.1
60	95	18	22.7	110	22	36.7	48.1	130	31	62.6	77.4
65	100	18	23.5	120	23	38.9	49.9	140	33	71.3	93.5
70	110	20	29.3	125	24	42.6	54.6	150	35	80.0	105
75	115	20	30.2	130	25	46.4	59.8	160	37	87.1	115
80	125	22	36.6	140	26	54.9	70.7	170	39	94.7	128
85	130	22	38.0	150	28	63.9	82.1	180	41	102	138

Source. Compiled by R. J. Erisman.

18-10 ADDITIONAL FACTORS AFFECTING LIFE (BALL BEARINGS)

Bearings Operating with Combined Radial and Thrust Loads

Many bearing applications will have imposed thrust loads as well as the radial loads just discussed. Thrust loads derive from helical, bevel, or worm gears, spring-loaded actuating devices, or bearing preloads. External loading may also apply thrust loads that must ultimately be absorbed by the bearings. Vehicles operating on uneven ground, on slopes, or on curves create such loads. Similarly, many machining operations impose thrust loads on mounting spindles of machine tools. Thus bearings are often operated with some combination of radial and thrust loads.

Except for the special case of pure thrust bearings, bearing ratings shown in manufacturers' catalogs are for radial loads. When thrust is present, an *equivalent* radial load must be determined before consulting the catalog for bearing selection. This is defined as the radial load producing the same theoretical fatigue life as the combined radial and thrust loads. Most bearing manufacturers give methods of combining thrust and radial loads in accordance with ANSI standards to obtain the equivalent load. Thus

$$RE = RF_{th} \tag{18-6}$$

where

RE = equivalent radial load
R = radial load
F_{th} = thrust factor

A bearing catalog will display separate tables

TABLE 18-2 Thrust Factor

T/R[a]	F_{th}	T/R	F_{th}
0.25	1.0	1.00	1.63
0.30	1.025	1.05	1.68
0.35	1.06	1.10	1.73
0.40	1.095	1.15	1.79
0.45	1.13	1.20	1.83
0.50	1.17	1.25	1.88
0.55	1.21	1.30	1.94
0.60	1.25	1.35	1.99
0.65	1.29	1.40	2.03
0.70	1.33	1.45	2.08
0.75	1.38	1.50	2.12
0.80	1.43		
0.85	1.48		
0.90	1.54		
0.95	1.58		

Source. International Harvester Corporation.

[a]Thrust-to-radial load ratio. For T/R values < 0.25, thrust loads can be neglected. Loading-groove bearings should not be used if $T/R > 0.6$.

of values to cover single-row, double-row, and angular-contact variations. Table 18-2 applies to single-row, Conrad bearings. These values are also suitable for the Max type when limited to T/R values no greater than 0.6.

EXAMPLE 18-3

(*a*) A given bearing has a radial load of 2700 lb and a thrust load of 1460 lb. What is the equivalent radial load? (*b*) What is the bearing life if operated at 1750 rpm? Catalog rating is 68.1 kN. (*c*) What is the life with thrust load removed?

SOLUTION

$$(a)\ T/R = \frac{1460 \text{ lb}}{2700 \text{ lb}} = 0.541;$$

$$F_{th} = 1.20 \text{ (from Table 18-2)}$$

$$RE = RF_{th} = (2700 \text{ lb})(1.2) = 3240 \text{ lb}$$

$$= \frac{3240 \text{ lb}}{1000}\left(4.448\frac{\text{N}}{\text{lb}}\right) = \mathbf{14.4\ kN}$$

$$(b)\ \text{Life} = \frac{10^6}{60n}\left(\frac{\text{catalog rating}}{\text{applied load}}\right)^3$$

$$= \frac{10^6}{(60 \text{ min/h})(1750 \text{ rpm})}\left(\frac{68.1 \text{ kN}}{14.4 \text{ kN}}\right)^3$$

$$= \mathbf{1007\ h}$$

$$(c)\ 2700 \text{ lb} = \frac{2700 \text{ lb}}{1000}\left(4.448\frac{\text{N}}{\text{lb}}\right) = 12.0 \text{ kN}$$

$$\text{Life} = \frac{10^6}{(60 \text{ min/h})(1750 \text{ rpm})}\left(\frac{68.1 \text{ kN}}{12 \text{ kN}}\right)^3$$

$$= \mathbf{1741\ h}$$

Outer Race Rotation

The normal mounting practice is to place the outer race in the stationary member of the mechanism and the inner race around the rotating member. This corresponds to the way bearing manufacturers actually perform their rating tests and typifies the vast majority of applications. When the mounting practice is reversed, the actual bearing life will be less than indicated by calculations.

To understand the reason for this, consider the typical force diagrams shown in Fig. 18-19. Note that the resultant force in each case is stationary, even though the shafts are rotating. This means that the stationary member of the bearing will have one localized zone that will sustain the maximum loads.

The rotating part of the bearing, on the other hand, will have all of its surface exposed to the fatigue cycles, since it must continually rotate through the maximum load zone. Obviously, fatigue cycles will accumulate more rapidly when all are concentrated in one place instead of being distributed around the whole circumference. *Because contact stresses are greater at the inner race (diverging surfaces), it follows that this race should rotate to distribute the stresses as much as possible.*

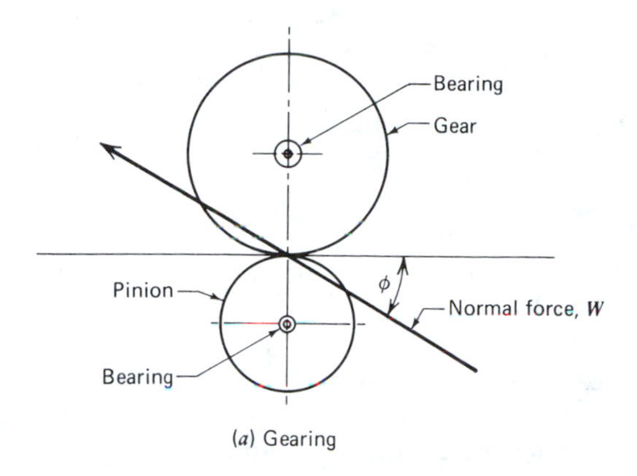

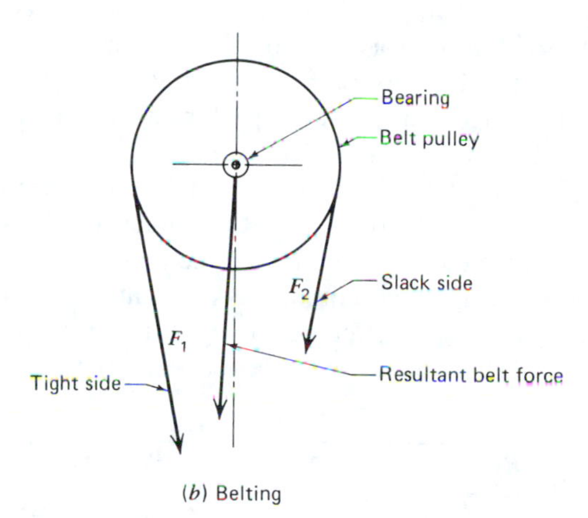

Figure 18-19 Force diagrams showing stationary force placement.

When the design makes it impossible to follow this rule, a larger bearing should be used to achieve the design life criterion. Mathematically, this is accomplished by dividing the bearing capacity by some factor. Thus a larger bearing must be selected so that the desired final capacity will be obtained.

The factor most commonly quoted for this phenomenon is 1.2, which was probably derived empirically. Some bearing suppliers choose to ignore the outer race rotation factor, claiming that there is enough bearing creep in the mounting to provide the desired load distribution. This may be true, but it would be good to verify this

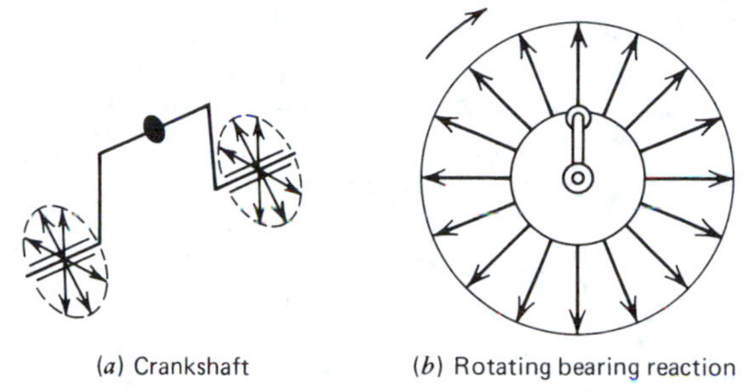

Figure 18-20 Force diagram showing rotating forces.

experimentally because the life reduction factor is significant:

$$\frac{1}{(1.2)^3} = 0.579 \text{ or } 42\% \text{ reduction in bearing life}$$

Notice also that not all uses of outer race rotation are subject to this phenomenon. A bearing supporting an unbalanced rotating crank is one exception (Fig. 18-20). In this case, the load is rotating and a stationary inner race actually provides the desired load distribution. The designer, of course, must make the proper analysis in each situation.

Reliability Other Than 90%

The catalog rating of bearings is usually on the basis of 90% reliability (10% failure). There are, of course, many applications where a greater reliability must be assured. Since it would be inconvenient to have several rating tables for different levels of reliability, a series of multiplying factors provides the same result. Thus

$$L_e = L_d F_{\text{rel}} \qquad (18\text{-}7)$$

where

L_e = equivalent life (90% reliability); h
L_d = desired life (reliability other than 90%); h
F_{rel} = reliability factor

A set of values for F_{rel} is shown in Table 18-3. To use this table, a designer would select the multiplying factor corresponding to the desired reliability, modify the hours of life accordingly,

TABLE 18-3 Life Adjustment Factor for Varying Reliabilities

Reliability	F_{rel}
0.5	0.200
0.6	0.260
0.7	0.353
0.8	0.527
0.90	1.00
0.95	1.85
0.96	2.25
0.97	2.89
0.98	4.10
0.99	7.46
0.995	13.51
0.999	53.56

and choose the appropriate bearing in the usual manner.

EXAMPLE 18-4

What catalog rating is required for a ball bearing to obtain 5000 h of life with a reliability of 95%? The bearing has an applied load of 8 kN and operates at 2100 rpm.

SOLUTION

By substituting into Eq. 18-7,

$$L_d = 5000 \text{ h}$$

$$F_{\text{rel}} = 1.85 \text{ (from Table 18-3)}$$

the designer obtains

$$L_e = (5000 \text{ h})(1.85) = 9250 \text{ h}$$

$$\text{Life} = \frac{10^6}{60n}\left(\frac{\text{catalog rating}}{\text{applied load}}\right)^3$$

$$9250 \text{ h} = \frac{10^6}{60 \text{ min/h}(2100 \text{ rpm})}\left(\frac{\text{catalog rating kN}}{8 \text{ kN}}\right)^3$$

$$\text{Catalog rating} = (8)\left[\frac{(9250)(60)(2100)}{10^6}\right]^{1/3}$$

$$= \mathbf{84.2\ kN}$$

Weighted Life Determination[3]

In many applications a bearing does not run at constant load and speed but has a well-defined pattern of varied loading conditions. It is convenient to express the life of a bearing that operates in this manner as the *weighted life*, in which each of several operating conditions has been taken into account. An automobile transmission is a good example of this kind of variable loading. The three forward speeds and the one reverse speed are used for varying amounts of time that, totaled, represent the performance demand of the vehicle.

To calculate the weighted life of a bearing under several loading conditions, one must first determine the indicated life in each mode. These lives are then combined, taking into consideration the various percentages of time for each mode. The general formula for weighted life is

$$L_w = \frac{100}{(P_1/L_1) + (P_2/L_2) + \cdots + (P_n/L_n)} \tag{18-8}$$

where

L_w = weighted life; h

P_1, P_2, P_3, P_n = the percentage of times in each loading mode

L_1, L_2, L_3, L_n = the individual lives in each loading mode

---

EXAMPLE 18-5

A bearing operates under four distinct loading conditions. The individual lives for each loading condition are 35, 130, 525, and 2000 h.

---

[3]*Weighted* is a *statistical* expression. It refers to a constant assigned to a single item in a frequency distribution indicative of the item's relative importance.

What is the weighted life if (*a*) the bearing operates equally in each loading mode, and (*b*) the time pattern for the four load conditions is 5, 10, 20, and 65%, respectively?

SOLUTION

$$L_w = \frac{100}{(P_1/L_1) + (P_2/L_2) + (P_3/L_3) + (P_4/L_4)}$$

$$(a)\ L_w = \frac{100}{\dfrac{25}{35\ h} + \dfrac{25}{130\ h} + \dfrac{25}{525\ h} + \dfrac{25}{2000\ h}}$$

$$= \mathbf{103\ h}$$

$$(b)\ L_w = \frac{100}{\dfrac{5}{35\ h} + \dfrac{10}{130\ h} + \dfrac{20}{525\ h} + \dfrac{65}{2000\ h}}$$

$$= \mathbf{344\ h}$$

18-11 BEARING SELECTION—OTHER THAN BALL TYPES

Designers may wish to become equally familiar with the design techniques for other types of bearings available for mechanical equipment. To discuss them all in detail would be needless repetition, since they follow similar patterns, as outlined earlier. Designers should refer to individual catalogs readily available from manufacturers. Some of the more noticeable differences are discussed in the following paragraphs.

Needle Bearings

These bearings are very susceptible to shaft distortion and are not recommended if the deflection angle of the shaft at the bearing is greater than 0.5 deg. In order to reduce deflection, needle bearings are sometimes used without inner races, thus maximizing shaft diameter (Fig. 18-13). This practice, however, presents another problem. The rated capacity of the bearing is only achieved when the shaft hardness is at least 58 on the Rockwell C scale. When shaft hardness is below this value, the calculated life must be reduced.

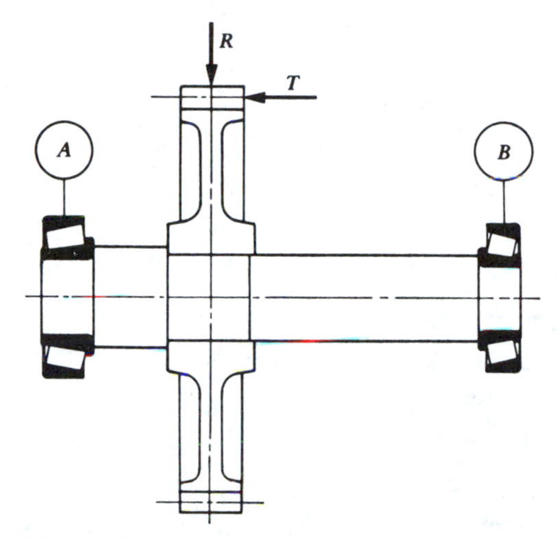

Figure 18-21 Tapered roller bearing in a back-to-back mounting.

Tapered Roller Bearings

Determination of bearing reaction loads is handled in a unique manner. Bearing centerlines projected to the shaft are located at the so-called *effective center*. This is found by a line directed towards the shaft centerline perpendicularly from the roller midpoint of contact with the outer race. This has the effect of shortening the bearing span when bearings are mounted back-to-back and lengthening it for a front-to-front mounting (Figs. 18-21 and 18-22). Combining radial and thrust loads is complex; refer to manufacturers' catalogs.

18-12 ADDITIONAL INFORMATION AVAILABLE FROM BEARING CATALOGS

After a bearing has been chosen that will satisfactorily meet load-space demands, the catalog should be further consulted in finishing the design and compiling final detailed specifications. The design itself should contain the recom-

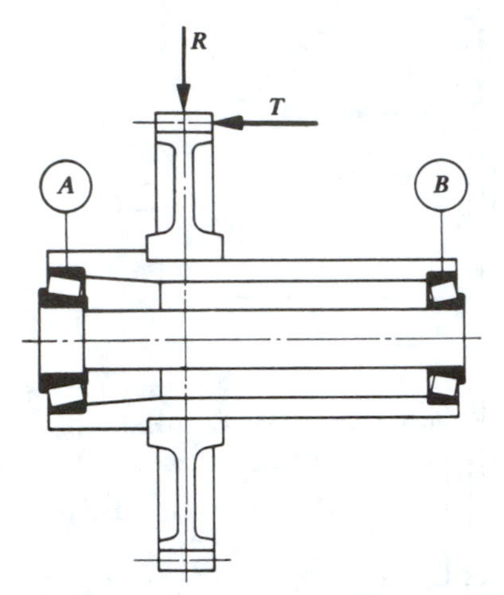

Figure 18-22 Tapered roller bearings in a front-to-front mounting.

mended shoulder heights for shaft and housing as well as connecting fillet radii. These dimensions provide the bearing's proper seating and alignment. Dimensions are also given for standard locknuts and preload springs. The use of these items, when required, offers the obvious advantages of economy in standardized parts and availability of assembly and maintenance tools.

Two of the most important dimensions for successful operation of rolling element bearings are the mounting diameters of the shaft and housing. These can be found in the vendor catalog for the various classes of fit encountered. Likewise, if the bearing is to be fitted with a snap ring for axial location, the matching shoulder in the housing must be accurately dimensioned to secure the proper position. These dimensions can also be obtained from the vendor catalog.

Finally, the bearing identification must be properly specified. Because seals and shields may be used singly, in pairs, or in combination for bearings with or without snap rings, proper identification is very important. A standard bearing is available in 12 configurations, and each has its own distinctive identification. Furthermore, each bearing manufacturer has a different identification system, thereby increasing confusion. Designers must become familiar with the various bearing catalogs, both for their own convenience and to aid in communicating with others concerning the selected bearings.

18-13 MOUNTINGS FOR ROLLING BEARINGS

Rolling bearings must be mounted so that their shape is not distorted, the rolling elements are not bound, and the inner and outer races are aligned. Self-aligning bearings can be used when the last condition cannot be met. In addition, when two or more bearings are mounted on a common shaft, the axes of all must have a common alignment.

Rolling bearings are usually mounted with one race a press-fit (interference) and the other a push-fit (zero to minimal clearance). Normally, the press-fit race is used with the *rotating* part and the push-fit race with the *stationary* part. However, where large bearings encounter heavy loads and high speeds, both races may be *pressed* into place. This is particularly true for tapered roller bearings.

Some of the most common methods of mounting roller bearings are shown in Fig. 18-23. Tapered roller bearings require an axial adjustment to obtain the proper running clearance. This is accomplished by turning the nut as in Fig. 18-23*a*, matching the cone widths as in Fig. 18-23*b*, and using shims as in Fig. 18-23*c*.

Typical methods of mounting ball bearings are shown in Fig. 18-24. Notice the various means used to maintain the axial location of the bearing on the shaft and in the housing.

18-14 LUBRICATION

Rolling element bearings require less lubrication than plain bearings because of the absence of severe sliding contact. However, proper selection of both lubrication method and lubricant is essential to good bearing performance.

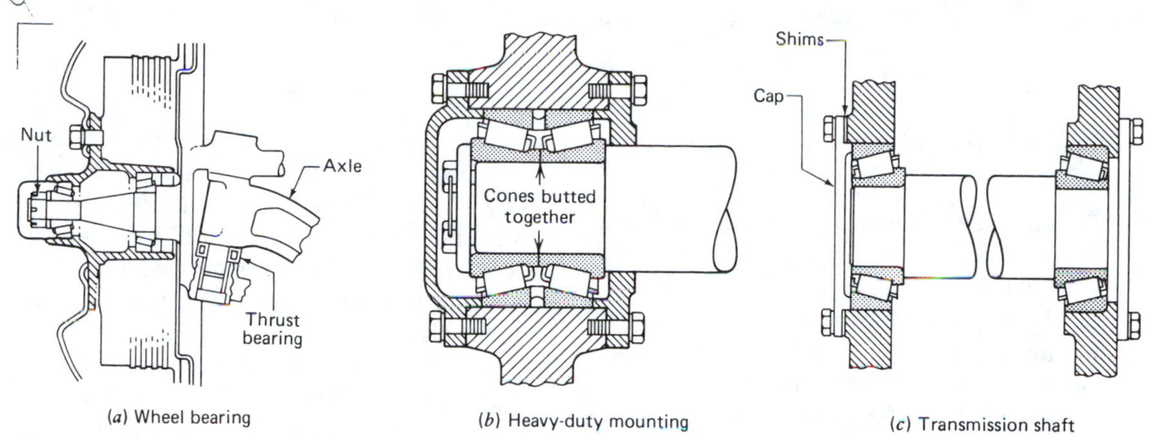

(*a*) Wheel bearing (*b*) Heavy-duty mounting (*c*) Transmission shaft

Figure 18-23 Typical roller bearing mountings, showing different means of adjustment. (Courtesy The Timken Company.)

Figure 18-24 Typical ball bearing mountings, showing different means of axial location. (Courtesy Deere and Company Technical Services.)

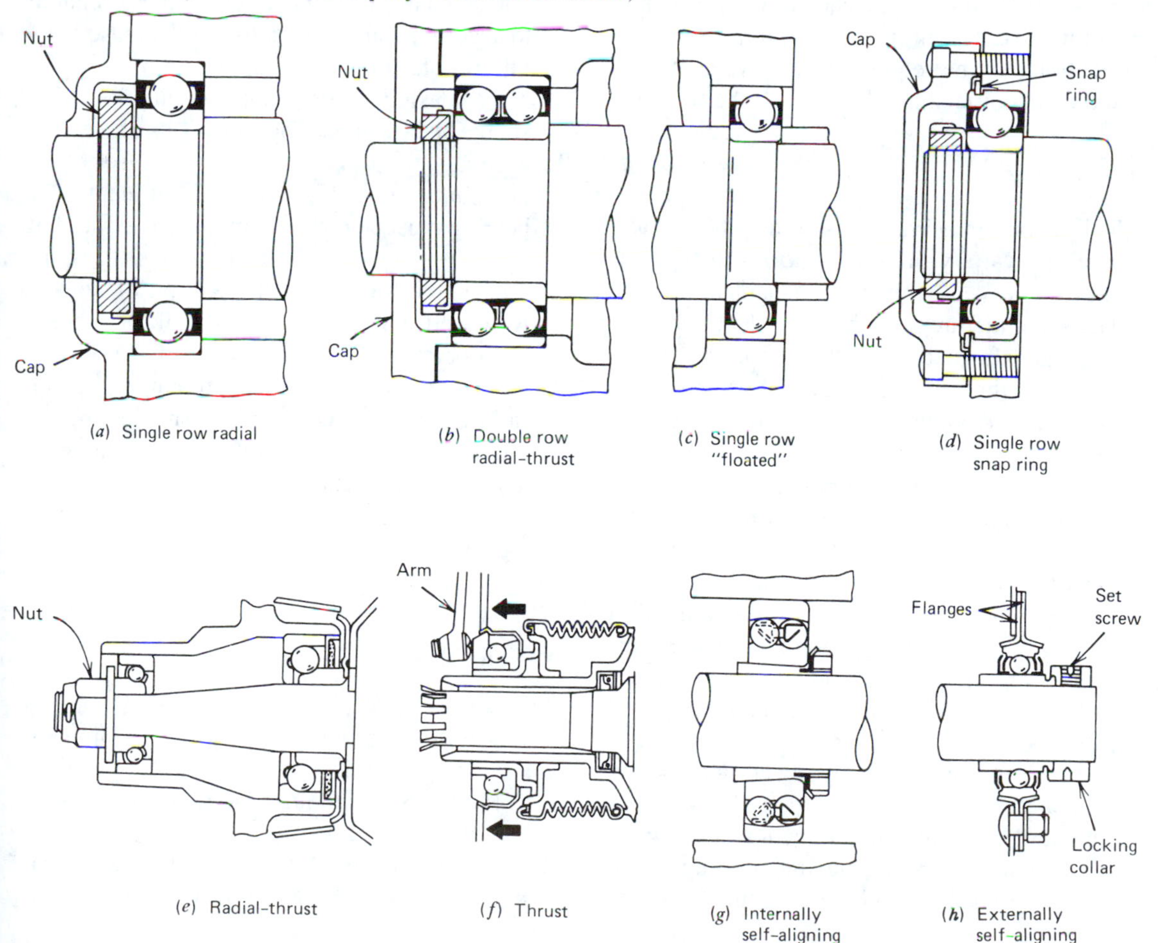

(*a*) Single row radial (*b*) Double row radial-thrust (*c*) Single row "floated" (*d*) Single row snap ring

(*e*) Radial-thrust (*f*) Thrust (*g*) Internally self-aligning (*h*) Externally self-aligning

With rolling bearings the function of the lubricant is to:

1. Reduce metal-to-metal contact in the load zone. Maximum bearing life is realized only when the rollers and race are separated by a continuous film of lubricant.
2. Lubricate the rubbing and sliding contacts between rolling elements and cages.
3. Protect finely polished surfaces from corrosion.
4. Dissipate heat when necessary.

Lubricants can be supplied to bearings in several ways. Common oil lubrication methods include oil-air mist, recirculation, oil jet, bath, splash, drip, and wick feed. Space and economics usually dictate the choice. Oil-air mist is recommended for high-speed applications. Jet lubrication may be required to remove heat generated from some source other than the bearing.

SUMMARY

Rolling bearings are precision members available in a variety of configurations to adapt to almost every conceivable mechanical device. The manufacturers of these products provide detailed information on their proper use and sizing in the catalogs they publish. To use these data, designers must determine the total loads and speeds that are present at the shaft supports of the application being considered. Gears, pulleys, sprockets, brakes, and clutches—the common machine members mounted on shafts—all generate or are actuated by forces that are, in turn, transmitted to the supporting bearings. Values for the individual force magnitudes are first determined, and the bearing reactions are then calculated from the principles of statics.

The criterion for bearing selection is usually some acceptable time basis instead of the consideration of stress or deflection. Designers must thus have available the rotating speed and the expected daily usage time and the number of years that the product must survive to be acceptable. In many products, there is no single set of operating conditions, but several. A method of combining these different conditions into a single weighted average from which bearing selection can be made is given in this chapter.

Bearing failure is a statistical phenomenon. A group of bearings operating under identical conditions will show a pattern of failures in which any two failures may vary in a ratio of 20:1 on a time scale. Bearing manufacturers usually base their ratings on a 10% failure rate. That is, out of a group of bearings operating at the catalog rating, 10% will fail within the stated time frame, and 90% will exceed that standard. Factors are given in this chapter to accommodate an increase or decrease in the failure rate for a given application.

The design job does not end with the proper sizing of bearings. The design itself must be adequate to protect the bearings from contaminants and to retain the lubricant. The bearing must be properly mounted on the shaft and in the housing. Structural integrity is also a prime consideration. Deflection of housing walls and shafts must be minimal for successful operation. Consideration of heat dissipation is also important. Premature bearing failure is often encountered by neglecting this factor. The volume of lubricant in the sump and the area of housing walls are two additional items of consideration to proficient designers.

REFERENCES

18-1 *Bearing Technical Journal*. Indianapolis: FMC Corporation, Bearing Division, 1973. (Well suited for study of rolling bearings, although beyond the scope of this chapter.)

18-2 *Engineering Data*. King of Prussia, Pa.: SKF, Inc., Bearing Group, 1980. (Contains 128 pages of engineering data. Can be obtained from a local dealer.)

18-3 *Engineering Journal,* Vol. I. Canton, Ohio: The Timken Company, 1972. (Contains engineering data and bearing capacities.)

PROBLEMS

P18-1 A bearing rotates at a constant speed of 1600 rpm 8 h/day and 5 days/wk. How many load cycles are accumulated in 12 yr?

P18-2 An astronomical observatory uses a protecting dome having a mass of 3600 kg. It is supported by 16 steel balls, each having a diameter of 100 mm. The balls are confined between two steel races with a 4.5 m pitch diameter. The coefficient of rolling friction (f_r) is $0.21/R$. What force at the race periphery is required to rotate the dome? What motor rating is required to rotate the dome at a constant speed of 0.5 rpm?

P18-3 What is the change in life of a ball bearing when:

(*a*) The load is doubled?
(*b*) The speed is halved?
(*c*) Both (*a*) and (*b*) occur simultaneously?

P18-4 A bearing is to carry a radial load of 750 lb and a thrust load of 450 lb. The bearing will be in use 40 h/wk for 3 yr. The speed of the shaft is 1000 rpm. What catalog rating is required to achieve the desired life?

P18-5 Two symmetrically located, light-series Conrad bearings of 20 mm bore support a flywheel having a mass of 200 kg. If the mounting shaft rotates at 1800 rpm, what is the B-10 life rating of the bearings? The shaft is driven by an electric motor.

P18-6 Solve P18-5 when the entire unit is placed on a 30 deg incline.

P18-7 In redesigning a transmission, a bearing is found to have an increased speed of 16% and an increased load of 22%. What increase in capacity is required to maintain the same life hours?

P18-8 What reduction in life hours would result if the bearing in P18-7 were not changed?

P18-9 A 35 mm bore, medium-series, Conrad ball bearing is operating under the following conditions: 1800 rpm, 1125 lb radial load, 900 lb thrust load. What is the life expectancy considering (*a*) 10% failure rate, and (*b*) 5% failure rate?

P18-10 Select a bearing to support a 6.0 kN radial load at 750 rpm. The thrust component is 2.5 kN. The bearing is to operate 8 h/day, 6 days/wk for 10 yr. The reliability is not critical and may be assumed as 70%. The limits of the bore size are 45 ± 5 mm.

P18-11 The front wheel (nondriving) of an automobile is mounted on two ball bearings spaced 88 mm apart (Fig. 18-23*a*). The rolling radius of the tire is 305 mm. The camber and geometry of the wheel are such that the inner bearing is directly in line with the ground contact. The ground thrust reaction due to cornering is found to be 0.3 of the weight reaction and can act either inward or outward. When the thrust is directed inward, the inner bearing sustains the thrust; when directed outward, the outer bearing sustains it. What catalog ratings should be chosen to achieve 1000 h of life at 10% failure rate? Design speed is at 100 km/h and the vertical reaction at the wheel is 7 kN.

P18-12 A transmission drive shaft is shown schematically in Fig. 18-25. The

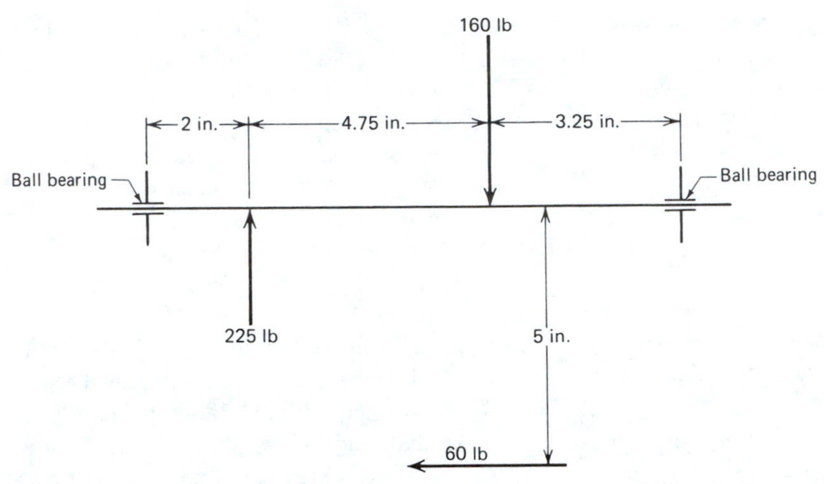

Figure 18-25 Transmission drive shaft. Illustration for Problem 18-12.

forces are all in one plane. What are the bearing lives (95% reliability) if the shaft rotates at 3600 rpm and the bearings are 30 mm bore extralight series? Assume thrust is applied on the least radially loaded bearing.

P18-13 A jackshaft in a tractor gear box is supported by a pair of bearings with the following radial loads. One of the bearings also carries the indicated centrally directed thrust loads. What is the B-10 life for a 70 mm, medium-series Conrad ball bearing for equal time distribution? Calculate the most heavily loaded bearing only.

Radial Load, kN	Thrust Load, kN	Speed, rpm
30.3	10.1	225
21.6	7.21	315
15.5	5.15	441
11.0	3.68	617
7.89	2.63	864

P18-14 The shaft in Fig. 18-26 has two V-belt pulleys mounted as shown. The tangential force of the 100 mm pulley is 760 N. The value of $e^{f'\theta}$ for each pulley is 3.5. The belt pulls are directed vertically downward for the small pulley and horizontally into the paper for the large pulley. Select the proper size light-series ball bearings to be used for an operating speed of 3000 rpm and a life expectancy of 5000 h at 80% reliability.

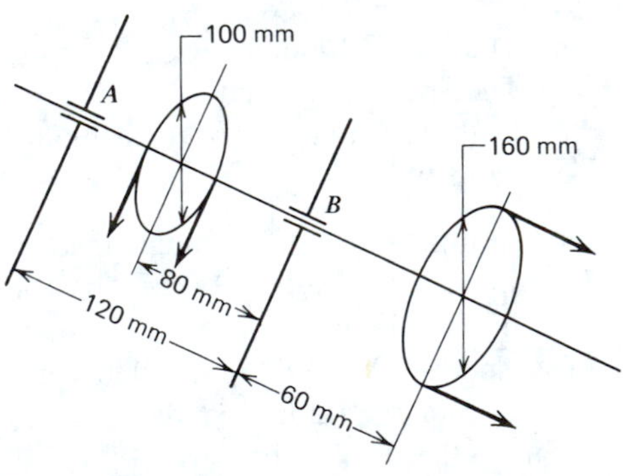

Figure 18-26 Shaft carrying two V-belt pulleys. Illustration for Problem 18-14.

P18-15 The support for a tractor final drive gear is a pair of 60 mm bore, medium-series, Max type ball bearings. Assuming 90% reliability, what is the weighted life for the following speed, load, and time usage conditions?

	Load, kN	Speed, rpm	Use, %
First	61.2	51.1	5
Second	40.8	76.7	10
Third	27.2	115	30
Fourth	18.1	172	25
Fifth	12.1	259	15
Sixth	8.05	388	10
Seventh	5.37	582	3
Eighth	3.58	874	2

P18-16 A drill press uses a 3.75 kW motor operating at 1725 rpm (Fig. 18-27). It has a pair of four-step pulleys whose diameters are 125, 100, 75, and 50 mm, respectively. The pulleys are mounted in opposition on 400 mm centers and provide four speed ratios. The vertical spindle is mounted on two ball bearings of the same size spaced 75 mm apart. The pulley is mounted outboard of the bearings, with the centerline of the largest groove 25 mm away from the centerline of the top bearing. The remaining grooves are spaced in increments of 15 mm further away from the top bearing. What size ball bearings are required to achieve 10,000 h of life at 95% reliability? Assume operation time is equally divided between the four speeds and at the rated capacity of the motor, and $f' = 0.42$.

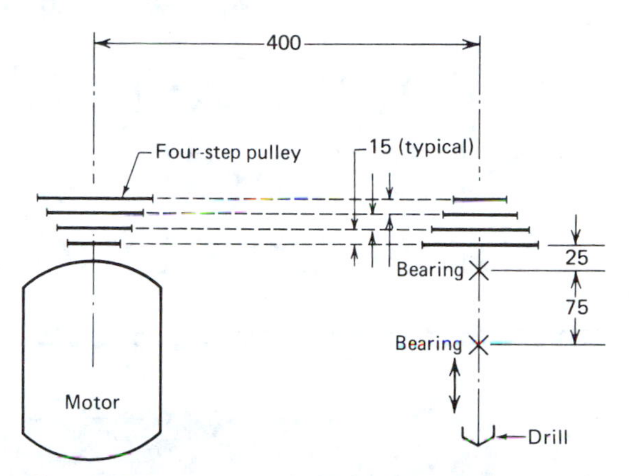

Figure 18-27 Drill press with four speeds. Illustration for Problem 18-16.

P18-17 A single-row ball bearing supports a 3000 lb radial load at 1800 rpm. If a thrust load of 1350 lb is added, find the speed change required to produce equivalent life with the same bearing.

P18-18 A bearing has a radial load of 600 lb. Which types of bearings should be used for a rated life of 30,000 h?

P18-19 A bearing has a radial load of 26 kN. Which types of bearings should be used for a rated life of 30,000 h?

P18-20 A shaft is to rotate at 100 rpm and have a rated life of 30,000 h. Which bearing parameter determines the type of bearing to be used?

PART 5

Machine Elements for Transmitting and Controlling Rotary Power and for Sealing and Enclosing Fluids

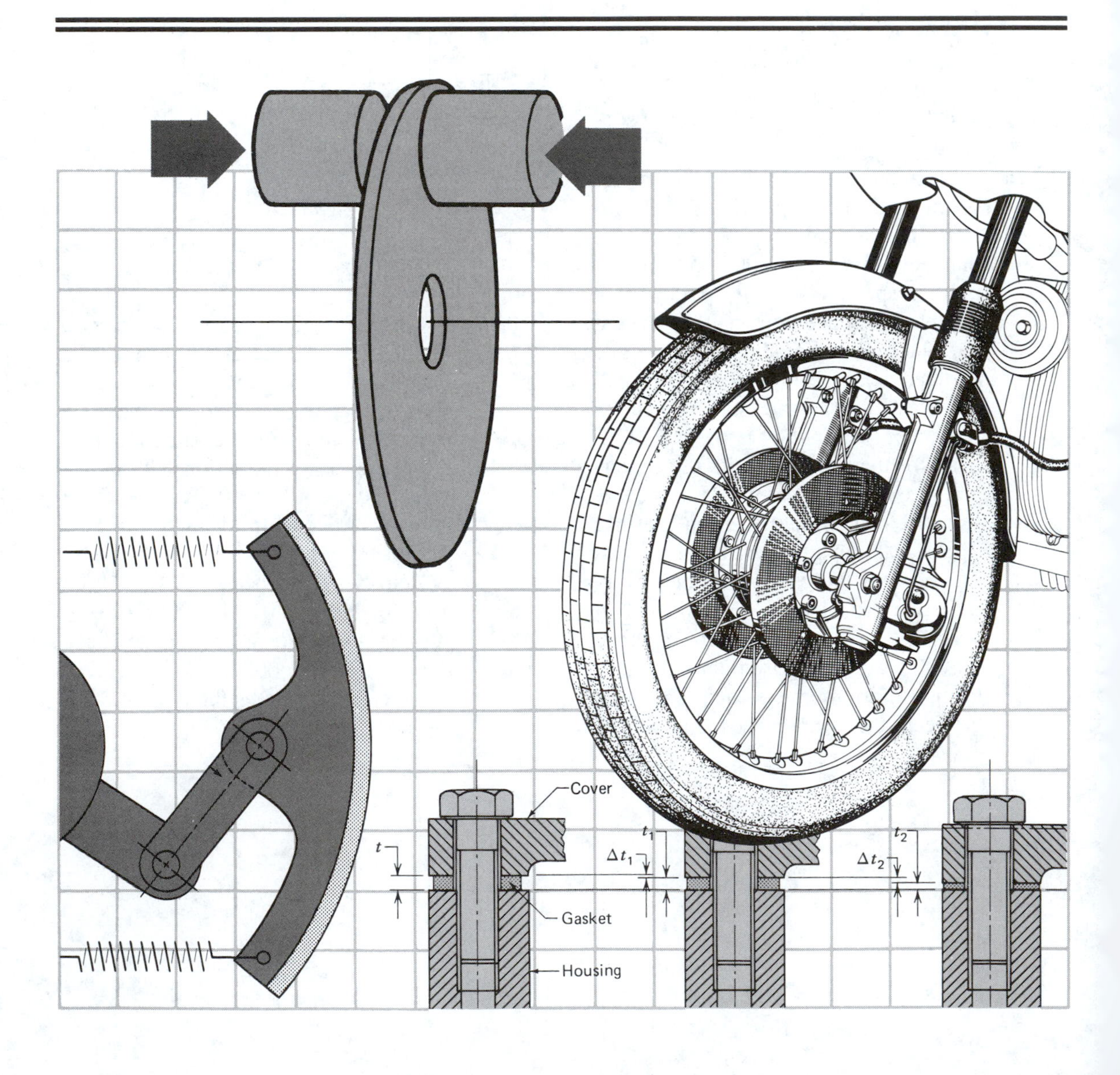

Chapter 19
Mechanical Clutches

Experience is the best of school masters but the fees are high.

THOMAS CARLYLE

NOMENCLATURE

d = inside diameter of clutch facing
D = outside diameter of clutch facing
f = coefficient of friction
F_a = actuating force
F_n = contact force
F_D = drive factor
F_L = load factor
F_S = starting factor
K_s = service factor
m = mass of shoe
n = rotational speed
n_e = engagement speed
N_p = number of pairs of contacting faces
p_{all} = allowable contact pressure
p_{ave} = average contact pressure
p_{max} = maximum contact pressure
P = tension in garter spring
r = radius to center of gravity of shoe
R = inner radius of drum
R_e = effective friction radius
T = nominal torque capacity
T_{des} = design torque capacity
T_{total} = total torque capacity

Clutches control the flow of power to ensure the safe and effective operation of machinery. The most familiar use of the clutch is to connect a loaded shaft to the output shaft of a prime mover to bring the load up to speed smoothly and gradually. The mechanical clutch is largely a product of the automobile. Internal combustion engines will not start under load. They lack power at low speed and need a device that will delay power takeoff until the engine has gained speed. Friction clutches are the type of clutch most frequently used. The friction clutch slips until the load has been brought up to speed; then it acts simply as a coupling.

Transmitting power is not the only reason for using clutches. They appear as important parts of the steering systems of tracked vehicles such as army tanks or crawler tractors. In industry they are employed to achieve smooth engagement and disengagement of tools and work pieces.

Although they are important, frictional clutches are not the only type of clutch. In the synchromesh device in a manual shift transmission, for example, a friction clutch is used to bring two shafts to be connected to approximately the same speed. Then a positive-contact clutch, typically a jaw clutch, is engaged to provide a nonslip coupling. In automatic transmissions, on the other hand, overrunning clutches are essential parts of the gear-changing mechanisms. They permit two rotating members to run independently in one direction but lock them together if the relative motion is reversed.

19-1 FUNCTION

A clutch consists principally of two main sections that are engaged or disengaged either at will by a hand- or foot-operated controlling device or by some power-driven device. In modern production machinery, however, clutch activation without power assistance is rare. The actuating forces are generally provided by air or oil under pressure or by magnetic fields. Safety as well as a continuing trend toward automation has recently focused attention on clutch-brake systems that respond to electrical commands. For emergency stops, an electrical system responds much faster than hydraulic or pneumatic systems.

In clutches power transmission is achieved through (1) interlocking, (2) friction, (3) wedging, (4) magnetic fields, or (5) moving fluids. Clutch design is thus based on strength and wear considerations. The kinematics involved are generally very simple.

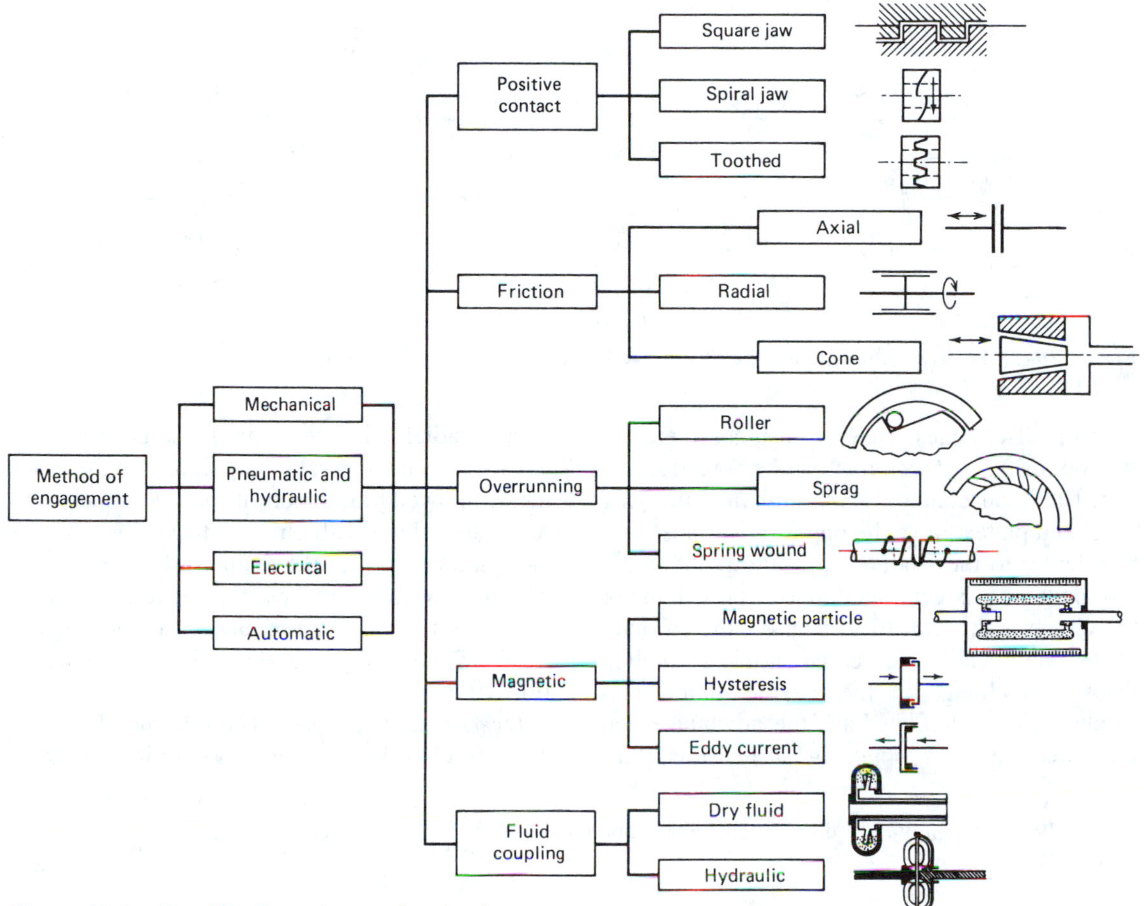

Figure 19-1 Classification schemes for clutches.

Special-purpose clutches perform a variety of jobs such as maintaining constant speed or constant torque, imposing limitations on power and torque, and achieving quick starts and stops.

19-2 CLUTCH TYPES[1]

A classification scheme for clutches must reflect the answers to at least these two questions.

1. What physical principle is used to transmit torque from one member to another?
2. What is the means by which the members are engaged or by which their relative speed is controlled?

A classification scheme responsive to these two questions is shown in Fig. 19-1.

Coupling Methods

Positive-contact clutches. The engaging surfaces interlock to form a rigid mechanical junction. Commercially available (Fig. 19-2) are *square-jaw clutches, spiral-jaw clutches* with intermeshing ratchet teeth that permit operation in only one direction, and *tooth clutches* that use a variety of gear-tooth designs.

Frictional clutches. These are the most com-

[1]Based largely on an article by John Proctor (Reference 11).

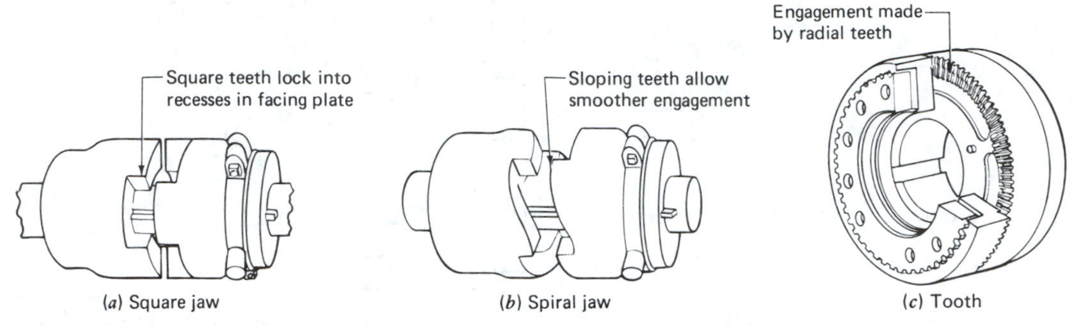

Figure 19-2 Positive contact clutches. (Courtesy Penton Publishing Company.)

mon of all clutches. In each instance there are two opposing surfaces that can be forced into a firm frictional contact. *Axial clutches,* utilizing disks and plates, have the opposing surfaces perpendicular to the shafts. Up to the point where too much heat is generated in a given volume or the clutch is difficult to disengage, the capacity of the clutch can be increased simply by adding disks. No change in the outside diameter is needed. *Cone clutches* have the advantage that the force between opposing surfaces is principally radial. Thus the axial force necessary to engage the clutch is relatively light. On the other hand, the wedging action of the two cones may make the clutch difficult to disengage. In the earliest automobiles the main clutch was a cone clutch. Now disk clutches are used almost exclusively as the main automotive clutches in the United States. A typical disk clutch is shown in Fig. 19-3.

Overrunning clutches. The two members can move freely relative to one another in one direc-

Figure 19-3 A heavy-duty disk clutch for dry operation. (Courtesy John Deere Service Publications.)

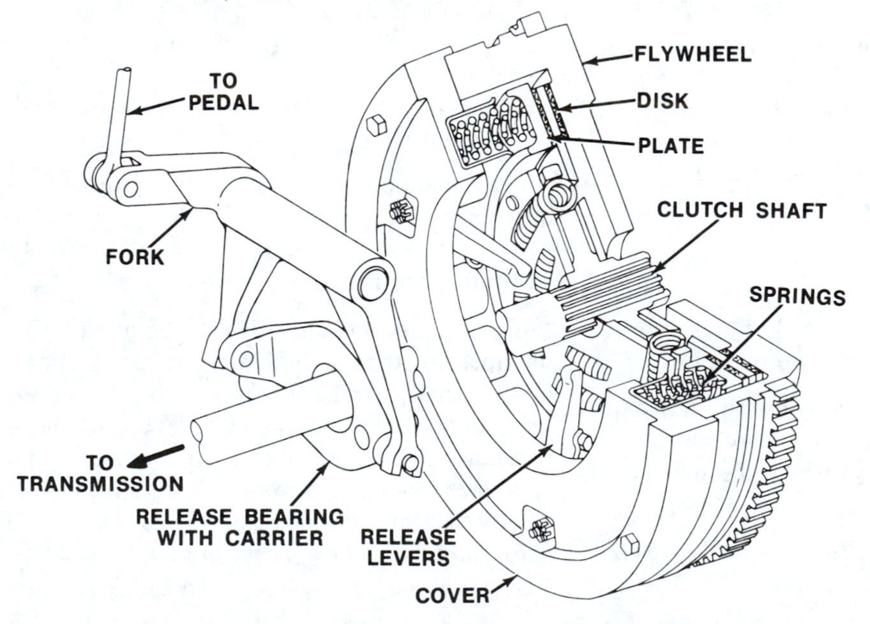

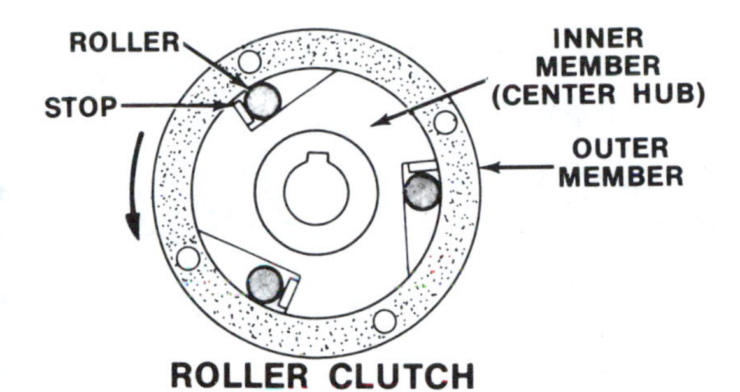

Figure 19-4 A roller/ramp type of overrunning clutch. (Courtesy Deere and Company Technical Services.)

tion but are locked together in the other. In *roller-ramp clutches* (Fig. 19-4) the members are wedged together by concentric rollers (or balls) riding on a race with a slight cam profile. In *sprag clutches* (Fig. 19-5) eccentric cams are pinched between concentric races. A large torque can be developed in *spring-wound clutches* by the gripping action of a helical spring when rotation tends to wind up the helix.

Magnetic clutches. In this instance magnetic forces are used to couple the clutch members. *Magnetic particle clutches* (Fig. 19-6*a*) make use of a ferromagnetic powder in a magnetic field to transmit the torque from one member to the other. The powder, mixed with a lubricant, partially fills an annular gap between the members. When a magnetic field is induced by a direct-current coil, the iron particles, by their mutual attraction, form chains that provide a torque capability between the drive and driven members. *Hysteresis clutches* (Fig. 19-6*b*) directly couple

Figure 19-5 A sprag or cam type of overrunning clutch. (Courtesy Deere and Company Technical Services.)

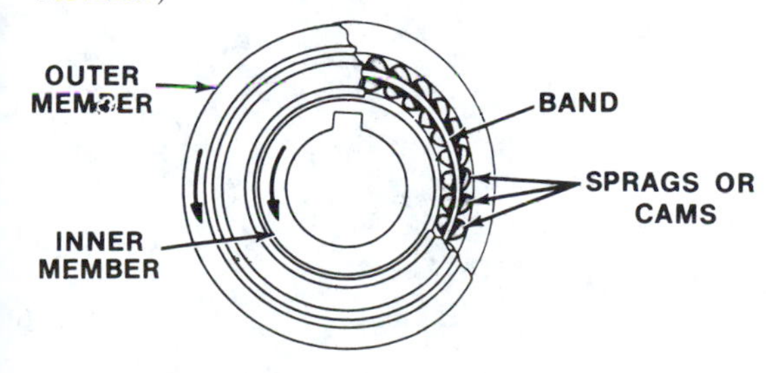

two members provided the load does not exceed the torque rating. They also can slip continuously to maintain a constant torque output independent of speed. Although similar in construction to the hysteresis clutch, the *eddy-current clutch* (Fig. 19-6*c*) requires slip to develop torque, which is the result of dissipation of eddy currents through the rotor ring's electrical resistance. If an electromagnetic unit is used, the output torque can be varied over a wide range by varying the excitation current.

Fluid couplings. In *dry-fluid couplings* a quantity of heat-treated steel shot is thrown centrifugally to the perimeter housing (keyed to the input shaft) when the input shaft starts to turn. It is packed against the rotor, which transmits power to the load. At a predetermined full speed, the shot charge is packed solid and the housing and rotor lock together with no slip unless the torque becomes excessive. The *hydraulic coupling* employs oil to transfer energy.

Control Methods

If the primary function of the clutch is to operate, in effect, as an on-off switch—to transmit full power if engaged, none if disengaged—the only control needed is the control of engagement itself. But if the clutch is to serve as a variable operating control or as an automatic safety device, the control must be more subtle. It requires control of the rate of slip.

Mechanical control. A variety of linkages, balls or rollers working over cams, rocking wedges and sliding wedges, or other mechanical devices are used to achieve a mechanical advantage. The clutch can be actuated manually by means of a lever, or the shifting can be done by solenoid, electric motor, air cylinder, or hydraulic ram.

Electrical control. Standard electromagnetically operated friction and tooth clutches are engaged electrically and spring-released. That makes them fail-safe because, if power fails, the clutch is disengaged. However, spring-engaged and electrically disengaged clutches are also

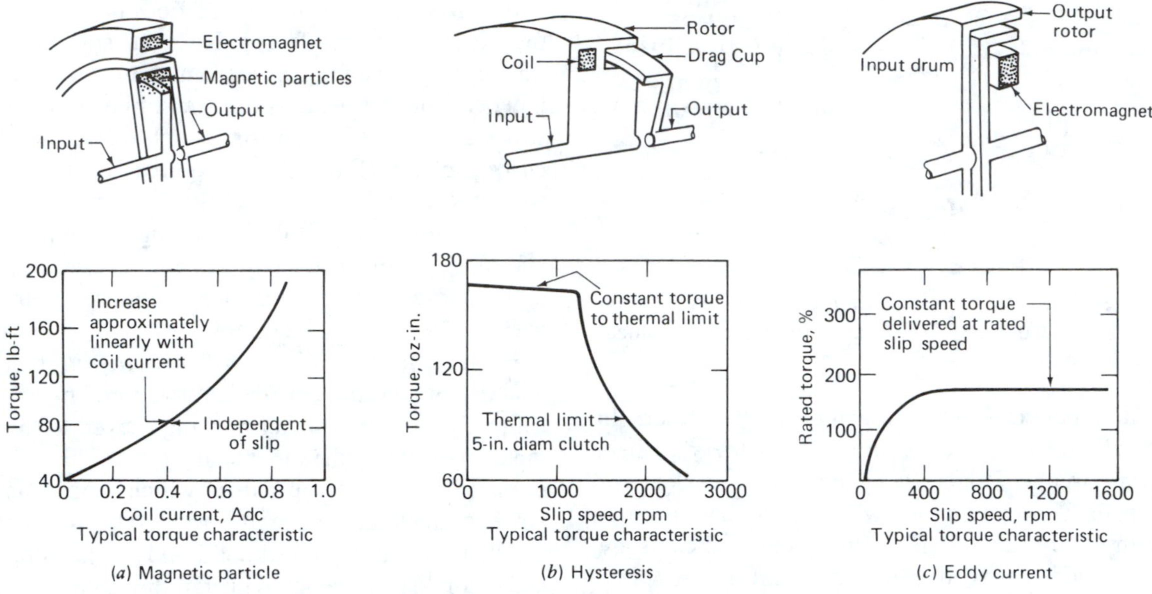

Figure 19-6 Magnetic clutches. (Courtesy Penton Publishing Company.)

available for applications where shafts are coupled for much longer periods than they are uncoupled.

Pneumatic or hydraulic control. Some clutches make use of pistons or pressure plates to move the actuating linkage. In other cases, direct air actuation is used. Some clutches use an inflatable tube or gland to apply the pressure needed for engagement. An example is shown in Fig. 19-7. These clutches offer close control of the gripping force, and they are also fail-safe, since an air or hydraulic failure causes the clutch to disengage. Multiple-disk wet clutches in automatic transmissions generally have hydraulically actuated pistons that press directly against the end disk.

Automatic control. Instead of responding directly to an external signal, they are "programmed" to react to predetermined conditions as they arise. Sensitivity to speed is characteristic of friction and dry-fluid couplings that work on centrifugal force. With both hydraulic couplings and eddy-current clutches, torque is automatically regulated by the slip.

19-3 SERVICE FACTORS

In order to start a load from rest and accelerate it, a clutch should have a torque capacity substantially higher than the nominal torque require-

Figure 19-7 An air-tube actuated drum clutch. (Courtesy Eaton Corporation, Industrial Drive Division.)

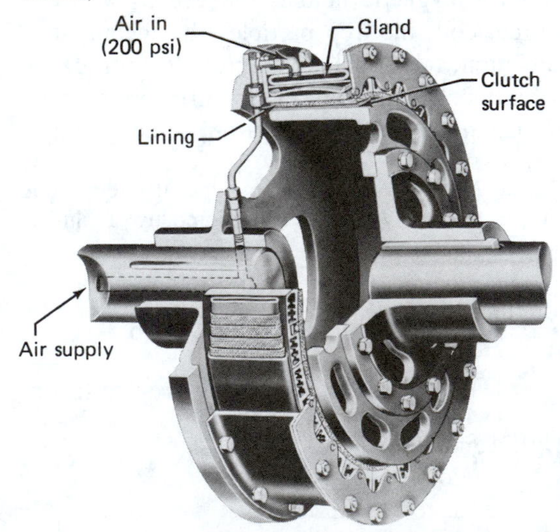

TABLE 19-1 Torque Factors for Friction Clutches

Starting Factor F_S		Drive Factor F_D		Load Factor F_L	
Free start with no load	1.0	Nonpulsating (e.g., three-phase motor)	1.0	Nonpulsating (e.g., blower, centrifugal pump, generators under steady load)	1.0
Average inertia load	2.0	Moderate pulsation (e.g., single-phase motor or multicylinder engine)	1.5	Moderate shock (e.g., multicylinder pump)	1.5–1.75
High inertia load	3.0	Severe pulsation (e.g., single-cylinder gas engine)	2.0	Severe shock (e.g., crane, shovel, single-cylinder compressor, rock crusher)	2.0–3.0

Source. A. F. Gagne, Jr., "Torque Capacity and Design of Cone and Disk Clutches," *Product Engineering*, December 1953.

ment so that the clutch can accelerate that load without excessive slip. If either the prime mover torque or the load torque fluctuates, the torque capacity of the clutch must be made greater than the nominal value so that it can cope with momentary peaks.

A. F. Gagne, Jr. (Reference 4) has recommended the following technique for calculating clutch capacity for design purposes. His equation is

$$T_{des} = K_s T \qquad (19\text{-}1)$$

where K_s is the service factor that takes into account starting and torque fluctuations. The service factor K_s is calculated by the following equation.

$$K_s = \sqrt{F_S^2 + F_D^2 + F_L^2 - 2} \qquad (19\text{-}2)$$

where F_S, F_D, and F_L are the starting, prime mover, and load factors. Note that if all three factors are unity, the service factor is also. Recommended values of F_S, F_D, and F_L are presented in Table 19-1.

Gagne devised this expression on the assumption that it is unlikely that the prime mover torque and the load torque will peak at the same time. In Eq. 19-1, T is the average torque capacity needed to handle the load once it has been brought up to speed. But T_{des} is the torque capacity on which the clutch design is based.

EXAMPLE 19-1

An eight-cylinder diesel engine developing a full-load power of 185 hp at 1350 rpm is being used to drive a rock crusher in a limestone quarry. Although the engine will be brought up to speed before any rocks are fed in, the crusher itself is a high inertia load. What should be the design torque for a friction clutch located between the engine and the rock crusher?

SOLUTION

(*a*) The nominal torque corresponding to 185 hp at 1350 rpm is found by using Eq. 1-10. Hence

$$T = \frac{63{,}000P}{n}$$

$$= \frac{(63{,}000)(185 \text{ hp})}{(1350 \text{ rpm})} = 8637 \text{ lb-in.}$$

(*b*) Using Table 19-1, these values for the torque factors were selected.

$$F_S = 3.0$$

$$F_D = 1.5$$

$$F_L = 3.0$$

(*c*) Equation 19-2 is applied to find the service factor.

$$K_s = \sqrt{F_S^2 + F_D^2 + F_L^2 - 2}$$

$$= \sqrt{(3)^2 + (1.5)^2 + (3)^2 - 2} = 4.272$$

(*d*) The design torque is, by Eq. 19-1,

$$T_{des} = K_s T$$

$$= (4.272)(8637 \text{ lb-in.})$$

$$= \mathbf{36{,}900\ lb\text{-}in.}$$

19-4 FRICTION CLUTCHES

Most clutches make use of a friction torque transmitted between mating surfaces. In the typical friction clutch, contact occurs between a *replaceable* lining and a *metallic* surface. Torque is developed due to friction forces acting some distance from the axis of rotation. These friction forces are proportional to the applied normal forces pressing the disks together. For maximum capacity and minimum wear, the normal forces must be carefully controlled. Many designs make use of coil springs to clamp the friction elements together and an activating mechanism for temporary disengagement. By a gradual increase of the normal force, engagement can be made smoothly. When engaged, the clutch members tend to rotate as a single unit; but they are free to slip when the input and output torques are not in equilibrium. The ability to slip during engagement or under shock loading is a primary advantage of friction clutches, but heat generated by slip must be carefully anticipated to avoid any burning of friction faces or distortion of mounting faces.

Types

There are three major types of friction clutches, based on the plane in which the normal or actuating forces operate. They are the *axial, radial,* and *cone* types.

Axial types. These include all single- and multiple-disk or plate clutches. The contact pressure is applied axially, while the friction forces act in a plane (or planes) perpendicular to the shaft axis. The simplest form has two flanges mounted on the end of in-line shafts, with some means of forcing the faces into contact. Standard models are generally more complex. Here one or more friction disks are clamped between driving and driven plates. The torque capacity of such clutches can be readily multiplied by increasing the number of plates. Multidisk clutches are ideal for static or low-speed applications where high static torque is required. In clutches for high-speed dynamic use, the fewer the friction surfaces, the better because of fewer cooling problems (e.g.; in cars and tractors).

Disk clutches are used extensively with engine and transmission units to facilitate changing gears. A compact design is achieved by utilizing one side of the engine flywheel as the friction surface of a single-plate disk clutch. Figure 19-8 shows such an arrangement schematically. A commercial dry disk clutch was shown in Fig. 19-3.

Radial types. These include all drum, rim, shoe, and band clutches. They operate on a common axis with the shaft. The friction surface is cylindrical, and the opposing mechanisms move into action radially, either by expanding outward toward the rim or by contracting inward toward the hub. Clutches of this type are often built directly into pulleys and flywheels.

Cone types. These are of a composite type. The clutch surfaces are slightly tapered, so the load between the surfaces is primarily radial. However, the two members are engaged by pushing them together axially—rather like pushing a tapered cork into a tapered bottleneck. A

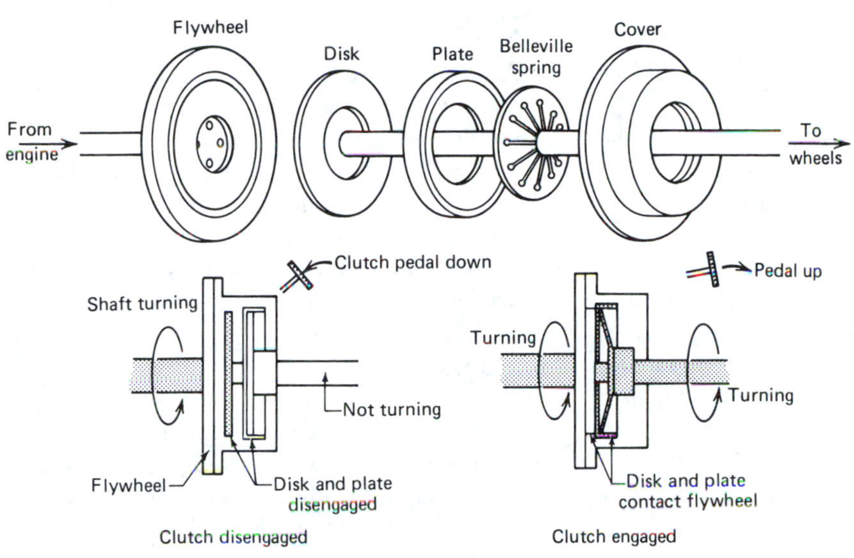

Figure 19-8 Operation of a friction clutch. (Courtesy Deere and Company Technical Services.)

fairly heavy torque capacity can be achieved with a relatively light axial force, but the clutch has a tendency to grab and to resist disengagement.

19-5 DESIGN OF DISK CLUTCHES

A schematic drawing of a disk clutch with just one pair of mating surfaces is shown in Fig. 19-9. The disks are pressed together with a force F_a so that a torque capacity T is developed. This capacity is shown in SI units and in English units.

$$T = \frac{fF_aR_e}{1000}\ \text{N}\cdot\text{m} \tag{19-3}$$

$$T = fF_aR_e\ \text{lb-in.} \tag{19-4}$$

where

T = torque capacity; N · m, lb-in.
f = coefficient of friction
F_a = actuating force; N, lb
R_e = effective friction radius; mm, in.

When a disk clutch is new, the pressure distribution across the lining is fairly uniform. Once the clutch has worn in, the pressure distribution changes to produce a uniform rate of wear across the face of the lining. Then the maximum contact pressure occurs at the inside radius.

The effective friction radius is simply the average radius

$$R_e = \frac{D + d}{4} \tag{19-5}$$

Figure 19-9 Schematic drawings of a disk clutch.

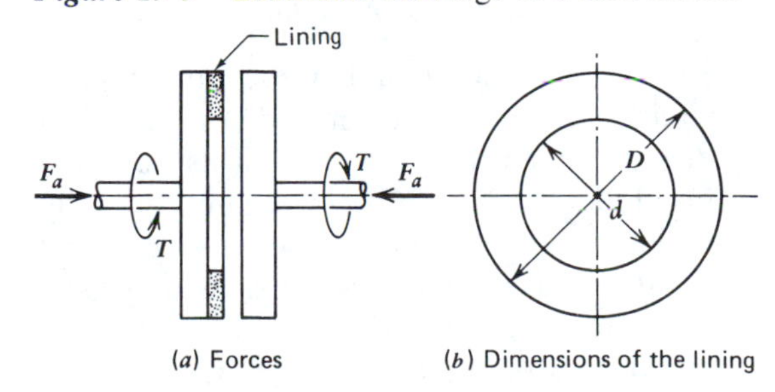

where

D = outside diameter of contact area; mm, in.
d = inside diameter of contact area; mm, in.

For any selected outside diameter D, the torque capacity is *maximized* for $D/d = 1.732$. Good practice is to use a D/d ratio between 1.5 and 2.

To keep clutch wear to an economical level, manufacturers recommend values for the maximum contact pressure. Table 19-2 lists manufacturers' recommendations. The maximum contact pressure at the inside diameter of the disk should not exceed the allowable pressure p_{all} selected from the table.

The actuating force F_a depends on the disk dimensions and the allowable contact pressure. It is

$$F_a = 0.5\pi d(D - d)p_{all} \qquad (19\text{-}6)$$

Some manufacturers of clutch linings recommend an average contact pressure, so it can be useful in design to have a relationship between the maximum and the average contact pressures.

$$\frac{p_{max}}{p_{ave}} = 0.5\left(\frac{D}{d} + 1\right) \qquad (19\text{-}7)$$

For example, for $D/d = 1.732$, $p_{max}/p_{ave} = 1.37$.

If there are N_p pairs of contacting faces, the torque capacity is

$$T = \frac{fF_aR_eN_p}{1000}\ \text{N}\cdot\text{m} \qquad (19\text{-}8)$$

$$T = fF_aR_eN_p\ \text{lb-in.} \qquad (19\text{-}9)$$

There is a subtle problem, however. In a typical multidisk clutch the mating disks slide on splines. Friction between the disks and the splines reduces the effective actuating force, by as much as 10% per pair of successive disks. Thus, after four or five pairs, the actuating force has been reduced seriously.

Derivations of these basic equations will be found in Section 19-10.

EXAMPLE 19-2

A clutch is needed between a 24-kW, three-phase motor running at 1800 rpm and a large multicylinder pump. A disk clutch operating dry has been selected. Space limitations dictate a maximum outside diameter of 195 mm for the disks. Determine the principal dimensions of the clutch and the actuating force needed.

SOLUTION

(*a*) Some initial design decisions must be made. Let us use molded phenolic plastic lining operating dry with $D = 195$ mm and $d = 195/1.732 = 133$ mm. Table 19-2 provides values for coefficient of friction and maximum contact pressure. The tabulated values are $f = 0.25$ and $p_{all} = 0.69$ MPa.

(*b*) The nominal torque is found by using Eq. 1-7 and solving for T.

$$T = \frac{9550P}{n}$$

$$= \frac{(9550)(24)}{1800} = 127.3\ \text{N}\cdot\text{m}$$

(*c*) Gagne's method can be used to find the design torque. Table 19-1 is used to select appropriate starting, drive, and load factors. Use $F_s = 2.0$, $F_D = 1.0$, and $F_L = 1.75$. The service factor is, by Eq. 19-2,

$$K_s = \sqrt{F_S^2 + F_D^2 + F_L^2 - 2}$$

$$= \sqrt{(2.0)^2 + (1.0)^2 + (1.75)^2 - 2} = 2.46$$

The design torque is, by Eq. 19-1,

$$T_{des} = K_sT$$

$$= (2.46)(127.3\ \text{N}\cdot\text{m}) = 313.4\ \text{N}\cdot\text{m}$$

(*d*) The effective friction radius R_e is found by use of Eq. (19-5).

TABLE 19-2 Friction Materials for Clutches

Contact Surfaces		Friction Coefficient[a]					
Wearing	Opposing	Wet	Dry	Maximum temperature, °C	Allowable pressure,[b] MPa	Relative cost	Comment
Cast bronze	Cast iron or steel	0.05	—	149	0.55–0.83	Low	Subject to seizing
Cast iron	Cast iron	0.05	0.15–0.2	316	1.03–1.72	Very low	Good at low speeds
Cast iron	Steel	0.06	—	260	0.83–1.38	Very low	Fair at low speeds
Hard steel	Hard steel	0.05	—	260	0.69	Moderate	Subject to galling
Hard steel	Hard steel, chromium plated	0.03	—	360	1.38	High	Durable combination
Hard-drawn phosphor bronze	Hard steel, chromium plated	0.03	—	260	1.03	High	Good wearing qualities
Sintered metal	Cast iron or steel	0.05–0.1	0.1–0.4	538	1.03	High	Good wearing qualities
Sintered metal	Hard steel, chromium plated	0.05–0.1	0.1–0.3	538	2.07	Very high	High energy absorption
Wood	Cast iron or steel	0.16	0.2–0.35	149	0.41–0.62	Lowest	Unsuitable at high speed
Leather	Cast iron or steel	0.12–0.15	0.3–0.5	90.3	0.07–0.27	Very low	Subject to glazing
Cork	Cast iron or steel	0.15–0.25	0.3–0.5	90.3	0.05–0.1	Very low	Cork-insert type preferred
Felt	Cast iron or steel	0.18	0.22	138	0.03–0.07	Low	Resilient engagement
Vulcanized fiber or paper	Cast iron or steel	—	0.3–0.5	90.3	0.07–0.27	Very low	Low speeds, light duty
Carbon graphite	Steel	0.05–0.1	0.25	359–538	2.07	High	For critical requirements
Molded phenolic plastic, macerated cloth base	Cast iron	0.1–0.15	0.25	149	0.69	Low	For light special service

[a]Conservative values should be used to allow for possible glazing of clutch surfaces in service and for adverse operating conditions.
[b]Where a range of values is given for allowable pressure, use the lower end of the range for dry operation and the higher end for wet operation.

$$R_e = \frac{D + d}{4}$$

$$= \frac{195\text{ mm} + 113\text{ mm}}{4} = \mathbf{77\text{ mm}}$$

(*e*) Limiting the maximum contact pressure implies limiting the actuating force. For $p_{\text{all}} = 0.69$ MPa and by Eq. 19-6,

$$F_a = 0.5\pi d(D - d)p_{\text{all}}$$

$$= 0.5\pi(113\text{ mm})$$

$$(195\text{ mm} - 113\text{ mm})(0.69\text{ N/mm}^2)$$

$$= \mathbf{10\ 040\ N}$$

(*f*) Finally, the number of pairs of mating surfaces needed can be found by using Eq. 19-8.

$$N_p = \frac{1000T}{fF_aR_e}$$

$$= \frac{(1000)(313.4\text{ N}\cdot\text{m})}{(0.25)(10\ 040\text{ N})(77\text{ mm})}$$

$$= \mathbf{1.62\ pairs}$$

Two pairs should suffice, even taking into account some loss of actuating force due to friction between splines and disks.

19-6 FRICTION MATERIALS FOR CLUTCHES

A friction material (e.g., sintered metal) is placed on only *one* of the two mating surfaces having relative motion. The reason is *heat dissipation*. If both surfaces were lined with a friction material, heat could not be removed fast enough. Thus one of the surfaces is usually steel or cast iron.

A friction material suitable for clutches will provide a high coefficient of friction, resist seizing of the mating surfaces, and have a low enough wear rate to guarantee an economically acceptable period of operation between relinings.

Dry Operation

Operating a clutch dry implies that the coefficient of friction is *high* and the torque capacity is high. It is then essential to maintain the dry operation. Any contamination will reduce torque capacity and cause erratic behavior. If a dry clutch is operating near well-lubricated machinery, sealing can be a problem. Of course, the absence of a lubricant makes heat dissipation more difficult.

Wet Operation

This implies a *lower* coefficient of friction and a lower torque capacity. There are also some advantages. Engagement is smoother. Heat dissipation is less difficult, since the lubricant will carry away much of the heat. Wear rates are far less, about 1% of the rate expected for dry operation.

Since the coefficient of friction is reduced, there must be a larger actuating force, more clutch plates, or larger clutch plates to provide an adequate torque capacity. Fortunately, wet operation eases the problems of wear and heat dissipation enough that a much larger actuating force can be used than would be feasible in dry operation.

Wet clutch facings are grooved to provide for the escape of lubricant. Thus any design calculations involving face area should be based on the net area. Subtract the groove area from the disk area to get the net area.

Material Properties

Table 19-2 provides data suitable for use in preliminary design calculations. They are representative only. Designers should consult manufacturers of friction materials for more precise data. The maximum pressure listed should not be exceeded, both to avoid excessive buildup of temperature in the lining material and to keep down the rate of wear.

19-7 PLACEMENT OF THE CLUTCH

Consider a paper mill driven by a multicylinder diesel engine. At optimum operating speed, the engine runs too fast for direct drive to the mill. A gear reduction unit is thus used to reduce the speed. A clutch is also needed so that the engine can be brought up to speed before taking over the load. Where should it be located, at the engine side or at the paper mill side of the reduction unit?

Since power is the product of torque and speed, for any given power to be transmitted, the greater the speed the lower the torque needed. It seems logical to place the clutch at the high-speed end, between the engine and the gear reduction unit. Indeed, this is where it would usually be located. However, higher speed, combined with the smaller physical size of the clutch, will tend to increase the problems of heating and wear. So various manufacturers have suggested that at times it is worth the greater initial cost of a larger clutch at the low-speed end of the gear unit to achieve lower maintenance costs and energy losses.

The locations of the clutches for an earth compactor and a recreational vehicle (a go-cart) are shown in Fig. 19-10.

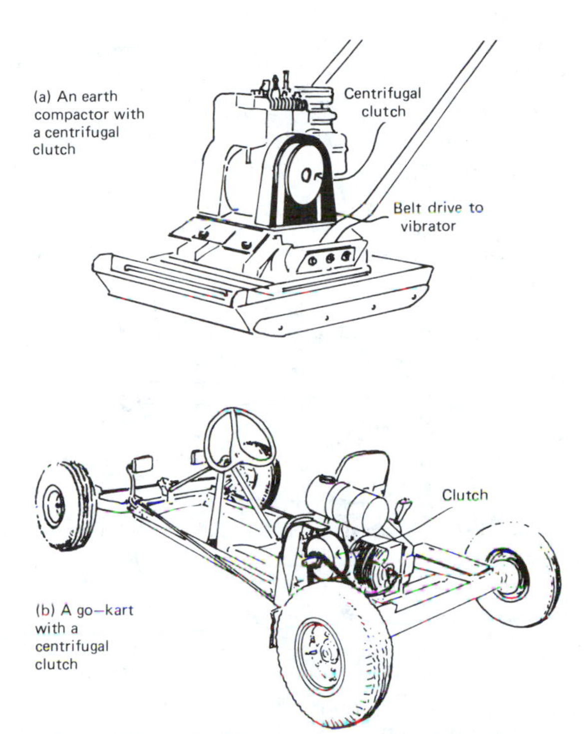

Figure 19-10 Typical locations of the clutch. (Courtesy Mercury Products, a Division of Dyneer Corp.)

19-8 CENTRIFUGAL CLUTCHES

Suppose you have to select a clutch for a golf cart. Frequent starts and stops are the rule, so it would be convenient if a single floor pedal sufficed to control speed and switch the motor on and off. A centrifugal clutch is a practical and economical choice. When the golfer's foot leaves the pedal and the motor slows down, disengagement is automatic. When the pedal is pressed to start the motor and move on, the engagement is very smooth because the motor has the chance to accelerate to near operating speed before it has to take up the load. Chain saws, riding lawn mowers, and small recreational vehicles are other familiar applications of centrifugal clutches.

The centrifugal clutch is practical also in a variety of heavy-duty applications where a high-inertia load must be brought up to speed. By providing a *time delay* sufficient to permit the prime mover to gain momentum before taking over the load, centrifugal clutches provide smooth engagement. And, since a large torque output at low speed is unnecessary, a substantially smaller prime mover is needed. Furthermore, the speed at which engagement occurs can be controlled by adjusting the tension of clutch springs. Heavy mobile equipment such as cranes, cement kilns, and ball mills have made good use of centrifugal clutches. Indeed, these clutches have been used so frequently that for starting applications they are often referred to generically as *starting clutches*.

The simplest design is the free-shoe type shown in Fig. 19-11. It has three members. The input member, or *spider*, delivers power from

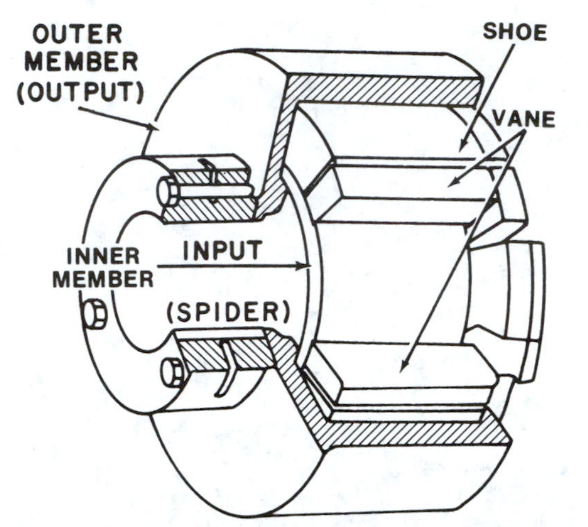

Figure 19-11 Free-shoe centrifugal clutch. (Courtesy Deere and Company Technical Services.)

the prime mover. Weights are thrown radially outward as the input shaft rotates. They bear against the inner surface of a drum connected to the output pulley and transfer torque through friction. When the drum reaches operating speed, slip ceases and the clutch acts simply as a coupling. By adding a garter spring in circumferential grooves in the outer portions of the weights (Fig. 19-12), designers can preset engagement speed by providing the correct tension at assembly.

Figure 19-12 Free-shoe centrifugal clutch with garter spring to regulate engagement speed.

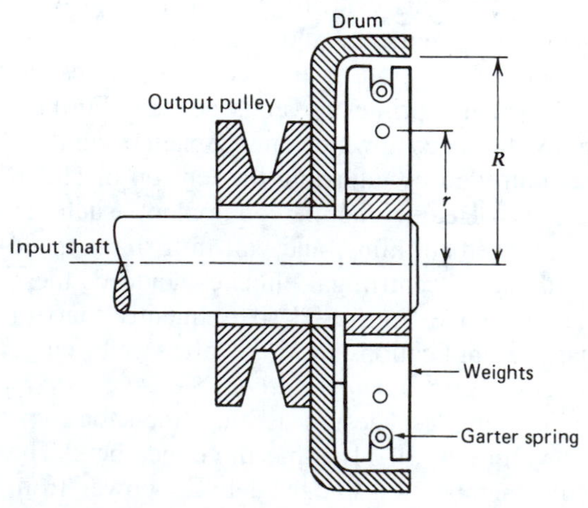

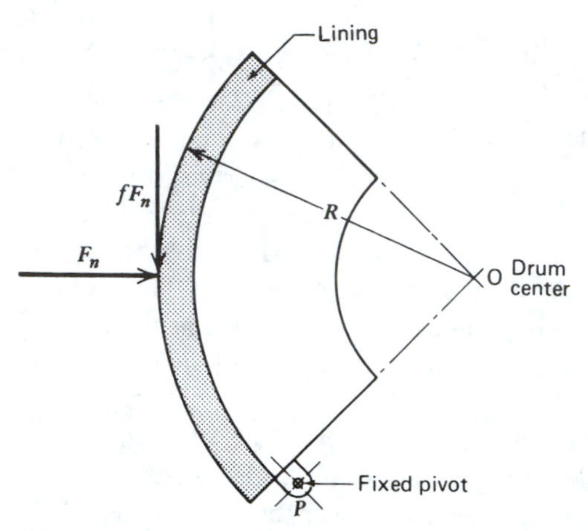

Figure 19-13 Fixed-pivot shoe for a centrifugal clutch.

In somewhat more complex designs, the shoe may be connected to the input member by a fixed pivot (Fig. 19-13) or by a floating link (Fig. 19-14). Design equations are given next for the free-shoe configuration with a garter spring. An article by Spotts (Reference 13) shows how to deal with the fixed-pivot shoe and the floating-link designs.

Figure 19-14 Floating-link design for a centrifugal clutch.

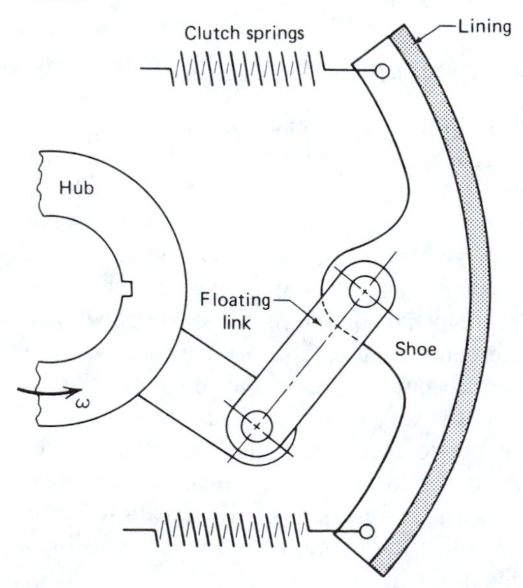

Design of Free-Shoe Clutch with Garter Spring

The torque capacity of the clutch is

$$T = \frac{fF_n R N_s}{1000} \text{ N}\cdot\text{m} \qquad (19\text{-}10)$$

$$T = fF_n R N_s \text{ lb-in.} \qquad (19\text{-}11)$$

where

T = total frictional torque for N_s shoes; N · m, lb-in.
f = coefficient of friction
F_n = force between each shoe and the inner surface of the drum; N, lb
R = radius of inner (contact) surface of the drum; mm, in.
N_s = number of shoes used

The product fF_n is, of course, the frictional force developed between a shoe and the drum. When multiplied by the radius of the contact surface of the drum, the product is the friction torque developed per shoe.

The normal force F_n that is exerted on the drum's inner surface by each shoe is principally due to the centrifugal force, but there is some inward force as a result of the garter spring's tension. The net normal force F_n is

$$F_n = \frac{mrn^2}{C} - 2P\cos\left(90 \text{ deg} - \frac{180 \text{ deg}}{N_s}\right) \qquad (19\text{-}12)$$

where

F_n = normal force exerted by each shoe on the drum's inner surface; N, lb
m = mass of each shoe; kg, lbm
r = radius from center of drum to center of gravity of shoe; mm, in.
n = rotational speed of input shaft; rpm
P = total initial tension in garter spring; N, lb
N_s = number of shoes used

$$C = \begin{cases} 91\,200 \text{ for SI units} \\ 35{,}200 \text{ for English units} \end{cases}$$

Engagement begins when the input shaft reaches a speed n_e such that F_n becomes *positive*. The value of n_e, then, is the value of n that corresponds to $F_n = 0$. But the designer's problem is to select the garter spring tension P such that the engagement will occur at some selected speed n_e (often 75% of running speed). So by setting $n = n_e$ and $F_n = 0$ in Eq. 19-12, one can readily solve for the corresponding value of P.

$$P = \frac{mrn_e^2}{2C\cos[90 \text{ deg} - (180 \text{ deg}/N_s)]} \qquad (19\text{-}13)$$

EXAMPLE 19-3

A free-shoe centrifugal clutch with a circumferential garter spring makes use of four shoes. Each has a mass of 0.31 lbm and a radius of 1 in. from the drum's center to the shoe's center of gravity. The drum's contact diameter (inner surface) is 3 in. The coefficient of friction is 0.30. The service factor is 2.5 for the application.

(*a*) What should be the tension at assembly of the garter spring for an engagement at 1000 rpm?
(*b*) What is the torque capacity at the operating speed of 3000 rpm?
(*c*) What is the power rating?

SOLUTION

(*a*) To find the garter spring initial tension, P, Eq. 19-13 is used with $C = 35{,}200$.

$$P = \frac{mrn_e^2}{2C\cos[90 \text{ deg} - (180 \text{ deg}/N_s)]}$$

$$= \frac{(0.31 \text{ lbm})(1 \text{ in.})(1000 \text{ rpm})^2}{2(35{,}200)\cos[90 \text{ deg} - (180 \text{ deg}/4)]}$$

$$= \mathbf{6.23 \text{ lb}}$$

(*b*) Equations 19-12 and 19-11 are used to determine the torque capacity T.

$$F_n = \frac{mrn^2}{C} - 2P\cos\left(90 \text{ deg} - \frac{180 \text{ deg}}{N_s}\right)$$

$$= \frac{(0.31 \text{ lbm})(1 \text{ in.})(3000 \text{ rpm})^2}{(35{,}200)} - 2(6.23 \text{ lb}) \cos\left(90 \text{ deg} - \frac{180 \text{ deg}}{4}\right)$$

$$= \mathbf{70.5\ lb}$$

$$T = fF_nRN_s$$

$$= (0.3)(70.5 \text{ lb})(3/2 \text{ in.})(4)$$

$$= \mathbf{126.9\ lb\text{-}in.}$$

This is the design torque capacity. The nominal torque capacity is the design torque divided by the service factor.

$$T = \frac{T_{des}}{K_s} = \frac{126.9 \text{ lb-in.}}{2.5}$$

$$= \mathbf{50.8\ lb\text{-}in.}$$

(*c*) The nominal power capacity is

$$P = \frac{Tn}{63{,}000}$$

$$= \frac{(50.8 \text{ lb-in.})(3000 \text{ rpm})}{63{,}000} = \mathbf{2.42\ hp}$$

19-9 SELECTING THE RIGHT CLUTCH

The starting point is to select the general type of clutch needed. Table 19-3 offers a starting point. Designers may want to refer to the article itself to narrow down their choice further. Four additional tables are given there comparing different types of general-utility, self-actuating, continuous-slip, and overrunning clutches with respect to speed, physical size, capacity, and type of load. Designers may also want to consult the manufacturers of that clutch type.

19-10 DERIVATION OF BASIC DESIGN EQUATIONS FOR DISK CLUTCHES

This section amplifies Section 19-5 by supplying the derivations for the basic design equations introduced there.

Pressure Distribution

One contact face of a disk clutch is sketched in Fig. 19-15. An elementary annular region of width dr at radius r is shown. The uniform wear assumption supposes that the wear rate is uniform across the face of the disk from $r = d/2$ to

TABLE 19-3 Selection Chart for Clutches

Function	General Utility	Centrifugal and Fluid Self-Actuating	Continuous Slip		Overrunning or Freewheeling
			Automatic	Variable	
No-load start Manual or externally controlled	√				√
Automatic		√			√
Smooth load pickup Normal load	√	√	√		
High-inertia load		√		√	

TABLE 19-3 (Continued)

Function	General Utility	Centrifugal and Fluid Self-Actuating	Continuous Slip		Overrunning or Freewheeling
			Automatic	Variable	
High-breakaway load (more than 100% running torque)		√		√	
Automatic delayed pickup		√		√	
Extended acceleration		√		√	
Auxiliary starter					√
Running operation Normal load (no slip at full load, full speed)	√	√	√		
Control variable-torque load				√	
Control constant-torque load			√	√	
Control constant-tension load				√	
Overload protection and stopping General protection: transient and infrequent overloads	√	√			
Limit speed (prevent runaway load)		√			√
Limit torque		√	√		√
Automatic overload release	√				√
Dynamic braking		√			√
Back stopping		√			√
Intermittent operation On-off, with driver at speed	√				√
Inching and jogging	√				
Indexing and load positioning					√
Dual-drive and standby operation		√			√

Source. John Proctor, "Selecting Clutches for Mechanical Drives," *Product Engineering*, June 19, 1961. Courtesy of McGraw-Hill Book Company.

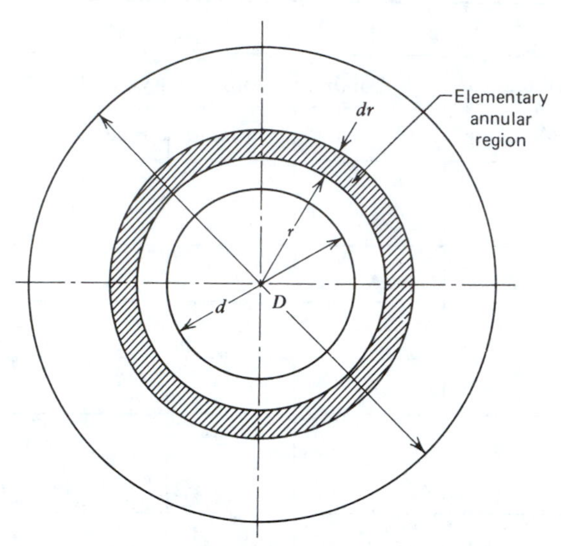

Figure 19-15 Contact surface of a disk clutch.

$r = D/2$. Furthermore, the wear rate is assumed to be proportional to the product of contact pressure and rubbing velocity. Since at any instant the rubbing velocity is itself proportional to the radius r, it follows that pr = constant. But then the maximum contact pressure must occur at the inner radius. Thus

$$pr = \frac{p_{\max} d}{2} = \text{constant}$$

or

$$p = \frac{p_{\max} d}{2r} \tag{19-14}$$

Actuating Force

The area of the elementary annular region cross-hatched in Fig. 19-15 is

$$dA = 2\pi r\, dr$$

The normal force on this region is the product of pressure and area.

$$dF_a = p\, dA = 2\pi pr\, dr$$

The total actuating force is found by integration after substituting for p from Eq. 19-14.

$$F_a = \int_{d/2}^{D/2} dF_a = \pi d p_{\max} \int_{d/2}^{D/2} dr$$

$$= \frac{\pi}{2} d(D - d)p_{\max} \tag{19-15}$$

Torque Capacity

The friction force on the elementary annular region in Fig. 19-15 is

$$dF = f\, dF_a = 2\pi f pr\, dr$$

The corresponding frictional torque is

$$dT = r\, dF = 2\pi f p r^2\, dr$$

This can be integrated to find the total actuating torque after substituting for p from Eq. 19-14.

$$T = \int_{d/2}^{D/2} dT = \pi f p_{\max}\, d \int_{d/2}^{D/2} r\, dr$$

$$T = \frac{\pi}{2} f p_{\max} \frac{(D^2 - d^2)d}{4} \tag{19-16}$$

Effective Friction Radius

The equation for torque capacity is more useful if written in terms of the actuating force. To do this, first define an effective friction radius R_e such that

$$T = fF_a R_e \tag{19-17}$$

The value to be used for R_e in design is needed. By setting equal the two expressions for T found in Eqs. 19-16 and 19-17 and substituting for F_a from Eq. 19-15, this result is obtained.

$$R_e = \frac{D + d}{4} \tag{19-18}$$

For the uniform rate-of-wear assumption, the effective friction radius turns out to be simply the average radius of the contact area.

Average Pressure

The actuating force can also be written in terms of an average contact pressure

$$F_a = \frac{\pi}{4}(D^2 - d^2)p_{ave} \qquad (19\text{-}19)$$

Then, by equating the two expressions for F_a, it is found that

$$\frac{p_{max}}{p_{ave}} = \frac{1}{2}\left(\frac{D}{d} + 1\right) \qquad (19\text{-}20)$$

SUMMARY

A clutch is usually used to connect a loaded shaft to the output shaft of a prime mover in order to bring the load up to speed smoothly and gradually. The clutch type depends on the physical principle used to transmit torque from one member to another and on the means of controlling engagement.

When designing a clutch, the design value of the torque capacity should be established by applying a service factor to the nominal torque requirement. The service factor is affected by the starting inertia of the load, the nature of the prime mover's output torque, and the character of the load.

To keep wear within bounds and to prevent overheating, the design of a disk clutch is based on a maximum value of contact pressure recommended by the manufacturer of the friction material. Multidisk clutches increase capacity, but friction forces between the disks and the splines they slide on will reduce the actuating force available. Friction materials suitable for clutches provide a high coefficient of friction, yet resist seizing of the mating surfaces and have a low wear rate.

Usually a clutch is placed between the prime mover and the speed reduction unit. However, on some applications a larger clutch at the low-speed end may be justified by lower energy loss and lower maintenance cost.

Centrifugal clutches are practical where a time delay is necessary so that a prime mover can be brought up to speed before taking over a high-inertia load. The free-shoe clutch with a garter spring is easy to design for a specified engagement speed.

For guidance in selecting the right clutch, selection tables are available to narrow down the choice quickly.

REFERENCES

19-1 Beach, Karl. "Try These Formulas for Centrifugal Clutches." *Product Engineering,* July 9, 1962.

19-2 Bette, A. J. "Basic Properties and Selection of Friction Materials." *Machine Design,* September 29, 1960.

19-3 *Design and Product Manual: Friction Materials.* Detroit: D. A. B. Industries, Inc.

19-4 Gagne, A. F., Jr. "Torque Capacity and Design of Cone and Disk Clutches." *Product Engineering,* December 1953.

19-5 Greenwood, Douglas C. "Clutches." *Mechanical Power Transmission.* New York: McGraw-Hill, 1962.

19-6 Jania, Z. J., and David Sinclair. "Friction Clutches and Brakes." *Mechanical Design and Systems Handbook.* Harold A. Rothbart (ed.). New York: McGraw-Hill, 1964.

19-7 Jensen, W. R. "Friction Materials." *Machine Design,* January 27, 1972.

19-8 Kaplan, Joseph, and Donald Marshal. "Design Equations and Nomographs for Self-Energizing Types of Spring Clutches." *Machine Design,* April 19, 1956.

19-9 Machen, James F. "Seven Low-cost Designs for Overrunning Clutches." *Product Engineering,* February 8, 1960.

19-10 *Machine Design. Mechanical Drives Reference Issue,* 1976.

19-11 Proctor, John. "Selecting Clutches for Mechanical Drives." *Product Engineering,* June 19, 1961.

19-12 "Selection of Friction Clutches." *Tribology Handbook.* M. J. Neale (ed.). New York: Wiley, 1973.

19-13 Spotts, M. F. "New Design Formulas for Long-shoe Centrifugal Clutches." *Product Engineering,* February 17, 1964.

PROBLEMS

P19-1 A three-phase induction motor is used to drive an automobile engine on a test stand to determine friction losses in the engine. This is called a motoring test. A disk clutch is used between the electrical motor and the engine. What is a suitable service factor for the clutch?

P19-2 A disk clutch is to be used in a railroad crane. A multicylinder diesel engine is the power source. A high-inertia load must be brought up to speed each time the clutch is engaged. The engine produces 90 kW at 1520 rpm. What service factor should be used? What should the design torque be?

P19-3 A disk clutch operated wet is needed to connect a single-cylinder gasoline engine to a portable generator. One available clutch is designed for a torque capacity of 3200 lb-in. What is a reasonable nominal power rating at 1200 rpm?

P19-4 Large drying drums of a paper making machine are driven by a three-phase motor by means of a helical gear reduction unit. The clutch is between the motor and the gear reduction unit. The motor develops 112 kW at 1800 rpm. Why does this motor need a clutch? (Usually motors do not need a clutch.) Is a clutch design for a torque capacity of 1100 N · m adequate for this application?

P19-5 A dry disk clutch has one pair of contacting surfaces. The dimensions are $D = 12$ in. and $d = 8$ in. The wearing material is molded phenolic plastic opposed by cast steel. For an allowable contact pressure of 70 psi, what should the actuating force be?

P19-6 For the clutch in P19-5, what is the average contact pressure?

P19-7 A disk clutch for wet operation uses hard-drawn phosphor bronze as the wearing material and chromium plated steel as the opposing material. What is the maximum actuating force that should be used in operation for dimensions of $D = 335$ mm and $d = 280$ mm?

P19-8 For the clutch described in P19-7, what is the average contact pressure if an allowable contact pressure of 0.85 MPa is used for design purposes? Calculate the actuating force based on average contact pressure and on maximum contact pressure.

P19-9 A wet disk clutch has two pairs of mating surfaces, each with the dimension $D = 300$ mm and $d = 200$ mm. The linings are molded phenolic plastic opposed by cast iron. A service factor of 2.85 is to be used. What is the nominal power rating for 580 rpm?

P19-10 A dry disk clutch has mating surfaces with the dimensions $D = 13.8$ in. and $d = 5.9$ in. The friction lining is sintered metal. The opposing disk is cast steel. There is only one pair of mating surfaces. For a coefficient of friction assumed to be $f = 0.15$ and an average contact pressure of 50 psi, what is the design torque capacity? What is the nominal power capacity at 1200 rpm for a service factor of 2.75?

P19-11 A 45-kW, 3600-rpm, single-phase electric motor is to be equipped with a centrifugal starting clutch designed with eight equally spaced shoes inside a 150-mm inside diameter drum. Engagement is to occur at 2700 rpm. Assume a coefficient of friction of 0.35 and a 36-deg arc of contact between each shoe and the drum. The center of gravity of each shoe is 65 mm from the drum's axis. Operating conditions are moderate shock load and 24-h service. Find the required mass for each cast iron shoe and calculate its width for an allowable average pressure of 1.5 MPa. *Hint:* Find the initial garter spring tension P in terms of shoe mass m.

P19-12 A 0.5-hp chain saw is to be equipped with a centrifugal starting clutch designed to have four equally spaced steel shoes inside a 3.8-in. inside diameter drum. Engagement is to occur at 1480 rpm. Running speed is 1755 rpm. Assume a coefficient of friction of $f = 0.32$ and a 60-deg arc of contact between each shoe and the drum. The center of gravity of each shoe is 1.45 in. from the drum's axis. Find the required mass of each shoe and calculate its width for an allowable average pressure of 200 psi. Use a service factor of 2.5.

P19-13 A dry disk clutch is to carry 134 kW at 400 rpm. The outside disk diameter is not to be larger than 610 mm. The service factor is 2.15. Complete the design of the clutch.

P19-14 Design a disk clutch for dry operation. The nominal power capacity is 18.5 kW at 500 rpm. An average inertia load has to be brought up to speed by a multicylinder engine. The load is nonpulsating.

P19-15 Design a clutch for wet operation with a nominal capacity of 25 kW at 1000 rpm. Space limits the outside disk diameter to no more than 275 mm.

P19-16 Design a disk clutch for dry operation using two disks only. The nominal power capacity is 20 hp at 500 rpm. An average inertia load has to be brought up to speed by a multicylinder engine. The load is nonpulsating. Use an inexpensive friction material. The clutch is to be mass produced.

Chapter 20
Mechanical Brakes

Life is the art of drawing sufficient conclusions from insufficient premises.

SAMUEL BUTLER

NOMENCLATURE

a = length of band brake lever

a = acceleration; m/s², ft/s²

A = exposed area of brake disk

A_p = piston area at wheel cylinder

A_{pad} = area of brake pad

b = distance from "taut" side of band to lever pivot

b = thickness of brake disk; mm, in.

c = distance from "slack" side of band to lever pivot

c = specific heat; J/kg · °C, Btu/lbm-°F

d = brake disk diameter

D = brake drum diameter

e = distance from shaft center to center of pad

E = heat energy to be dissipated in a brake application; J, Btu

f = coefficient of friction

f' = design value of coefficient of friction to avoid self-locking

f_R = tire-road coefficient of adhesion

f_v = correction factor for air velocity

F_a = actuating force

F_1 = larger band tension

F_2 = smaller band tension

F_{bf} = braking force at front axle

F_{br} = braking force at rear axle

$F_{bf\max}$ = maximum braking force at front axle

$F_{br\max}$ = maximum braking force at rear axle

$F_{b\max}$ = maximum braking force for vehicle

g = local acceleration of gravity; m/s², ft/s² (standard gravity = 9.807 m/s² or 32.174 ft/s²)

h = height of vehicle's center of gravity above ground; m, ft

h = overall heat transfer coefficient; W/m² · °C, Btu/s-in.²-°F

h_r = radiation coefficient; W/m² · °C, Btu/s-in.²-°F

h_c = convection coefficient in still air; W/m² · °C, Btu/s-in.²-°F

$H_{\max}$ = initial rate of heat generation in braking; W, Btu/s

H_{ave} = average rate of heat generation in braking; W, Btu/s

H_{diss} = rate of heat dissipation; W, Btu/s

J = mass moment of inertia; kg · m², lb-in.²

l_1 = horizontal distance from front axle to vehicle's center of gravity

l_2 = horizontal distance from rear axle to vehicle's center of gravity

L = wheelbase

m = vehicle mass; kg, slugs

m_d = mass of brake disk; kg, lbm

n = rotational speed

n_f = final rotational speed

n_o = initial rotational speed

p_{ave} = average pressure

p_h = hydraulic pressure

r_f = effective rolling radius of front tire

r_r = effective rolling radius of rear tire

R = radius of circular brake pad

R_e = effective friction radius

R_i = inside radius of annular brake pad

R_o = outside radius of annular brake pad

s = distance traveled by vehicle

$s_{\min}$ = minimum stopping distance

S = stops per hour by an industrial brake

t_s = slowdown time

$T_{f\max}$ = maximum braking torque for front axle

$T_{r\max}$ = maximum braking torque for rear axle

W = axle load

W_f = front axle load

W_r = rear axle load
v_f = final vehicle velocity
v_o = initial vehicle velocity
δ = effective friction radius factor
θ = angular extent of annular brake pad; rad
θ = angle of wrap in a band brake; rad
ρ = density; kg/m^3, lbm/in.^{3}
ω_f = final rotational velocity; rad/s
ω_o = initial rotational velocity; rad/s

Safe operation of a vehicle requires full control over it. An essential ingredient in that control is a set of well-designed and properly functioning brakes. They diminish speed or bring the vehicle completely to a halt safely. In vehicles as light and delicate as a 10-speed bicycle or as huge and ponderous as an earth mover, the rule is the same: *safe operation demands dependable brakes*.

This rule also applies to industrial equipment. Brakes are needed to protect the operator and to avoid damage to the equipment itself. The brakes also provide necessary control during routine operation. In hoists and elevators the rate of descent is controlled by harmlessly dissipating some of the potential energy. In production equipment brakes are used, for example, to hold parts in rest positions or to prevent reversal of the direction of rotation.

This chapter covers mechanical brakes, with an emphasis on disk and band brakes. It also provides an introduction to the methods for determining the braking needs of both vehicles and stationary industrial equipment.

20-1 BRAKING SYSTEMS

The invention of braking systems occurred almost simultaneously with the invention of the wheel. Of course, stopping a vehicle has less appeal to the adventurous than setting it in motion in the first place, but safety requires control of speed.

The first modern braking system appeared on ore wagons in Germany in the sixteenth century. It involved a remotely controlled wooden post brake. The invention of the steam locomotive and the rapid growth of railroads in the nineteenth century created serious braking problems unlike anything previously encountered. In 1869 George Westinghouse patented the basic air brake and founded the Westinghouse Air Brake Company. In 1872 he invented the automatic air brake, which was quickly adopted by American railroads and, gradually, by European railroads. The automatic air brake is applied on *release* of air from the system, either intentionally released or accidentally released by a broken pipe or other failure. Thus it is *fail-safe*. A refined version of the Westinghouse automatic air brake is now standard equipment on rail cars.

Early automobile and truck brakes simply imitated the external block brakes used on horse-drawn carriages and wagons. Soon the internal shoe brake was developed. In it rigid shoes carrying friction linings are pressed outward against the inner surface of a drum by a linkage and cam or by the pistons of a hydraulic cylinder. In modern air brakes for heavy trucks, the linkage that actuates the shoes is actuated by an air cylinder. By 1930, the internal shoe drum brake was almost universally used on trucks and automobiles.

Although Dr. F. Lanchester patented the caliper disk brake in 1902 and used it on his 25-hp Lanchester car in 1906, the automotive industry did not adopt it generally until the 1950s. Early disk brakes were noisy, and the pads wore excessively. After a period of intensive development, disk brakes were used successfully on aircraft and in racing cars. Caliper disk brakes are now used widely for automobiles, especially for the front wheels, because caliper brakes are relatively insensitive to changes in the friction coefficient, whether from heat or from water. Modern disk brakes also have a better heat-dissipation capacity than drum brakes.

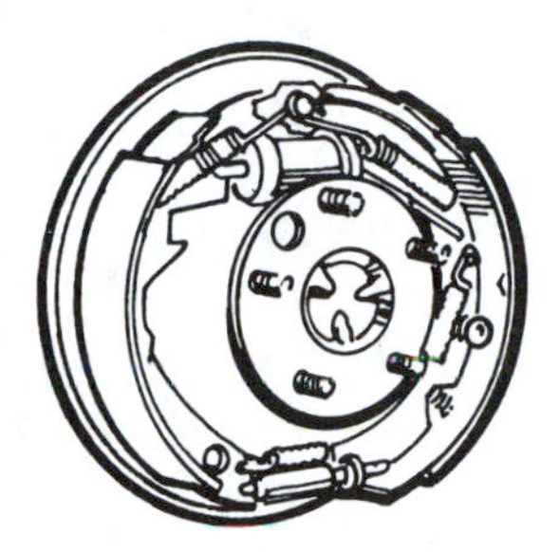

Drum Brake System

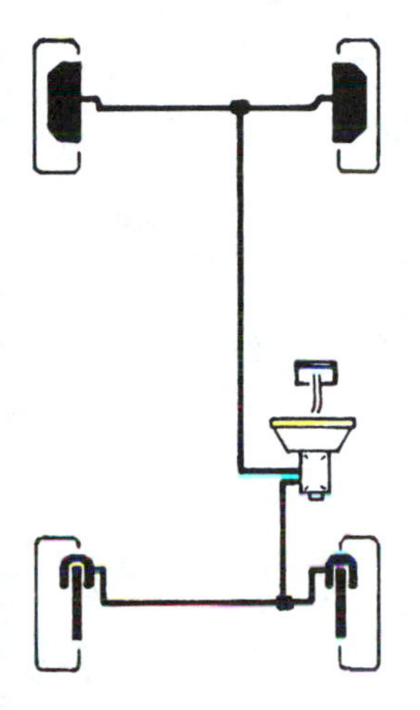

Disc Brake System

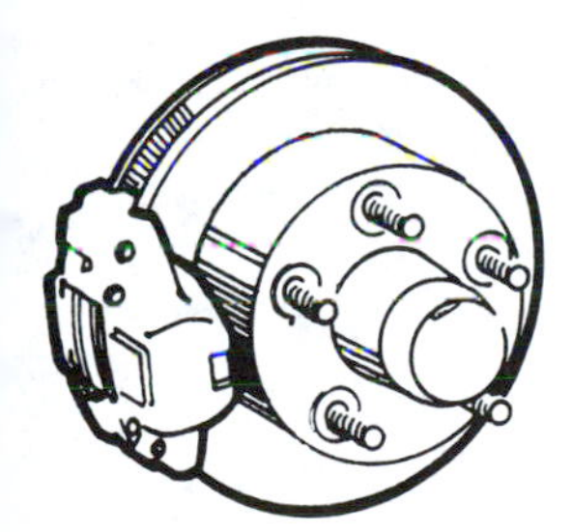

Figure 20-1 Automotive brake systems. (Copyright ©1981 by Amoco Enterprises, Inc., *Adventure Road Magazine*.)

One approach to passenger car design that has proved effective is to use disk brakes on the front wheels and drum brakes on the rear wheels, as shown in Fig. 20-1. However, in many instances disk brakes are used for all four wheels. Figures 20-2 and 20-3 show the characteristics of disk and drum brakes, respectively.

In *industrial* applications a wide variety of brake types is encountered. Drum brakes using external blocks, usually two pivoted shoes gripping the drum, have been widely used. Internal drum brakes are also used, especially multishoe types designed so that the shoes travel radially from the drum's center, thus ensuring even engagement and lining wear. Where space is at a premium, plate or disk brakes are used, operated dry or wet. As in clutch design, energy dissipation in an electromagnetic field offers an alternative to friction as the operating principle. Eddy-current, hysteresis, and magnetic particle brakes are all available commercially.

20-2 DISK BRAKES

Capable of a *high torque capacity in a small volume and easily controlled,* disk brakes are rapidly becoming the single most widely used brake for vehicular and industrial applications. In vehicles the relative insensitivity of the disk brake to changes in coefficient of friction, the ability of the brake to dissipate heat rapidly, and the absence of mechanical fade are the three basic reasons for the widespread use of disk brakes, especially for the front axles of passenger cars.

The analysis of plate disk brakes is basically the same as for multiple disk clutches, so no further details are given here.

Caliper Disk Brakes

Caliper disk brakes are now used in many applications. The dual disk brake on the front of the BMW motorcycle (Fig. 20-4) is hydraulically actuated. The Tol-o-matic mechanical caliper disk brake in Fig. 20-5 has been used for recreational vehicles, lift trucks, farm machinery, and other light mobile equipment.

A sectional view of a hydraulically operated disk brake appears in Fig. 20-6*a*. The hydraulic

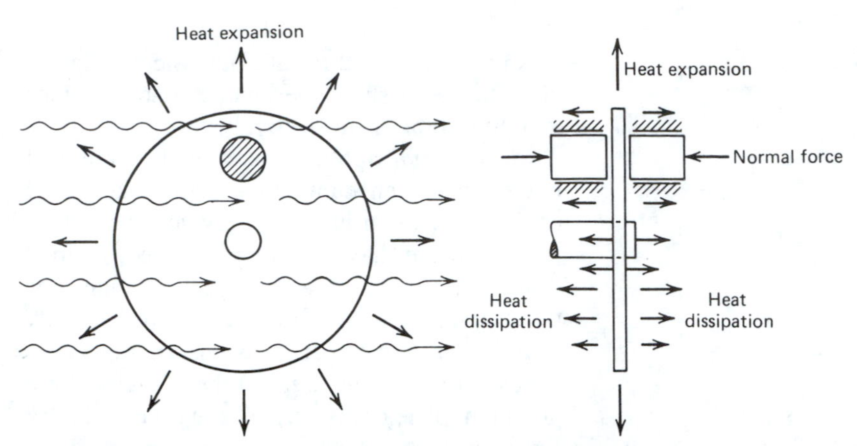

Figure 20-2 *Disk brakes*. The capacity for heat dissipation is great because most of the disk surface is continuously exposed to and cooled by a moving air stream created by vehicle motion. Heat expansion is no problem.

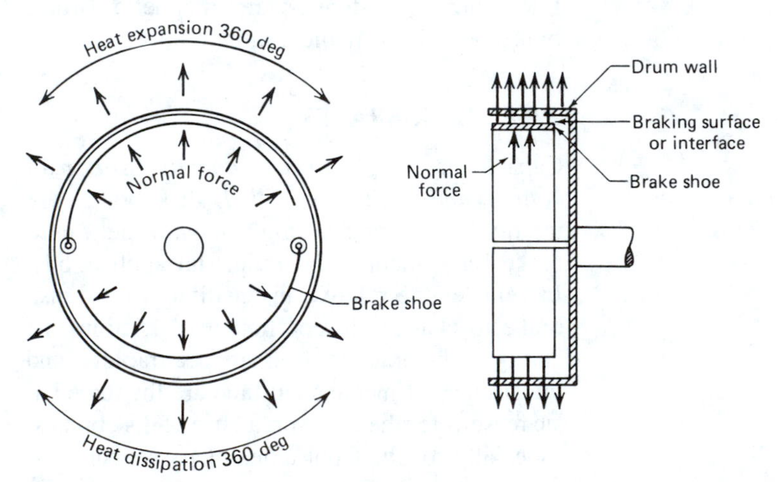

Figure 20-3 *Drum brakes*. Heat generated at the braking surface must be transmitted through the drum wall before it can be dissipated to the atmosphere. Heat tends to build up in the system, thus progressively reducing the available coefficient of friction. Limited radial expansion contributes to a "thermal fade effect."

pressure from the master cylinder applies a force to the caliper's piston and causes the inboard shoe and lining (on the right) to move against the rotor. The reaction from the piston forces the outboard shoe and lining (on the left) into similar contact with the rotor. Thus the caliper grips the rotor in a pincerslike manner, as shown in Fig. 20-6*b*.

Pad shapes. The pressure pads in caliper disk brakes are usually either of two shapes. The most commonly used pads are of the annular shape shown in Fig. 20-7. Three dimensions suffice to

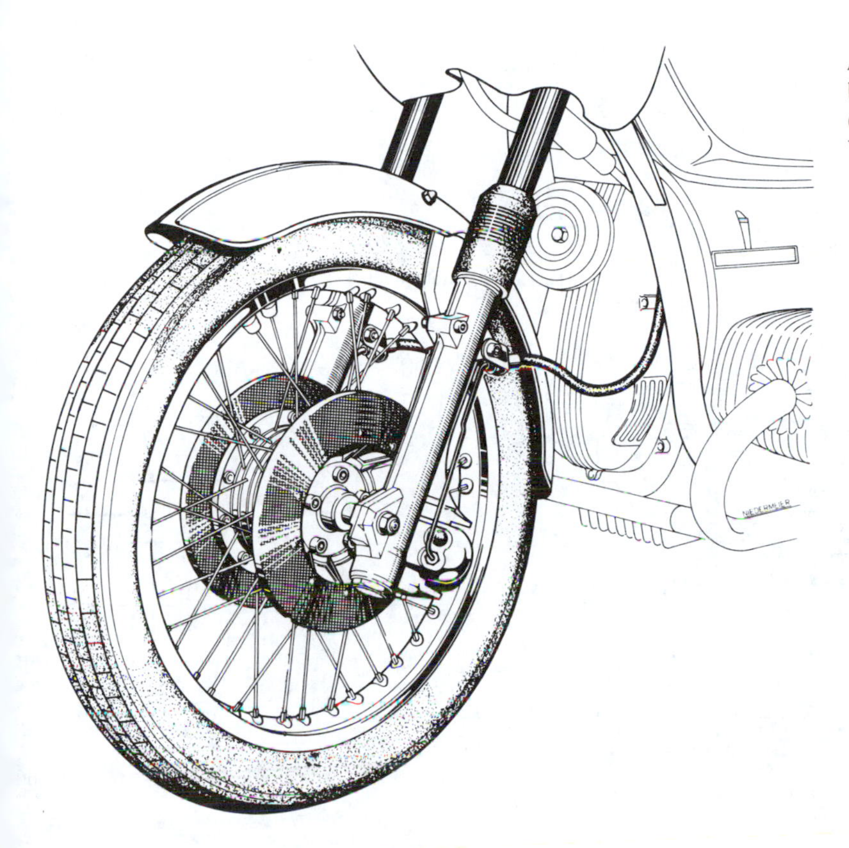

Figure 20-4 Dual disk brake on a motorcycle. (Courtesy Bavarian Motor Works.)

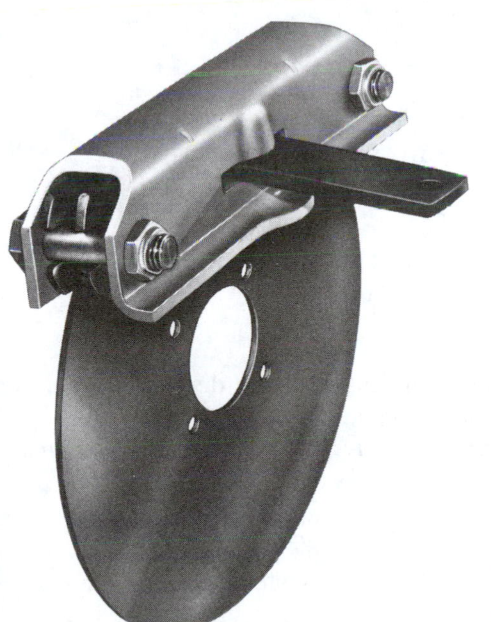

Figure 20-5 A mechanically operated disk brake. (Courtesy Auto Specialties Manufacturing Company, St. Joseph, Michigan.)

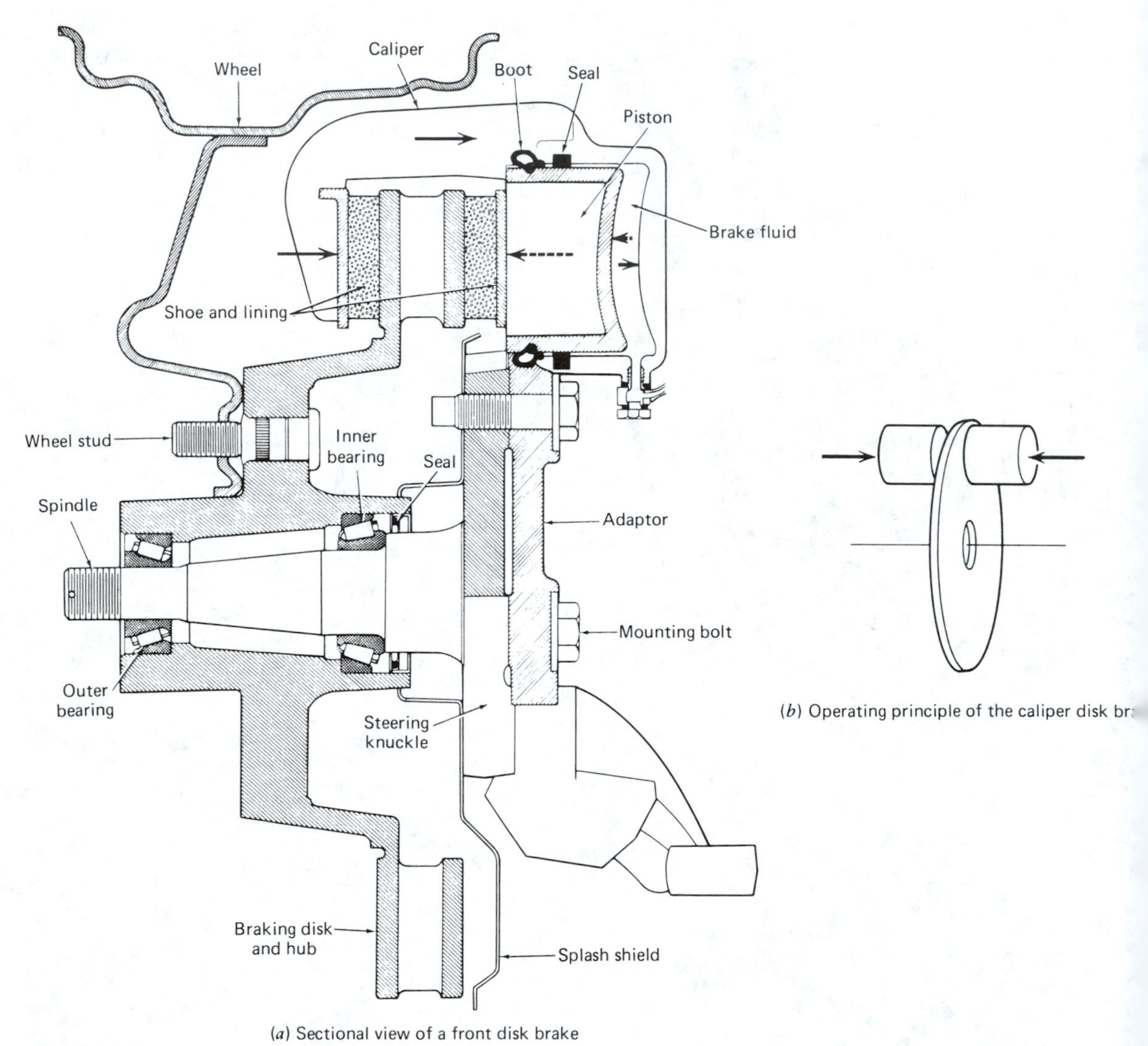

(*b*) Operating principle of the caliper disk br

(*a*) Sectional view of a front disk brake

Figure 20-6 Automotive disk brake. (Courtesy of Chrysler Corporation.)

describe its size and shape. The inside and outside radii are R_i and R_o. Its angular width is angle θ. The location of the pad is given by dimension e, the distance from the center of the disk to the middle of the pad.

$$e = 0.5(R_i + R_o) \tag{20-1}$$

The circular pad is shown in Fig. 20-8. Only one dimension is needed, the pad radius R. Its location is dimension e, the distance from disk center to pad center.

Torque capacity. The torque capacity *per pad* is

$$T = \frac{fF_aR_e}{1000}\ \text{N}\cdot\text{m} \tag{20-2}$$

$$T = fF_aR_e\ \text{lb-in.} \tag{20-3}$$

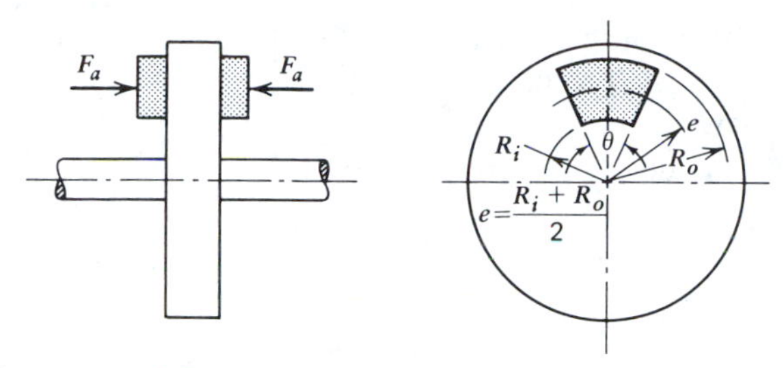

Figure 20-7 Disk brake with two annular opposing pads.

where

T = torque capacity; N · m, lb-in.
f = coefficient of friction for pad and disk
F_a = actuating force; N, lb
R_e = effective friction radius; mm, in.

Effective friction radius for the annular pad. The effective friction radius R_e for the *annular* pad is

$$R_e = \frac{2}{3}\frac{(R_o^3 - R_i^3)}{(R_o^2 - R_i^2)} \qquad (20\text{-}4)$$

The pressure varies little over the annular pad, so a recommended average pressure is used in calculating the allowable actuating force.

For the *circular pad,* the friction radius R_e, is

$$R_e = \delta e \qquad (20\text{-}5)$$

where the δ values are tabulated in Table 20-1.

Actuating force. The actuating force is the product of the average pressure and the pad's contact area.

$$F_a = p_{ave} A_{pad} \qquad (20\text{-}6)$$

Figure 20-8 Disk brake with two opposed circular pads.

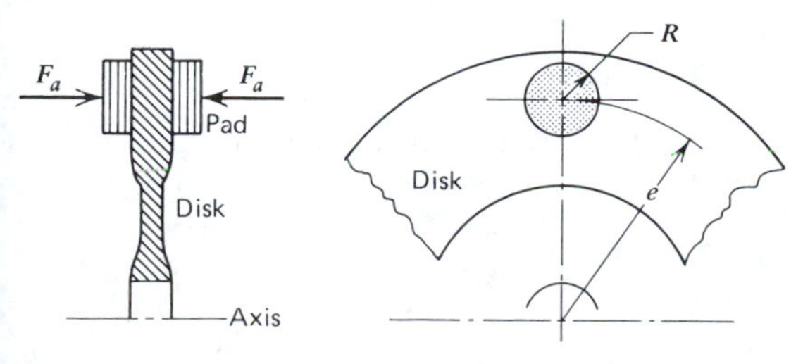

TABLE 20-1 Design Factors for Caliper Disk Brakes with Circular Pads

R/e	$\delta = R_e/e$	p_{max}/p_{ave}
0.0	1.0000	1.0000
0.1	0.9833	1.0926
0.2	0.9693	1.2116
0.3	0.9572	1.3674
0.4	0.9467	1.5779
0.5	0.9375	1.8751

For an annular pad, the area of the pad is

$$A_{pad} = 0.5\theta(R_o^2 - R_i^2) \qquad (20\text{-}7)$$

For a circular pad, however, it is simply the area of a circle of radius R. In these last two equations

F_a = actuating force; N, lb
p_{ave} = average contact pressure; MPa, psi
A_{pad} = pad's contact area with the disk; mm², in.²
θ = angular dimension of pad; rad
R_i, R_o = inside and outside radii of an annular pad; mm, in.

Pressure distribution. If the pad is relatively large, the maximum contact pressure will be significantly higher than the average pressure. Tables 20-1 and 20-2 list the ratios of maximum

TABLE 20-2 Pressure Distribution Factor for Caliper Disk Brakes with Annular Pads

R_i/R_o	p_{max}/p_{ave}
0.00	∞
0.10	5.5
0.20	3.0
0.30	2.17
0.40	1.75
0.50	1.5
0.60	1.33
0.70	1.21
0.80	1.13
0.90	1.06
1.00	1.00

to average pressures for the two pad shapes. Suppliers of pad materials will recommend average pressures to use in design, but they will often also list a maximum contact pressure that should not be exceeded in normal operation. Designers should check the design to be sure this maximum pressure is not exceeded.

EXAMPLE 20-1

For an American sports car, front disk brakes are to be used with two pads per wheel. Space limitations permit dimensions of $R_i = 98$ mm, $R_o = 141$ mm, and $\theta = 108$ deg. The friction material supplier guarantees a minimum coefficient of friction of 0.37. Each pad will be actuated by two hydraulic pistons of 38-mm diameter. A front axle braking capacity of 2944 N · m is needed. What hydraulic pressure is needed at the wheel cylinders? What are the average and maximum contact pressures?

SOLUTION

Since the front axle braking capacity is to total 2944 N · m, the torque capacity per pad should be 2944/4 = 736 N · m.

The mean friction radius for each pad is, by Eq. 20-4,

$$R_e = \frac{2}{3}\frac{(R_o^3 - R_i^3)}{(R_o^2 - R_i^2)}$$

$$= \frac{2}{3}\frac{(141^3 - 98^3)}{(141^2 - 98^2)}$$

$$= 120.8 \text{ mm}$$

The actuating force needed for each pad is found by solving Eq. 20-2 for F_a.

$$F_a = \frac{1000T}{fR_e}$$

$$= \frac{(1000)(736 \text{ N}\cdot\text{m})}{(0.37)(120.8 \text{ mm})} = 16\,467 \text{ N}$$

Each of the two pistons actuating the pad must supply a force of 16 467/2 = 8234 N, so the hydraulic pressure must be the actuating force divided by the piston area.

$$p_h = \frac{F_a}{A_p} = \frac{4F_a}{\pi d_p^2}$$

$$= \frac{(4)(8234\ N)}{(\pi)(38 \text{ mm})^2} = \mathbf{7.26\ MPa}$$

This is a high pressure. A power-assist system will be needed.

The average contact pressure is, by Eqs. 20-6 and 20-7,

$$p_{\text{ave}} = \frac{2F_a}{\theta(R_o^2 - R_i^2)}$$

$$= \frac{(2)(16\,467 \text{ N})}{(108/57.296)[(141 \text{ mm})^2 - (98 \text{ mm})^2]}$$

$$= \mathbf{1.70\ MPa}$$

The maximum contact pressure is found by use of Table 20-2. Since $R_i/R_o = 98/141 = 0.7$, $p_{\max}/p_{\text{ave}} = 1.21$. The maximum contact pressure is 21% higher than the average. It is **2.06 MPa.**

EXAMPLE 20-2

Suppose circular pads are used in the front disk brakes of the American sports car in Example 20-1. Three pads of 38-mm diameter are to be used on each side of each disk. Each pad is actuated by a 35-mm diameter piston. What will the hydraulic pressure at these wheel cylinders have to be? Assume $e = 120$ mm and $f = 0.37$. What will the maximum contact pressure be?

SOLUTION

There will be a total of 12 pads in the two front brakes. Each must contribute a braking torque capacity of 2944/12 = 245 N · m.

The ratio R/e is

$$\frac{R}{e} = \frac{38/2}{120} = 0.158$$

From Table 20-1, the value of δ is estimated to be 0.975. Thus the effective friction radius is

$$R_e = \delta e = (0.975)(120 \text{ mm}) = 117 \text{ mm}$$

By Eq. 20-2, the necessary actuating force per pad is

$$F_a = \frac{1000T}{fR_e} = \frac{(1000)(245 \text{ N}\cdot\text{m})}{(0.37)(117 \text{ mm})} = 5660 \text{ N}$$

The hydraulic pressure is therefore the actuating force divided by the piston area.

$$p_h = \frac{F_a}{A_p} = \frac{4F_a}{\pi d_p^2} = \frac{(4)(5660 \text{ N})}{(\pi)(35 \text{ mm})^2} = \mathbf{5.88\ MPa}$$

The average contact pressure on each brake pad is

$$p_{\text{ave}} = \frac{F_a}{A_{\text{pad}}} = \frac{F_a}{\pi R^2} = \frac{5660 \text{ N}}{(\pi)(38 \text{ mm}/2)^2} = 4.99 \text{ MPa}$$

In Table 20-1, for $R/e = 19/120 = 0.158$, the value of $p_{\max}/p_{\text{ave}}$ is estimated to be 1.16. So the maximum contact pressure is (1.16)(4.99 MPa) = **5.8 MPa.** Caliper disk brakes make severe demands on friction materials.

20-3 FRICTION MATERIALS FOR BRAKES

A brake's friction lining or pad is used to convert mechanical energy to heat energy, which must then be dissipated by the brake unit. Three factors are involved in the choice of a friction material: (1) friction coefficient, (2) fade resistance (ability of the material to maintain its friction coefficient at fairly high temperatures), and (3) wear resistance. More than any other single phenomenon, temperature rise complicates the choice of the brake's lining or pad material.

Types of Brake Friction Material

Three main groups of friction materials for brakes are woven cotton, sintered metal, and cermets (ceramic and metal mixtures). *Woven cotton* linings are made of layers of closely woven cotton impregnated with resins. They are used in industrial drum brakes where high temperatures are not a problem. A high coefficient of friction and low cost are their chief merits. *Sintered metal* friction materials are also used in heavy-duty applications. Iron and copper powders are molded in shape and heated under pressure ("sintered"). *Cermets* are mixtures of ceramic and metallic materials, molded and sintered. The sintered materials can be operated at higher pressures than the organic materials. Their thermal conductivity is 10 to 20 times that of the organic materials, so they dissipate the heat generated more easily.

Typical properties of common friction materials for brakes are listed in Table 20-3. For the operating pressure in each instance, a wide range is given. The higher the rubbing velocity, the lower the contact pressure should be to avoid excessive temperature rise and wear. For any given torque capacity, of course, the average contact pressure is dependent on the area of the friction material.

In design, therefore, a good starting point is to estimate the total rubbing area of friction material required for the average braking power. Table 20-4 lists conservative estimates for three levels of duty.

TABLE 20-3 Typical Operating Conditions and Properties for Friction Materials Used in Brakes

Material	f	Wear rate, mm^3/J at 100°C ($in.^3/Btu$ at 212°F)	Maximum operating temperature, °C (°F)	Maximum temperature, °C (°F)	Working pressure, MPa (psi)	Maximum pressure, MPa (psi)
Woven cotton	0.50	$(12.2)(10^{-6})$ $(7.85)(10^{-7})$	100 (212)	150 (300)	0.07–0.7 (10–100)	1.5 (220)
Sintered metal	0.30	Used at higher temperatures	300 (572)	650 (1200)	0.35–3.5 (50–500)	5.5 (800)
Cermet	0.32	Used at higher temperatures	400 (750)	800 (1470)	0.35–1.05 (50–150)	6.9 (1000)

Source. M. J. Neale (editor) *The Tribology Handbook,* C7. Halsted Press, John Wiley and Sons, New York, 1973.

EXAMPLE 20-3

A caliper disk brake is being developed as a safety brake for a printing press. It is used infrequently. A braking torque of 4100 lb-in. is required to stop the press in an acceptable time limit. The initial disk speed is 1200 rpm. Two pads are to be used, one on each side of the disk. What should be the minimum contact area per pad?

SOLUTION

The initial braking power is found using the power equation (Eq. 1-10).

$$P = \frac{Tn}{63{,}000}\ \text{hp}$$

$$= \frac{(4100\ \text{lb-in.})(1200\ \text{rpm})}{63{,}000} = 78.1\ \text{hp}$$

The average braking power is half of this: 39.1 hp or 27.6 Btu/s.

Table 20-4 suggests an area to braking power ratio of 0.28 $in.^2$-s/Btu for a caliper disk brake used infrequently. The total contact area should be

$$(0.28\ \text{in.}^2\text{-s/Btu})(27.6\ \text{Btu/s}) = 7.73\ \text{in.}^2$$

Each pad should have a contact area of 7.73/2

TABLE 20-4 Required Friction Material Area for a Given Average Braking Power.

Duty	Typical applications	Area/average braking power, mm^2/W ($in.^2/Btu/s$)		
		Band and drum brakes	Plate disk brakes	Caliper disk brakes
Infrequent	Emergency brakes	0.5 (0.82)	1.7 (2.78)	0.17 (0.28)
Intermittent	Elevators, cranes, winches	1.7 (2.78)	4.4 (7.2)	0.43 (0.70)
Heavy duty	Excavators, presses	3.4–4.2 (5.56–6.87)	8.3 (13.6)	0.86 (1.41)

Source. M. J. Neale (editor) *The Tribology Handbook,* A51, Halsted Press, John Wiley and Sons, New York, 1973; and *Friction Materials for Engineers,* 2nd ed., Ferrodo, Ltd.

in.^{2} = **3.86 in.^{2}** If a circular pad is used, a diameter of at least 2.2 in. will suffice.

20-4 BAND BRAKES

Braking action can also be obtained by pulling a band tightly against a drum, as in Fig. 20-9. When a belt or band is wrapped on a pulley or drum, the ratio of the band tensions is, from Eq. 11-10,

$$\frac{F_1}{F_2} = e^{f\theta} \quad (20\text{-}8)$$

and the net torque applied to the drum by the band is

$$T = \frac{(F_1 - F_2)D}{2} \quad (20\text{-}9)$$

where

F_1, F_2 = the larger and smaller band tensions; N, lb
f = coefficient of friction
θ = angle of wrap; rad
T = torque; N · m, lb-in.
D = drum diameter; m, in.

If the desired braking torque is known, Eqs. 20-8 and 20-9 can be solved simultaneously to determine the band tensions F_1 and F_2.

Figure 20-9 Differential band brake.

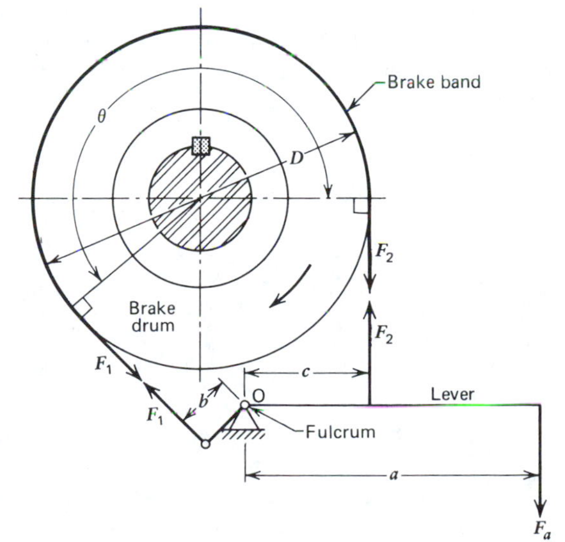

For the braking lever (Fig. 20-9), the sum of moments about the pivot point O must be zero.

$$\sum M_O = F_2 c - F_1 b - F_a a = 0 \quad (20\text{-}10)$$

where

F_a = the force applied to the lever; N, lb

A simultaneous solution of these three equations yields a useful design equation for the actuating force F_a necessary to achieve any braking torque T.

$$F_a = \frac{2T}{Da}\frac{(c - be^{f\theta})}{(e^{f\theta} - 1)} \quad (20\text{-}11)$$

This brake is said to be *self-energizing* for $b > 0$. What is meant by this can best be seen by inspecting Fig. 20-9. To apply the brake, a clockwise moment has to be applied to the brake lever. This is done by applying a brake force F_a downward, since it is necessary to overcome the counterclockwise resisting moment exerted by the band tension F_2. Notice that band tension F_1 exerts a clockwise moment on the lever and assists in applying the brake. Friction forces developed in the brake assist the brake operation.

This also shows up in Eq. 20-11. The term $be^{f\theta}$ in the numerator reduces the actuating force needed for any braking torque.

However, if the designer is careless, there is the danger that the brake will be *self-locking* because if b is large enough, the numerator in Eq. 20-11 can be made zero or even negative. The slightest touch on the lever would cause the brake to seize. To avoid self-locking, select the dimension b such that

$$b \leq \frac{c}{e^{f'\theta}} \quad (20\text{-}12)$$

where f' is substantially greater than (say 50% greater than) the largest coefficient of friction the manufacturer of the friction material expects.

EXAMPLE 20-4

A differential band brake is to have a braking torque capacity of 1050 N · m. Some basic dimensions have already been selected: $D = 1100$ mm, $a = 1700$ mm, $c = 500$ mm, $\theta = 200$ deg. The manufacturer of the band material lists a coefficient of friction of $f = 0.40 \pm 10\%$. Select a value for the dimension b and calculate the maximum value of the applied force F_a necessary to achieve the desired braking capacity.

SOLUTION

The first task is to select a value for b. The maximum value of friction coefficient quoted is 0.44 (0.40 plus 10%). Let us use $f' = 1.5(0.44) = 0.66$. The angle of wrap is $\theta = 200$ deg/57.3 deg/rad $= 3.49$ rad. The value of b can now be selected.

$$b \leq \frac{c}{e^{f'\theta}} = \frac{500 \text{ mm}}{e^{(0.66)(3.49)}} = \mathbf{50\ mm}$$

Dimension b should not exceed 50 mm. That dimension will be used.

To calculate the maximum actuating force needed to achieve a torque capacity of 1050 N · m, use the smallest friction coefficient to be expected, 0.36 (0.40 less 10%). For this coefficient the ratio of band tensions is

$$e^{f\theta} = e^{(0.36)(3.49)} = 3.51$$

Then, by Eq. 20-11, the actuating force is

$$F_a = \frac{2T}{Da}\frac{(c - be^{f\theta})}{(e^{f\theta} - 1)}$$

$$= \frac{(2)(1050 \text{ N} \cdot \text{m})[500 \text{ mm} - (50 \text{ mm})(3.51)]}{(1.1 \text{ m})(1700 \text{ mm})(3.51 - 1)}$$

$$= \mathbf{145\ N}$$

Making use of the self-energizing character of this brake produces an economy in the size of actuating force needed. If one were to use $b = 0$, the brake would not be self-energizing and the necessary actuating force would be

$$F_a = \frac{(2)(1050 \text{ mm})(500 \text{ mm} - 0)}{(1.1 \text{ m})(1700 \text{ mm})(3.51 - 1)} = 224 \text{ N}$$

This difference in actuating forces shows the practical advantage of the self-energizing design.

20-5 DRUM BRAKES

Although rapidly being displaced by disk brakes in both vehicular and industrial applications, drum brakes are still important and widely used where *high* braking capacity is needed, as in trucks. A long-shoe internal drum brake is shown in Fig. 20-10. In this case the internal shoes are moved against the inner surface of the brake drum by the double-acting wheel cylinder at the top, itself activated by hydraulic pressure. For air brakes in heavy vehicles, on the other hand, the shoes are moved outward by a cam or a wedge that is actuated by an air cylinder and a linkage. The air brake in Fig. 20-11 is of suitable design for heavy-duty trucks.

For industrial applications, air or hydraulic fluid also is often used for actuating brakes but,

Figure 20-10 A leading-trailing shoe brake. (Reprinted from Newcomb and Spurr, *Automobile Brakes and Braking Systems*, courtesy Robert Bentley, Inc.)

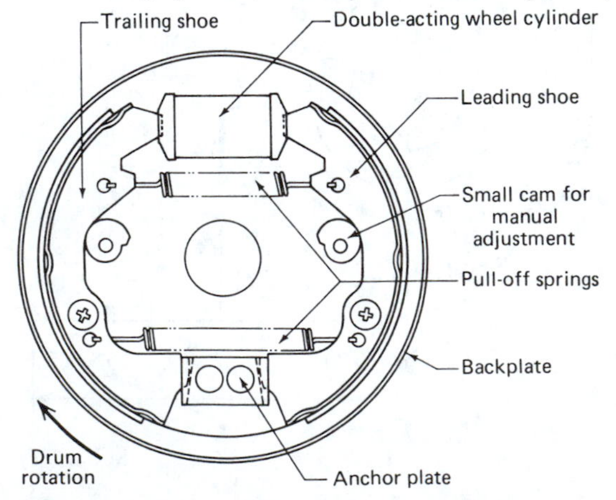

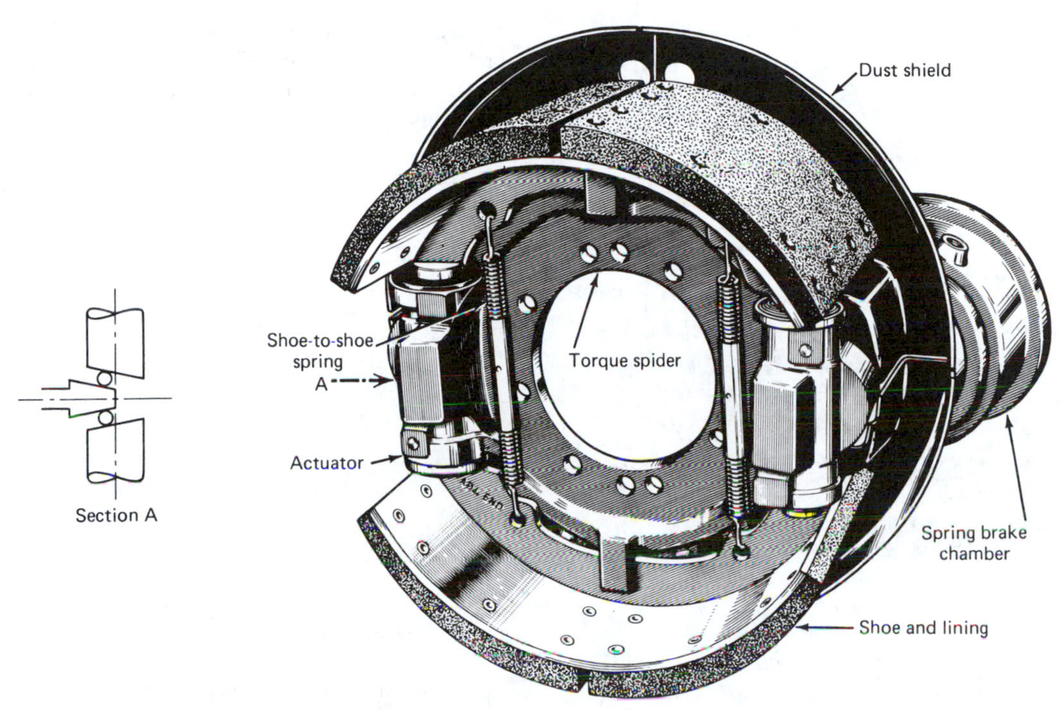

Figure 20-11 A wedge operated air brake for trucks. (Courtesy International Harvester Company.)

in these applications, electrical actuation is common. The shoes of the industrial drum brake in Fig. 20-12 are forced outward by the actuating cam, which is turned by the actuating lever. When current is applied to the electromagnet, the magnet contacts an armature surface on the brake drum and operates the actuating lever.

Design of a drum brake is not easy. A major complicating factor is the sharp variation in contact pressure along a shoe's lining. References 1 and 9 will be helpful to designers seeking to gain competence in drum brake design.

20-6 BRAKING REQUIREMENTS FOR VEHICLES

A two-axle vehicle during braking is shown in Fig. 20-13. For simplicity, the aerodynamic resistance and the tire-rolling resistance have been neglected. The only forces shown are the vehicle's weight mg, the front and rear axle loads W_f and W_r, and the front and rear braking forces F_{bf} and F_{br}. Even with this simple model of a vehicle, we can deal with two important questions. What braking torque capacity should be assigned to each axle? What is the minimum stopping distance?

Assignment of Braking Capacity

During braking, there is a load shift from the rear to the front axle. This suggests that additional braking capacity should be assigned to the front axle. The maximum braking force that can be supported at either axle depends on both the tire-road coefficient of adhesion f_R and the axle load W. It is usual to assign braking capacities to front and rear axles in such a way that both brakes can lock the wheels at the same time. Then the maximum braking forces are

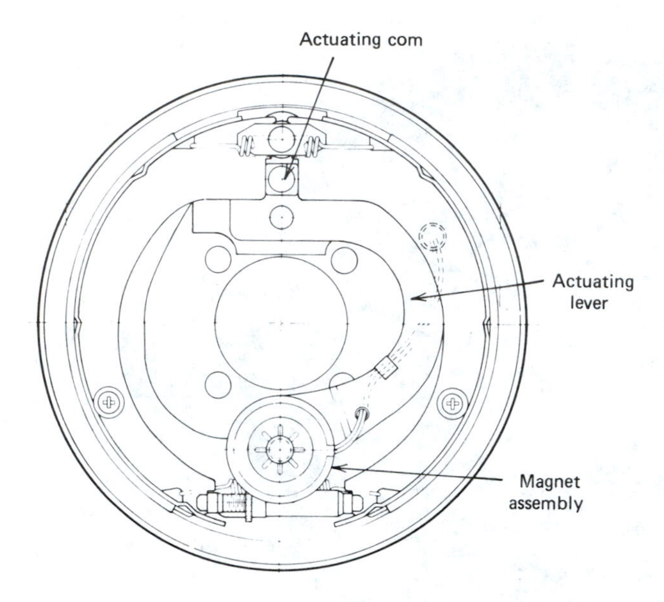

Figure 20-12 An electrically actuated drum brake for industrial or vehicular applications. (Courtesy Warner Electric Brake and Clutch Company.)

$$F_{bf\max} = \frac{f_R mg}{L}(l_2 + hf_R) \qquad (20\text{-}13)$$

$$F_{br\max} = \frac{f_R mg}{L}(l_1 - hf_R) \qquad (20\text{-}14)$$

where

$F_{bf\max}$, $F_{br\max}$ = maximum braking forces at front and rear axles; N, lb
m = vehicle mass; kg or slugs
g = local acceleration of gravity; m/s^2, ft/s^2
f_R = tire-road coefficient of adhesion

Figure 20-13 Forces acting on a two-axle vehicle during braking on a level road.

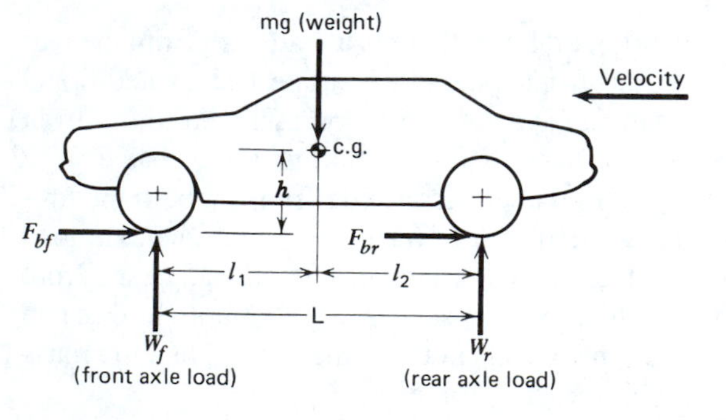

L = length of wheel base (Fig. 20-13); m, ft
l_1, l_2, h = location of center of gravity (c.g.) (Fig. 20-13); m, ft

The corresponding brake torque capacities should be

$$T_{f\max} = r_f F_{bf\max} \qquad (20\text{-}15)$$

$$T_{r\max} = r_r F_{br\max} \qquad (20\text{-}16)$$

where

$T_{f\max}$, $T_{r\max}$ = assigned braking torques to front and rear axles; N · m, lb-in.
r_f, r_r = front and rear effective tire rolling radii; m, in.

Tires and roads vary so much that there is no single value for f_R on which brake designers can rely. Table 20-5 lists some typical f_R values for pneumatic tires on various road surfaces. To design brakes for commercial vehicles, a common practice is to provide for a braking force at any braked wheel equal to 60% of the vehicle load at that wheel. Hence, to achieve this braking force, the coefficient of friction must be at least 0.60. In Great Britain the Ministry of Transport requires that a vehicle have sufficient braking ca-

TABLE 20-5 Friction Coefficients for Pneumatic Tires on a Variety of Road Surfaces

Road surface	f_R
Dry concrete pavement	0.8–0.9
Wet concrete pavement	0.8
Wet asphalt	0.5–0.7
Gravel	0.6
Dry packed earth	0.68
Wet packed earth	0.55
Packed snow	0.2
Ice	0.1

pacity to achieve a deceleration rate of at least 0.5g on dry, level highway. Most passenger vehicles have sufficient braking capacity to obtain a deceleration rate of 0.8g or above on level, dry, concrete highways in good condition.

EXAMPLE 20-5

The 1965 Corvette had a test mass (car, driver, and gasoline) of 3480 lbm (108.16 slugs). The wheelbase was 8.17 ft. The center of gravity was 3.9 ft forward of the rear axle and 1.48 ft above ground. The effective tire rolling radius was 13 in., both in front and rear. Assign braking capacities to the front and rear axles for an assumed tire-road coefficient of 0.9. Local g = 32.174 ft/s^2.

SOLUTION

(a) The braking force capacities for front and rear axles are found by using Eqs. 20-13 and 20-14.

$$F_{bf\max} = \frac{f_R mg}{L}(l_2 + hf_R)$$

$$= \frac{(0.9)(108.16 \text{ slugs})(32.174 \text{ ft/s}^2)}{(8.17 \text{ ft})}$$

$$[(3.9 \text{ ft} + 1.48 \text{ ft})(0.9)]$$

$$= 2006 \text{ lb}$$

$$F_{br\max} = \frac{f_R mg}{L}(l_1 - hf_R)$$

$$= \left[\frac{(0.9)(108.16 \text{ slugs})(32.174 \text{ ft/s}^2)}{(8.17 \text{ ft})}\right]$$

$$[(8.17 \text{ ft} - 3.9 \text{ ft}) - (1.48 \text{ ft})(0.9)]$$

$$= 1126 \text{ lb}$$

(b) The brake torque capacities are, by Eqs. 20-15 and 20-16,

$$T_{f\max} = r_f F_{bf\max} = (13 \text{ in.})(2006 \text{ lb})$$

$$= 26{,}080 \text{ lb-in.}$$

$$T_{r\max} = r_r F_{br\max} = (13 \text{ in.})(1126 \text{ lb})$$

$$= 14{,}640 \text{ lb-in.}$$

For each axle the torque capacity just calculated is the total braking capacity for the axle. Thus each front wheel brake should have a capacity of 26,080 lb-in./2 = **13,040 lb-in.,** and each rear axle brake should have a capacity of 14,640 lb-in./2 = **7320 lb-in.** Note that for an assumed value of f_R = 0.9, 64% of the vehicle's braking capacity is assigned to the front axle. When the vehicle is not decelerating (or accelerating), only 48% of the vehicle's weight is on the front axle.

Stopping Distance

The basic kinematic relationship for motion with constant acceleration is

$$v_f{}^2 = v_o^2 + 2as \tag{20-17}$$

where

v_f = final velocity; m/s, ft/s
v_o = initial velocity; m/s, ft/s
a = acceleration; m/s^2, ft/s^2
s = distance traveled during acceleration period; m, ft

To find the minimum stopping distance, it is assumed that the maximum braking force is being used at each axle. From Eqs. 20-13 and

20-14, or by inspecting Fig. 20-13, the maximum braking force is

$$F_{b\max} = F_{bf\max} + F_{br\max} = f_R mg \quad (20\text{-}18)$$

Therefore, by Newton's second law, the maximum acceleration rate is

$$a = -\frac{F_{b\max}}{m} = -\frac{f_R mg}{m} = -f_R g \quad (20\text{-}19)$$

The acceleration is negative, since the vehicle is slowing down. By combining Eqs. 20-17 and 20-19 and by setting $v_f = 0$, the minimum possible stopping distance is

$$s_{\min} = \frac{v_o^2}{2f_R g} \quad (20\text{-}20)$$

EXAMPLE 20-6

What is the minimum stopping distance from 100 km/h if $f_R = 0.9$ for the 1965 Corvette?

SOLUTION

The initial velocity is

$$v_o = \left(100\,\frac{\text{km}}{\text{h}}\right)\left(\frac{1000\text{ m}}{\text{km}}\right) + \frac{\text{h}}{3600\text{ s}}$$
$$= 27.78\text{ m/s}$$

So by Eq. 20-20, the minimum stopping distance is

$$s_{\min} = \frac{v_o^2}{2f_R g}$$
$$= \frac{(27.78\text{ m/s})^2}{2(0.9)(9.807\text{ m/s}^2)} = \mathbf{43.72\text{ m}}$$

20-7 BRAKING REQUIREMENTS FOR INDUSTRIAL BRAKES

In industrial applications brakes usually bring the rotating members of a machine to a stop in a specified time or slow them down to a lower speed in some desired time. From elementary mechanics, the torque necessary to decelerate a rotating body is

$$T = \frac{2\pi J(n_o - n_f)}{t_s}\ \text{N}\cdot\text{m} \quad (20\text{-}21)$$

$$T = \frac{J(n_o - n_f)}{61.43t_s}\ \text{lb-in.} \quad (20\text{-}22)$$

where

T = braking torque; N · m, lb-in.
J = mass moment of inertia about the axis of rotation; kg · m^2, lbm-in.^{2}
n_o, n_f = initial and final angular speeds; rps
t_s = slowdown time; s

There is usually more than one rotating mass. Gearing or belt-and-pulley drives connect shafts rotating at different speeds. Consequently, it is necessary to use the *equivalent inertia* for the whole system.

EXAMPLE 20-7

A small packaging machine has an equivalent mass moment of inertia of 19.3 kg · m^2. What braking torque is necessary to slow it down from 15 rps to 3 rps in 12 s?

SOLUTION

Equation 20-21 is all that is needed.

$$T = \frac{2\pi J(n_o - n_f)}{t_s}$$
$$= \frac{2\pi(19.3\text{ kg}\cdot\text{m}^2)(15\text{ rps} - 3\text{ rps})}{12\text{ s}}$$
$$= \mathbf{121\text{ N}\cdot\text{m}}$$

20-8 ENERGY GENERATION

One of the greatest challenges in brake design is to produce a brake that has ample braking torque and the ability to dissipate harmlessly the large quantities of heat developed by friction.

Vehicle Brakes

If a vehicle is slowed down from an initial velocity of v_o to a final velocity of v_f, the heat energy E that the brakes must dissipate is estimated to be the decrease in the vehicle's kinetic energy. Of course, some of the energy is dissipated in rolling resistance, air resistance, and driveline losses, so this is a conservative approach. The change in kinetic energy is

$$E = \frac{m}{2}(v_o^2 - v_f^2) \text{ J} \qquad (20\text{-}23)$$

$$E = \frac{m}{1556}(v_o^2 - v_f^2) \text{ Btu} \qquad (20\text{-}24)$$

where

E = kinetic energy change; J, Btu
m = vehicle mass; kg, slugs
v_o, v_f = initial and final vehicle velocities; m/s, ft/s

The maximum rate at which heat energy is generated by the brake occurs at the beginning of the braking effort. It is

$$H_{max} = f_R m g v_o \text{W} \qquad (20\text{-}25)$$

$$H_{max} = \frac{f_R m g v_o}{778} \text{ Btu/s} \qquad (20\text{-}26)$$

where

H_{max} = maximum rate of energy generation; W, Btu/s

The other quantities have already been defined. To select friction materials or to estimate temperature rise, vehicle brake designers often use the average rate of heat generation during braking.

$$H_{ave} = 0.5H_{max} \qquad (20\text{-}27)$$

Once the average rate of heat generation has been estimated, Table 20-4 helps to determine the friction surface needed.

Industrial Brakes

The heat energy generated per application of the brake is the change in rotational kinetic energy of the machine. It is

$$E = 19.74J(n_o^2 - n_f^2) \text{ J} \qquad (20\text{-}28)$$

$$E = \frac{J(n_o^2 - n_f^2)}{182{,}565} \text{ Btu} \qquad (20\text{-}29)$$

where

E = energy generated; J, Btu
J = mass moment of inertia; kg · m^2, lbm-in.^{2}
n_o, n_f = initial and final rotational velocities; rps

In many industrial applications the brakes are applied frequently. The average rate of heat generation is

$$H_{ave} = \frac{ES}{3600} \qquad (20\text{-}30)$$

where

H_{ave} = average rate of heat generation; W, Btu/s
E = heat energy generated per application; J, Btu
S = stops per hour

20-9 HEAT DISSIPATION BY INDUSTRIAL CALIPER DISK BRAKES

Two approaches can be taken. If braking applications are frequent, the average rate at which heat is generated by the braking operations must be matched by the average rate at which heat is dissipated from the disk by convection and radiation. If the brake is infrequently used, the disk can be regarded as a temporary heat-storage device. It is then assumed (conservatively) that all of the heat generated by the braking operation is absorbed by the disk alone. It is necessary in this case only to be sure that the disk mass is large enough so that the temperature rise of the disk is not excessive. These two approaches are suitable for preliminary design estimates.

Frequent Application

The average rate at which heat must be dissipated is given by Eq. 20-30. The rate at which the disk is capable of dissipating heat by convection and radiation is estimated by Eq. 20-31.

$$H_{\text{diss}} = hA(t_d - t_a) \tag{20-31}$$

where

H_{diss} = rate of heat dissipation; W, Btu/s
h = overall heat transfer rate; W/m² · °C, Btu/in.²-s-°F
t_d = disk temperature; °C, °F
t_a = temperature of surrounding air; °C, °F
A = exposed surface of disk; m², in.²

The heat transfer coefficient h expresses the combined effect of convection and radiation. However, the convection heat transfer is very dependent on the average velocity of the air passing over the disk. The heat transfer coefficient is

$$h = h_r + f_v h_c \tag{20-32}$$

where

h_r = radiation coefficient; W/m² · °C, Btu/in.²-s-°F
h_c = convection coefficient for still air; W/m² · °C, Btu/in.²-s-°F
f_v = factor to correct for the velocity of moving air

The quantities h_r and h_c are plotted in Figs. 20-14 and 20-15, and the values of f_v appear in Fig. 20-16.

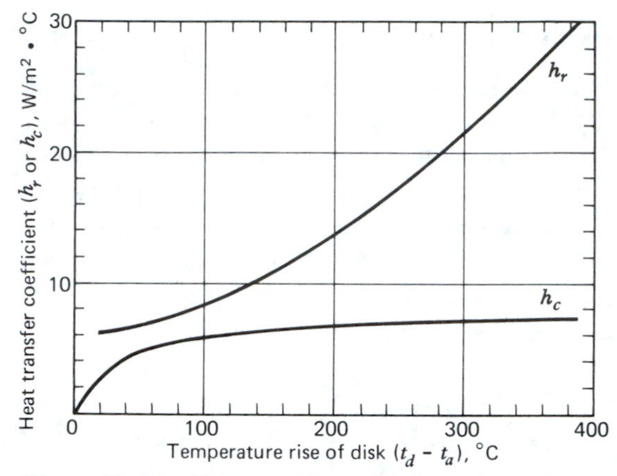

Figure 20-14 Heat transfer coefficients for radiation and convection in still air (SI units).

EXAMPLE 20-8

An industrial caliper disk brake is used an average of 23 times per hour to stop a machine with an equivalent mass moment of inertia of 126,436 lbm-in.² from a speed of 285 rpm. What minimum exposed area on the disk is needed to limit the temperature rise of the disk to 284°F? The mean air velocity is 33 ft/s. The disk cannot be more than 11.6 in. in diameter.

SOLUTION

The energy the brake must dissipate per stop is found by applying Eq. 20-29. The initial speed $n_o = 285/60 = 4.75$ rps. The energy converted into heat per braking operation is

$$E = \frac{J(n_o^2 - n_f^2)}{182{,}565}$$

$$= \frac{(126{,}436\ \text{lb-in.}^2)[(4.75\ \text{rps})^2 - 0]}{182{,}565}$$

$$= 15.6\ \text{Btu}$$

So the average rate at which heat energy must be dissipated by the brake is, by Eq. 20-30,

Figure 20-15 Heat transfer coefficients for radiation and convection in still air (English units).

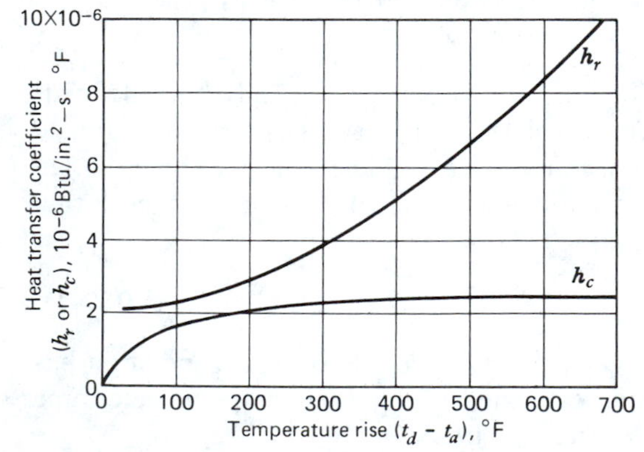

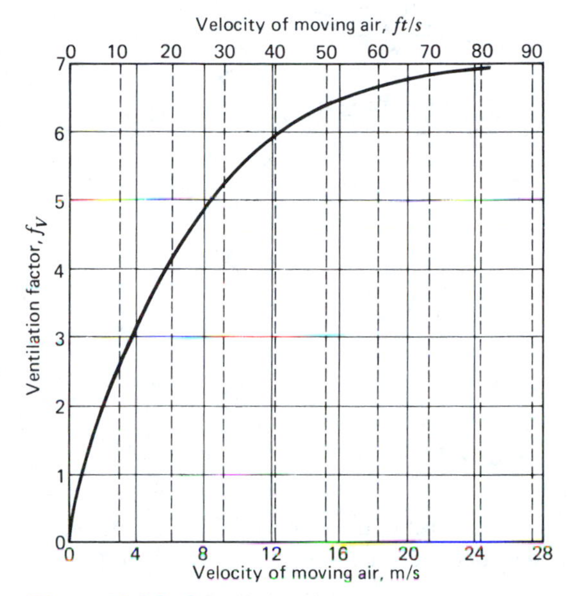

Figure 20-16 Ventilation factors.

$$H_{\text{ave}} = \frac{ES}{3600 \text{ s/h}}$$

$$= \frac{(15.6 \text{ Btu/oper})(23 \text{ oper/h})}{3600 \text{ s/h}}$$

$$= 0.1 \text{ Btu/s}$$

The overall heat transfer coefficient is found by use of Eq. 20-32 and Figs. 20-15 and 20-16.

$$h = h_r + f_v h_c$$

$$= (3.53)(10^{-6} \text{ Btu/in.}^2\text{-s-°F})$$

$$+ (5.5)(2.14)(10^{-6} \text{ Btu/in.}^2\text{-s-°F})$$

$$= (1.529)(10^{-5} \text{ Btu/in.}^2\text{-s-°F})$$

So the required exposed area on the disk is, by Eq. 20-31,

$$A = \frac{H_{\text{diss}}}{h(t_d - t_a)}$$

$$= \frac{0.1 \text{ Btu/s}}{(1.529)(10^{-5} \text{ Btu/in.}^2\text{-s-°F})(284\text{°F} - 32\text{°F})}$$

$$= \mathbf{26 \text{ in.}^2}$$

Infrequent Stops

In this case the disk is assumed to absorb all of the heat generated during braking, but there is adequate time between braking operations for the disk to cool down. The design criterion is to limit the temperature rise of the disk to a safe value by making sure it has sufficient mass. The energy the disk is assumed to absorb is

$$E_{\text{abs}} = m_d c(t_d - t_a) \qquad (20\text{-}33)$$

where

E_{abs} = absorbed energy; J, Btu
c = specific heat of the disk material; J/kg · °C, Btu/lbm-°F
t_d = final disk temperature; °C, °F
t_a = surrounding air temperature, assumed to be initial disk temperature as well; °C, °F
m_d = mass of the disk; kg, lbm

This equation is based on the assumption that the disk is initially at the temperature of the surrounding air and has time to return to that temperature before the next braking operation.

For a simple cylindrical disk, this equation becomes

$$E_{\text{abs}} = \frac{\pi}{4} d^2 b \rho c(t_d - t_a) \qquad (20\text{-}34)$$

where

d = disk diameter; m, in.
b = disk thickness; m, in.
ρ = density of disk material; kg/m^3, lbm/in.^{3}
c = specific heat of disk material; J/kg · °C, Btu/lbm-°F

Typical values of the density and specific heat of steel are ρ = 7833 kg/m^3 or 0.283 lbm/in.^{3} and c = 490 J/kg · °C or 0.117 Btu/lbm-°F.

EXAMPLE 20-9

Suppose the brake described in Example 20-8 were used only three times per hour. What would the minimum thickness of the disk have to be to limit the temperature rise to 248°F for a disk diameter of 11.6 in.?

SOLUTION

The energy to be dealt with per braking operation was found to be 15.6 Btu in Example 20-8. This is the energy to be absorbed by the disk. By Eq. 20-34, the minimum disk thickness is

$$b = \frac{4E_{\text{abs}}}{\pi d^2 \rho c (t_d - t_a)}$$

$$= \frac{(4)(15.6\ \text{Btu})}{\pi (11.6\ \text{in.})^2 (0.283\ \text{lbm/in.}^3) c (t_d - t_a)}$$

$$= \frac{62.4\ \text{Btu}}{(119.633\ \text{lbm})(0.117\ \text{Btu/lbm-°F})(248\text{°F})}$$

$$= \mathbf{0.018\ in.}$$

Since any practical brake disk of this diameter will be much thicker than 0.02 in., there will be no problem with using this brake infrequently.

20-10 DERIVATIONS OF DESIGN EQUATIONS

In the earlier discussions of disk brake design (Section 20-2) and braking requirements for vehicles and industrial machines (Sections 20-6 and 20-7), a few basic equations were presented without proof. Here are the necessary derivations.

Effective Radius of Annular Pad

An annular pad for a disk brake is shown in Fig. 20-17. On it is identified an elementary region of radial width *dr*. In this derivation the contact pressure is assumed to be constant across the pad.

The area of the elementary region is

$$dA = r\theta\, dr$$

The corresponding portion of the actuating force is

$$dF_a = p\, dA = p\theta r\, dr$$

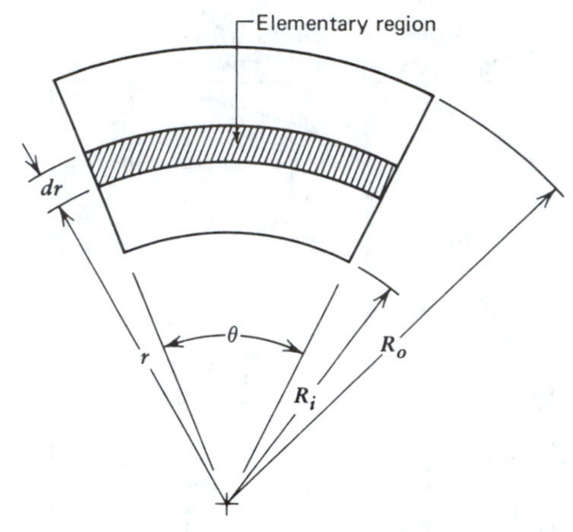

Figure 20-17 Annular pad for a disk brake.

By integrating radially across the pad, an equation is obtained for the total actuating force.

$$F_a = \int_{R_i}^{R_o} dF_a = p\theta \int_{R_i}^{R_o} r\, dr$$

$$F_a = \frac{p\theta(R_o^2 - R_i^2)}{2} \qquad (20\text{-}35)$$

The friction torque developed in the elementary region is

$$dT = f\, dF_a r = fp\theta r^2\, dr$$

The total friction torque capacity is, by integration,

$$T = \int_{R_i}^{R_o} dT = fp\theta \int_{R_i}^{R_o} r^2\, dr$$

$$T = \frac{fp\theta(R_o^3 - R_i^3)}{3} \qquad (20\text{-}36)$$

It is more convenient to express the torque capacity in terms of the actuating force. (This was also done for disk clutches in Section 19-10.) An effective radius R_e is defined by

$$R_e = \frac{T}{fF_a} \qquad (20\text{-}37)$$

An expression for R_e in terms of the pad dimensions is found by solving simultaneously Eqs. 20-35, 20-36, and 20-37. The result is

$$R_e = \frac{2}{3}\frac{(R_o^3 - R_i^3)}{(R_o^2 - R_i^2)} \qquad (20\text{-}38)$$

This is the same as Eq. 20-4.

In the analysis of disk clutches in Section 19-10, a uniform rate of wear was assumed. Here the assumption has been that the pressure, not the wear rate, is uniform over the pad. To make a comparison, consider a pad with the dimensions $R_i = 180$ mm and $R_o = 235$ mm. For the assumption of uniform wear, $R_e = 207.5$ mm but, for the uniform pressure assumption, $R_e = 208.7$ mm. For R_i/R_o greater than 0.6, the difference between the two estimates of R_e is 2.1% or less. For annular brake pads, R_i/R_e is usually greater than 0.6.

Assignment of Braking Capacity

Here Eqs. 20-13 and 20-14 for the maximum braking forces are derived. Figure 20-18 is Fig. 20-13 redrawn with two additions. An inertia force ma has been shown, and the center of the rear tire patch has been labeled as O. No inertia couple is shown. Any angular acceleration during braking is neglected.

The sum of the horizontal forces must be zero, and the sum of the vertical forces must also be zero.

$$F_{bf\max} + F_{br\max} + ma_{\max} = 0 \qquad (20\text{-}39)$$

$$W_r + W_f - mg = 0 \qquad (20\text{-}40)$$

The maximum braking forces at each axle depend on the tire-road coefficient of adhesion.

$$F_{bf\max} = f_R W_f \qquad (20\text{-}41)$$

$$F_{br\max} = f_R W_r \qquad (20\text{-}42)$$

Simultaneous solution of these equations yields

$$a_{\max} = -f_R g \qquad (20\text{-}43)$$

The acceleration is negative because the vehicle is slowing down.

To apportion braking capacity to the two axles, the axle loads W_f and W_r must be known. Summing moments about point O, the center of the rear tire patch,

$$l_2 mg - LW_f - mah = 0 \qquad (20\text{-}44)$$

W_r and W_f can be found by simultaneous solution of Eqs. 20-40, 20-43, and 20-44.

$$W_f = \frac{mg}{L}(l_2 + hf_R) \qquad (20\text{-}45)$$

$$W_r = \frac{mg}{L}(l_1 - hf_R) \qquad (20\text{-}46)$$

Therefore the maximum braking forces are

$$F_{bf\max} = f_R W_f$$

$$F_{br\max} = f_R W_r$$

Equations 20-13 and 20-14, which we are to derive, follow immediately by substituting for W_f and W_r from Eqs. 20-45 and 20-46. Thus

$$F_{bf\max} = \frac{f_R mg}{L}(l_2 + hf_R) \qquad (20\text{-}13)$$

$$F_{br\max} = \frac{f_R mg}{L}(l_1 - hf_R) \qquad (20\text{-}14)$$

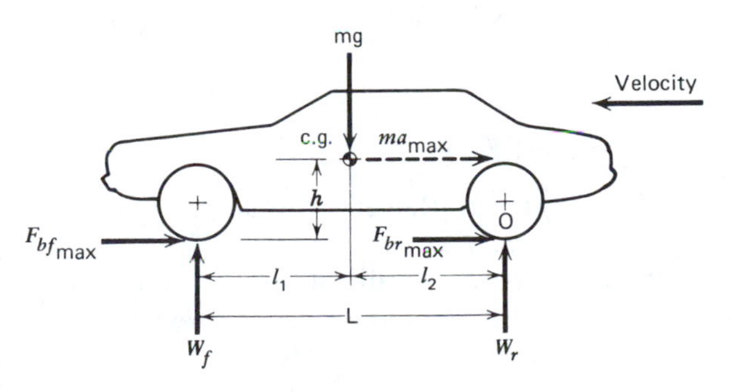

Figure 20-18 Forces acting on a two-axle vehicle during braking on a level road.

Torque to Decelerate a Rotating Body

Equations 20-21 and 20-22 are the last equations to be derived. From basic mechanics, the torque to decelerate a rotating mass is

$$T = J\alpha \tag{20-47}$$

where

T = torque; N · m, lb-in.
J = mass moment of inertia; kg · m^2, lbm-in.^{2}
α = angular deceleration; rad/s^2

The angular deceleration in a short time interval averages

$$\alpha = \frac{\omega_o - \omega_f}{t_s}$$

where

ω_o, ω_f = initial and final angular velocities; rad/s

However,

$$\omega = 2\pi n$$

where

n = angular velocity; rps

Hence

$$\alpha = \frac{2\pi}{t_s}(n_o - n_f)$$

Substituting for α in Eq. 20-47

$$T = \frac{2\pi J}{t_s}(n_o - n_f) \text{ N} \cdot \text{m} \tag{20-21}$$

$$T = \frac{J(n_o - n_f)}{61.43 t_s} \text{ lb-in.} \tag{20-22}$$

This is Eq. 20-21. Equation 20-22 is derived by inserting conversion factors to convert from metric to English units.

SUMMARY

Brakes are essential in vehicles and in many industrial machines to control speed and thus protect the operators and the equipment.

Caliper disk brakes, using either annular or circular pads, are now being applied to vehicles of all sizes and to many industrial machines. Their design is based principally on practical limits for pressure, temperature, and rate of wear.

Friction materials are available in a variety of forms, from woven cotton to ceramic-metal mixtures. The choice is influenced by coefficient of friction, fade resistance, and wear resistance.

A band brake is relatively simple to design and manufacture for use in hoists and other types of industrial machinery. It can be designed to be self-energizing.

Internal shoe drum brakes are used where very high capacity is needed, as in heavy trucks. They, too, can be designed to be self-energizing.

Braking capacity is assigned to the front and rear axles of a vehicle in such a way that both sets of brakes can lock at the same time. Since there is substantial weight shift to the front axle during braking, the larger part of the braking capacity will be put there. During braking, the vehicle's kinetic energy is being changed to heat energy; consequently, the brakes must be able to dissipate large quantities of heat without overheating.

Brakes for industrial machines slow down or stop a rotating mass in some specified time. For frequent braking operations, industrial brakes must be able to dissipate heat continuously. For infrequent braking, the disk must simply be heavy enough to absorb all of the heat generated without itself overheating. It can dissipate the heat between braking operations.

REFERENCES

20-1 Auman, Reiner J. "Brake Design." *Machine Design,* June 14, 1956.

20-2 Crouse, William H. "Automotive Brakes." *Automotive Chassis and Body,* Fourth Edition. New York: McGraw-Hill, 1971.

20-3 Fazekas, G. A. "On Circular Spot Brakes." *ASME Transactions: Journal of Engineering for Industry,* August 1972.

20-4 Fenwick, Keld. "Industrial Clutches and Brakes." *Chartered Mechanical Engineer,* September 1976.

20-5 *Machine Design. Mechanical Drives Reference Issue,* 1976.

20-6 Mathews, G. P. "Art and Science of Braking Heavy Duty Vehicles." Special Publication SP-251. Warrendale, Pa.: Society of Automotive Engineers, 1964.

20-7 Newcomb, T. P., and R. T. Spurr. *Automobile Brakes and Braking Systems.* Cambridge, Mass.: Robert Bentley, 1969.

20-8 Sinclair, David. "Friction Brakes." *Mechanical Design and Systems Handbook.* New York: McGraw-Hill, 1964.

20-9 Spotts, M. F. "Design Formulas for Floating-Shoe Brakes and Clutches." *Product Engineering,* September 28, 1964.

20-10 Wong, J. Y. *Theory of Ground Vehicles.* New York: Wiley, 1978.

PROBLEMS

P20-1 The front caliper disk brakes of a passenger car are each to have a torque capacity of 1238 N · m. There is room for a 250-mm diameter disk. It has been decided tentatively to use two annular pads per brake, one on each side, with R_o = 120 mm, R_i = 80 mm, and θ = 40 deg. The pad manufacturer guarantees a coefficient of friction of at least 0.36. Each pad will be actuated by a 35-mm diameter piston.

(*a*) What is the effective friction radius?

(*b*) What actuating force is needed (remember that there are two pads)?

(*c*) What hydraulic pressure is needed at the wheel cylinders?

P20-2 What is the torque capacity of a caliper disk brake that uses two circular pads, one on each side of the disk? The dimensions are: R = 0.7 in., e = 3.5 in. The coefficient of friction will be at least 0.32. Express the torque as a multiple of the actuating force. What actuating force is needed for a torque capacity of 4340 lb-in.?

P20-3 The band brake shown in Fig. 20-19 is used in well-drilling operations and is operated by compressed air. What should the piston diameter in the air cylinder be to achieve a torque in braking of 17 430 N · m? All dimensions are in millimeters.

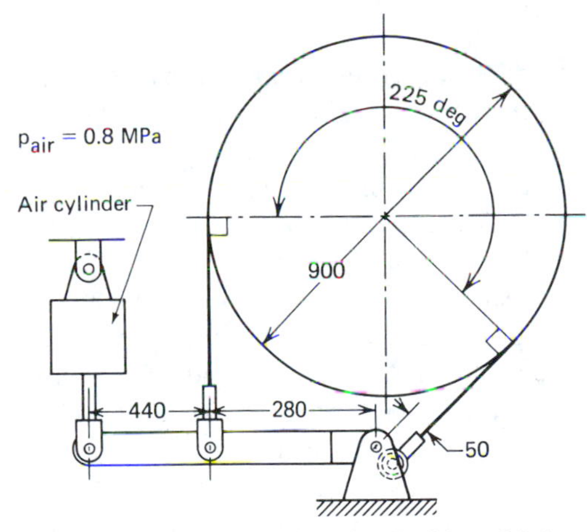

Figure 20-19 Illustration for Problem 20-3.

P20-4 A backstop is shown in Fig. 20-20. It is used to prevent accidental descent of a load. For counterclockwise rotation, the drum turns freely. However, if the shaft should attempt to turn the drum clockwise, the brake is self-locking. (*a*) What is the minimum coefficient of friction between band and drum for the brake to operate properly? (*b*) If f = 0.33 and the band tension is lim-

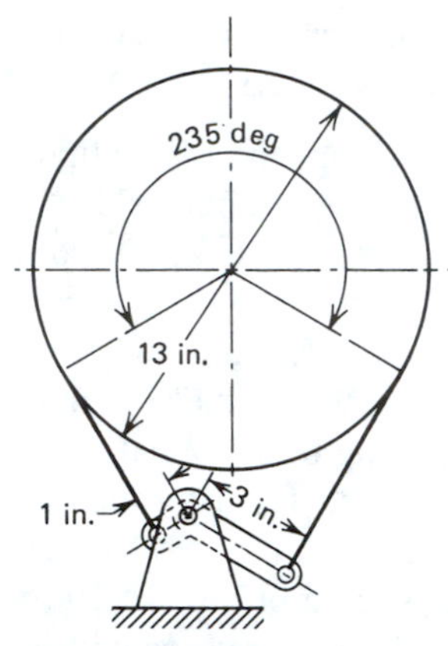

Figure 20-20 Illustration for Problem 20-4.

ited to 1070 lb, what torque capacity does the brake have?

P20-5 A sports car has a mass (with driver, one passenger, and fuel) of 1790 kg and a wheelbase of 2.71 m. The center of gravity is 1.22 m behind the front axle and 0.39 m above ground. The tire rolling radius is 0.31 m. Assign braking capacity to the front and rear brakes assuming that $f_R = 0.90$.

P20-6 A two-axle straight truck has a wheelbase of 12.17 ft and a total mass, when loaded, of 28,000 lbm. Its center of gravity is 7.82 ft behind the front axle and 3.02 ft above ground when fully loaded. The tire rolling radius is 1.65 ft. For an assumed tire-road coefficient of 0.65, assign braking capacity to the front and rear axles.

P20-7 Find the minimum braking distances for a vehicle traveling at 30 km/h, 45 km/h, 60 km/h, and 90 km/h if the tire-road coefficient of friction is 0.8. Repeat the calculations under the assumption that a light rain has reduced the coefficient to 0.48. Estimate the total stopping distance from 90 km/h if the coefficient is 0.8 and the reaction time is 0.3 second.

P20-8 What total energy must be dissipated by the brakes of a 9900-lbm school bus (fully loaded) when it is stopped from 45 mph?

P20-9 A school bus with a mass of 4491 kg fully loaded is to be stopped from a maximum speed of 80 km/h. It is assumed that the coefficient of friction is 0.70 between tire and road. What is the average braking power?

P20-10 For the school bus described in P20-9, how much lining area should be used at a minimum?

P20-11 A plate disk brake is attached to the shaft of a press whose equivalent mass moment of inertia is $(1.09)(10^6)$ lbm-in.^{2} What braking torque is needed for a complete stop from 475 rpm in 0.9 s?

P20-12 For the press described in P20-11, what total energy must be dissipated in a complete stop from 475 rpm? What is the average rate of heat dissipation required for 20 stops per hour?

P20-13 Estimate the total pad area needed for a caliper disk brake used in a hoist. The use is intermittent. The average braking power is 48 hp during braking.

P20-14 A caliper disk brake is used to brake a load in an industrial machine with an equivalent mass moment of inertia of $J = 28\ \text{kg} \cdot \text{m}^2$ from 650 rpm to rest. The disk diameter is 195 mm. What should be the disk thickness to limit the disk temperature to 150°C when the surrounding air is 24°C?

P20-15 An intermittently used caliper disk brake is used 17 times per hour to stop a load of $J = (1.78)(10^5)$ lbm-in.^{2} from 985 rpm. Is a single disk of 24-in. diameter sufficient for a maximum disk temperature of 300°F? The surrounding air temperature is 68°F. Average air velocity over the disk is 33 ft/s.

Chapter 21
Gaskets and Seals

Everything in nature is a cause, from which there flows some effect.

SPINOZA

NOMENCLATURE

A = gasket contact area
b = effective gasket width
b_o = basic gasket width
C = cross-sectional width of rectangular sealing ring
D = bolt diameter; diameter of gasket load reaction
d_i = O-ring inside diameter
d_o = outside diameter of gasket sealing surface, defined in Table 21-5
E = depth of groove for seal ring
F = width of groove for seal ring; tensile force in bolt
l = effective bolt length
Δl = change in bolt length
m = gasket maintenance factor
n = number of bolts
p = pressure
p_a = apparent flange pressure
p_f = minimum flange pressure
T = bolt torque
t = gasket thickness
Δt = change in gasket thickness
W = diameter of O-ring cross section; depth of cross section of rectangular section ring

The need for seals arose with the discovery that lubricants reduce bearing friction. An early form of seal was a leather strap wrapped around a vehicle axle to retain grease in the bearing. Later, packings and rope were used. These performed adequately despite some leakage.

Use of reciprocating motion in the steam engine and in pumps led to a need for seals on sliding members. The piston ring was developed for pistons under high temperature and high pressure. For low-temperature and low-pressure conditions the leather cup, used in bicycle tire pumps, was found to be adequate. At about the same time, packings were used to seal piston rods.

Gaskets were developed to prevent leakage of gases and liquids between metal surfaces that are bolted together. The early cork and paper gaskets have now been largely replaced by composite materials.

Today many designs of seals, packings, and gaskets are used to prevent leakage of all the liquids and gases used in industry. All types may be divided into two major classes: static seals and dynamic seals. Static seals are used with permanent or semipermanent connections. Dynamic seals are used with connections involving moving parts.

21-1 SEALS FOR STATIC CONNECTIONS: GASKETS

Gaskets are the most commonly used device for providing a seal between metal surfaces where motion is not intended to occur. A gasket is defined by ASTM Standard F118-72 as "a deformable material, which when clamped between essentially stationary faces, prevents the passage of matter through an opening or joint." Thus a gasket is a metal or a nonmetallic material used to contain or exclude gases, liquids, or solids.

A gasket must meet the following requirements.

1. Compatibility with the fluid to be sealed.
2. Stability at the maximum operating temperature.
3. Provision of the necessary seal.
4. Maintenance of the seal without excessive loss of bolt tension.

The first two requirements are met through material selection. The other two, however, involve the joint design as well as the selection of gasket material. This holds true except when the joint design is fixed as with standard pipe flanges, or when a new gasket is installed in existing equipment. In such cases, a gasket must be selected to meet the conditions encountered.

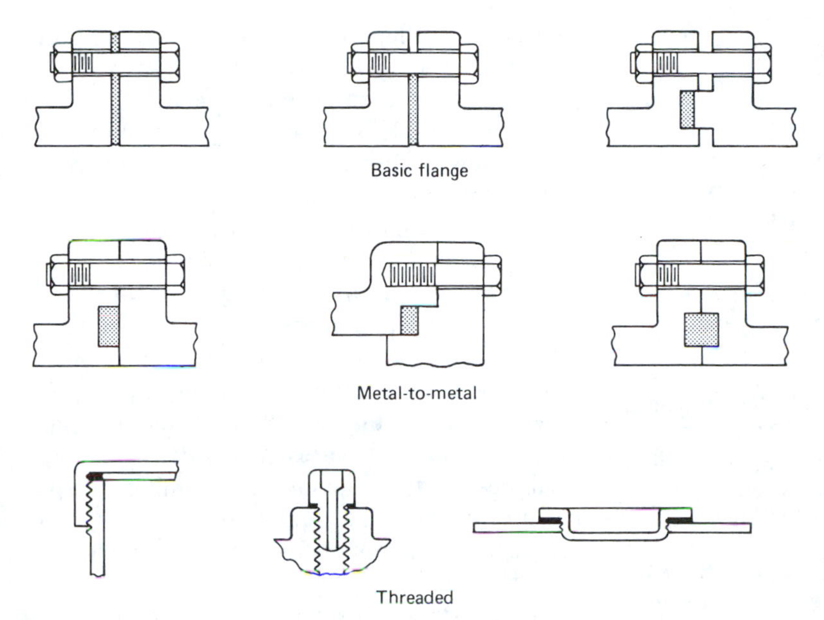

Figure 21-1 Examples of gasketed joints. (From Earl M. Smoley, "Sealing with Gaskets," *Machine Design*, Vol. 38, No. 25, pp. 172-187, 1966.)

Separable static joints may be sealed with a flat gasket, with a nonmetallic or metallic cylindrical ring, or with a sealant. Each of these methods will be discussed separately.

21-2 GASKETS FOR FLAT FLANGES

Gaskets (Fig. 21-1) are intended for use between essentially *parallel* flanges. In the basic flange arrangement the amount of gasket compression is determined by the gasket characteristics and the tension in the bolts. In the metal-to-metal configuration the amount of gasket compression is determined by the difference between the initial gasket thickness and the depth of the cavity.

The basic flange, with the gasket extending beyond the bolts, is the most frequently used arrangement in machinery. The flat flange can have almost any outline and requires a minimum of machining. In addition, the bolt holes provide a method of locating the gasket between the flanges during assembly.

The basic materials from which flat gaskets are made are asbestos, cellulose fibers (papers and vulcanized fiber), cork, metals, plastics, and rubbers. Solid metal gaskets of aluminum, copper, or low-carbon steel are sometimes used, especially in threaded connections (Fig. 21-1). More often, however, metal is used in combination with asbestos. For example, many cylinder head gaskets for internal combustion engines are made with a sheet of asbestos between two thin sheets of steel.

Some of the nonmetals, especially cork, paper, and rubber, are used without the addition of other materials. However, adding various binders and elastomers[1] to asbestos, cork, or paper makes it possible to produce an almost infinite number of gasket compositions.

Table 21-1 lists the general categories of gasket materials that can be used for various ranges of operating temperatures. The temperature of the flanges in contact with the gasket is of greater importance than the temperature of the fluid to be sealed because only an edge of the gasket is exposed to the fluid.

The categories of materials that can be used

[1]An elastomer is an elastic substance resembling rubber. These are manufactured materials of which fluorocarbon, nitrile-butadiene, polyacrylate, polychloroprene, and silicone represent some of the elastomers used in gaskets.

TABLE 21-1 Gasket Material Selection Based on Temperature

Temperature Range	Material Categories
Up to 149°C (300°F)	Cellulose or asbestos compositions, cork materials, or conventional elastomers
149–260°C (300–500°F)	Asbestos compositions or special elastomers
Above 260°C (500°F)	Metal-reinforced asbestos compositions or metallic constructions

Source. Alexander L. Gordon, "Performance Criteria of Gasket Materials," Society of Automotive Engineers Paper 770070 (1977).

under various ranges of fluid pressure are listed in Table 21-2. In general, materials satisfactory for low-temperature applications are also satisfactory for low pressures.

Pressure must be applied to a gasket to provide and maintain the seal between flanges. Without sufficient clamping pressure, leakage will occur through porous gasket materials. Highly compressible materials such as cork, elastomers, and soft asbestos compositions require relatively low flange pressures. Other materials such as compressed asbestos require high flange pressures to effect a seal.

In general, the highly compressible gasket materials are less expensive. They also permit the use of smaller bolts and thus provide an additional cost saving. Consequently, these materials are ordinarily used where pressure and temperature conditions permit. High pressures and temperatures require material of low compressibility and high density.

Sometimes, however, it is necessary to use a gasket material of low compressibility for structural reasons. Sufficient flange pressure must then be provided to create a seal. In mobile machinery, for example, gaskets of low or moderate compressibility must be used in joints between castings that are part of the machine structure. In such cases a low-density (high-compressibility) material may permit excessive motion in the joint. Joint motion can lead to extrusion of the gasket from the joint and loss of bolt tension.

The cost of a gasket can be minimized by attention to detail when designing a gasketed joint. Some of the common faults and suggested remedies are shown in Fig. 21-2. Designers should always remember that most gasket materials are low in strength and are also compressible.

21-3 FLANGE PRESSURE

The *apparent* flange pressure is the tensile force in the bolts holding a joint together divided by the area of contact between gasket and flanges. If

TABLE 21-2 Gasket Material Selection Based on Internal Pressure

Pressure Range	Material Categories
0–0.7 N/mm² (0–100 psi)	Wide choice, including low-density cellulose or asbestos compositions
0.7–3.5 N/mm² (100–500 psi)	Medium- to high-density cellulose and asbestos compositions
3.5–7.0 N/mm² (500–1000 psi)	High-density compressed asbestos
Above 7.0 N/mm² (1000 psi)	Metal-reinforced composites and metallic constructions

Source. Alexander L. Gordon, "Performance Criteria of Gasket Materials," Society of Automotive Engineers Paper 770070 (1977).

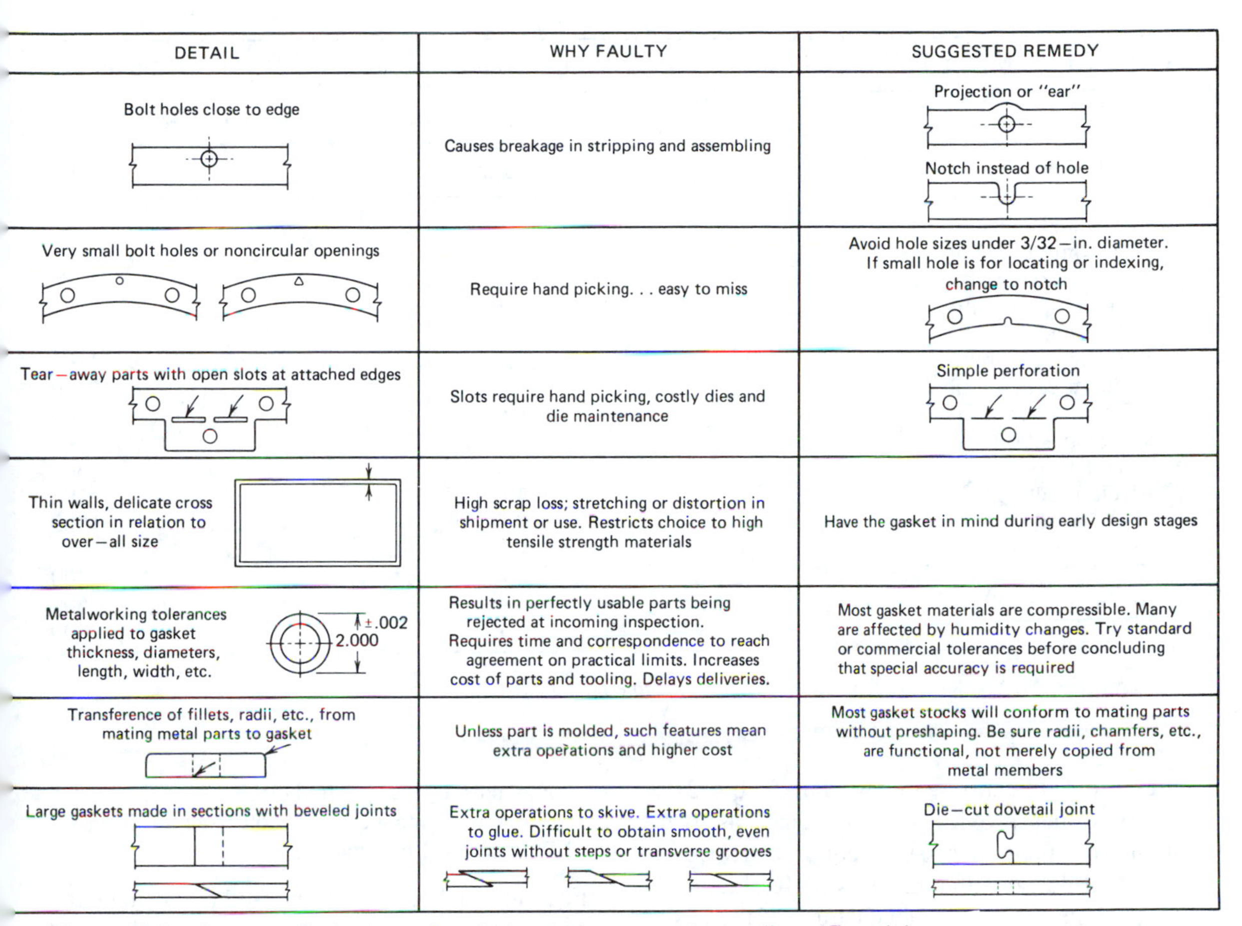

DETAIL	WHY FAULTY	SUGGESTED REMEDY
Bolt holes close to edge	Causes breakage in stripping and assembling	Projection or "ear" Notch instead of hole
Very small bolt holes or noncircular openings	Require hand picking. . . easy to miss	Avoid hole sizes under 3/32–in. diameter. If small hole is for locating or indexing, change to notch
Tear–away parts with open slots at attached edges	Slots require hand picking, costly dies and die maintenance	Simple perforation
Thin walls, delicate cross section in relation to over–all size	High scrap loss; stretching or distortion in shipment or use. Restricts choice to high tensile strength materials	Have the gasket in mind during early design stages
Metalworking tolerances applied to gasket thickness, diameters, length, width, etc. ±.002 2.000	Results in perfectly usable parts being rejected at incoming inspection. Requires time and correspondence to reach agreement on practical limits. Increases cost of parts and tooling. Delays deliveries.	Most gasket materials are compressible. Many are affected by humidity changes. Try standard or commercial tolerances before concluding that special accuracy is required
Transference of fillets, radii, etc., from mating metal parts to gasket	Unless part is molded, such features mean extra operations and higher cost	Most gasket stocks will conform to mating parts without preshaping. Be sure radii, chamfers, etc., are functional, not merely copied from metal members
Large gaskets made in sections with beveled joints	Extra operations to skive. Extra operations to glue. Difficult to obtain smooth, even joints without steps or transverse grooves	Die–cut dovetail joint

Figure 21-2 Common faults in gasket design and suggested remedies. (Copyright, Penton Publishing Co., reproduced with permission.)

all bolts are of the same size, this can be expressed as

$$p_a = \frac{nF}{A} \qquad (21\text{-}1)$$

where

p_a = apparent flange pressure; N/mm^2, psi
n = number of bolts
F = tensile force in each bolt; N, lb
A = contact area between gasket and flanges; mm^2, in.^{2}

Each gasket material has a *minimum* required flange pressure[2] for prevention of leakage. The standard method of measuring flange pressure is defined in ASTM F401-74, in which heavy steel flanges are used, and bolt forces are determined within 1% of the true value. Under these conditions the flange pressure is more uniform than that in many actual situations. The minimum flange pressures for some materials used in low

[2]This quantity is designated by y in the ASTM standard and in the ASME Boiler and Pressure Vessel Code. It is also known as yield factor or minimum seating stress.

TABLE 21-3 Minimum Flange Pressures for Gasket Materials Used in Low-Pressure Applications

Type of Material	Minimum Flange Pressure N/mm^2	psi
Asbestos and rubber		
Low compressibility	5.5	800
Medium compressibility	10.3	1500
High compressibility	13.8	2000
Cellulose and rubber		
Low compressibility	10.3	1500
Medium compressibility	13.8	2000
High compressibility	6.9	1000
Cork composition		
Low density	3.4	500
Medium density	4.8	700
High density	6.9	1000
Cork and rubber		
Soft	2.8	400
Medium	3.4	500
Firm	4.1	600

Source. Earl M. Smoley, "Sealing with Gaskets," *Machine Design*, Oct. 27, 1966.

pressure applications are listed in Table 21-3. These values provide some guidance as to the minimum flange pressure required. However, in actual design situations it is preferable to obtain a value from the manufacturer of the gasket material.

Most flanges in machines with low internal pressures are more flexible than those used for measurement of minimum flange pressures by the ASTM method mentioned earlier. In addition, calculated bolt tensions may deviate greatly from the actual tensions because of assembly conditions. A study of actual gasketed joints resulted in the curve in Fig. 21-3 as a recommendation for design purposes. Here the minimum flange pressure is the value from standard measurements. The apparent flange pressure is calculated from Eq. 21-1, with F defined as

$$F = \frac{T}{0.2D} \qquad (21\text{-}2)$$

where

F = tensile force in the bolt; N, lb
T = torque applied to the bolt; N · mm, lb-in.
D = nominal bolt diameter; mm, in.

Equation 21-2 is frequently used to calculate the approximate tension induced in an unlubricated bolt by the applied torque.

EXAMPLE 21-1

The standard dimensions for the mounting pad for a power takeoff from a truck transmission according to SAE Standard J704b are shown in Fig. 21-4 (dimensions in inches). This is for a gear-driven power takeoff for which a low-compressibility gasket is needed. Low compressibility is required to maintain control of the center distance between the gears in the transmission case and the power takeoff. What is a suitable choice of gasket material? Should Grade 5 bolts torqued to 35 lb-ft or Grade 8 bolts torqued to 49 lb-ft (Table 7-8) be used?

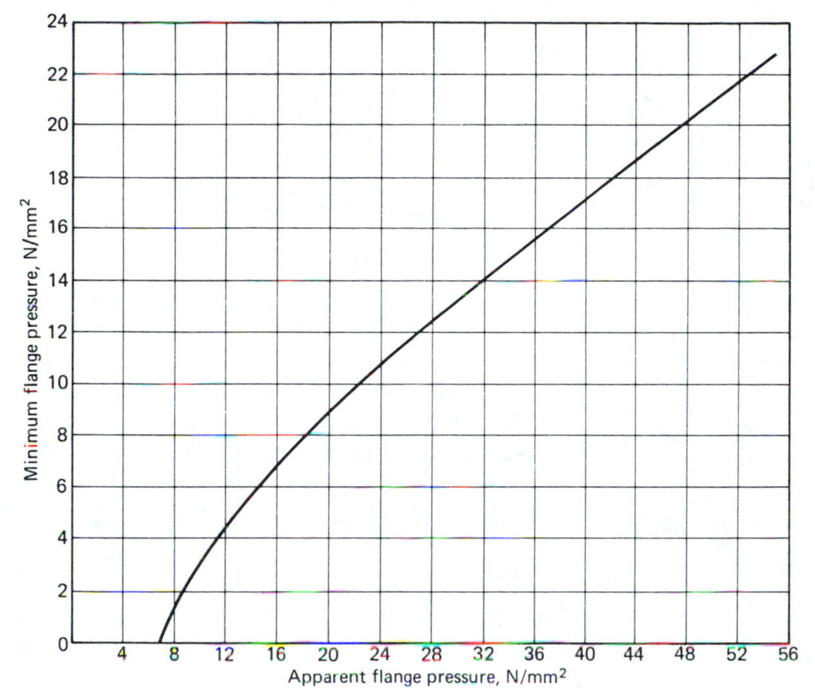

Figure 21-3 Relationship between minimum flange pressure and apparent flange pressure for low-pressure gasketed joints. (From I. G. Nolt and E. M. Smoley, "Gasket Loads in Flanged Joints," *Machine Design,* Vol. 33, No. 20, pp. 128-134, 1961.)

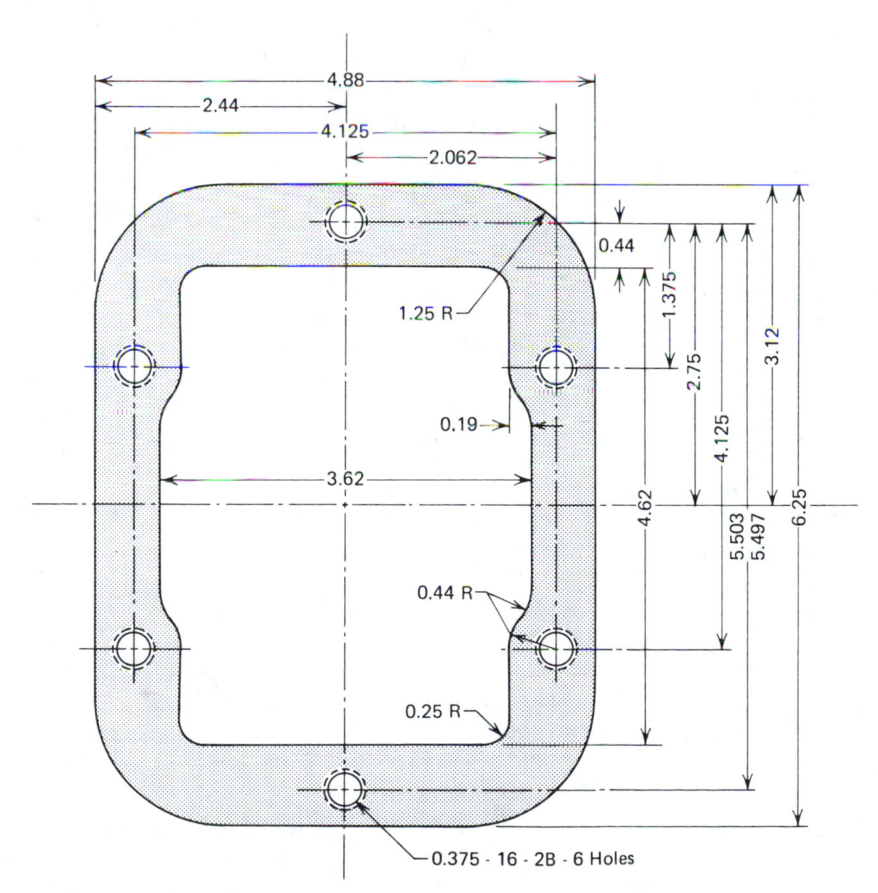

Figure 21-4 Dimensions of mounting pad for power takeoff from a truck transmission—SAE J704b. (From *SAE Handbook,* Society of Automotive Engineers, Inc., Warrendale, Pa., 1980.)

SOLUTION

The temperature of a truck transmission housing probably should not exceed 93°C (200°F). Thus, according to Table 21-1, temperature is not a limiting factor in the selection of a gasket material.

The pressure in a truck transmission is very low, probably less than 1 psig (0.007 N/mm^2 gauge). Table 21-2 indicates a wide choice of materials for this condition. In this instance, the selection of gasket material depends on material compressibility and cost. A cellulose and rubber gasket of medium compressibility seems to be a good choice. Table 21-3 indicates that the minimum flange pressure should be 1500 psi (10.3 N/mm^2).

Assuming that the contact area between gasket and flange will be the same as the area of the mounting pad, $A = 11.80$ in.^{2} (calculated).

For the Grade 5 bolts torqued to 35 lb-ft (420 lb-in.), the bolt tension, according to Eq. 21-2, is

$$F = \frac{T}{0.2D} = \frac{420 \text{ lb}}{(0.2)(0.375 \text{ in.})} = 5600 \text{ lb}$$

The apparent flange pressure can now be determined with Eq. 21-1 for six bolts and a contact area of 11.80 in.^{2}.

$$p_a = \frac{nF}{A} = \frac{(6)(5600 \text{ lb})}{11.80 \text{ in.}^2} = 2847 \text{ psi}$$

The gasket requires a minimum flange pressure of 1500 psi or 10.34 N/mm^2. Figure 21-3 indicates that this requires an apparent flange pressure of 23.2 N/mm^2 or 3364 psi, which is more than the value obtained with Grade 5 bolts. Thus, there may be gasket leakage if Grade 5 bolts are used.

It is necessary to repeat the computations with Grade 8 bolts torqued to 49 lb-ft (588 lb-in.).

$$F = \frac{T}{0.2D} = \frac{588 \text{ lb-in.}}{(0.2)(0.375 \text{ in.})} = 7840 \text{ lb}$$

and

$$p_a = \frac{nF}{A} = \frac{(6)(7840 \text{ lb})}{11.80 \text{ in.}^2} = 3986 \text{ psi}$$

The apparent flange pressure of 3986 psi exceeds the value of 3364 psi required for the gasket material. Consequently, oil leakage should not occur with Grade 8 bolts. It is possible, however, that a review of gasket manufacturers' literature would reveal a material of sufficiently low compressibility and minimum sealing pressure such that Grade 5 bolts could be used.

In order to obtain a high flange pressure to ensure that a gasket seals the joint, why not use a narrow gasket? Although a narrow gasket will certainly have higher flange pressure for the same bolt tension, flange pressure is not the only consideration. Because of rough flange and gasket surfaces, as well as the porosity of some gasket materials, *the width should be at least twice the thickness*. Gasket width is not considered in Fig. 21-3 or Table 21-3; however, narrow gaskets actually require higher flange pressures. Gasket widths of *six* times the gasket thickness or more are common in joints where the gasket is not recessed in a flange.

The minimum flange pressures in Table 21-3 are usually sufficient to seal the joint with typical machined flanges. However, gasket materials of low compressibility require smoother flanges than the more compressible materials do.

21-4 GASKET CREEP

A gasket material must be compressible in order to seal a joint. There is a possibility, however, that the gasket may creep or relax after the bolts are tightened. This results in reduced thickness of the gasket material and reduced bolt tension, which can cause a leaking joint. Figure 21-5*a* shows a joint with a flat face gasket of thickness t before the bolts are tightened. Torquing the bolts compresses the gasket to thickness t_1 with

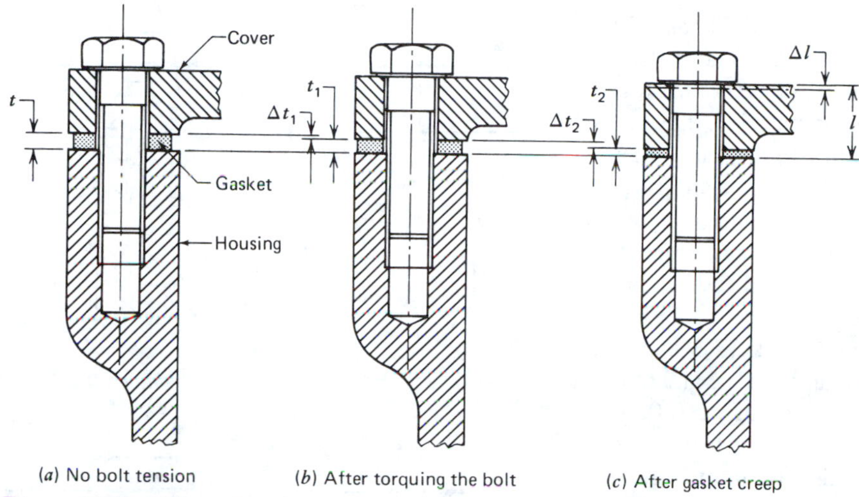

Figure 21-5 Effect of gasket creep on a gasketed joint.

a gasket compression of Δt_1 (Fig. 21-5*b*). Over a period of time, and especially with exposure to higher temperatures, the gasket will creep and become permanently thinner, leading to a new thickness t_2 (Fig. 21-5*c*). This results in a change in gasket thickness of Δt_2 with the bolts tightened (exaggerated in the drawing for clarity).

When the bolt is tightened, the length between the housing and the bolt head l is increased[3] by an amount Δl. When the gasket creeps, the bolt becomes shorter and tension is lost. If $\Delta t_2 = \Delta l$, the bolt has lost all of its tension and is loose. This condition, of course, cannot be tolerated.

The amount of creep that occurs in a gasket depends on the gasket material and the gasket thickness. Gasket materials are rated in order of increasing creep, as follows: asbestos and rubber, cellulose and rubber, cork and rubber, cork composition, and rubber. Thus a gasket of asbestos and rubber is the best choice for minimizing creep, and a gasket of plain rubber is the poorest choice. A standard method[4] of measuring creep relaxation of gasket materials is frequently used. Creep values may be as high as 60%.

Thin gaskets creep less than thick gaskets. Consequently, thin gaskets are preferred as long as the thickness is sufficient to provide a seal. If the machined surfaces on the housing and cover in Fig. 21-5 are very nearly flat, a thin gasket of low-compressibility material will provide a seal. If, however, the cover is warped, a thick gasket of high compressibility may be required to compensate for this warping. Lack of flatness in flanges tends to be a problem, particularly with stamped metal parts. For this reason, cork composition or cork and rubber gaskets are frequently used with stamped parts.

After selecting the material and the thickness for a gasket, the designer should consult manufacturers' literature to determine the available thicknesses and tolerances.

Manufacturers of gasket materials have their own standards for gasket thickness. These vary according to material type. Materials containing cork or asbestos usually range from 0.031 to

---

[3] $\Delta l = \dfrac{Fl}{AE}$

where

F = bolt tension
l = length with cross-sectional area A
E = modulus of elasticity of bolt material

[4] The detailed procedure is provided in ASTM F 38.

0.125 in. (0.79 to 3.18 mm) in thickness. The increments in thickness for these materials are usually 0.016 in. (0.41 mm) or 0.031 in. (0.79 mm). Some cellulose gasket materials may be as thin as 0.003 in. (0.08 mm). In any case, materials for flat gaskets in machine joints are usually no thicker than 0.125 in. (3.18 mm).

Tolerances for the thickness of gasket materials are specified in the Society of Automotive Engineers Recommended Practice SAE J90b, Nonmetallic Automotive Gasket Materials. These vary according to material type and thickness but are mostly in the range of $\pm$ 0.005 to $\pm$ 0.015 in.

There are other methods of minimizing the loss of bolt tension caused by gasket creep. One method, which tends to be expensive, is to place conical washers (Belleville springs) under the bolt heads. If the washers are compressed flat when the bolts are tightened, they regain their conical shape when gasket creep occurs. Thus the washers compensate for the change in gasket thickness.

Another method of minimizing the loss of bolt tension caused by gasket creep is to provide a long, effective bolt length in the joint. In Figs. 21-5 and 21-6*a*, the effective length l is equal to the combined thickness of the cover and the compressed gasket. If through bolts instead of tap bolts were used, the entire length between the bolt head and nut would be in tension (Fig. 21-6*b*). Then the change in bolt length resulting from gasket creep would be a smaller percentage of the elongation due to tightening. The lower the percentage of lost elongation is, the greater the tension remaining in the bolt. Even greater effective length can be obtained by adding raised bosses to the flanges of castings or forgings, as in Fig. 21-6*c*.

Much of the lost bolt tension due to gasket creep or relaxation can be regained by *retorquing* the bolts after a period of operation at normal operating temperature. Tests indicate that nearly all of the creep will occur within the *first 24 h* of exposure to higher temperature. The retorquing is, of course, an additional expense and may also be an inconvenience.

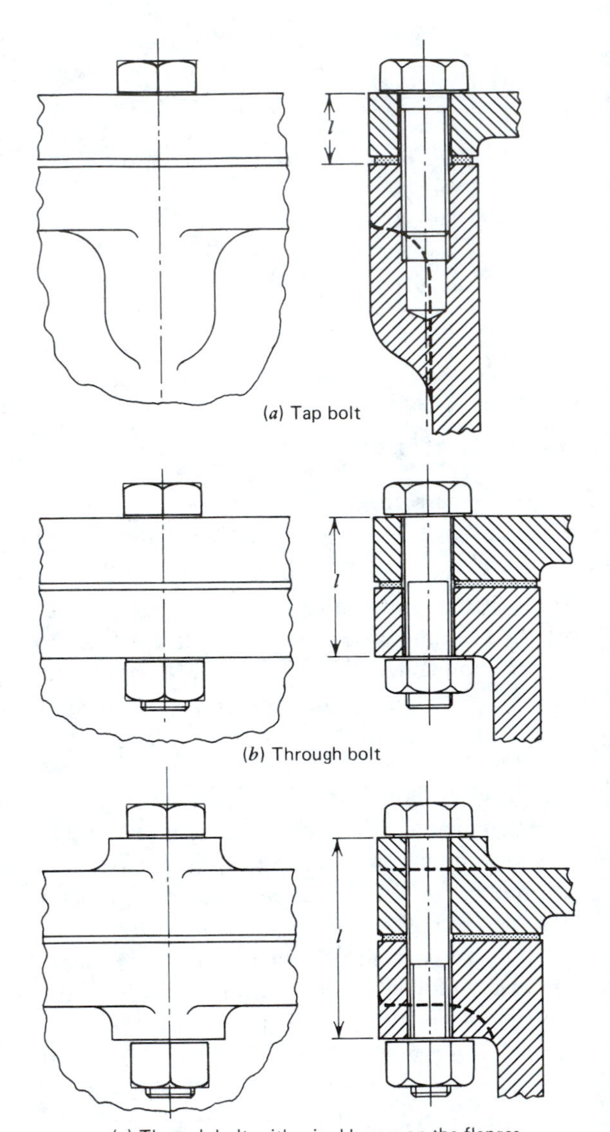

Figure 21-6 Effective length of bolts.

21-5 FLANGE THICKNESS AND BOLT SPACING

Bolted flanges deflect *between* the bolts. The greatest deflection and lowest flange pressure occur midway between adjacent bolts. Thick

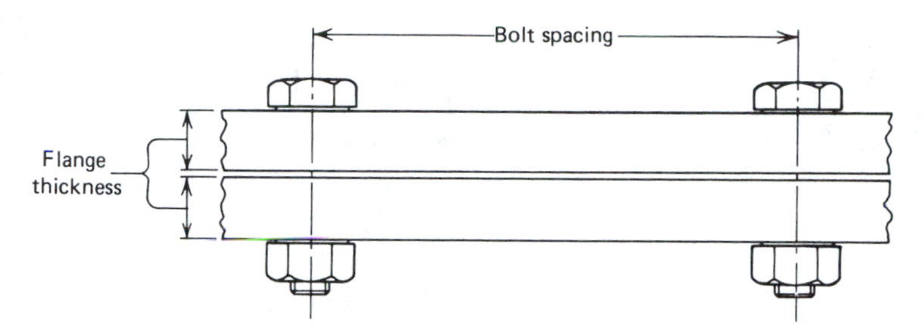

Figure 21-7 Definition of bolt spacing.

flanges are stiffer and will deflect less than thin flanges. Thus fewer bolts can be used with a thick flange because flange thickness and bolt spacing are interrelated. Experiments[5] with flat steel covers indicate that *the bolt spacing should not exceed 12 times the cover thickness* (Fig. 21-7). Bolt spacings in the range of 8 to 12 times the flange or cover thickness are good practice with steel parts. Because gray cast iron has a lower modulus of elasticity than steel, somewhat closer bolt spacing is required with that material.

There is a relatively simple technique for checking flange deflection. It consists of punching holes in a gasket, placing pieces of soft solder in the holes, assembling the parts, and tightening the bolts. When the gasket is removed, the thickness of each piece of solder is checked with a micrometer. Because the solder is soft, the thickness equals that of the surrounding gasket at that point. Comparison of the thickness of each solder plug to the original gasket thickness shows the amount of gasket compression that occurred at each point.

Bolt spacings of 12 times the metal thickness are not obtainable with parts formed from sheet metal. Other methods must then be used to increase the effective flange thickness. A turned-up lip, as shown in Fig. 21-8*a*, increases flange stiffness. Spot welding a strip to the flange through which the bolts pass (Fig. 21-8*b*) also adds stiffness to the flange. An alternative to the added strip is the formation of a bead in the flange between bolts. This adds stiffness to the flange and also increases the flange pressure.

21-6 GASKETED JOINTS FOR PIPING AND PRESSURE VESSELS

The design of gasketed joints for piping and pressure vessels has received more attention over the years than the design of gasketed joints in machinery. This disproportion undoubtedly results from the hazards caused by the escape of high-temperature, pressurized fluids. In addition, the circular shapes used in piping and pressure vessels are much easier to analyze than the odd shapes that frequently occur with machine housings and covers.

When a fluid under pressure must be sealed, a higher flange pressure is sometimes required because of the pressure acting on the exposed edge of the gasket, which tends to blow the gasket out of the joint. The additional bolt load that is required is calculated with a "maintenance factor."

Gasket materials used to contain pressurized fluids tend to be somewhat different from those suitable for gasketed joints in machinery. The minimum flange pressure p_f and maintenance factor m for some of the pressure vessel gasket materials are listed in Table 21-4. The material listed as "vegetable fiber sheet" includes various papers that are used for gaskets. Values of p_f and m for other gasket materials are available in the ASME code.

A method of computing the force required to obtain and maintain a seal in a gasketed joint is provided in the ASME code. This method in-

[5] Murl W. Clark, "We Need Better Gasket Materials," *Product Engineering*, Aug. 30, 1965, pp. 43-48.

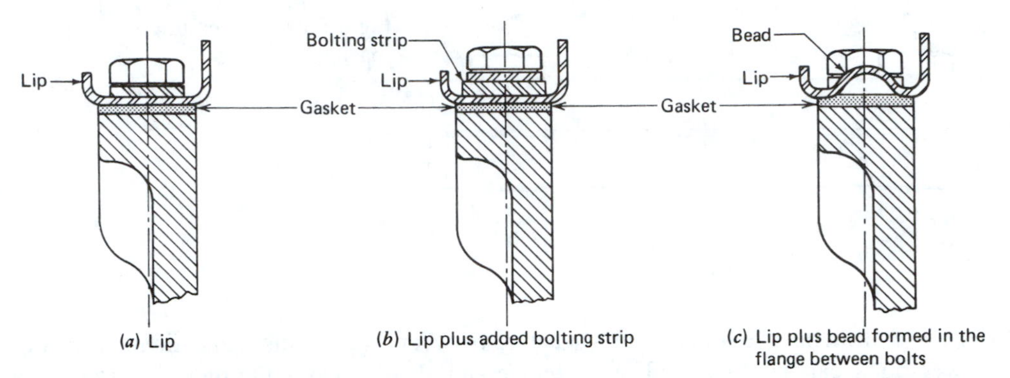

Figure 21-8 Methods of improving the sealing of thin metal flanges.

volves the gasket properties, gasket dimensions, and pressure to be sealed. The effect of gasket thickness is accounted for in the values of p_f and m. The effective seating width and the diameter at which the clamping force acts on the gasket are calculated from the formulas provided in the code.

Table 21-5 provides formulas for calculating the basic gasket seating width for some common flange conditions. These formulas apply only when the entire gasket is contained inside the inner edges of the bolt holes in the flanges. Additional formulas for conditions involving serrated flanges and ring-shaped metal gaskets are provided in the code.

A basic gasket seating width b_o from a formula in Table 21-5 must be converted into an effective seating width b. When $b_o > 0.25$ in.,

TABLE 21-4 Minimum Flange Pressures and Maintenance Factors for Gaskets in Pressure Vessels

Material	Thickness		Minimum Flange Pressure, p_f		Maintenance Factor, m
	mm	in.	N/mm²	psi	
Asbestos, compressed	0.4	0.016	24.1	3,500	
	0.8	0.031	17.2	2,500	3.50
	1.6	0.062	12.4	1,800	2.75
	3.2	0.125	9.7	1,400	2.00
Asbestos cloth, impregnated	2.4	0.093	11.0	1,600	
Metals, solid flat					
Aluminum, soft	0.8 to 1.6	0.031–0.062	137.9	20,000	4.00
Copper, soft			310.3	45,000	4.75
Monel			586.1	85,000	6.00
Steel, soft			482.6	70,000	5.50
Steel, stainless			655.0	95,000	6.50
Rubber					
75 durometer	0.8	0.031	1.4	200	1.00
60 durometer	minimum	minimum	1.2	175	0.50
Vegetable fiber sheet	All	All	5.2	750	1.00

Source. Boiler and Pressure Vessel Code Section III-1, Appendix XI, Table XI-3221.1-1, American Society of Mechanical Engineers, 1977.

TABLE 21-5 Formulas for Basic Gasket Seating Width

Gasket and Flange Situation	Formula
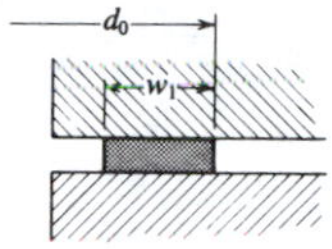 d_0, w_1	$b_o = \frac{w_1}{2}$
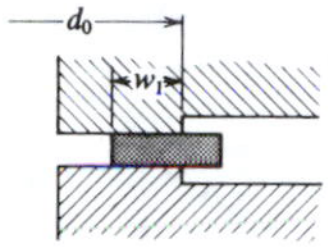 d_0, w_1	$b_o = \frac{w_1}{2}$
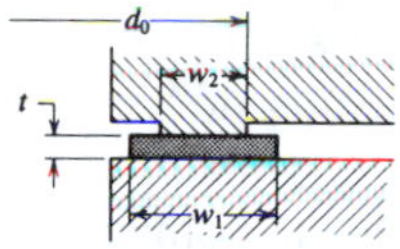 d_0, w_2, t, w_1	When $w_2 \leq w_1$ $b_o = \frac{w_2 + t}{2}$, max $b_o = \frac{w_1 + w_2}{4}$ When $w_2 \leq \frac{w_1}{2}$ $b_o = \frac{w_1 + w_2}{4}$ for solid metal gaskets $b_o = \frac{3w_1 + w_2}{8}$ for other materials
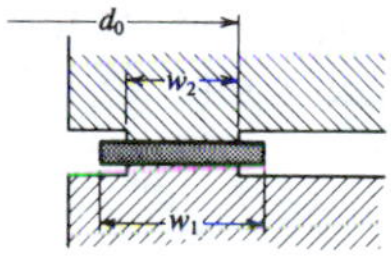 d_0, w_2, w_1	For solid metal gaskets $b_o = \frac{w_2}{2}$, min. $b_o = \frac{w_1}{4}$ For other gasket materials $b_o = \frac{w_1 + w_2}{4}$, min. $b_o = \frac{3w_1}{8}$

Source. Boiler and Pressure Vessel Code Section III-1, Appendix XI, Table XI-3221.1-2, American Society of Mechanical Engineers, 1977.

$$b = \frac{1}{2}(b_o)^{1/2} \qquad (21\text{-}3a)$$

when b and b_o are in inches, and

$$b = 12.7\left(\frac{b_o}{25.4}\right)^{1/2} \qquad (21\text{-}3b)$$

when b and b_o are in millimeters. The diameter D at which the gasket load reaction occurs is calculated, when $b_o > 0.25$ in., from

$$D = d_o - 2b \qquad (21\text{-}4)$$

where d_o is defined in the figures in Table 21-5.

A pressurized joint must be checked for two conditions: seating of the gasket at assembly and maintenance of a tight joint under fluid pressure.

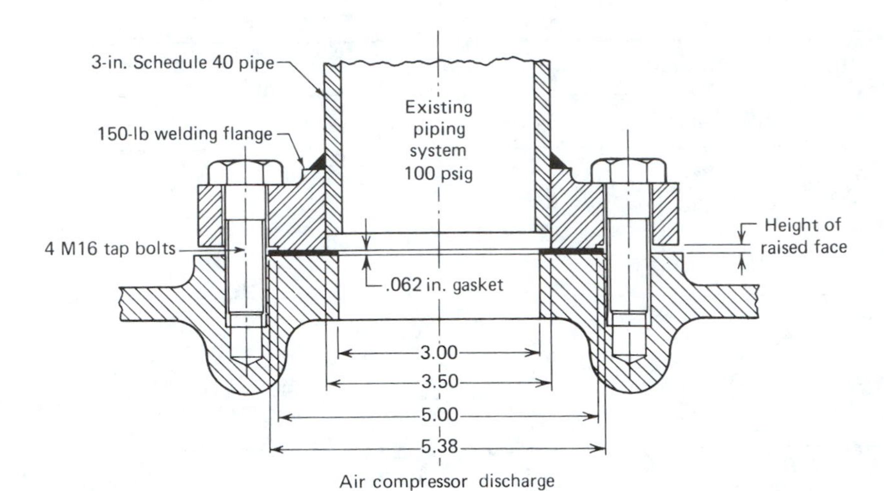

Figure 21-9 Section through gasketed joint.

Seating of the gasket is computed with the formula

$$F = \frac{\pi}{n} p_f D b \qquad (21\text{-}5)$$

where

F = required tensile force per bolt; N, lb
n = number of bolts in the joint
p_f = minimum flange pressure; N/mm², psi
D = diameter from Eq. 21-4; mm, in.
b = width from Eq. 21-3; mm, in.

The second condition, maintaining a tight joint under fluid pressure, is computed from the formula

$$F = \frac{\pi p D}{n}\left(\frac{D}{4} + 2mb\right) \qquad (21\text{-}6)$$

where F, n, D, and b are defined the same as for Eq. 21-5, p is the internal fluid gauge pressure (N/mm² or psi), and m is the maintenance factor for the gasket from Table 21-4.

EXAMPLE 21-2

An air compressor is to be replaced by a new machine connected to the existing piping system consisting of 3-in. pipe and 150-lb[6] welding flanges with raised faces. The joint between the air compressor discharge and the piping system is shown in Fig. 21-9, in which the dimensions are in inches. The tapped holes in the air compressor are for M16 tap bolts, not the $\frac{5}{8}$-in. tap bolts specified by the welding flange standard.

The system pressure of 100 psig (0.69 N/mm² gauge) and the temperature of the fluid do not necessitate an asbestos gasket. However, the plans are to use a 0.062-in. (1.6-mm) thick gasket that is on hand for other piping in the power plant. The flange requires four bolts. If M16 property class 4.8 bolts are used and torqued to 125 N · m, will the gasket be satisfactory?

SOLUTION

The tensile force developed in each 16-mm diameter bolt by applying 125-N · m (125 000-N · mm) torque can be computed from Eq. 21-2. Thus

---

[6]Although it is called "150-lb flange," what is actually meant is "150-psi flange."

$$F = \frac{T}{0.2D} = \frac{125\,000\text{ N}\cdot\text{mm}}{(0.2)(16\text{ mm})} = 39\,062\text{ N/bolt}$$

The basic gasket width is obtained from the third situation in Table 21-5. The gasket width is $w_1 = (5.38 - 3.00)/2 = 1.19$ in., or 30.2 mm. The width of the flange face is $w_2 = (5.00 - 3.50)/2 = 0.75$ in., or 19.0 mm. Since w_2 is less than w_1 but greater than $w_1/2$, we use the formula (with $t = 1.6$ mm)

$$b_o = \frac{w_2 + t}{2} = \frac{19.0 + 1.6}{2} = 10.3\text{ mm}$$

However, there is the restriction

$$\text{Max } b_o = \frac{w_1 + w_2}{4} = \frac{30.2 + 19.0}{4} = 12.3\text{ mm}$$

Using the smaller of these two values, $b_o = 10.3$ mm. Substituting this value into Eq. 21-3b to calculate the effective seating width,

$$b = 12.7\left(\frac{b_o}{25.4}\right)^{1/2} = 12.7\left(\frac{10.3}{25.4}\right)^{1/2} = 8.09\text{ mm}$$

Substituting values into Eq. 21-4, with $d_o = 5.0$ in. or 127.0 mm,

$$D = d_o - 2b = 127.0 - (2)(8.09) = 110.8\text{ mm}$$

The force required to seat the gasket can be computed with Eq. 21-5 after obtaining $p_f = 12.4\text{ N/mm}^2$ from Table 21-4.

$$F = \frac{\pi}{n} p_f D b$$

$$= \frac{\pi(12.4\text{ N/mm}^2)(110.8\text{ mm})(8.09\text{ mm})}{4}$$

$$= 8730\text{ N/bolt}$$

Since the tensile force in each bolt (39 062 N) exceeds the force required to seat the gasket, the solution can be continued.

The force per bolt required to maintain a tight joint under a pressure of 0.69 N/mm^2 is obtained from Eq. 21-6. Substituting values, after obtaining $m = 2.75$ from Table 21-4,

$$F = \frac{\pi p D}{n}\left(\frac{D}{4} + 2mb\right)$$

$$= \frac{\pi(0.69\text{ N/mm}^2)(110.8\text{ mm})}{4}$$

$$\left[\frac{110.8\text{ mm}}{4} + (2)(2.75)(8.09\text{ mm})\right]$$

$$= \mathbf{4335\text{ N/bolt}}$$

This force is less than that required to seat the gasket. Since the tensile force in each bolt (39 062 N) is 4.47 times the force required to seat the gasket (8730 N), the gasket and bolt selections are satisfactory.

The basic flange portion of Fig. 21-1 shows a flat gasket recessed in a flange as one example. A gasket for this situation would be analyzed in the same manner as in Example 21-2. The gasket should cover the bottom of the recess with minimum clearances while allowing for tolerances on the inside and outside diameters of the recess.

Minimum flange pressures and maintenance factors for some newer gasket materials have been reported.[7] This source also provides formulas for calculating the tensile force remaining in the bolts after some gasket materials have been exposed to operating temperature. A method of computing the maximum pressure that a gasketed joint will seal has also been cited.[8] The gasket data that are required for use of this method have been reported for only a very limited number of materials.

21-7 GASKETS FOR METAL-TO-METAL JOINTS

The metal-to-metal joints in Fig. 21-1 can be sealed with O-rings or rectangular section rings of elastomeric material (Fig. 21-10). O-rings have a circular cross section and are molded to

---

[7]Robert H. Swick, "Designing the Leakproof Gasket," *Machine Design*, January 22, 1976, pp. 100-103.

[8]Robert H. Swick, "How Much Pressure Will That Gasket Take?" *Machine Design*, Nov. 24, 1977, pp. 86-89.

Figure 21-10 Seal rings.

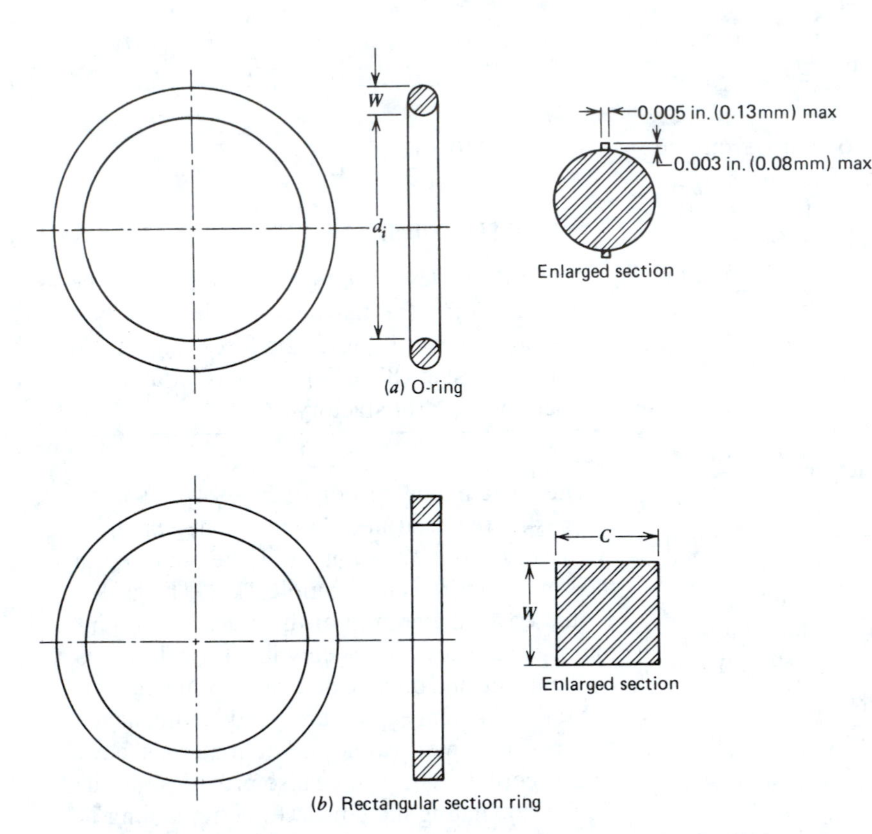

size. The "size" of an O-ring is designated by the inside diameter and the cross-sectional diameter. The standard nominal cross sections are 0.070 in., 0.103 in., 0.139 in., 0.210 in., and 0.274 in. They are available in a wide range of diameters and in several different families of materials.

Rectangular section rings are manufactured by a different method. A tube of material is molded, and the rings are then made by cutting off portions to the desired length. Because of this method of manufacture, they are also known as lathe-cut rings. The Recommended Practice SAE J120a, Rubber Rings for Automotive Applications, specifies nominal square cross sections of 0.066 in., 0.099 in., 0.134 in., 0.203 in., and 0.265 in. The sizes of these rings can be specified by listing SAE specification numbers.

The most commonly used materials for elastomeric sealing rings are nitrile (Buna N) compounds. These compounds are compatible with many fluids, including lubricating oils, hydraulic fluids, gasoline, alcohol, and water. Nitrile compounds are available in a range of hardness, and they will withstand a considerable range of temperatures. Other compounds, such as neoprene, butyl, fluorocarbon, and silicone, are used in special cases to obtain compatibility with fluids and resistance to higher temperatures. Data on fluid compatibility and temperature limits are provided in manufacturers' literature.

The standard shape of the groove for elastomeric sealing rings is shown in Fig. 21-11. The groove dimensions depend on the type of ring cross section, size of ring cross section, and type of seal (static or dynamic). With static seals the gap between the metal parts shown in Fig. 21-11 is not present. These rings seal only when compressed; thus the groove dimensions must be selected to provide "squeeze." Groove dimensions

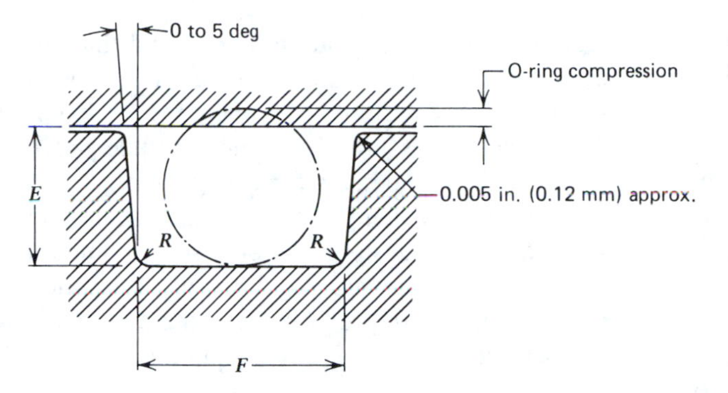

Figure 21-11 Shape of groove for seal rings.

for sealing are provided in manufacturers' literature and in SAE J120a, mentioned previously.

An O-ring static installation in which the joint is subject to internal pressure only is shown in Fig. 21-12*a*. This is usually less expensive to machine than the groove in Fig. 21-12*b*, which is for a joint subject to external pressure (internal vacuum). A rectangular section ring would serve equally well in these applications. It is generally advisable to use as large a ring cross section as will fit in the available space. The tolerance on groove depth is greater with larger cross sections. This results in less careful control of machining dimensions and generally lower machining costs.

O-rings are also used to seal hydraulic tube fittings screwed into tapped holes. Dimensions are specified in SAE Standard J514j, Hydraulic Tube Fittings.

Elastomeric materials are almost incompressible. This makes it essential to provide sufficient volume for the compressed ring in the groove for the ring. Otherwise, a metal-to-metal contact will not be obtained in a static joint.

21-8 FORMED-IN-PLACE GASKETS

Gaskets for flat flanges are cut from sheets of gasket material. The trimmings must be used to make smaller gaskets or must be discarded. Hence cut gaskets may require considerably more material than appears in the actual part, thereby increasing the cost of the gasket. These gaskets must then be shipped from the gasket

Figure 21-12 Static O-ring seals.

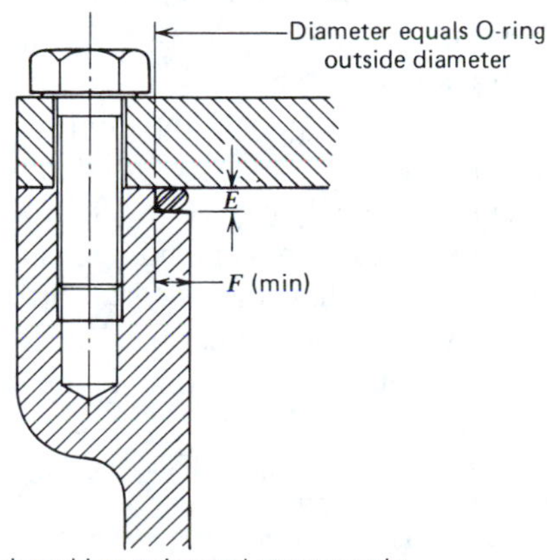

(*a*) Joint subject to internal pressure only

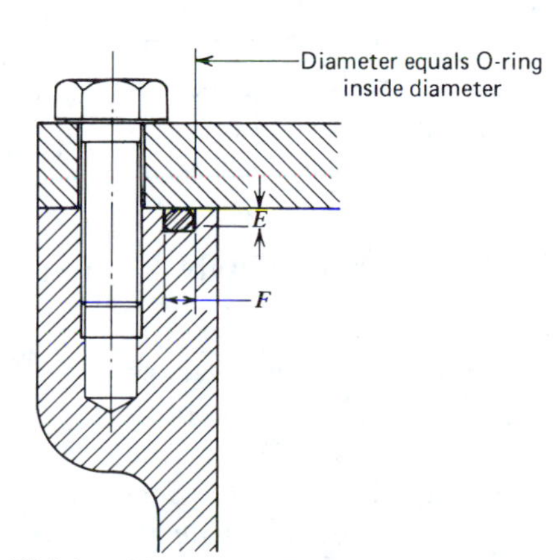

(*b*) Joint subject to external pressure

manufacturer to the machinery manufacturer, inspected, stored, and inventoried. In addition to these disadvantages, flat gaskets may not provide a tight joint.

The formed-in-place gasket is an attractive alternative to the flat gasket. Applying a paste or gel to a flange before assembly provides a seal while permitting metal-to-metal contact of the flanges. The material fills in the rough surfaces and any small gaps between mating flanges.

The two most commonly used types of compounds are silicone rubber, which vulcanizes at room temperature, and anaerobic materials. "Anaerobic" means life in the absence of oxygen. Anaerobic materials remain liquid until confined between two close-fitting surfaces. These materials start curing in the center and harden outward to form a tough plastic that will tolerate temperatures up to 204°C (400°F).

The room-temperature, -vulcanizing silicone rubber gasket material cures as a result of humidity in the atmosphere. Thus the curing begins at the exposed surfaces and progresses inward. When curing is complete, the gasket is like rubber and will tolerate temperatures up to 232°C (450°F). Silicone compounds usually require more time to cure than anaerobic compounds.

Machine methods are available for uniformly applying the formed-in-place gasket material to a flange prior to assembly. These methods are adapted to mass-production situations. For other situations and for machine repair, the material can be applied manually from a tube. Designs utilizing formed-in-place gaskets should be discussed with manufacturers' representatives.

21-9 SEALS FOR ROTARY MOTION

Shafts in machinery are usually supported by bearings lubricated with oil or grease. Seals are frequently required to prevent the escape of lubricant to the outside of the machine or into another compartment. Seals also prevent the entry of dust and dirt that may contaminate the lubricant. Abrasive particles mixed with oil or grease can create a grinding compound that accelerates wear. There are still other situations in which seals are required to prevent the escape of fluids under pressure. Examples include valve stem packings and seals for turbine shafts.

There are so many types of shaft seals and such varied seal designs that only some common seals can be discussed. Additional information can be found in the references at the end of the chapter and in manufacturers' literature.

Radial Lip Seals

Radial lip seals (Fig. 21-13) are usually placed near bearings to retain lubricant and exclude dirt. They are held in place by an interference fit with the housing. These seals were first produced about 1927 and used leather as the sealing element. Later, a garter spring[9] was added to provide a uniform force around the sealing element. Most radial lip seals are now made with elastomeric sealing elements instead of leather. These seals are effective, inexpensive, compact, and easily installed.

Leakage is less likely to occur if hydrodynamic sealing lips are used. These lips have very shallow grooves molded into the primary sealing lip that pump oil out of the contact area. Hydrodynamic seals are made for rotation in one direction only or for rotation in either direction.

Various elastomers are used for sealing lips, but nitrile (Buna N) rubber compounds are most commonly used. These compounds are compatible with lubricating oils, greases, and hydraulic fluids. They have fair to poor compatibility with the extreme pressure (*EP*) additives used in some gear lubricants, so a polyacrylate or fluoroelastomer is a better choice in such situations.

The portion of the shaft that the seal lip contacts should have a minimum hardness of Rockwell C 30 to prevent shaft scoring. If the shaft

[9]A helical spring with the ends joined so that the spring forms a circle. When the spring is expanded to a larger diameter, it exerts a uniform radial force around the inner circumference.

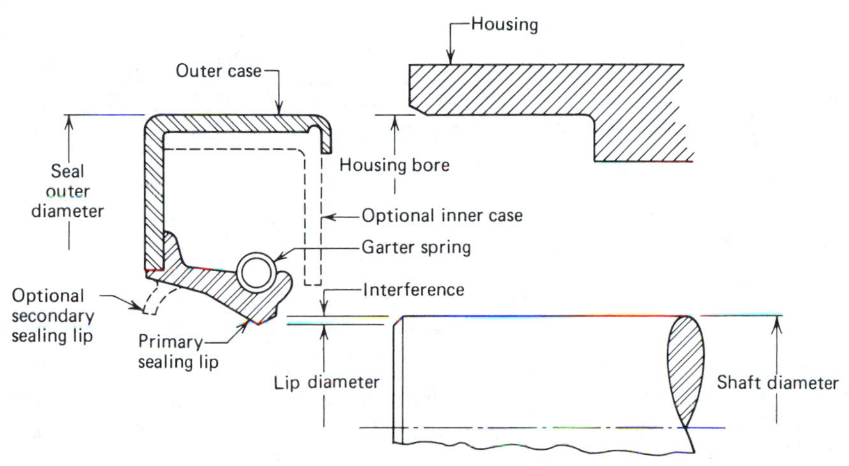

Figure 21-13 Radial lip seal terminology.

may be damaged in handling, a minimum hardness of Rockwell C 45 will provide protection against damage. A hard surface can be provided for soft shafts by using a hardened wear sleeve of thin steel that is held in place by an interference fit.

The radial lip seals perform best with carbon, alloy, or stainless steel shafts. Nickel-plated surfaces are also satisfactory. Use of brass, bronze, aluminum alloys, magnesium, zinc, or similar metals is not recommended.

Some of the common terminology for radial lip seals is shown in Fig. 21-13. The sealing lips are smaller in diameter than the shafts for which they are intended. This interference provides the force to maintain contact betwen the seal and shaft. A thin film of lubricant is required under the sealing lip to reduce friction and prevent overheating of the sealing element.

The use of radial lip seals is limited to sealing pressures of 3 to 7 psig (0.02 to 0.05 N/mm^2 gauge) depending on the shaft speed. When higher pressures are to be sealed, a mechanical face seal is preferable.

Face Seals

The radial lip seal makes contact with the shaft circumference. In contrast, a face seal acts against a surface that is perpendicular to the shaft axis, such as a shoulder or collar on a shaft. The seal can be stationary like a radial lip seal or be mounted on the shaft and seal against the housing.

Elastomeric face seals (Fig. 21-14*a*) are used where a suitable sealing surface for a radial lip seal is unavailable. Essentially the same elastomers are used in face seals as in radial lip seals. The elastomeric face seal has the disadvantage of having to be located precisely with respect to the sealing surface in order to provide the proper force on the seal. The mating surface for an elastomeric face seal must be flat and smooth—10 to 20 μin. (0.25 to 0.50 μm) surface finish.

Mechanical face seals are used where a radial lip seal or an elastomeric face seal would be unsatisfactory. The automotive engine water pump seal (Fig. 21-14*b*) is an application where pressure is relatively low. In this application a stationary, spring-loaded seal ring contacts a flat surface that rotates with the shaft.

If high pressure acts on the seal ring in Fig. 21-14*b*, a high axial sealing force is created. This results in loss of energy and possible overheating due to friction. Consequently, the balanced mechanical face seal (Fig. 21-14*c*) is used for higher pressures. These seals are proportioned so that most of the force due to pressure on the opposing surfaces is balanced. This leaves

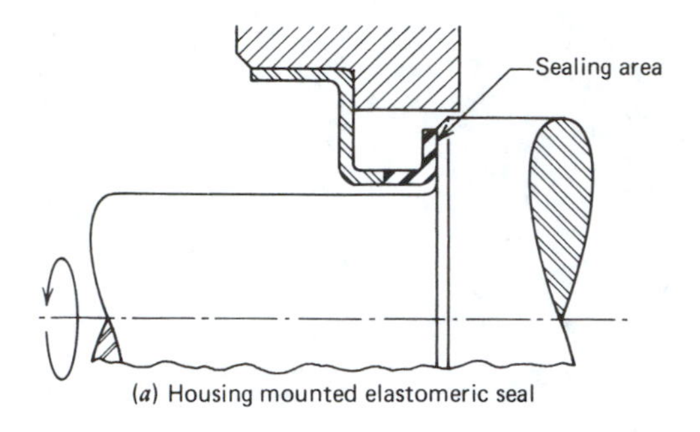

(*a*) Housing mounted elastomeric seal

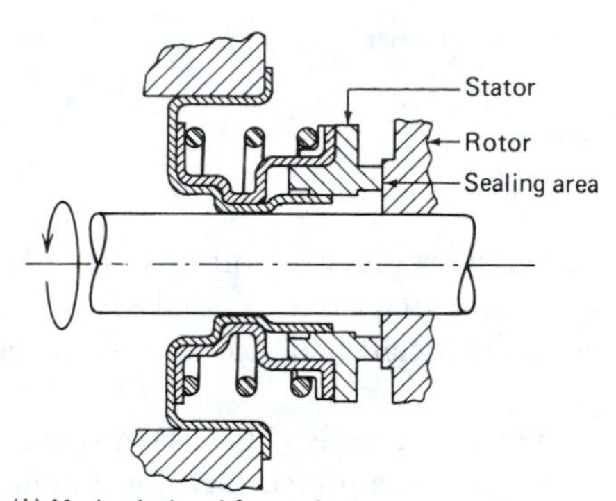

(*b*) Mechanical seal for engine water pump

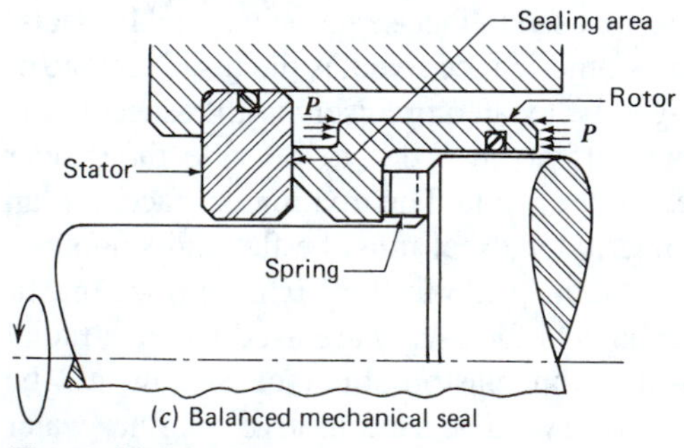

(*c*) Balanced mechanical seal

Figure 21-14 Face seals.

a small net force holding the rotor against the stator.

Compression Packings

The stuffing box (Fig. 21-15) was developed for sealing either rotating or reciprocating shafts. It is now used primarily for sealing fluids under pressure. Sealing action is obtained by expansion of the packing against the shaft caused by axial gland movement when the nuts are tight-

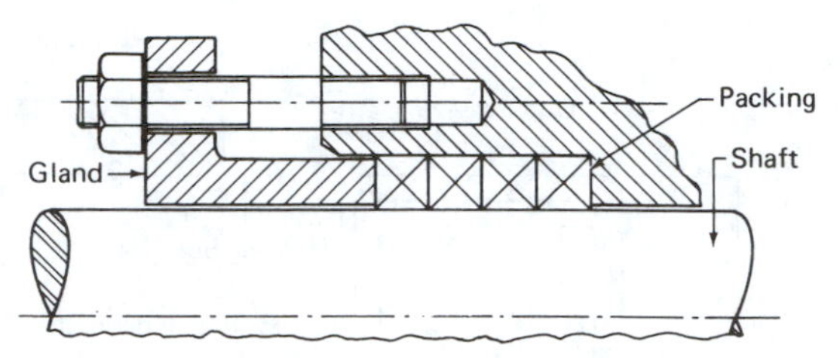

Figure 21-15 Stuffing box for a rotating shaft.

ened. Friction between the shaft and packing causes wear, which is compensated for by periodic tightening of the nuts.

Compression packings are usually provided in a square or rectangular cross section in straight lengths. Pieces are cut off and formed into rings that fit the stuffing box dimensions. A wide range of materials is used for packings, and the material selection depends on the fluid to be sealed. These materials include plant fibers such as jute and cotton, asbestos, leather, artificial fibers, graphite, and metals. The metals include aluminum, copper, lead, and nickel. Metal packings are formed from foil that is then compressed into the packing shape. Metal packings are used for temperature conditions where other materials would be inadequate.

For soft packings, there are recommended stuffing box dimensions. In the design of a stuffing box, small clearances are used between the shaft and the housing and between the shaft and the gland. A low clearance between the shaft and housing reduces the pressure acting on the packing. The clearances of the gland with the shaft and the housing should also be small in order to minimize extrusion of the packing around the gland.

Valve stems have a helical rather than a pure rotary motion when a valve is opened or closed. Research into the prevention of leakage and wear of valve stems resulted in a procedure for designing valve stuffing boxes.[10] This procedure could probably also be applied to rotating shafts.

[10] L. I. Ezekoye and J. A. George, "Valve Packings That Don't Leak," *Machine Design*, Jan. 20, 1977, pp. 142–143.

O-Rings

Some attempts to use O-rings to seal shafts have been unsuccessful. These failures are due to a characteristic of elastomers: they shrink when heated. If an O-ring is stretched around a rotating shaft, friction between the O-ring and shaft generates heat that makes the ring shrink. Contraction of the ring causes more heating due to friction and additional contraction. This leads to rapid failure of the seal if the O-ring is installed under tension.

The key to using O-rings to seal rotating shafts is to install the ring in a smaller than normal diameter groove in the housing so that the ring remains under compression when it shrinks due to heating. O-rings can then be used to seal pressures up to 1.38 N/mm^2 (200 psi) and surface speeds up to 3.8 m/s (750 ft/min).

Noncontacting Seals

The loss of energy due to friction that occurs with contacting seals can be prevented by using a seal that does not make physical contact. A noncontacting seal cannot, however, prevent leakage entirely but instead reduces it to a tolerable level. One method of achieving this low leakage rate is to provide a low clearance between the shaft and the housing or a bushing. The longer the low clearance passage is, the greater the reduction in fluid flow.

Labyrinth seals (Fig. 21-16) are used in machines such as blowers and turbines. This type of noncontacting seal can be used for retaining lubricant or for sealing the working fluid. An advantage of labyrinth seals is that they eliminate seal friction and can therefore be used at high peripheral shaft speeds. The seal effectiveness depends on low clearance between the shaft and seal. Consequently, labyrinth seals are usually made of a relatively soft metal such as aluminum or bronze so that the shaft will not be damaged if contact does occur. The type shown in Fig. 21-16 is the simplest, but other types are also used.

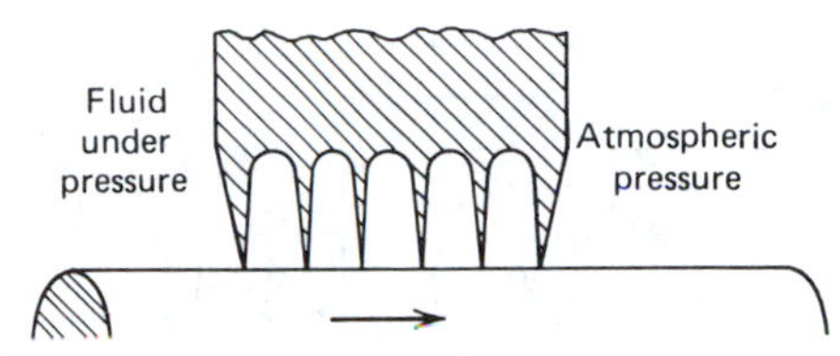

Figure 21-16 Labyrinth type of noncontacting seal. A succession of sealing lips creates a series of pressure drops and thus reduces leakage.

21-10 SEALS FOR RECIPROCATING MOTION

Some, but not all, of the seals used for rotary motion are also used for reciprocating motion. In particular, O-rings and compression packings, as well as some other types that will be discussed, are used to seal reciprocating rods, shafts, and pistons.

O-Rings

The O-ring (Fig. 21-17) is used in many applications because of its low installed cost and effectiveness as a seal. In addition to its use for static seals, and to some extent for sealing rotary motion, it is used extensively as a seal for reciprocating motion. Many of these applications are in hydraulic equipment, where the hydraulic oil acts as a lubricant for the O-ring. Figure 21-17 shows applications as a static seal and as seals for a piston and piston rod. Rectangular section rings (Fig. 21-10*b*) are used only for static applications because they are not suited to reciprocating motion.

The shape of the O-ring groove (Fig. 21-11) is the same for sealing reciprocating motion as for a static seal. The recommended groove depth E, however, is slightly different for a reciprocating seal. Dimensional recommendations are provided in SAE Recommended Practice J120a and in manufacturers' literature.

The O-rings for reciprocating motion in Fig. 21-17 seal the leakage paths between the piston and cylinder and between the piston rod and end

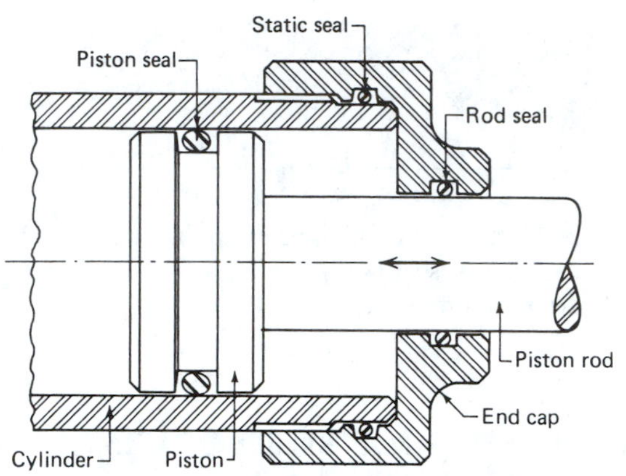

Figure 21-17 Application of O-rings as a static seal and as seals for reciprocating motion.

cap. The size of the gap that an O-ring can seal depends on the hardness of the elastomer and the fluid pressure. If the clearance is too large, the O-ring is extruded into the clearance space (Fig. 21-18*a*). The reciprocating motion then tears away small pieces of the O-ring. This action results in a leaking seal and contamination of the working fluid by the O-ring particles.

At least one O-ring manufacturer recommends a maximum pressure of 10.34 N/mm^2 (1500 psi) when the radial clearance approaches zero and a maximum pressure of 0.69 N/mm^2 (100 psi) for a clearance of 0.705 mm (0.028 in.). These values are for a hardness of 60 durometer.[11] The permissible pressure increases with higher hardness of the O-ring material.

The pressure limitations of O-rings are overcome by using backup rings (Fig. 21-18*b*) or other devices that prevent extrusion of the O-ring into the clearance space. Some backup rings are made of a material that deforms under pressure to seal the gap. Backup rings are usually made of leather, plastic, or metal. Metal backup rings are split like a piston ring so that they can be compressed radially for assembly.

[11]Shore durometer is a measure of the hardness of an elastomer and an approximate measure of the modulus of elasticity.

Because an O-ring is squeezed radially to seal the working fluid, friction is present when motion occurs. The friction force in a particular installation can vary considerably due to the tolerances on the dimensions of the parts involved. This variation can be troublesome when O-rings

Figure 21-18 O-rings.

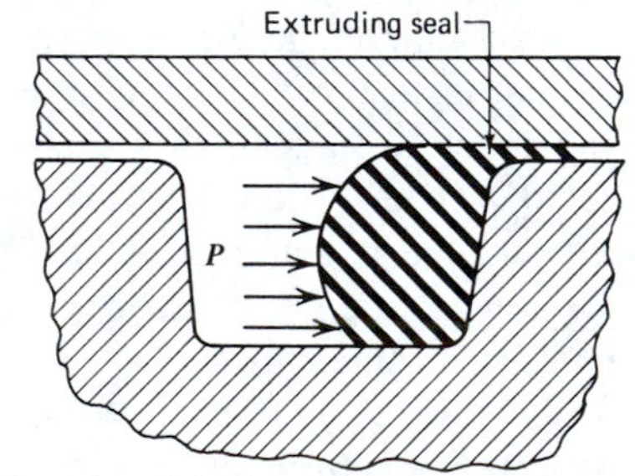

(*a*) Extrusion of ring into clearance space due to pressure

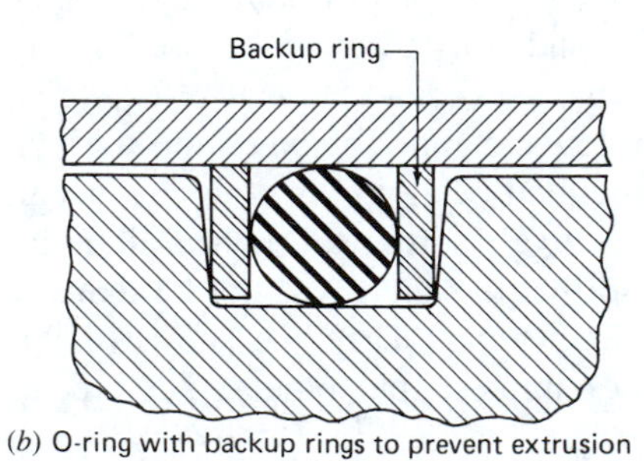

(*b*) O-ring with backup rings to prevent extrusion

are used as dynamic seals in mechanisms that require low friction.[12] It is advisable to check with the O-ring supplier for friction data on the particular elastomer used.

O-rings may be damaged if forced over sharp edges during assembly. Adding chamfers to corners is an inexpensive method of reducing damage in assembly. In addition, rubbing surfaces must be smooth in order to prevent excessive wear due to abrasion. A serious source of trouble is to have an O-ring pass over a hole in a cylinder wall while under pressure. If this occurs, the ring expands into the hole and must later be forced back into the groove. This action tends to nibble away the O-ring and thus destroy its sealing ability.

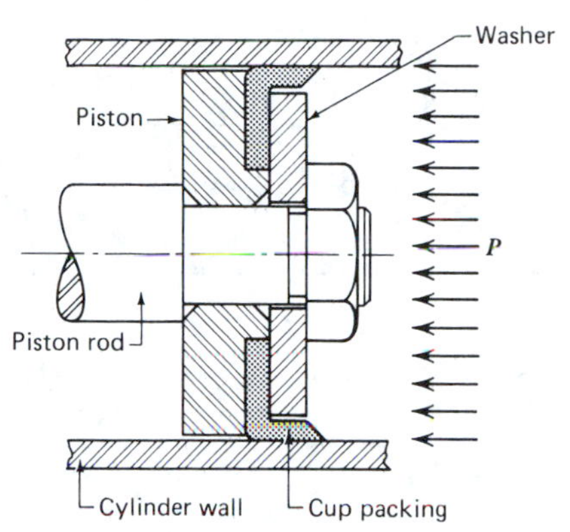

Figure 21-19 Cup packing for single-acting cylinder.

Lip Packings

Lip packings, such as cup packings, U-seals, and V-ring packings, are used primarily for reciprocating motion. They are usually made of leather, solid rubber, or fabric-reinforced rubber with special compounds available for difficult situations. Leather packings possess the advantage of a low coefficient of friction on the order of 0.006 to 0.008, depending on the tanning process. Life is increased when friction is low because less heat is generated.

Cup packings (Fig. 21-19) have long been used as a piston seal for pneumatic and hydraulic applications. Sealing is provided by the fluid pressure, which expands the lip of the cup outward against the cylinder wall. Because of this feature, a double-acting cylinder requires two packings in order to seal against pressure in both directions. A boss is shown on the inner portion of the piston in Fig. 21-19 to prevent excessive tightening of the washer against the cup. Otherwise, the cup may be crushed against the piston, and poor sealing will result.

Elastomeric U-seals are shown on a double-acting piston in Fig. 21-20. This type can also be used to seal between the piston rod and housing. When U-seals are made of leather, a filler is required between the lips in order to prevent collapse of the seal. Elastomeric U-seals have approximately the same pressure limitations as O-rings; therefore backup rings are required for higher pressures.

The V-ring packing (Fig. 21-21) is used primarily for sealing shafts, although it can also be used to seal pistons. The pressure-sealing capability depends on the type of material and num-

Figure 21-20 U-seals for a double-acting piston.

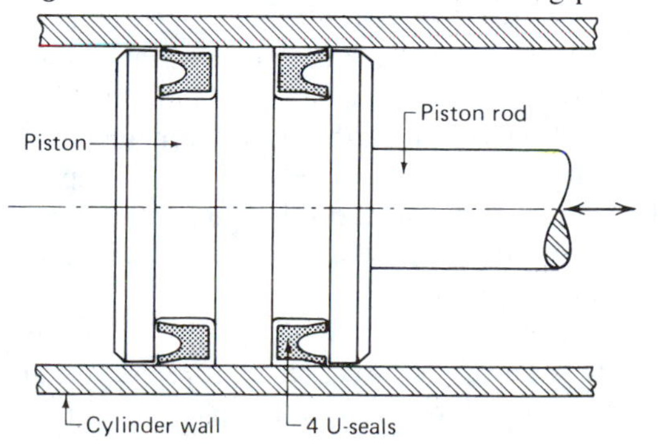

[12] A method of calculating the frictional force due to pressure on an O-ring is given by David R. Pearl, "O-Ring Seals in the Design of Hydraulic Mechanisms," *SAE Transactions*, *55* (1947), pp. 602-611.

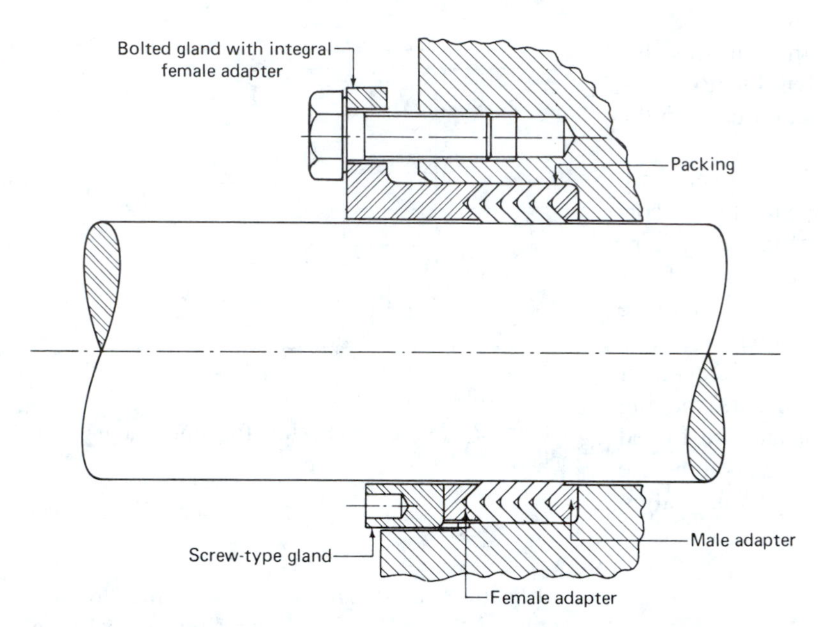

Figure 21-21 V-ring packing for a reciprocating shaft.

ber of individual packings. This type of packing is considered superior to other lip types at high pressures, especially above 345 N/mm^2 (50,000 psi).

Figure 21-21 shows the V-shape of the packing established by support obtained from V-shaped adapters. The pressure of the working fluid effectively expands the packings against the shaft and housing to provide the seal. Better sealing is obtained if the rings are continuous rather than split for ease of installation and removal.

SUMMARY

Seals of various types are used to contain fluids and exclude contaminants. Static seals are used with permanent or semipermanent connections such as those between machine housings and covers or between pipe flanges. Dynamic seals, on the other hand, are used with moving parts such as pistons, reciprocating rods, and rotating shafts.

Gaskets are used exclusively for static sealing. The selection of an appropriate gasket material involves the gasket material characteristics and the operating temperature and pressure. It is necessary to apply sufficient pressure to the gasket both to seal the gasket and to maintain a seal. The sealing effectiveness of a gasketed joint is influenced by bolting flange thickness and bolt spacing.

Gasket creep is detrimental to sealing effectiveness because it results in a loss of bolt tension. The effect of creep can be reduced by designing joints with long, effective bolt lengths.

O-rings and rectangular section rings are used for both static and dynamic sealing. A multitude of seal types is available for sealing rotary and reciprocating motions. The selection of an appropriate type of seal and seal material depends on the nature of the motion, operating pressure, shaft velocity, and the fluid to be sealed.

REFERENCES

21-1 "Designing with Leather Seals." *Design News,* January 17, 1977.

21-2 Fazekas, Thomas W., and William R. Kosec. "Where Balanced Face Seals

Pay Off." *Machine Design,* June 7, 1977.

21-3 Gordon, Alexander L. "Performance Criteria of Gasket Materials." Society of Automotive Engineers Paper 770070 (1977).

21-4 *Handbook of Mechanical Packings and Gasket Materials*. Mechanical Packing Association, 1960.

21-5 Horve, Les. "How to Select Radial Lip Seals, Part 1: Principles of Operation," *Power Transmission Design,* November 1976; Part 2: "Sealing System Requirements and Recommendations," December 1976; Part 3: "Shaft Seal Materials," February 1977; Part 4: "Seal Design Evolution," April 1977.

21-6 *Machine Design*. Seals Reference Issue, September 13, 1973.

21-7 Martini, Leonard J. "Sealing Rotary Shafts with O-Rings." *Machine Design,* May 26, 1977.

21-8 Nolt, I. G., and E. M. Smoley. "Gasket Loads in Flanged Joints." *Machine Design,* September 28, 1961.

21-9 Otto, Dennis Lee. "Sealing Considerations for Tapered Roller Bearings." Society of Automotive Engineers Paper 780401 (1978).

21-10 Smoley, Earl M. "Sealing with Gaskets." *Machine Design,* October 27, 1966.

21-11 *SAE Handbook,* Warrendale, Pa.: Society of Automotive Engineers, Inc., 1980.

21-12 Tokarski, Joseph. "Formed-in-Place Gaskets: Concept vs. Reality." Society of Automotive Engineers Paper 770073 (1977).

PROBLEMS

P21-1 A pipe contains steam at 200°C and 4 N/mm^2 pressure. What type of gasket material is suitable for use in the pipe joint?

P21-2 Why is a cellulose gasket unsuitable as a cylinder head gasket for a diesel engine?

P21-3 A gasketed joint has a contact area between the flanges and gasket of 4000 mm^2. The joint is held together by four bolts with a tensile force of 10 000 N in each.

(*a*) What is the apparent flange pressure?

(*b*) What is the minimum flange pressure?

P21-4 If an asbestos and rubber gasket of medium compressibility were used in Example 21-1, what grade of bolts would be satisfactory? If a low-compressibility asbestos and rubber gasket were used?

P21-5 If the gasket in Example 21-1 were made with the 6.25-in. dimension in Fig. 21-4 reduced to 6.12 in. and the 4.88-in. dimension reduced to 4.75 in., would the Grade 5 bolts be satisfactory?

P21-6 A cork gasket is 0.125 in. thick. Should a gasket width of 0.50 in. be satisfactory?

P21-7 Bolts in a gasketed joint are tightened to produce an elongation of 0.1 mm. If the gasket creeps to the extent that the thickness is reduced from 1.6 mm to 1.5 mm, what tension remains in the bolts?

P21-8 The bolts in Example 21-1 have an effective length of 0.56 in., of which 0.42 in. is threaded. What is the elongation of the bolts due to 50 lb-ft torque? If the gasket loses 0.002 in. of compressed thickness due to creep, what is the remaining bolt tension?

P21-9 A gray cast iron cover on a housing has a bolt spacing of six times the cover thickness. If it is replaced by a steel cover of the same thickness, will the bolt spacing be satisfactory?

P21-10 If the ratio of bolt spacing to flange thickness for the joint in Fig. 21-4 should not exceed eight, what is the minimum flange thickness?

P21-11 If the outside diameter of the asbestos gasket in Example 21-2 were reduced from 5.38 in. to 4.75 in., would the gasket be satisfactory?

P21-12 If the gasket in Example 21-2 is replaced by a soft aluminum gasket, what is the lowest property class of bolt that can be used if the tightening torques are: class 4.8—156 N · m; class 8.8—245 N · m; class 10.9—335 N · m; and class 12.9—400 N · m?

P21-13 A rubber O-ring has the dimensions (Fig. 21-10) of $d_i = 1.739 \pm 0.010$ in. and $w = 0.070 \pm 0.003$ in. It is used as a dynamic seal in a cylinder bore of $1.875\ ^{+0.002}_{-0.000}$ in. and fits in a groove of $1.765\ ^{+0.000}_{-0.002}$-in. diameter. What is the minimum compression of the cross section? What is the maximum compression of the cross section?

P21-14 An O-ring with a cross section diameter $w = 0.139 \pm 0.004$ in. is to be placed in a nonstandard rectangular section groove that is 0.185 ± 0.005 in. wide and 0.140 ± 0.005 in. deep. Will the groove accept the O-ring?

P21-15 Would the O-ring groove dimensions for P21-14 be suitable for a rotating shaft? Why or why not?

P21-16 A radial lip seal is to be used on a shaft made from SAE 1020 hot rolled steel that is not to be heat treated in any manner. Is the shaft likely to be scored by the seal?

P21-17 A radial lip seal is proposed for use on a shaft where the lubricating oil level will be 250 mm above the shaft center. Is the seal likely to perform satisfactorily?

P21-18 A radial lip seal is to be used on the pinion shaft of an automobile differential that is lubricated by oil that contains an extreme pressure additive. Will a nitrile rubber sealing lip be satisfactory? If nitrile rubber is not satisfactory, what material should be specified for the sealing lip?

P21-19 A face seal of the type shown in Fig. 21-14 seals a fluid under a pressure of 1.5 N/mm^2. The outside diameter of the rotor is 200 mm. The outside diameter of the portion that contacts the stator is 175 mm, and the outside diameter of the shaft at the opposite end of the rotor is 172 mm. What is the force due to fluid pressure that holds the stator against the rotor?

PART 6
Design Projects and Auxiliary Problems

Appendixes

Appendix A Simple Machines

Appendix B Review of Statics and Strength of Materials

Appendix C Auxiliary Problems

Appendix D Design Projects

Appendix E Associations of Interest to Designers

Appendix F Tables and Graphs

Appendix G Summary of Equations and Formulas

Appendix H Answers to Selected Problems

Appendix A
Simple Machines[1]

A-1 INTRODUCTION

A simple machine is a device with at least *one* mechanically actuated member (lever, screw, or piston). Although eight simple machines are generally recognized, physicists see only three basic principles in simple machines.

1. Lever action.
2. The inclined plane.
3. Pascal's principle of equal pressure distribution in fluids.

This is an important distinction because mechanisms based on lever action generate *little* friction and therefore have *high* efficiency. In contrast, those based on the inclined plane have high friction and low efficiency. The hydraulic press lies somewhere between the two with regard to friction and efficiency.

Friction plays a *dual* role in the three simple machines based on the inclined plane. It lowers efficiency but compensates by adding a useful self-locking feature not found in those based on the lever.

Simple machines are useful because of the *mechanical advantage* (*MA*) they yield,

$$MA = \frac{\text{resistance}}{\text{effort}} \tag{A-1}$$

A mechanical advantage greater than *one* indicates an *increase* of force at the expense of speed. A fractional mechanical advantage ($MA < 1$) is really a mechanical disadvantage, since effort is greater than resistance; speed is increased at the expense of force. Resistance to motion invariably includes friction, a quantity that usually cannot be measured with accuracy. When resistance due to friction is omitted, we can determine the *theoretical* mechanical advantage (*TMA*).

$$TMA = \frac{\text{distance effort moves}}{\text{distance resistance moves}} \tag{A-2}$$

The *efficiency e* of simple machines is defined as the mechanical advantage divided by the theoretical mechanical advantage:

$$e = \frac{MA}{TMA} \tag{A-3}$$

Because of friction $MA < TMA$ and, hence, $e < 1$. Note that for all simple machines the mechanical advantage and efficiency are *independent* of speed and load within wide limits.

A-2 LEVER ACTION

The four simple machines based on lever action—lever, wheel and axle, pulley, and gearing—will now be discussed in detail.

Lever

A lever is a rigid bar that is free to pivot about an axis through a point called the *fulcrum*. The applied force is called *effort*, and the load to be overcome is termed *resistance*. Three classes of levers are generally recognized, as shown in Fig. A-1. The location of the pivot point in relation to the resistance determines the lever class. In each class there is an advantage to be gained in either force or distance. In *first*-class levers (e.g., crowbars, shears, and pliers), the fulcrum is located *between* the effort and resistance. The

[1] Also referred to as basic machines or mechanical powers.

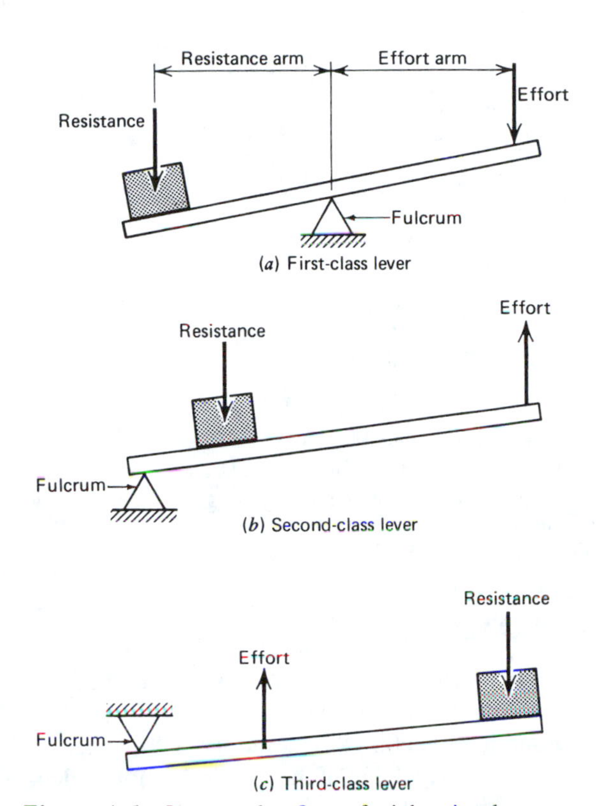

Figure A-1 Lever—the first of eight simple machines.

second-class lever (e.g., wheelbarrow) has the fulcrum at one end and the effort at the other, with the resistance somewhere between these two points. Both first- and second-class levers are used to help overcome big resistances with a relatively small effort. Herein lies the main importance of levers and simple machines in general.

Third-class levers apply the effort on the *same* side of the fulcrum as the resistance, but on a much *shorter* arm. They are thus used primarily to gain speed or distance (reach, as in a cherry picker) at the expense of effort. Both move in the same direction. The human arm is a third-class lever. Development of powerful hydraulic and pneumatic cylinders has vastly increased the use of this third class.

Levers must often be angled, as in bell cranks (Fig. A-2*a*), curved, as in pump handles (Fig. A-2*b*), or offset, as in rocker arms (Fig. A-2*c*), to satisfy space requirements and/or a need for changing the force's direction.

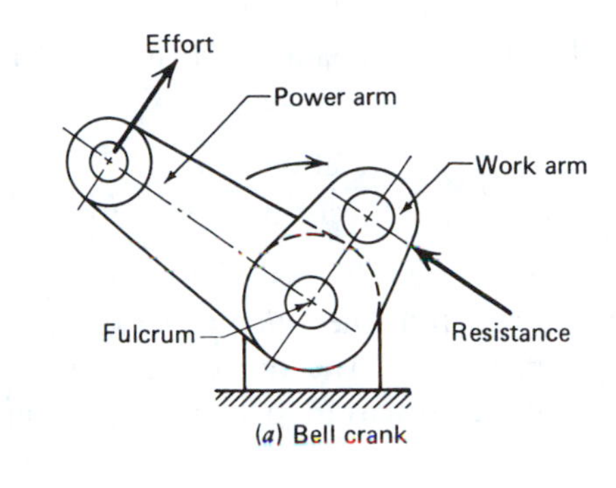

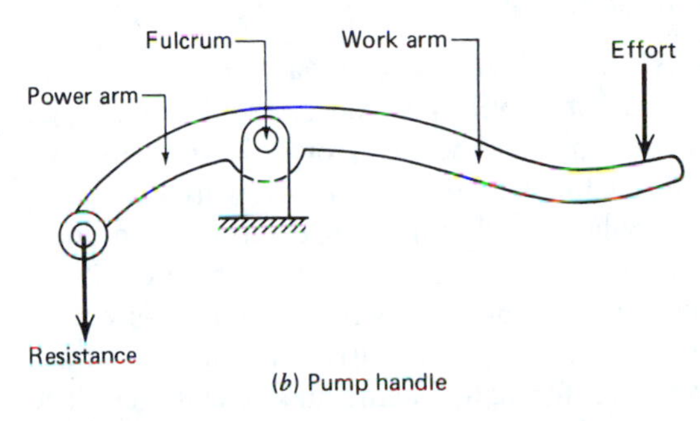

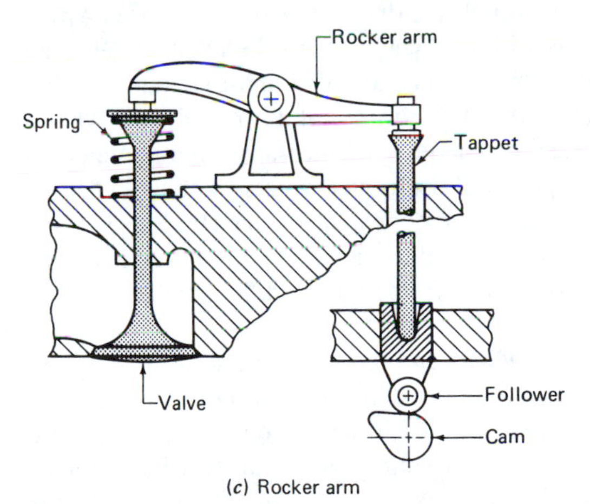

Figure A-2 Angled and curved levers. The lever appears in a wide variety of disguises because the moving parts of many mechanical contrivances can be reduced to levers—in structure or in principle.

The theoretical mechanical advantage of levers is

$$TMA = \frac{\text{effort arm}}{\text{resistance arm}} \tag{A-4}$$

The effort arm is the distance (length of the perpendicular) from the fulcrum to the line of action of the effort. The resistance arm is the distance from the fulcrum to the line of action of the resistance.

Wheel and Axle

This simple device is essentially a *circular* lever capable of rotating instead of oscillating around its fulcrum. It is the basis for tools, mechanisms, and machines as diverse as screwdrivers, keys, waterwheels, windmills, and turbines. In addition to its rolling function, as with automobiles and railroad cars, the wheel-and-axle assembly performs an important drive-and-ratio function, the basis for belt, chain, and gear drives. The wheel-axle assembly consists of a circular member or crank rigidly attached to the axle, which turns with the wheel. Thus the front wheel of an automobile is not a wheel-and-axle machine because the axle does not turn with the wheel. Wheel-and-axle machines are often depicted as composed of a grooved wheel fixed to a shaft or drum and two ropes, or cables (Fig. A-3). One rope is attached to the wheel circumference, and the other is wrapped around the smaller shaft or drum.

The wheel and axle may be used to magnify either the applied force (effort) or the motion. When the applied force acts on the wheel, a large mechanical advantage is obtained (drum hoist). Acceleration takes place when the input effort is applied to the axle. This is a fundamental design principle in vehicle drives.

The mechanical advantage, from Eq. A-1, is

$$MA = \frac{\text{load}}{\text{effort}} \tag{A-5}$$

$$TMA = \frac{\text{effort arm}}{\text{load arm}} \tag{A-6}$$

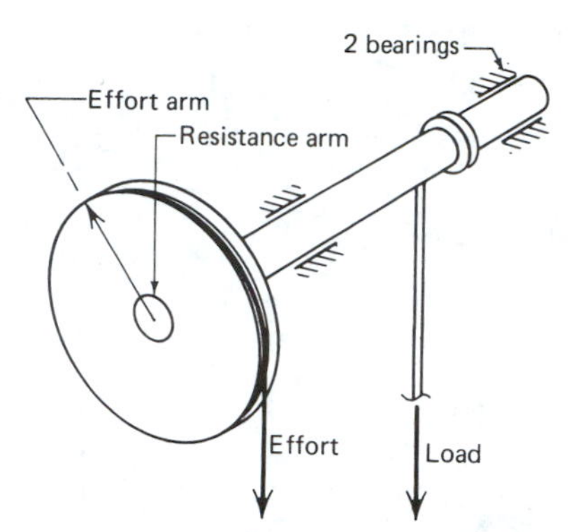

Figure A-3 Wheel and axle. Essentially a "round" lever, it has hundreds of applications, from small toys to giant turbines.

EXAMPLE A-1

The effort on a wheel-and-axle device, as shown in Fig. A-3, is 62 N. The corresponding load is 600 N. The radius of the wheel is 250 mm, and that of the axle is 25 mm. Calculate the mechanical advantage, theoretical mechanical advantage, and efficiency for this simple machine.

SOLUTION

$$MA = \frac{600\text{ N}}{62\text{ N}} = \mathbf{9.68}$$

$$TMA = \frac{250\text{ mm}}{25\text{ mm}} = \mathbf{10}$$

$$e = \frac{MA}{TMA} = \frac{9.68}{10} = \mathbf{0.97}$$

The efficiency is very high, as is to be expected in a simple machine based on lever action.

Pulley

In its simplest form the pulley consists of a single grooved wheel or sheave turned by means of a rope or chain partially confined to the groove. A

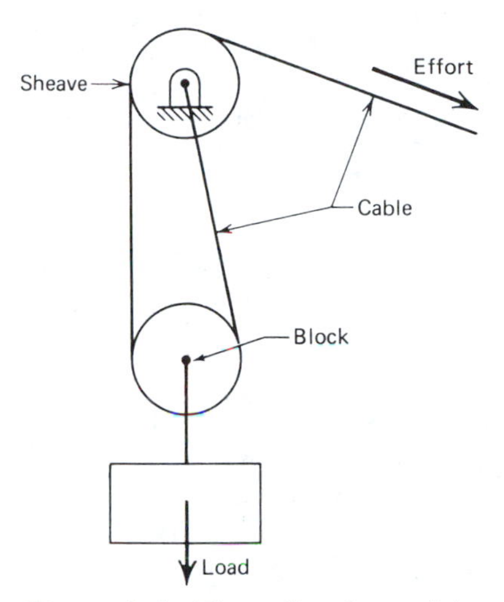

Figure A-4 The pulley is used in countless applications from small hoists to giant cranes.

single pulley yields no mechanical advantage but conveniently directs the applied force downward when a burden is lifted. When two pulleys are interconnected, as in Fig. A-4, a force can be generated as much as twice that applied. For each pulley added the applied force is cut according to Eg. A-7.

$$TMA = \text{number of sheaves} \qquad \text{(A-7)}$$

Pulley systems are therefore a means of changing the effort and speed of the load. They can also be used to change the direction of force to make the force more useful. Note that the mechanical advantage is *independent* of the pulley diameter, except for very small pulleys when resistance to bending of the cable adversely affects efficiency.

EXAMPLE A-2

Find the efficiency of a pulley system in which the load moves 0.5 m while the applied force covers a distance of 2 m. Furthermore, a load of 100 N corresponds to an effort of 26 lb. How many pulleys are required?

SOLUTION

From Eq. A-1 $MA = \frac{100 \text{ N}}{26 \text{ N}} = \mathbf{3.85}$

From Eq. A-2 $TMA = \frac{2 \text{ m}}{0.5 \text{ m}} = \mathbf{4.0}$

$$e = \frac{3.85}{4.0} = \mathbf{0.96} \text{ (very high)}$$

Four sheaves are required according to Eq. A-7.

Gears

Gears are wheels with mating teeth cut in the rim or surface so that one can turn the other without slippage. The lever is basic to gearing (Fig. A-5). A pair of gears is essentially a set of spinning levers each with two equal arms acting in turn. Effective lever action is possible, however, only because a pair of mating gear-tooth profiles act against each other to produce the relative motion desired. Gears can be mounted on parallel, intersecting, or nonintersecting shafts. Consequently, they can:

1. Change the plane of rotation.
2. Increase or decrease the speed of applied motion.
3. Magnify or reduce the applied force.
4. Provide a drive without slippage.

The theoretical mechanical advantage of any gear train is the product of the number of teeth on the driven wheels (gears) divided by the product of the number of teeth on the driver gears (pinions).

$$TMA = \frac{N_G}{N_p} \qquad \text{(A-8)}$$

A-3 THE INCLINED PLANE

The three simple machines based on the principle of the inclined plane will now be discussed in detail.

(*a*) A gear is essentially a first-class lever with arms of equal length

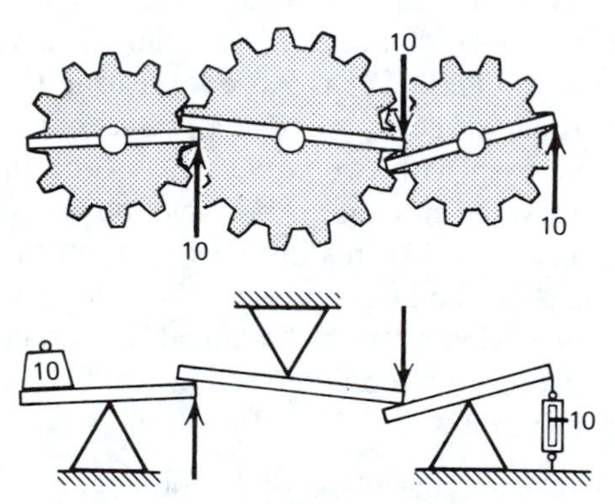

(*b*) A pair of gears is essentially a set of spinning levers acting in turn

Figure A-5 The lever is basic to gearing. Gearing is essential to devices ranging from watches to drawbridges. (Courtesy Caterpillar Tractor Company.)

Inclined Plane

A plane surface set at an angle other than a right angle against a horizontal surface (Fig. A-6) is an inclined plane. The inclined plane permits one to overcome a *large* resistance by applying a relatively *small* force through a *longer* distance, L, than the load is to be raised, H. Note that F_t, the component parallel to the plane, is much smaller than the weight, W, itself. The applied force F parallel to the plane (effort) needed to move the load *up* the plane is equal to F_t plus any *frictional* force but will generally be less than the

Figure A-6 Inclined plane. For the same height H, any decrease in force F_t is accompanied by a reciprocal increase in distance L, and vice versa.

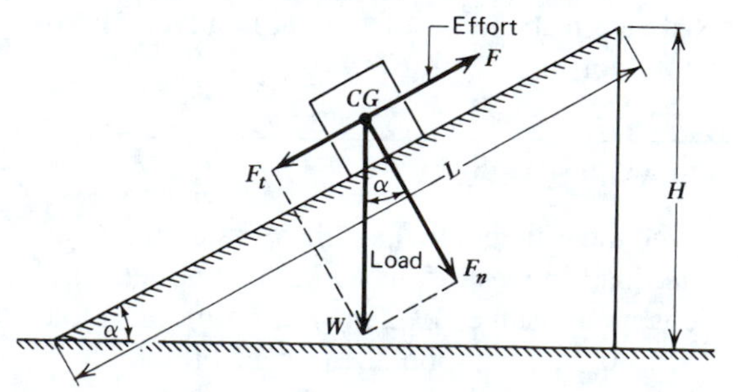

weight, W. Thus human or mechanical effort can be greatly economized.

The mechanical advantage for motion *up* the plane is

$$MA = \frac{W}{F} = \frac{1}{f \cos \alpha + \sin \alpha} \qquad \text{(A-9)}$$

$$TMA = \frac{1}{\sin \alpha} \qquad (f = 0) \qquad \text{(A-10)}$$

where

f = coefficient of friction
α = plane angle; deg

For movement *down* the plane

$$MA = \frac{W}{F} = \frac{1}{f \cos \alpha - \sin \alpha} \qquad \text{(A-11)}$$

If the applied force acts *parallel* to the base of the plane

$$MA = \cot(\alpha + \theta) \qquad \text{(A-12)}$$

where

θ = arctan f
= friction angle

The friction angle is a convenient alternative to the coefficient of friction. It is defined as shown in Fig. A-7*a*. The true meaning of friction angle, however, is best seen in Fig. A-7*b*. If the hinged plane is tilted progressively, the force F will steadily decrease until a position is reached for which F is zero. The gravity component $W \sin \alpha$ will simultaneously increase until it becomes sufficiently large to move the body down the plane. The plane angle for which motion is impending is $\theta = \arctan f$. The friction angle is thus a *limiting* angle, *a criterion for impending motion*, and herein lies its importance.

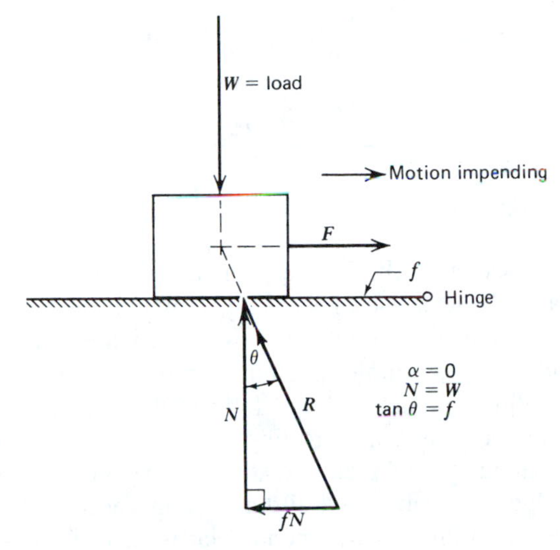

(*a*) For impending motion, the friction angle θ, is the angle between the normal force N and the resultant R of the normal force and the friction force.

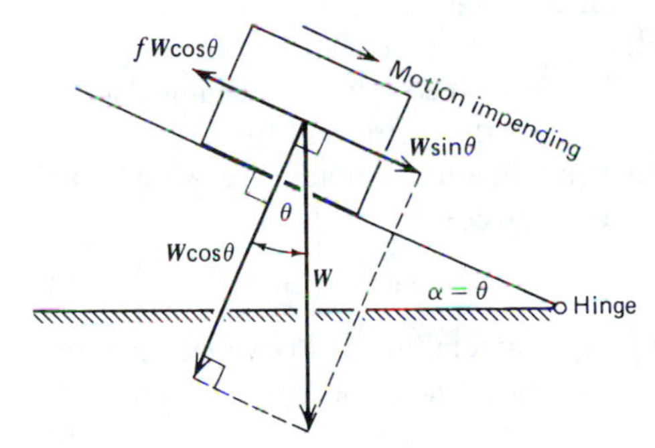

(*b*) For impending downward motion $F = 0$ when $\alpha = \theta$ and $\tan \theta = f$. The friction angle is still the angle between the normal and the resultant.

Figure A-7 Definition and true meaning of friction angle, θ.

EXAMPLE A-3

Find the mechanical advantage and the efficiency of a plane inclined at an angle of 30 deg when the coefficient of friction is 0.10.

SOLUTION

From Eq. A-9,

$$MA = \frac{1}{f \cos \alpha + \sin \alpha}$$

$$= \frac{1}{0.1 \cos 30 \text{ deg} + \sin 30 \text{ deg}} = \mathbf{1.71}$$

From Eq. A-10,

$$TMA = \frac{1}{\sin \alpha} = \frac{1}{\sin 30 \text{ deg}} = \mathbf{2.0}$$

$$e = \frac{1.71}{2.0} = \mathbf{0.86}$$

Wedge

The wedge is the *active* twin of the inclined plane. It does useful work by moving; the inclined plane, in contrast, always remains stationary. This simple machine consists of a *pair* of inclined planes, set face to face, that can sustain relative sliding or rolling motion (Fig. A-8). By moving one plane relative to the other, a wedge is capable of building up an enormous force in a direction perpendicular to that of the moving wedge. Force multiplication varies inversely with the size of the wedge angle; a sharp wedge (small included angle) yields a large force. With adequate friction at the interfaces, the wedge becomes a separating, holding, and stopping device with countless mechanical applications. *Sprag* is the term used for the wedge in many such applications, notably unidirectional clutches. For a wedge,

$$MA = \cot (\alpha + \theta) \tag{A-13}$$

The *cam* is a special form of wedge that has either a rotary or a linear motion (Figs. A-9 and A-2*c*).

Figure A-8 Wedge. This device consists of a pair of inclined planes that can sustain relative sliding or rolling motion. The wedge can support heavy loads at self-locking angles.

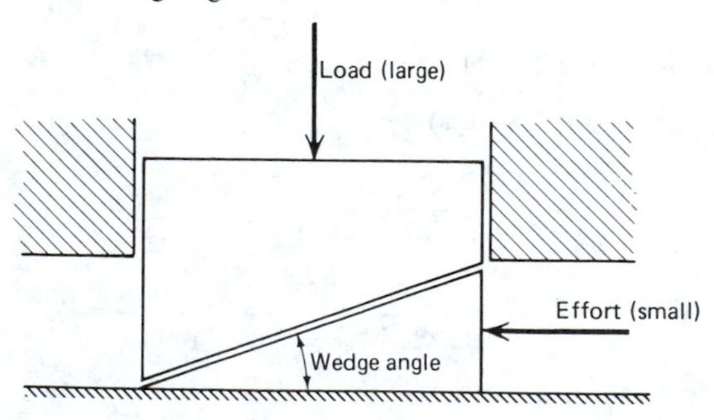

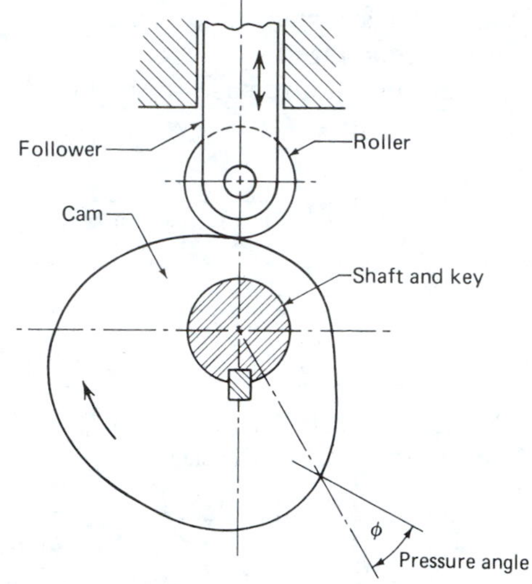

(*a*) Cams are in essence rotary wedges, in which the pressure angle varies between zero and a maximum value.

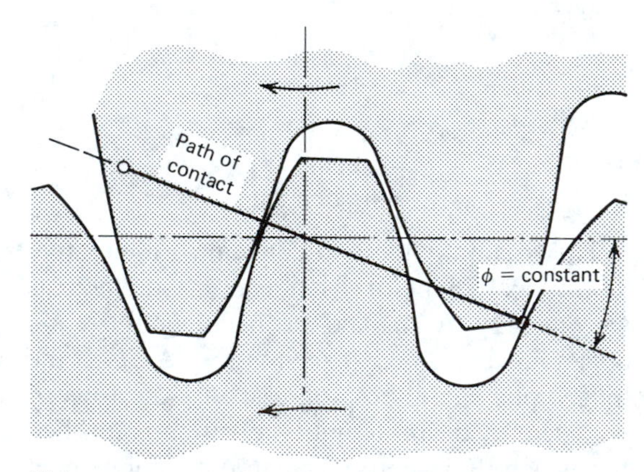

(*b*) In essence, a pair of mating gear tooth profiles are cams, the one acting against the other, to produce the relative motion desired. For involute profiles the pressure angle remains constant throughout mesh, a feature of great practical importance.

Figure A-9 The cam is a special form of a wedge.

Screw

This simple machine is a *modification* of the wedge designed to yield a very *large* mechanical advantage in *minimal* space. The principle of the screw can be easily understood by cutting a right triangle out of paper and wrapping it around a pencil (Fig. A-10*a*). The triangle's hypotenuse

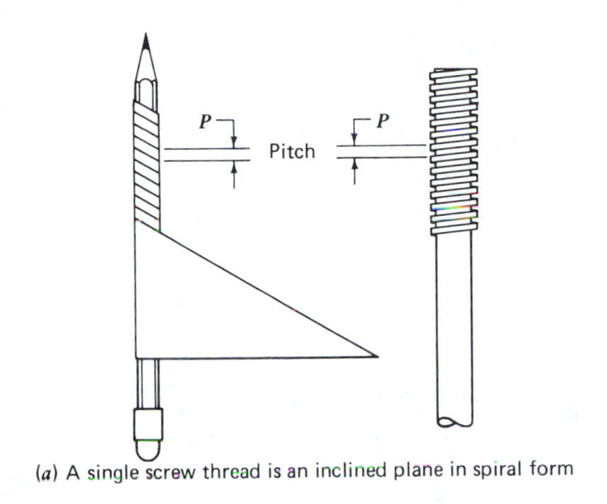

(*a*) A single screw thread is an inclined plane in spiral form

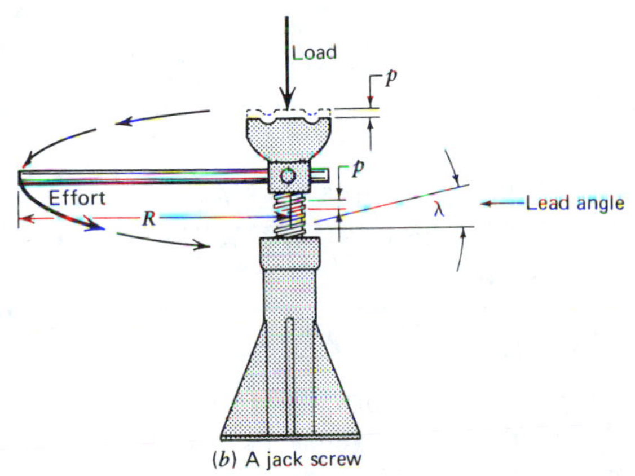

(*b*) A jack screw

Figure A-10 The screw—complexity in simplicity—is a modification of the wedge. (Courtesy The Bureau of Naval Personnel.)

becomes the inclined plane of a single thread of the screw. In one sense, a screw is *not* a simple machine at all, since it depends on a lever (wrench) or wheel and axle (screwdriver) for its operation. Figure A-10*b* shows that when the lever arm makes a full turn in the direction shown, the screw is elevated a distance equal to the pitch p. The screw is the basis of threaded fasteners, the propeller, the auger, screw gearing, and screws for linear motion. A screw and nut are opposed twins, the nut being an inversion of the solid screw. The screw and nut combination is essentially a wedge guided along a helical track. Consequently, large amounts of *circular* motion are reduced to very small amounts of *straight-line* motion, either with high or low friction. If axial resistance is encountered, the opposite occurs: a *small* applied circumferential force is transformed into a *large* axial force.

For motion opposite load, which *opposes* the load motion in that direction,

$$MA = \frac{2\pi r - fp}{2\pi fr + p}\left(\frac{R}{r}\right) \tag{A-14}$$

where

r = pitch radius of screw
f = coefficient of friction
p = lead of screw (lead equals pitch for single threads, not for multiple threads)
R = moment of applied force

For motion opposite load, which *assists* it

$$MA = \frac{2\pi r + fp}{2\pi fr - p}\left(\frac{R}{r}\right) \tag{A-15}$$

When $f = 0$,

$$TMA = \frac{2\pi R}{p} \tag{A-16}$$

The theoretical mechanical advantage of screws is of limited interest because of the large friction forces present in most screws. Whenever it serves a useful purpose, much friction can be built into a screw.

The screw is essentially a *transfer device* for motion and/or force. The *lead angle* is the deciding parameter (Fig. A-11). For *small* lead (helix) angles (less than 6 deg), a large mechanical advantage makes the screw well suited as a fastener—a device for transferring and maintaining a large force. This function is facilitated by self-locking, a trait inherent in all screws with small helix angles. Self-locking means that no matter how large the axial force becomes, it cannot cause the nut to rotate (backdrive). Self-locking is enhanced by friction at the screw-nut interface.

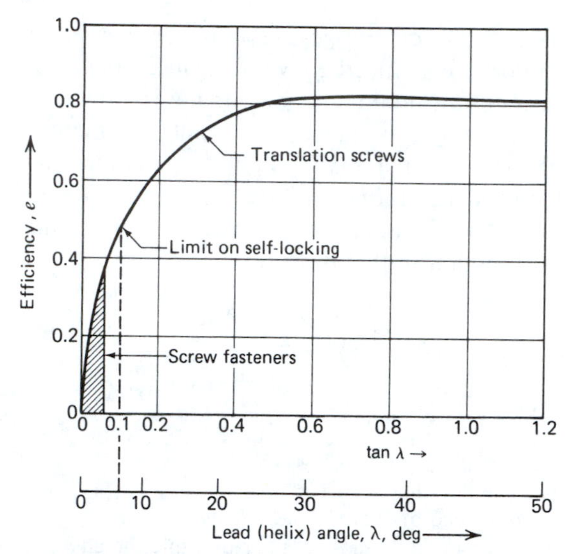

Figure A-11 Efficiency of a standard metric screw as a function of the lead (helix) angle. This curve was based on a coefficient of friction $f = \tan\theta = 0.01$; $\theta = 5.71$ deg. (From O. Niemann, *Maschinenelemente,* I., p. 203, Springer-Verlag, Berlin, 1975.)

For *large* lead (helix) angles, the efficiency has increased at the expense of the mechanical advantage, generating a combination of motion and force required by power screws.

EXAMPLE A-4

Calculate the mechanical advantage and the efficiency of a jack screw based on these data.

$r = 48$ mm $\quad p = 5$ mm

$f = 0.07$ mm $\quad R = 300$ mm

What effort is required to lower and elevate a load of 1000 N?

SOLUTION

For motion *opposite* load, Eq. A-14 gives

$$MA = \frac{2\pi(48 \text{ mm}) - 0.07(5 \text{ mm})}{2\pi(0.07)(48 \text{ mm}) + 5 \text{ mm}}\left(\frac{300 \text{ mm}}{48 \text{ mm}}\right)$$

$$= \mathbf{72}$$

From Eq. A-16 $\quad TMA = \dfrac{2\pi(300 \text{ mm})}{5 \text{ mm}} = \mathbf{377}$

From Eq. A-3 $\quad e = \dfrac{72}{377} = \mathbf{0.19}$

From Eq. A-1 $\quad \text{Effort} = \dfrac{1000 \text{ N}}{72} = \mathbf{13.9 \text{ N}}$

For motion *assisted* by the load, Eq. A-15 gives

$$MA = \frac{2\pi(48 \text{ mm}) + 0.07(5 \text{ mm})}{2\pi(0.07)(48 \text{ mm}) - 5 \text{ mm}}\left(\frac{300 \text{ mm}}{48 \text{ mm}}\right)$$

$$= \mathbf{117}$$

From Eq. A-3 $\quad e = \dfrac{117}{377} = \mathbf{0.31}$

From Eq. A-1 $\quad \text{Effort} = \dfrac{1000 \text{ N}}{117} = \mathbf{8.5 \text{ N}}$

A-4 PASCAL'S PRINCIPLE OF EQUAL PRESSURE DISTRIBUTION IN FLUIDS

Hydraulic Press[2]

The outstanding feature of this simple machine, based on the law of hydrostatics, is its potential for:

1. A *tremendous* mechanical advantage limited only by the strength of its own components.
2. An ability to operate power units in *remote* or inaccessible places, thereby eliminating the necessity for complicated mechanical devices that would otherwise be required.
3. An ability to generate *smooth* linear motion, free of harmful vibration.

According to the law of hydrostatics, external pressure exerted on a confined fluid is trans-

[2]Invented by Joseph Bramah (1784–1814), England (the only simple machine with a known inventor).

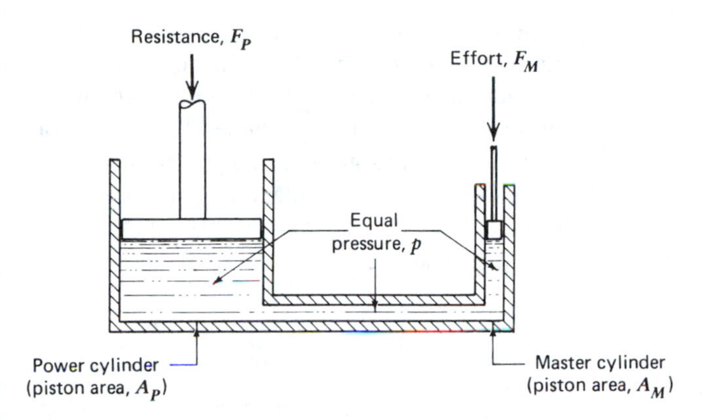

Figure A-12 The hydraulic press—powerful, versatile, yet smooth and gentle in action.

ferred, *undiminished,* to all surfaces in contact with the fluid. The hydraulic press is designed to take advantage of this fundamental principle. It has a confined body of fluid connecting two cylinders of *different* areas, each fitted with a piston (Fig. A-12). As a result of *equal* fluid pressure per unit area on the pistons of different areas, a greater force will be exerted on the larger piston than on the smaller one, or

$$\frac{F_M}{A_M} = \frac{F_P}{A_P} \qquad \text{(A-17)}$$

Thus a *large* force F_P results on the piston of the power cylinder when a *small* force F_M is applied to the piston of the master cylinder. The theoretical mechanical advantage is therefore the ratio of the area of the power cylinder to the area of the master cylinder.

$$TMA = \frac{A_P}{A_M} \qquad \text{(A-18)}$$

If levers or linkages are used to actuate the master cylinder, their mechanical advantage must be included in the overall mechanical advantage.

The hydraulic press also facilitates remote control of large forces in places where levers, linkages, screws, and wedges could not be accommodated. Moreover, numerous cylinders can be actuated singly or in unison by the master cylinder (Fig. A-13). Consequently, designers are using third-class levers to increase speed and force in applications including machine tools, construction equipment, farm machinery, loading platforms, and cranes.

EXAMPLE A-5

In a hydraulic press, find the resistance that corresponds to an effort of 100 N. The master cylinder has a diameter of 50 mm, and the power cylinder has a diameter of 500 mm. What is the theoretical mechanical advantage?

SOLUTION

From Eq. A-17, one obtains,

Figure A-13 Numerous cylinders can be actuated in unison by a single cylinder—the master cylinder. (Courtesy The Bureau of Naval Personnel.)

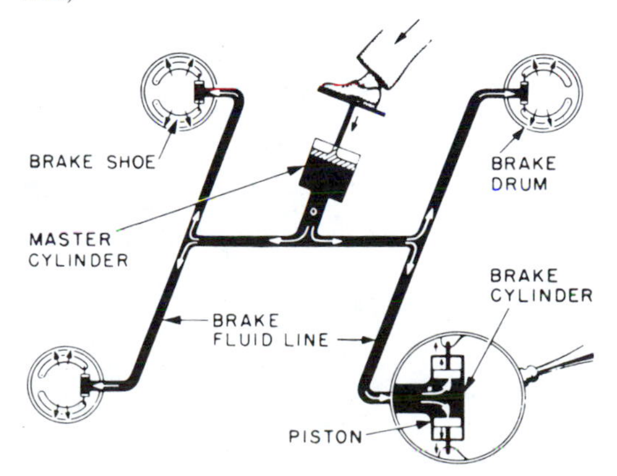

$$F_P = F_M \frac{A_P}{A_M}$$

$$= (100\ \text{N}) \frac{0.25\pi(500\ \text{mm})^2}{0.25\pi(50\ \text{mm})^2} = \mathbf{10\ kN}$$

$$TMA = \frac{A_P}{A_M} = \frac{500^2}{50^2} = \mathbf{100}$$

The concept of simple machines is useful in obtaining a *fundamental* knowledge of machines but inadequate in the realm of higher mechanics. Not all machines can be reduced to simple machines. Spring design is based on Hooke's law, not on any of the simple machines. The theory of sliding bearings is based on hydrodynamics, not on simple machines.

Appendix B
Review of Statics and Strength of Materials

B-1 MECHANICS OF MATERIALS APPLIED TO MACHINE DESIGN

Calculations for strength are based on the fundamentals of *statics* and *material properties*. Statics as a discipline is concerned with *equilibrium* of bodies; it assumes that all bodies acted on by external forces (loads or moments) are rigid and remain so; yet, to a varying degree, all real bodies are elastic and deform under the influence of external forces (roughly in proportion to the load).

Under load, a body responds by setting up *internal reactions* (pressures, tensions, and shears), the effect of which is equal and opposite to the load.[1] Thus, if a structural member does not move or deform excessively, equilibrium prevails. The magnitude of the internal reactions depends on the amount and distribution of material in the structural member that is capable of resisting the external loads. Internal forces are, for the most part, *passive* by nature; that is, they react only when acted on. The magnitude of the limits of their resistance to external load is termed material strength. Thus we may conclude that *equilibrium is an indication that the material has adequate strength to support the imposed loading*. However, in design one must be aware of service conditions other than equilibrium that enter into the proper sizing of structural members.

Under load, a body may deform *elastically* or *plastically* (permanent deformation). Machine design is concerned mainly with elastic deformations of small magnitude so that accuracy of all (moving) parts is maintained.

As a discipline, strength of materials involves three major concerns: *load, material,* and *size*. Three major types of calculations are therefore common.

1. Calculation of relevant loads for given materials and sizes.
2. Establishment of adequate overall sizes based on given loads and given materials.
3. Calculation of deformations based on given loads and given materials.

B-2 BASIC ASSUMPTIONS FOR STRENGTH CALCULATIONS

Calculations are based on certain *idealized* conditions such as isotropic materials, that is, materials whose properties are independent of the directions along which they are measured. Calculations are also limited to areas not coincident with the point of application of the load. The initial effect of applying a load is ignored in machine design (unless it is applied suddenly), and equilibrium is assumed.[2]

Machine design calculations aim to keep all deformations well within the elastic range. When elastic deformations are small, the geometry of any structural member is essentially retained and the body, for most practical purposes, is rigid. Thus the principles of statics apply to strength-of-materials calculations. Struc-

[1] Internal reactions are called stresses.

[2] For suddenly applied loads, impact derating factors are used (see Chapter 3).

tural components, which have large deflections, are notable exceptions.

In design deflections must remain so small that malfunction is clearly prevented. This requirement leads to the concept of factors of safety. A *factor of safety* can be defined as the ratio of the limit of material strength to the extreme value of the internal reactions. The limit of material strength, established by testing, should never be reached in service.

Finally, it is assumed that all loads are *static;* that is, they increase slowly from zero to constant values at which they remain. This means that a structural part is free of all but static loads. These assumptions, then, aid in establishing suitable mathematical models of actual conditions. They enable us to assign to a part the appropriate combination of load, material, and size.

B-3 FREE BODY DIAGRAM

Before we can assess the effect of a load on a structural member, the internal forces and moments must be linked to the external forces and moments. The *free-body* concept specifically serves this purpose. Essentially a means of obtaining a realistic mathematical model, free-body considerations involve four steps.

1. Pass an imaginary plane (or set of planes) through the body at a section of interest.
2. Remove the parts of least interest.
3. Show on the drawing all pertinent external forces and moments and indicate the internal forces and moments. Include all possible nonzero internal forces and moments.
4. Apply the equations of static equilibrium.

The laws of static equilibrium in a single plane require that two conditions be fulfilled.

1. The sum (Σ, sigma) of all forces and reactions must be zero.

$$\sum F = 0$$

2. The sum of all moments with respect to any point in the plane must be zero.

$$\sum M = 0$$

The first condition yields *two* equations at the most. When applied to a Cartesian coordinate system, these equations are $\Sigma F_x = 0$ and $\Sigma F_y = 0$. The second condition ($\Sigma M = 0$) theoretically yields an *infinite* number of equations, but they are *not* independent of each other. In essence, they simply say the same thing in many different ways. In practice, however, it should be applied first to *points of concurrency*, since this procedure reduces the number of unknowns in the equation (zero moment arm), or to other points where the moment arm lengths are known.

B-4 STRESS, STRENGTH, AND STRAIN

A stress is an internal reaction that exists within a material or machine part and resists deformation due to external forces. Stress is expressed as resistance per unit area of cross section. Simple stresses are produced by constant loading conditions on members of uniform cross section. Steady lateral or axial loads on beams, rods, or bars are examples. Compound stresses develop when two or more simple stresses act simultaneously at the same point.

Regardless of type of stress (simple or compound), an *upper stress limit* exists that should never be exceeded in service. This limit is the allowable or working stress, which is a function of the material strength and is usually obtained by dividing the material strength by a safety factor. Thus we arrive at a fundamental rule of design

Induced stress $\leq$ working stress

The induced stress is caused by loads and is obtained largely through the application of static equilibrium. By contrast, the working stress is a function of material strength, a measure of the ability of the material to resist the application of load without rupture or appreciable deflection. The factor of safety provides a margin of safety that depends first on the nature of loading (static or dynamic) and material (brittle or ductile). Other considerations are duration of load (intermittent or continuous) and safety. If failure endangers human lives, the safety factors must be of such magnitude that failures are precluded.[3] In general, strength is based on

1. *Yield* strength for *ductile* materials.
2. *Ultimate* strength for *brittle* materials.
3. *Fatigue* strength for parts subjected to *cyclic* stresses.

Fatigue aspects of design are discussed in Chapter 3. General recommendations for values of factors of safety are given in Chapter 3 and in *Machinery's Handbook*. In accordance with the preceding statements, we have:

$$fs = \frac{\text{ultimate strength}}{\text{working stress}} \quad \text{(brittle material)}$$

$$fs = \frac{\text{yield strength}}{\text{working stress}} \quad \text{(ductile material)}$$

Strain, the constant companion of stress, is a measure of the change in size or shape of a body as compared to its original size or shape. *Linear* strain, the deformation (elongation) per unit length in a given direction, is a practical concept. Maximum strain is nearly always in the direction of the applied load. As unit values, stress and strain provide a basis for comparing engineering materials using stress-strain curves.

Within a broad range of conditions, strain is proportional to stress.

$$\sigma = E\epsilon \tag{B-1}$$

where

σ (sigma) = stress (tensile and compressive)
E = modulus of elasticity
ϵ (epsilon) = strain

[3]See Chapter 3, Section 3-4, Working Stresses.

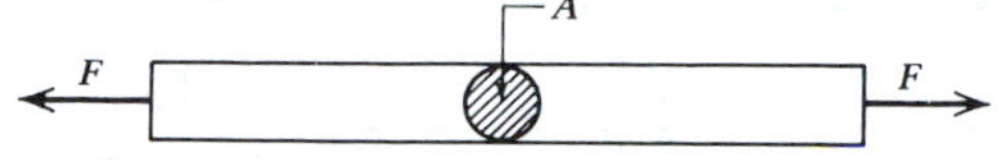

Figure B-1 Uniform bar in tension.

This relationship, known as *Hooke's law, says that both stress and deflection are proportional to the applied load*. This simple relationship, which holds true for all materials that exhibit elastic behavior, is of enormous practical importance in the design of machinery.

B-5 TENSION AND COMPRESSION IN DESIGN

Simple tension and compression occur when the applied force or load is in line with the axis of the member (Fig. B-1). Simple or direct stress develops under direct loading conditions.

For a bar of uniform cross section, the design is guided by:

$$\sigma_t = \frac{F}{A} \leq \sigma_{tw} \tag{B-2}$$

$$\sigma_c = -\frac{F}{A} \leq -\sigma_{cw} \tag{B-3}$$

where

σ_t = induced tensile stress
σ_c = induced compressive stress
F = tensile or compressive force
A = cross-sectional area
σ_{tw} = working stress in tension
σ_{cw} = working stress in compression

Several practical concepts and associated formulas will now be presented:

For a uniform bar in tension, total elongation ΔL is (Fig. B-2):

$$\Delta L = \frac{FL_0}{EA} \tag{B-4}$$

where

ΔL = elongation
L_0 = initial length
A = initial cross-section

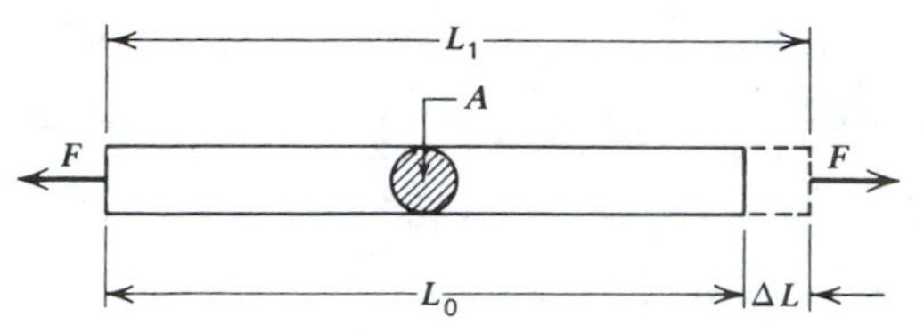

Figure B-2 Elongation of cylindrical bar.

Also, by definition,

$$\epsilon = \frac{\Delta L}{L_0} = \frac{L_1 - L_0}{L_0} \tag{B-5}$$

Bearing Stress

Bearing stress or *crushing* is a *limiting* condition for journal bearings, pins, rivets, and shear bolts when loaded *transversely*. What these diverse machine members have in common is an intimate contact between two semicylindrical surfaces and their fabrication from ductile materials (Fig. B-3). When subjected to high pressure, ductile materials will yield and deform permanently. Since this cannot be tolerated, the lower yield strength of the two materials constitutes a limiting condition.

When two semicylindrical surfaces are in close contact and loaded transversely, the contact stress p may be approximated by the formula

$$p = \frac{F}{dL} = \frac{F}{A} \tag{B-6}$$

Figure B-3 Bearing stress.

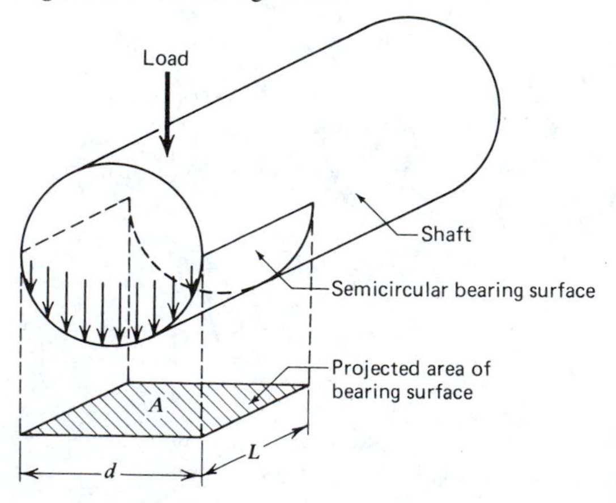

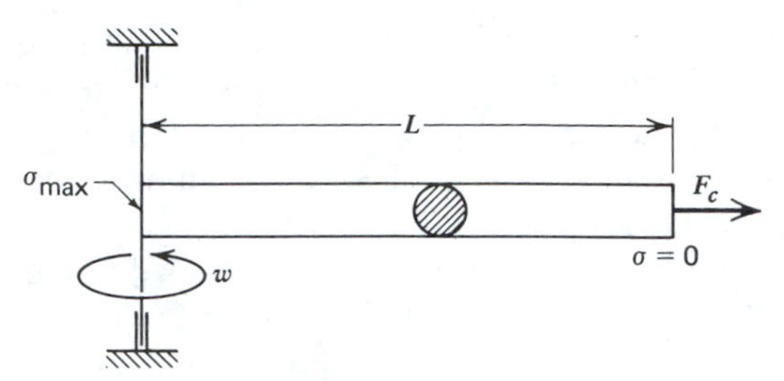

Figure B-4 Centrifugal force on a straight bar.

where

d = common diameter
L = length of bearing surface
A = projected area (rectangle)

The bearing stress naturally should not exceed the allowable compressive stress, which is the compressive yield strength divided by a suitable factor of safety. If p is greater than the allowable or working pressure, another material may be selected or the bearing surface may be enlarged.

Centrifugal Stresses

Tensile loads due to centrifugal forces play a major role in machinery, where the prevailing motion is rotary. For the cylindrical, radially oriented bar shown in Fig. B-4, the maximum tensile stress occurring at the inner end is

$$\sigma_{\max} = 0.5\rho L^2\omega^2 \tag{B-7}$$

where

ρ (rho) = mass per unit of volume
L = total length
ω (omega) = angular velocity; rad/s

For a ring or hoop (Fig. B-5), the tension is

$$\sigma_{\max} = \rho r^2\omega^2 \tag{B-8}$$

B-6 SHEAR LOADING

Direct shear loading is generated by opposing forces acting perpendicular to a rigid member (Fig. B-6). Transverse loading is another name for such forces. Shear loading is closely related to bending. Most transverse loads will generate

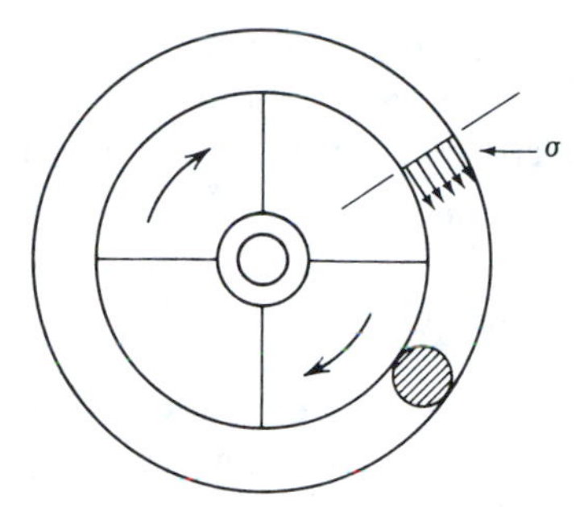

Figure B-5 Hoop stress due to rotation.

both shear and bending. However, when the two shear loads are close together, maximum stress occurs as direct shear and not as a result of bending.

In machine design we have

$$\tau = \frac{F}{A} < \tau_w = \frac{S_s}{fs} \tag{B-9}$$

where

τ_w = working stress in shear
S_s = shear strength of material
fs = factor of safety

When engineering materials are loaded in direct shear, the resulting shear strain is also proportional to stress. Thus we have

$$\tau = G\gamma \tag{B-10}$$

where

G = shear modulus
γ (gamma) = shear strain

Figure B-6 Direct shear.

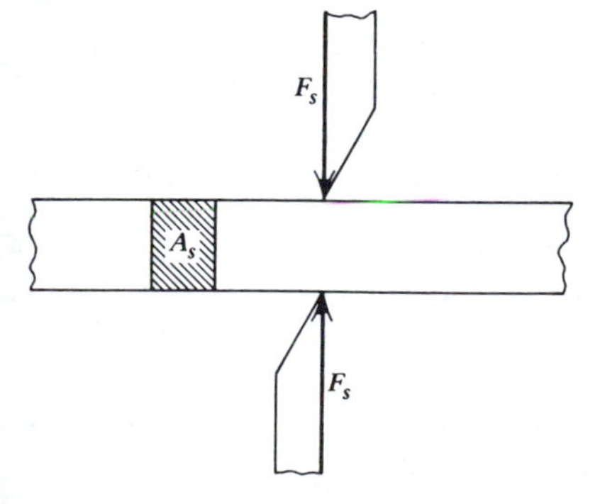

Since steel is much weaker in shear than in tension, G, although related to E, is smaller for steel by a factor of roughly 2.5. The exact relationship is

$$G = \frac{E}{2(1 + \nu)} \tag{B-11}$$

For most metals the constant μ (Greek mu), known as Poisson's ratio, has a value of roughly 0.30. It is used for transverse strain effects such as those that occur in interference fits between machine parts. Values for Poisson's ratio are available in *Machinery's Handbook,* p. 351.

B-7 BENDING OF STRAIGHT BEAMS

A beam is a structural member that is *long* relative to its width and height; its function is to carry *lateral* loads and *bending* moments. A transverse force acting on a beam between supports (Fig. B-7) generates both shear and bending stresses. Since the induced stresses should always be checked against allowable stresses, the question arises as to which is the larger. For a sizable beam length, maximum bending stresses are usually more critical than shear stresses. Only for short beams does it become necessary to check shear as well as bending.[4]

The first step toward computing stresses is to find the beam reactions. The equations of static equilibrium ($\Sigma F_y = 0$ and $\Sigma M = 0$) serve this purpose.

The second step is to find the bending moment in the beam. It can be defined as the sum of all the moments acting to the left of the section considered. In this case "the section considered" refers to that section along the beam where the bending moment is desired.

As an exercise, the reactions and maximum

---

[4]For a typical cantilever beam loaded at the end, the shear stress is approximately equal to the bending stress when the beam's length is equal to its cross-sectional depth. However, when such a beam is 10 times as long as it is deep, the bending stress is roughly 100 times the shear stress.

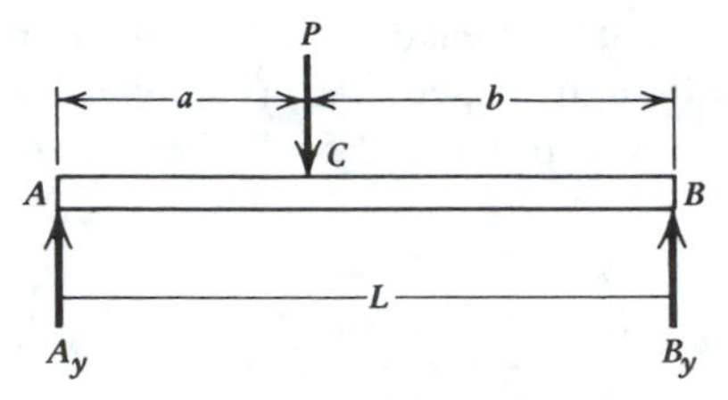

Figure B-7 Transverse force acting on a beam between supports.

bending moment of a simple beam will be computed.

EXAMPLE B-1

Find reactions A_y, B_y, and the bending moment at C, for the simple supported beam shown in Fig. B-7.

SOLUTION

$$\sum M_B = 0 \Rightarrow \quad A_yL - Pb + B_y(0) = 0$$

$$A_y = P\frac{b}{L}$$

$$\sum M_A = 0 \Rightarrow \quad B_yL - Pa + A_y(0) = 0$$

$$B_y = P\frac{a}{L}$$

Always check reactions by means of $\Sigma F_y = 0$.

$$A_y + B_y = P\frac{b}{L} + P\frac{a}{L} = P\left(\frac{a+b}{L}\right) = P$$

The moment at C is the sum of the moments to the left of C—in this case, one contribution only.

$$M_c = aA_y = P\frac{ab}{L}$$

Note that for a single load the maximum moment is always found at the load. If several single loads are present, the maximum moment is found under one of them. For a uniform load, it becomes necessary to draw the moment diagram in order to find the maximum moment.[5] The same is true for a combination of uniform and single loads.

B-8 SIZING OF BEAMS

The basic method for sizing beams is to compute the load-induced bending stresses and compare them to the allowable or working stress. The bending stresses in a beam are tensile and compressive and are computed by the formula

$$\sigma_b = \pm\frac{Mc}{I} = \pm\frac{M}{Z} \tag{B-12}$$

where

σ_b = bending stress
M = bending moment
c = distance from neutral axis to outer fibers
I = moment of inertia about the neutral axis
Z = section modulus

Note the similarity of form between $\sigma_b = M/Z$ and $\sigma_t = F/A$, as well as the difference. Z is a measure of beam strength based not on the square of the area but on the distribution of the beam cross section relative to the neutral axis. It recognizes that the usefulness of material increases with its distance from the neutral axis squared, as exemplified by the flanges on an I-beam. Figure B-8 gives values for I and Z for common sections. Since the section modulus is not a measure of material strength, by itself it can only be used to compare beams of *like* materials. Yield strength, in contrast, can only indicate *relative* strength of identical beams. But the *product* of the two enables designers to compare beams of different materials and different sections. This product is the *resisting bending moment* (*RBM*). Thus

$$RBM = ZS_y$$

---

[5]*Machinery's Handbook* provides formulas for maximum moments due to uniform loads.

Figure B-8 Elements of sections.

Section	I	Z	J	Z_p
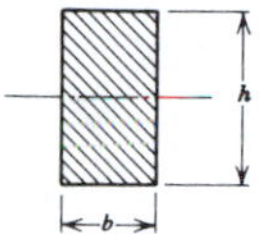 Rectangle (b, h)	$\frac{bh^3}{12}$	$\frac{bh^2}{6}$	—	—
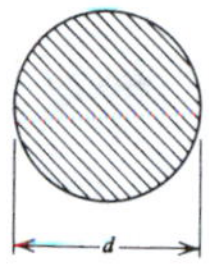 Solid circle (d)	$\frac{\pi d^4}{64}$	$\frac{\pi d^3}{32}$	$\frac{\pi d^4}{32}$	$\frac{\pi d^3}{16}$
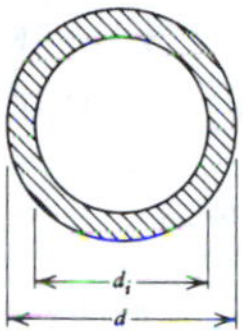 Hollow circle (d_i, d)	$\frac{\pi(d^4 - d_i^4)}{64}$	$\frac{\pi(d^4 - d_i^4)}{32d}$	$\frac{\pi(d^4 - d_i^4)}{32}$	$\frac{\pi(d^4 - d_i^4)}{16d}$

Since the induced stress should not exceed the working stress, the applicable relationship is

$$\sigma_b = \frac{M}{Z} \leq \frac{RBM}{Z} = \frac{S_y}{fs} = \sigma_w$$

This measurement shows that a beam with a smaller-section modulus of a higher-strength material can be as strong (in terms of maximum induced stress) as a larger-section modulus beam of a lower-strength material.[6]

It is common practice in beam calculations to compare the induced bending stress with the working stress rather than the induced moment with the resisting bending moment. Thus

$$\sigma_b = \frac{M}{Z} \leq \frac{S_y}{fs} = \sigma_w \qquad \text{(B-13)}$$

Again, the working stress is obtained from the material strength divided by a suitable factor of safety.

[6]In general, increasing the "strength" of a beam by changing to a stronger material will not reduce deflections to smaller values.

B-9 DEFLECTION

Machine members designed for strength usually have adequate rigidity if they are short in proportion to their own width, depth, or diameter. Shafts are notable exceptions. There are two types of bending deflections to consider: maximum deviation of the neutral axis, and the angular deflection at each end relative to the undeflected position of the neutral axis (Fig. B-9).

The deflection curve for an entire beam is rarely needed. Only the maximum deflection is generally needed in order to anticipate possible adverse effects of deflection, such as interference of machine members. Tables of maximum

Figure B-9 Linear and angular deflection due to bending.

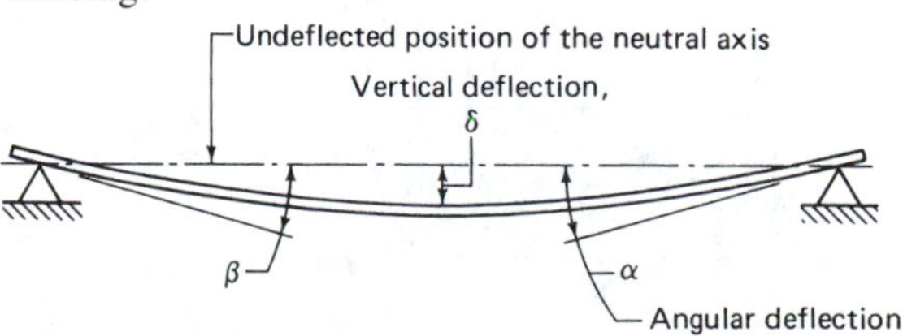

deflections for many commonly occurring cases are given in *Machinery's Handbook*.

B-10 TORSION OF CYLINDRICAL SHAFTS

Since rotary motion prevails in machinery, shear stress and angular deflection obviously become limiting conditions on common machine members such as shafts. Manufacturing considerations, with only a few exceptions, require the use of shafts with circular cross sections.

Strength-of-materials formulas for torsional stress and deflection assume that *circular cross sections perpendicular to the axis remain circular* and perpendicular to the axis, and that the radii of every cross section remain straight. Parallel cross sections, therefore, rotate slightly relative to each other, generating shear stresses and a slight angular deflection—assuming, of course, that the elastic limit of the material is not exceeded. The shear stresses increase linearly from zero at the center to a maximum value at the surface, as shown in Fig. B-10. Only maximum stresses are of interest, and they are computed from a formula analogous to that of bending.

$$\tau_{\max} = \frac{Tr}{J} = \frac{T}{Z_p} \tag{B-14}$$

where

r = radius of shaft
J = polar moment of inertia
Z_p = polar section modulus

Figure B-10 Torsional shear stress in circular shaft.

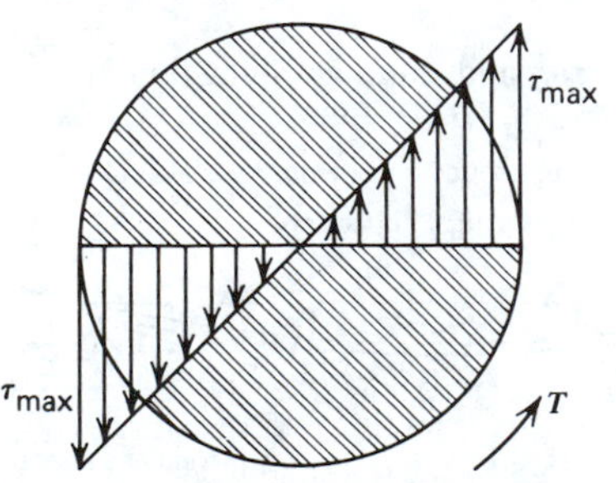

For a shaft of circular cross section,

$$J = \frac{\pi d^4}{32} \quad \text{and} \quad Z_p = \frac{\pi d^3}{16}$$

The corresponding maximum stress is

$$\tau = \frac{16T}{\pi d^3} \leq \tau_w = \frac{S_s}{fs}$$

For a hollow circular shaft of inside diameter d_i and outside diameter d_o, the maximum stress is

$$\tau = \frac{16Td_o}{\pi(d_o^4 - d_i^4)} \leq \tau_w$$

The most common application of these two formulas is finding the shaft size for given torque and material.

B-11 ANGLE OF TWIST

The angle of twist θ is a limiting condition in shafts carrying precision gearing. Since the design of gears is based on zero angular displacement of their respective shafts, the amount of displacement must clearly be limited to prevent excessive wear. Angular displacement combined with manufacturing errors limits gear width. Torsional vibration, a topic beyond the scope of this book, is another undesirable effect of angular displacement.

Figure B-11 shows the angle of twist of a circular shaft, greatly exaggerated for clarity. The formula for calculating θ is

Figure B-11 Angle of twist of a circular shaft.

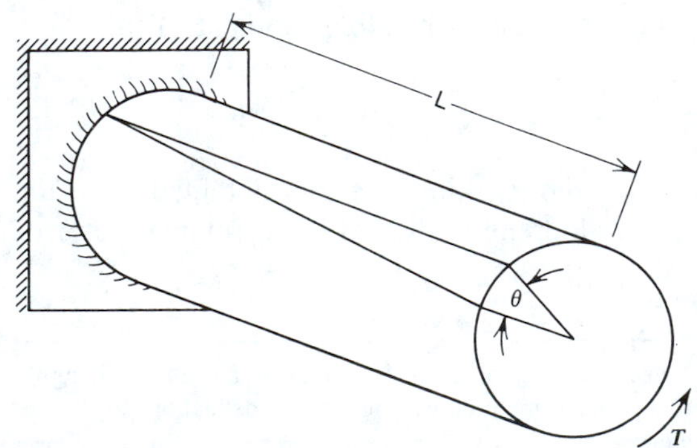

$$\theta = \frac{TL}{JG} \qquad \text{(B-15)}$$

where

θ = angle of twist; rad
L = length of shaft subjected to torque
G = modulus of rigidity

The product JG is a measure of torsional rigidity.

B-12 COMBINED STRESSES

Although some machine members are exposed to simple stresses only (tension, torsion, or compression), machines are generally too complicated to yield anything but combined stresses. When two different types of stresses, normal or shear, act at the same point simultaneously, the stress state is referred to as combined stress. Stresses of the same type may be added algebraically, provided they are collinear. If shear stresses on a single plane are not collinear, they can be combined vectorially.

Often, however, both normal and shear stresses act at a point. In that case, stress parameters termed *equivalent stresses* are used to evaluate strength. A first step, however, prior to evaluating strength for multiple stresses of the same type, is to use superposition to determine resultant stresses.

Superposition of Stresses

One situation is shown in Fig. B-12 where axial (normal) stresses are produced by bending and tension. The resulting tensile stress (σ_{res}) is determined as follows:

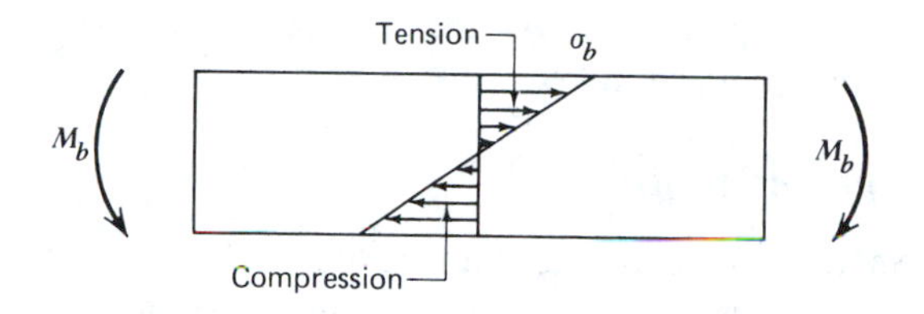

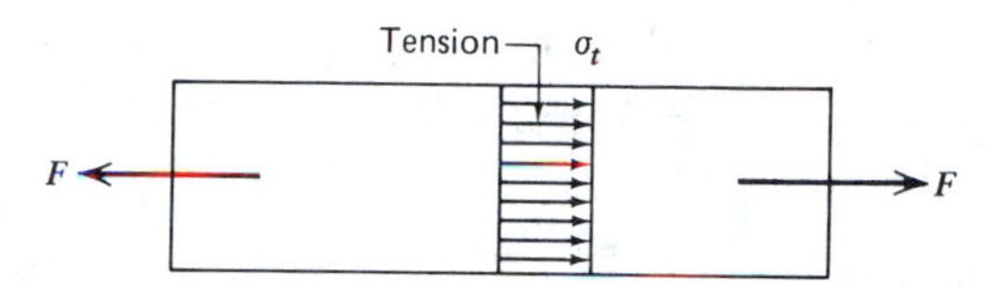

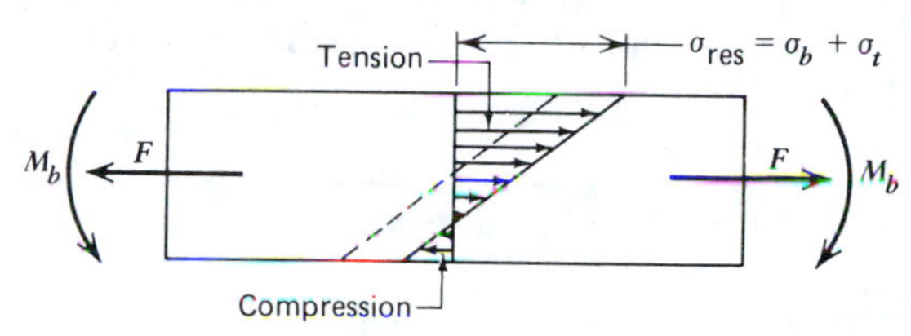

Figure B-12 Superposition of stress: compression and bending combined.

$$\sigma_{res} = \sigma_t + \sigma_b$$

$$\sigma_{res} = \frac{F}{A} + \frac{Mc}{I} \qquad \text{(B-16)}$$

For bending and compression (eccentric loading) (Fig. B-13), the resulting compressive strength is given by

$$\sigma_{res} = \sigma_c + \sigma_b$$

$$\sigma_{res} = \frac{F}{A} + \frac{Mc}{I} \qquad \text{(B-17)}$$

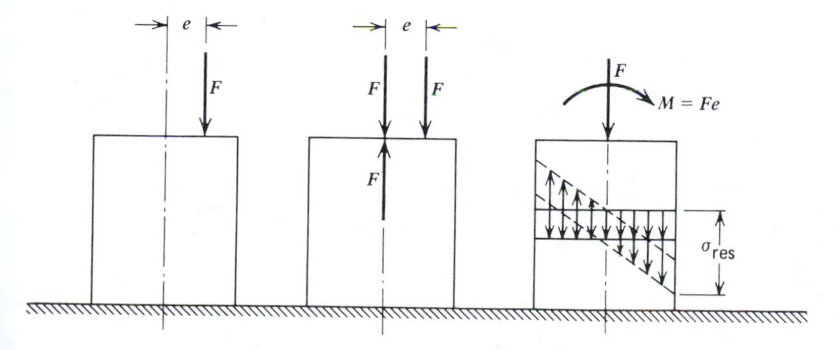

Figure B-13 Combined stress of bending and compression (eccentric loading).

which applies to simple machine frames, clamps, and the like.

Equivalent Stresses

Shafts that are an integral part of belt, chain, and gear drives are often loaded transversely and in torsion (Fig. B-14). For ductile materials, the equivalent stress is the equivalent octahedral induced normal stress.

$$\sigma_i = \sqrt{\sigma_i^2 + 3\tau_i^2} \qquad \text{(B-18)}$$

The use of this equation is discussed in Chapter 3. Useful simple stress equations are given in Fig. B-15.

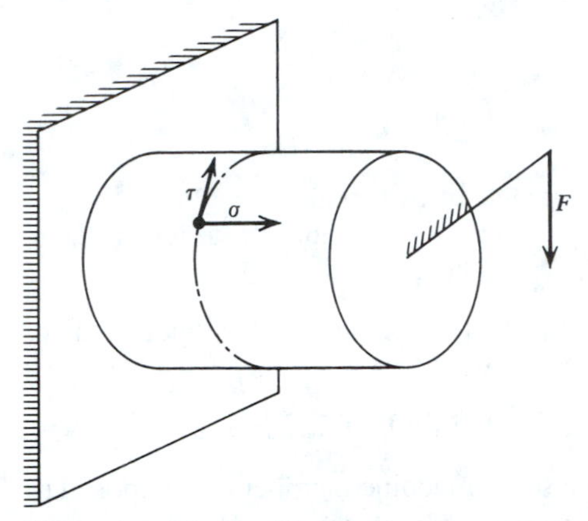

Figure B-14 Bending and shear stress combined.

Figure B-15 Simple stresses.

Type of Loading	Illustration	Type of Stress	Stress Equations
Direct Tension	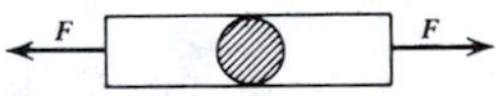	Tensile	$\sigma_t = \dfrac{F}{A}$
Direct Compression		Compressive	$\sigma_c = -\dfrac{F}{A}$
Bending	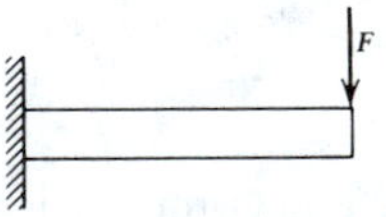	Bending	$\sigma_b = \pm\dfrac{M}{Z} = \pm\dfrac{Mc}{I}$
Direct Shear	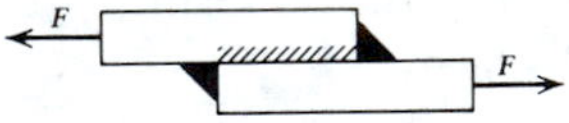	Shear	$\tau = \dfrac{F}{A}$
Direct Shear	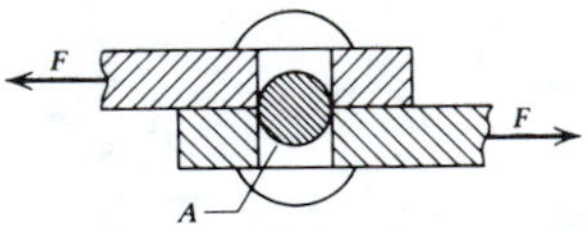	Shear	$\tau = \dfrac{F}{A}$
Torsion	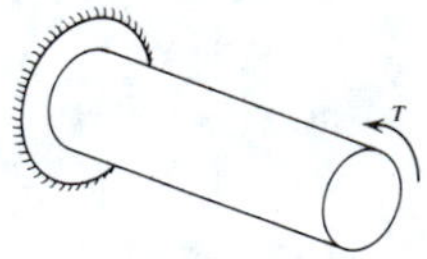	Shear	$\tau = \dfrac{T}{Z_p} = \dfrac{Tc}{J}$
Crushing or Bearing	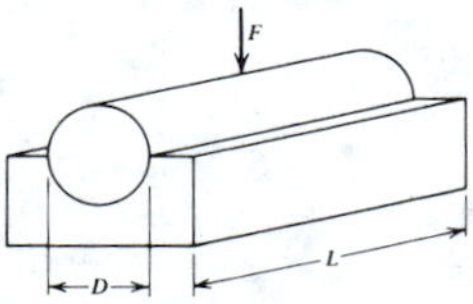	Compressive or Bearing	$\sigma_c = \dfrac{F}{LD}$

Appendix C
Auxiliary Problems

These problems are an exercise in integrated design. For their solution they require the use of information given in *several* chapters. In some cases students may seek information from *outside* sources; in other cases they must use their own judgment to add *missing* information. In this respect, the problems bear some resemblance to those given at professional engineers' examinations. *The instructor should feel free to change the problems to suit his level of teaching.* All dimensions are in millimeters unless stated otherwise.

C-1 In order to simplify low-speed drives, electric motors are built integrally with a speed reduction gear (gearhead) as, for example, in electric drills. Assume a gearhead motor runs at 1800 rpm and has an output of approximately 1.10 kW through a double (reverted) gear reduction unit of 3 : 1 ratio per step. Motor efficiency is 0.75, and each gear set has an efficiency of 0.97. Assume a life of 7000 h for the unit. Calculate (*a*) overall efficiency, (*b*) output torque, (*c*) power input; (*d*) select a suitable set of gears using the method outlined in *Machinery's Handbook;* (*e*) make a full-scale layout of the gears and shafts; (*f*) calculate the size of the jackshaft; and (*g*) select a set of ball bearings for the jackshaft.

C-2 A milling cutter is 100 mm in diameter and 150 mm wide (Fig. C-1). At a tip or peripheral speed of 1 m/s, in a certain material, it is capable of a feed of 1.25 mm per revolution while cutting to a depth of 10 mm.

In the course of machining, a material undergoes severe plastic deformation that consumes large amounts of energy. If we assume the energy consumption is 0.023 W/mm^3 and the cutter efficiency is 0.60, calculate (*a*) feed in millimeters per second, (*b*) material removed in cubic millimeters per second, and (*c*) output of drive motor.

Figure C-1

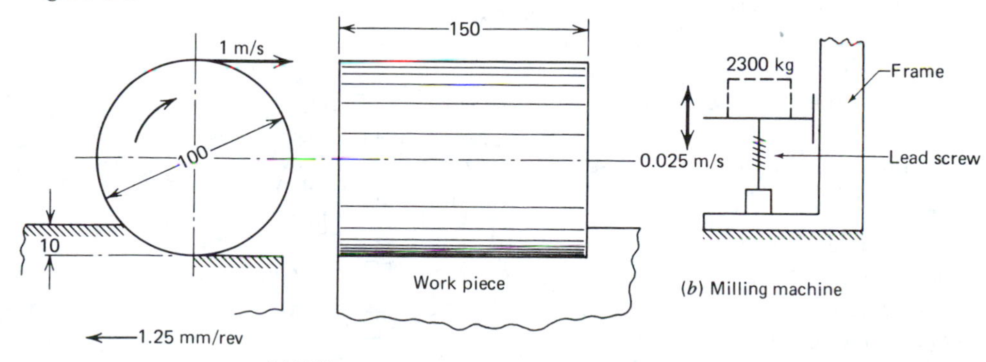

The combined mass of table, fixture, and work piece does not exceed 2300 kg. For this load the motor must have the capacity to elevate the table (rapid traverse) at a speed of 0.025 m/s. Assuming an efficiency of 0.50 for this combined lead-screw gear mechanism, calculate the necessary power. Will the drive motor suffice for this operation? Select a standard motor adequate for both operations.

C-3 An overhead crane with a projected lifting capacity of 450 kN and a travel speed of 1.25 m/s has a mass of 6800 kg. This motor-driven, bridgelike steel structure runs on two elevated rails and is supported by four steel wheels. A safe coefficient of rolling friction is 0.015. Assume an efficiency of the drive mechanism of 0.75. For full load, answer the following.

(*a*) What power is needed to maintain the required crane travel?
(*b*) If the crane is to reach a travel speed of 1.25 m/s in 7 s, what overload must the travel motor be able to develop?
(*c*) In what distance will the crane reach its projected travel speed?
(*d*) If the crane is to be brought to a stop from full speed in 10 s, how much energy must the brakes dissipate as heat and what will the stopping distance be?
(*e*) Where should the brake or brakes be located for maximum efficiency and minimum size?
(*f*) What should be the advantages and disadvantages of using:

(*1*) Two separate drive motors instead of one?
(*2*) A hydraulic drive?
(*3*) A pneumatic drive?
(*4*) Magnetic brakes?

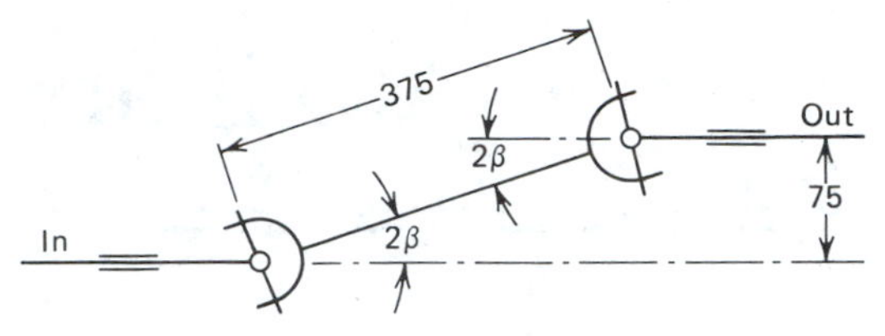

Figure C-2

(*5*) Disk brakes?
(*6*) Block brakes on the wheels?

C-4 Two parallel shafts spaced 75 mm apart are connected by means of an intermediate shaft 375 mm long (Fig. C-2). For an input speed of 375 rpm calculate the following.

(*a*) The shaft angle 2β.
(*b*) The limiting speeds of the intermediate shaft.
(*c*) Angular acceleration of the intermediate shaft.
(*d*) If the shaft speed were greatly increased, how would you expect the intermediate shaft to fail?
(*e*) How can failure be minimized?
(*f*) For a shaft speed of 500 rpm and a power input of 20 kW, estimate the size of the two parallel (input-output) shafts and the intermediate shaft.

C-5 A reciprocating, two-cylinder refrigeration compressor is designed to operate intermittently for 18 h a day at a speed of 500 rpm while consuming 7.5 kW (Fig.

Figure C-3

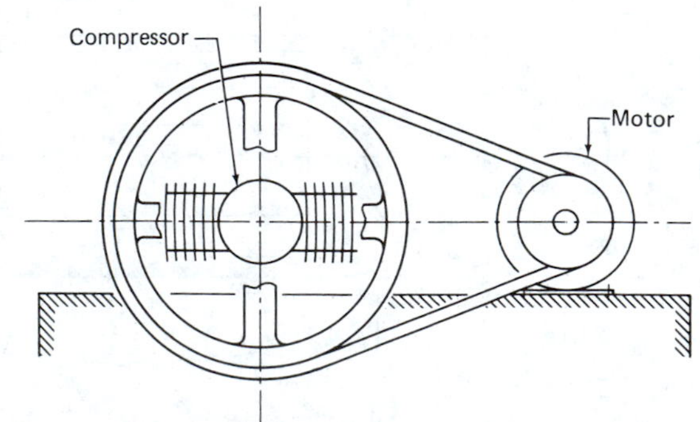

C-3). Power will be transmitted from a 1200-rpm motor to the compressor flywheel, which has a minimum diameter of 600 mm.

(*a*) Select a suitable belt drive for this compressor.
(*b*) Select a suitable chain drive.
(*c*) Design a gear reducer for this unit.
(*d*) Which drive is preferable?

C-6 The last step of a rubber belt conveyor drive is a roller chain (Fig. C-4). The drive sprocket revolves at 125 rpm and transmits 3.1 kW. The driven shaft, which also contains the conveyor belt pulley, turns at 50 rpm. The shaft center distance should be roughly 360 mm and must be inclined at an angle of 35 deg to horizontal. Operating conditions are light shock under full load for 8 h a day, five days a week. Assume a chain life of 10 000 h and deficient lubrication.

(*a*) Select a suitable chain drive.
(*b*) Calculate the transverse shaft load.
(*c*) What type of drive would you suggest using to precede the chain drive?
(*d*) Why did the assignment suggest a chain and not a belt drive for the final drive?
(*e*) How would you prevent the conveyor from "backdriving"?

Figure C-4

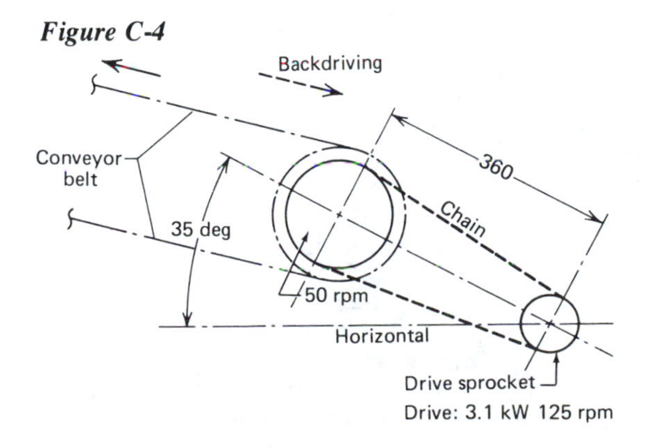

(*f*) How can the drive system be protected against overload?
(*g*) Which power source do you consider best suited for this type of conveyor?

C-7 A 7.5-kW, 1800-rpm generator is to be driven by a "low-head" water turbine with a regulated optimum speed of approximately 30 rpm.

(*a*) Discuss possible means of power transmission for vertical and horizontal shaft positions.
(*b*) For a vertical turbine shaft, design a suitable belt drive.
(*c*) For a horizontal turbine shaft, design a combined chain and belt drive.
(*d*) Can the same set of bearings be used for both designs?
(*e*) Make an assembly drawing of drive as specified by your instructor.

C-8 (*a*) The spur pinion shown in Fig. C-5 belongs to a machine tool. It transmits a torque of 180 N · m. For the following data:

$$\phi = 20 \text{ deg} \qquad a = m$$
$$N = 25 \qquad \sigma_{bw} = 110 \text{ MPa}$$
$$m_P = 1.70 \qquad F = 14 \text{ m}$$

calculate (*1*) module m, (*2*) pitch diameter d, and (*3*) outside diameter d_o.

(*b*) The pinion is mounted on a cone of length 35 mm. The end diameters are 35 and 31.5 mm. Assume a coefficient of friction between cone and gear $f = 0.10$. Calculate the axial force F_a necessary to transmit the torque of 180 N · m. The necessary axial force F_a is obtained by means of the nut and thread shown at the end of the shaft. The thread is M30, and the coefficient of friction $f_t =$ 0.15. The average diameter d_m of the nut contact surface is 36 mm, and the

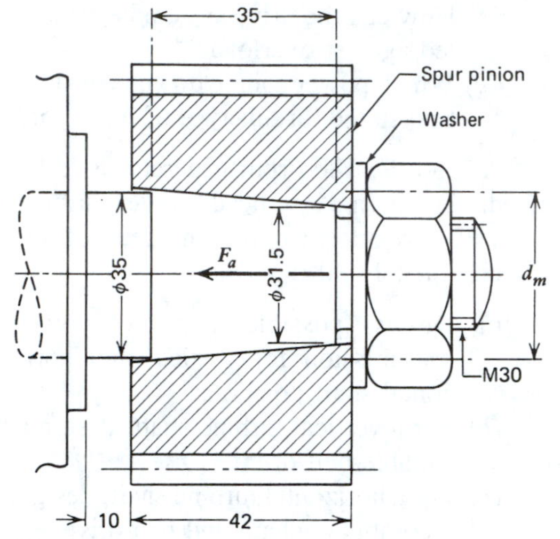

Figure C-5

coefficient of friction between nut and washer is $f_a = 0.15$.

(*c*) Calculate the tightening torque on the nut necessary to obtain F_a [the friction angle θ can be determined by $\theta = f_t / \cos(0.5\beta)$].

(*d*) Select a suitable bearing to carry this pinion (use *Machinery's Handbook*).

(*e*) What are (*1*) the maximum bending stress in this cantilever shaft, and (*2*) the maximum combined stress?

(*f*) How would you expect this shaft to fail if the load is increased enough to cause failure?

(*g*) What other types of failure could make the system inoperative?

(*h*) Replace this module based gear with a pitch based gear. Use *Machinery's Handbook* to calculate this gear in accordance with AGMA standards.

(Courtesy Odense Institute of Technology)

C-9 Figure C-6 shows a schematic outline of a motor-driven mechanical hoist. A simple reduction gear reduces motor speed to that of the drum. Rotation of the drum, in turn, moves the load either up or down. Maximum lifting capacity is 49 kN. Drum diameter is 350 mm, total

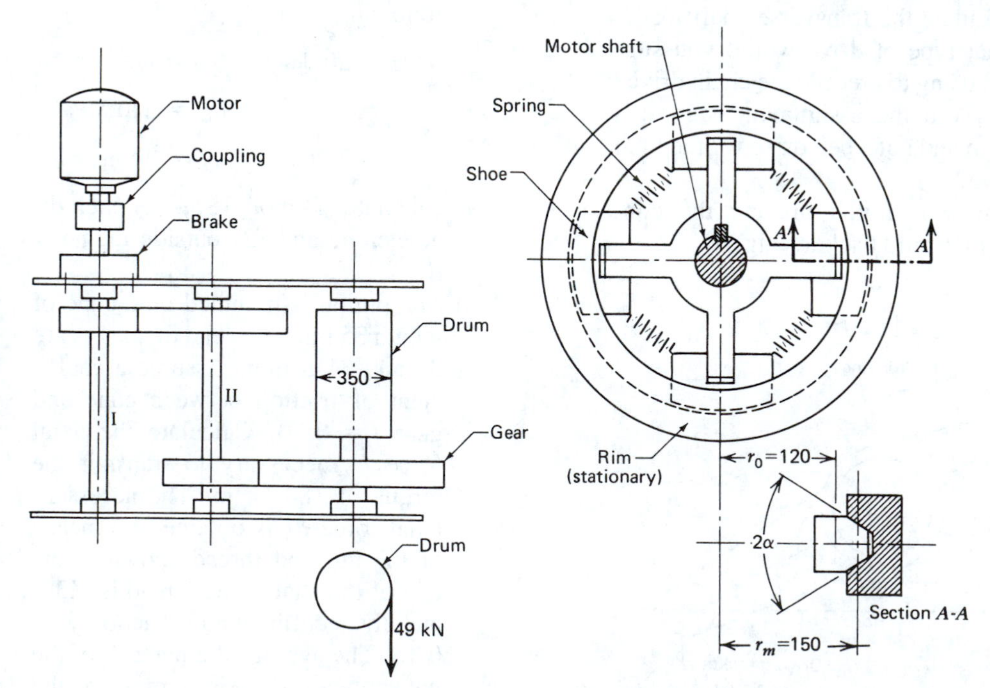

Figure C-6

efficiency 0.78, and motor torque 200 N · m.

The safety brake on the right has four "shoes" for power input, each of mass 0.13 kg, guided radially, and balanced by four springs. When the four shoes simultaneously make contact with the outer rim (output), each spring force is 320 N, and the distance r_o to the center of gravity of each shoe is 120 mm. The brake must keep the speed below 15 m/s. Use $f = 0.15$ for the brake shoe interface.

Part A

(*a*) Why is the brake placed on the motor shaft and not on the drum shaft?
(*b*) Why is the brake shoe wedge shaped?
(*c*) Specify the type of (*1*) coupling, (*2*) gearing, and (*3*) bearing most likely to be used in this arrangement. Give reasons for your answer.

Part B. Calculate:

(*d*) Total speed reduction of the two gear sets.
(*e*) Speed of motor shaft corresponding to a descent of 15 m/min.
(*f*) Braking torque needed to assure a speed of descent *not* exceeding 15 m/min.
(*g*) Wedge angle 2α of the brake shoe.
(*h*) The coil springs required.

Part C

If the instructor so desires, also calculate the gears needed and make layouts of specific parts or of the entire hoist.

(Courtesy Odense Institute of Technology)

C-10 Figure C-7 shows a platform that can be elevated and lowered mechanically while carrying a heavy load *F*. The weight of table and load is 30 kN, concentrated at the center of table shown. The platform, assumed rigid, is to be lifted by two vertical spindles at a speed $v = 5$ m/min. The spindles are turned by two sets of bevel gears for which $N_p = 24$ and $N_G = 48$. The following data are available.

Bevel gears: $e = 0.96$

Spindle bearings: $e = 0.97$

Threads: $e = \dfrac{\tan \alpha}{\tan(\theta + \alpha)}$

α = plane angle

θ = friction angle

Nut and spindle: $f = 0.05$

Service factor: k_s. Use Fig. 1-6.

Shear modulus: $G = 80\,000$ MPa

Calculate the following.

(*a*) The necessary speed of the drive shaft.
(*b*) The power necessary to elevate the table at constant speed.
(*c*) The diameter of the drive shaft between the two bevel pinions. Assume $\tau_{\text{all}} = 15$ MPa. Specify a standard size.
(*d*) The combined stress in the spindle during lifting.
(*e*) The difference in elevation of the two sides due to angular displacement of the horizontal drive shaft. Assume a length of 2500 mm.
(*f*) What sort of power source would you suggest for this platform? Give reasons for your choice. Give alternate power sources.
(*g*) What type of coupling would you recommend?
(*h*) Would it be advantageous to specify ball screws? Give reasons for your answer.
(*i*) What is the primary mode of failure of the spindles? How safe are the

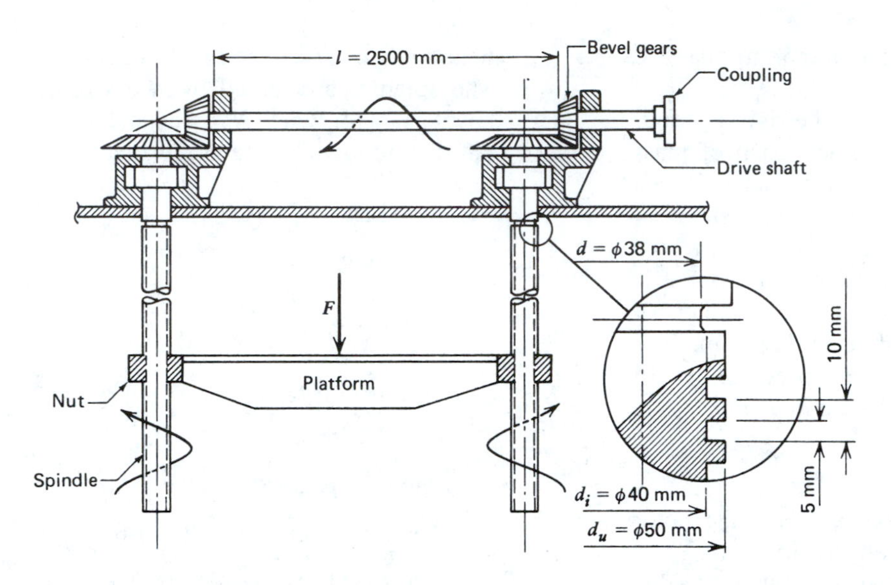

Figure C-7

spindles, assuming this mode of failure?

(*j*) Which simple machine is the basis for this lifting platform? Redesign the platform based on another and fundamentally different simple machine.

(Courtesy Odense Institute of Technology)

C-11 A typical snow blower operates on the principle of the auger. Two power-driven augers of the same diameter and length, but of opposite hand, mounted on the same horizontal shaft pull the snow in from both sides and discharge it radially with great force (Fig. C-8). The engine rated 5.5 kW at 2700 rpm has its speed reduced in two steps. First, it is reduced by 1.5 : 1 in a belt drive; next, a chain cuts speed by a factor of 2.25 : 1.

(*a*) Select a suitable belt drive.
(*b*) Specify a practical chain drive.
(*c*) Calculate the intermediary shaft.
(*d*) Select suitable bearings.
(*e*) Make suitable layouts as specified by your instructor.

Figure C-8

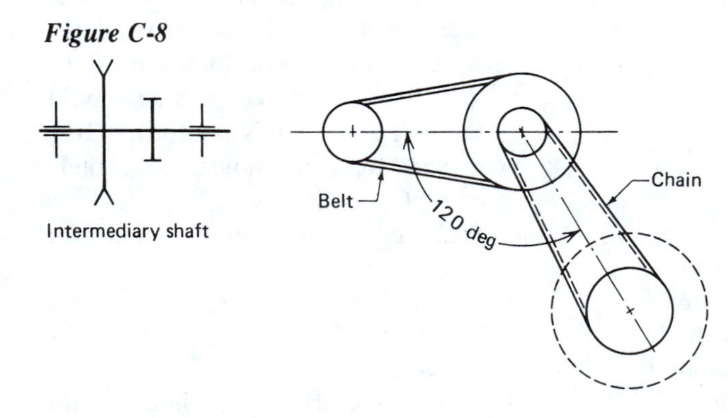

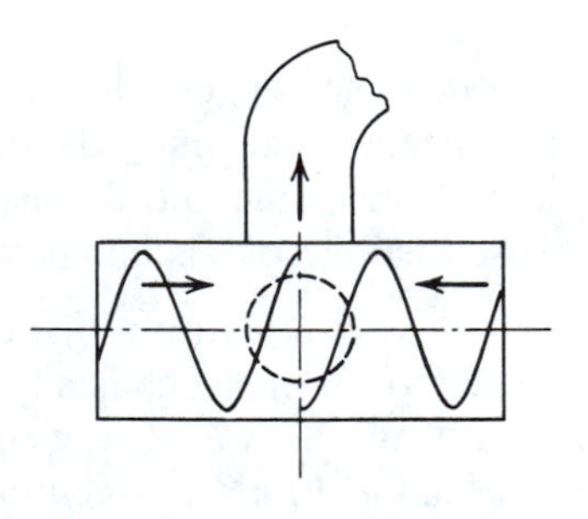

Appendix D
Design Projects

These sources are suggested for practice in applying the principles of the preceding chapters in more comprehensive and practical situations. Other sources are:

Theory and Problems of Machine Design, by A. S. Hall, A. R. Holowenko, and H. G. Laughlin. Schaum Publishing Company, 1961. (Contains 36 projects.)

Design Engineering Projects, by M. F. Spotts. Prentice-Hall, Inc., 1968. (Contains over 100 design projects geared to actual machines and devices now in production.)

Design Ideas for Weldments I & II. The James F. Lincoln Arc Welding Foundation, P. O. Box 3035, Cleveland, Ohio, 44117.

SOURCES FOR DESIGN PROJECTS (AND AUXILIARY PROBLEMS)

1. Linear actuator for aircraft control.
2. Conveyor problems.
3. Problems involving ball screws.
4. Hydraulic actuators involving cylinders, tanks, bolts, and the like.
5. Mechanical hoists.
6. Major parts of machines such as snowmobiles; snow blowers; lawn mowers; rotor tillers; street sweepers; pumps; paper machinery; food machinery elevators; oil-drilling equipment; motorcycles; racing cars; water, gas, and steam turbines; cement machinery; rock crushers; military equipment of all kinds; and machine tools.

GUIDELINES FOR DESIGN CALCULATIONS

1. The purpose of calculations is either to *confirm* or *predict* the necessary dimensions, the expected effects, or the advantage of one solution over another.

Aside from numerical errors, the main obstacle to useful results lies in the basis or prerequisites for the calculations. Even the most careful calculations are of little use when the basis for them is incorrect or when material specifications are not fulfilled. It is therefore important to provide basic quantities as accurately as possible and to specify all prerequisites accurately. This is particularly true when new conditions prevail, in which case empirical values and equations may have changed. The following questions therefore arise.

(a) Are all "known" figures applicable to the "new" conditions?
(b) Which quantities of importance appear in the equations?
(c) Does this simulate real conditions?
(d) Are there other quantities of importance not contained in the equations?
(e) What do the equations yield under boundary conditions?

Designers must, at an early stage, decide what the equations really contain and evaluate the expected results. In this case, designers should guard against self-deception. What one expects and what one gets are often two different things. At this stage, preliminary calculations and the use of models can be very helpful.

Dimensional errors are common. Designers must be careful to substitute the right numerical values in all equations. As a check, they should evaluate their size through rough calculations using figures from previous calculations or figures based on tests or experience. When extensive calculations are performed, it is wise to tabulate them and show the results graphically.

2. It may be possible to vary the properties of the system's elements or the inputs or some aspects of the environment so as to affect the properties of the system. Where this is the case, it may be possible to choose a combination of variables that yields the best system performance. Finding this combination is termed *optimization*.

Example 1. A doctoral candidate designs a series of fatigue tests for bolts but forgets to account for temperature and humidity effects. The dissertation is rejected for publication.

Example 2. The axle of a small tractor fails after 6 to 12 months in the field. The designer has erroneously assumed a unidirectional torque. Under certain conditions a reverse torque is induced, causing fatigue failure.

Note to instructor: All dimensions are in millimeters unless stated otherwise. Also, feel free to change the problems to suit your level of teaching.

D-1 A mine-shaft elevator is shown schematically in Fig. D-1. It is driven by a diesel engine through a reduction gear (not shown). The counterweight has a mass of 2000 kg. The cage has a mass of 1000 kg empty; when loaded, the total mass is not to exceed 3000 kg. The reduction drive, including the drive pulley, has an efficiency of 0.90. The driven pulleys each have an efficiency of 0.98.

The cable has a mass of 0.3 kg/m and is inelastic. The overhanging cable length, and that part only, should be included in any calculations. The overhanging length does not exceed 200 m. The hoisting speed is 5 m/s.

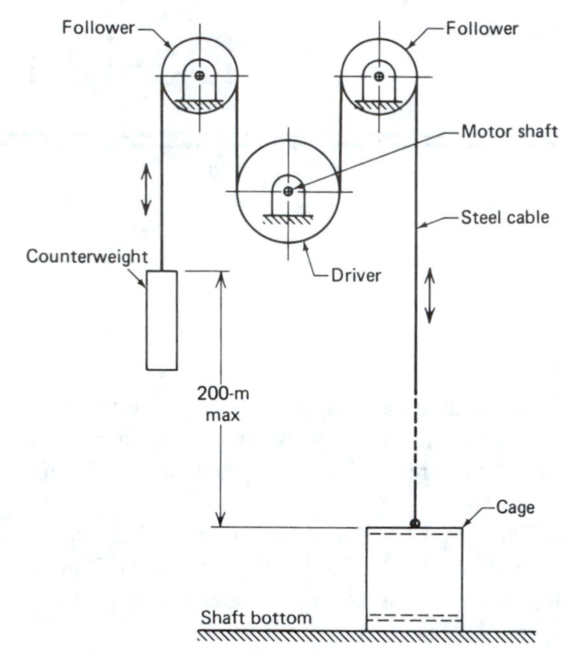

Figure D-1

(*a*) Calculate the motor power in kilowatts required for continuous operation.

(*b*) Calculate the motor power required for accelerated hoisting at 2 m/s^2.

The drive sheave has a semicircular groove with an inlay of rubber in the bottom for increased friction. The coefficient of friction between the inlay and the cable varies between 0.4 and 0.5.

(*c*) Calculate the maximum time for descent of the cage, in case the drive sheave is blocked.

(*d*) Calculate maximum cable tension created through blocking of the drive sheave during descent.

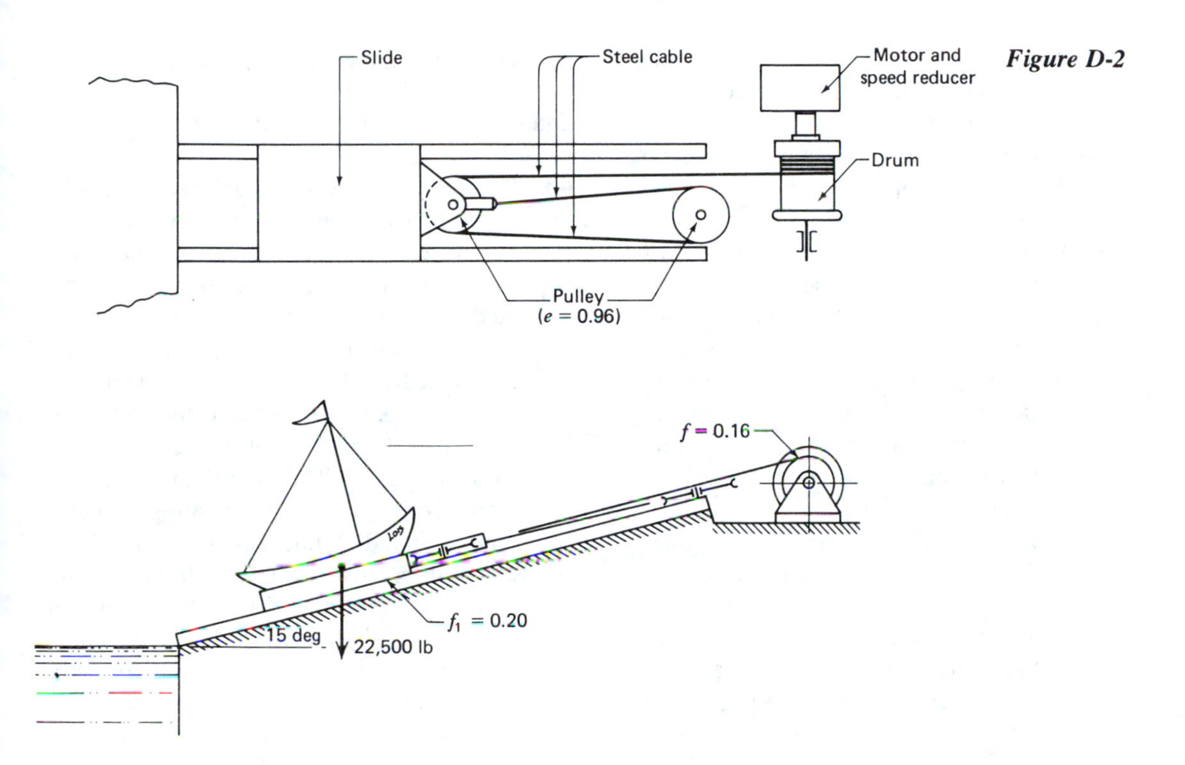

Figure D-2

(*e*) Will the cable "slip" or the motor stall if the cage is blocked in its bottom position?

(*f*) Calculate maximum cable tension in part (*e*).

(*g*) Calculate the necessary ultimate tensile strength of the cable for a safety factor of 6. Select a suitable cable from *Machinery's Handbook*.

(*h*) Make the necessary layouts from which detailed drawings can be made.

(Courtesy Technical University of Denmark)

D-2 Figure D-2 shows a motorized hoisting tackle for pulling boats up a ramp onto land. The boat is mounted on a slide with a load capacity of 22,500 lb. The incline of the ramp is 15 deg, and the maximum hoisting speed is 20 ft/min. The hoisting tackle uses steel rope. Each pulley has an efficiency of 0.96. The coefficient of friction between slide and ramp is $f_1 = 0.2$. Between rope and drum, the coefficient of friction is $f_2 = 0.16$. Disregard inertia forces in all calculations. (*Note to instructor:* This assignment can be expanded to include frames, anchors, and bolts.)

(*a*) Use data given in *Machinery's Handbook*, pp. 485–489, to select or calculate the cable size and drum diameter. Specify mean drum diameter in inches, rounded off to the nearest larger whole number.

(*b*) The rope is clamped onto the drum surface by means of a suitable clamping device. The maximum pull in the rope occurs when only one-and-a-half windings are left on the drum. Specify a suitable drum material.

(*c*) The hoisting drum is driven by an electric motor through a reduction gear with three or four sets of spur gears and an efficiency of 0.96 per gear set. What should be the speed ratio of the gear train? Calculate the design power needed and select a motor accordingly.

(*d*) Specify pulley diameters. Select a suitable steel pin for mounting each pulley.

(*e*) Calculate the approximate length of the drum for a towing length of 100 ft.

(*f*) Calculate and specify a shaft of uniform diameter for the hoisting drum.

(*g*) Specify (*1*) bearings for the hoisting drum and (*2*) one coupling to connect speed reducer and drum shaft (assume a base-mounted speed reducer and motor).

(*h*) Calculate the spur gears in the reduction gear.

(*i*) Select a speed reducer from a manufacturer's catalog.

(*j*) Estimate the size of the bolts needed to hold the drum assembly in place.

(*k*) Replace the slide with a platform supported by rollers.

(Courtesy Odense Institute of Technology)

D-3 Figure D-3 shows a draw-bench, a machine that can be designed to perform a variety of jobs, such as tube, shaft, and wire drawing. The work stroke is generated by a horizontal, centrally located, motor-driven spindle that can be retracted at a speed of 0.11 m/s while exerting a pull of 30 kN. The spindle is prevented from turning by means of a key and a suitable guide. The axial movement is therefore produced by a rotating, square, double-threaded nut engaging matching external threads on the spindle. The "nut" is also the hub of a 70-tooth gear wheel revolving on ball bearings and driven by a 20-tooth helical pinion mounted directly on the motor shaft.

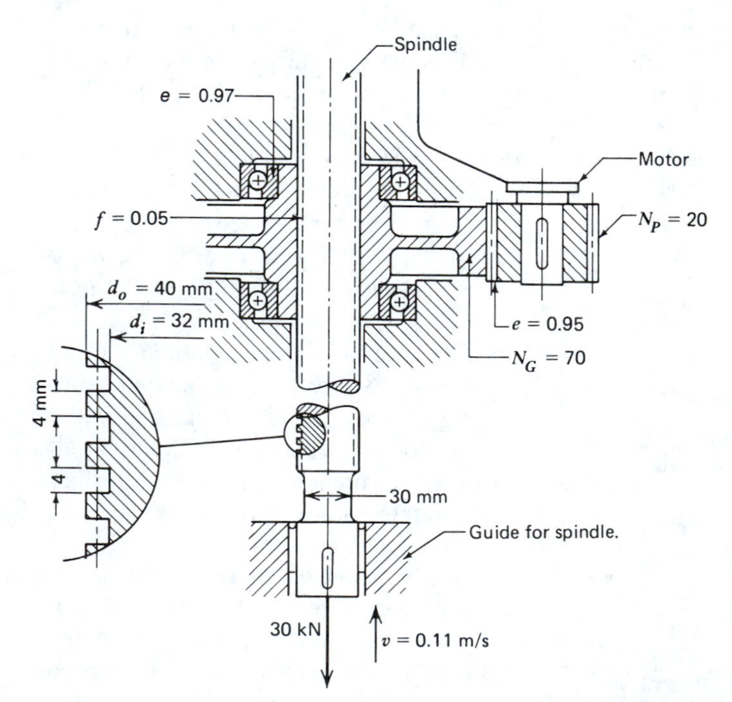

Figure D-3

Figure D-4

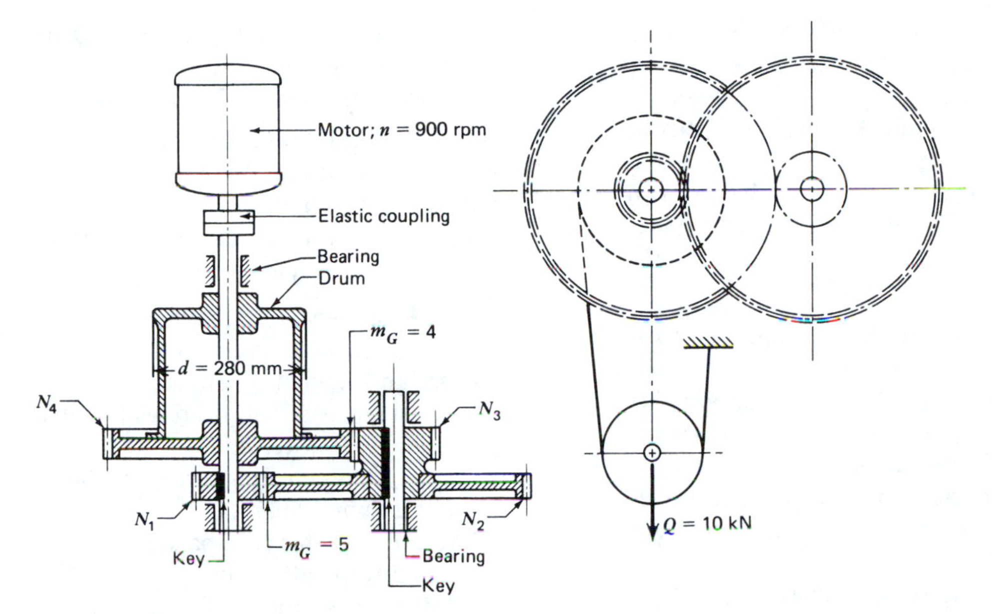

Efficiencies and coefficient of friction are as follows.

Guide:	$e = 0.98$
Gearing:	$e = 0.95$
Spindle and nut:	$f = 0.05$
Ball bearings:	$e = 0.97$

(*a*) If the efficiency of the spindle and nut combination is determined by

$$e_{sp} = \frac{\tan \lambda}{\tan(\lambda + \theta)}$$

where

λ = lead angle

θ = friction angle

estimate

(*1*) Motor power; kW
(*2*) Motor speed; rpm
(*3*) Running torque; N · m
(*4*) Motor size

(*b*) Calculate the tangential, normal, and axial forces generated by the tooth mesh. Which way should the normal force be directed? How is this achieved?

(*c*) Find the nominal stress at section *A-A*, diameter = 30 mm, of the spindle by means of

$$\sigma_n = (\sigma_t^2 + 3\tau^2)^{1/2}$$

(*d*) Select suitable ball bearings.

(*e*) Are there alternate and perhaps simpler solutions to the design of a drawbench? Name at least two.

(*f*) Draw assembly and detail drawings according to directions by your instructor.

(Courtesy Odense Institute of Technology)

D-4 Figure D-4 shows a motor-driven hoist intended to lift a burden weighing 10 kN by means of a simple drum and pulley arrangement. A compact design was obtained (*1*) by mounting the drum concentrically with an extended motor shaft, (*2*) by using a reverted gear train for speed reduction, and (*3*) by mounting the input pinion on the motor shaft extension. The first set of spur gears has a reduction ratio of 5; the second set of spur gears has a reduction ratio of 4. Gear No. 1 has 20 teeth; gear No. 3 has 25 teeth. The motor speed of 900 rpm is transferred directly to the motor shaft extension by means of a flexible coupling. The drum has a pitch diameter of 280 mm. Include service factors in all calculations but ignore efficiency. Determine the following:

(*a*) The velocity of hoisting and the power required by the motor.
(*b*) Pitch and center distance for both sets of gears (assuming only pitch-based gears are available).
(*c*) Specify the type of coupling needed. Refer to the types discussed in Chapter 10.
(*d*) Select a suitable wire rope from *Machinery's Handbook*.
(*e*) Calculate the shaft diameters needed.
(*f*) How would you secure the gears to their respective shafts? What size connectors would you use?
(*g*) The drum is bolted to one gear. Estimate the number and size of bolts needed to hold by friction, not shear.
(*h*) Select suitable bearings from a manufacturer's catalog.
(*i*) Draw assembly and detail drawings as directed by your instructor.

(Courtesy Odense Institute of Technology)

D-5 In use, the band brake shown in Fig. D-5 operates as follows. Braking takes place with constant retardation from $n = 1400$ rpm to $n = 0$ in 20 s. The braking torque is constant and equal to 300 N · m.

(*a*) Find the resulting bearing reaction during the braking operation.
(*b*) Design a suitable shaft and key.
(*c*) Select a pair of ball bearings that will endure 5000 cycles.
(*d*) What is the actuating force for a brake arm 200 mm long?
(*e*) Select a suitable steel band and design the corresponding pins.
(*f*) Assume cast iron for the brake wheel and select a suitable friction material and width of wheel.
(*g*) Replace the key with a frictional shaft connector or a spline.

(Courtesy Technical University of Denmark)

D-6 For the hoist shown in Fig. D-6, determine the following.

(*a*) Velocity of hoisting for the given motor speed.

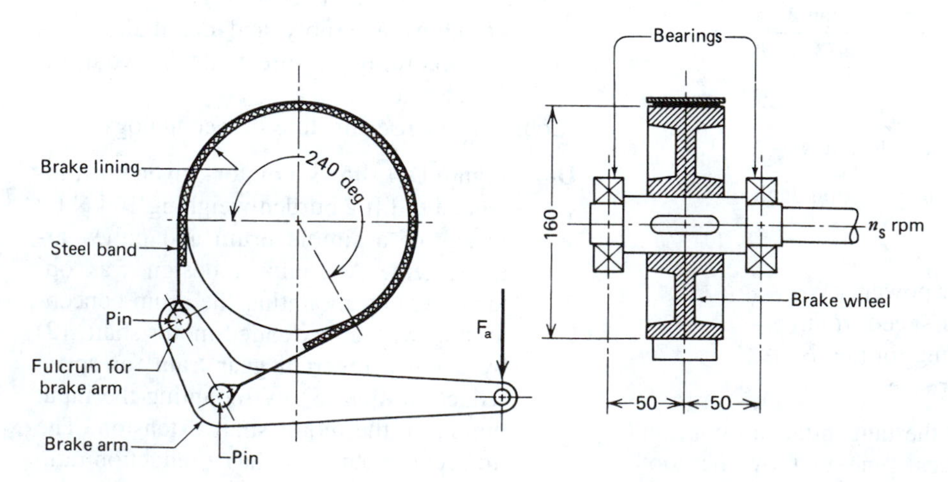

Figure D-5

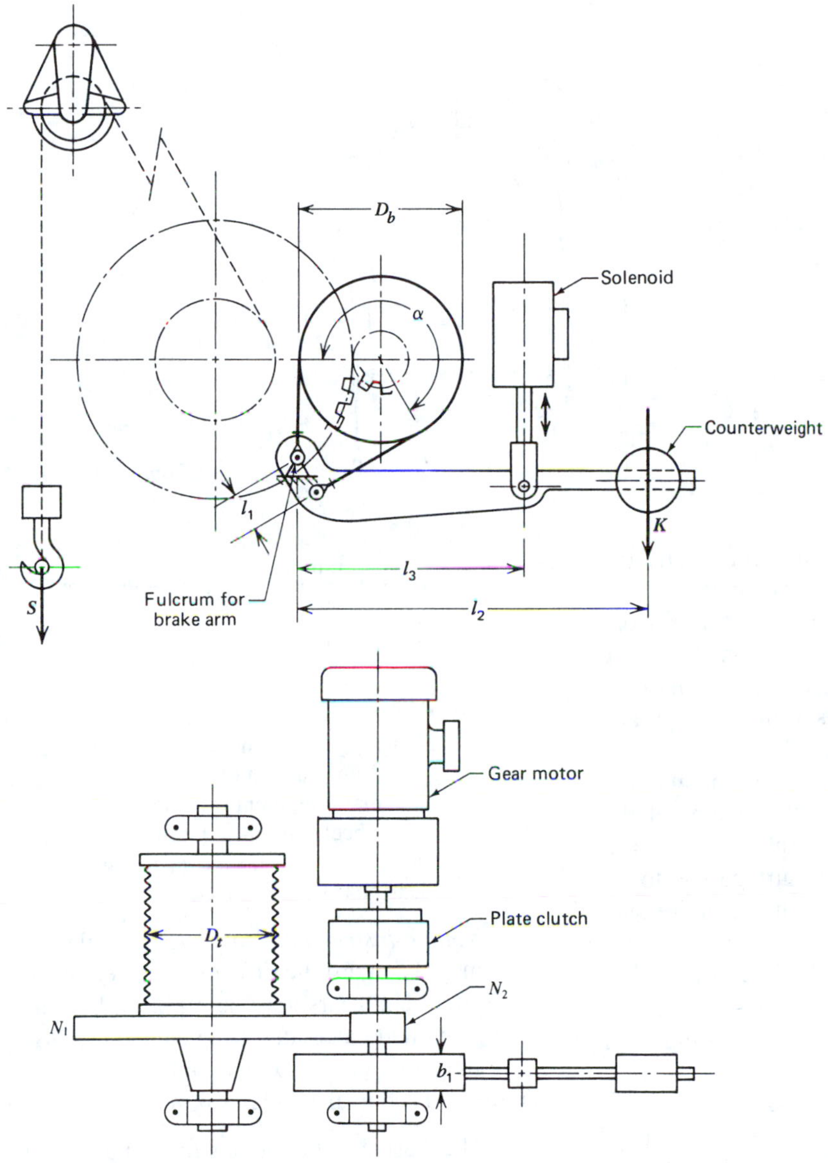

Figure D-6

(*b*) Maximum load for the given motor power when hoisting takes place at constant velocity. The efficiency is 0.98 for each of the following machine members: sheave for rope; drum with bearings; and open gear with bearings and the enclosed gearing attached to the motor.

(*c*) The load on the brake lining (in newtons per square millimeter) when the load is lowered with a constant velocity of 2 m/s through a distance of 10 m with a load of 10 kN. Lowering of the load is established by disengaging and lifting the counterweight. The efficiency is 1.0 for lowering the loads.

(*d*) The necessary brake counterweight

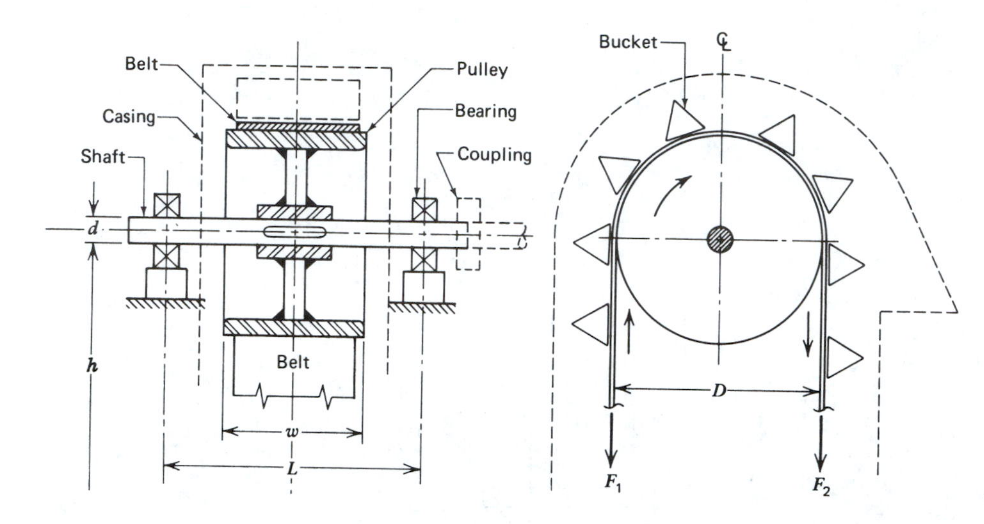

Figure D-7

K when the hoisting conditions described in part (c) take place in 2 s. Make your calculations for a worn surface, that is, steel on cast iron ($f = 0.2$). Disregard the weight of the magnet anchor as well as the brake arm.

(e) The minimum necessary height and force of lifting for the anchor of the brake magnet and brake arm when a band without wear surface has to be lifted 1.5 mm to clear the entire surface.

(f) How would you obtain an even more compact design by means of different components such as gearing and brake?

(g) Discuss bearings to be used.

(h) Select or calculate a suitable clutch. The data are as follows (Fig. D-6).

$D_b = \phi\ 350$ mm $\qquad b_1 = 100$ mm

$\alpha = 240$ deg $\qquad l_1 = 100$ mm

$l_2 = 800$ mm $\qquad l_3 = 500$ mm

$D_t = \phi\ 250$ mm

$N_1 = 105$ teeth $\qquad N_2 = 21$ teeth

$m = 5\ \dfrac{\text{mm}}{\text{tooth}}$ $\qquad F = 60$ mm

$I_1 = 0.12\ \text{kgm}^2$ $\qquad I_2 = 0.025\ \text{kgm}^2$

where

I_1 = mass moment of inertia of gear, drum, and shaft

I_2 = mass moment of inertia of brake wheel, pinion, coupling and shaft

Motor: $P = 7.5$ hp $\qquad n_s = 1400$ rpm

Gear box: $m_G = 7.4$

Note to instructor: Simplified calculations may disregard the effect of moment of inertia. The gears may be replaced by pitch gears and calculated by means of AGMA standards or the Lewis equation.

(Courtesy Technical University of Denmark)

D-7 The "drive" end of a simple bucket elevator for transporting granular material (grain, sand, etc.) is shown in Fig. D-7. An endless, oversize flat belt with buckets attached scoops up material at the bottom and discharges as it moves around its top position. Basic data are:

Capacity: 5000 kg/h

Shaft-to-shaft center distance: $h = 30$ m

Pulley speed: $v = 3.35$ m/s approx.

Pulley width: $w = 300$ mm
Pulley diameter: $D = 800$ mm
Shaft support distance: $L = 575$ mm approx.

(*a*) Design the pulley as a weldment.
(*b*) Calculate shaft and key.
(*c*) Select suitable bearings.
(*d*) Specify a suitable coupling, speed reduction unit, and motor. Alternately, the reduction unit may be designed instead of selected.
(*e*) Draw assembly and detail drawings as specified by your instructor.

Note to instructor: A variety of assignments can be obtained by varying the tonnage at increments of, for instance, 10 tons, while maintaining belt speed.

D-8 DESIGN OF A WIDE-TRACK RAILROAD

The nation's railroads have been in financial trouble for more than two decades. The reasons for this situation are well known because they have been discussed in the press and on television.

The fundamental reason, however, may be *technical* in nature. Retention of the present track width of 4 ft, $8\frac{1}{2}$ in. (less than most trucks) lends an instability to rail cars and locomotives out of proportion to the loads they carry and much greater than that of vehicles. This *inherent* limitation on stability and size may have prevented the railroads from increasing speed, capacity, economy, and safety and thereby counteracting the ever growing competition from vehicle, motor, and air transportation.

The objective is therefore to study the overall effect of a much wider track, 5 to 6 m (18 to 20 ft). This size track would provide a width comparable to jumbo jets with eight passenger seats across and a load capacity six to eight times that of present freight cars. Flat cars would be four times greater, facilitating piggy-backs and automatic loading and unloading. A new type of locomotive should be designed.

Is a wide-track train a radical new idea? Not at all! The following excerpt from *Strange Stories, Amazing Facts,* published by the Reader's Digest Association, Inc., testifies to the soundness of the wide-track concept.

On the Right Lines

"The railway track that showed the way to safety"

Right from its opening in 1841, the Great Western Railway was different from other British railways—thanks to its brilliant, but unorthodox, engineer Isambard Kingdom Brunel (1806–1859).

While other railway companies laid their lines with the rails 4 ft, $8\frac{1}{2}$ in. apart, Brunel had other ideas. He set the Great Western's tracks *seven* feet apart, wide enough for engines and carriages to be *slung between the wheels,* giving them much greater stability. This *broad gauge track* served the company well, and for half a century it ran expresses at top speed without a hitch.

Even when mishaps did occur the broad gauge proved its worth. In 1847 a driving wheel on the Exeter to London express collapsed at top speed on an open viaduct near Southall, yet not a single wheel jumped the rails.

Lack of foresight among railroad builders, disregard for safety, and lack of compatibility with domestic and continental railroads were probably the main reasons for abandoning the wide track after 50 years of flawless service. One can only speculate as to how many lives might have been saved and how much material damage avoided if the British had adopted this superior design. Equally strange is the fact that American railroad builders, for all their great visions of spanning a continent with steel tentacles, apparently never considered a much broader gauge track that would have permitted greater speeds with greater safety at moderate, if any, additional cost.

Not until 1964 was the wide-track concept considered for rail transportation in the United States. The innovator was the General American Transportation Corporation of Chicago, Illinois. The concept, termed RRollway, was planned for the Chicago-St. Louis route.

Resistance to motion in trains and vehicles was discussed in Chapter 4, Section 4-9. *The key factor is the coefficient of rolling friction f_r*, which has the following values.

Railroad cars: $f_r = 0.0015 - 0.0035$

Vehicles on asphalt and concrete:

$$f_r = 0.015 - 0.025$$

For *trucks, the main competitor to railroads, resistance to rolling is almost 10 times greater than that of rail cars.* In a world faced with energy conservation this fundamental fact, if applied to transportation *assures huge energy savings in national transportation.*

Presently, neither cars nor aircraft are really energy efficient. Only trains and water transportation are. The wide-track design would add capacity and speed to the rail concept.

Figure D-8 shows the comparative size of standard and Wide Track Train (WTT) rail cars. Proper relative size makes the two systems compatible. The large-size car can accommodate automated loading and unloading devices.

Figure D-9 shows the WTT flat car accommodating four truck trailers riding "piggyback."

Figure D-10 shows major WTT lines in the United States. Most of these will make large-capacity means of transportation available to areas not served by inland waterways.

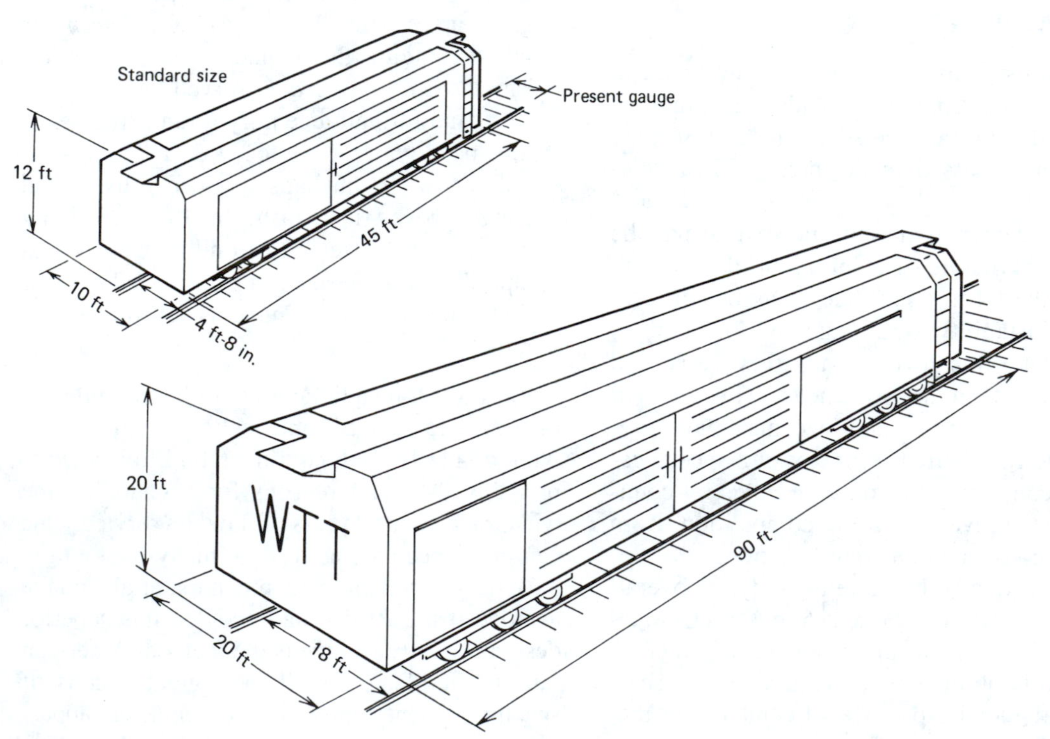

Figure D-8 Comparison of WTT with a standard-size railcar.

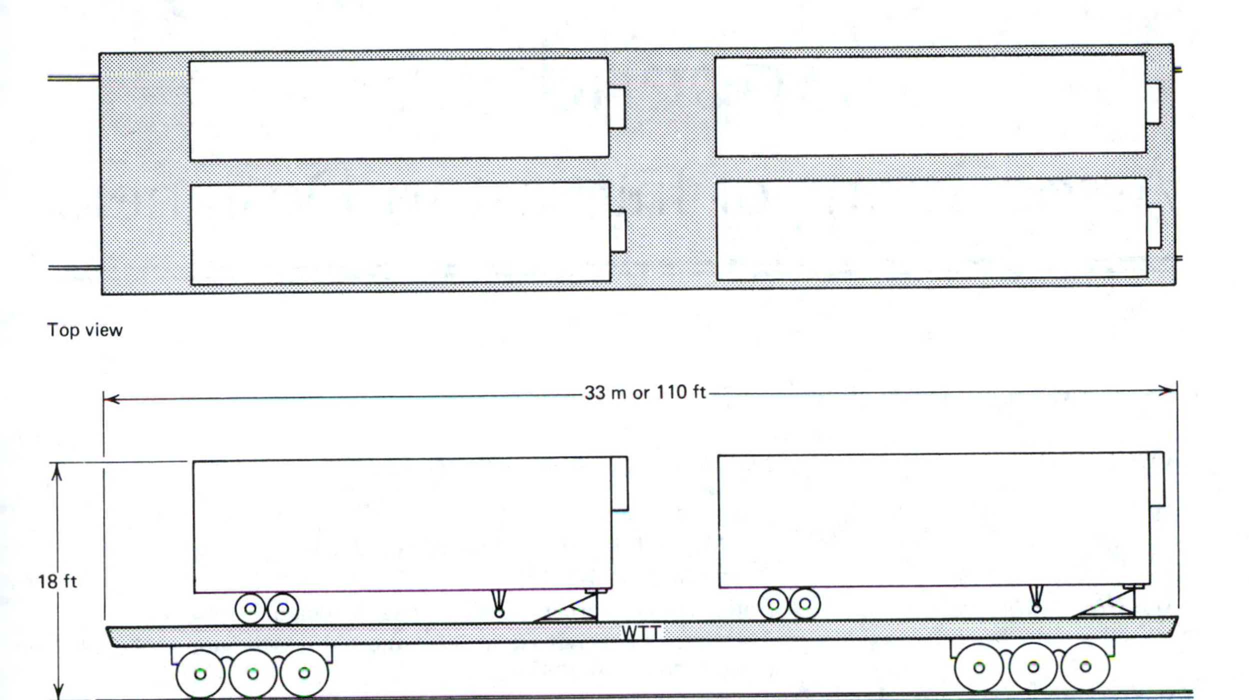

Figure D-9 One WTT flat car will accommodate four trailers riding "piggyback."

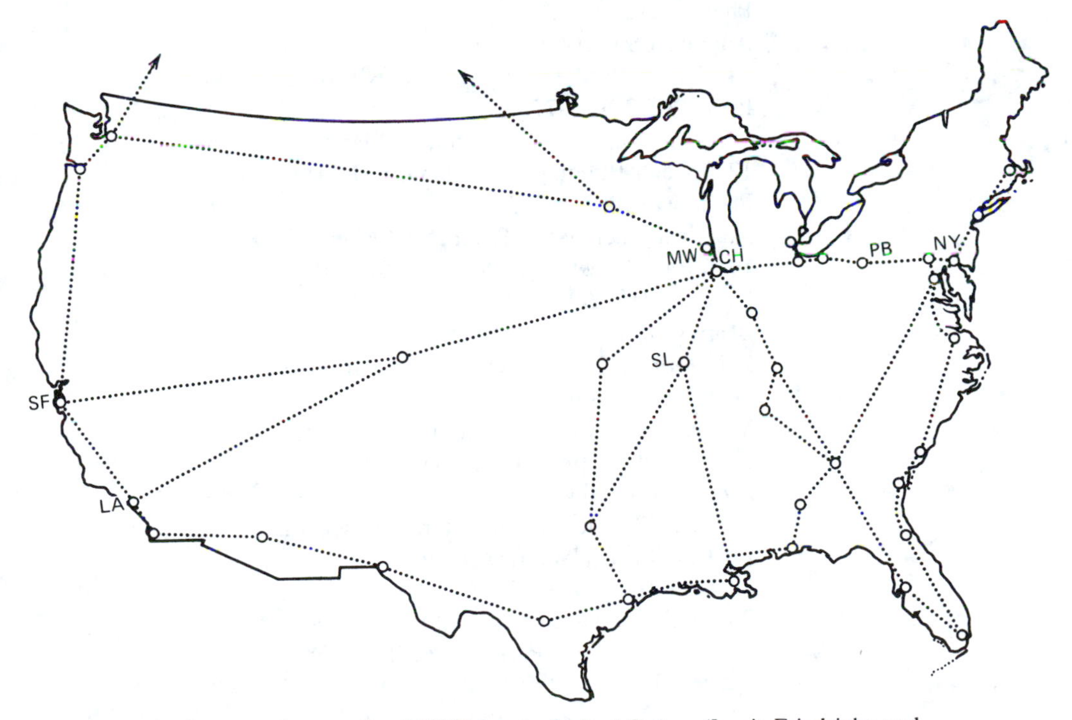

Figure D-10 Suggested network of WTT for the United States. (Louis Friedrichs and Richard Bachta.)

Appendix E
Associations of Interest to Designers

LISTED ALPHABETICALLY BY ABBREVIATION

Abbreviation	*Full Name and Headquarters Location*
ACA	American Chain Association 160 Meredith Drive, Englewood, FL 33533 Phone: (813) 474-7013
AFBMA	Anti-Friction Bearing Manufacturers Association 2341 Jefferson Davis Highway, Suite 1015, Arlington, VA 22202 Phone: (703) 979-1261
AGMA	American Gear Manufacturers Association 1901 Ft. Myer Drive, Suite 1000, Arlington, VA 22209 Phone: (703) 525-6000
ANSI	American National Standards Institute, Inc. 1430 Broadway, New York, NY 10017 Phone: (212) 354-3300
ASME	American Society of Mechanical Engineers 345 East 47th Street, New York, NY 10017 Phone: (212) 644-7722
ASMMA	American Supply and Machinery Manufacturers Association 1230 Keith Building, Cleveland, OH 44115 Phone: (216) 241-7333
ASTM	American Society for Testing and Materials 1916 Race Street, Philadelphia, PA 19103 Phone: (215) 299-5400
BSA	Bearing Specialists Association 221 North LaSalle Street, Chicago, IL 60601 Phone: (312) 346-1862
MPTA	Mechanical Power Transmission Association 1717 Howard Street, Evanston, IL 60202 Phone: (312) 869-6983
NEMA	National Electrical Manufacturers Association 2101 L Street, NW, Washington, DC 20037 Phone: (202) 457-8452
PTDA	Power Transmission Distributors Association 100 Higgins Road, Park Ridge, IL 60068 Phone: (312) 852-2000
SAE	Society of Automotive Engineers, Inc. 400 Commonwealth Drive, Warrendale, PA 15096 Phone: (412) 776-4841

Appendix F
Tables and Graphs

TABLE F-1 Conversion Factors[a]

Angle	1 rad = 57.296 deg
Length	1 in. = 25.4 mm = 2.54 cm 1 ft = 0.3048 m
Area	1 in.^{2} = 645.16 mm^2 = 6.4516 cm^2 1 cm^2 = 0.155 in.^{2}
Volume	1 in.^{3} = 16.387 cm^3 1 cm^3 = 0.061024 in.^{3}
Mass	1 lbm = 0.45359 kg 1 kg = 2.2046 lbm
Velocity	1 ft/s = 0.3048 m/s 1 m/s = 3.2808 ft/s 1 mph = 1.6093 km/h
Acceleration	1 ft/s^2 = 0.3048 m/s^2
Standard gravity	g = 386.09 in./s^2 = 32.174 ft/s^2 = 9.8066 m/s^2
Force	1 lbf = 4.4482 N 1 N = 0.22481 lbf
Pressure and stress	1 Pa = 1 N/m^2 1 N/mm^2 = 1 MPa 1 lbf/in.^{2} (psi) = (6.8948)(10^{-3}) N/mm^2 (MPa) 1 N/mm^2 (MPa) = 145.04 psi 1 atm = 0.10133 N/mm^2
Work and energy	1 J = 1 N · m 1 Btu = 778.17 ft-lbf = 1055 J 1 ft-lbf = 1.3558 J 1 J = 0.73757 ft-lbf 1 J = (9.4782)(10^{-4}) Btu
Power	1 W = 1 J/s 1 hp = 550 ft-lbf/s = 33,000 ft-lbf/min 1 kW = 1.34 hp 1 hp = 0.7457 kW
Bending moment or torque	1 lbf-in. = 0.11298 N · m 1 lbf-ft = 1.3558 N · m 1 N · m = 8.8507 lbf-in. 1 N · m = 0.73757 lbf-ft
Temperature	$t_C = (t_F - 32)/1.8$ $t_F = 1.8t_C + 32$ $t_K = t_C + 273.15$

[a]To five significant figures.

TABLE F-2 Section Properties for Bending and Torsion

Type of Section	Properties for Bending: Moment of Inertia	Properties for Bending: Section Modulus	Properties for Torsion: Moment of Inertia or Torsion Constant*	Properties for Torsion: Section Modulus or Stress Formula*
1. Solid circle	$I_x = I_y = \frac{\pi d^4}{64}$	$Z_x = Z_y = \frac{\pi d^3}{32}$	$J_p = \frac{\pi d^4}{32}$	$Z_p = \frac{\pi d^3}{16}$
2. Hollow circle	$I_x = I_y = \frac{\pi(d_o^4 - d_i^4)}{64}$	$Z_x = Z_y = \frac{\pi(d_o^4 - d_i^4)}{32d_o}$	$J_p = \frac{\pi(d_o^4 - d_i^4)}{32}$	$Z_p = \frac{\pi(d_o^4 - d_i^4)}{16d_o}$
3. Square	$I_x = I_y = \frac{a^4}{12}$	$Z_x = Z_y = \frac{a^3}{6}$	$K = 0.1406\, a^4$	Max $\tau = \frac{T}{0.208a^3}$ Maximum stress occurs at the midpoint of each side
4. Square 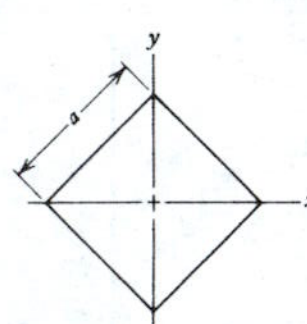	$I_x = I_y = \frac{a^4}{12}$	$Z_x = Z_y = \frac{a^3}{6\sqrt{2}}$	Same as for Case 3	Same as for Case 3

5. Rectangle	$I_x = \frac{bh^3}{12}$ $I_y = \frac{hb^3}{12}$	$Z_x = \frac{bh^2}{6}$ $Z_y = \frac{hb^2}{6}$	$K = \frac{bh^3}{16}\left[\frac{16}{3} - 3.36\frac{b}{a}\left(1 - \frac{b^4}{12a^4}\right)\right]$	Max $\tau = \frac{2T(1.5b + 0.9h)}{b^2h^2}$
6. Hollow Rectangle	$I_x = \frac{BH^3 - bh^3}{12}$ $I_y = \frac{HB^3 - hb^3}{12}$	$Z_x = \frac{BH^3 - bh^3}{6H}$ $Z_y = \frac{HB^3 - hb^3}{6B}$	$K = \frac{2tt_1(a - t)^2(b - t)^2}{at + bt_1 - t^2 - t_1^2}$ When $t = t_1$ $K = \frac{2t(a - t)^2(b - t)^2}{a + b - 2t}$	At midpoint of short sides $\tau = \frac{T}{2t(a - t)(b - t_1)}$ At midpoint of long sides $\tau = \frac{T}{2t_1(a - t)(b - t_1)}$ The stress will be greater at inner corners unless fillets are fairly large

TABLE F-3 Formulas for Bending Moment and Deflection of Beams

Loading, Support, and Reference Number	Maximum Bending Moment	Maximum Deflection of Uniform Beam
1. Cantilever, end load 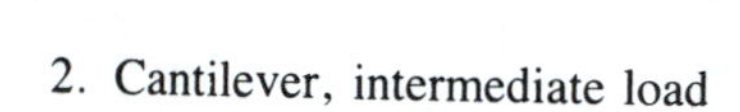	$M = -Wl$ at B	$y = \dfrac{-Wl^3}{3EI}$ at A
2. Cantilever, intermediate load 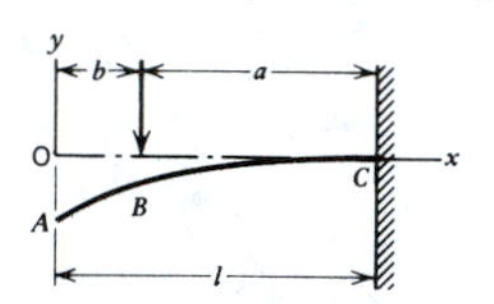	$M = -Wa$ at C	$y = \dfrac{-W(3a^2l - a^3)}{6EI}$ at A
3. Cantilever, end moment 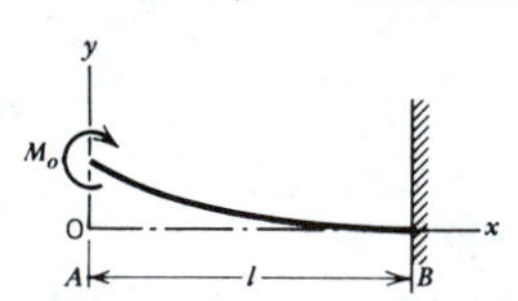	$M = M_o$ from A to B	$y = \dfrac{M_o l^2}{2EI}$ at A
4. Cantilever, uniform load 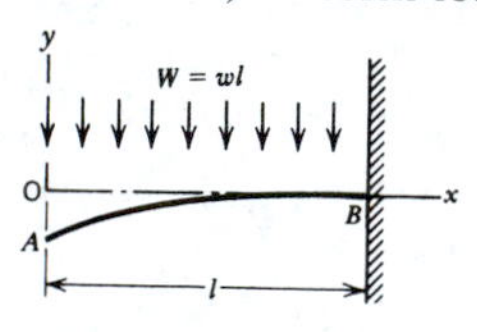	$M = \dfrac{-Wl}{2}$ at B	$y = \dfrac{-Wl^3}{8EI}$ at A
5. End supports, center load 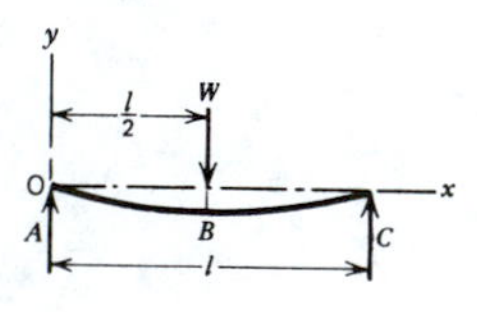	$M = \dfrac{Wl}{4}$ at B	$y = \dfrac{-Wl^3}{48EI}$ at B
6. End supports, intermediate load 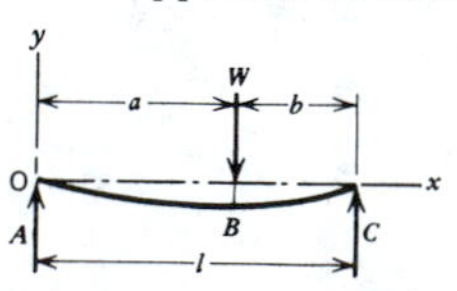	$M = \dfrac{W\,ab}{l}$ at B	$y = \dfrac{-W\,ab(a+2b)[3a(a+2b)]^{1/2}}{27EIl}$ at $x = \left[\dfrac{a(a+2b)}{3}\right]^{1/2}$ when $a > b$

TABLE F-3 Formulas for Bending Moment and Deflection of Beams

Loading, Support, and Reference Number	Maximum Bending Moment	Maximum Deflection of Uniform Beam
7. End supports, two symmetrical loads	$M = Wa$ from B to C	$y = \dfrac{-Wa(3l^2 - 4a^2)}{24EI}$ at $x = \dfrac{l}{2}$
8. Equal overhangs, and loads	$M = -Wc$	$y = \dfrac{-Wc^2(2c + 3d)}{6EI}$ at A and D
9. End supports, uniform load 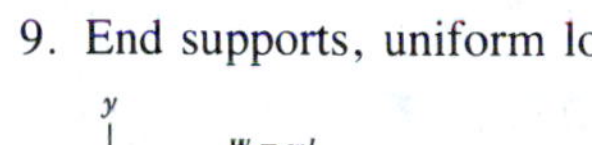 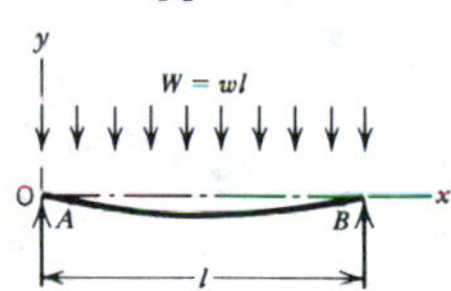	$M = \dfrac{Wl}{8}$ at $x = \dfrac{l}{2}$	$y = \dfrac{-5Wl^3}{384EI}$ at $x = \dfrac{l}{2}$

TABLE F-4 Properties of Metals and Alloys at Room Temperature

Metal or alloy	Density[a] g/mm³	Density[a] lb/in.³	E (tension) N/mm²	E (tension) psi	G N/mm²	G psi	α mm/mm °C	α in./in. °F
	$\times 10^{-3}$		$\times 10^{3}$	$\times 10^{6}$	$\times 10^{3}$	$\times 10^{6}$	$\times 10^{-6}$	$\times 10^{-6}$
Aluminum, pure	2.70	0.0975	62.1	9.0			22.21	12.34
Aluminum alloys, cast	2.57–2.96	0.093–0.107	65.5–80.0	9.5–11.6	24.5–30.0	3.55–4.35	20.7–25.6	11.5–14.2
Aluminum alloys, wrought	2.63–2.82	0.095–0.102	69.0–78.6	10.0–11.4	25.9–29.7	3.75–4.30	19.8–24.3	11.0–13.5
Beryllium copper	8.19–8.25	0.296–0.298	117.2–131.0	17.0–19.0	44.8–50.3	6.5–7.3	16.74	9.30
Brass	8.41–8.89	0.304–0.321	96.6–117.2	14.0–17.0	36.6–44.1	5.3–6.4	17.23	9.57
Bronze	8.41–8.75	0.304–0.316	96.6–124.1	14.0–18.0	36.6–46.2	5.3–6.7	17.75	9.86
Copper, pure	8.97	0.324	117.2	17.0			15.97	8.87
Iron, gray cast, No. 20	6.92–7.75	0.25–0.28	66.2–96.6	9.6–14.0	26.9–38.6	3.9–5.6		
Iron, gray cast, No. 30	6.92–7.75	0.25–0.28	89.7–113.1	13.0–16.4	35.9–45.5	5.2–6.6		
Iron, gray cast, No. 40	6.92–7.75	0.25–0.28	110.3–137.9	16.0–20.0	44.1–53.8	6.4–7.8	10.01	5.56
Iron, gray cast, No. 50	6.92–7.75	0.25–0.28	129.7–157.2	18.8–22.8	49.7–55.2	7.2–8.0		
Iron, gray cast, No. 60	6.92–7.75	0.25–0.28	140.7–162.1	20.4–23.5	53.8–58.6	7.8–8.5		
Iron, cast, ferritic malleable	7.20–7.45	0.260–0.269	151.7–172.4	22.0–25.0	65.5–75.9	9.5–11.0		
Iron, cast, pearlite malleable	7.20–7.45	0.260–0.269	175.9–193.1	25.5–28.0	66.9–69.0	9.7–10.0		
Iron, cast, nodular	7.20	0.26	144.8–172.4	21.0–25.0	65.5–68.3	9.5–9.9	11.52	6.40
Iron, pure	7.88	0.2845	193.1–200.0	28.0–29.0	78.9–81.7	11.44–11.84	11.70	6.50
Iron, sintered	4.51–7.50	0.163–0.271	48.3–206.9	7.0–30.0				
Magnesium, pure	1.74	0.0628	39.8–43.0	5.77–6.24			25.74	14.30
Magnesium alloys, cast	1.80–1.86	0.065–0.067	44.8	6.5	16.6–17.2	2.4–2.5	25.92	14.40
Magnesium alloys, wrought	1.74–1.83	0.063–0.066	44.8	6.5	16.6	2.4	25.92	14.40
Monel, wrought	8.47–8.55	0.306–0.309	179.3	26.0	65.5	9.5	13.99	7.77
Nickel, pure	8.90	0.3216	206.9	30.0			12.51	6.95
Steel, cast	7.83	0.283	206.9	30.0	77.2	11.2		
Steel, wrought, alloy	7.83–8.00	0.283–0.289	196.6–211.0	28.5–30.6	77.9–84.1	11.3–12.2	11.45	6.36
Steel, wrought, carbon	7.83–7.86	0.283–0.284	194.5–209.7	28.2–30.4	74.5–82.8	10.8–12.0		
Steel, wrought, stainless	7.47–8.03	0.27–0.29	193.1–200.0	28.0–29.0	80.7–86.2	11.7–12.5	18.00	10.00
Titanium, commercially pure	4.51	0.163	106.9	15.5	44.8	6.5	7.13	3.96
Titanium, wrought, alloy	4.43–4.73	0.160–0.171	103.4–120.7	15.0–17.5	42.8–46.2	6.2–6.7		
Zinc, pure	7.14	0.258	82.8	12.0			26.46	14.70
Zinc alloys (for die castings)	6.64	0.24	62.1[b]	9.0[b]				

[a]Density in the SI system is in terms of mass per unit volume. The values for density in the English system of measurements are in terms of weight per unit volume. The English system values are specific weight, not density.

[b]Value for instantaneous loading at 45.5 N/mm² (6600 psi) stress, the approximate fatigue limit. Because of creep, the apparent modulus of elasticity may be much less when loaded over an extended period of time.

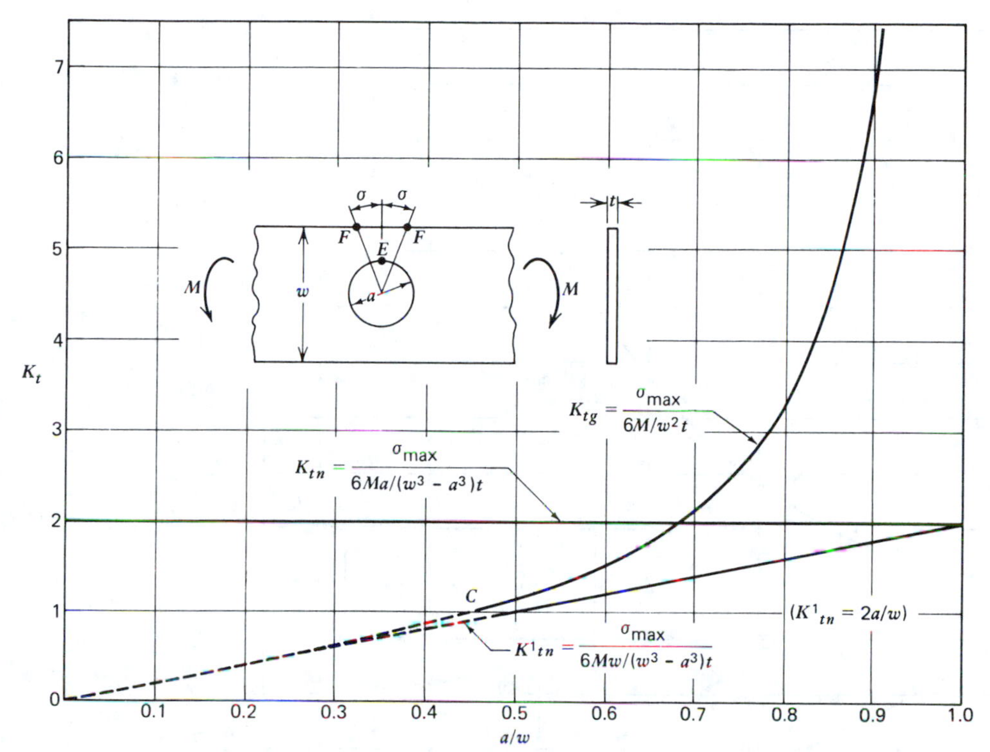

Figure F-1 Stress concentration factor for bending of a thin rectangular beam with a central circular hole.

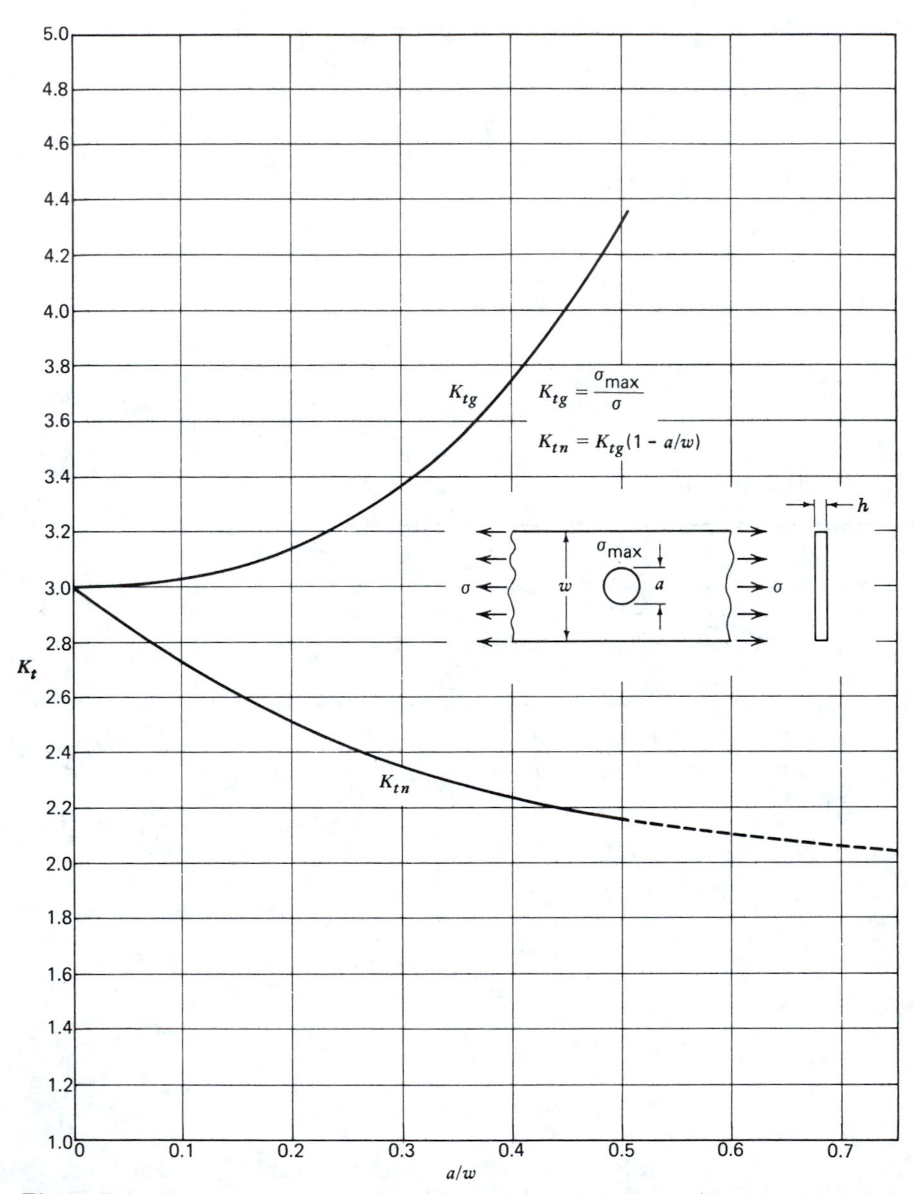

Figure F-2 Stress concentration factor for a finite-width plate with a circular hole in direct tension.

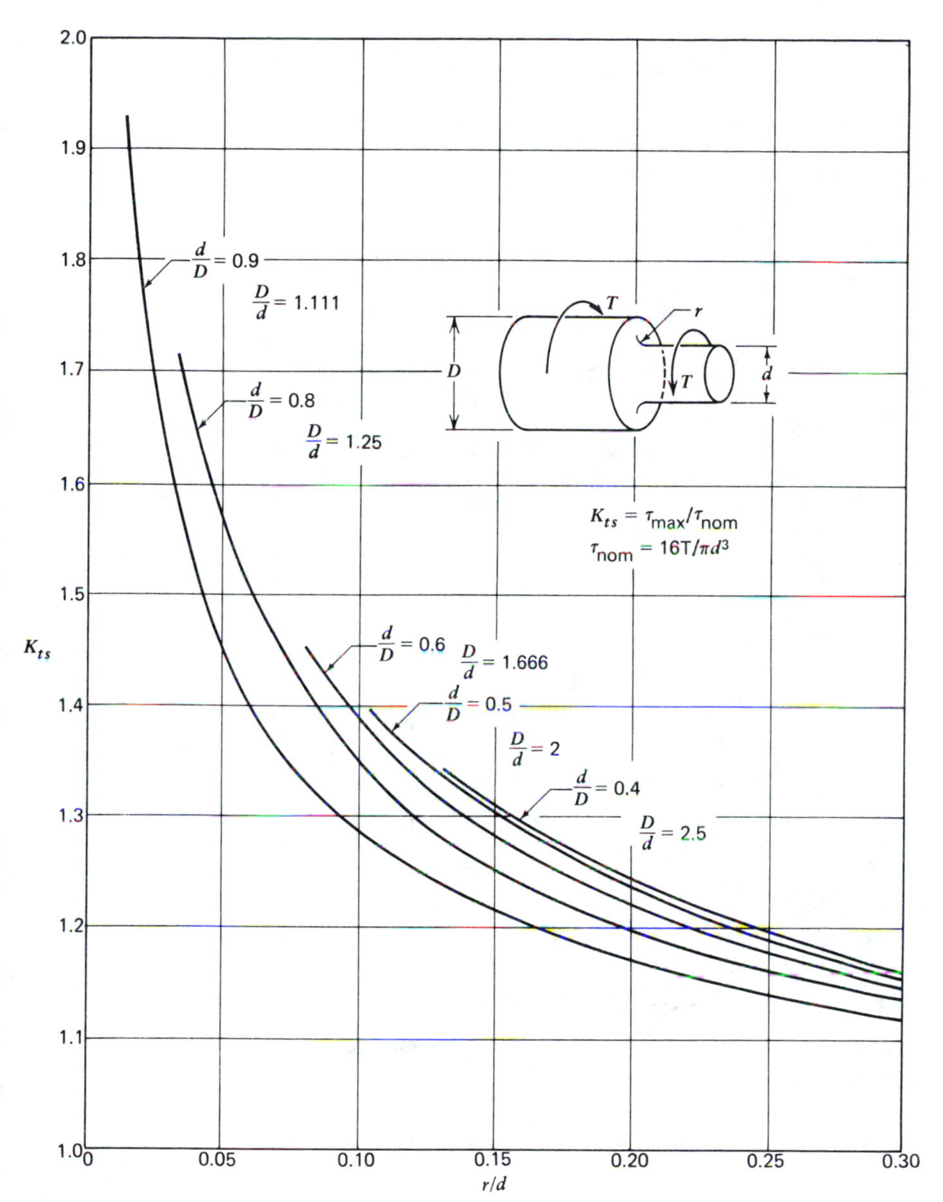

Figure F-3 Stress concentration factor for a stepped round bar in torsion.

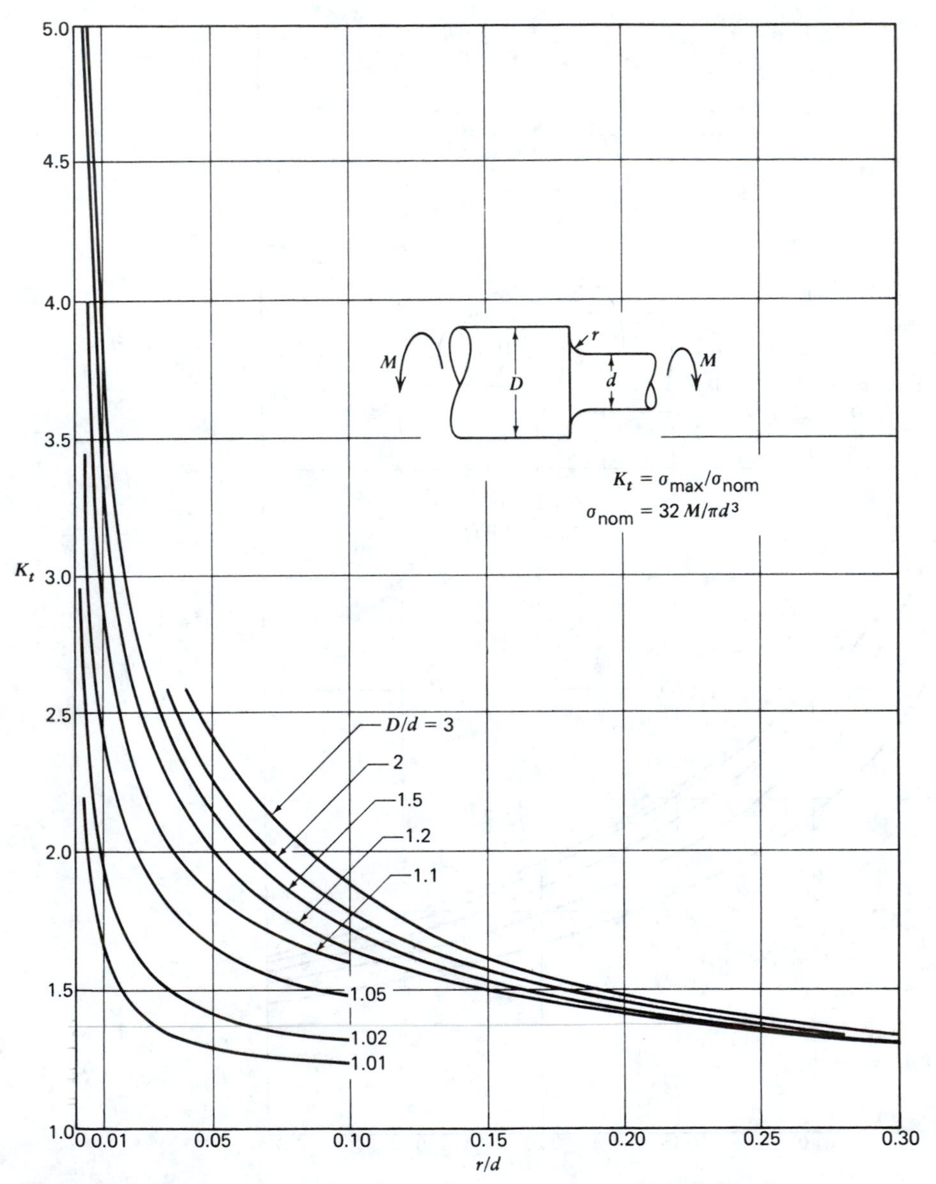

Figure F-4 Stress concentration factor for a stepped round bar in bending.

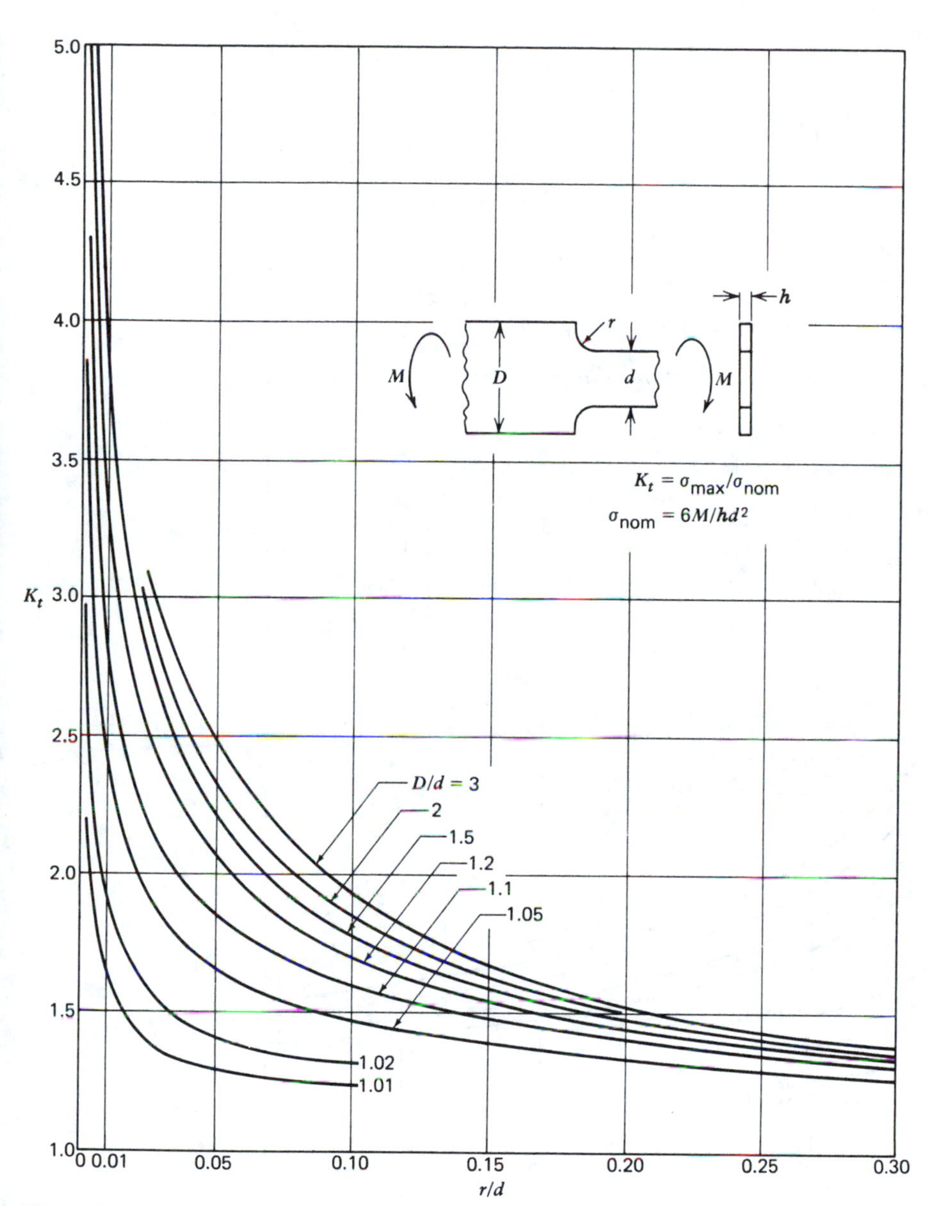

Figure F-5 Stress concentration factor for a stepped flat bar in bending.

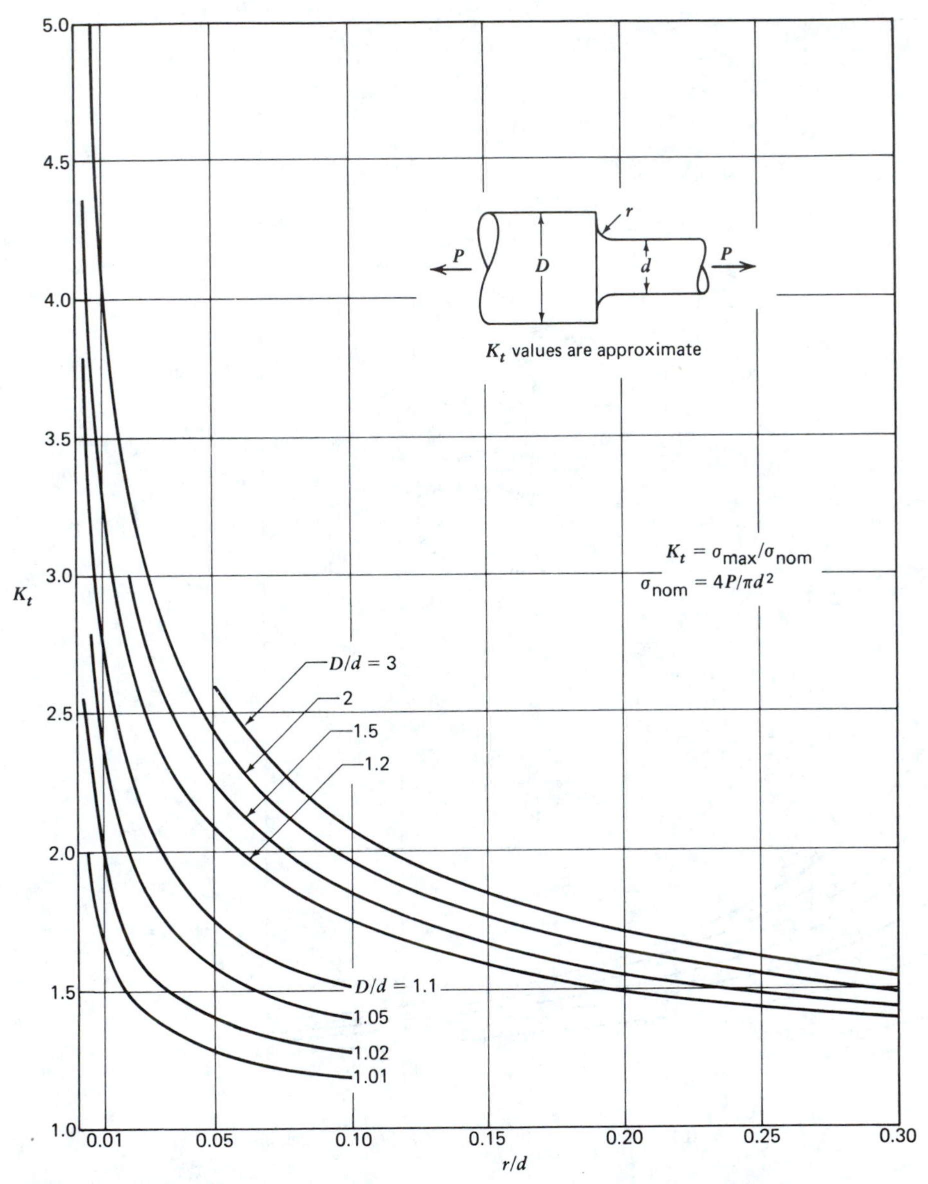

Figure F-6 Stress concentration factor for a stepped round bar in direct tension.

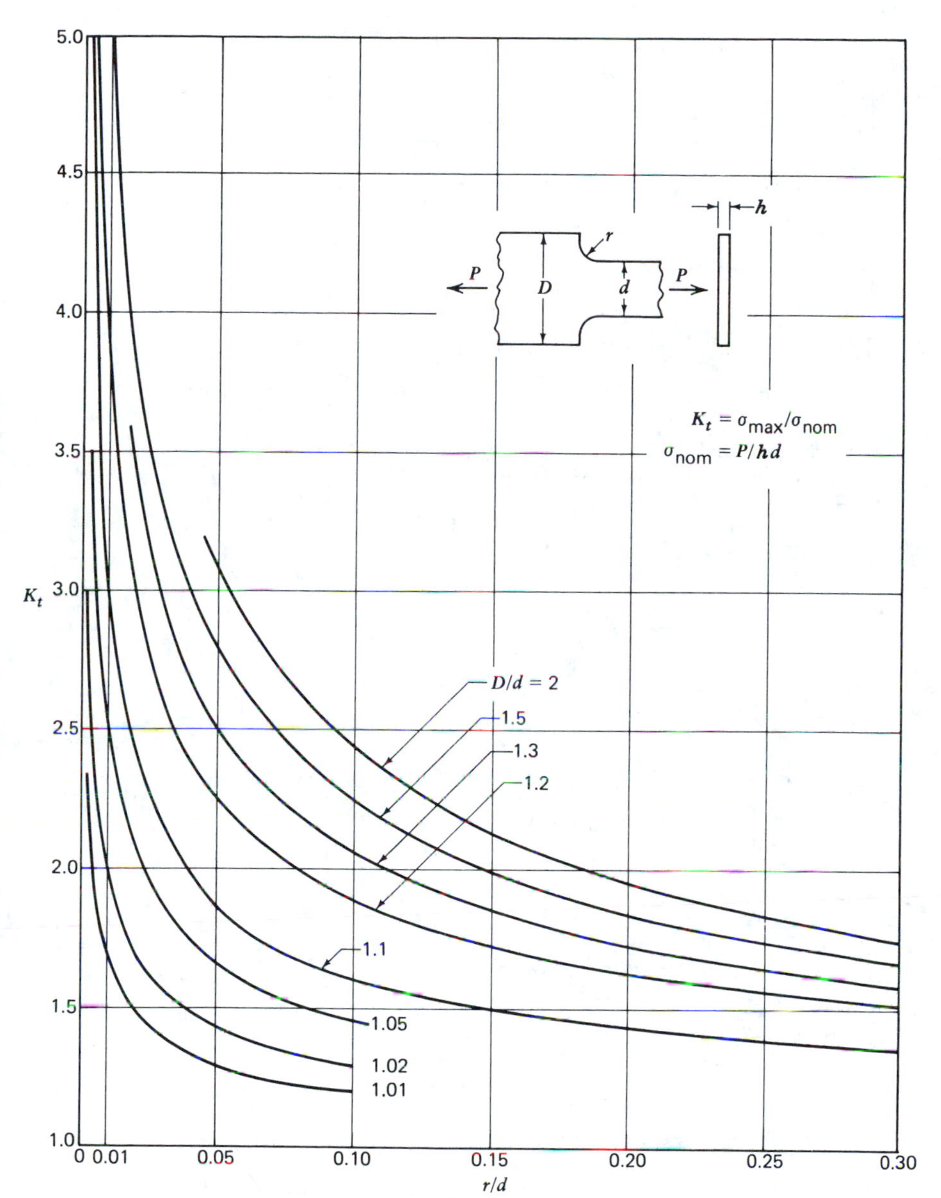

Figure F-7 Stress concentration factor for a stepped flat bar in direct tension.

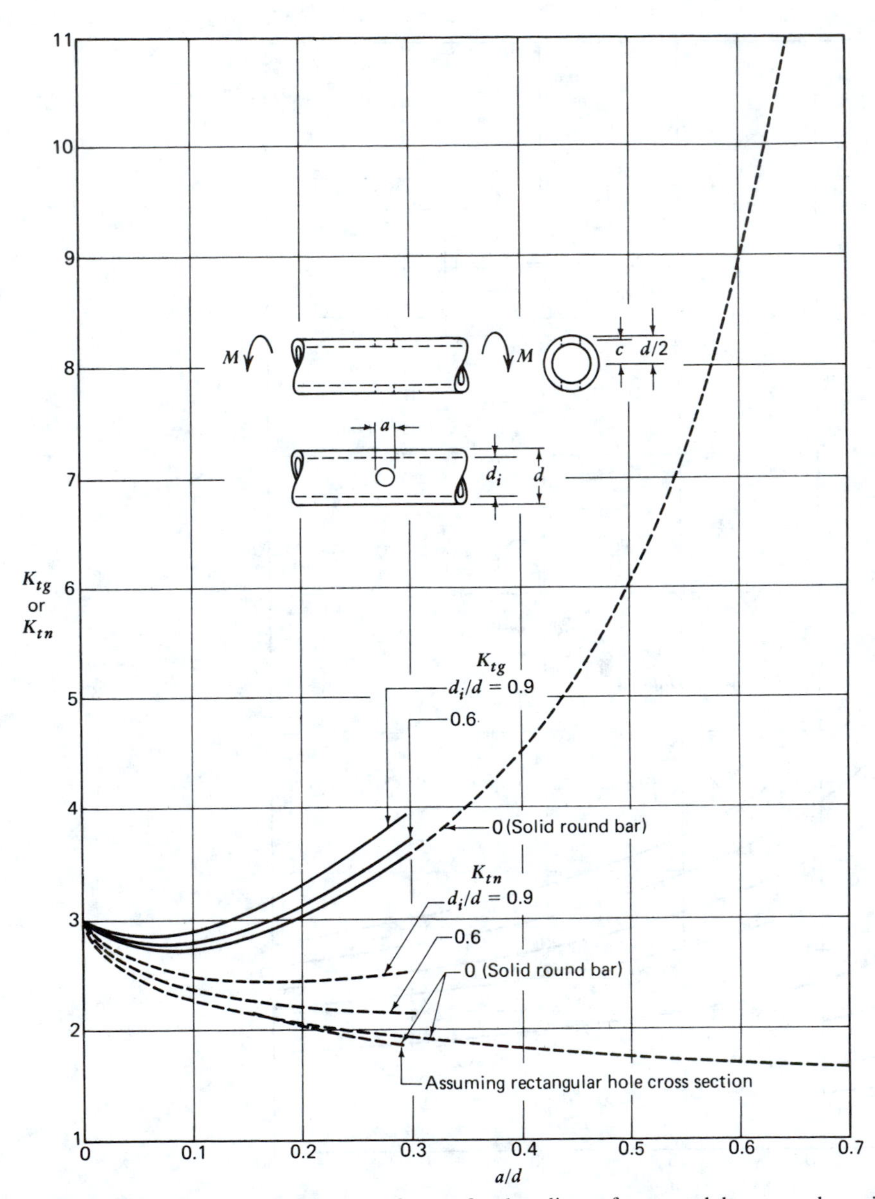

Figure F-8 Stress concentration factor for bending of a round bar or tube with a transverse hole.

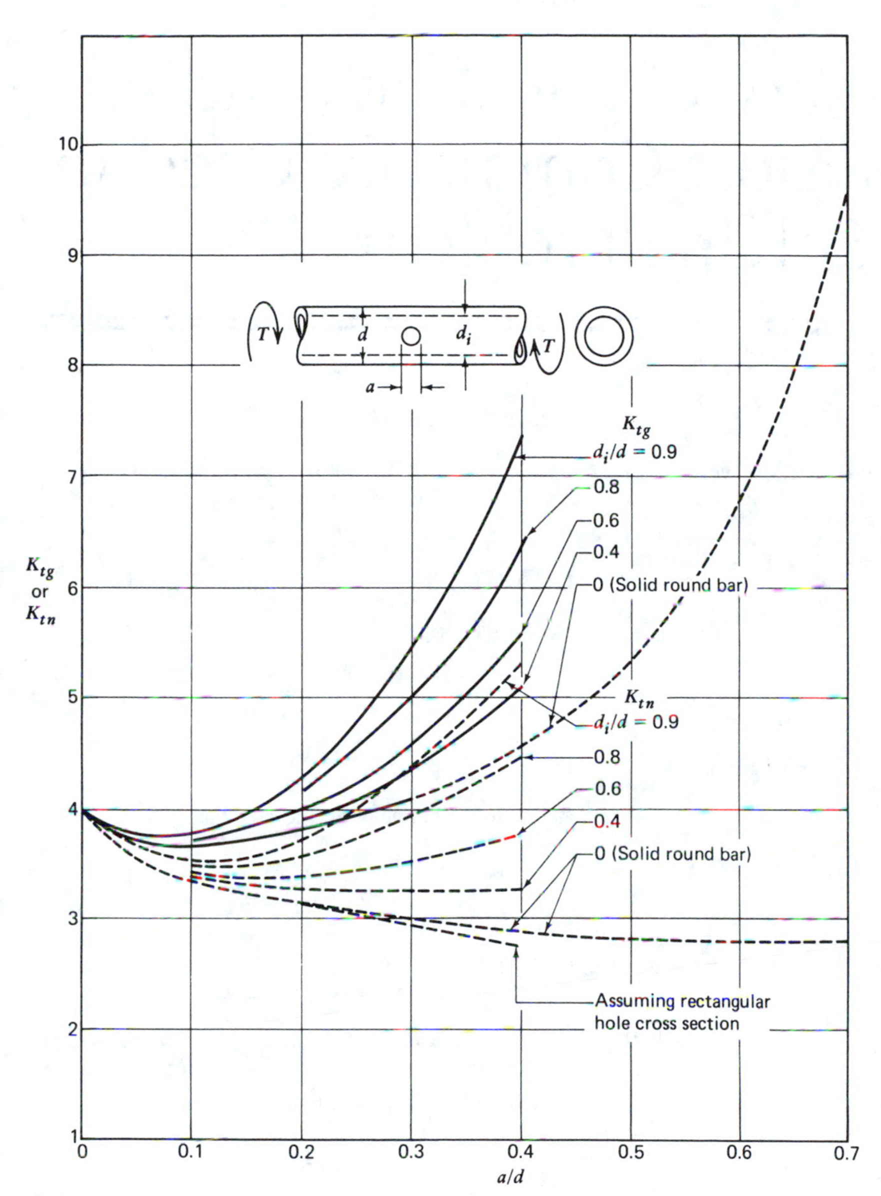

Figure F-9 Stress concentration factor for a round bar or tube with a transverse hole in torsion.

Appendix G
Summary of Equations and Formulas Most Commonly Used in Solving Problems

CHAPTER 1

$$\text{Work input} = \text{work output} + \text{lost work} \quad (1\text{-}2)$$

$$\text{Efficiency} = \frac{\text{work output}}{\text{work input}} = \frac{\text{energy delivered}}{\text{energy received}} \quad (1\text{-}3)$$

Energy: $$E_0 = eE_i \quad (1\text{-}4)$$

Efficiency:

$$e = (e_1)(e_2)\ .\ .\ .\ (e_{n-1})(e_n) \quad (1\text{-}5)$$

Power: $$P = \frac{Fv}{1000}\ \text{kW} \quad (1\text{-}6)$$

Power: $$P = \frac{Tn}{9550}\ \text{kW} \quad (1\text{-}7)$$

Power: $$P = \frac{T\omega}{1000}\ \text{kW} \quad (1\text{-}8)$$

Power: $$P = \frac{Fv}{33{,}000}\ \text{hp} \quad (1\text{-}9)$$

Power: $$P = \frac{Tn}{63{,}000}\ \text{hp} \quad (1\text{-}10)$$

Starting torque: $$T_s = k_r T_r \quad (1\text{-}11)$$

Design torque: $$P_d = k_s P_r \quad (1\text{-}12)$$

CHAPTER 2

Absolute error

$$= \text{measured value} - \text{exact value} \quad (2\text{-}1)$$

$$\text{Relative error} = \frac{\text{absolute error}}{\text{exact value}} \quad (2\text{-}2)$$

Relative change (RC)

$$= \frac{\text{new value} - \text{old value}}{\text{old value}} \quad (2\text{-}3)$$

CHAPTER 3

Hooke's law:

$$\sigma = \pm E\epsilon \quad (3\text{-}1)$$

$$E = \tan\theta \quad (3\text{-}2)$$

$$\tau = G\gamma \quad (3\text{-}3)$$

Relationship between elastic constants:

$$G = \frac{E}{2(1 + \nu)} \quad (3\text{-}4)$$

Poisson's ratio:

$$\mu = \frac{\text{transverse strain}}{\text{axial strain}} \quad (3\text{-}5)$$

Ultimate strength of steel approximation:

$$S_u(\text{MPa}) = 3.45\ \text{Bhn}$$

$$S_u(\text{psi}) = 500\ \text{Bhn} \quad (3\text{-}6)$$

Derating factors:

$$\text{Working stress} = (\text{derating factors})(\text{material strength}) \quad (3\text{-}7)$$

Factor of safety:

$$\text{Working stress} = \frac{\text{material strength}}{\text{factor of safety}} \quad (3\text{-}8)$$

Derating factors and factor of safety:

$$\text{Working stress} = \frac{\text{derating factors}}{\text{factor of safety}} \text{(material strength)} \quad (3\text{-}9)$$

Equivalent induced normal stress:

$$\sigma_i' = \sqrt{\sigma_i^2 + 3\tau_i^2} \quad (3\text{-}10)$$

Working stress for yielding:

$$\sigma_w = \frac{S_y}{fs} \quad (3\text{-}11)$$

Maximum induced (principal) normal stress:

$$\sigma_{i1} = \frac{\sigma}{2} + \sqrt{\left(\frac{\sigma}{2}\right)^2 + \tau^2} \quad (3\text{-}12)$$

Minimum induced (principal) normal stress:

$$\sigma_{i2} = \frac{\sigma}{2} - \sqrt{\left(\frac{\sigma}{2}\right)^2 + \tau^2} \quad (3\text{-}13)$$

Tensile ultimate working stress:

$$\sigma_{wt} = \frac{S_{ut}}{fs} \quad (3\text{-}14)$$

Compressive ultimate working stress:

$$\sigma_{wc} = \frac{S_{uc}}{fs} \quad (3\text{-}15)$$

Mean endurance limit:

For steel,

$$S_e' = 0.5S_{ut}; \qquad S_{ut} \leq 1400 \text{ MPa (200 Ksi)}$$

$$S_e' = 700 \text{ MPa (100 Ksi)};$$

$$S_{ut} > 1400 \text{ MPa (200 Ksi)} \quad (3\text{-}16)$$

For cast iron,

$$S_e' = 0.4S_{ut} \quad (3\text{-}17)$$

Mean fatigue strength of aluminum for the range 100 million to 500 million cycles:

For wrought aluminum alloys,

$$S_e' = 0.4S_{ut} \quad (3\text{-}18)$$

For cast aluminum alloys,

$$S_e' = 0.3S_{ut} \quad (3\text{-}19)$$

Torsional fatigue strength:

$$S_{se}' = 0.57S_e' \quad (3\text{-}20)$$

Endurance of a machine part:

$$S_e = k_a k_b k_c k_d k_e k_f S_e' \quad (3\text{-}21)$$

Surface finish derating factor, k_a:

For steels,

$$k_a = \frac{k_a S_e'}{S_e'} = \frac{k_a S_e'}{0.5S_{ut}}; \qquad S_{ut} < 1400 \text{ MPa}$$

$$= \frac{k_a S_e'}{700 \text{ MPa}}; \qquad S_{ut} > 1400 \text{ MPa} \quad (3\text{-}22)$$

For cast iron,

$$k_a = \frac{k_a S_e'}{0.4S_{ut}} \quad (3\text{-}23)$$

Temperature derating factor, k_d:

$$k_d = \frac{620}{460 + T} \quad \text{(Fahrenheit)}$$

$$k_d = \frac{344}{273 + T} \quad \text{(Celsius)} \quad (3\text{-}24)$$

Maximum stress at a discontinuity:

Direct load: $$\sigma = K_t \frac{F}{A} \quad (3\text{-}25)$$

Bending: $$\sigma = K_t \frac{Mc}{I} \quad (3\text{-}26)$$

Torsion: $$\tau = K_t \frac{Tr}{J} \quad (3\text{-}27)$$

Actual stress concentration factor, K_f:

$$K_f = 1 + q(K_t - 1) \quad (3\text{-}28)$$

Modifying (derating) factor for stress concentration, k_e:

$$k_e = \frac{1}{K_f} \quad (3\text{-}29)$$

Cyclic mean stress:

$$\sigma_m = 0.5(\sigma_{\max} + \sigma_{\min}) \qquad (3\text{-}30)$$

Cyclic amplitude (alternating) stress:

$$\sigma_a = 0.5(\sigma_{\max} - \sigma_{\min}) \qquad (3\text{-}31)$$

Soderberg formula:

$$\sigma_a = S_e\left(1 - \frac{\sigma_m}{S_{yt}}\right) \qquad (3\text{-}32)$$

Goodman formula:

$$\sigma_a = S_e\left(1 - \frac{\sigma_m}{S_{ut}}\right) \qquad (3\text{-}33)$$

Cyclic amplitude (alternating) shear stress:

$$\tau_a = S_{se} \qquad (3\text{-}34)$$

Shear failure by yielding:

$$\tau_a + \tau_m = 0.57S_y \qquad (3\text{-}35)$$

Cyclic shear mean stress:

$$\tau_m = 0.5(\tau_{\max} + \tau_{\min}) \qquad (3\text{-}36)$$

Cyclic shear amplitude (alternating) stress:

$$\tau_a = 0.5(\tau_{\max} - \tau_{\min}) \qquad (3\text{-}37)$$

Equivalent mean octahedral induced normal stress:

$$\sigma'_{im} = \sqrt{\sigma_{im}^2 + 3\tau_{im}^2} \qquad (3\text{-}38)$$

Equivalent amplitude (alternating) octahedral induced normal stress:

$$\sigma'_{ia} = \sqrt{\sigma_{ia}^2 + 3\tau_{ia}^2} \qquad (3\text{-}39)$$

CHAPTER 4

Columns:

The critical load of a *short* column is determined by the J. B. Johnson formula, when $Q/r^2 < 2$

$$P_{cr} = AS_y\left(1 - \frac{Q}{4r^2}\right) \qquad (4\text{-}1)$$

where

$$Q = \frac{S_y L^2}{n\pi^2 E} \quad \text{and} \quad r = \sqrt{\frac{I}{A}}$$

Often r is not known. Thus $Q/r^2 < 2$ must be confirmed *afterward*. When $Q/r^2 > 2$, use the *Euler* equation.

Euler: $$P_{cr} = \frac{S_y A r^2}{Q} \qquad (4\text{-}2)$$

Friction:

Coulomb's law: $$F_f = fN \qquad (4\text{-}3)$$

Friction angle: $$\theta = \arctan f \qquad (4\text{-}4)$$

Rolling friction: $$F_R = f_R W \qquad (4\text{-}5)$$

Resistance to rolling friction:

$$F_R = f_v W \qquad (4\text{-}6)$$

CHAPTER 6

Welding:

Butt weld: $$\sigma_{t,i} = \frac{P}{wt} \qquad (6\text{-}1)$$

Fillet welds:

Direct loading:

$$\tau = \frac{P}{0.707\omega(2L)} \qquad (6\text{-}2)$$

$$P = 1000fL \text{ lb} \qquad (6\text{-}3)$$

Axial loading: $$f = \frac{P}{\pi d} \qquad (6\text{-}4)$$

Torsion: $$f = \frac{F}{\pi d} = \frac{2T}{\pi d^2} \qquad (6\text{-}5)$$

Bending:

$$\sigma_b = \frac{32M(d + 2t_e)}{\pi[(d + 2t_e)^4 - d^4]} \qquad (6\text{-}6)$$

$$\sigma_{\max} = \frac{4M}{\pi d^2 t_e} \quad \text{for} \quad t_e \leq 0.1d \qquad (6\text{-}7)$$

Eccentric Loading:

$$\tau_1 = \frac{P}{2t_eL} \qquad (6\text{-}8)$$

$$\tau_2 = \frac{3T\sqrt{L^2 + w^2}}{t_e(L^3 - 3Lw^2)} \qquad (6\text{-}9)$$

$$\tau_{max} = \sqrt{\tau_1^2 + 2\tau_1\tau_2 \cos \emptyset + \tau^2} \qquad (6\text{-}10)$$

$$\emptyset = \arctan \frac{w}{L} \qquad (6\text{-}11)$$

Riveting:

Shear:

$$\tau = \frac{P}{A_s} = \frac{4P}{\pi d^2} \quad \text{(Code: 60.7 MPa)} \qquad (6\text{-}12)$$

Compression:

$$\sigma_c = \frac{P}{td} \quad \text{(Code: 131 MPa)} \qquad (6\text{-}13)$$

Plate:

$$\sigma_t = \frac{P}{(w - d)t} \quad \text{(Code: 75.8 MPa)} \qquad (6\text{-}14)$$

CHAPTER 7

Tensile stress: $\sigma_t = \dfrac{F}{A_s}$ (7-4)

Shear stress in bolt thread:

$$\tau = \frac{F}{(AS_s)h} \quad \text{(external)} \qquad (7\text{-}5)$$

$$\tau = \frac{F}{(AS_n)h} \quad \text{(internal)} \qquad (7\text{-}6)$$

Torque-tension coefficient:

$$C = \frac{T}{F_iD} \quad \text{(Fig. 7-27)} \qquad (7\text{-}7)$$

CHAPTER 8

$$k = \frac{P_2 - P_1}{\delta_2 - \delta_1} \qquad (8\text{-}2)$$

$L = (N + 1)d$ Unground Ends Fig. (8-10)

$L = Nd$ Ground Ends Fig. (8-10)

$$\delta = \frac{P}{k} \qquad (8\text{-}3)$$

$$\tau = \frac{8PD}{\pi d^3} \qquad (8\text{-}5)$$

$$\tau = \left(\frac{8PD}{\pi d^3}\right)\left(1 + \frac{1}{2C}\right) \qquad (8\text{-}7)$$

$$\tau_2 = \frac{8PD}{\pi d^3} K \quad (K\text{: Fig. 8-12}) \qquad (8\text{-}9)$$

$$k = \frac{Gd^4}{8D^3n} \qquad (8\text{-}12)$$

$$k = \frac{P_e}{fn} \qquad (8\text{-}14)$$

$$IT = P_e\left(\frac{d}{D}\right) \qquad (8\text{-}16)$$

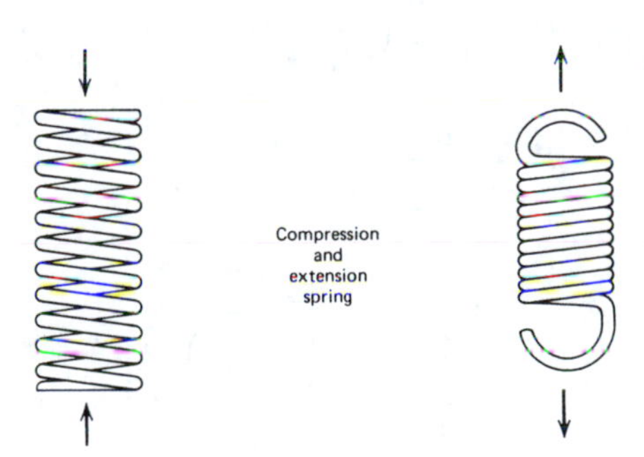

$$\sigma = \frac{10.2T}{d^3} \qquad (8\text{-}17)$$

$$T = \frac{Ed^4}{10.2Dn} \quad \text{360 deg deflection} \qquad (8\text{-}18)$$

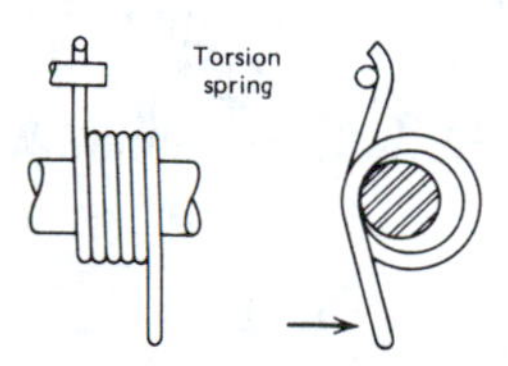

$$\sigma_b = \frac{6PL}{bt^2} \tag{8-19}$$

$$\delta = \frac{4PL^3K}{Ebt^3} \tag{8-20}$$

$$k = \frac{Ebt^3}{4L^3K} \tag{8-21}$$

$$t = \frac{L^2\sigma_b K}{1.5\ \delta E} \tag{8-22}$$

$$b = \frac{6PL}{\sigma_b t^2} \tag{8-23}$$

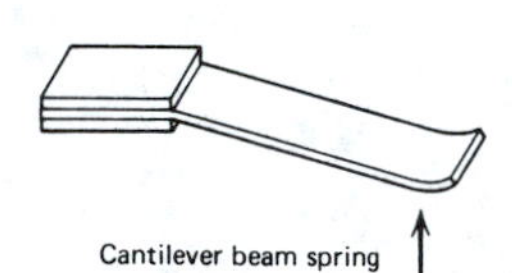

$$\sigma = \frac{-E\emptyset}{1\text{-}\upsilon^2}\left(\frac{t}{d}\right)^2 \tag{8-24}$$

$$P = \frac{-\sigma t^2}{\beta} \tag{8-25}$$

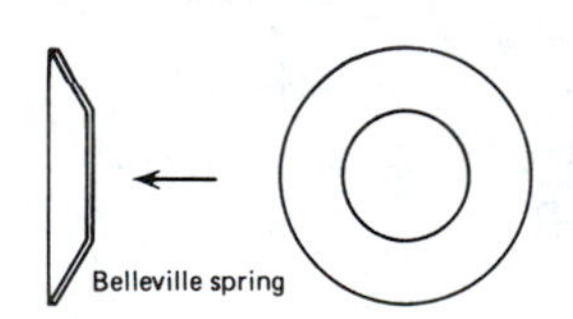

$$k = k_1 + k_2 + k_3 + \cdots + k_n \tag{8-26}$$

$$\frac{1}{k} = \frac{1}{k_1} + \frac{1}{k_2} + \frac{1}{k_3} + \cdots + \frac{1}{k_n} \tag{8-27}$$

Coupling of springs

Parallel series

Useful Conversions
Metric to English

$$\text{Load, lb} = \frac{\text{load, N}}{4.45}$$

$$\text{Rate, lb/in.} = \text{rate}\frac{N}{\text{mm}}(5.71)$$

$$\text{Stress, psi} = \text{stress}\frac{N}{\text{mm}^2}(145)$$

Common Wire Gauges:
Inch Values

1—0.2830 in.	11—0.1205 in.
2—0.2625 in.	12—0.1055 in.
3—0.2437 in.	13—0.0915 in.
4—0.2253 in.	14—0.0800 in.
5—0.2070 in.	15—0.0720 in.
6—0.1920 in.	16—0.0625 in.
7—0.1770 in.	17—0.0540 in.
8—0.1620 in.	18—0.0475 in.
9—0.1483 in.	19—0.0410 in.
10—0.1350 in.	20—0.0348 in.

CHAPTER 9

Flywheel, energy transfer:

$$E = 0.5k^2m(\omega_1^2 - \omega_2^2) \tag{9-1}$$

Transverse load on a shaft:

$$F_t = k_t\frac{19\ 100P}{Dn}\ \text{N} \tag{9-2}$$

$$F_t = k_t\frac{126{,}000P}{Dn}\ \text{lb} \tag{9-3}$$

Single chain:	$k_t = 1.0$
Gears:	$k_t = 1.25$
Double chain:	$k_t = 1.25$
V-belts:	$k_t = 1.50$
Flat belts:	$k_t = 2.5$

CHAPTER 11

Angle of contact:

$$\theta = \pi - \frac{D - d}{C} \text{ rad, approximately} \quad (11\text{-}1)$$

Center distance:

$$C = 0.0625\{b + [b^2 - 32(D - d)^2]\}^{1/2} \quad (11\text{-}2)$$

Belt length:

$$L = 2C + 1.57(D + d) + \frac{(D - d)^2}{4C} \quad (11\text{-}3)$$

Speed ratio:

$$i = \frac{\text{speed of drive shaft}}{\text{speed of driven shaft}} \quad (11\text{-}4)$$

Speed ratio:

$$i = \frac{\text{diameter of larger pulley}}{\text{diameter of smaller pulley}} \quad (11\text{-}5)$$

Belt speed: $v = \pi D n_D = \pi d n_d$ (11-6)

Speed of driven sheave without slip:

$$n_D = \frac{n_d}{i} \quad (11\text{-}7)$$

Speed of driven sheave:

$$n_D = n_d \frac{1 - S}{i} \quad (11\text{-}8)$$

$$n_D = \frac{n_d}{i}(1 - S_1)(1 - S_2) \quad (11\text{-}9)$$

Static forces: $F_1 = F_2 e^{f\theta}$ (11-10)

V-belt: $F_1 = F_2 e^{f'\theta}$ (11-11)

$$f' = \frac{f}{\sin \beta} \quad (11\text{-}12)$$

Total normal force:

$$2N = \frac{F_n}{\sin \beta} \quad (11\text{-}13)$$

Friction force:

$$F_f = \frac{F_n f}{\sin \beta} = F_n f' \quad (11\text{-}14)$$

Tangential force:

$$F_t = F_1 - F_2 = F_1(1 - e^{f'\theta}) \quad (11\text{-}15)$$

Power transmitted by static forces:

$$P = F_1(1 - e^{-f'\theta})v\,10^{-3} \text{ kW} \quad (11\text{-}16)$$

$$P = F_1(1 - e^{-f'\theta})v\,(550)^{-1} \text{ hp} \quad (11\text{-}17)$$

Centrifugal force:

$$F_c = \rho(1.0m)Av^2 \text{ N} \quad (11\text{-}18)$$

Mean tensile stress:

$$\sigma_t = \frac{F_c}{A} \text{ MPa} \quad (11\text{-}19)$$

$$\sigma_1 = \sigma_{max} - \sigma_c \quad (11\text{-}20)$$

Power transmission, including dynamic effects:

$$P = (F_{max} - F_c)(1 - e^{-f'\theta})v\,10^{-3} \text{ kW} \quad (11\text{-}21)$$

$$P = (F_{max} - F_c)(1 - e^{-f'\theta})v\,(550)^{-1} \text{ hp} \quad (11\text{-}22)$$

CHAPTER 12

Pitch diameter:

$$d = \frac{p}{\sin (180\ deg/n)} \quad (12\text{-}1)$$

Center distance:

$$C = D + 0.5d \quad \text{for} \quad D \gg d \quad (12\text{-}2)$$

Angle of wrap:

$$\theta = \pi - 2\alpha = \pi - \frac{D - d}{C} \text{ approximately} \quad (12\text{-}3)$$

Chain length:

$$L = \frac{N + n}{2} + \frac{2C}{p} + \frac{p(N3n)^2}{39.5C} \text{ pitches} \quad (12\text{-}4)$$

Chain speed:

$$v = \pi d n_s 10^{-3} \text{ m/s} \quad (12\text{-}5)$$

$$v = \pi n_s\left(\frac{d}{12}\right) \text{ fps} \tag{12-6}$$

$$v = (pn)n_s 10^{-3} \text{ m/s} \tag{12-7}$$

$$v = nn_s\left(\frac{p}{12}\right) \text{ fps} \tag{12-8}$$

Speed variation:

$$v = \pi\left(\frac{R}{12}\right)n_s(1 - \cos\theta)$$

for $\quad 0 < \theta < 180 \text{ deg}/n \qquad (12\text{-}9)$

Chordal rise:

$$\Delta r = \frac{R}{12}(1 - \cos\theta) \tag{12-10}$$

Chain tension: $\quad F = F_t + F_c \qquad (12\text{-}11)$

$$F_t = \frac{2T}{d} \tag{12-12}$$

$$F_c = ma \tag{12-13}$$

Basic equations for power transmission:

$$P = (F - mv^2)(v)10^{-3} \text{ kW} \tag{12-14}$$

$$P = (F - mv^2)(v)(550^{-1}) \text{ hp} \tag{12-15}$$

Maximum speed:

$$n = \left(\frac{2000}{p}\right)\left(\frac{A}{Wp}\right) \text{ rpm} \tag{12-16}$$

CHAPTER 13

Spur Gears:

Pressure angle:

$$\cos\phi = \frac{r_b}{r} = \frac{R_b}{R} \tag{13-1}$$

$$\cos\phi = \frac{d_b}{d} = \frac{D_b}{D} \tag{13-2}$$

Center distance:

$$C = 0.5(D + d) \tag{13-3}$$

Speed ratio:

$$m_G = \frac{n_p}{n_G} = \frac{D}{d} = \frac{N_G}{N_p} \tag{13-4}$$

Base pitch: $\quad p_b = \dfrac{\pi D_b}{N} \qquad (13\text{-}5)$

Circular pitch: $\quad p_c = \dfrac{\pi D}{N} \qquad (13\text{-}6)$

Module: $\quad m = \dfrac{D}{N} \text{ mm/tooth} \qquad (13\text{-}7)$

Diametral pitch:

$$P_d = \frac{N}{D} \text{ teeth/in.} \tag{13-8}$$

Base pitch: $\quad p_b = p_c \cos\phi \qquad (13\text{-}9)$

Pitch and module:

$$mP_d = 25.4 \tag{13-10}$$

Circular pitch: $\quad p_c = \pi m \qquad (13\text{-}11)$

Diametral and circular pitch:

$$P_d p_c = \pi \tag{13-12}$$

Contact ratio:

$$m_p = \frac{Z}{p_b} = \frac{\sqrt{R_o^2 - R_b^2} + \sqrt{r_o^2 - r_b^2} - C\sin\phi}{p_c\cos\phi} \tag{13-13}$$

N_p (minimum): $\quad N_c = \dfrac{2}{\sin^2\phi} \qquad (13\text{-}14)$

Interference can be avoided if:

$$R_o \le \sqrt{R_b^2 + C^2\sin^2\phi} \tag{13-15}$$

$$r_o \le \sqrt{r_b^2 + C^2\sin^2\phi} \tag{13-16}$$

Tangential force:

$$W_t = W\cos\phi \tag{13-17}$$

$$W_r = W_t\tan\phi \tag{13-18}$$

$$W_t = \frac{19\,100P}{dn} \text{ N} \tag{13-19}$$

$$W_t = \frac{126{,}000P}{dn} \text{ lb} \quad (13\text{-}20)$$

Tooth size for known center distance:

$$P_d^2 = \frac{S_e C_v k \pi^2 y_p}{W_t} \quad (13\text{-}26)$$

Tooth size for unknown center distance:

$$P_d^3 = \frac{S_e C_v k \pi^2 N_p y_p}{2T_p} \quad (13\text{-}27)$$

Total tooth load:

$$W_T = W_t + W_d \quad (13\text{-}28)$$

Helical Gears:

Normal circular pitch:

$$p_n = p_c \cos \psi \quad (13\text{-}30)$$

Normal diametral pitch:

$$P_n = \frac{P_d}{\cos \psi} \quad (13\text{-}31)$$

Axial pitch: $$p_a = \frac{p_c}{\tan \psi} = \frac{p_n}{\sin \psi} \quad (13\text{-}32)$$

Velocity ratio:

$$m_G = \frac{n_p}{n_G} = \frac{N_G}{N_p} = \frac{D}{d} \quad (13\text{-}33)$$

Center distance:

$$C = \frac{N_G + N_p}{2P_d} \quad (13\text{-}34)$$

Lead: $$L = \frac{\pi D}{\tan \psi} \quad (13\text{-}35)$$

Pressure angle: $$\tan \phi = \frac{\tan \phi}{\cos \psi} n \quad (13\text{-}36)$$

Axial load: $$W_a = W_t \tan \psi \quad (13\text{-}39)$$

Total load: $$W = \frac{W_t}{\cos \phi_n \cos \psi} \quad (13\text{-}40)$$

Radial load: $$W_r = W_t \frac{\tan \phi_n}{\cos \psi} \quad (13\text{-}41)$$

Total contact ratio:

$$m_t = m_p + m_f \quad (13\text{-}42)$$

Face contact ratio:

$$m_f = \frac{F \tan \psi}{p_c} \quad (13\text{-}43)$$

Face width: $$F \geq \frac{1.15 p_c}{\tan \psi} \quad (13\text{-}45)$$

CHAPTER 14

Bevel gears: $$\sum = \Gamma_p + \Gamma_G \quad (14\text{-}1)$$

For $\sum = 90$

$$\tan \Gamma_p = \frac{d}{D} = \frac{N_p}{N_G} \quad (14\text{-}2)$$

$$\tan \Gamma_G = \frac{D}{d} = \frac{N_G}{N_p} \quad (14\text{-}3)$$

Worm gearing:

Lead: $$L = N_w p_a \quad (14\text{-}4)$$

Speed ratio of worm gearing:

$$m_G = \frac{P_a N_G}{L} = \frac{N_G}{N_w} = \frac{n_w}{n_G} = \frac{D_G}{D_w \tan \lambda_w} \quad (14\text{-}5)$$

Axial and circular pitch:

$$p_a = p_c = \frac{\pi D_G}{N_G} \quad (14\text{-}6)$$

Speed: $$n_G = \frac{L}{\pi D_G} n_w \quad (14\text{-}7)$$

Lead angle:

$$\tan \lambda_w = \frac{L}{\pi D_w} \quad (14\text{-}8)$$

Center distance:

$$C = \frac{L}{2\pi}(m_G + \cot \lambda_w) \quad (14\text{-}9)$$

Efficiency: $$e = \frac{\tan \lambda}{\tan (\lambda + \theta)} \quad (14\text{-}10)$$

CHAPTER 15

Steady torque:

$$d = \left(\frac{8.935Tfs}{S_{yt}}\right)^{1/3} \tag{15-3}$$

Fluctuating torque:

$$d = \left\{\frac{8.935[T_m + (S_{sy}/S_{se})T_a]fs}{S_{yt}}\right\}^{1/3} \tag{15-4}$$

Steady bending loads (shaft rotating):

$$d = \left(\frac{10.186M_a fs}{S_e}\right)^{1/3} \tag{15-5}$$

Design for strength—combined loading:

$$d = \left[\frac{10.186fs}{S_{yt}}\sqrt{\left(\frac{S_{yt}}{S_e}M\right)^2 + \frac{3}{4}T^2}\right]^{1/3} \tag{15-10}$$

Hollow shaft:

$$C = \frac{d_i}{d_o} \tag{15-13}$$

$$d_o = \left[\frac{10.186fs}{S_{yt}(1 - C^4)}\sqrt{\left(\frac{S_{yt}}{S_e}M\right)^2 + \frac{3}{4}T^2}\right]^{1/3} \tag{15-16}$$

CHAPTER 16

Rectangular key:

$$\tau = \frac{F}{bL} \tag{16-4}$$

$$\tau = \frac{2T}{bLD} \tag{16-5}$$

$$\sigma_c = \frac{4T}{tLD} \tag{16-6}$$

Round key:

$$\tau = \frac{2T}{dLD} \tag{16-7}$$

Dowel pin:

$$\tau = \frac{4T}{\pi D d^2} \tag{16-8}$$

Tapered shaft:

$$T = 0.5fF_n D_m \tag{16-9}$$

$$F_n = \frac{F_a \cos\theta}{\sin(0.5\alpha + \theta)} \tag{16-10}$$

Maximum torque:

$$T = \frac{D_m F_a \sin\theta}{2\sin(0.5\alpha + \theta)} \tag{16-11}$$

Compressive stress:

$$\sigma_c = \frac{2T}{fD_m^2 L} \tag{16-12}$$

CHAPTER 17

Dynamic viscosity:

$$\mu = \frac{Fh}{AV} \tag{17-3}$$

Bearing characteristic number:

$$P = \frac{W}{LD} \tag{17-4}$$

Rubbing velocity:

$$v = \frac{\pi Dn}{1000} \text{ m/s} \tag{17-5}$$

Oscillating shaft:

$$v = \frac{\pi D\theta f}{360\,000} \text{ m/s} \tag{17-6}$$

Rubbing velocity:

$$v = 5\pi Dn \text{ fpm} \tag{17-7}$$

Oscillating shaft:

$$v = \frac{\pi D\theta f}{72} \text{ fpm} \tag{17-8}$$

Oil film thickness:

$$h = C + e\cos\theta \tag{17-9}$$

Min. oil film thickness:

$$h_o = C - e \tag{17-10}$$

$$h_o = C(1 - \epsilon) \tag{17-11}$$

Sommerfeld number:

$$S = \left(\frac{R}{C}\right)^2\left(\frac{\mu n}{P}\right) \tag{17-12}$$

Dimensionless clearance factor:

$$m = 1000\frac{C}{R} \qquad (17\text{-}13)$$

$$S = \frac{\mu n}{1000 m^2 P}\ \text{MPa}\cdot\text{s} \qquad (17\text{-}14)$$

$$S = \frac{\mu n}{m^2 P}\ \text{microreyns} \qquad (17\text{-}15)$$

Rate of heat generation:

$$H_f = \left(\frac{WnC}{159.15}\right)\left(\frac{R}{C}f\right)\ \text{W} \qquad (17\text{-}16)$$

$$H_f = \left(\frac{WnC}{1485.9}\right)\left(\frac{R}{C}f\right)\ \text{Btu/s} \qquad (17\text{-}17)$$

Flow rate; cm^3/s, or $\text{in.}^3/\text{s}$:

$$Q_s = \frac{k_g m^3 D^3(1 + 1.5\epsilon^2)p_s}{\mu}10^{-6} \qquad (17\text{-}18)$$

Flow factor:

$$k_g = \frac{32.725}{L'/D} \qquad (17\text{-}19)$$

Rate of dissipation:

$$H_d = C\rho Q_s\frac{\Delta t}{2} \qquad (17\text{-}20)$$

CHAPTER 18

Life:

$$L_{\text{rev}} = K\left(\frac{C}{R}\right)^p \qquad (18\text{-}1)$$

Ball bearings:

$$L = \left(\frac{10^6}{60n}\right)\left(\frac{\text{catalog rating}}{\text{applied load}}\right)^3 h \qquad (18\text{-}2)$$

Equivalent radial load:

$$RE = RF_{th} \qquad (18\text{-}6)$$

Equivalent life:

$$L_e = L_d F_{\text{rel}} \qquad (18\text{-}7)$$

Weighted life:

$$L_w = \frac{100}{(P_1/L_1) + (P_2/L_2) + \cdots + (P_n/L_n)} \qquad (18\text{-}8)$$

CHAPTER 19

Design torque:

$$T_{\text{des}} = K_s T \qquad (19\text{-}1)$$

Service factor:

$$K_s = \sqrt{F_s^2 + F_D^2 + F_L^2 - 2} \qquad (19\text{-}2)$$

Design of disk clutches based on uniform pressure distribution:

Torque capacity:

$$T = fF_aR_e10^{-3}\ \text{N}\cdot\text{m} \qquad (19\text{-}3)$$

$$T = fF_aR_e\ \text{lb-in.} \qquad (19\text{-}4)$$

Effective friction radius:

$$R_e = 0.25(D + d) \qquad (19\text{-}5)$$

Actuating force:

$$F_a = 0.5\pi d(D - d)p_{\text{all}} \qquad (19\text{-}6)$$

Design of disk clutches based on average contact pressure:

$$\frac{P_{\max}}{P_{\text{ave}}} = 0.5\left(\frac{D}{d} + 1\right) \qquad (19\text{-}7)$$

Torque capacity:

$$T = fF_aR_eN_p10^{-3}\ \text{N}\cdot\text{m} \qquad (19\text{-}8)$$

$$T = fF_aR_eN_p\ \text{lb-in.} \qquad (19\text{-}9)$$

Design of free-shoe clutch with garter spring:

Torque capacity:

$$T = fF_nRN_s10^{-3}\ \text{N}\cdot\text{m} \qquad (19\text{-}10)$$

$$T = fF_nRN_s\ \text{lb-in.} \qquad (19\text{-}11)$$

The net normal force:

$$F_n = \frac{mrn^2}{C} - 2P\cos\left(90\text{ deg} - \frac{180\text{ deg}}{N_s}\right) \qquad (19\text{-}12)$$

$$C = \begin{cases} 91\,200 & \text{for SI} \\ 35{,}200 & \text{for EU} \end{cases}$$

The tension P in the garter spring:

$$P = \frac{mrn_e^2}{2\,C\cos[90\text{ deg} - (180\text{ deg}/N_s)]} \qquad (19\text{-}13)$$

CHAPTER 20

Disk brakes:

$$e = 0.5(R_i + R_o) \quad (20\text{-}1)$$

Torque capacity:

$$T = fF_aR_e 10^{-3}\ \text{N}\cdot\text{m} \quad (20\text{-}2)$$

$$T = fF_aR_e\ \text{lb-in.} \quad (20\text{-}3)$$

Effective friction radius R_e for the annular pad:

$$R_e = \frac{2}{3}\frac{R_o^3 - R_i^3}{R_o^2 - R_i^2} \quad (20\text{-}4)$$

Circular pad: $R_e = \delta e$ (20-5)

Actuating force:

$$F_a = p_{\text{ave}} A_{\text{pad}} \quad (20\text{-}6)$$

Annular pad:

$$A_{\text{pad}} = 0.5\theta(R_o^2 - R_i^2) \quad (20\text{-}7)$$

Band brakes:

Band tensions: $\dfrac{F_1}{F_2} = e^{f\theta}$ (20-8)

Torque: $T = \dfrac{(F_1 - F_2)D}{2}$ (20-9)

Actuating force:

$$F_a = \frac{2T(c - be^{f\theta})}{Da(e^{f\theta} - 1)} \quad (20\text{-}11)$$

Self-energizing: $b > 0$

Self-locking: $b \le \dfrac{c}{e^{f'\theta}}$ (20-12)

CHAPTER 21

Apparent flange pressure:

$$p_a = \frac{nF}{A} \quad (21\text{-}1)$$

Tensile force in bolt:

$$F = \frac{T}{0.2D} \quad (21\text{-}2)$$

Gasket seating width:

$$b = 0.5\sqrt{b_0}\ \text{in.} \quad (21\text{-}3a)$$

For $b_0 > 0.25$ in.

$$b = 12.7\sqrt{\frac{b_0}{25.4}}\ \text{mm} \quad (21\text{-}3b)$$

Diameter for gasket load:

$$D = d_0 - 2b \quad (21\text{-}4)$$

Tensile bolt force:

$$F = \frac{\pi}{n} p_f D b \quad (21\text{-}5)$$

$$F = \frac{\pi p D}{n}\left(\frac{D}{4} + 2mb\right) \quad (21\text{-}6)$$

Appendix H
Answers to Selected Problems

CHAPTER 1

P1-4 (*a*) 3 hp (*b*) 2.24 kW
P1-5 $T_r = 19.23$ to 20.83 N · m
$P = 2.42$ to 2.62 kW
P1-9 $T_r = 96$ to 150 N · m
$T_R = -150$ N · m
P1-12 $P = 86$ kW
P1-13 $k_s = 1.3$ approx. 15-hp motor
P1-15 (*a*) $e = 0.3042$ (*b*) 82.18 J
P1-16 (*a*) $D = 170$ mm (*b*) 7.5-hp motor
(*c*) $F_c = 266$ N approx.
P1-18 $k_s = 1.65$ approx.
P1-20 $k_s = 2.5$ approx.
P1-22 (*a*) $k_s = 2.6$ approx. (*b*) $k_s = 1.5$ approx.

CHAPTER 2

P2-2 (*a*) 0.5% (*b*) 0.31% (*c*) 0.03%
P2-3 (*a*) 0.25% (*b*) 0.40% (*c*) 0.65%
P2-4 (*a*) 6% (*b*) 1.5%
P2-6 3.3%; a change may not be necessary
P2-8 (*a*) 20% (*b*) 15% (*c*) 5.25%
P2-10 (*a*) $n = 710$ rpm approx.
(*b*) $n = 706$ rpm

CHAPTER 3

P3-1 $F = 548$ lb
P3-2 A: $k_t = 1.51$ B: $k_t = 2.33$
C: $k_t = 1.42$
Section B will fail first.
P3-3 (*a*) $d = 1.3$ in.; $d = 1.5$ in. (*b*) 68%
P3-4 $d = 1.7$ in.
P3-5 (*a*) $d = 44.9$ mm; (*b*) $r = 11.2$ mm (*c*) $D = 67.35$ mm
P3-6 $fs = 1.73$
P3-7 $r = 11$ mm; $d = 43.88$ mm
P3-8 $d = 1.24$ in.; $r = 0.37$ in.;
$D = 1.99$ in.
P3-9 The stress is acceptable
P3-10 Adequate strength is available: $fs = 1.8$
P3-11 (*a*) $S_e = 115$ MPa
(*b*) No
P3-12 (*a*) $d = 2.10$ in. (*b*) $r = 0.26$ in.
P3-13 (*a*) $d = 1.45$ in. (*b*) $d = 1.37$ in.
(*c*) 5.5%
P3-14 (*a*) $t = 3.07$ mm (*b*) $t = 7.97$ mm
(*c*) 160%
P3-15 (*a*) $t = 4.29$ mm and 3.28 mm
(*b*) $t = 5.5$ mm and 4.8 mm
(*c*) $t = 6.74$ mm and 6.40 mm
P3-17 $fs = 1.52$
P3-18 (*a*) $fs = 2.29$ (*b*) $fs = 9$
P3-19 Infinite life
P3-20 Infinite life
P3-21 $d = 1.42$ in.
P3-23 (*a*) $t = 5.9$ mm (*b*) $t = 21.6$ mm
(*c*) $t = 11.6$ mm (*d*) 266%; 97%; fatigue is sensitive to the magnitude of the stress amplitudes

CHAPTER 4

P4-1 $L_1 = 50$ mm; $L_{0.25} = 25$ mm; $L_2 = 71$ mm; $L_3 = 86.6$ mm; $L_4 = 100$ mm
P4-3 Interpolation suggests a 1.5-in. dia. rod
P4-5 Reduction in mass = 11.2%
P4-6 A 25% reduction in mass
P4-8 Use $d_i = 1.0$ in.; $d_o = 2$ in.
P4-9 4-in. standard pipe; $3\frac{1}{2}$ in. extra strong in. double strong
P4-10 $d \geq 18$ mm; Chart: $d = 31$ mm approximately

P4-11 Use $d = 16$ mm
P4-12 $t = 72$ mm
P4-13 $P = 134\,867$ N (Euler) approximately
P4-14 Motion for $\alpha > 5.71$ deg
P4-16 (*a*) 3600 N (*b*) 1800 N
(*c*) $T = 540$ N · m
P4-17 $P_{max} = 5.63$ hp; use a 7.5-hp motor

CHAPTER 5

P5-1 $F = 11.2$ kN approximately
P5-2 (*a*) $t = 6.5$-mm rigidity controls
(*b*) 57% (*c*) $R = 1.97\%$
P5-3 $F = 21.8$ kN approximately

CHAPTER 6

P6-1 $P = 225{,}000$ lb
P6-2 $P = 33{,}360$ lb
P6-3 $L = 6.50$ lb
P6-4 $\frac{3}{16}$-in. weld
P6-5 (*a*) $P = 118{,}720$ lb
(*b*) $P = 79{,}147$ lb
P6-6 (*a*) $P_{st} = 14{,}840$ lb
(*b*) $P_{dyn} = 7{,}420$ lb
P6-7 (*a*) $F_{st} = 10{,}390$ lb
(*b*) $F_{dyn} = 5{,}195$ lb
P6-8 (*a*) $P_{st} = 66{,}780$ lb
(*b*) $P_{dyn} = 44{,}520$ lb
P6-10 $d = 12$ mm
P6-12 $d = 6$ mm $w = 44$ m
P6-13 5%
P6-17 $P = 13{,}760$ lb
P6-18 (*a*) 0.0087 in. (*b*) $T = 1380$°F

CHAPTER 7

P7-1 $R = 34$ in.
P7-2 Pull = 1844 N; σ_t (max) = 264 MPa
P7-3 (*a*) Proof stress = 310 MPa
(*b*) $A_s = 171$ mm^2
(*c*) M 20 × 2.5
(*d*) $fs = 1.43$
P7-5 M 14 × 2; $fs = 1.50$
P7-6 (*a*) $\sigma_t = 466$ MPa (*b*) 459 deg
P7-7 (*b*) $\frac{5}{8}$ − 11 UNC (*c*) $h = 0.7$ in.
P7-9 Eye bolt M24 × 3, 20-mm engagement
P7-10 (*a*) $F = 33\,852$ N (*b*) $F = 40\,816$ N
(*c*) $F = 46\,992$ N
P7-12 $P = 226$ N
P7-16 $h = 0.40D$
P7-17 (*a*) $F_i = 29.58$ kN (*b*) $T = 65$ N · m
(*c*) $\rho = 1.66$ mm (*d*) 398 deg

CHAPTER 8

P8-1 $P = 600$ N
P8-2 $n = 14.5$
P8-3 Maximum stress = 112,682 psi; short life because $fs \sim 1.036$
P8-4 $\tau_1 = 603$; $n = 14$ coils
P8-5 $P = 65.7$ lb
P8-6 (*a*) $k = 8.52$ N/mm (*b*) $P = 613$ N
(*c*) $\tau_1 = 805$ N/mm^2 (*d*) $L_1 = 78$ mm
P8-7 $OD = 17$ mm; $d = 2.25$ mm;
$N = 12.5$; $L_0 = 63.2$ mm
P8-11 $d = 12.5$ mm; $N = 12.7$;
$L_0 = 310.75$ mm
P8-12 $OD = 1.5 \pm 0.03$ in.; $d = 0.177 \pm 0.003$ in.; $N = 12 \pm 0.6$ in.;
$L_0 = 4.12 \pm 0.12$ in.; $L_s = 2.12$ in.
P8-14 $P = 4418$ N; $k = 19.6$ N;
$k_s = 176$ N/mm
P8-16 $n = 11$; $k = 34.9$ N/mm;
$L_0 = 116.8$ mm
P8-18 $t = 0.0129$ in.; $\sigma_b = 36{,}282$ psi
P8-20 $D = 90$ mm; $d = 56$ mm; $h = t = 2$ mm

CHAPTER 9

P9-4 At the work station
P9-6 12.53%
P9-7 $D > 86.2$ mm
P9-9 (*a*) $T_A = 1272$ N · m (*b*) $F_t = 1669$ N
(*c*) $n_w = 501$ rpm (*d*) 69.5 kW
(*e*) 368 kN (*f*) 37 kN
(*g*) 815 m and 8150 m
P9-10 (*a*) 13 415 N (*b*) 74.65 rpm
(*c*) 12.5 kW (*d*) 13.16 kW
(*e*) No

CHAPTER 10

P10-6 $T_s = 239$ N · m <600N · m for $f_s = 2.5$

P10-7 28 folds

P10-8 (*a*) $\tau_{max} = 3\ F_s\ (a/b)R \cos \alpha$
(*b*) $P = 3.45$ kW
(*c*) (1) Vary the length of arm
(2) Change of spring constant

P10-9 $d = 7.98$ mm

P10-10 $n_{max} = 1015$ rpm; $n_{min} = 984$ rpm

CHAPTER 11

P11-2 (*a*) 226 rpm (*b*) 14.4 hp

P11-3 $i_{max} = 1.71$

P11-4 $P = 4.85$ hp

P11-5 $P = 7.77$ hp; 60% increase

P11-6 (*a*) 0.93 (*b*) 6.16 m/s (*c*) 2.023 m
(*d*) 262 N (*e*) 358 N (*f*) 25 mm

P11-7 (*b*) $n_d = 16.24$ rps $n_D = 3.95$ rps
(*c*) 28.81 kW
(*d*) 2.14
(*e*) 4.9 kN

P11-8 68.5%

P11-10 7 belts—size C162; $L = 164.9$ in.; $C = 51.65$ in.

P11-12 Use 4 belts; C162 designation, $C = 51.14$ in.

P11-14 $d_1 = 225$; $d_2 = 179$; $d_3 = 148$; $d_4 = 225$; $d_5 = 269$; $d_6 = 296$

CHAPTER 12

P12-3 (*a*) 47 (or 46)
(*b*) $d = 2.721$ in.; $D = 7.486$ in.
(*c*) 8.847 in.
(*d*) $L = 70$ pitches
(*e*) 641 fpm, 638 fpm
(*f*) 4.98 hp
(*g*) $W = \frac{5}{16}$ in.; $D_r = 0.312$ in.; outside width = 0.5525 in.
(*h*) Inside clearance: use 0.875 in.

P12-4 (*a*) 43
(*b*) $d = 3.134$ in.; $D = 10.275$ in.
(*c*) 6.83 fps
(*d*) 807 lb.

P12-5 (*a*) RC 50; use 21 and 95 teeth in sprockets
(*b*) $d = 4.193$ in.; $D = 18.903$ in.; $C = 21$ in.; $L = 132p$ C (corrected) = 21.19 in.
(*c*) $\theta = 140$ deg
(*d*) Maximum bore = $2\frac{9}{32}$ in. Maximum hub diameter = $3\frac{31}{64}$ in.
(*e*) $C = 14.04$ in.

P12-6 (*a*) RC 40, $n = 19$, $N = 57$
(*b*) $d = 3.038$ in.; $D = 9.076$
(*c*) $C = 10.60$ in.; $\theta = 147$ deg
(*d*) $L = 84$ links; $C_a = 10.57$ in. (for 82 links)
(*e*) 11.93 fps
(*f*) Type B lubrication

P12-7 (*a*) RC 80, four-strand $n = 17$, type B lubrication (*b*) $N = 51$

P12-8 RC 50, 17 and 11 teeth in sprockets; $D = 3.401$ in.; $d = 2.612$ in.; $C = 3.5$ in. Type B lubrication

CHAPTER 13

P13-2 $N_p = 50$; $N_G = 127$; $m = 1$

P13-3 $m = 2.5$; $p_c = 7.854$ mm; $p_b = 7.118$ mm; $h = 5.125$ mm; $t = 3.927$ mm

P13-4 (*a*) External contact: $N_p = 18$; $N_G = 45$
(*b*) Internal contact: $N_p = 42$; $N_G = 105$

P13-5 $n_G = 271.43$ rpm; $h_t = 9$ mm; $t_c = 6.28$ mm

P13-6 (*a*) $m = 4$
(*b*) $d = 72$ mm; $D = 112$ mm
(*c*) 92 mm
(*d*) $r_b = 33.83$ mm; $R_b = 52.62$ mm

P13-7 $m_p = 1.58$; no interference

P13-8 External contact: $C = 4.5$ in.; $n_p = 945$ rpm. Internal contact: $C = 2.5$ in.; 44% reduction

P13-9 $m_p = 1.58$

P13-10 $m = 20$; $b = 25$ mm; $t_c = 31.416$ mm; $p_c = 62.832$ mm $p_b = 59.043$ mm

P13-11 (*a*) Exact center distance: $m = 4$; $N_p = 25$; $N_G = 102$
(*b*) Exact speed ratio: $m = 4$
$N_p = 25$; $N_G = 100$; $Np = 26$; $N_G = 104$
$m = 4.5$: $N_p = 23$; $N_G = 92$
$N_p = 22$; $N_G = 88$

P13-12 $m_p = 1.50$ acceptable; no interference

P13-16 $P_d = 12$; $N_G = 144$; $N_p = 24$; $C = 7$ in.; $d = 2$ in.; $D = 12$ in.; $F = 0.875$ in.

P13-17 (*a*) $P_d = 4$ (*b*) $N_p = 30$
(*c*) $N_G = 90$ (*d*) $F = 2.50$ in.

P13-20 $P_d = 5$; $N_p = 18$; $N_G = 54$; $C = 7.2$ in. $F = 1.10$ in.

P13-21 9 hp; $F = 1.25$ in.

P13-30 16.04 deg $< \psi <$ 29.89 deg

P13-32 $P_n = 4$

P13-33 $m = 8.465$ mm

P13-35 $L = 467.24$ mm

P13-36 $m_n = 12$ mm; $p_n = 37.70$ mm; $\phi_n = 18.52$ deg

P13-37 $m_t = \begin{cases} 3.96 \text{ max.} \\ 2.76 \text{ min.} \end{cases}$

CHAPTER 14

P14-1 33.18 deg; 56.82 deg

P14-2 Γ_1 33.23 deg; Γ_2 56.77 deg; $d = 228$ mm; $D = 348$ mm; $a = 12$ mm; $b = 15$ mm; $w = 108$ mm; $H = 59.18$ mm; $h = 90.34$ mm

P14-4 (*a*) $D_G = 276.93$ mm
(*b*) $\lambda = 46.69$ deg
(*c*) $C = 144.47$ mm
(*d*) $m_G = 21.75$
(*e*) $e = 0.94$
(*f*) $P = 18.8$ kW

P14-5 $\lambda_w = 34.30$; $\psi_G = 55.70$; $D_G = 17.11$ in.; $C = 9.43$ in.; $m_G = 14.33$

CHAPTER 15

P15-1 $fs = 23.15$

P15-2 $d = 0.71$

P15-3 $fs = 3.30$

P15-4 $fs = 1.80$

P15-5 $d = 31.6$ mm

P15-6 $d = 93.41$ mm

P15-10 $d = 3.56$ in.

P15-11 11,900 lb-in. at pulley

P15-17 (*a*) 12.3 MPa (*b*) 154 MPa
(*c*) 158 MPa (*d*) $fs = 3.53$

P15-21 $\theta = 0.11$

P15-23 $y = 0.0048$ in. at 180-lb load; $y = 0.0069$ in. at 90-lb load

CHAPTER 16

P16-1 M14: 496.2 N · m; M16: 474.2 N · m

P16-2 M4 setscrew

P16-3 Use $\frac{1}{4}$-in. setscrew

P16-5 0.1875 in. × 0.1875 in. × 1.0 in. key

P16-7 Use two keys $\frac{3}{4}$ in. × $\frac{7}{8}$ in. × 2 $\frac{1}{2}$ in. spaced 90 deg apart.

P16-8 8 mm × 8 mm × 35 mm key

P16-9 (*a*) AISI 4140 (*b*) 47 kW
(*c*) $\tau = 106$ MPa < 129 MPa

P16-11 Key: $b \times t \times L = 0.75$ in. × 0.5 in. × 1.875 in.

P16-12 $d \times L = \frac{5}{8}$ in. × 3 $\frac{1}{2}$ in.

P16-15 $\tau = 176$ MPa; $\sigma_c = 44$ MPa

P16-18 A 4 mm standard duty AISI 1420 pin. $fs = 2.3$ approximately

P16-19 T = 429 N · m; σ_c 341 MPa

P16-20 T = 169 N · m

P16-22 T = 3000 lb-in.

CHAPTER 17

P17-1 SAE 40

P17-3 2.2 microreyns

P17-4 60 mm

P17-7 (*a*) $h_o = 0.02$ mm (*b*) $\varepsilon = 0.6$
(*c*) $h = 0.029$ mm at 135 deg
P17-8 (*a*) $h_o = 0.0039$ mm
(*b*) $\emptyset = 31$ deg (*c*) $H_f = 12.9$
P17-10 $h_o = 324\ \mu$in.
P17-11 $L_{min} = 48$ mm

CHAPTER 18

P18-1 (2.396) 10^6 cycles
P18-2 148 N; 17.4 W
P18-3 (*a*) 1/8 (*b*) 2/1 (*c*) 1/4
P18-4 RE = 4170 N; CR = 30.06 kN
P18-5 9430 h
P18-6 73% reduction
P18-7 27%
P18-8 52.5% reduction
P18-9 (*a*) $L_{10} = 425$ h (*b*) $L_5 = 230$ h
P18-10 ID = 45 mm Max type; CR = 50.2 kN
P18-12 $L_{95} = 7644$ h; $L_{95} = 17\,440$ h
P18-15 $L_w = 3597$ h
P18-16 Medium series: Bore 20 mm Max type
Bore 25 mm Conrad
Light series: Bore 30 mm Max type
Bore 35 mm Conrad
Extra light: Bore 45 mm

CHAPTER 19

P19-2 2226 N · m
P19-4 Clutch inadequate; $P_{nom} = 65$ k W
P19-5 35,191 lb
P19-7 24 900 N
P19-10 $T_{des} = 4520$ lb-in.; $P_{nom} = 31$ hp
P19-11 Each shoe must have a mass of 0.296 kg and a width of at least 2.1 mm

CHAPTER 20

P20-1 (*a*) $R_e = 101.3$ mm
(*b*) $F_a = 16\,980$ N
(*c*) $p_h = 17.6$ MPa
P20-3 49 mm
P20-5 $T_{fmax} = 3327$ N · m;
$T_{rmax} = 1570$ N · m
P20-7 Stopping distances for $f_R = 0.8$ are 4.4, 10, 17.7, 39.8 m; Stopping distances for $f_R = 0.48$ are 7.4, 16.6, 29.5, 66.4 m; 47.3 m (assuming vehicle travels at 90 km/hr during the 0.3-s reaction time)
P20-9 343 kW
P20-13 24 in.^{2}
P20-15 Required disk diameter is 21.2 in. The 24-in. diameter is sufficient.

CHAPTER 21

P21-1 Asbestos compositions
P21-2 Unable to withstand the cylinder pressure
P21-3 (*a*) Apparent flange pressure = 10 N/mm^2
(*b*) Minimum flange pressure = 3 N/mm^2
P21-4 Asbestos and rubber gasket - Grade 8 bolts; low compressibility asbestos and rubber gasket - Grade 5 bolts
P21-5 No
P21-6 Yes
P21-7 None
P21-8 0.0018 in. elongation due to torque; no bolt tension remains after gasket creep
P21-9 Yes
P21-10 0.34 in.

Index